T0140374

Advances in Intelligent Systems and Computing

Volume 800

Series editors

Janusz Kacprzyk, Systems Research Institute, Polish Academy of Sciences, Warsaw, Poland
e-mail: kacprzyk@ibspan.waw.pl

The series "Advances in Intelligent Systems and Computing" contains publications on theory, applications, and design methods of Intelligent Systems and Intelligent Computing. Virtually all disciplines such as engineering, natural sciences, computer and information science, ICT, economics, business, e-commerce, environment, healthcare, life science are covered. The list of topics spans all the areas of modern intelligent systems and computing such as: computational intelligence, soft computing including neural networks, fuzzy systems, evolutionary computing and the fusion of these paradigms, social intelligence, ambient intelligence, computational neuroscience, artificial life, virtual worlds and society, cognitive science and systems, Perception and Vision, DNA and immune based systems, self-organizing and adaptive systems, e-Learning and teaching, human-centered and human-centric computing, recommender systems, intelligent control, robotics and mechatronics including human-machine teaming, knowledge-based paradigms, learning paradigms, machine ethics, intelligent data analysis, knowledge management, intelligent agents, intelligent decision making and support, intelligent network security, trust management, interactive entertainment, Web intelligence and multimedia.

The publications within "Advances in Intelligent Systems and Computing" are primarily proceedings of important conferences, symposia and congresses. They cover significant recent developments in the field, both of a foundational and applicable character. An important characteristic feature of the series is the short publication time and world-wide distribution. This permits a rapid and broad dissemination of research results.

** Indexing: The books of this series are submitted to ISI Proceedings, EI-Compendex, DBLP, SCOPUS, Google Scholar and Springerlink **

More information about this series at http://www.springer.com/series/11156

Shahram Latifi

Editor

16th International Conference on Information Technology-New Generations (ITNG 2019)

 Springer

Editor
Shahram Latifi
Department of Electrical and Computer Engineering
University of Nevada
Las Vegas, NV, USA

ISSN 2194-5357   ISSN 2194-5365   (electronic)
Advances in Intelligent Systems and Computing
ISBN 978-3-030-14072-4   ISBN 978-3-030-14070-0   (eBook)
https://doi.org/10.1007/978-3-030-14070-0

This Springer imprint is published by the registered company Springer Nature Switzerland AG.
The registered company address is: Gewerbestrasse 11, 6330 Cham, Switzerland

Contents

Part IX Computer Vision, Image Processing/Analysis, Tracking

Part X Potpourri

Chair Message

Welcome to the 16th International Conference on Information Technology, New Generations, ITNG 2019. It is a pleasure to report that we have another successful year for the ITNG 2019. Gaining popularity and recognition in the IT community around the globe, the conference was able to attract many papers from authors worldwide. The papers were reviewed for their technical soundness, originality, clarity, and relevance to the conference. The conference enjoyed expert opinion of over 100 author and non-author scientists who participated in the review process. Each paper was reviewed by at least two independent reviewers. A total of 80 articles were accepted as regular papers, and 10 were accepted as short papers (posters).

The articles in this book of chapters address the most recent advances in such areas as Wireless Communications and Networking, Software Engineering, Information Security, Data Mining and Machine Learning, Informatics, High-Performance Computing Architectures, the Internet, and Image and Video Processing.

As customary, the conference features two keynote speakers on Monday and Tuesday. There will be a short tutorial session Wednesday morning on Deep Learning. The presentations for Monday, Tuesday, and Wednesday are organized in 2 meeting rooms simultaneously, covering a total of 20 technical sessions. Poster presentations are scheduled for the morning and afternoon of these days. The award ceremony, conference reception, and dinner are scheduled for Tuesday evening.

Many people contributed to the success of this year's conference by organizing symposia or technical tracks for the ITNG. Dr. Doina Bein served in the capacity of conference vice chair. We benefited from the professional and timely services of Dr. Ping Wang who not only organized the Security Track but helped in shaping the students' book of abstracts to be published online at the ITNG website. Dr. Yenumula Reddy deserves much credit for spearheading the review process and running a symposium on Wireless Communications and Networking. My sincere thanks go to other major track organizers and associate editors, namely, Glauco Carneiro, Luiz Alberto Vieira Dias, Fred Harris, Ray Hashemi, Thomas Jell, Atsushi Kawaguchi, Teruya Minamoto, and Fangyan Shen.

Others who were responsible for solicitation, review, and handling the papers submitted to their respective tracks/sessions include Drs. Azita Bahrami, Wolfgang Bein, and Mei Yang.

The help and support of the Springer team in preparing the ITNG proceedings are specially appreciated. Many thanks are due to Michael Luby, the Senior Editor, and Nicole Lowary, the Assistant Editor of the Springer Supervisor of Publications, for the timely handling of our publication order. We also appreciate the hard work by Menas Kiran, the Springer Project Coordinator, who looked very closely at every single article to make sure they are formatted correctly according to the publisher guidelines. Finally, the great efforts of the Conference Secretary, Ms. Mary Roberts, who dealt with the day-to-day conference affairs, including timely handling volumes of emails, are acknowledged.

The conference venue is Tuscany Suites Hotel. The hotel, conveniently located within half a mile of Las Vegas Strip, provides an easy access to other major resorts and recreational centers. I hope and trust that you have an academically and socially fulfilling stay in Las Vegas.

Shahram Latifi
The ITNG General Chair

Reviewers List

Abbas, Haider	Figueiredo, Eduardo	Nwaigwe, Adaeze
Abreu, Fernando	Fong, Andy	Paiva, Ana
Aguiar, Ademar	Ford, George	Pang, Les
Ahmed, Adel	França, Joyce	Peiper, Les
Amâncio, José	Fujinoki, Kensuke	Peiper, Cad
Andro-Vasko, James	Garcia, Vinicius	Pirouz, Matin
Anikeev, Maxim	Girma, Anteneh	Popa, Vlad
Araújo, Marco	Gofman, Mikhail	Resende, Antonio Maria
Avelino, Guilherme	Gueron, Shay	Roccosalvo, Janine
Baguda, Yaqoub	Guo, Kendra	Rossi, Gustavo
Bahrami, Azita	Hashemi, Ray	Santos, Katyusco
Barford, Lee	Jell, Thomas	Sbeit, Raed
Bein, Doina	Kannan, Sudesh	Sharma, Sharad
Bein, Wolfgang	Kayano, Mitsunori	Shen, Fangyang
Cagnin, Maria	Kelley, Richard	Silva, Bruno
Canedo, Edna	Khan, Zahoor	Silva, Paulo Caetano
Carneiro, Glauco	Kinyua, Johson	Stuart, Jeff
Chapman, Matthew	Koch, Fernando	Takano, Shigeru
Christos, Kalloniatis	Laskar, John	Tayeb, Shahab
Colaço Jr., Methanias	Lee, Byoeng Kil	Terra, Ricardo
Daniels, Jeff	Maciel, Rita	Tsetse, Anthony
Darwish, Marwan	Mahto, Rakesh	Ueki, Masao
Dascalu, Sergiu	Maia, Paulo	Wang, Jau-Hwang
David, José	Mascarenhas, Ana	Wang, Yi
Dawson, Maurice	Mateos, Cristian	Wang, Ping
De La Santa, Kimberly	Matsui, Kota	Williams, Kenneth
Ventura, Michele	Mendonça, Manoel	Wu, Rui
Durelli, Rafael	Mernik, Marjan	Xu, Frank
Eler, Marcelo	Monteiro, Miguel	Yamamoto, Michio
El-Ziq, Yacoub	Morimoto, Akira	Yi, Beifang
Faria, João	Novais, Renato	Zhang, Zhong

Part I
Cybersecurity

Nir Drucker and Shay Gueron

1.1 Introduction

The Multiply and Accumulate (MAC) operation consumes three inputs a, b, c, and computes $a = a + b \cdot c$. It is a fundamental step in many floating-point and integer computations. Examples are dot product calculations, matrix multiplications, and modular arithmetic. Modern processors offer instructions for performing MAC over floating-point inputs, e.g., AMD Bulldozer's Fused Multiply-Add (FMA), and Intel's Single Instruction Multiple Data (SIMD)-FMA (starting with the microarchitecture Codename Haswell). Here, we focus on Intel's AVX512IFMA instructions [1] that compute MAC on unsigned integers.

The AVX512IFMA instructions are defined, but are not yet widely available. However, a demonstration of their capabilities is already given in [11], showing a 2x potential speedup for 1024-bit integer multiplication (and more for larger operands). Another example is [6], where we showed a 6x potential speedup over OpenSSL's Montgomery Multiplication (MM). Additional code examples [5, 10] contributed to OpenSSL, include optimized 1024/1536/2048-bit MM. These demonstrations did not optimize modular squaring specifically; rather, they used a multiplication routine for squaring as well. Here, we show how to use the AVX512-IFMA instructions for optimizing modular squaring. Our developments build on top of the AMS optimization of [7] (other squaring methods can be found in [4, 9, 13]).

This work was done prior to joining Amazon.

N. Drucker · S. Gueron (✉)
University of Haifa, Haifa, Israel

Amazon Web Services Inc, Seattle, WA, USA
e-mail: shay@math.haifa.ac.il

The paper is organized as follows. Section 1.2 discusses some preliminaries. Section 1.3 deals with implementing the AMS algorithm with the AVX512IFMA instructions. In Sect. 1.4, we propose a potential improvement to the definition of AVX512IFMA. Finally, we show our experimental results in Sect. 1.5, and provide our conclusions in Sect. 1.6.

1.2 Preliminaries and Notation

Hereafter, we use lower case letters to represent scalars (64-bit integers), and upper case letters to represent 512-bit wide register. We denote zero extension of a 64 bits variable x by $ZE(x)$.

1.2.1 The AVX512IFMA Instructions

Intel's Software Developer Manual [1] introduces two instructions called AVX512IFMA: VPMADD52LUQ and VPMADD52HUQ. Their functionality is illustrated in Algorithm 1.1. These instructions multiply eight 52-bit unsigned integers residing in wide 512-bit registers, produce the low (VPMADD52LUQ) and high (VPMADD52HUQ) halves of the 104-bit products, and add the results to 64-bit accumulators (i.e., SIMD elements), placing them in the destination register. They are designed for supporting big number multiplications, when the inputs are stored in a "redundant representation" using radix 2^{52} (as explained in [8]).

The AVX512IFMA instructions build on the existence of other instructions called SIMD-FMA, which are designed to support IEEE standard Floating-Point Arithmetic [12]. The SIMD-FMA instructions handle double-precision floating-point numbers ($x[63 : 0]$), where the bits are viewed as: (a)

© Springer Nature Switzerland AG 2019
S. Latifi (ed.), *16th International Conference on Information Technology-New Generations (ITNG 2019)*,
Advances in Intelligent Systems and Computing 800,
https://doi.org/10.1007/978-3-030-14070-0_1

Algorithm 1.1 DST = VPMADD52(A,B,C) [1]

    Inputs: A,B,C (512-bit wide registers)
    Outputs: DST (a 512-bit wide register)
1: **procedure** VPMADD52LUQ(A, B, C)
2:    **for** j := 0 to 7 **do**
3:        i := j × 64
4:        TMP[127 : 0] := ZE(B[i+51:i]) × ZE(C[i+51:i])
5:        DST[i+63:i] := A[i+63:i] + ZE(TMP[**51 : 0**])
6: **procedure** VPMADD52HUQ(A, B, C)
7:    **for** j := 0 to 7 **do**
8:        i := j × 64
9:        TMP[127 : 0] := ZE(B[i+51:i]) × ZE(C[i+51:i])
10:      DST[i+63:i] := A[i+63:i] + ZE(TMP[**103 : 52**])

---

fraction $x[51 : 0]$ (53 bits where only 52 bits are explicitly stored); (b) exponent $x[62 : 52]$; (c) sign bit $x[63]$.

1.2.2 Almost Montgomery Multiplication

MM is an efficient technique for computing modular multiplications [14]. Let t be a positive integer, k an odd modulus and $0 \leq a, b < k$ integers. We denote the MM by $MM(a, b) = a \cdot b \cdot 2^{-t} \pmod{k}$, where 2^t is the Montgomery parameter. A variant of MM, called Almost Montgomery Multiplication (AMM) [7], is defined as follows. Let k and t be defined as above, and $0 \leq a, b < B$ integers, then $AMM(a, b)$ is an integer U that satisfies: (1) $U \pmod{m} = a \cdot b \cdot 2^{-t} \pmod{k}$; (2) $U \leq B$.

The advantage of AMM over MM is that the former does not require a (conditional) "final reduction" step. This allows using the output of one invocation as the input to a subsequent invocation. The relation between AMM and MM is the following. If $0 \leq a, b < B$, $RR = 2^{2t} \pmod{k}$, $a' = AMM(a, RR)$, $b' = AMM(b, RR)$, $u' = AMM(a', b')$ and $u = AMM(u', 1)$, then $u = a \cdot b \pmod{k}$.

1.3 Implementing AMS with AVX512IFMA

One of the common squaring algorithms [4] is the following. Let $A = \sum_{i=0}^{n} B^i a_i$ be an n digits integer in base B, $a_i \geq 0$. Then,

$$
\begin{aligned}
A^2 &= \sum_{i=0}^{n} \sum_{j=0}^{n} B^{i+j} a_i a_j \\
&= \sum_{i=0}^{n} B^{2i} a_i^2 + 2 \sum_{i=0}^{n} \sum_{j=i+1}^{n} B^{i+j} a_i a_j
\end{aligned}
\tag{1.1}
$$

where the last multiplication by 2 can be carried out by a series of left shift operations [9]. This reduces about half of the single-precision multiplications (compared to

regular multiplication). Additional improvement is achieved by using vectorization. For example, [8] shows squaring implementations that use Intel's Advanced Vector Extensions (AVX) and AVX2 instructions. In these implementations, integers are stored in a "redundant representation" with radix $B = 2^{28}$ (each of the n digits is placed in a 32-bit container, padded from above with 4 zero bits). Each of the AVX2 256-bit wide registers (ymm) can hold up to eight 32-bit containers. This allows for (left) shifting of 8 digits in parallel, without losing their carry bit.

Algorithm 1.2 describes an implementation of AMS= AMM(a,a) that uses the AVX512IFMA instructions. Let the input (a), the modulus (m) and the result (x) be n-digit integers in radix $B = 2^{52}$, where each digit is placed in a 64-bit container (padded with 12 zero bits from above). Let $z = \lceil n/8 \rceil$ be the total number of wide registers needed for holding an n-digit number, and denote $k_0 = -m^{-1} \pmod{2^{52}}$. The final step of Algorithm 1.2 returns the result to the radix $B = 2^{52}$ format, by rearranging the carry bits. An illustration of a simple AMS flow is given in Fig. 1.1 that shows how ~20% of the VPMADD52 calls (left as blank spaces in the figure) are saved, compared to an AMM. The algorithm applies the left shift optimization of [9] to the AVX512IFMA AMM implementation of [11]. This can be done through either Eq. 1.1 (perform all MAC calculations and then shift the result by one), or according to:

Algorithm 1.2 x = AMS52(a, m, k_0)

    Inputs: a,m (n-digit unsigned integers), k_0 (52-bit unsigned integer)
    Outputs: x (n-digit unsigned integers)
1: **procedure** MULA[L/H]PART(i)
2:    X_i := VPMADD52[L/H]UQ(X_i, A_{curr}, A_i)
3:    **for** j := i + 1 to z **do**
4:        T := VPMADD52[L/H]UQ($ZERO, A_{curr}, A_j$)
5:        X_j := $X_j + (T \ll 1)$
1: **procedure** AMS52(a, m, k_0)
2:    load a into $A_0 \ldots A_z$ and m into $M_0 \ldots M_z$
3:    zero($X_0 \ldots X_z, ZERO$)
4:    **for** i := 0 to z **do**
5:        **for** j := 0 to min{8, n − (8 · i)} **do**
6:            A_{curr} = broadcast($a[8 \cdot i + j]$)
7:            MulALPart(i)
8:            $y[127 : 0]$:= $k_0 \cdot X_0[63 : 0]$
9:            Y := broadcast($y[52 : 0]$)
10:          **for** l := 0 to z **do**
11:            X_l := VPMADD52LUQ(X_l, M_l, Y)
12:          x_0 := $X_0[63 : 0] \gg 52$
13:          X := $X \gg 64$
14:          $X_0[63 : 0]$ = $X_0[63 : 0] + x_0$
15:          MulAHPart(i)
16:          **for** l := 0 to z **do**
17:            X_l := VPMADD52HUQ(X_l, M_l, Y)
18:        FixRedundantRepresentation(X)
19:    **return** X

Fig. 1.1 Flow illustration of $x = \mathrm{SQR}(a, m, k_0)$, where a, m and x are 16-digit operands, each one is accommodated in two *zmm* registers

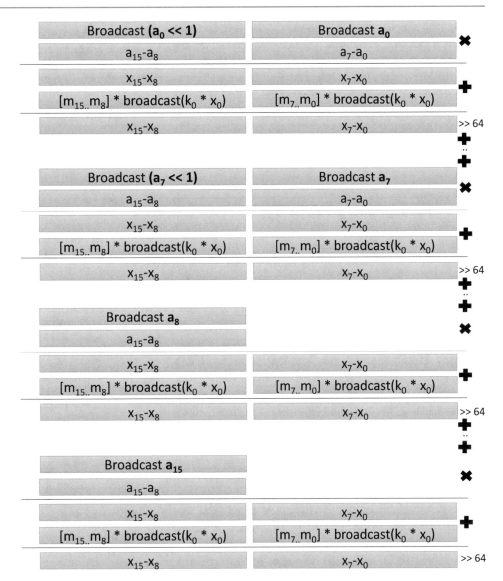

$$A^2 = \sum_{i=0}^{n} B^{2i} a_i^2 + \sum_{i=0}^{n} \sum_{j=i+1}^{n} B^{i+j} a_i a_j' \quad (1.2)$$

where $a' = a \ll 1$. An efficient implementation of the first approach requires to accommodate a, m, and x in wide registers (not in memory), while an implementation of the second approach requires accommodating a' in wide registers as well. Consequently, the AVX512, which has only 32 wide registers, can hold n-digit integers up to $n \leq 85$ with the first approach, or up to $n \leq 64$ with the second approach. For example, 4096-bit modular squaring (part of a 4096-bit exponentiation, e.g., for Paillier encryption) has $n = 80$-digits operands (written in radix $B = 2^{52}$). It requires 40 wide registers with the second approach (but there are not enough). With the first approach, only 30 wide registers are needed (there are 32). This situation seems better, but in practice, it is not good enough.

Performing left shifting of an n-digit number requires some extra wide registers. These are not necessarily available for use with the above two approaches. Thus, we propose Algorithm 1.2, that is based on the following identity:

$$A^2 = \sum_{i=0}^{n} B^{2i} a_i^2 + \sum_{i=0}^{n} \sum_{j=i+1}^{n} 2(B^{i+j} a_i a_j) \quad (1.3)$$

Here, the left shifts are performed on-the-fly, and free some wide registers for supporting other operations.

Identifying an additional bottleneck On-the-fly left shifting can be implemented in three ways, but unfortunately, all three do not go along well with the AVX512IFMA architecture. The first alternative is to multiply, accumulate and shift the result. This may double shift some of the previously accumulated data. The second alternative is to shift one of the

VPMADD52's input operands. This may lead to a set carry bit in position 53, which would be (erroneously) ignored during the multiplication (see Algorithm 1.1). The third alternative splits the MAC operation, to inject the shift between. This is not feasible with the atomic operation of VPMADD52, but can be resolved by performing the Multiply-Shift-Accumulate operation in two steps, with an extra temporary (zeroed) wide register. Indeed, Algorithm 1.2, MulA[L/H]Part (steps 4, 5) executes this flow.

1.4   Is Using Radix 2^{51} Better?

In this section, we discuss the selection of the radix. The AVX512IFMA instructions leverage hardware that is needed anyhow, for the FMA unit (floating-point operations need 53-bit multiplication for a 106-bit mantissa). Obviously, given AVX512IFMA, it is natural to work with radix $B = 2^{52}$. Using a larger radix (e.g., $B = 2^{58}$) could be better in theory, but will incur too many costly conversions to allow for using VPMADD52. We also note that no native SIMD instructions for a larger radix are available. A smaller radix (e.g., 2^{51}) is, however, possible to choose. This allows to cut about half of the serialized instructions in steps 3–5 of $MulA[L/H]Part$, by left shifting one of the operands before   the multiplication.

Algorithm 1.3 is a modification of Algorithm 1.2, operating in radix 2^{51}. While it avoids the shift operations before the VPMADD52LUQ, it still needs to perform the shifting before the VPMADD52HUQ instruction. For example, Let a, b, c_1, c_2 be 51-bit integers. After performing $c_1 = \text{VPMADD52LUQ}(0, a, b) = (a \times b)[51 : 0]$ and $c_2 = \text{VPMADD52HUQ}(0, a, b) = (a \times b)[102 : 52]$, c_1 and c_2 are no longer in (pure) radix 2^{51}. Propagating the carry bit in c_1 can be delayed to step 22 of Algorithm 1.3. In contrary, c_2 must be shifted prior to the accumulation step. As we show in Sect. 1.5, Algorithm 1.3 does not lead to faster squaring, with the current architecture. This suggests a possible improvement to the architectural definition.

1.4.1   A Possible Improvement for AVX512IFMA

Algorithm 1.3 offers better parallelization compared to Algorithm 1.2, but still includes serialized steps (e.g., the function MulHighPart). A completely parallelized algorithm requires hardware support. To this end, we suggest a new instruction that we call Fused Multiply-Shift-Add (FMSA), and describe in Algorithm 1.4. It shifts the multiplication

---

Algorithm 1.3 x = AMS51(a, m, k_0)

Inputs: a,m (n-digit unsigned integers), k_0 (52-bit unsigned integer)

Outputs: x (n-digit unsigned integers)

1: **procedure** MulHighPart(SRC1, SRC2, DEST)
2:     TMP := VPMADD52HUQ(ZERO, SRC1, SRC2)
3:     DEST := DEST + (TMP $\ll$ 1)
1: **procedure** AMS51(a, m, k_0)
2:     load a into $A_0 \dots A_z$ and m into $M_0 \dots M_z$
3:     zero($X_0 \dots X_z$, ZERO)
4:     **for** $i := 0$ to z **do**
5:         **for** $j := 0$ to $\min\{8, n - (8 \cdot i)\}$ **do**
6:             $A_{curr} = \text{broadcast}(a[8 \cdot i + j])$
7:             $A_{shifted} = A_{curr} \ll 1$
8:             $X_i := \text{VPMADD52LUQ}(X_i, A_{curr}, A_i)$
9:             **for** $j := i + 1$ to z **do**
10:                 $X_i := \text{VPMADD52LUQ}(X_i, A_{shifted}, A_i)$
11:         $y[127 : 0] := k_0 \cdot X_0[63 : 0]$
12:         $Y := \text{broadcast}(y[51 : 0])$
13:         **for** $l := 0$ to z **do**
14:             $X_l := \text{VPMADD52LUQ}(X_l, M_l, Y)$
15:         $x_0 := X_0[63 : 0] \gg 51$
16:         $X := X \gg 64$
17:         $X_0[63 : 0] := X_0[63 : 0] + x_0$
18:         MulAHighPart(X_i, A_{curr}, A_i)
19:         **for** $l := i + 1$ to z **do**
20:             MulAHighPart($X_l, A_{shifted}, A_l$)
21:         **for** $l := 0$ to z **do**
22:             MulAHighPart(X_l, M_l, Y)
23:         FixRedundantRepresentation(X)
24:     **return** X

Algorithm 1.4 DST=FMSA(A,B,C,imm8)

1: **for** j := 0 to 7 **do**
2:     i := j*64
3:     TMP[127 : 0] := ZE(B[i+51:i]) $\times$ ZE(C[i+51:i])
4:     DST[i+63:i] := A[i+63:i] +
                    ZE(**TMP[103 : 52]** $\ll$ **imm8**)

---

result by an immediate value ($imm8$) before accumulating it. This instruction can be based on the same hardware that supports FMA (just as AVX512IFMA). Note that when $imm8 = 0$ then this instruction is exactly VPMADD52HUQ.

1.5   Results

1.5.1   Results for the Current Architecture

This section provides our performance results. For this study, we wrote new optimized code for all the algorithms discussed above, and measured them with the following methodology.

Currently, there are only a limited series of processors with VPMADD52, which we currently don't have. Therefore, to predict the potential improvement on future Intel architectures we used the Intel Software Developer Emulator

(SDE) [3]. This tool allows us to count the number of instructions executed during each of the tested functions. We marked the start/end boundaries of each function with "SSC marks" 1 and 2, respectively. This is done by executing "movl ssc_mark, %ebx; .byte 0x64, 0x67, 0x90" and invoking the SDE with the flags "-start_ssc_mark 1 -stop_ssc_mark 2 -mix -cnl". The rationale is that a reduced number of instructions typically indicates improved performance that will be observed on a real processor (although the exact relation between the instructions count and the eventual cycles count is not known in advanced).

Our measurements show that the overall number of instructions in our AMM and AMS implementations (in radix 2^{52}) is almost identical. However, the number of occurrences per instruction varies between the two algorithms. The most noticeable change was for the VPADDQ, VPMADD52, VPSLLQ, and VPXORQ instructions. Let u_{AMS}/u_{AMM} be the number of occurrences of the instruction u in AMS/AMM code, and let t_{AMM} be the total number of instructions in the AMM code. We write $r_u = (u_{AMS} - u_{AMM})/t_{AMM}$. Table 1.1 compares the r_u values for different u and operands sizes. It shows that reducing the number of VPMADD52 instructions is achieved through increasing the number of other instruction (e.g., VPADDQ, VPSLLQ, and VPXORQ).

To assess the impact of the above trade-off, we note that the latency of VPADDQ, VPSLLQ, and VPXORQ is 1 cycle, the throughput of VPADDQ and VPXORQ is 0.33 cycles, and the throughput of VPSLLQ is 1 cycle [2]. By comparison, we can expect that the latency/throughput of a future VPMADD52 would be similar to VPMADDWD (i.e., 5/1), or to VFMA$\star$ (i.e., 4/0.5). It appears that trading one VPMADD52 for 4 other instructions (which is worse than the trade-off we have to our AMS implementation) could still be faster than the AMM implementation.

To study the effects at the higher scale of the modular exponentiation code, we define the following notation. Let $u_{ModExpAMS}/u_{ModExp}$ be the number of occurrences of the instruction u in the modular exponentiation code, with and without AMS, respectively, and let t_{ModExp} be the overall number of instructions in this code (w/o AMS). We write $s_u = (u_{ModExpAMS} - u_{ModExp})/t_{ModExp}$. Table 1.2 shows the values s_u.

Table 1.1 Values of r_u for different u instructions and different operands sizes

Size	VPADDQ	VPMADD52	VPSLLQ	VPXORQ
1024	0.06	−0.05	0.06	0.06
1536	0.13	−0.07	0.07	0.07
2048	0.05	−0.09	0.08	0.08
3072	0.05	−0.13	0.15	0.15
4096	0.12	−0.12	0.12	0.12

Table 1.2 Values of s_u for different instructions (u) and different operands sizes

Size	VPADDQ	VPMADD52	VPSLLQ	VPXORQ
1024	0.01	−0.01	0.02	0.01
1536	0.02	−0.01	0.02	0.02
2048	0.02	−0.03	0.02	0.02
3072	0.04	−0.04	0.04	0.04

Table 1.3 Values of w_u^{AMM}, w_u^{AMS}, and w_u^{ModExp} for different instructions (u) and different operands sizes

Function name	VPADDQ	VPMADD52	VPSLLQ	VPXORQ
AMM3072	0.32	0.01	0.32	0.32
AMM4080	0.28	0.01	0.28	0.28
AMM4096	0.28	0.01	0.28	0.27
AMS3072	0.07	0.00	0.09	0.07
AMS4080	0.07	0.00	0.10	0.07
AMS4096	0.08	0.01	0.10	0.06
ModExp3072	0.05	0.00	0.06	0.05

We use the following notation for evaluating the radix 2^{51} technique. Let $u_{AMS51}/u_{AMM51}/u_{ModExpAMS51}$ be the number of occurrences of the instruction u in radix 2^{51} code. We write

$$w_u^{AMM} = (u_{AMM} - u_{AMM51})/t_{AMM}$$

$$w_u^{AMS} = (u_{AMS} - u_{AMS51})/t_{AMS}$$

$$w_u^{ModExp} = (u_{ModExpAMS} - u_{ModExpAMS51})/t_{ModExp}$$

Table 1.3 shows the values w_u^{AMM}, w_u^{AMS}, and w_u^{ModExp}. Here, we see that the number of VPMADD52 instructions is almost unchanged, but the number of VPADDQ, VPXORQ, and VPSLLQ was increased. Therefore, we predict that implementations with operands in radix 2^{51} will be slower than those in radix 2^{52}.

1.5.2 A "what if" Question: The Potential of FMSA

Table 1.4 is similar to Table 1.3, where we replace the instructions in the MulHighPart with only one VPMADD52HUQ instruction, emulating our new FMSA instruction. Here, the added number of VPADDQ, VPSLLQ, and VPXORQ instructions is no longer needed, and the full power of our AMS can be seen.

Table 1.4 Values of w_u^{AMM}, w_u^{AMS}, and w_u^{ModExp}, when using the FMSA instruction, for different instructions (u) and different operands sizes

Function name	VPADDQ	VPMADD52	VPSLLQ	VPXORQ
AMM3072	0.00	0.01	0.00	0.00
AMM4080	0.00	0.01	0.00	0.00
AMM4096	0.00	0.01	0.00	−0.01
AMS3072	−0.03	0.00	−0.09	−0.10
AMS4080	−0.10	0.00	−0.08	−0.10
AMS4096	−0.10	0.01	−0.08	−0.11
ModExp3072	−0.04	0.00	−0.03	−0.04

1.6   Conclusion

This paper showed a method to use Intel's new AVX512-IFMA instructions, for optimizing software that computes AMS on modern processor architectures. Section 1.5 motivates our prediction that the proposed implementation would further improve the implementations of modular exponentiation described in [7]. As a future research we are aiming on measuring our code on a real processor once it will be widely available.

In addition, we analyzed the hypothetical benefit of using a different radix: 2^{51} instead of 2^{52}. This can significantly improve the AMS algorithm (only) if a new instruction, which we call FMSA, is also added to the architecture. We note that FMSA requires only a small tweak over the current AVX512IFMA, and no new hardware.

Acknowledgements This research was supported by: The Israel Science Foundation (grant No. 1018/16); The Center for Cyber Law and Policy at the University of Haifa in conjunction with the Israel National Cyber Directorate in the Prime Ministers Office.

References

1. Intel ® 64 and IA-32 Architectures Software Developers Manual, Sept 2015
2. Intel ® 64 and IA-32 Architectures Optimization Reference Manual, June 2016
3. Intel® Software Development Emulator, version 8.12.0. https://software.intel.com/en-us/articles/intel-software-development-emulator, Jan 2017
4. Brent, R.P., Zimmermann, P.: Modern Computer Arithmetic, vol. 18. Cambridge University Press, Leiden (2010)
5. Drucker, N., Gueron, S.: [openssl patch] Fast 1536-bit modular exponentiation with the new VPMADD52 instructions. http://openssl.6102.n7.nabble.com/openssl-org-4032-PATCH-Fast-1536-bit-modular-exponentiation-with-the-new-VPMADD52-instructions-td60082.html, Sept 2015
6. Drucker, N., Gueron, S.: Paillier-encrypted databases with fast aggregated queries. In: 2017 14th IEEE Annual Consumer Communications Networking Conference (CCNC), Las Vegas, pp. 848–853, Jan 2017
7. Gueron, S.: Efficient software implementations of modular exponentiation. J. Cryptogr. Eng. 2(1), 31–43 (2012)
8. Gueron, S., Krasnov, V.: Software implementation of modular exponentiation, using advanced vector instructions architectures. WAIFI 12, 119–135 (2012)
9. Gueron, S., Krasnov, V.: Speeding up big-numbers squaring. In: 2012 Ninth International Conference on Information Technology: New Generations (ITNG), Las Vegas, pp. 821–823. IEEE (2012)
10. Gueron, S., Krasnov, V.: [openssl patch] Fast modular exponentiation with the new VPMADD52 instructions. https://rt.openssl.org/Ticket/Display.html?id=3590, Nov 2014
11. Gueron, S., Krasnov, V.: Accelerating big integer arithmetic using intel IFMA extensions. In: 2016 IEEE 23rd Symposium on Computer Arithmetic (ARITH), Silicon Valley, pp. 32–38. IEEE (2016)
12. IEEE Standard for Binary Floating-Point Arithmetic. ANSI/IEEE Std 754–1985 (1985)
13. Menezes, A.J., Van Oorschot, P.C., Vanstone, S.A.: Handbook of Applied Cryptography. CRC Press, Boca Raton/London/New York (1996)
14. Montgomery, P.L.: Modular multiplication without trial division. Math. Comput. 44(170), 519–521 (1985)

Password Security in Organizations: User Attitudes and Behaviors Regarding Password Strength

Tahani Almehmadi and Fahad Alsolami

2.1 Introduction

With the world gradually becoming a global village, cyber security is no longer an alternative; it is a necessity. In Saudi Arabia for instance, the Shamoon virus attack in 2012 that targeted Saudi Aramco, a state-owned oil corporation, and which resurfaced again in late 2016 disrupting operations in several state corporations, forced Riyadh to invest substantially in updating its information technology (IT) infrastructure [1]. Moreover, according to the security organization Symantec, it exceeded the global averages of email incidents of spam and phishing, according to the March 26 Internet security report [2]. In response, the country established the National Authority for Cyber Security as a first step towards fulfilling the aims of Vision 2030 a body meant to enhance data, network, and IT/operating system protection in light of the country threat-laden cyber landscape [3]. In essence, the prioritization of a nationwide cyber security strategy undoubtedly points to passwords as one of the strategies basic elements. Saudi Arabia's Vision 2030 targets diversification of the economy [3]; technology is intended to be a significant contributor regarding enabling the vision. As such, digitization of nearly all of the government's data is certain to increase the country's exposure to hackers. With such important data stored digitally, the importance of passwords as one mode of protecting government data warehouses only increases.

T. Almehmadi (✉)
Technical College for Girls in Jeddah, Technical and Vocational Training Corporation, Jeddah, KSA
e-mail: talmehmadi1@tvtc.gov.sa

F. Alsolami
Faculty of Computing and Information Technology,
King Abdulaziz University, Jeddah, KSA

With the expansion of government digital services to lower bureaucracy, confidentiality and reliability of information systems are critical to the success of government Saudi Arabia as the case in point. Along these lines, passwords are considered as the most feasible method of authenticating the access of individuals. Unfortunately, with the increased sophistication of cyber attacks, such as the spear phishing campaigns like Shamoon in Saudi Arabia that were intended to steal sensitive data, password use as a predominant method of user authentication has been shown to be ineffective thanks to hacker intrusions [4]. This is to say that the choice of password characters, a source of password strength, is a troubling issue. A lot has been said regarding password vulnerabilities, but passwords are and in the near future will be ubiquitous in user authentication. Inherent in passwords is a trade-off between security and usability; where a strong password is difficult for a hacker to guess, it is conversely generally hard for its user to memorize [4]. This presents a dilemma for user-created passwords. Typically, a password ought to be hard to guess and easy to memorize [4]. For passwords guessing to be difficult, its selection should be from a broad domain. Strong passwords are typically random character strings, but users generally cannot memorize them, even though guessing of arbitrary, long, random-character passwords is hard for hackers. The majority of users deliberately opt for weak passwords or those with (very) few characters, due in part to the perception that adhering to best practices in security impedes their workflow. Therefore, identification of the reasons for this insecure behavior among password users and examination of the factors that trigger the behaviors are imperative. These endeavors help inform organizations on effective password policies. Organizational policies on passwords should ideally find a middle ground between safety and usability.

In light of the fact that passwords are extensively used in modern authentication processes, users are expected to select passwords that are hard to guess but easy to remember [4]. However, most users circumvent the rule and use meaningful particulars in their choice of characters, such as their dates of birth and nicknames [4]. As such, this study attempts to evaluate the status quo among password users via a comprehensive questionnaire made up of assorted questions targeting user attitudes and practices as regards password choices.

2.2   Related Work

As Saudi Arabia example shows, a large number of organizations across the globe have been and are continuously relying on IT systems for internal operations. At the same time, the systems interconnectivity has increased so much that the possibility of defacement, theft, intrusion, and other forms of loss has increased. Despite the fact that a number of studies demonstrate that a significant number of security incidents emerge from the internal aspects, organizations tend to be more concerned with vulnerability to external threats. Yan, Blackwell, Anderson, and Grant [5] observe that a lot of the weaknesses of password authentication systems are owed to limitations of human memory. To justify this, the researchers glean from cognitive psychology studies on the minds password remembrance to determine that authentication of passwords involves a trade-off. Specifically, certain passwords are easily remembered but can be easily guessed via searching the dictionary. Other passwords may be guess-proof but hard to remember. Here, human limitation jeopardizes the security of used passwords since a user may write a password somewhere insecure or opt for unsafe procedures for backup authentication in case the password is forgotten. In an empirical investigation of the balance between security and memorability in practice, the researchers compare the effects of offering three different forms of advice on password selection to separate user groups [3]. Contrary to the advice that many sizable organizations give new users on good password selection (being reasonably long with a reasonably large set of characters and being easy to memorize), special characters were not used by anybody in two of the groups. The only group in which they were used was the pass phrase group [3]; it is assumed this group used special characters based on the guidance provided, as they were given examples of punctuation-containing passwords.

Zviran and Haga [4] note that most of the user-selected passwords are made up of meaningful personal information are moderately short, hardly change, have alpha-numeric characteristics and are typically written down. The study's findings further affirm that password selection methods influence password memorizability; recurrent password alteration hinders password recall. In response to such behavior, organizations need to enhance the effectiveness of their in-structive endeavors to raise system users security awareness and put in place guidelines on choice and use of strong passwords. Adams and Sasse [6] further argue that it is very important for security departments to implement a user-centered design method and communicate more with the users to obtain information on their behavior (s) and perceptions with regards to password mechanisms. The researchers note that it is not the user who is the enemy of security; rather, the user is a collaborator in need of applicable information to help preserve system security.

The Martinson study [7] investigates if users adhere to guidelines pertaining to password use. To substantiate his argument, he provides survey questions targeting exploration of password memorization methods often used by (system) users. Findings indicate that most users are irritated by organizational guidelines and prefer to write down passwords. Interestingly, the study finds that users can, on average, memorize five passwords Still, several users use the same passwords for multiple applications. Likewise, Florencio and Herley [8] underscore the significance of passwords for IT system users. Most web users, according to their findings, use poor-quality passwords for account protection, such as in the case of email and social networks. Also highlighted was this critical concern: most web users disremember their passwords, making their accounts susceptible to hacker attacks. Furthermore, knowledge of the users has insufficient knowledge regarding what institutes a secure password. It was also found that the majority of the users are unlikely to change their passwords.

Unlike the above-discussed studies, the study by Gaw and Felten [9] focuses on determining how technology use can improve password use. The researchers adopted a unique approach to data collection by measuring the results of actual login tries. Its findings confirm that users do not rely on technology but their memory when it comes to password remembrance. The findings reveal that password reuse rates increase because of the build-up of more accounts and avoidance of creating new passwords. Even so, users defend the behavior by maintaining that reuse of the same passwords makes password management easy. The contribution of human factors, as Hoonakker, Bornoe, and Carayon [10] note, ought to be considered since end-user behavior can escalate information systems vulnerability. Users either use the same passwords all the time or use passwords that are comparatively uncomplicated.

By deduction, the choice of the right passwords by the users is essential for ensuring the security of systems. Nevertheless, issues such as social engineering, brute force attacks and dictionary attacks [7] are some potential ways to break the passwords. The intruders typically attack the passwords that are simple and consist of words that can be commonly found in the dictionary. In addition, the intruders can also introduce a program in the system that can easily guess the passwords. These programs consist of a large number of words in different languages that belong to distinctive cultures [3, 5]. These threats have compelled the organizations to devise complex passwords.

Several criteria are used for ensuring different levels of password security, including password ownership, password lifetime and password composition [7]. As per the National Infrastructure Protection Center [7], it is comparatively difficult to memorize long passwords, but there are few techniques that can be integrated to ease the difficulty. For instance, the individuals can link the passwords to something that is familiar to them. Selection of passwords is a crucial step in computer security; however, human fallibility makes it almost impossible to follow stipulated policies. Normally, the passwords selected by the individuals are based on meaningful information including pets, places, and people. On the other hand, past research notes that users do not adhere to the policy recommendations as regards passwords. As Adams and Sasse [6] point out, the restrictions imposed by the companies to devise strong and secure passwords may lead to a generation of passwords that are less memorable because most of the users perceive the security mechanism is complex and needless. Therefore, it is imperative for organizations to identify the undesirable behaviors of users with respect to passwords that compromise the security of information systems. It is essential to conduct an analysis of passwords that takes into account both human factors and security aspects. This particular study analyzes the undesirable behaviors of the users with respect to the passwords.

2.3   Methodology

The survey questions are designed to identify the attitudes and behavior of the user regarding the password strength. The survey consists of an assortment of questions that are categorized according to the themes pertaining to undesirable behaviors. Survey questions are categorized into one of six pertinent areas: the respondent's information; password features; password use and re-use; how often new passwords are created and changed; password sharing; and responsiveness to organizations security policies. The questions were reviewed to ensure correct understanding by the respondents by fielding questions to three of the reviewers and adjusting according to the instructions given.

2.3.1   Data Collection

In this particular study, the data is collected through a survey with questions intended for self-completion, which is published via Google Forms. It consists of 36 close-ended questions intended to obtain information regarding unde-

sirable behavior of the users regarding passwords. Initially, we identify the potential undesirable behavior of the users and devise questions according to the themes. Additionally, before the questionnaire is issued, an introduction explains the purpose of the study and emphasizes that all information, including the demographic data and responses, is to be kept confidential. Additionally, the respondents are informed that none of their responses are manipulated in any way.

2.3.2   Research Sample

For this research, the survey is distributed on the random sample through email and social media. 233 participants complete the questionnaires and give their viewpoints regarding undesirable user behavior with passwords.

2.3.3   Data Analysis

In this study, quantitative data were analyzed using frequency distribution and statistical techniques. Microsoft Excel was used to analyze the data gathered from survey questionnaires. In addition, Microsoft Excel was used to process the viewpoints into the statistical format and then represent it in a graphical form, for instance, bar charts or frequency tables (Tables 2.1 and 2.2).

2.4   Results

Table 2.1 Password use practices

1	37% of the participants have 5 accounts or less. 47% of the participants have 6–10 accounts
2	93% of the total population have a set of passwords and reuse them
3	90% of the participants have not used any password management software and may not even be aware of such programs and their relevance
4	64% of respondents prefer using biometrics (i.e., fingerprint or face recognition) instead of a password
5	67% of participants used the same password in their personal, study, or work accounts
6	72% of the total populations use personal information in the composition of the password
7	More than 70% of the participants change their passwords willingly so that they can easily recall it
8	48% of the participants write down their passwords 1–3 times in a day
9	74% of the surveyed populace deny sharing passwords with the workgroup
10	49% of the participants have shared their password with friends, family, co-workers or others
11	Majority of participants indicate that they never change their password

Table 2.2 Password security measures

1	Majority of the participants find it comparatively hard to create a password that adheres to the password rules
2	89% of the participants find creating new passwords continuously is annoying
3	63% of the participants agreed that changing passwords make them more secure
4	53% of the total population find password policies of organizations burdensome
5	51% of the participants find password procedures and instructions to be a nuisance
6	57% of the surveyed follow the password procedures based on the organization's guidelines

2.5   Discussion

Among the different security measures, passwords are deemed to be the most effective way of authenticating the users. Along these lines, the security and effectiveness of the passwords typically rely on the selection of the passwords made by the users, as well as their use and reuse habits. A number of investigations in the past, for example, Duggan, Johnson, and Grawemeyer [11] have noted that employees are the weakest link in the security of the passwords. Even though organizations incorporate the latest and cutting-edge technologies for securing their employee's sensitive information, the actions and behaviors of the users can lead to numerous security failures. This study intends to analyze the behaviors of the end-users towards passwords through an action research. Several interesting findings have emerged during this research. For instance, one of the findings of this study revealed that most of the users have 1–3 passwords for different accounts. Along these lines, the outcomes of the survey highlight that a significant number of the participants write down their passwords nearly 1–3 times per day. Length of passwords plays a crucial role in ensuring their safety. Nevertheless, the responses of the survey indicated that majority of the users create passwords that use a maximum of eight characters with a minimum of four characters, and most of the passwords include letters, numbers, and special symbols. The preceding finding supports the research conducted by Rhodes-Ousley [12], which found that users develop passwords that are shorter in length and comprise characters and words that are readily available in dictionaries. In contrast, passwords that are longer in length and have more characters can be characterized as strong passwords. In the same vein, Herley and Van Oorschot [13] note that strong passwords ought not to be derived from personal information including security numbers and name. If users use information that is not personal, they cannot memorize long or complex passwords, instead choosing to write them down.

One of the findings of the research notes that people have an explicit set of passwords which they reuse for different accounts. These passwords are utilized for 2–5 different accounts. Another finding affirms this as nearly 67% of participants emphasize that they use the same password in their personal, study, or work accounts.

The findings of the survey further noted that most of the participants devise passwords in their native language, and they do not use special symbols while creating passwords for distinctive accounts. This finding of the research is in line with the study by Gulenko [14], which shows that, when users have the choice of selecting their passwords, they typically devise passwords that can be memorized easily and comprise of predictable patterns. The author concludes that this approach makes the passwords vulnerable to several attacks, especially dictionary attacks.

The outcomes of the survey also reveal some astonishing findings. For instance, 44% of the participants have an account with a unique password that is not used in another account. However, these unique passwords are used only in important calculations from the point of view of participants. In this, we find a clear indication of participants' awareness of the importance and sensitivity of passwords.

Another observation is that most of the users find it relatively difficult to create passwords that are in accordance with rules and policies. They also find the password development procedure annoying. In fact, as per the majority of the participants in the survey, the procedures and instructions for creating the passwords are troublesome.

The outcomes of the study reveal that participants change their passwords so that they can easily recall them; however, the participants are not mindful of the negative consequences of not changing their passwords regularly. One of the crucial findings of the research is that most of the users had the propensity to share passwords with their co-workers, relatives, and friends. This finding is in line with the research by Charoen, Raman, and Olfman [15]; the study points out that the practice of sharing passwords may be attributed to a situation where individuals are compelled to share passwords with co-workers. However, this practice makes the passwords vulnerable to external threats. Thus, a password policy requiring frequent modification(s) of passwords could significantly facilitate safeguarding of user's passwords.

2.6   Conclusion

Over the years, most of the attention to enhancing computer and information system security has been placed on software and hardware solutions with a very little emphasis on the people aspect. According to various studies in the past, individuals and the way in which they interact with computer systems can be considered the weakest link in computer information security. The chief goal of this research is to assess the attitudes of end users towards passwords. To achieve this aim, a quantitative research design is adopted based on a survey. The survey questionnaire comprises an array of questions that are categorized according to different user behaviors. In light of the findings of the survey, we conclude that most of the users create weak, relatively simple passwords with short lengths and in native languages for use with work computers. We also found that a significant population of users are likely to use identical passwords for numerous accounts since they want to lessen the burden of recalling different passwords. We also found that the participants are not aware of the threats associated with their attitudes and behaviors toward passwords. By deduction, it is essential for organizations to implement policies and methods to enhance password security, urge users to avoid reusing the same passwords for multiple accounts and prevent the sharing of passwords with others. Policies such as an implementation of two-factor authentication provide an additional layer of security in effect safeguarding authenticated user logins and preventing hackers from backdoor login using stolen credentials.

References

1. Perlroth, N., Krauss, C.: A cyber attack in Saudi Arabia failed to cause carnage, but the next could be deadly. The Independent, 21 Mar 2018 [Online]. Available: https://www.independent.co.uk/news/long_reads/cyber-warfare-saudi-arabia-petrochemical-security-america-a8258636.html. [Accessed 26 Aug 2018]
2. Symantec Corporation: Internet Security Threat Report (ISTR) 2018 | Symantec, Symantec, Mar 2018 [Online]. Available: https://www.symantec.com/security-center/threat-report. [Accessed 28 Aug 2018]
3. A. U. is an analyst for government et al.: Saudi Arabia's new cyber security authority faces an uphill battle | GRI, Global Risk Insights, 27 Nov 2017 [Online]. Available: https://globalriskinsights.com/2017/11/saudi-arabia-cyber-security-authority-challenges/. [Accessed 26 Aug 2018]
4. Zviran, M., Haga, W.J.: Password security: an empirical study. J. Manag. Inf. Syst. 15(4), 161–185 (1999)
5. Yan, J., Blackwell, A., Anderson, R., Grant, A.: Password memorability and security: empirical results. IEEE Secur. Priv. 2(5), 25–31 (2004)
6. Adams, A., Sasse, M.A.: Users are not the enemy. Commun. ACM 42(12), 40–46 (1999)
7. Martinson, K.W.: Passwords: a survey on usage and policy. Technical report, AIR Force INST of Tech Wright-Patterson AFB OH School of Engineering and Management (2005)
8. Florencio, D., Herley, C.: A large-scale study of web password habits. In: Proceedings of the 16th International Conference on World Wide Web, Banff, pp. 657–666. ACM (2007)
9. Gaw, S., Felten, E.W.: Password management strategies for online accounts. In: Proceedings of the Second Symposium on Usable Privacy and Security, Pittsburgh, pp. 44–55. ACM (2006)
10. Hoonakker, P., Bornoe, N., Carayon, P.: Password authentication from a human factors perspective: results of a survey among end-users. In: Proceedings of the Human Factors and Ergonomics Society Annual Meeting, vol. 53, pp. 459–463. SAGE Publications, Sage/Los Angeles (2009)
11. Duggan, G.B., Johnson, H., Grawemeyer, B.: Rational security: modelling everyday password use. Int. J. Hum-Comput. Stud. 70(6), 415–431 (2012)
12. Sasse, M.A., Brostoff, S., Weirich, D.: Transforming the 'weakest link'—a human/computer interaction approach to usable and effective security. BT Technol. J. 19(3), 122–131 (2001)
13. Herley, C., Van Oorschot, P.: A research agenda acknowledging the persistence of passwords. IEEE Secur. Priv. 10(1), 28–36 (2012)
14. Gulenko, I.: Improving passwords: influence of emotions on security behaviour. Inf. Manag. Comput. Secur. 22(2), 167–178 (2014)
15. Charoen, D., Raman, M., Olfman, L.: Improving end user behaviour in password utilization: an action research initiative. Syst. Pract. Action Res. 21(1), 55–72 (2008)

Detecting and Preventing File Alterations in the Cloud Using a Distributed Collaborative Approach

José Antonio Cárdenas-Haro and Maurice Dawson Jr.

3.1 Introduction

With the continued development and improvement of new communication and computing technologies, we are now in a situation where most people and companies have most of their digital data in the Cloud. It is not only the data but now also the application software which is stored and run in the Cloud. This is a game changer and new security challenges have arisen, it is relatively easier now for hackers to compromise large interconnected databases and systems [1, 2, 8, 9, 20]. Malicious users could gain access to files in the Cloud and alter the contents of them to influence the system on their behalf; there are plenty of examples of this kind of attacks, the Democratic party database system was hacked and altered [5], many bank systems have been hacked [3], as well as companies as Facebook, Amazon, Yahoo, among others have had their systems compromised.

The management of the data in the databases, the application software and the verification of their integrity cannot be delegated to only one entity, that is too risky. Here we propose a public protocol to verify and ensure the integrity of the application software and the data in the Cloud. Cloud computing has plenty of advantages, it is not only about creating and/or storing data in remote servers, it also saves a lot of resources at the client side not only in memory but also in processing power or CPU cycles. The Cloud technology provides a lot of services via the Internet, some with a fee and others totally free. This is appealing not only for personal use, all these advantages attract companies and organizations of any size around the world, since it saves time and huge amounts of managing and maintenance costs, and overall Cloud services tend to be more efficient every time. This new model and all these advantages, on the other hand, constitutes a big challenge in data security and privacy. Security breaches are usually a lot harder to be spotted and cause more damage, due to the fact that it represents the access to more data and users; moreover there is the risk of an inside attack from the service provider company. All this and more represent new research challenges in security, privacy, reliability and interoperability [4, 12, 17]. Plenty of vulnerabilities have been detected in the diverse Cloud services, as an example Mulazzani et al. [15] expose the weaknesses of the Dropbox Cloud service. The security breaches of the Apache Hadoop technology used to handle vast amounts of computation and data are also exposed, analyzed and discussed by Hamlen et al. [11] and Moreno et al. [14]. Many research lines are under development to deal with different problems related to the Cloud computing. Here we present a novel algorithm developed to ensure the integrity of the data and applications.

3.2 Previous Related Work

Several algorithms have been developed and implemented to ensure that the access to the data and applications in the Cloud is granted only to the authorized users of that data or applications [19–22]. Some data could be under the shared mode or it could be users private data. Only the authorized accesses must be possible at all times, this one is considered the core or most important aspect of the Cloud computing and services. For that reason research related to this topic has been always under development. Eranna et al. [8] developed an algorithm for integrity verification, their work is based in the Merkle Hash Tree (MHT) construction for block tag

J. A. Cárdenas-Haro (✉)
Department of Computer & Electrical Engineering & Computer Science (CEE/CS), California State University Bakersfield, Bakersfield, CA, USA
e-mail: jcardenas23@csub.edu; jacarde1@asu.edu

M. Dawson Jr.
Illinois Institute of Technology, Chicago, IL, USA

© Springer Nature Switzerland AG 2019
S. Latifi (ed.), *16th International Conference on Information Technology-New Generations (ITNG 2019)*, Advances in Intelligent Systems and Computing 800, https://doi.org/10.1007/978-3-030-14070-0_3

authentication. They need for that multiple auditing tasks but the reliability is not ensured. Qian Wang et al. [20] in their model consider a third party auditor (TPA) to verify the integrity of the data, and they also use the MHT for authentication. The use of a TPA means extra costs and overhead, and there is the risk of attacks from inside, the use of the MHT on top means extra overhead. Other works rely only in encryption but that approach is insufficient in live systems, all the data is decoded on the fly and in a compromised system the hacker would have access to all the data in live mode. Encryption is a very important ingredient here but not enough by itself. Other works base their algorithms in the Trusted Platform Module (TPM) which uses the Kerberos systems to authenticate the users [18].

3.3 Our Algorithm to Avoid the Hash Value Manipulation Attack

In order to detect when a file have been illegally altered, we use here a collaborative pool of nodes in which every node build a reputation through the interactions with its peers. The nodes gain reputation through the honest reporting of the hash values, which proves to be a Nash Equilibrium for correct reporting [6, 10].

We developed an algorithm to address the problem of ensuring the data integrity in Cloud systems. First we need to keep a table with the hash values of all those files that we want to protect, they will be under monitoring. The hash generator could be the SHA-256 or the SHA-512 algorithms; there is always the risk of a coincidental value collisions but the probabilities for this to happen are so minor, insignificant. The verification of the hash values is in a collaborative manner using a distributed authority system. We keep track of the authorized updates of the files, and subsequently the corresponding hash values are recalculated and updated on the tables.

Before the update is authorized it is verified first the consistency of the previous file hash value collaboratively, not every peer has to participate in every verification, the minimum number of peers needed depends on a constant value set as a part of the algorithm.

$$k_v \leqslant V_p \leqslant N_p \qquad (3.1)$$

Where k_v is an integer constant value set as the minimum number of peers required for the collaborative verification; V_p is the number of peer nodes participating in the collaborative verification; whereas N_p represent the total number of peers in the system. This approach prevents the HVMA (Hash Value Manipulation Attack) which is one variation of the Man-in-the-middle attack to replace hash values.

Since there is a lack of a certifying authority, which in this case is an advantage, the policies and rules used are enforced by the nodes through the interaction with their peers. Honest nodes, as a way of protection, need a mechanism to detect malicious users (Sybil nodes) and protect the system against a possible attack.

3.3.1 Our Algorithm for the Verification of Hash Values and for the Reputation System

Our algorithm never restricts a node in any way in participating in the collaborative system for the verification of the hash values and for the analysis of the behavior of the other nodes, which is another of the advantages. We consider every new node trustable enough to participate in the system; although, without enough reputation earned the damage that it may cause is so restricted because that node is not trustable enough. In other algorithms a node first has to be certified as trustable to be admitted into the network, and that may cause a big problem later since the certifier authority may be compromised already. Randomly a pair of nodes are picked to perform the verification of a file hash value, randomly picked as well. Two arrays are used here, one with the hashes of all the files, and the other with the reputation points of all the nodes participating. When the interaction between peers for the verification of certain hash value is over, each node reports the results of that participation to the authority node; Then the authority node or certification node broadcast a report that contains not only the result of the verification of the corresponding file hash value, but also the reputation points earned through that interaction which is based in the game theory algorithm. The report is broadcast to all the other peer nodes which keep track of all the results. Even though the earned reputation points could be different, i.e. asymmetric, the hash value is one, either a new one if the file has been updated by an authorized writer, or the same previous hash if the file has not been altered. Every peer updates the data in their reputation array. There is a set of nodes, from those with the higher reputation, that also play the role of authority nodes. Any anomaly detected is handled by them as explained in the Sect. 3.3.4.

3.3.2 The Arrays Used

We need to build two arrays for our system. First we take the hash values of all the files that need to be secured or protected and we create an array of size M where M corresponds to the total number of files under custody. The values in there are updated only after a verified value of the previous hash value, and the corresponding verification that the related file has been modified by an authorized user. In order to keep track of the reputation of all the nodes we need a second array, in this case of size N where N represents the total number of nodes participating in the system. When a new node is added it gets a copy of these arrays, when a new file is added or discarded, the information is broadcast so every node can update the array.

3.3.3 The Selection Process for the Verification

The authority nodes are a set of nodes selected from those with the higher reputation points. They do the same as all the other nodes but they are also in charge of some extra responsibilities. They randomly have to select one of the files under custody and a pair of nodes that will verify the integrity of that file through the corresponding hash value stored in the array described in Sect. 3.3.2. Then after verifying that the file remains unaltered, they broadcast their results independently to all the other peers in the system as explained in Sect. 3.3.1. Every node keeps copy of the arrays described in Sect. 3.3.2 and they update those correspondingly with the new hash values and the reputation points are updated for the corresponding nodes. The nodes with the higher reputation belong to the group of certification nodes, these perform the same work as the regular nodes plus some extra responsibilities; these are in charge of some administrative duties. They randomly are selecting the files to be checked and the nodes that will verify the hash value of it. Both nodes give the result to the corresponding certification node which will assign the earned reputation points by those nodes. Truthful reporting is proven to be Nash equilibrium [13]. In case of a mismatch in the value reported, the certification node would select another pair of nodes to run the hash value checking, once there is a match, reputation points will be assigned to the previous pair of nodes based on the outcome. Through the broadcast reports the peers obtain the information necessary for the updating of the corresponding values in the arrays.

3.3.4 Detecting and Reporting Inconsistencies in the Files Under Custody

Files are under validation periodically by the authority nodes; they are in charge of randomly selecting the files for verification, as well as the pair of nodes that will run the verification. The average time from validation to validation can be adjusted accordingly to the requirements in the system. Once an inconsistency is detected, i.e. unauthorized alteration or modification of a file, that authority node will report that to all the group of authority or certification nodes, then another of the authority nodes randomly selected will run the verification algorithm all over again. The result will be broadcast to all the authority group; if the inconsistency is real, then the anomaly would be reported to the users of the system as a form of intrusion detection.

3.3.5 Authorized File Updates

When a file has been updated the system check first that it has been done by an authorized user, and at an authorized time; then the authority nodes update the corresponding hash value and broadcast the new value so that all the peers in the system can update their arrays with the new authorized value.

3.3.6 Advantages of Our Algorithm Over Other Solutions

A blockchain model would also work in protecting the integrity of the files but there are some inconveniences with it. First of all it requires a lot of computational power, it has to solve complex algorithms that are of the order $O(n^2)$; also the computations take long time, it could even take several hours to finalize. The scalability is another problem in the blockchain algorithm, the block size limits the number of operations allowed per unit of time, also adding more nodes increases the complexity of the system [7, 16]. Using just a plain checksum comparison is so fragile, it would not be enough, and it is susceptible to the Hash Value Manipulation Attack that is one kind of Man-in-the-middle attack used to replace hash values. Our algorithm is light, it requires $O(n)$ hashing computations in total, and every node will perform just a fraction of that $O(n)$. The scalability is not a problem, more nodes and/or files to be resguarded can be added at any time; it is only about increasing the size of the arrays held by the nodes and giving copies of them to all the new nodes.

3.4 Conclusions

Our algorithm works well as a tool to ensure the integrity of the data, i.e. it detects the unauthorized alteration or modification of the data. The random selection of the file that will be checked as well as the pair of nodes that will perform the verification is performed in $O(1)$ time. Every update in the information in the arrays can also be performed in a $O(1)$ time. The asymptotic time complexity of the hash algorithms, as it is the case of the SHA-256, is known to be $O(n)$ with a space of $O(1)$, so we can say that our algorithm is light and supports good scalability.

The limitations here is that still other parties may gain reading access to the data and remain undetected as long as they do not perform any modification to any of the files under custody; This is still an open problem in this case.

References

1. Ateniese, G., Burns, R., Curtmola, R., Herring, J., Kissner, L., Peterson, Z., Song, D.: Provable data possession at untrusted stores. In: Proceedings of the 14th ACM Conference on Computer and Communications Security, CCS 07, pp. 598–609. ACM (2007)

2. Ateniese, G., Di Pietro, R., Mancini, L., Tsudik, G.: Scalable and efficient provable data possession. In: Proceedings of the 4th International Conference on Security and Privacy in Communication Networks, p. 110. ACM (2008)

3. BBC News: Russian man charged over 'massive' US hack attacks. 10 Sept 2018

4. Bowers, K., Juels, A., Oprea, A.: Proofs of retrievability: theory and implementation. In: Proceedings of the 2009 ACM Workshop on Cloud Computing Security, pp. 43–54. ACM (2009)

5. Bump, P.: Timeline: how Russian agents allegedly hacked the DNC and Clinton's campaign. The Washington Post. 13 July 2018

6. Cárdenas-Haro, J.A., Konjevod, G.: Detecting Sybil nodes in static and dynamic networks. In: Meersman, R., Dillon, T., Herrero, P. (eds.) On the Move to Meaningful Internet Systems, OTM 2010. OTM 2010. Lecture Notes in Computer Science, vol. 6427. Springer, Berlin/Heidelberg

7. Crosby, M., et al.: Blockchain technology: beyond bitcoin. scet.berkeley.edu Applied Innovation Review; Issue No. 2, June 2016

8. Eranna, M., Muarali Krishna, S.: Ensuring the integrity of data storage security in cloud computing. Int. J. Comput. Appl. 4(2) (2012). ISSN:2250-1797

9. Erway, C., Küpkü, A., Papamanthou, C., Tamassia, R.: Dynamic provable data possession. In: Proceedings of the 16th ACM Conference on Computer and Communications Security, pp. 213–222. ACM (2009)

10. Fellman, P.V.: The nash equilibrium revisited: chaos and complexity hidden in simplicity. In: Minai, A.A., Braha, D., Bar-Yam, Y. (eds.) Unifying Themes in Complex Systems, Chap. 13, pp. 105–112. Springer, Berlin/Heidelberg (2011)

11. Hamlen, K., Kantarcioglu, M., Khan, L., Thuraisingham, B.: Security issues for cloud computing. Int. J. Inf. Secur. Priv. 4(2), 39–51 (2010)

12. Juels, A., Kaliski, B., Jr.: PORs: proofs of retrievability for large files. Future Internet 8(3), 1–16. 16p. (2016). In: Proceedings of the 14th ACM Conference on Computer and Communications Security, pp. 584–597. ACM (2007)

13. Kreps, D.M.: Nash equilibrium. In: Game Theory, pp. 167–177. Palgrave Macmillan, London (1989)

14. Moreno, J., Serrano, M.A., Fernndez-Medina, E.: Main issues in big data security. Future Internet 8(3), 1–16 16p. (2016)

15. Mulazzani, M., Schrittwieser, S., Leithner, M., Huber, M., Weippl, E.: Dark clouds on the horizon: using cloud storage as attack vector and online slack space. In: Proceedings of the 20th USENIX Conference on Security; SEC'11, San Francisco, CA. USENIX Association, Berkeley

16. Nakamoto, S.: Bitcoin: a peer-to-peer electronic cash system. academia.edu (2009)

17. Padhy, R.P., Patra, M.R., Satapathy, S.C.: Cloud computing: security issues and research challenges. IRACST – Int. J. Comput. Sci. Inf. Technol. Secur. (IJCSITS) 1(2), 136–146 (2011)

18. Sharma, S., Chugh, A.: Survey paper on cloud storage security. Int. J. Innov. Res. Comput. Commun. Eng. 1(2), 208–213 (2013)

19. Wang, S.-H., Chen, D.-W., Wang, Z.-W., Chang, S.-Q.: Public auditing for ensuring cloud data storage security with zero knowledge privacy. College of Computer, Nanjing University of Posts and Telecommunications, China (2009)

20. Wang, Q., Wang, C., Li, J., Ren, K., Lou, W.: Enabling public verifiability and data dynamics for storage security in cloud computing. In: Proceedings of the 14th European Conference on Research in Computer Security, ESORICS'09. Springer (2009)

21. Wang, C., Wang, Q., Ren, K., Lou, W.: Ensuring data storage security in cloud computing. In: Proceedings of IWQoS09 (2009)

22. Xu, C., He, X., Abraha-Weldemariam, D.: Cryptanalysis of Wang's auditing protocol for data storage security in cloud computing. In: Information Computing and Applications. ICICA 2012. Communications in Computer and Information Science, vol. 308. Springer, Berlin/Heidelberg

Comparing Black Ridge Transport Access Control (TAC), Brain Waves Authentication Technology and Secure Sockets Layer Visibility Appliance (SSL-VA)

Hossein Zare, Mohammad J. Zare, Peter Olsen, and Mojgan Azadi

4.1 Introduction

It has not been more than four decades since the early 1990s when the Internet was privatized and opened to commercial traffic. Nowadays, the Internet has become an essential part of people's daily lives. The Internet became "quietly and unobtrusively an integral component of our home life and our jobs" [24]. Statistics show that on average more than 51% of the world population had access to the Internet in 2017. With this huge development, one can expect to see different sources of crimes in this virtual environment. The authors of a new report published by Cisco [7] claimed that one in five organizations lost customers lost 30% of their revenue and their customers because of an attack. To protect organizational assets and information from these malicious hackers, IT experts have developed sophisticated methods to improve the security of networks and cyberspace. In this paper, we discuss in detail three new technologies: BlackRidge Transport Access Control (TAC), brain wave authentication technology (BAT), and Secure Sockets Layer Visibility Appliance (SSL-VA).

4.2 BlackRidge Transport Access Control Computer Layer and Internet Communication

Before analyzing TAC, we need to review the five computer layers for Internet communication. The physical layer moves "the actual bits between the nodes of networks, on a best effort basis" ([14], p. 224). The ability to transmit bits between a pair of network nodes is an abstraction of this level to the next highest level. The link layer transfers data "between a pair of network nodes or between nodes in a local area network" ([14], p. 224). The network layer, also known as the Internet layer, provides the best effort basis for passing packets between any two hosts. The main protocol provided by this layer is the Internet protocol (IP). The main task of the transport layer protocol (TCP) layer is to "support communication and connection between applications based on IP addresses and ports" ([14], p. 224). A virtual connection established by TCP delivers "all packets guaranteed between a client and server" in an ordered fashion ([14], p. 225). There is no guarantee of delivering data as quickly as possible; for this purpose, another layer is necessary. The fifth layer is the application layer. Based on the services provided by TCP, the task of this layer is to support useful functions; "HTTP is an example that uses TCP and support web browser" ([14], p. 225).

H. Zare (✉)
Department of Health Services Management, University of Maryland University College (UMUC), Adelphi, MD, USA

The Johns Hopkins University, Department of Health Policy and Management, Baltimore, MD, USA
e-mail: Hossein.Zare@faculty.umuc.edu

M. J. Zare
Department of Computer Science and Engineering, Azad University, Yazd, Iran

P. Olsen
Department of Computer Science and Electrical Engineering, University of Maryland Baltimore County, Baltimore, MD, USA
e-mail: MDolsen@sigmaxi.net

M. Azadi
University of Maryland University College, Adelphi, MD, USA

The Johns Hopkins University School of Nursing, Baltimore, MD, USA
e-mail: mazadi2@jhmi.edu

© Springer Nature Switzerland AG 2019
S. Latifi (ed.), *16th International Conference on Information Technology-New Generations (ITNG 2019)*,
Advances in Intelligent Systems and Computing 800,
https://doi.org/10.1007/978-3-030-14070-0_4

4.2.1   Concept and Development

TAC was introduced in 2010 by Black Ridge Technology (BRT), a company located in the state of Nevada [21]. The BlackRidge Technology defense system defends against cyber-attacks by blocking unauthorized access by isolating and cloaking servers and clouds, segmenting networks, and providing identity attribution. TAC stops attackers by identifying authentication on the first packet of network sessions; "First Packet Authentication provides low and deterministic latency." [3].

Using this technique, TAC uses a new, real-time protection system to block or redirect unidentified traffic, including port scanning and reconnaissance [3]. TAC uses "end-to-end across network boundaries with multiple policy enforcement points" with high compatibility with existing network security technologies, such as "IPv4, IPV6 as well as client-server, server-server, cloud and mesh networks" [4]. Here are the main steps of TAC's process (See Fig. 4.1):

- TAC sets up authentication on the first packet of the network, defining a new real time;
- TAC inserts a cryptographically secure code into the first packet of a TCP connection request;
- After receiving the TCP connection request on the server side, TAC extracts, checks, and authenticates the packet;
- Packet is delivered, redirected, or discarded based on the results of the previous step, and connection is permitted or denied;
- TAC records logs for each policy action with information about real-time, unauthorized access, and unidentified requests. It also collects information for early detection from insiders, outsiders, and third parties.

4.2.2   Main Features of the Technology

There are several privileges of having TAC as a defense system for a network, including its ease of use, deployment, and maintenance. TAC operates in a "set it and forget it" mode [3]. It is also compatible with different types of networks, virtual servers, and even cloud configurations as well as physical appliances. The other privilege of TAC is the flexibility of deploying it in front of the existing security technologies to filter unwanted, unauthorized traffic. It can be used inside the network to protect servers or isolate a specific part of a network.

Improving Cybersecurity

TAC isolates and protects services and offers network segmentation and remote office protection. As described, TAC provides a new level of defense to protect and isolate critical services across the enterprise and hybrid clouds. In addition, using a software-based approach, TAC can define access or block network connections by using identity-based access controls. Finally, by using end-to-end security architecture technology and by extending identity across virtual private networks (VPN), TAC can reduce risks by applying policies, permitting, rejecting, or redirecting from a remote office (third parties) into an organization's network. See Fig. 4.2 for a visual representation of the general privilege of TAC [3].

Dynamic Identity Integration

TAC gateways can not only configure static identities but can also perform as a dynamic learning system by integrating with Microsoft Active Directory (MAD) to implement specific policies and directions.

Fig. 4.1 First packet authentication using TAC [4]

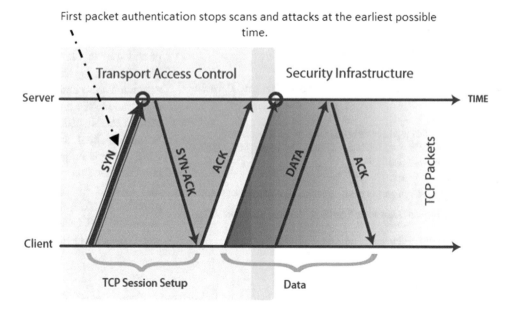

Fig. 4.2 Using BATC
technology to protect, isolate and
segment a cloud [3]

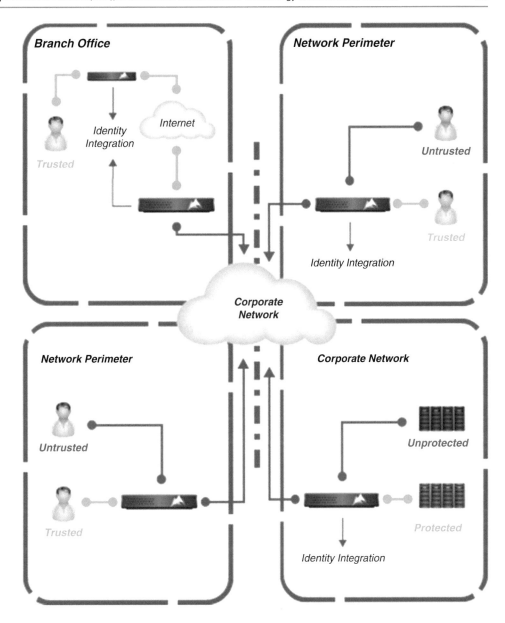

Hiding Networks by Blocking Network Scanning and Identification

The general rule is "when you cannot see, you cannot attack." Using this technique, TAC blocks all servers and networks from unauthorized/unidentified users.

End-to-End Protection and Identity Attribution to Improve the Security of Networks

"BlackRidge works across LAN and router boundaries and automatically adjusts to changing network topologies, ensuring that systems are secure end-to-end" [5]. Moreover, while using identity attribution as the earliest possible time, TAC protects systems from spoofing TCP/IP [3]. Additionally,

each TAC identity generates individually, expires after a short period of time, and cannot be reused [4].

Using the segment implementation and isolating the protected resources, TAC can block WannaCry (the most harmful ransomware ever). It can also protect a network from Denial of Service (DOS) attacks by hiding a network from unauthorized users. TAC also covers vulnerabilities of servers—such as DOS—by adding one more security level [3]. A paper by DeCusatis, Liengtiraphan, Sager and Pinelli [9] showed that TAC "can provide enhanced security in enterprise computing and cloud environments as part of defense-in-depth strategy" [9].

4.2.3 Organization Use and Real-World Example

As BlackRidge was founded in 2010, a report published by BlackRidge shows that BRT and the Cloud Computing Analytics Center (CCAC) are working on the Marist cloud that hosts 25 organizations. After proof-of-concept with support from the New York State Regional Economic Development Council, TAC is in a good position for more progress. Additionally, TAC provides more value to organizations by reducing network traffic, reducing system cost, and reducing compliance costs [2].

4.2.4 Government Role

One of the main responsibilities of the federal government is supporting local products, and BlackRidge is an American organization located in Nevada. With increasing tension, especially in cyberspace, between the United States and Russia and possibilities of Russia using data gathered by Kaspersky, supporting technologies such as TAC is so crucial. In view of the economic impact and new job creation, the state of New York raised more than $3.3 million in capital in 2017 [22]. By spreading this technology nationwide and into international markets, one can expect more economic impact from TAC on the American job market.

4.3 Brain Wave Authentication Technology Concept and Development

Just recently, more than 143 million individual identities were stolen in the United States [13], and more than 500-million compromised in a major security attack on Yahoo. These recently published attacks raised a lot of questions about cybersecurity and authentication technologies. At the same time, most of the stolen personal information is someplace in cyberspace. It requires more sophisticated technology to protect individual information. This is not only a personal issue. These attacks encourage organizations and people to use biometric authentication such as fingerprints, retina scans, and iris identification, but the concern still exists that this information can be stolen or replicated, and hence require certain level of protection [7]. What is the solution?

The idea of using BAT is to "keep the system aware of the person who is using it" [28]. Using these brain patterns, a system can keep tracking a confidential metric about a user and "immediately prompt for reentry of the password whenever the confidence metric falls below a certain threshold" [28]. A National Science Foundation research grant was used to develop a model that uses electroencephalogram (EEG) as a new authentication technology.

Brain wave technologies such as ones using EEG use several unique techniques and privileges including but not limited to the following:

- A brain wave tells more about a user in comparison with a single password. For example, "it could reveal medical, behavioral or emotional aspects of a person" [28].
- In comparison with a few years ago, the EEG devices are "becoming much more affordable, accurate and portable" [28], and by spending less than $100, a user can find an EEG-equipped device that fits on his/her head.
- There are new apps such as brain-sensing apps to capture brain signals, and there is no need to go to clinics to capture EEG; anyone with a cellphone and an app can do it.

4.3.1 Main Features of the Technology

Although number of published reports have suggested that there is enough depth in the EEG recording, this technology is still not available on the market, still needs more research. Palaniappan and Mandic [25] performed a personal identification experiment and reported 95–98% accuracy. Marcel and Millán [18] and Millán et al. [19] reported a 93.4% accuracy rate for personal verification [16]. Claimed that the accuracy of using EEG is still not high enough to allow for a direct application [17].

In recent years, authors of a few published studies have shown that this technique can be used accurately for individual identification. In research performed by Min et al. [20], EEG signals were produced "by a steady-state visually evoked potential-inducing grid-shaped top-down paradigm" and by using "top-down cognitive features analyzed by individuals' differently characterized neuro-dynamic causal connectivity" [20]. They reached a maximum accuracy of 98.6% using 16 brain regions with 5-second intervals of EEG signals. Min et al. believe that the system accurately diagnosed the EEG signals (See Fig. 4.3).

4.3.2 Brain Waves and Cybersecurity

As described, the EEG could potentially provide maximum authentication accuracy by using several variables, such as the following:

- User templates
- Signal frequency ranges

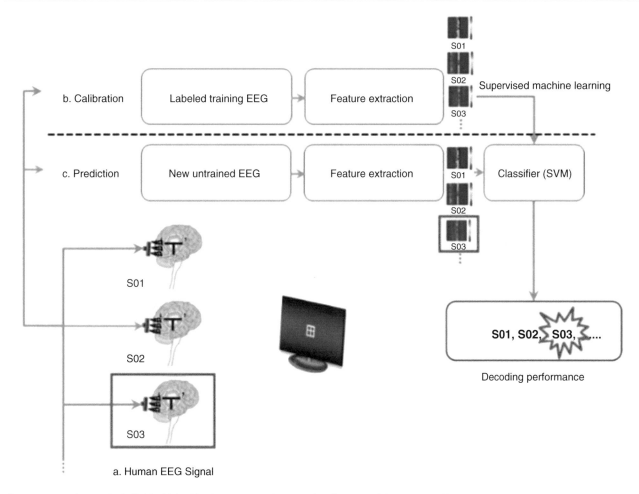

Fig. 4.3 Overview of the individual identification system using causal brain connectivity patterns [20]

- Different parts of the brain can generate different ranges of waves
- This technique could be merged with fingerprint "ridges and valleys on a finger" and work as a unique individual identity [28].

In addition, the EEG technology "can be applied for forensic and security devices" for everyday activities such as opening a door lock [20]. It also can be connected to remote control to turn on a TV or even a car. In advanced models, by adding a number of rows and columns, the sensitivity of the system can potentially be increased for organizations such as the FBI, military, or people who work with highly classified information.

This system requires further improvement with more research and development, such as using different technology to record brain waves, for example, functional near-infrared spectroscopy, which has a higher signal-to-noise ratio than EEG [28]. The most recently released technology—steady-state visual evoked potential—was developed in the Korean language; more research is needed to develop this technology in English [20].

4.3.3 Organizational Use and Real-World Example

Some researchers suggest using multimodal interaction using brain waves. This model benefits from three components of brain–computer interface using a combination of EEG, MEG, NIRS, fMRI, ECOG, and MEA [15]. As presented in Table 4.1, each of the brain activity measurements has some advantages and disadvantages, and more research is needed to find the most accurate, cost-effective model.

4.3.4 Government Role

One of the main roles of the federal government is to pay attention to the features of the mentioned techniques, how these technological devices are performing, on which computing environment should it be placed, and how they are the most accurate and easy-to-use technologies.

In comparison with other technologies, such as finger-prints, voice recognition, and facial recognition, which or-

Table 4.1 Comparing brain activity measurement methods [15]

	EEC	MEG	NIRS	fMRI	ECoG	MEA
Deployment	Noninvasive	Noninvasive	Noninvasive	Noninvasive	Invasive	Invasive
Measured activity	Electrical	Magnetic	Hemodynamic	Hemodynamic	Electrical	Electrical
Temporal resolution	Good	Good	Low	Low	High	High
Spatial resolution	Low	Low	Low	Good	Good	High
Portability	High	Low	High	Low	High	High
Cost	Low	High	Low	High	High	High

Note: EEG electroencephalography, *MEG* magnetoencephalography, *NIRS* near infrared spectroscopy, *fMRI* functional magnetic resonance imaging, *ECoG* electrocorticography, *MEA* microelectrode (or multielectrode) array

ganizations already use for forensic investigations or in the computer industry (for example, a new version of ASUS computers uses facial recognition as an authentication system), brain wave technology needs more time and research to be commercialized. Most of the activities in the real world are presented as research studies in laboratory environments. Here are a few examples: (a) using EEG to move a robot between several rooms using brain wave technology [19], (b) using technology (steady-state visually evoked potential) as an authentication technology with 98.6% accuracy [20], and (c) using EEG as a biometric signature [16].

4.4 Secure Sockets Layer Visibility Appliance Concept and Development

As described in the introduction section of this paper, Transport Layer Security (TLS) "support[s] communication and connection between applications based on IP addresses and ports" ([14], p. 224). TLS and Secure Sockets Layer (SSL) are frequently called SSL. This standard security technology establishes an encrypted link between a server and a client. A client is a Web server and browser or a mail server and mail client, such as Outlook.

Sensitive information such as personal information (e.g., social security number, credit card numbers, etc.) are transmitted insecurely with a plain text format. This plain text is vulnerable to several types of cyber-attacks and could be compromised.

Basically, SSL works as a security protocol. "This protocol determines variables of the encryption for both the link and the data being transmitted" [23]. SSL requires having a secure connection browser and the server's required SSL certificate. In this section, we discuss SSL-VA in detail and ways this application provides a secure environment for transferring information.

Symantec Corporation developed SSL-VA models SV3800, SV3800B, and SV3800B-20. According to a Symantec [26] report, this product provides two main functions after being deployed within a network. The first function is called *SSL inspection*. This function enables

"other security appliances to see a non-encrypted version of SSL/TLS traffic that is crossing the network" [8]. The second function acts as a policy control point. This function enables "explicit control over what SSL/TLS traffic is and is not allowed across the network" [8].

SSL-VA provides a non-encrypted version of SSL/TLS but keeps the SSL/TLS end-to-end connection between a client and the involved server. The SSL-VA appliances are also compatible with existing security devices, such as the following:

- Intrusion prevention systems,
- Intrusion detection systems,
- Data loss prevention systems, and
- The Network Forensic appliance [8].

SSL-VA is capable of working in three modes: passive-inline, active-inline, and passive-tap.

4.4.1 Main Features of the Technology

SSL-VA is a newly offered technology by SSL, a leader in enterprise security. This application was developed by *Brian Capital* in March 2015 [11]. The Blue Coat SSL (BC-SSL) Visibility Appliance "decrypts multiple streams of SSL content across all network ports to provide intrusion detection and prevention (IDS/IPS), logging, forensics, and data loss prevention" [6]. The following are the main benefits and features of this new technology.

Improving Line-Rate Network Performance
The BC-SSL sends non-SSL flows to the attached security appliance and minimizes delay because of network traffic. Additionally, SSL-VA could potentially be set up to analyze up to 6,000,000 simultaneous TCP flows. Both functions reduce network traffic and improve network performance.

Application Preservation
SSL-VA generates a TCP stream with the packet headers and delivers intercepted plain text. This function helps intrusion

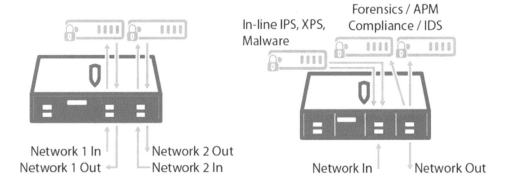

Fig. 4.4 Multiple segment support and port mirroring in Blue Coat Secure Sockets Layer [7]

A. Multiple Segment Support   **B. Port Mirroring (Decrypt once, feed many)**

detection systems and intrusion protect systems to provide better services for SSL-encrypted traffic [6]. This function not only cuts the complexity but also increases the SSL visibility (see Fig. 4.4; compare application delivery controllers and Blue Coat). As described in the previous section, by using multiple segment support and port mirroring, Blue Coat cuts the complexity [1, 6].

By using input aggregation technology, the SSL "allows aggregation of traffic from multiple network taps onto a single passive-tap segment for inspection" [26], and the output mirroring functions allow SSL-VA to feed up to two attached passive security appliances in addition to one primary security appliance.

BC-SSL Visibility Appliance has a powerful Web-based management user interface, with the capability of using SSL policies and e-mail alerts. It is also flexible enough to work with inline and tap models. Finally, the main capacity of the BC-SSL Visibility Appliance is "recording session log details of all SSL flows," [27], the log record's start time, segment ID, send-receive IP port, destination port, domain name, certification status, cipher status, action, and status. It helps IT administrators trace suspicious activity trends and patterns to recognize types of cyber-attacks [27].

4.4.2 Improves Cybersecurity

A combination of control and visibility of outbound encrypted traffic, using other state-of-the-art security devices for encrypting data center traffic, and being "the world's best web traffic classification system" [6], placed the BC-SSL Visibility Appliance as one of the most secure and easy-to-use providers. A new article by Durumeric et al. [10] categorizes Blue Coat as having the second largest fingerprint after Avast Antivirus. At the same time, Blue Coat received the highest grade for using mirror client ciphers with the highest secure TLS interception middlebox performance [6, 10].

4.4.3 Organization Use and Real-World Example

SSL-encrypted traffic is a widely spread encryption technology with enough capacity to quickly grow in the market in the United States and at the international level. In spite of all the advantages of SSL, however, SSL "presents a blind spot for current security tools and applications as malware and advanced threats use SSL to hide from detection" [12]. Addressing this dilemma effectively is an essential key to reducing the risk of damage due to cyber-attacks.

4.5 Comparison and Requirements

In accepting cybersecurity as a public good, governments need to follow different policies and principles to fix market failures and to enhance national security. The first and most important part is law, regulations, and policies. Research shows that most policies create more barriers than laws and regulations and need more clarification. The second important issue involves changing behaviors in private companies, establishing minimum standards, giving tax credits to private companies, requiring liability insurance, and establishing cyber shelters for small and medium-size companies. Finally, establishing rules and regulations nationwide is essential, but an international agreement will be required to have more secure networks. The United States government can lead this effort to create a safe and more secure world to live in.

The world needs a new agreement on how to use the Internet. Any agreements might potentially reduce private or business use of the World Wide Web, but these malicious activities and even spying on citizens must be stopped. In spite of international agreements and regulations, there is an essential need for local regulations.

4.6 Conclusion

Cyberspace allows for the possibility of attacks from anywhere around the world and from any nation. We used several main indicators to compare BlackRidge TAC, brain waves technology, and SSL-VA technologies. TAC uses the first packet and defines a new real time to identify authorized users. Several ongoing projects use brain waves with grand-averaged technology and EEG to develop an authentication mechanism. SSL-VA uses advanced technology to decrypt packets between servers; it also benefits from input aggregation and output-mirroring technology to decrypt packets (See Tables 4.2 and 4.3 for more details).

We used CIA indicators to categorize these technologies and found that SSL provides a much more secure environment than other technologies.

Table 4.2 Comparing BlackRidge TAC, Brain Waves Technology, and SSL-VA

Tools	Developer	Features
BlackRidge Transport Access Control	BlackRidge Technology, a company in Nevada [2]	1. Sets up authentication on the first packet of the network and defines a new real-time. 2. Inserts a cryptographically secure into the first packet of a TCP connection request. 3. After receiving the TCP connection request on the server side, BTAC extracts, checks, and authenticates the packet. 4. The packet is delivered, redirected, or discarded based on the results of the previous step, and the connection is permitted or denied.
Brain Wave Authentication Technology	There are several products, such as grand-averaged topographies developed by a Korean researcher in 2017. Electroencephalogram (EEG), as a new authentication technology, is still being developed by a group of researchers at Texas University.	1. EEG tells more about a user in comparison with a single password; for example, "it could reveal medical, behavioral or emotional aspects of a person" [28]. 2. "This technology is becoming much more affordable, accurate and portable" [28]. 3. The grand-averaged topographies "can be applied for forensic and security devices" and for every-day activities such as opening a door lock [20].
Secure Sockets Layer (SSL) Visibility Appliance	Symantec Corporation developed Secure Sockets Layer Visibility Appliance models SV3800, SV3800B, and SV3800B-20.	"Decrypts multiple streams of SSL content across all network ports to provide intrusion detection and prevention (IDS/IPS), logging, forensics, and data loss prevention" [6]. Main features: *Improving line-rate network performance* *Application preservation* *Input aggregation and output mirroring* *Improving data security center*

Table 4.3 Comparing BlackRidge TAC, Brain Waves Technology, and SSL-VA, in viewpoint of cybersecurity

	Cybersecurity				
	Confidentiality	Integrity	Availability	Use by organization	Rank
BlackRidge Transport Access Control	++ (2)	+ (1)	+ (1)	Still in development process. Approved by Cloud Computing Analytics Center by the state of NY.	3
Brain wave Authentication Technology	+++ (3)	+ (1)	+ (1)	Most activities in the real world presented as research studies in laboratory environ-ments, e.g.: 1. Using EEG to move robot [19] 2. EEG as a password 3. EEG to open home door	2
Secure Sockets Layer (SSL) Visibility Appliance	+++ (3)	++ (2)	++ (1)	Blue Coat Systemic is one of the well-known technology companies and is used at the local and international level.	1

Note: To compare the impact of the selected technologies on cybersecurity, I scored them by "confidentiality," "integrity," and "availability" with +++ for the highest score and + as the lowest impact. These scores are based on my understanding of the technology. Based on these scores, the most secure technology is Secure Sockets Layer, then brain waves, and finally BlackRidge Transport Access Control.

References

1. Appliance and Gemalto SafeNet Luna SP: Gemalto.com. Retrieved September 30, 2017 from https://safenet.gemalto.com/resources/solution-brief/data-protection/Blue_Coat_SSL_Visibility_Appliance_-_Luna_SP_-_Solution_Brief/?langtype=1033

2. Black Ridge Technology: Transport Access Control, revision 12. blackridge.us. Retrieved September 30, 2017 from https://webcache.googleusercontent.com/search?q=cache:rMJpKmkXG0UJ:https://wikileaks.org/hbgary-emails//fileid/15265/5402+&cd=8&hl=en&ct=clnk&gl=us (2010)

3. Black Ridge Technology: Identity-based network and cyber defense. blackridge.us. Retrieved September 30, 2017 from https://www.blackridge.us/images/site/page-content/BlackRidge_Feature_Datasheet.pdf (2017)

4. Black Ridge Technology: Black Ridge Technology Transport Access Control: overview. blackridge.us. Retrieved September 30, 2017 from https://www.blackridge.us/images/site/page-content/BlackRidge_TAC_Overview.pdf (2017)

5. Black Ridge Technology: BlackRidge Enterprise Manager Datasheet. blackridge.us. Retrieved September 30, 2017 from https://www.blackridge.us/images/site/page-content/BlackRidge_Feature_Datasheet.pdf (2017)

6. Blue Coat System Inc: Blue Coat SSL Visibility Appliances SV1800 / SV2800 / SV3800. Retrieved September 30, 2017 from http://webcache.googleusercontent.com/search?q=cache:hHQxAMIfeEEJ:me.westcon.com/documents/48993/bcs_ds_SSL_Visibility_SV1800-SV2800-SV3800_EN_v2e.pdf+&cd=10&hl=en&ct=clnk&gl=us (2013)

7. Cisco: The Zettabyte Era: trends and analysis. Cisco Visual Networking Index™. Retrieved September 13, 2017 from https://www.cisco.com/c/en/us/solutions/collateral/service-provider/visual-networking-index-vni/vni-hyperconnectivity-wp.pdf (2017)

8. Corporation Symantec: Symantec Corporation SSL Visibility Appliance, document revision 12/22/2016. Retrieved September 30, 2017 from https://csrc.nist.gov/csrc/media/projects/cryptographic-module-validation-program/documents/security-policies/140sp2821.pdf (2016)

9. DeCusatis, C., Liengtiraphan, P., Sager, A., Pinelli, M.: Implementing zero trust cloud networks with transport access control and first packet authentication. In: 2016 IEEE International Conference on Smart Cloud (SmartCloud), pp. 5–10. IEEE (2016)

10. Durumeric, Z., Ma, Z., Springall, D., Barnes, R., Sullivan, N., Bursztein, E., et al.: The security impact of HTTPS interception. In Network and Distributed Systems Symposium. Retrieved September 30, 2017 from https://jhalderm.com/pub/papers/interception-ndss17.pdf (2017)

11. Esentire: Blue Coat expand security industry's largest collaboration for encrypted traffic. Esentire.com. Retrieved September 30, 2017 from https://www.esentire.com/news-and-events/coverage/blue-coat-expands-security-industrys-largest-collaboration-for-encrypted-traffic/ (October 2015)

12. Gemalto: Ensuring Trust in SSL-encrypted Networks: Blue Coat SSL Visibility https://www.bluecoat.com/ko/documents/download/baa2863d-c45d-4338-9db8-043ceda0cda3 (2017)

13. Gleason, S.: Equifax Shares Tank After Company Announces Data Breach of 143 Million Customers. TheStreet.com. Retrieved September 13, 2017 from https://www.thestreet.com/story/14298310/1/equinox-shares-tank-after-company-announces-data-breach-of-143-million-customers.html (September 2017)

14. Goodrich, M., Tammasia, R.: Introduction to Computer Security. Pearson Education Inc, Gambrills (2011)

15. Gürkök, H., Nijholt, A.: Brain–computer interfaces for multimodal interaction: a survey and principles. Int. J. Hum. Comput. Int. **28**(5), 292–307 (2012)

16. Hu, J.F.: New biometric approach based on motor imagery EEG signals. In: 2009 International Conference on Future BioMedical Information Engineering (2009)

17. Khalifa, W., Salem, A., Roushdy, M., Revett, K.: A survey of EEG based user authentication schemes. In: 8th International Conference on Informatics and Systems (INFOS), p. BIO-55. IEEE (2012)

18. Marcel, S., Millán, J.D.R.: Person authentication using brainwaves (EEG) and maximum a posteriori model adaptation. IEEE Trans. Pattern Anal. Mach. Intell. **29**(4), 743–752 (2007)

19. Millán, J.R., Renkens, F., Mourino, J., Gerstner, W.: Noninvasive brain-actuated control of a mobile robot by human EEG. IEEE Trans. Biomed. Eng. **51**(6), 1026–1033 (2004)

20. Min, B.K., Suk, H.I., Ahn, M.H., Lee, M.H., Lee, S.W.: Individual identification using cognitive electroencephalographic neurodynamics. IEEE Trans. Inf. Forensic Secur. **12**, 2159–2167 (2017)

21. Mirace, M.: BlackRidge Technology: Beyond the Firewall. Blackridge.us. Retrieved September 30, 2017 from https://www.blackridge.us/images/site/page-content/BlackRidge_and_HPCNY_Success_Story.pdf (2017)

22. Miracle, M.: BlackRidge Technology. Hpc-nyy.org. Retrieved September 28, 2017 from https://hpc-ny.org/collaboration/blackridge-technology/ (February 2017)

23. Odigicert: What is an SSL certificate and how does it work? digicert.com. Retrieved September, 2017 from https://www.digicert.com/ssl/ (2017)

24. Okin, J.R.: The Internet Revolution: The Not-for-Dummies Guide to the History, Technology, and Use of the Internet. Omnigraphics, Detroit (2005)

25. Palaniappan, R., Mandic, D.P.: EEG based biometric framework for automatic identity verification. J. VLSI Signal Process. Syst. Signal Image Video Technol. **49**, 243–250 (2007)

26. Symantect Corporation: Symantec SSL Visibility SV800/SV1800B/SV2800B/ SV3800B/SV3800B-20 Remove the Security Blind Spots Created by Encrypted Traffic. Retrieved September 30, 2017 from http://www.netfos.com.tw/symantec/Datasheet%20to%20partner/Blue%20Coat%20products/EN%20version/SSL/SSL%20Visibility%20Data%20Sheet.pdf (2016)

27. Symantect Corporation: Seven Reasons to Deploy Blue Coat SSL Visibility with Application Delivery Controllers. Retrieved September 30, 2017 from: https://www.symantec.com/content/dam/symantec/docs/solution-briefs/seven-reasons-to-deploy-blue-coat-ssl-visibility-with-application-delivery-controllers-en.pdf (2017)

28. Watson, G.: Could brainwaves replace your passwords? Texas Tech University. Retrieved September 28, 2017 from http://today.ttu.edu/posts/2016/09/brain-waves (September 2016)

Comparing Cellphones, Global Positioning Systems (GPSs), Email and Network and Cyber-Forensics

5

Hossein Zare, Peter Olsen, Mohammad J. Zare, and Mojgan Azadi

5.1 Introduction

"Computer forensics implies a connection between computers, the scientific method, and crime detection" [16]. The main objective in the field of computer forensics is to catch or recover any possible, seen information using all available pieces of evidence. Specialized software and skill/knowledge of using computer forensics techniques are two important factors behind successful analysis. Professional investigators using appropriate and approved software technologies can analyze a computer system, hard drive, clouds or other storage to detect hidden or deleted folders or files [18].

Today, investigators have varieties of devices and storage media at their disposal. There are also other running wireless devices such as cell phones, tablets and laptops up for consideration. People able to use internal (e.g. 4 TB home cloud) or external professional clouds (e.g. Apple clouds, Google drive, Dropbox or institutional clouds). Besides magnetic media, there are also optical drive such as DVD, CD, Flash drive and SD cards are available for investigation. There are hundreds of internet communication channels such as "WhatsApp", "Skype", "WeChat", etc., with streams of available communication data to a network. Each of these sources of evidence has their own challenges and required unique techniques for investigation. If a wireless device is connected to the internet, there is possibility of transferring or deleting evidence. They should be inaccessible at the early minutes of any investigation. For any desktop or laptop, the router or modem must be disconnected to keep evidence inaccessible for any potential criminal interventions [19].

For this paper, we focused on four sources of data which could be used in a digital forensics investigation: Cellphone, Global Positioning System (GPS), Email and Network.

H. Zare (✉)
Department of Health Services Management, University of Maryland University College (UMUC), Adelphi, MD, USA

The Johns Hopkins University, Department of Health Policy and Management, Baltimore, MD, USA
e-mail: Hossein.Zare@faculty.umuc.edu

P. Olsen
Department of Computer Science and Electrical Engineering, University of Maryland Baltimore County, Baltimore, MD, USA
e-mail: MDolsen@sigmaxi.net

M. J. Zare
Department of Computer Science and Engineering, Azad University, Yazd, Iran

M. Azadi
University of Maryland University College, Adelphi, MD, USA

The Johns Hopkins University School of Nursing, Baltimore, MD, USA
e-mail: mazadi2@jhmi.edu

5.2 Cellphone and Cellular Network

Evidence and data can be stored not only in "cellphone" or "memory card" but also on a cellular provider's network. One needs to see the layout of a cellular network. Individual cells are backbone of any cellular network, "each cell uses a predetermined range of frequencies to provide service to distinct geographical area" [19]. Cells have varieties of sizes and shapes and can cover a city block or specific square miles (See Fig. 5.1).

5.2.1 Cellular Network Component and Their Types

With Global System of Mobile Communication (GSM) or Code Division Multiple Access (CDMA) customers can

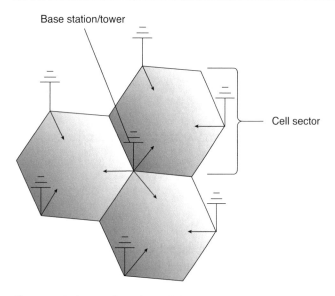

Base station/tower

Cell sector

Fig. 5.1 The layout of a typical cellular network [19]

talk. Service (SMS) and Multimedia Messaging Service (MMS), and they can send/receive text messages with limited characters (e.g. 160 characters for SMS). As described, cellular networks will use four different ways to transmit data; CDMA, GSM, Integrated Digitally Enhanced Network (iDEN), and Prepaid Cell Phones. Knowing this information, an investigator can identify a usage area received and data held by subscribers and providers [8, 20].

5.2.2   Cellphone Operating System

Forensics examination are significantly impacted by phone's operating system (OS). An OS shows that how data is stored and created. Most popular operating systems include Apples iOS, Blackberry OS, Google Android and Windows CE. Each has its own way to perform storing and generating data [9]. A forensic investigator needs to consider what specific criteria of an OS can do best during forensic data collection phase [19].

5.2.3   Cellphone Evidence

Cellphone information can be stored at multiple locations within a handset and its related network. Table 5.1 presents some of the potential items that may be considered as evidence [19].

The most important indicators any forensic investigator needs to be considered is encryption technology and cell PIN's (personal Identification Number). Using a wrong PIN number can result in a phone being locked out. Unlocking a cellphone requires a Personal Unlock Key (PUK) code that is only known by the an owner or a SIM card's provider [2].

"Predictive text" is another interesting technology in use by cellphones. This technology guesses words that most likely will be used with previous-used words stored in user's data dictionary (mobile phone, dictionry.org, 2009, [27]). A forensic expert can use this "learning capability of mobile dictionary" for finding slangs, abbreviations and words not limited to just abbreviations, email addresses, URL addresses, GPS and locations that has been searched by cellphone users. Predictive text is especially important for investigating specific concern and sue only for that type of concern (such as drug trafficking), where there are specific abbreviations [11].

5.2.4   Call Detail Records (CRD)

The CRD is another important piece of evidence for a forensic investigation. Providers use cell phone Call Detail Records to check system performance and for troubleshooting. CRD data include;

- Call duration, data and time of each call;
- Incoming and outgoing calls;
- Identity of connecting towers (originating tower and terminating tower) and possible locations of call participants.

An investigator needs to consider the difference between a CRD and a subscriber's information. Individual information (address, name, payment system, internet account, email address, services, social security number and bank account) are registered as subscriber information, available only during specifics period depending upon a provider's policy. For instance, most providers keep SMS for just 7 days but maintain a subscriber's records (especially billing and payment) longer [22]. This large stock of billing data can become a valuable source of detailed information during an investigation [9]. Using this technology in combination of Human activity-travel behavior (ATB) an investigator is able to predict a user's habitual daily, weekly, monthly, and seasonal routines [10].

5.2.5   Collecting and Handling Cellphone Evidence

Cellphone evidence must be carefully preserved. This means protecting the phone from both physical and electronic access. Phones may be very vulnerable Many phones can be accessed from the mobile network, some even when the phone appears to be turned off.

"*Faraday bags*" or "*Arson cans*" can isolate the phone from the network, preventing malicious remote-based hackers or providers from "wiping" the phone. Investigators may

Table 5.1 Sources of cellphone evidence

Call history	Text message	Email	Application
Video Picture Contact numbers Cell calendar	Voice memo Text message including deleted text message GPS information	Any documents Chat (text and history) Browser history	Web-based phone applications (WeChat, What's App, etc.) Social media application (download/upload files and phots, etc.) Connection time and GPS

also want to turn off the phone, but this is not always the case. This is especially so if the phone requires a PIN to restart. For example, the IMEL can be used to find the phone manufacturer.

Depending on the phone, an investigator may need forensic tools to acquire the data and document details such as text, voicemails, photos, contact numbers and internet-based phone applications [9].

5.2.6 Subscription Identity Modules (SIMs)

Subscription Identity Modules are another type of evidence that include valuable information such as International Mobile Subscriber identity (IMSI) [7], type of provider, Integrated Circuit Card Identifier (ICC-ID), user's language preferences, phone location, user's contact numbers, user-dialed phone numbers, and SMS (even deleted SMS from provider).

5.2.7 Cell Phone Acquisition

Cell phone acquisition could be performed physically or logically by the logical model of the phone's own software. Using forensic tools an investigator can capture and copy bit-for-bit data that includes deleted files. However, in using the logical model (without a forensic tool) only existing files and folders can be copied. If time is critical (for example looking for a missing child) each method can be prioritized by an investigation team [19].

5.2.8 Cellphone Forensic Tools

In case of forensic investigation, cellphone forensic tools depend on cell phones model and vendors (e.g. Apple, Android Samsung, LG). Tools available to investigator export include [9]:

BitPim, is a useful tool for CDMA phones such as LG and Samsung [1];
Oxygen Forensic Suite, a powerful forensic program for cell phones that can collect any type of phone data including SIM card, phonebook, video, photo, contact list, Java application, SMS, and deleted SMS [13];

Cellebrite Universal Forensic Extraction Device (UFED), is a stand-alone self-contained; hardware device that supports more than 2500 phones, and available in most cell phone stores, used for transferring data from an old cellphone to a new one [15];
EnCase, is yet another tool that collects and reviews data from a cell phone and a tablet.

5.3 Global Positioning Systems (GPSs)

The Global Positioning Systems (GPSs) can be a tremendous source of information and possible evidence. GPS can be used to show the location and device activities. GPS also can be used to identify possible "suspect destination[s]". Depending on the type, following information can be captured by GPS [19]:

- Cell phone logs;
- SMS message (recently most GPS can connect to the cell phone and read messages);
- Photos and image;
- Address book;
- Previous address used by driver;
- Home address;
- Parking address;
- Car dealership.

5.3.1 Types of GPS

From the 27 GPS satellites encircling around the US only 24 of them are in use at any one time, three are held in reserve. GPS satellites using mathematical models called trilateration [4]. There are four types of GPS and each of them store some type of information.

Simple GPS

This type of GPS is designed to guide a driver/user from one point to another point. Most of these kinds of GPS store information such as [14]:

- TrackPoint;
- Waypoint;
- Track logs.

Smart GPS

These units are categorized in two different storage device styles; "automotive and USB mass storage device". Traditionally GPS devices usually have either internal storage or external storage on an SD card or USB drive. In addition, many GPSs can "save" favorite places, view photos and play MP3.

Hybrid GPS

These GPSs are a great source of evidence. In addition of their GPS functionality, these devices can connect to cellphones using Bluetooth technology. This GPS style can also be considered as a secondary source of cellphone data, including:

- Call logs;
- Cellphone address books;
- Cellphone SMS records;
- "MAC address of up to 10 last phones that have been connected to the unit [14].

Connected GPS

This is a more advanced system of GPS, providing all the functionality of hybrid GPS, but can be connected to google maps and traffic information centers. Using a traffic information center requires a subscription. Car owners' subscriptions can be valuable sources of information. Some new connected GPS systems have voice recognizer functions that can record sounds, including speech, from inside the vehicle. Users can control the general system parameters, but once those are set, these systems collect information without any further specific user authorization. Data, once collected, cannot be modified by the user.

Track Log

A track log retraces a user path and stores comprehensive a list of TrackPoints. This log is a great source of evidence for forensic investigations.

Waypoint

Waypoints are user-generated data that specify intermediate points along a route.

5.3.2 Forensic Investigation and GPSs

As discussed previously, GPSs resemble cellphone forensic investigation. As with mobile phones, investigators should take care to protect external memory devices such as SD cards or USB drives. A real-world example of GPS forensic investigation is the case of Las-Vegas dancer Debbie Flores-Narvaez, where police found her body using GPS evidence from the GPS in a rented U-Haul truck [6].

5.4 Emails and Internet

Emails and the Internet are two of the most important sources of information for evidence collection. By default, email is only for an intended recipient. However, emails can be retrieved from multiple Internet locations even, away from a specific user's computer.

Email systems employ several design paradigms. Two of the most common are host-based systems such as Windows Outlook and Apple's AppleMail, web-based systems such as Hotmail, Gmail and Yahoo. Host-based systems are often easier to analyze, especially if investigators can get physical access to the host's storage devices. Email uses different protocols to send and receive messages. Three popular protocols are: Simple Mail Transfer Protocol (SMTP), Post Office Protocol (POP), and Internet Message Access Protocol (IMAP). SMTP is used to send messages between computers; POP is used for downloading messages to hosts from servers, and IMAP is used to synchronize mailboxes on hosts and servers [19].

Email messages have three main parts: a header, a body, and (optional) attachments. Headers give the address of the originator, the intended recipient, and the path the message took between the two, among other things. These details can make headers particularly valuable for forensic investigation.

Figure 5.2 shows an example of an email header.

The message id 445-243-800-00AA is a unique number assigned by mailer01.example.net [24].

Fig. 5.2 Example of an email header

```
▪ Date: Sat, 24 Nov 2035 11:45:15- 0500

▪ Received: from mailer01.example.net (mailer01.example.net
  [192.168.01])

      by ixde-df8.example.com (Internet Inbound) with ESMTP id
  44524380000AA   for <recipient@example.com>; Sat, 24 Nov 2035
  11:45:15 -0500 (EST)

▪ Resent-Date: Sat, 24 Nov 2035 11:45:15 -500
```

Specifications for these protocols and formatting of email messages themselves are given in documents called "Requests for Comments" (RFC)s. These are available on the Internet.

5.4.1 Email as a Source of Evidence

Any email is a potentially unique source of information. Potential evidence that can be recovered by any email includes:

- Contents;
- Sender's and receivers' email addresses;
- Sender's and receivers' organizational and domain identification (".gov", ".edu", etc)
- Time and date of origination and reception.
- The path taken between sender and receiver, including time stamps along the way.
- Computer IP addresses;
- Attached files (photos, documents, video and audio files);

Emails can be found in multiple locations, example include senders' and receivers' computers; servers; cellphones; logs, queues and backup media. Some systems save "deleted" messages until they are explicitly purged. Comparing these different sources with modern forensic tools can help investigators recover emails from a user's machine or account even after the user believes they have been deleted.

Email can provide a trove of information but recovering it may not be easy. There are several well-known challenges including (Spoofing, Shared Accounts, Onetime Accounts):

Spoofing
Applications can generate realistic faux emails. These can appear real enough to mislead untrained or inexperienced investigators. Header information on real emails can be forged. The correct information usually can be obtained from machine logs, but this can be time-consuming.

Shared Accounts
Users can pass information by editing "draft" emails. These are never sent, so they can't be screened by host or server filters.

One-Time Accounts
Correspondents can set up "one-time" accounts for limited numbers of emails or even single emails. Account IDs can be shared via key-generation algorithms or even one-time pads. "Richard Clarke — former US counterterrorism czar — recently says that one-time anonymous account is extremely difficult to monitor" [3].

5.5   Network Evidence and Investigation

Network investigations are often harder than investigations of a standalone computer. Evidence can spread across multiple devices, organizations, or geographical areas. This increased complexity gives hackers more targets to attack. Hackers will usually follow a specific path through a network to attack a system. Retracing this path is a critical step in tracking an attack to identify its source.

To track an attack, investigator must "hop backwards" from device to device. Usually these devices are routers (although there may be other types of devices, such as firewalls). Routers thus become critical points in an investigation. Routers pose a challenge because of their volatile memory Powering down a router may erase potentially critical evidence.

5.5.1   Log Files

Log files are primary sources of network forensic evidence. There are several types of log files including "authentication, application, operating system and firewall logs" [19]. Each of them stores some specific types of evidence and information.

Application Log
Application logs records some critical information such as:

- Date
- Time
- Application identifier

Using this log, a forensic investigator could identify the date, time, and length of time of an application's execution.

Operating System Logs
Operating Systems control the use of system devices, including the execution of application software. In addition, all local network abnormalities are stored in OS logs [12]. Evidence stored in OS logs include:

- Running devices;
- System starts, stops, crashes, and reboots;
- Abnormal system operation or resource usage– this is especially important in discovering malwarecattacks;
- Command history;
- Recently accessed and reviewed files;
- Audit records.

Forensic evidence can be found in the most surprising places. In "The Cookoo's Egg", Cliff Stoll explains how he discovered a system penetration by noticing a 25-cent discrepancy in system cost-accounting logs.

Investigators must be careful when relying on OS because some information is available only when a system is running and is lost when it shuts down.

Firewall and Router Logs

A firewall is another part of a computer security system for preventing unauthorized access [25]. Firewalls may be incorporated into a system's own software, or it may be a separate device. Firewalls fit between a system (or its operating system) and the networks to which it connects. Systems may have firewalls on all of their network interfaces or just on some of them.

Messages entering or leaving protected networks pass through their firewalls, which enforce specific security policies [26]. Firewall and router logs are uniquely valuable source of information for a system under any attack.

Routers direct the paths data transmissions take between systems. Router logs can contain such valuable information as:

- Source and destination IP Addresses and domain names;
- Transmission paths and time stamps;
- Source and destination processes and port numbers.

Caution When Working with Logs

Working with log files always requires professional skills.

- Investigators must take control of systems as soon as possible. Dynamic system data can be over-written or otherwise lost. Log files can be changed, modified, or deleted when a system stops, starts, or reboots or they can be truncated to control their size.
- Investigators often require system administrator access to read system logs. They must take care not to cause any side effects that might destroy or obfuscate evidence [19].

5.5.2 Investigation Tools for Network

Capturing and analyzing network traffic are the *sine qua non* of network investigation tools. The most famous sniffing tools are Snort and Wireshark. Netwitness Investigator and NetIntersept are also popular tools. Investigators must consider legal and organizational issues before using any sniffer. Casey states that "monitoring a network traffic in certain instance can be considered wiretapping" [5].

Table 5.2 Risk assessment analysis of four types of evidence

	Cellphone and Cellular network	Global Positioning Systems	Emails	Network
Motivation of threat source	3	2	4	4
Vulnerabilities	1	1	4	5
Possibility of using external attack	2	1	5	5
Possibility of insider attack	1	1	4	5
Possibility of modifying files/evidence	3	2	4	3
Total score	7	5	17	19
Risk ratio	0.35	0.25	0.85	0.95
Confidentiality ratio	0.6	0.4	0.8	0.6

Note: Scores moves between more likely with score 5 and less likely with score 1

Table 5.3 Easy to recovery data and evidence

	Cellphone and cellular network	Global positioning systems	Emails	Network
Availability of data	3	2	4	5
Easy to recovery	5	4	3	2
Located in more than one location	3	2	4	5
Availability of Forensic tools	3	2	3	5
Requires professional skills	3	2	4	5
Total score	17	12	18	22
Availability ratio	0.68	0.48	0.72	0.88

5.6 Comparison and Prioritization

To compare integrity and availability of data we used risk assessment analysis model [23]. We used five indicators, four to measure integrity and one to measure confidentiality. We also used five indicators to measure the availability and ease-of-recovery of data using forensic lab.

As presented in Tables 5.2 and 5.3, network forensics receive the highest risk with 0.95 risk ratio and GPS the lowest risk with 0.25 risk ratio. Ease-of-recovery ratio showed that Network data with 0.88 is the most available source of data.

5.7 Conclusion

Evidence collection of digital devices is the most important step of forensic investigation. Collection and analysis require the use of appropriate tools and techniques to systematically search digital devices for pertinent evidence [17]. We discussed the sources of evidence on Cellphones, Global Positioning Systems (GPSs), email systems, and networks. We also discussed aspects of operating systems, storage devices, and logs as evidence sources. We considered how these sources could help forensic investigators gather the information they need.

Cellphone evidence sources include SD card content, contact numbers, SMS, video, image, call logs, call location and duration, Call Detail Records (CDRs), time/date of originating and terminating towers. Collecting evidence requires professional skills, with Cellphones isolated and disconnected from the network to protect any proof to evidence from being wiped up and overwritten.

Global Positioning Systems (GPSs) evidence sources include TrackPoints, WayPoints and log files. Hybrid and connected models are valuable source of cellphone information including video, image and call logs. Voice-activated GPS systems may provide a record of sounds from inside the vehicle. GPSs can guide an investigator to identify "where the unit has been and where a user intended to go". The main challenge of using GPSs evidence is isolating and disconnecting them from network to protect evidence from overwritten (Jansen and Ayers).

Email evidence sources include contents, dates, times, senders, receivers, computer IPs, transmission paths, and account holder information.

Network evidence sources include router logs, firewall logs, and operating system logs including user, password, data and time of using a system, application performance, configuration settings and last visited files with using volatile and non-volatile evidence.

Professional skills are needed to protect evidence from overwritten in and modification. Network sniffers – Snort, WireShark and NetIntercep – are technical tools for investigation of Network activities. Investigators must consider the legal and organizational implications of network sniffing, which may be considered wiretapping [21, 28].

Table 5.4 Evaluation of four evidence sources based on CIA indicators

	Loss of integrity	Loss of availability	Loss of confidentiality	CIA score
Cellphone and cellular network	0.65	0.32	0.4	1.37
Global positioning systems	0.75	0.52	0.6	1.87
Emails	0.15	0.28	0.2	0.63
Network	0.05	0.12	0.4	0.57

Source: The study calculation using CIA risk assessment analysis
Loss of integrity (1-Risk Ratio)
Loss of Availability (1-Availability Ratio)
Loss of Confidentiality (1-Confidentiality Ratio)

Finally using the CIA's model, we prioritize these four sources of evidence (See Table 5.4). As focuses; Email and network received the lowest rank; Cellphone and GPSs the highest. Additionally, Integrity and tracking evidence from Email – especially with using anonymous and shared account – cannot be guaranteed. Malware can protect some external attackers but there are potential risks of insider hackers.

References

1. Ayers, R., Jansen, W., Cilleros, N., Daniellou, R.: Cell Phone Forensic Tools. Retrieved 10 Oct 2018 from https://ws680.nist.gov/publication/get_pdf.cfm?pub_id=150375 (2005)
2. Barbara, J.J: SIM Forensics: Part 1. Retrieved 10 Mar 2017 from: https://www.forensicmag.com/article/2011/04/sim-forensics-part-1 (2011)
3. Bailey, T.D., Grimaila, M.R.: Running the blockade: information technology, terrorism, and the transformation of Islamic mass culture. Terrorism Polit. Viol. **18**(4), 523–543 (2006)
4. Brian, M., Harris, T.: How GPS Receivers Work. Retrieved 10 Mar 2017 from http://electronics.howstuffworks.com/gadgets/travel/gps.htm/printable (2011)
5. Casey, E.: Digital Evidence and Computer Crime: Forensic Science, Computers and the Internet. Elsevier Inc, Waltham (2011)
6. Hartenstein, M., Sheridan, M.: Missing Vegas Showgirl Debbie Flores-Narvaez was Pregnant, Beaten by her ex, According to Police. Retrieved 10 Mar 2017 from: http://www.nydailynews.com/news/national/missing-vegas-showgirl-debbie-flores-narvaez-pregnant-beaten-police-article-1.149194 (2010)
7. He, S., Paar, I.C.: SIM Card Security. Chair for Communication Security. Retrieved 10 Mar 2017 from: https://pdfs.semanticscholar.org/9adb/d6044393c8f0fb0ab6329d286e4ad64cae6c.pdf (2007)
8. Hillebrand, F. (ed.): GSM and UMTS: The Creation of Global Mobile Communication, p. 371. Wiley, Chichester (2002)
9. Jansen, W., Ayers, R.: Guidelines on cell phone forensics. NIST Special Publication, 800, 101. Retrieved 10 Mar 2017 from: http://www.4law.co.il/cell1.pdf (2007)
10. Järv, O., Ahas, R., Witlox, F.: Understanding monthly variability in human activity spaces: a twelve-month study using mobile phone call detail records. Trans. Res. Part C: Emerg. Technol. **38**, 122–135 (2014)
11. Kessler, G.: Cell Phone Analysis: Technology, Tools, and Processes. Mobile Forensics World. Purdue University, Chicago (2010)
12. Kizza, J.M.: Guide to Computer Network Security. Springer, Swindon, UK (2015)
13. Lee, X., Yang, C., Chen, S., Wu, J.: Design and implementation of forensic system in Android smart phone. In: The 5th Joint Workshop on Information Security (2009)
14. LeMere, B.: Enhancing Investigations with GPS Evidence. Retrieved 10 Mar 2017 from http://www.forensicmag.com/article/enhancing-investigations-gps-evidence (2011)
15. Mahajan, A., Dahiya, M.S., Sanghvi, H.P.: Forensic analysis of instant messenger applications on Android devices. arXiv preprint arXiv:1304.4915 (2013)
16. Oluwasegun, S., David, O.E., Esther, E., Victor, O.: Computer forensics for law enforcement. J. Emerg. Trend. Eng. Appl. Sci. (JETEAS). **5**(1), 35–38 (2014)

17. Reith, M., Carr, C., Gunsch, G.: An examination of digital forensic models. Int. J. Digit. Evid. **1**(3), 1–12 (2002). Retrieved 10 Mar 2017 from http://www.di-srv.unisa.it/~ads/corso-security/www/CORSO-9900/a5/gsmreport/gsmreport.pdf

18. Ryder, S., Le-Khac, N.A.: The end of effective law enforcement in the cloud? To encypt, or not to encrypt. arXiv preprint arXiv:1609.07602 (2016). Retrieved 10 Mar 2017 from https://arxiv.org/ftp/arxiv/papers/1609/1609.07602.pdf

19. Sammons, J.: The Basics of Digital Forensics-2nd Edition: the Primer for Getting Started in Digital Forensics. Elsevier Inc, Waltham (2015)

20. Scourias, J.: Overview of the global system for mobile communications. University of Waterloo, 4 (1995)

21. Snort: Snort User's Manual 2.9.9. Retrieved March 10, 2017 from http://manual-snort-org.s3-website-us-east-1.amazonaws.com/ (2016)

22. Steeh, C., Buskirk, T.D., Callegaro, M.: Using text messages in US mobile phone surveys. Field Methods. **19**(1), 59–75 (2007)

23. Stoneburner, G., Goguen, A.Y., Feringa, A.: Sp 800–30. Risk management guide for information technology systems (2002)

24. Tschabitscher, H.: How to Understand Date and Time in Email Headers. Retrieved 10 Mar 2017 from https://www.lifewire.com/what-is-an-email-header-1171127 (24 Aug 2016)

25. UMUC: Enterprise Network Intrusion Prevention Systems, CSEC 630 Module 1, Document posted in University of Maryland University College prevention and protection strategies in cybersecurity-CSE630 online classroom. Archived at: https://leoprdws.umuc.edu/CSEC630/1306/csec630_01/assets/csec630_01.pdf (2016)

26. Valacich, J., Schneider, C.: Information System Today: Managing in the Digital World, 6th edn. Pearson, New Jersy (2014)

27. Weiss, S.M., Indurkhya, N., Zhang, T., Damerau, F.: Text Mining: Predictive Methods for Analyzing Unstructured Information. Springer Science & Business Media, New York (2010)

28. Wireshark: Wireshark User's Guide. Retrieved 10 Mar 2017 from https://www.wireshark.org/docs/wsug_html_chunked/ (2016)

Making AES Great Again: The Forthcoming Vectorized AES Instruction

Nir Drucker, Shay Gueron, and Vlad Krasnov

6.1 Introduction

AES is the most ubiquitous symmetric cipher, used in many applications and scenarios. A prominent example is the exponentially growing volume of encrypted online data. Evidence for this growth, which is strongly supported by the industry (e.g., Intel's new AES-NI instructions [1–3], and Google's announcement [4] on favoring sites that use HTTPS) can be observed, for example, in [5] showing that more than 70% of online websites today use encryption.

This makes the performance of AES a major target for optimization in software and hardware. Dedicated hardware solutions were presented (e.g., [6,7]) and via the introduction of the AES-NI instructions that were added to x86 general purpose CPUs (and other architectures). These instructions, together with the progress made in processors' microarchitectures, allow software to run the Authenticated Encryption with Additional Authentication Data (AEAD) scheme AES-GCM at 0.64 cycles per byte (C/B hereafter), approaching the theoretical performance of encryption only, 0.625 C/B, on such CPUs. Other software optimizations, written in OpenCL or CUDA that aim for the Graphical Processor Unit (GPU) [8, 9] achieve the performance of 0.56 C/B and 0.44 C/B, respectively. Last year, AMD introduced the new "Zen" processor that has two AES units [10], and this reduces the theoretical throughput of AES encryption to 0.31 C/B.

Recently, Intel has announced [11] that its future architecture, microarchitecture codename "Ice Lake", will

add vectorized capabilities to the existing AES-NI instructions, namely VAESENC, VAESENCLAST, VAESDEC, and VAESDECLAST (VAES* for short). These instructions are intended to push the performance of AES software further down, to a new theoretical throughput of 0.16 C/B.

This can directly speed up AES modes such as AES-CTR and AES-CBC, and also more elaborate schemes such as AES-GCM and AES-GCM-SIV [12, 13] (a nonce misuse resistant AEAD). These two schemes require fast computations of the almost XOR-universal hash functions GHASH and POLYVAL, which are significantly sped up with dedicated "carry-less multiplication" instruction PCLMULQDQ [2, 14]. Indeed, fast AES-GCM(-SIV) implementations can be achieved by using the new instruction that vectorizes the PCLMULQDQ instruction (VPCLMULQDQ) (see [15]).

In this paper, we demonstrate how to write software that efficiently uses the new VAES* and VPCLMULQDQ instructions. While the correctness of our algorithms (and code) can be verified with existing public tools, the actual performance measurements require a real CPU, which is currently unavailable. To address this difficulty, we give predictions based on instructions' count of current and new implementations.

The paper is organized as follows. Section 6.2 describes the new VAES* and VPCLMULQDQ instructions. Section 6.3 describes our implementations of AES encryption modes AES-CTR and AES-CBC. Section 6.4 focuses on the AEAD schemes AES-GCM and AES-GCM-SIV. In Sect. 6.5, we explain our results, and we conclude in Sect. 6.6.

6.2 Preliminaries

We use AES to refer to AES128. The xor operation is denoted by $\oplus$, and concatenation is denoted by $||$ (e.g., $00100111||10101100 = 0010011110101100$, which, in

This work was done prior to joining Amazon.

N. Drucker · S. Gueron (✉)
University of Haifa, Haifa, Israel

Amazon Web Services Inc, Seattle, WA, USA
e-mail: shay@math.haifa.ac.il

V. Krasnov
CloudFlare, Inc., San Francisco, CA, USA

S. Latifi (ed.), *16th International Conference on Information Technology-New Generations (ITNG 2019)*,
Advances in Intelligent Systems and Computing 800,
https://doi.org/10.1007/978-3-030-14070-0_6

hexadecimal notation, is the same as $\mathtt{0x27}$ || $\mathtt{0xac}$ = $\mathtt{0x27ac}$). The notation $X[j : i]$, $j > i$ refers to the values of an array X between positions i and j (included). The case $i = j$ degenerates to $X[i]$. Here, X can be an array of bits or of bytes, depending on the context. For an array of bytes X, we denote by $\overline{X}$ the corresponding byte swapped array (e.g., $X = \mathtt{0x1234}$, $\overline{X} = \mathtt{0x3412}$). The two new vectorized AES-NI and $\mathtt{PCLMULQDQ}$ instructions are described next. The description of other assembly instructions can be found in [16].

6.2.1   Vectorized AES-NI

Intel's AES-NI instructions ($\mathtt{AES*}$) include $\mathtt{AESKEYGENA}$ $\mathtt{SSIST}$ and $\mathtt{AESIMC}$ to support AES key expansion and $\mathtt{AESENC/DEC(LAST)}$ to support the AES encryption/decryption, respectively. Algorithm 6.1 illustrates the new $\mathtt{VAES*}$ instructions. These are able to perform one round of AES encryption/decryption on $KL = 1/2/4$ 128-bit operands (two qwords), having both register-memory and register-register variant (we use only the latter here). The inputs are two source operands, which are 128/256/512-bit registers (named xmm, ymm, zmm, respectively), that (presumably) represent the round key and the state (plaintext/ciphertext). The special case $KL = 1$ using xmm registers degenerates to the current version of $\mathtt{AES*}$.

Algorithm 6.1 $\mathtt{VAES*}$, and $\mathtt{VPCLMULQDQ}$ instructions [11]

Inputs: SRC1, SRC2 (wide registers)
Outputs: DST (a wide register)
1: **procedure** $\mathtt{VAES*}$(SRC1, SRC2)
2:    **for** $i := 0$ to $KL - 1$ **do**
3:        $j = 128i$
4:        RoundKey[127 : 0] = SRC2[$j + 127 : j$]
5:        T[127 : 0] = (Inv)ShiftRows(SRC1[$j + 127 : j$])
6:        T[127 : 0] = (Inv)SubBytes(T[127 : 0])
7:        T[127 : 0] = (Inv)MixColumns(T[127 : 0])
                                    ▷ Only on $\mathtt{VAESENC/VAESDEC}$.
8:        DST[$j + 127 : j$] = T[127 : 0] $\oplus$ RoundKey[127 : 0]
9:    **return** DST

Inputs: SRC1, SRC2 (wide registers) Imm8 (8 bits)
Outputs: DST (a wide register)
1: **procedure** $\mathtt{VPCLMULQDQ}$(SRC1, SRC2, Imm8)
2:    **for** $i := 0$ to $KL - 1$ **do**
3:        $j_1 = 2i + Imm8[0]$
4:        $j_2 = 2i + Imm8[4]$
5:        T1[63 : 0] = SRC1[$64(j_1 + 1) - 1 : 64j_1$]
6:        T2[63 : 0] = SRC2[$64(j_2 + 1) - 1 : 64j_2$]
7:        DST[$128(i + 1) - 1 : 128i$] = PCLMULQDQ(T1, T2)
8:    **return** DST

6.2.2   Vectorized $\mathtt{VPCLMULQDQ}$

Algorithm 6.1 (bottom) illustrate the functionality of the new vectorized $\mathtt{VPCLMULQDQ}$ instruction. It vectorizes

polynomial (carry-less) multiplication, and is able to perform $KL = 1/2/4$ multiplications of two qwords in parallel. The 64-bit multiplicands are selected from two source operands and are determined by the value of the immediate byte. The case $KL = 1$ degenerates to the current $\mathtt{VPCLMULQDQ}$ instruction.

6.3   Accelerating AES with $\mathtt{VAES*}$

The use of the $\mathtt{VAES*}$ instructions for optimizing the various uses of AES is straightforward for some cases (e.g., AES-ECB, AES-CTR, AES-CBC decryption). For example, to optimize AES-CTR, which is a naturally parallelizable mode, we only need to replace each xmm with zmm register and handle the counter in a vectorized form. In some other case, using the new instruction is more elaborate (e.g., optimizing AES-CBC encryption, AES-GCM, or AES-GCM-SIV).

Figure 6.1 compares legacy (Panel a) and vectorized (Panel b) codes of AES-CTR. In both cases, the counter is loaded and incremented first (Steps 6–8 and 7–8, respectively). In Panel b, Steps 9–11, the key schedule is duplicated 4 times in 11-zmm registers (zmm0-zmm10). The encryption is executed in Steps 9–13, and 12–16, of Panels a and b, respectively. Finally, the plaintext is xored and the results are stored.

A mode like AES-CBC encryption is serial by nature, and cannot be parallelized. However, we note that the $\mathtt{VAES*}$ instructions encrypt 2/4 independent plaintext streams in parallel. To do this, we need to rearrange ("transpose") the inputs/outputs in order to make them suitable for vectorized code such as in Fig. 6.1.

Figure 6.2 illustrates how to handle four independent $4 * 128$-bit plaintext streams (A, B, C, D). We first load the four 512-bit values into four zmm registers (red upper left vectors), then use the $\mathtt{VPERMI2Q}$ instruction to permute the qwords of each two vectors A, B and C, D (orange vectors). $\mathtt{VPERMI2Q}$ receives two source operands and a mask operand, which is also the destination operand (all are wide registers). Therefore, the mask must be re-set before each $\mathtt{VPERMI2Q}$ execution. Finally, we use the $\mathtt{VPERMI2Q}$ instruction to calculate the final results (green right bottom vectors). The flow requires 4 loads 8 permutations and 8 mask preparations, with total of 20 instructions per 256-bytes of processed data (e.g., plaintexts). We find this method to be very efficient. Other transposing methods can use the $\mathtt{VPGATHERQQ/VPSCATTERQQ}$ or the $\mathtt{VPUNPCK}$ instructions, but suffer from high instructions' latency, or need to use more instructions for the same task. Note that the $\mathtt{VPUNPCK}$ instruction is recommended for transposing a matrix of size 4×4 with elements of size $4 * 64$-bit, but this is not the case here.

(a)

```
1  .set  t,  %xmm12
2  .set  ctrReg,  %xmm11
3  inc_mask:
4  .long  0,0,0,0x01000000
5
6  vmovdqu  (ctr),  ctrReg
7  .irp  j,1,2,3,4
8   vpadd  inc_mask(%rip),ctrReg
9   vpxor  (key),ctrReg,t
10   .irp  i,1,2,3,4,5,6,7,8,9
11    vaesenc  \i*0x10(key),t,t
12   .endr
13   vaesenclast  10*0x10(key),t,t
14   vpxor  (pt),t,t
15   vmovdqu  tmp,  (ct)
16   lea  0x10(pt),  pt
17   lea  0x10(ct),  ct
18  .endr
```

(b)

```
1  .set  t,  %zmm12
2  .set  ctrReg,  %zmm11
3  inc_mask:
4  .long  0,0,0,0x01000000,
5  .long  0,0,0,0x02000000,
6  .long  0,0,0,0x03000000,
7  .long  0,0,0,0x04000000
8
9  vbroadcasti64x2  (ctr),ctrReg
10  vpadd  inc_mask(%rip),ctrReg
11  .irp  i,0,1,2,3,4,5,6,7,8,9,10,11
12   vbroadcasti64x2  \i*0x10(key),%zmm\i
13  .endr
14  vpxorq  %zmm0,ctrReg,t
15  .irp  i,1,2,3,4,5,6,7,8,9
16   vaesenc  %zmm\i,t,t
17  .endr
18  vaesenclast  %zmm10,t,t
19  vpxorq  (pt),t,t
20  vmovdqu64  t,(ct)
```

Fig. 6.1 AES-CTR sample (AT&T assembly syntax); ct[511 : 0]=AES-CTR(pt[511 : 0], key). (**a**) Legacy AES-CTR. (**b**) Vectorized AES-CTR

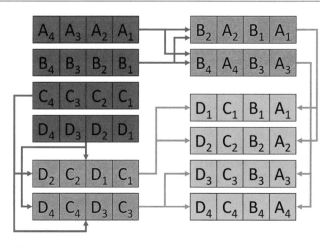

Fig. 6.2 Transposing a 4×4 128-bit (input in red; output in green) by executing eight VPERMI2Q instructions. Every group of four instructions can run in parallel

Leveraging the pipeline capabilities efficiently Fast AES computations need to operate on multiple independent blocks in parallel [3], in order to hide the latency of the instructions and make the flow depend only on their throughput. The optimal number of blocks is determined by the latency and throughput of the VAES* instructions [17], and the number of available registers. The latency of VAES* is 4 cycles on architecture codename "Skylake" (was 7 cycles on earlier processor generations), and their throughput is 1 cycle. In addition, the AVX512 architecture has 32 zmm registers, which can be used together with the VAES* instructions. For example, in AES-CTR we can allocate 11 registers for the AES round keys, and split the rest among the counters and their the plaintext/ciphertext states ($\sim$10 each). Consequently, it is possible to process 10 packets of 4 blocks in

parallel, instead of only 8 packets of 1 block, as with the current instructions.

6.4   AES-GCM and AES-GCM-SIV

AES-GCM [18, 19] and AES-GCM-SIV [12, 13] are AEAD schemes (AES-GCM-SIV is nonce-misuse resistant). Their encryption flows are outlined in Algorithms 6.2 and 6.3. Both modes include AES-CTR encryption (the code is already shown in Fig. 6.1).

Algorithm 6.2 AES-GCM encryption [18, 19]

Inputs: K (128 bits), IV (96 bits), A (AAD), M (message)
Outputs: T (tag, 128 bits), C (ciphertext)

1: **procedure** AES-GCM(K, IV, A, M)
2:     $H = \text{AES}(K, 0^{128})$
3:     $CTR_0 = IV||0^{31}1$
4:     **for** $i = 0, 1, \ldots, v - 1$ **do**
5:         $CTR_i = CTR[127 : 32]||((CTR[31 : 0] + i) \pmod{2^{32}})$
6:         $C_i = \text{AES}(K, CTR_i) \oplus M_i$
7:     $T = \text{GHASH}(H, A, C) \oplus \text{AES}(K, CTR_0)$
8:     **return** C, T

We focus on optimizing the universal hashing parts of these algorithms, which are not identical: AES-GCM uses GHASH and AES-GCM-SIV uses POLYVAL. Both hash functions operate in $\mathbb{F} = \mathbb{F}_{2^{128}}$ (but with different reduction polynomials) and evaluate a polynomial with coefficients $X_1, X_2, \ldots, X_s$ (for some s) in $\mathbb{F}$ at some point $H \in \mathbb{F}$ (which is the hash key). As shown in [13]:

$$POLYVAL(H, X_1, X_2, \ldots, X_s) =$$
$$\overline{(GHASH((\overline{H \otimes x}), (\overline{X_1}), (\overline{X_2}), \ldots, (\overline{X_s})))}$$

so if suffices to demonstrate the implementation of POLY-VAL.

Algorithm 6.3 AES-GCM-SIV encryption [12, 13]

Inputs: K_1 (128 bits), K_2 (128 or 256 bits), N (96 bits), A (AAD),
L_A (A length in bytes), M (message), L_M (M length in bytes)
Outputs: T (tag, 128 bits), C (ciphertext)

1: **procedure** AES-GCM-SIV($K_1, K_2, N, A, L_A, M, L_M$)
2:     $Tmp = \text{POLYVAL}(K_1, A||M||L_A||L_M)$
3:     $T = \text{AES}(K_2, 0||(Tmp \oplus N)[126:0])$
4:     **for** $i = 0, 1, \ldots, v - 1$ **do**
5:         $CTR_i = 1||T[126:32]||((T[31:0] + i) \pmod{2^{32}})$
6:         $C_i = AES(K_2, CTR_i) \oplus M_i$
7:     **return** $C = (C_1, \ldots, C_{v-1}), T$

The "Aggregated Reduction" method (see [14]) replaces Hoeren's method with a per-block reduction ($T_i = ((X_i \oplus T_{i-1}) \otimes H) \pmod{Q(x)}$), with a deferred reduction based on pre-computing $t > 0$ powers of H stored in a table (Htbl).

$$T_i = ((X_i \otimes H) \oplus (X_{i-1} \otimes H^2) \oplus \cdots \oplus$$
$$(X_{i-(t-1)} + T_{i-t}) \otimes H^t) \pmod{Q(x)}$$

($\otimes$ is field multiplication; $Q(x)$ is the reduction polynomial).

Figure 6.3 compares codes for initializing Htbl. Panel (a) describes the legacy implementation with $t = 8$. Panel (b) describes a vectorized implementation for calculating

(a) Legacy Htbl-init(8)

```
1  vmovdqu    (H) , %xmm0
2  vmovdqu    %xmm0, %xmm1
3  .irp  i,0,1,2,3,4,5,6
4    vmovdqu %xmm0,\ i*0x10(Htbl)
5    call       GFMUL
6  .endr
7  vmovdqu    %xmm0,  7*0x10(Htbl)
8  ret
```

(b) Vectorized Htbl-init(32)

```
1  vmovdqu    (H) , %xmm0
2  vmovdqu    %xmm0, %xmm1
3  vmovdqu    %xmm0, (Htbl)
4  call       GFMUL
5  vmovdqu    %xmm0, 0x10(Htbl)
6  call       GFMUL
7  vbroadcasti64x2  0x10(Htbl),%ymm1
8  vmovdqu64  (Htbl) , %ymm0
9  call       GFMUL2
10 vmovdqu64  %ymm0, 0x20(Htbl)
11 vbroadcasti64x2  0x30(Htbl),%zmm1
12 vmovdqu64  (Htbl) , %zmm0
13 .irp  i,1,2,3,4,5,6,7
14   call       GFMUL4
15   vmovdqu64 %zmm0, \ i*0x40(Htbl)
16 .endr
17 ret
```

Fig. 6.3 Initializing the HTBL. (**a**) legacy $t = 8$, (**b**) vectorized $t = 8 * 4$

```
1  vpclmulqdq  $0x00,  %zmm1, %zmm0, %zmm2
2  vpclmulqdq  $0x11,  %zmm1, %zmm0, %zmm5
3  vpclmulqdq  $0x10,  %zmm1, %zmm0, %zmm3
4  vpclmulqdq  $0x01,  %zmm1, %zmm0, %zmm4
5  vpxorq      %zmm4, %zmm3, %zmm3
6
7  vpslldq  $8,  %zmm3, %zmm4
8  vpsrldq  $8,  %zmm3, %zmm3
9  vpxorq      %zmm4, %zmm2, %zmm2
10 vpxorq      %zmm3, %zmm5, %zmm5
11
12 vpclmulqdq  $0x10,  poly(%rip), %zmm2, %zmm3
13 vpshufd   $78 ,  %zmm2, %zmm4
14 vpxorq      %zmm4, %zmm3, %zmm2
15
16 vpclmulqdq  $0x10,  poly(%rip), %zmm2, %zmm3
17 vpshufd   $78 ,  %zmm2, %zmm4
18 vpxorq      %zmm4, %zmm3, %zmm2
19
20 vpxorq      %zmm5, %zmm2, %zmm0
21 ret
```

Fig. 6.4 GFMUL4 function, performs vectorized $A_1 \otimes A_2$ (mod $Q(x)$)

```
1  vextracti64x4  $1,  %zmm0, %ymm1
2  vpxor          %ymm1, %ymm0, %ymm0
3  vextracti128   $1,  %ymm0, %xmm1
4  vpxor          %ymm1, %xmm0, %xmm0
```

Fig. 6.5 Vectorized aggregated reduction method – final aggregation

$t = 4 * 8$ powers of H. Both snippets use the GFMUL function for the field multiplication. Figure 6.4 presents GFMUL4 that performs 4 multiplications in parallel. The same code is used for GFMUL and GFMUL2, but over different registers (xmm/ymm). Steps 1–10 perform "Schoolbook" multiplication, and Steps 12–20 perform the reduction (see [14]). An implementation that uses "Aggregated Reduction" should first perform $H \otimes X_t$ as in Fig. 6.4, Steps 1–5. Then process $H^i \otimes X_{t-i}, i = 1, \ldots, t-1$ in parallel and accumulate the results. Subsequently, perform the reduction steps 7–20.

In the vectorized implementation we load 4 values from Htbl into each zmm register e.g., zmm(i)= $\{H^{4i}, H^{4i+1}, H^{4i+2}, H^{4i+3}\}$. To multiply the matching values (H^i, X_{t-i}), we first need to reverse their order e.g., zmm(i)= $\{H^{4i+3}, H^{4i+2}, H^{4i+1}, H^{4i}\}$. We do this using the VSHUFI64X2 instruction: "vshufi64x2 0x1b, %zmm(i), %zmm(i), %zmm(i)". Eventually, we end with $T_j = \sum_{\substack{i \equiv j \pmod 4 \\ i=1,\ldots,t}} (H^i \otimes X_{t-i})$, $j = 1, \ldots, 4$.
Figure 6.5 shows the final aggregation step.

6.5   Results

We implemented x86 assembly code for AES-CTR and POLYVAL, using VAES* and VPCLMULQDQ instructions,

Table 6.1 Instructions count comparison (lower is better)

Algorithm	PT SIZE (bytes)	Legacy impl.	Vectorized impl.	Ratio
AES-CTR	512	608	178	3.42
AES-CTR	8,192	9,248	2,338	3.96
AES-CTRx8	512	493	150	3.29
AES-CTRx8	8,192	7,453	1,890	3.94
POLYVALx8	4,096	2,816	794	3.55
POLYVALx8	8,192	5,536	1,474	3.76
POLYVALx8	16,384	10,976	2,834	3.87

pipelining 1 or 8 streams in parallel (the suffix "x8" distinguishes the implementations). To predict the potential improvement on future architectures before real samples are available, we used the Intel SDE [20]. This tool allows us to count the number of instructions executed during each of the tested functions. We marked the start/end boundaries of each function with "SSC marks" 1 and 2, respectively. This is done by executing "`movl ssc_mark, %ebx; .byte 0x64, 0x67, 0x90`" and invoking the SDE with the flags "-start_ssc_mark 1 -stop_ssc_mark 2 -mix -icl". The rationale is that a reduced number of instructions typically indicates improved performance that will be observed on a real processor (although the exact relation between the instructions count and the eventual cycles count is not known in advanced).

Table 6.1 compares the instructions count in our implementations. The results confirm our prediction that AES algorithms can be sped up by a factor of $3-4\times$ and that better speedups are expected when operating on larger buffers.

6.6 Conclusion

This paper shows how to leverage Intel's new instruction `VAES*` and `VPCLMULQDQ` for accelerating encryption with AES. Our results predict that optimized vectorized AES code can approach the new theoretical bound of 0.16 C/B on forthcoming CPUs, about $4\times$ faster than current implementations. We demonstrated optimized AES-CTR and AES-GCM(-SIV) code snippets that can approach this limit. For serial mode such as AES-CBC, we showed how to optimize code by processing multiple message streams in parallel.

Acknowledgements This research was supported by the Center for Cyber Law and Policy at the University of Haifa in conjunction with the Israel National Cyber Directorate in the Prime Ministers Office

References

1. Gueron, S.: Intel® Advanced encryption standard (AES) new instructions set rev. 3.01. Intel Software Network (2010)

2. Gueron, S., Kounavis, M.: Efficient implementation of the Galois Counter Mode using a carry-less multiplier and a fast reduction algorithm. Inf. Process. Lett. **110**(14), 549–553 (2010)

3. Gueron, S.: Intel's new AES instructions for enhanced performance and security. In: FSE, vol. 5665, pp. 51–66. Springer (2009)

4. Bahajji, Z.A.: Indexing HTTPS pages by default, Dec 2015. https://security.googleblog.com/2015/12/indexing-https-pages-by-default.html

5. Bahajji, Z.A.: Percentage of web pages loaded by firefox using HTTPS, Jan 2018. https://letsencrypt.org/stats/#percent-pageloads

6. Hodjat, A., Verbauwhede, I.: Area-throughput trade-offs for fully pipelined 30 to 70 Gbits/s AES processors. IEEE Trans. Comput. **55**(4), 366–372 (2006)

7. Mathew, S., Satpathy, S., Suresh, V., Anders, M., Kaul, H., Agarwal, A., Hsu, S., Chen, G., Krishnamurthy, R.: 340 mV #x2013;1.1 V, 289 Gbps/W, 2090-gate nanoAES hardware accelerator with area-optimized encrypt/decrypt GF(2 4) 2 polynomials in 22 nm tri-gate CMOS. IEEE J. Solid-State Circuits **50**(4), 1048–1058 (2015)

8. Manavski, S.A.: CUDA Compatible GPU as an efficient hardware accelerator for AES cryptography. In: 2007 IEEE International Conference on Signal Processing and Communications, Nov 2007, pp. 65–68

9. Patchappen, M., Yassin, Y.M., Karuppiah, E.K.: Batch processing of multi-variant AES cipher with GPU. In: 2015 Second International Conference on Computing Technology and Information Management (ICCTIM), Apr 2015, pp. 32–36

10. Patchappen, M., Yassin, Y.M., Karuppiah, E.K.: The "Zen" core architecture, Jan 2018. http://www.amd.com/en/technologies/zen-core

11. Patchappen, M., Yassin, Y.M., Karuppiah, E.K.: Intel architecture instruction set extensions programming reference, Oct 2017. https://software.intel.com/sites/default/files/managed/c5/15/architecture-instruction-set-extensions-programming-reference.pdf

12. Gueron, S., Lindell, Y.: GCM-SIV: full nonce misuse-resistant authenticated encryption at under one cycle per byte. In: Proceedings of the 22nd ACM SIGSAC Conference on Computer and Communications Security, Ser. CCS '15, pp. 109–119. ACM, New York (2015)

13. Gueron, S., Langley, A., Lindell, Y.: AES-GCM-SIV: specification and analysis. Cryptology ePrint Archive, Report 2017/168 (2017). https://eprint.iacr.org/2017/168

14. Gueron, S., Kounavis, M.E.: Intel® carry-less multiplication instruction and its usage for computing the GCM mode. White Paper (2010)

15. Drucker, N., Gueron, S.: Fast multiplication of binary polynomials with the forthcoming vectorized VPCLMULQDQ instruction. In: 2018 IEEE 25th Symposium on Computer Arithmetic (ARITH), June 2018

16. Drucker, N., Gueron, S.: Intel ® 64 and IA-32 architectures software developers manual. Volume 3a: System Programming Guide, Sept 2015

17. Drucker, N., Gueron, S.: Intel ® 64 and IA-32 architectures optimization reference manual, June 2016

18. McGrew, D., Viega, J.: The Galois/counter mode of operation (GCM). Submission to NIST Modes of Operation Process, vol. 20 (2004)

19. McGrew, D.A., Viega, J.: The security and performance of the Galois/counter mode (GCM) of operation. In: Canteaut, A., Viswanathan, K. (eds.) Progress in Cryptology – INDOCRYPT 2004, pp. 343–355. Springer, Berlin/Heidelberg (2005)

20. McGrew, D.A., Viega, J.: Intel® Software Development Emulator, version 8.12.0, Jan 2017. https://software.intel.com/en-us/articles/intel-software-development-emulator

OntoCexp: A Proposal for Conceptual Formalization of Criminal Expressions

Ricardo Resende de Mendonça, Ferrucio de Franco Rosa, Antonio Carlos Theophilo Costa Jr., Rodrigo Bonacin, and Mario Jino

7.1   Introduction

Information security generally uses cryptography to make a message not understandable to unauthorized persons [1, 2]. People with low technological knowledge make use of simple strategies, such as social cryptography, by using expressions ciphered in a restrict idiom. Criminals use different types of speech such as group slang or marginal slang, used for communication among them. They change the meaning of words, turning slang into a "secret" language by using it in conjunction with data encryption (out of the scope of this paper) or not [3]. A study of 223 cases related to terrorism is presented in [4]; it points out that 61% of events contained evidence of their online activities. These cases included activities such as: (a) virtual communities and recruitment of new members; and (b) online communication among group members. This emphasizes the necessity of studies for supporting and automating the analysis of communication between criminals.

The fast increase of criminality [5], in countries such as Brazil, makes infeasible dealing with all criminal occurrences in a timely manner. Several criminal investigations are

R. R. de Mendonça (✉)
Faculty of Campo Limpo Paulista, Campo Limpo Paulista, SP, Brazil

F. de Franco Rosa · A. C. Theophilo Costa Jr.
Information Technology Center Renato Archer, Campinas, SP, Brazil

University of Campinas, Campinas, SP, Brazil
e-mail: ferrucio.rosa@cti.gov.br; antonio.theophilo@cti.gov.br

R. Bonacin (✉)
Faculty of Campo Limpo Paulista, Campo Limpo Paulista, SP, Brazil

Information Technology Center Renato Archer, Campinas, SP, Brazil
e-mail: rodrigo.bonacin@cti.gov.br

M. Jino
University of Campinas, Campinas, SP, Brazil
e-mail: jino@dca.fee.unicamp.br

inefficient, including those executed with Internet support. Research for development of tools and techniques is necessary to support the investigative and prevention processes. Automated services must be prioritized to increase their efficiency and provide fast response [6].

Methods of criminal investigations, as well as criminals' methods, are constantly changing. Nowadays, identifying and categorizing crimes executed with the support of Internet is a hard work; particularly, those with support of social networks. This task requires a lot of effort, especially when we consider the velocity and volume, for instance, for drug trafficking, terrorism, and other crimes [7].

Criminal Slang Expression (CSE) refers to an obscure dialect used for criminal and cyber criminal purposes. Various CSEs are already present in lexical dictionaries; however, CSEs are evolving fast, confusing police officers and security investigators. Thus, a precise and flexible representation of the key CSEs is needed.

The Semantic Web provides a web environment where data is understood by both humans and automated mechanisms [8]. Its technologies have shown to be very useful for establishing semantic relations between domain concepts. Requirements for understanding judicial aspects [6] can be achieved through the specification of semantic relations. This enables a better (semi)automatic interpretation of natural texts. Techniques of artificial intelligence are employed by security researchers, along with the adoption of ontologies. Use of ontologies has been intensified in the areas of artificial intelligence and security [9–11].

Although there are many methods of textual analysis in the literature, identification of slang (of criminal entourages) remains a major obstacle for (semi)automatic interpretation [12]. We propose the OntoCexp (Ontology of Criminal Expressions) as a common and extensible model for identifying the use of crime expressions in Internet. The initial version

S. Latifi (ed.), *16th International Conference on Information Technology-New Generations (ITNG 2019)*,
Advances in Intelligent Systems and Computing 800,
https://doi.org/10.1007/978-3-030-14070-0_7

of OntoCexp including its core elements are presented in this article; the complete ontology (OWL file) is available in [13].

The remainder of the paper is organized as follows: Section 7.2 presents a summary of a literature review and related work; Section 7.3 describes how OntoCexp was developed; Section 7.4 presents the core concepts of the ontology; Section 7.5 presents the instantiation and validation; finally, Section 7.6 draws conclusions.

7.2   Literature Review and Related Work

Our quasi-systematic review on ontologies and taxonomies was based on [14]. We summarized the papers, focusing on aspects we considered in the development of OntoCexp. Out of 63 papers of interest from the literature review, 17 papers were selected to be discussed here. Most of them aim to describe (sub-)domains of crime and legal cases.

Figure 7.1 illustrates a mapping including the focus and contributions of the 17 selected works. Development of criminal ontologies is the most frequent focus (*67.0%*). It is followed by work on methods for the definition of ontologies (*14.0%*) and on the development of criminal terminologies (*10.0%*). Other studies (*9%*) include: criminal confinement calculation, relevance of terms, classification of legal documents, data mining, and methods for knowledge formulation. Ontology is the main contribution of *76.0%* of the papers; the alternative approaches of crime investigation are addressed by *19%*. Other contributions (*5%*) include: crimes evidences, murders, legal terms, network security,

ontology improvement, extraction of criminal texts, online drug transactions, and knowledge extraction from police reports.

Methods for constructing ontologies are discussed in [7, 15]. A object-oriented view on ontology engineering process is presented in [10]. A conceptualization of the crime domain by means of conceptual graphs, named-entity, recognition and formal concept analysis is presented in [16].

A formalization of knowledge on crime is presented in [6, 9, 12, 17]. These provide a broad and general view of the criminal area, whereas other articles deal with specific aspects. A specific ontology on the murders domain is proposed in [18, 19] aimed at detailing knowledge of crime evidences. Identification of agents involved in a crime is proposed in [20]. Other studies focus on specific situations. Reference [21] describes the construction of an ontology on the events of the judicial case *Popov v. Hayashi*, 2002 in California. An ontology for the automatic classification of legal cases is presented in [22].

An analysis of legal cases through the relevance of terms is discussed in [23]; this is necessary for formalizing terms and documents hierarchy of the judicial area. An ontology focused on management of crimes against life, and criminal calculation, are presented in [19]. Reference [24] formalizes Lebanese laws knowledge and a set of logical rules to be used in applications. The studies presented in [17, 25] emphasize the need of an ontology on crimes in Brazil. Reference [25] presents further details on the use of UFO-B as a upper-ontology, whereas in [26] the use of DOLCE upper-ontology.

Differently from the previous approaches, our proposal focuses on the specification of a core model; where CSEs are

Fig. 7.1 Summary of mapping of related work

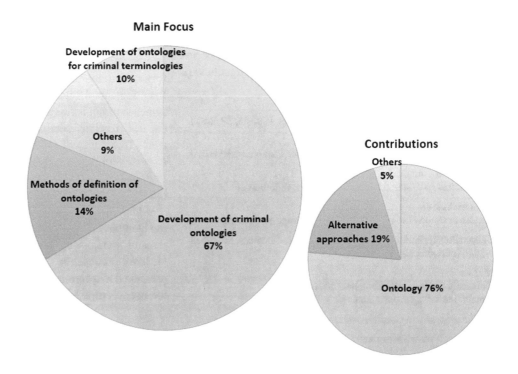

formalized to support the development of tools and methods for the detection and interpretation of encoded expressions by security researchers and police investigators.

7.3 The Engineering Process of OntoCexp

Our engineering process, inspired by reference [27], contains the following iterative steps:

(i) *Determination of the scope and reuse.* We defined as scope of our ontology the domain of crime, and performed the correlation between expressions and terms used by individuals in illicit activities. As proposed in [19, 24, 26], we evaluated the top-level ontologies SUMO, UFO-B and DOLCE. However, we chose not use them, due to the particularities of our scope.

(ii) *Enumerate terms, define classes, properties, and constraints.* We defined the concepts obtained based on [28]. This vocabulary is a result of 8 years of research cataloging of criminals' terms in Brazil. It contains 1009 frequently used expressions. Despite its importance, we need a more formal model such as an ontology. The definition of classes was performed hierarchically as a top-down model. The expressions were analyzed by three researchers, who made the modeling decisions. Table 7.1 has three columns, CSE Expression in Portuguese, CSE Expression with the literal translation to English, and CSE Meaning in English.

The expressions "*Bater*", "*Piar*", "*Vomitar*", "*X9*" and "*Xisnovear*" (Table 7.1) are very distant from their direct denotative meanings. For example, CSE Expression *Vomitar/vomit* (Portuguese/English) means betraying someone by reporting confidential information. We also investigated a dataset of tweets from criminal areas to analyze how the criminals use such expressions. We filtered tweets using terms from [28]. For 2 days of the first week of October 2018, we collected 5,896,549 tweets in the list of 1009 CSE expressions. Twitter standard API has an endpoint, which allows us to retrieve, for each invocation, at most 100 tweets, posted in the last 7 days containing an expression. For each CSE expression,

we iterated through the search API results to get all the tweets we could get until reaching the 7-day limit. We restricted our search to the Portuguese language to get more trustful data as well as we filtered out retweets to avoid repetitions. The code developed for this task is publicly available [29].

(iii) *Creation of instances and Validation.* We created a set of instances from the phrases extracted of our tweet dataset to validate the model. These phrases make references to the ontology's terms and relationships. We considered *Positive* (*i.e.*, criminals communications that use the modeled terms), *False Positive* (*i.e.*, usual communications that use the terms) and *False Negative* (*i.e.*, criminals communications that do not use the terms) to review OntoCexp. Several iterations were performed with the three researchers until the definition of the current version of OntoCexp. The ontology must be continually revised, given that criminal language is very dynamic. OntoCexp is available in the GitHub repository [13].

7.4 The Core Model of OntoCexp

We consider the core model containing the main concepts of the specific domain, allowing thus to structure and understand other related concepts.

7.4.1 OntoCexp Core Model Description

The core model description starts with the definition of 20 classes (Fig. 7.2). We structured the conceptualization into five groups of concepts, described in the following subsections.

Conceptualization: Human Beings

The *People* class allows the modeling of men and women, as well as specificities with respect to their sexuality, characteristics, professions, among other aspects. As shown in Fig. 7.3, the ontology can represent a person, a place and physical aggressions. Very common among criminals, the distinction with regard to sexual choice is present in various expressions. The class Sexuality represents this. Another relevant factor is the organization of groups of individuals for the execution of a large proportion of criminal acts. The ontology also represents the individual's profession.

Conceptualization: Crime Activities

There are various types of criminal activities (*e.g.*, obtain weapons, drugs, and money), including the development of cyber weapons and attacks. Crimes are categorized

Table 7.1 Criminal slang expressions (CSE)

CSE expression (Portuguese)	CSE expression (English)	CSE meaning (English)
Bater	Beat	To tattle on, to betray.
Piar (Piá)	Tweet	Show up, to betray.
Vomitar	Vomit	To report, to betray.
X9	–	Informer.
Xisnovear	–	To betray.

Fig. 7.2 Core classes of OntoCexp

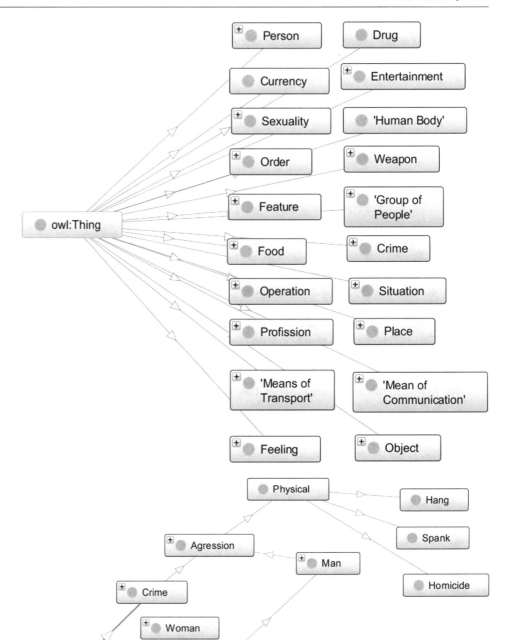

Fig. 7.3 Human beings related classes

as physical and verbal. The physical ones include beatings, murders, and hangings. The act of committing crimes can be still related with the type of weapon that was used.

Weapons are distinguished as firearms, cold, explosive, chemical, nuclear, and biological. This distinction is necessary, for example, to infer the scale of the crime. These acts can be, for instance, an armed robbery, a chemical attack in a public event, or even a massive cyber attack. Other criminal acts are associated with money, such as: purchasing drugs, bribery, and kidnappings.

Conceptualization: Transport and Communication

Police officers can intercept, understand, and make a better prediction of criminal acts when identify the transportation and communications means. For example, identification of a vehicle used by criminals to perform a crime is a key aspect in the investigation; the vehicle may have been negotiated using social networks.

Conceptualization: Entertainment and Subsistence

In the realm of our research, criminals usually participate in illegal gambling involving money; consumption of alcoholic beverages and drugs is also usual. Classes to represent these events are useful, since events they are frequently linked to other criminal activities.

Conceptualization: Criminal Operations

Place represents the geographic space where something related to a crime can be identified. Criminals have particular ways of referring to the locals where they act. Often it is necessary to identify what is happening in a particular place. For example, the expression "*Os coloniais vão berimbolar a gaiola*" does not make sense in standard Portuguese, but it is possible to extract three elements using the ontology: "*coloniais*", "*berimbolar*" and "*gaiola*", which, respectively, mean "*Prisoners of the Cândido Mendes Penitentiary*", "*Rebellion*" and "*Prison*". Thus, we can identify the intention of starting a rebellion of prisoners in a prison.

We also need to understand the execution of criminal (or police) operations; for example, ambush, lookout, invasion or escape operations. OntoCexp also models objects used by criminals. For example, the term "*Balancinho*" (child swing) is, in crime slang, a rope object that is used by detainees to climb to see through the jail.

7.4.2   Discussion on OntoCexp

OntoCexp describes the domain of Internet criminal communications. Criminals expressions were individually analyzed for conceptualize each term presented in [28]. With a well-defined scope, OntoCexp does not cover all concepts of

the crime expressions. The current version (1.0) of OntoCexp has 672 Axioms, 162 Logical Axioms, 133 Classes, 30 Object Properties, Object Property Domains, and Axioms. It attained ALCH(D) expressivity of the description logic.

OntoCexp represents a complex and dynamic domain; updating and evolution is a must. This implies continuous maintenance effort. It is also necessary to perform evaluation of its use in a tool supporting the task of monitoring and investigating crimes by means of Internet communication.

7.5   Instantiation and Validation of OntoCexp

We first filtered a set of candidates from the entire dataset (5,896,549 tweets) using criminal terms. Next, a subset of the twitter (274 messages) was analyzed. Finally, 14 selected "cases", as the most representative ones, were instantiated and used in the validation of OntoCexp. The following hypothetical scenario illustrates the utilization, validation and representativeness of the ontology.

This scenario starts with a criminal investigator searching Twitter messages related to a "kill by burning alive" crime. A search with "normal terms" may not return key messages related to the crime. Using OntoCexp, a supporting tool can highlight suspected messages. For instance, the following message is from our twitter dataset: *Soprar na unha, Fumo, Colocar no buraco, Ahaaaaa! Queimada. Apontar, Espremer* (Blow in the nail, Smoke, Put it in the hole, Ahaaaaa! Burned, To point, Squeeze). By using the ontology (Fig. 7.4), it is possible to identify key crime expressions: *Soprar* (Blow) means informing, *Unha* (Nail) means drug dealer, and *Colocar no buraco* (Put it in the hole) means to kill by burning alive.

Rules can be used to rank messages about criminal acts. This sentence gets a high-level of risk because three different terms related to the criminals' slang are identified in the same

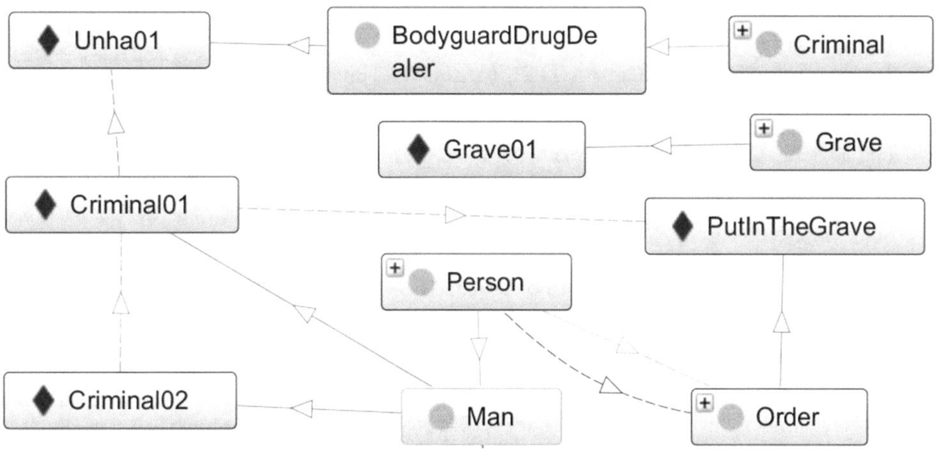

Fig. 7.4 Representation of the Tweet (example)

sentence. The risk indicator allows a more precise analysis, taking into account only messages with a high probability of being related to criminal acts. In these cases, OntoCexp act as a representation model to be used with other computational techniques such as search algorithms, data mining, as well as to support human based activities.

7.6   Conclusions

The OntoCexp is a conceptual proposal; it aims to provide a common and extensible model for identifying the use of crime expressions in the Internet. Its foundations come from an initial terminology and an analysis of written communication acts between criminals (from a Brazilian Twitter dataset). 17 papers on ontologies, out of 63 articles of interest, have been selected and used as input to our proposal. The initial version of OntoCexp containing its core elements is presented in this article. As future work, we intend to expand the universe of concepts to cover a larger number of terms, as well as, to incorporate automated systems to identify the intention in natural language texts. We also plan to extend the ontology to integrate criminal acts with the language used for planning Internet attacks.

References

1. Rupa, C., Avadhani, P.S.: Message encryption scheme using cheating text. ITNG 2009 – 6th Int. Conf. Inf. Technol. New Gener. **1**, 470–474 (2009)
2. Moghaddam, F.F., Karimi, O., Alrashdan, M.T.: A comparative study of applying real-time encryption in cloud computing environments. 2013 IEEE 2nd Int. Conf. Cloud Netw. **1**, 185–189 (2013)
3. De Matos, S.C.: A Lingua dos "Filhos errantes da Sociedade": Discurso, Poder e Discriminação nas gírias do sistema penitenciário do interior do Tocantins, no. 2004, pp. 559–570 (2013)
4. Gill, P., Corner, E., Conway, M., Thornton, A., Bloom, M., Horgan, J.: Terrorist use of the internet by the numbers: quantifying behaviors, patterns, and processes. Criminol. Public Policy. **16**(1), 99–117 (2017)
5. Cerqueira, D., Lima, R., Bueno, S., Neme, C., Ferreira, H., Coelho, D., Alves, P., Pinheiro, M., Astolfi, R., Marques, D., Reis, M., Merian, F.: Atlas da Violência. Ipea and Brazilian Forum on Public Security (2018)
6. Júnior, A.P.C., Veiga, E.F., Barbosa, J.L.F., Calixto, W.P., Silva, L.F.A., Campos, P.H.M., Gomes, V.M., Castro, L.L.O.P.: Ontology Applied in the Judicial Sentences. 2017 Chil. Conf. Electr. Electron. Eng. Inf. Commun. Technol. Proc. **2017**, 1–6 (2017)
7. Michel, M.C., Carvalho, M., Crawford, H., Esterline, A.C.: Cyber identity: salient trait ontology and computational framework to aid in solving cybercrime. 2018 17th IEEE Int. Conf. Trust. Secur. Priv. Comput. Commun. **1**, 1242–1249 (2018)
8. Jo, D.W., Kim, M.H.: Web-based semantic web retrieval service for law ontology. Proc. – 2013 IEEE Int. Conf. High Perform. Comput. Commun. **1**, 666–673 (2014)
9. Osathitporn, P., Soonthornphisaj, N., Vatanawood, W.: A scheme of criminal law knowledge acquisition using ontology. Proc.18th IEEE/ACIS Int. Conf. Softw. Eng. Artif. Intell. Netw. Parallel/Distributed Comput. **1**, 29–34 (2017)
10. Gang, L., Yingge, M., Kejun, W., Shaobin, H.: A domain security ontology network constructing and hardening technology. Proc. – 2014 4th Int. Conf. Instrum. Meas. Comput. Commun. Control. IMCCC. **2014**, 788–793 (2014)
11. Rosa, F.F., Jino, M., Bonacin, R.: Towards an ontology of security assessment: a Core model proposal. In: Latifi, S. (ed.) Information Technology – New Generations. Advances in Intelligent Systems and Computing, vol. 738, pp. 75–80. Springer, Cham/Las Vegas (2018)
12. Wu, L., Morstatter, F., Liu, H.: SlangSD: building, expanding and using a sentiment dictionary of slang words for short-text sentiment classification. Lang. Resour. Eval. **52**(3), 839–852 (2018)
13. Mendonça, R.R., Rosa, F.F., Bonacin, R.: OntoCexp – ontology of criminal expressions. Available: https://github.com/ricardoresende/OntoCexp (2018). Accessed 19 Oct 2018
14. Kitchenham, B.: Procedures for performing systematic reviews. Keele, UK, Keele Univ. **33**(TR/SE-0401), 28 (2004)
15. Santos, L.E., Girardi, R., Novais, P.: A case study on the construction of application ontologies. Proc. 10th Int. Conf. Inf. Technol. New Gener. ITNG. 619–624 (2013)
16. Orphanides, C., Akhgar, B., Bayerl, P.S.: Discovering knowledge in online drug transactions using conceptual graphs and formal concept analysis. Proc. – 2016 Eur. Intell. Secur. Informatics Conf. EISIC. **2016**, 100–103 (2017)
17. Rodrigues, C.M.O., de Azevedo, R.R., de Freitas, F.L.G., da Silva, E.P., Barros, P.V.S.: An ontological approach for simulating legal action in the Brazilian penal code. Proc. 30th Annu. ACM Symp. Appl. Comput. SAC '15. 376–381 (2015)
18. McDaniel, M., Sloan, E., Day, S., Mayes, J., Esterline, A., Roy, K., Nick, W.: Situation-based ontologies for a computational framework for identity focusing on crime scenes. 2017 IEEE Conf. Cogn. Comput. Asp. Situat. Manag. CogSIMA. **2017**, 1–7 (2017)
19. Rodrigues, C.M.O., De Freitas, F.L.G., Oliveira, I.J.S.: An ontological approach to the three-phase method of imposing penalties in the Brazilian criminal code. Proc. – 2017 Brazilian Conf. Intell. Syst. BRACIS 2017. **2018**, 414–419 (2018)
20. McDaniel, M., Sloan, E., Nick, W., Mayes, J., Esterline, A.: Ontologies for situation-based crime scene identities. SoutheastCon 2017, Charlotte, NC, 1–8 (2017)
21. Wyner, A., Hoekstra, R.: A legal case OWL ontology with an instantiation of Popov v. Hayashi. Artif. Intell. Law. **20**(1), 83–107 (2012)
22. Capuano, N., De Maio, C., Salerno, S., Toti, D.: A methodology based on commonsense knowledge and ontologies for the automatic classification of legal cases. Proc. 4th Int. Conf. Web Intell. Min. Semant. – WIMS '14. **1**, 1–6 (2014)
23. de Araujo, D.A., Rigo, S.J., Barbosa, J.L.V.: Ontology-based information extraction for juridical events with case studies in Brazilian legal realm. Artif. Intell. Law. **25**(4), 379–396 (2017)
24. El Ghosh, M., Naja, H., Abdulrab, H., Khalil, M.: Towards a legal rule-based system grounded on the integration of criminal domain ontology and rules. Procedia Comput. Sci. **112**, 632–642 (2017)
25. Rodrigues, C.M.O., De Freitas, F.L.G., De Azevedo, R.R.: An ontology for property crime based on events from UFO-B foundational ontology. Proc. – 2016 5th Brazilian Conf. Intell. Syst. **1**, 331–336 (2016)
26. Dhouib, K., Gargouri, F.: Legal application ontology in Arabic. 4th Int. Conf. Inf. Commun. Technol. Access. ICTA. 2013, 1–6, (2013)
27. Noy, N.F., Mcguinness, D.: Ontology development 101: a guide to creating your first ontology. Knowl. Syst. Lab. **32**, 1–25 (2001)
28. Mota, J.A.: Glossário de Palavras e Expressões Utilizada oor Facções Criminosas e Presos [Online]. Available: http://egp.seaprj.com/wp-content/uploads/2018/01/Palavras-e-Expressões-Facções.pdf (2016). Accessed 19 Oct 2018
29. Theophilo, A.: Twitter reader – python code [Online]. Available: https://github.com/theocjr/twitter-reader (2018). Accessed 29 Oct 2018

An Instrument for Measuring Privacy in IoT Environments

Bruno Lopes, Diego Roberto Gonçalves de Pontes,
and Sergio Donizetti Zorzo

8.1 Introduction

The number of applications using the Internet of Things concept has grown in recent years, where machines and objects use communication in a smart way. This growth allowed IoT to evolve into a global platform capable of processing and self-managing information [3]. The autonomy of the IoT devices connected to various applications allows the design of intelligent environments [2]. In this way the concern with the privacy of the user is increased, because in these intelligent environments the devices will be connected in network [7].

IoT applications are constantly collecting personal data, allowing data to be compared and new information to be found, such as cell phone numbers, documents and addresses. On the other hand, the regulations needed to help users protect their privacy did not develop at the same speed as IoT [10]. Thus, aggressive access and data transmission practices are used by applications, mobile operating systems, and other objects that make these concerns even more daunting [18].

The understanding of privacy in the behavior of individuals is dependent on the contextual nature where privacy presents itself [5]. Users' concerns about their privacy in the context of various Internet of Things applications scenarios may highlight the flexibility of permitting the collection, storage and use of their private data [11]. However, there is a lack of tools capable of measuring such information in an IoT environment, as well as the relationship with privacy issues [16].

In this context, this paper describes an instrument capable of measuring the users privacy concerns in IoT environments. This instrument aims at cooperative research, allowing other researchers to use and test in their own applications. This paper also describes a proposed privacy inference model for implementing a IOT trading privacy mechanism.

This paper is organized as follows. Section 8.2 presents background knowledge on Internet of Things, an IoT reference model and existing scales. The Sect. 8.3 explains how IoTPC was created and its features. Section 8.4 has an explanation of the methodology used in this work and describes the IoT data analysis step. Section 8.5 is dedicated to the inference module. The results and assessment of IoTPC and the inference models are shown in Sect. 8.6. Finally, in Sect. 8.7 our conclusions are presented and the limitations of this approach and future works.

8.2 Background Knowledge

8.2.1 Internet of Things

Defining the term Internet of Things can be considered a difficult task since its concept can change over time depending on the approach it is used for.

Generally speaking, Internet of Things can be defined as a new approach regarding the interconnection of technology and objects through a network of computers, making it possible to define the concept of a global network of devices [8]. However, the Internet of Things depends on technological process to keep evolving. That said, IoT cannot be treated as a new disruptive technology, but as a paradigm of computing that is in constant evolution [19].

A few authors agree that the term "things" does not refer only to physical objects, but also to living entities or virtual representations. This way, anything that is connected to the

B. Lopes (✉) · D. R. G. de Pontes · S. D. Zorzo
Computer Science Department, Federal University of São Carlos, São Carlos, Brazil
e-mail: zorzo@ufscar.br

© Springer Nature Switzerland AG 2019
S. Latifi (ed.), *16th International Conference on Information Technology-New Generations (ITNG 2019)*,
Advances in Intelligent Systems and Computing 800,
https://doi.org/10.1007/978-3-030-14070-0_8

internet and is able to transmit information can be considered an IoT device [12].

8.2.2 Reference Model

The great number of devices in IoT allows the creation of different scenarios for smart environments. These scenarios can be described both by entities and the information flow that happens in these environments. The IoT reference model that we take into consideration in this paper was proposed by [19], who based their visions in the International Union of Telecommunications and the IoT European Investigation Council, which say that anything or any person is interconnected anywhere, anytime, by a network of any kind of service.

An example of this model is shown in Fig. 8.1, indicating that "smart things" can be any day-to-day device that is able to collect, process, and transmit data in a certain environment. These devices collect data about users who use the infrastructure of IoT, getting as a final product the dissemination of information through IoT services [19]. The usage of this module served as base for investigating the privacy concerns in different IoT scenarios.

8.2.3 Existing Scales

As mentioned before, the rise of new technologies have increased the concerns with privacy. Some authors proposed instruments that are able to measure privacy empirically. In [16], the authors proposed an instrument that was able to measure the privacy of users and organizational practices called "Concern for Information Privacy", CFIP. The CFIP was made by a scale of 15 items and the following privacy concern dimensions: collection, unauthorized secondary usage, unauthorized access and mistakes. In [11], observing the lack of trust from consumers regarding e-commerce privacy, the authors conducted a study on how big was the users' concerns with privacy and created an instru-

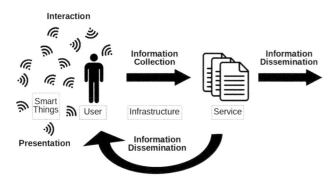

Fig. 8.1 IoT reference model. (Adapted from [19])

ment called "Internet Users' information Privacy Concerns", IUIPC.

IUIPC was developed based on CFIP, from which they extracted some scenarios that belonged to privacy concerns in e-commerce. IUIPC had a scale of 10 items and distributed them into three dimensions: collection, control and awareness of privacy practices.

Inspired by [11] and [16], the authors in [18] developed another instrument with a goal to represent users' privacy preferences in mobile devices called "Measuring Mobile Users' Concerns for Information Privacy", MUIPC. It had a scale of 9 items distributed across three dimensions: vigilance, intrusion and unauthorized secondary usage. Though it was able to measure the users' concerns with privacy in mobile devices, MUIPC did not reflect all relevant features in the concept of privacy concerns that exist in the literature, such as the dimensions of control and collection [5].

Observing the lack of these dimensions, in [5] the authors developed something similar to MUIPC that not only comprehended mobile devices but all smart devices that use apps as a technological interface. The instrument, called "App Information Privacy Concern", AIPC [5], was made with 17 items spreaded into 5 dimensions: collection, control, consciousness, unauthorized secondary usage and general concerns with privacy.

Though many instrument are able to measure privacy in certain contexts, none of them explore the Internet of Things scenario, being restricted only to mobile devices (MUIPC and AIPC), e-commerce (IUIPC) and a more general context (CFIP). For this reason, we proposed an instrument that is able to measure the concerns with privacy in IoT environments called "Internet of Things Privacy Concerns", IoTPC.

8.3 Internet of Things Privacy Concerns (IoTPC)

The developed instrument and presented in this paper is able to measure the degree of concerns related to privacy in IoT environments based on existing works in the literature such as IUIPC [11], MUIPC [18] and AIPC [5].

In the conception of the proposed instrument, first was needed to analyze and clearly define its dimensions so that afterwards its items could be organized into them. In this sense, the dimensions of IUIPC and MUIPC were selected as a starting point.

The dimensions of these two instruments were analyzed to check if they reflected some characteristics of concerns of privacy in IoT environments, such as the secondary data usage dimension that is present in MUIPC. This dimension usually reveals that users feel their privacy violated when third-parties use their data without their consent. This dimension can also comprehend privacy issues in IoT environments, since devices collect data and can eventually make them available for third-parties.

The other dimensions of IUIPC and MUIPC were analyzed and the dimensions used in IoTPC are described as:

1. IoT requests: in [11] the authors stated that the concept of acquisition is related to the degree that a certain individual is concerned about how much of their personal data is in the hands of third-parties by analyzing its cost-benefit. However, the heterogeneity of devices in the IoT scenario and the fact that the vast majority of these devices collect data from their users in an omnipresent way rise the concept of vigilance, which is also important to be considered for the definition of this dimension. In [13], the authors defined vigilance as the act of observing and listening to an individual's recordings. In a futuristic environment of IoT, the users' devices can eventually use the acquired data to track and spy on their own users.
2. Decision making: according to [11], the concept of control can be defined as an individual having interest in controlling or at least significantly influencing the use of their private data. Many times, in an IoT environment, users do not have absolute control of their private information, which creates certain discomfort and the sensation of having their privacy preferences violated. According to [13], intrusion can be defined as an invasive act that can disturb tranquility or loneliness. The concept of intrusion also applies to an IoT scenario, since devices can require data from their users, eventually disturbing or even taking them out of a calm state of mind.
3. Previous knowledge: this dimension is related to all knowledge about privacy practices that an IoT user has. We can use the concept of consciousness to define

this dimension, which is the degree that an individual is concerned with their knowledge about privacy practices related to private organizational information [11].

4. Secondary usage of information: sometimes data provided by users for a certain end is used in a different way, as opposed to what it was meant to and without the users' consent (for instance, profiling users and sending marketing messages to them) [16]. This practice reflects privacy concerns in IoT environments and was also taken into consideration to build the dimensions of IoTPC.
5. General privacy concerns: this dimension is responsible for representing items such as "I am concerned with threats to my privacy today" and "Compared to others, I am more concerned about the way IoT devices handle my personal data". This dimension aims at capturing the users' general concerns in an IoT environment [5].

Concerned with these items, the area that our instrument would act was limited in order to keep that it precise and accurate. It was analyzed 64 micro scenarios in IoT considering the research of [9]; for instance, "The fridge wants to have access to your bank records so it can buy food as soon as it is out of order" and 25 scenarios were collected to compose a general futuristic IoT scenario. The scenario exemplifies the routine of a university student interacting with the Internet of Things from their home to the university.

As well as the dimensions of IoTPC, the items related to each dimension were created based on the instruments mentioned in the previous section. The items of each instrument went through an assessment to check if they reflected any privacy characteristic in IoT environments, based on the scenario proposed in this research. The IoTPC items are shown in the Table 8.1.

Table 8.1 IoTPC items

N.º	Items
1	Privacy in IoT environments is a matter of users having the rightto control their decisions on how their data are collected, usedand shared
2	IoT devices that require data must indicate how they are collected,processed and used
3	It is very important to me to be aware and know about how mypersonal data is used
4	It usually bothers me when IoT devices ask for my personalinformation
5	When IoT devices ask for personal data, sometimes I think twicebefore allowing them to get it
6	It bothers me to provide personal data to so many IoT devices
7	I believe that the localization of my mobile device is monitoredat least part of the time
8	I am concerned that IoT devices are collecting too muchinformation about me
9	I am concerned that IoT devices can monitor my activities usingmy mobile device
10	I feel that, as a result of using IoT devices, other people knowmore about me than I would like to, making me uncomfortable
11	I believe that, as a result of using IoT devices, my private datais now more accessible to others than I would like to
12	I am concerned that IoT devices can use my personalinformation to other ends without notifying me or askingfor my authorization
13	When I provide personal information in IoT environments, I'mconcerned that IoT devices can use my data to other ends
14	I am concerned that IoT devices can share my personalinformation with third-parties without my authorization
15	Compared to other people, I am more sensible to the way IoTdevices handle my personal data
16	To me, the most important thing is to keep my privacy intactwhen using IoT devices
17	I am concerned about threats to my privacy nowadays

The fact that many scenarios are possible in a context of IoT makes it almost impossible to create an instrument that is able to cover all possible situations. In [9], the authors conducted a study on the factors that influence user privacy in IoT. Many different scenarios were created to try to exemplify and collect these factors. Among the scenarios, five main items were brought to light:

- "Where", the place where the data was collected;
- "What", the kind of data that was collected;
- "Who", the agent responsible for collecting such data;
- "Reason", the goal of collecting such data;
- "Persistence", the frequency with which such data is collected.

8.4 Instrument Validation

To identify the concerns with privacy, it was conducted a study focused on collecting privacy preferences of IoT users.

First, it was examined the nature of privacy during the literature review process as well as the definitions, principles, legal issues and their applications in different areas of study. From there, it was able to list possible privacy violations in an IoT environment. After this analysis, it was conducted a qualitative essay using an electronic questionnaire where IoTPC Items were measured using a Likert scale of five points, ranging from "Completely disagree" to "Completely agree", with the neutral point being "I have no opinion about it".

The essay was conducted between the first week of December 2017 and the last week of January 2018 and the participants were undergraduate students with ages ranging from 18 to 37 years of age. Around 60 participants (N = 61) contributed to our research, where 85.2% (N = 52) were male and 14.8% (N = 9) were female. There is a great number of statistical techniques in the literature that are able to assess efficacy of a certain instrument in a certain scenario. In [6], the authors stated that factor analysis is a technique used to identify groups or clusters of variables that can have three main uses: (a) to understand the structure of a set of variables; (b) to create a questionnaire that is able to assess a subjacent variable and (c) reduce a dataset to a more manageable size keeping as much of the original data as possible.

According to [4], factor analysis is usually understood as a set of models that are intimately related with exploring or establishing the structure of correlation between observed random variables. By trying to assess if our instrument was able to reflect the privacy preferences of IoT users, we used the exploratory factor analysis (EFA) to validate the 17 items of IoTPC. The technique used to extract latent variables was principal factor axis (PFA).

A Bartlett Sphericity Test was realized to check if the correlation of variables was high enough for the EFA as well as the Kaiser-Meyer-Olkin (KMO) measure, which is used to check if the sample size is big enough to conduct an analysis on. The Cronbach alpha measure was used to validate the trustability of our instrument; it was used the assessment criteria from [17], where values under 0.700 are considered insufficient and values above 0.800 are considered sufficient, whereas values above 0.300 are considered good for total-item correlation correction [6].

8.5 IoTPC Learning Inference Module

To use IoTPC as an instrument for privacy negotiation in the IoT environment, it was necessary to establish conversion rules to transpose the IoTPC responses in order to make inferences from IoT scenarios. The information and services required by the scenarios used in the IoTPC construction were classified as follows: (I) critical information – CI, (II) simple information – SI, (III) critical service – CS and (IV) simple service – SS.

Critical information regards scenarios that involve the collection of sensitive data (documents, bank records and others) such as in "The fridge wants access to your bank records in order to buy food as soon as it is out of order".

Simple information regards the collection of less sensitive data, such as musical preferences and eating preferences, as in "The air conditioner of your office wants to know your temperature preferences to adjust it accordingly".

In a similar way to information classification, some IoT scenarios not only ask for a particular kind of data, but they also state the purpose to which they will use the data. With that in mind, critical service regards scenarios that entail a cut in resources expenditure such as fuel, energy and others, as in "The thermostat of your house wants to know your day-by-day schedule to adjust the temperature accordingly and save energy".

Simple service is related to scenarios that use user data that are less important, such as "The gym wants to know your musical preferences to play songs that you like".

The answers provided by users were associated with a yes or no response, providing the required data or not, and thus determining the inference results for the scenario associated with that particular item.

In some cases, a single IoTPC item reflected the inference answer for more than one IoT scenario, being possible to get more than one service classification or more than one information for the scenarios. We built an inference module for IoT scenarios called IoTPC Learning. For this module, we generated a database with the inference results taken from the conversion rules, which led to the choice of a machine learning algorithm able to infer answers for these scenarios according to the privacy preferences given by the IoTPC. We decided to use a decision tree due to the small size of our dataset (N = 61) and the fact that the majority of our features were discrete, having only two possible outcomes (yes or no).

A decision tree is defined as a classification procedure that repetitively partitions a dataset into smaller subdivisions based on a set of defined tests for each branch of the tree [15]. Some preprocessing was conducted before generating the final decision trees. We applied a technique to avoid overfitting [1] and used a resample filter to balance the number of samples in each class. Finally, we used a 10-fold cross-validation during the training phase [1]. For each one of the 25 scenarios, a decision tree was created for it and the inference was based only on the personal features of the users, avoiding influence from the other 24 scenarios.

8.6 Results

8.6.1 Instrument Validation

The Cronbach alpha measure presented a result of (p = 0.911) and very high trustability [17]. The scores for individual items of IoTPC, mean, standard deviation, total-item correlation coefficient and Cronbach alpha measure for the excluded items are shown in the Table 8.2. The average score for individual items varied from 3.34 for item 15 ("Compared to other people, I am more sensible to the way IoT devices handle my personal data") to 4.74 for item 3 (It is very important to me to be aware and know about how my personal data is used). The total correlation of the corrected item was above 0.3, ranging from 0.33 for

item 2 ("IoT devices that require data must indicate how they are collected, processed and used") to 0.79 for item 8 ("I am concerned that IoT devices are collecting too much information about me").

The Sphericity Bartlett Test (p < 0.001) confirmed that there was no generation of an identity matrix, from which we conclude that the variables were significantly correlated.

The KMO analysis was enough to get a value of (p = 0.774) based on [6], indicating that the correlation matrix was adequate for the factor analysis. To maximize the extraction of factors was used an orthogonal rotation (Varimax) considering that the factors to be excluded did not have any kind of correlation [6]. The three extracted factors are shown in the Table 8.3.

In the Table 8.3 the Factor 1 is related to the dimension of IoT requests, since the items of this factor belong to this dimension. Factor 2 got variables from the IoT requests, decision making and general concerns of privacy dimensions. Finally, Factor 3 got items belonging to the dimensions of previous knowledge, secondary usage of information and decision making.

Factor 1 was composed by the items 14, 16, 8, 15 and 9. This new dimension of IoTPC clearly reflects the concerns of IoT users regarding the data collected by IoT devices. Given the fact that all items in this factor belong to the IoT requests, Factor 1 was named "IoT request".

Factor 2 was composed by the items 11, 15, 17, 16, 10 and 7. This new dimension is strongly related to two main aspects: the first one was the fact that IoT users are worried about controlling their private data and the second fact is related to questions about general privacy concerns, since all items that concern this dimension were comprehended in this

Table 8.2 Cronbach alpha analysis

Item	M	S.D.	Total correlation of the corrected item	Cronbach alpha if the item is deleted
1	4.44	0.958	0.498	0.909
2	4.41	1.131	0.330	0.913
3	4.74	0.705	0.502	0.909
4	3.82	1.232	0.620	0.905
5	4.11	1.212	0.562	0.907
6	3.87	1.245	0.634	0.905
7	4.28	1.280	0.332	0.914
8	3.85	1.352	0.790	0.900
9	3.87	1.372	0.644	0.905
10	3.80	1.364	0.624	0.905
11	3.97	1,211	0.536	0.908
12	4.30	1.022	0.702	0.904
13	4.21	1.112	0.697	0.903
14	4.34	1.031	0.559	0.907
15	3.34	1,569	0.757	0.901
16	3.59	1.383	0.444	0.911
17	3,89	1,380	0.773	0.900

Table 8.3 Result of principal axis analysis using varimax rotation

Item	Factor 1	Factor 2	Factor 3
14	**0.817**	0.177	0.066
16	**0.722**	0.116	0.291
8	**0.677**	0.379	0.336
15	**0.595**	0.130	0.230
9	**0.562**	0.403	0.190
11	0.181	**0.781**	0.041
15	0.556	**0.674**	0.127
17	0.493	**0.643**	0.227
16	0.133	**0.423**	0.280
10	0.398	**0.402**	0.320
7	0.045	**0.366**	0.218
2	0.098	−0.001	**0.601**
14	0.167	0.325	**0.586**
12	0.386	0.347	**0.585**
1	0.117	0.283	**0.569**
13	0.478	0.244	**0.568**
3	0.324	0.099	**0.491**

factor. Given that, the second factor extracted from IoTPC was named "Decision making".

Factor 3 was composed by the items 2, 14, 12, 1, 13 and 3. This new dimension reflected the IoT users' concerns with data collected by IoT devices and provided to third-parties without their authorization. It also reflected the understanding of good privacy practices. With that in mind, the third factor extracted from IoTPC was named "Caution".

8.6.2 IoTPC Learning Inference Module

The results of the 25 decision trees generated for IoT scenarios are presented in a grouped way. As can be seen in Fig. 8.2, from the 61 training samples used for learning, approximately 79% were correctly predicted, representing 48.32 answers. On the other hand, for wrong predictions, we got a ratio of 20.45%, which is equal to 12.48 answers. Generally speaking, the average accuracy level of 25 models was 79.20%, which is a good result.

A few interesting points can be noted during the analysis of each one of these models. In Fig. 8.3, we can clearly see that the level of correct predictions is higher than the number of incorrect predictions. Moreover, some scenarios got the same number of correct and incorrect inferences as well as the same level of accuracy, which was the case for scenarios 14 and 15. These similarities were due to the fact that among the 25 scenarios, some of them dealt with similar situations. In this example, scenario 14 ("Your car wants access to your final destination to calculate the most efficient route") and scenario 15 ("Your car wants information about your itinerary to determine if you have enough fuel for the route") try to save fuel while driving. In this sense, we expected similarities between the models of these two scenarios.

The IoTPC Learning structure was divided into three privacy factors extracted from the exploratory factor analysis:

IoT requests, Decision Making and Caution. This way, all 25 decision trees were distributed among these three factors, making it possible to control which dimension belonged the requests created in the inference module.

The results we got with IoTPC learning presented an accuracy level that is lower when compared to other existing mechanisms for privacy negotiation, such as privacy application [14]; nonetheless, there are a few points worth mentioning.

In privacy application, and the other privacy negotiation mechanisms, the learning process of their inference modules are given by the continuous use of the mechanism, making the user answer a series of questions until they are able to provide correct inferences. By integrating IoTPC learning to these instruments, the initial calibration would not be necessary, since together they would be able to tell which information is considered private or not in a particular environment.

Another point to observe is that the validations used in IoTPC Learning, such as pruning and resampling, ensure a higher level of trustability for the module. Since IoTPC Learning is in the area of privacy negotiation, its ability to learn becomes continuous. The more IoTPC Learning is used, the more it is able to gradually improve its predictions.

8.7   Limitations, Conclusions and Future Works

This paper presented the process of building a instrument to measure the concerns with privacy from users in a certain IoT scenario. It was also presented the developed inference module for IoT scenarios and the results were showed. The main contribution of this work was this new instrument, properly validated which is able to measure such concerns, and an option to inference models in the IoT scenarios that can to be useful in privacy negotiation mechanisms.

In this process, it was discovered three first order dimensions during the analysis of the instrument IoTPC and also checked the level of trustability of it. IoTPC can define the degree of concern with privacy of each user regarding IoT and how much each of the three dimensions is related to that particular concern.

The IoTPC Learning analysis presented 25 learning models, where each model represented the result of the inference corresponding to an IoT scenario. It also offered a starting point for inference and privacy negotiation mechanisms, being able to create a feedback about the most requested privacy dimensions in IoT devices.

This paper faced a few limitations. The fact that the privacy context in this paper was IoT made us develop an instrument for a very specific scenario, being unable to contemplate all privacy violation possibilities in an IoT environment.

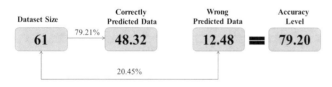

Fig. 8.2 Level of accuracy of the learning process

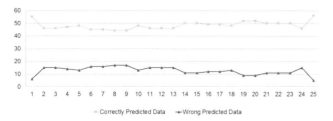

Fig. 8.3 Individual relationship of the number of correctly predicted data each scenario

The small dataset and small number of features, containing only age, sex and education level, also had an influence in the construction and validation of the inference module.

Future works aim to apply the Confirmatory Factor Analysis (CFA) to confirm the results got using EFA. Moreover, it is necessary to check if the inference module has a higher level of accuracy when compared to other modules from other privacy negotiation mechanisms.

References

1. Aggarwal, C.C.: Data Classification: Algorithms and Applications. CRC Press, London (2014)
2. Alaba, F.A., Othman, M., Hashem, I.A.T., Alotaibi, F.: Internet of Things security: a survey. J. Netw. Comput. Appl. **88**, 10–28 (2017)
3. Balte, A., Kashid, A., Patil, B.: Security issues in Internet of Things (IOT): a survey. Int. J. Advanced Res. Comput. Sci. Softw. Eng. **5**(4), (2015)
4. Basilevsky, A.T.: Statistical Factor Analysis and Related Methods: Theory and Applications, vol. 418. Wiley, New York (2009)
5. Buck, C., Burster, S.: App information privacy concerns. In: AMCIS – The Americas Conference on Information Systems (2017)
6. Field, A.: Discovering Statistics Using IBM SPSS Statistics: North American Edition. SAGE, London (2017)
7. Guo, K., Tang, Y., Zhang, P.: Csf: crowdsourcing semantic fusion for heterogeneous media big data in the Internet of Things. Inf. Fusion **37**, 77–85 (2017)
8. Koreshoff, T.L., Robertson, T., Leong, T.W.: Internet of Things: a review of literature and products. In: Proceedings of the 25th Australian Computer-Human Interaction Conference: Augmentation, Application, Innovation, Collaboration, pp. 335–344. ACM (2013)
9. Lee, H., Kobsa, A.: Understanding user privacy in Internet of Things environments. In: 2016 IEEE 3rd World Forum on Internet of Things (WF-IoT), pp. 407–412. IEEE (2016)
10. Lu, C.: Overview of security and privacy issues in the Internet of Things. In: Internet of Things (IoT): A vision, Architectural Elements, and Future Directions (2014)
11. Malhotra, N.K., Kim, S.S., Agarwal, J.: Internet users' information privacy concerns (IUIPC): the construct, the scale, and a causal model. Inf. Syst. Res. **15**(4), 336–355 (2004)
12. Oriwoh, E., Conrad, M.: "Things" in the Internet of Things: towards a definition. Int. J. Internet Things **4**(1), 1–5 (2015)
13. Peissl, W., Friedewald, M., Burgess, J.P., Bellanova, R., Čas, J.: Introduction: surveillance, privacy and security. In: Surveillance, Privacy and Security, Routledge, pp. 1–12 (2017)
14. Pereira Couto, F.R., Zorzo, S.: Privacy negotiation mechanism in Internet of Things environments. In: AMCIS – The Americas Conference on Information Systems (2018)
15. Rokach, L., Maimon, O.: Data Mining with Decision Trees: Theory and Applications. World Scientific, New Jersey (2014)
16. Smith, H.J., Milberg, S.J., Burke, S.J.: Information privacy: measuring individuals' concerns about organizational practices. MIS Q. **20**, 167–196 (1996)
17. Streiner, D.L.: Being inconsistent about consistency: when coefficient alpha does and doesn't matter. J. Pers. Assess. **80**(3), 217–222 (2003)
18. Xu, H., Gupta, S., Rosson, M.B., Carroll, J.M.: Measuring mobile users' concerns for information privacy. In: Thirty Third International Conference on Information Systems (2012)
19. Ziegeldorf, J.H., Morchon, O.G., Wehrle, K.: Privacy in the Internet of Things: threats and challenges. Secur. Commun. Netw. **7**(12), 2728–2742 (2014)

Forensic Analysis of LinkedIn's Desktop Application on Windows 10 OS

Saman Bashir, Haider Abbas, Narmeen Shafqat, Waseem Iqbal, and Kashif Saleem

9.1 Introduction

Digital forensics is defined as the implementation of scientific methods to find potential artefacts in a digital device that can be used as evidence in a forensic investigation [1, 2]. Digital forensics can be subdivided into three major domains; computer forensics, mobile forensics and network forensics. This paper pertains to computer forensics domain, and specifically, application forensics on Windows 10 Operating System (OS), since it is widely used today. Windows 10 was able to attract almost 14 million users on the initial day of launch in 2015 [3]. With 700 million users today [4], Windows 10 has eclipsed the usage of Windows 8 to become the second most used OS after Win 7 (Fig. 9.1).

Windows 10 has brought some novel features, such as Edge browser, Cortana Personal Assistant and Notification Database etc. [6]. From a forensic investigator's viewpoint, it brought changes in Jump lists [7], System Resource Usage Monitor (SRUM) [8] etc. Windows 10 is also available for mobile phones and hence its usage has considerably increased.

With the introduction of Windows AppStore and its ease of use, users now prefer to have both social media networks and work accounts on the same device. Subsequently, many users have downloaded desktop applications of WhatsApp, Linkedin, Facebook, Skype etc. from Windows AppStore on their PCs so that they can directly respond to notifications. Since social media applications provide a great deal of information, forensic investigators need to understand the forensic analysis of Windows based apps in order to locate maximum potential evidence.

LinkedIn, being one of the most widely used platforms, as depicted in Fig. 9.2, has been chosen for experimentation in this paper. LinkedIn, launched in 2003, now has around 562 million users around the globe [9]. It is a business-oriented application commonly used to find jobs, connect with people, share personal/ professional information and private messaging etc. All these private details can be used by criminals or hackers to their advantage [10], especially for masquerading and blackmailing etc. [11]. On the other hand, if any digital device of a criminal is found from the crime scene, then its forensic analysis can help reveal important details regarding the offender.

Criminal cases involving misuse of social media [13] are registered daily. In one incident [14], a job seeker was harassed by a recruiter through private messages and images over LinkedIn. Interestingly, the forensic analysis of RAM and hard disks of both parties revealed evidences of users' activities.

The forensic analysis of applications on the Windows Appstore is still in its infancy. The challenges include ever-developing nature of OS and the growing size of user's data [15]. Thus, a forensics investigator should be able to analyse the image files obtained from suspect's system either manually or using acquisition and analysis tools tabulated in Table 9.1.

The rest of the paper is structured as follows: Section 9.2 deals with related work done in this field and highlights shortcomings. Section 9.3 presents our proposed methodology followed by its evaluation in Sect. 9.4. Section 9.5 finally concludes the paper.

S. Bashir · H. Abbas (✉) · N. Shafqat · W. Iqbal
Department of Information Security, National University of Sciences and Technology, Islamabad, Pakistan
e-mail: sbashir.msis16mcs@students.mcs.edu.pk; haider@mcs.edu.pk; narmeen_shafqat@mcs.edu.pk; waseem.iqbal@mcs.edu.pk

K. Saleem
Center of Excellence in Information Assurance (CoEIA), King Saud University, Riyadh, Kingdom of Saudi Arabia
e-mail: ksaleem@ksu.edu.sa

© Springer Nature Switzerland AG 2019
S. Latifi (ed.), *16th International Conference on Information Technology-New Generations (ITNG 2019)*,
Advances in Intelligent Systems and Computing 800,
https://doi.org/10.1007/978-3-030-14070-0_9

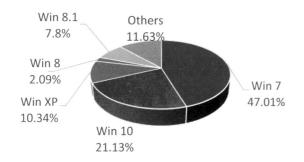

Fig. 9.1 Statistics of Windows OS [5]

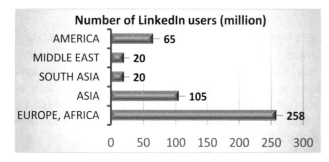

Fig. 9.2 Demographics of LinkedIn app [12]

9.2 Related Work

The applications available on Windows AppStore mimic the user-friendly mobile applications, while retaining traditional inputs from keyboard and mouse. These applications with APPX extension are found in %ProgramFiles\WindowsApps folder.

In 2011, N. Muttawa et al. tried to recover artefacts of Facebook from Win XP's hard disk [26]. They performed the test on Internet Explorer (IE), Chrome and Firefox and concluded that more artefacts can be recovered from IE.

In 2011, extensive research [27] focused on the forensic analysis of Windows registry from physical memory was carried out on Windows 7, XP and Vista. An algorithm for extracting the hives and pertinent files from memory was proposed.

In 2013, Saidi [28] defined ways to investigate illicit activities and unauthorized access by studying registry of Windows 7 using FTK Imager.

In 2014, Kumar, Majeed and Pundir [29] searched for sensitive information on physical memory of Windows server 2008 in various states like sleep, logoff, soft and hard reboot.

In 2015, Asma et al. [30] explored artefacts of Facebook, Skype and Viber on Windows 10 using an entire disk image.

Table 9.1 Acquisition and analysis tools

Name	Type	Issues
RAM image acquisition tools [16, 17]		
Dumpit	Open source	Sometimes de-allocates data
FTK imager	Open source	Cannot perform multi-tasking, deallocates memory data
Madiant memoryze	Open source	Difficult to use
Blekasoft live RAM capture	Open source	Tool needs to be reinstalled after image capture
RAM analysis tools [18–20]		
Hex editor	Free tool	Stability inconsistency
Blekasoft evidence center	Open source	Runs on a system that has python installed
Autopsy	Free tool	Little bit slow
Volatility	Open source	Crucial commands (cmd) can be missed
Storage image acquisition tools [21, 22]		
FTK imager	Open source	No multitasking, deallocates data
HDD raw copy tool	Freeware	It cannot lock a drive and close any open programs
Linux "dd"	Freeware	Ownership issues with dd cmd
Storage image analysis tools [23]		
FTK imager	Open source	No multitasking, search is laborious, deallocates data,
WinHex	Freeware	Manual search, time taking
Encase	Freeware	Often gives "job failed" error. Need to be reinstalled
X-ways	Free, paid (full)	Requires considerable experience to use
Autopsy	Free	Misses files sometimes
Registry analysis tools [24, 25]		
Reg edit	IIn windows	Often hangs on find cmd
Registry viewer	Free, paid (full)	Restarts after giving an error when lots of keys are open

Table 9.2 Summary of related work

Paper	OS	Memory	Summary
Muttawa et al. [26]	Win XP	Non-volatile	More Facebook artefacts recovered from IE than Chrome
Zhang [27]	Win XP, 7, Vista	Volatile	Recovered artefacts from hives of Windows registry
Saidi [28]	Win 7	Volatile, non volatile	Registry analysis for detection of malicious behaviour
Kumar [29]	Server 2008	Volatile, non- volatile	Recovered artefacts based on different modes of device
Majeed [30]	Win 10	Non-volatile	Recovered artefacts of Facebook, Viber, Skype
Lee [31]	Win 8	Non-volatile	Recovered artefacts from Viber and Line
Dija [32]	Win 7	Volatile	Framework proposed to recover.exe file
Yang [33]	Win 8.1	Volatile, non-volatile	Recovered artefacts and network analysis of Facebook and Skype
Choudhary et al. [34]	Win 10	Volatile	Recovered artefacts of Facebook (Didn't find deleted artefacts)
Ababneh [35]	Win 8	Volatile	Recovered artefacts from IMO
Meyers [36]	Win 10	Volatile	Propose automated tool for quick analysis of Windows 10 RAM

They were able to find various artefacts and their potential locations. Even deleted artefacts were recovered using Easeus software.

In 2015, Lee and Chung [31] studied the third-party Viber and Line apps on Windows 8 and identified that the package identifications (IDs) could be found by analysing the app caches. They also located records of account logins, contacts, chats, and transferred files. However, it was limited to dead analysis of hard disk.

In 2016, a research [32] focused on reconstruction of artefacts from physical memory of Windows 7 was carried out. This was challenging, since the sections reside in different pages and physical location.

In 2016, Yang et al. [33] conducted live and storage analysis of Facebook and Skype on Windows 8.1 and found contact lists, messages and conversations etc. The network analysis of these applications was also presented.

In 2016, Pankaj et al. [34] performed forensic analysis of Facebook on Windows 10 and found contact list, newsfeed, messages etc. in SQLite files. However, deleted artefacts were ignored.

In 2017, Ababneh et al. [35] carried out forensic study of IMO on mobile and Windows 8. Volatility and Windbg were used to capture volatile memory image of the OS and search for artefacts respectively.

In 2017, researchers [36] proposed an automated tool "Plugin for Autopsy" to extract image from volatile memory. All literature review carried out in this section has been summarized in Table 9.2.

The literature review done above is a blend of live and storage forensic analysis of Windows OS, artefacts recovery and their potential locations. The forensic analysis of Facebook, Skype and Viber [30, 33, 34] carried out on Windows 10 is limited in approach. Thus, it is concluded that minimal work has been done with regards to forensic analysis of applications on Windows AppStore. Also, major researches are conducted using tools for artefact's extraction but very few have been conducted manually.

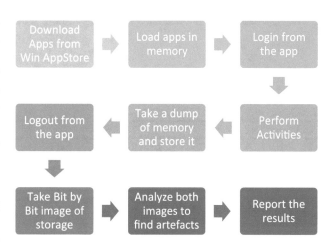

Fig. 9.3 Steps for Application Analysis

9.3 Proposed Methodology

In this section, we provide a generic forensic analysis methodology in order to analyse artefacts of APXX files downloaded from Windows AppStore. The suggested scenarios for this research are as follows.

- Tracking artefacts stored by live analysis of RAM,
- Analysing of artefacts stored in disk after logging in,
- Locating artefacts stored in the registry,
- Manually finding artefacts in the main folder where applications from the Windows AppStore and other folders are stored as well.

As per the proposed methodology, depicted in Fig. 9.3, the application is first downloaded from Windows Appstore on a laptop running Windows 10 OS. In our case, LinkedIn is downloaded and used for some time to generate sufficient user activity. When a user is signed in, the activities performed are stored in RAM. To analyse these activities, we need to take dump of memory using Dumpit tool. The image can be analysed using WinHex in order to gather artefacts and their timestamps.

Similarly, for storage analysis, the LinkedIn application was signed out and then memory dump was taken again using FTK imager and analysed carefully. The artefacts can also be recovered from registry using captured dumps.

9.4   Evaluation and Validation

The proposed technique follows manual analysis using Hex Editor and manual exploration of artefacts files in order to provide in-depth analysis of the selected application on Windows OS. Contrary to this, majority of the techniques mentioned in Sect. 9.2 use forensic tools to gather required artefacts, which may be incomplete or wrong. These tools

may not be compatible with all Operating systems and applications. Hence, the investigator needs to apply his intellect in order to find all potential evidences [37].

For validation of our proposed technique, we have tested our methodology on LinkedIn. Few dummy accounts were created and activities such as job search, messaging, invitation, comments and image exchange etc. were performed. Initial results of RAM, storage and registry artefacts obtained after analyzing LinkedIn app show that artefacts were stored in main folder i.e. DiskDrive/Users/System/AppData/Local/Packages/7EE7776C.LinkedInforWindows_w1wdnht996qgy.

After manual analysis of the images taken in WinHex and FTK imager, it was observed that artefacts were also found in various other folders outside main folder such as Microsoft edge browser folder, Microsoft\Windows\AppRepository,

Table 9.3 Artefacts and their potential locations

Artefacts	Location
RAM analysis	
Credentials	In Main folder found using WinHex
Image sent	Main folder\AC\INetCache\Y6HVKNWO
Private info	Main folder\LocalState\linkedIn.db
Notifications	AppData\local\Microsoft\Windows\Notifications\Wpndatabase file
Job search	INetCache folder, Microsoft Edge folder
Connections	Main folder\AC\INetCache\language.html
Install time	Microsoft\Windows\Apprepository\Linkedinforwindows.XML
DP of sender and receiver	AppData\Local\Microsoft\Windows\Notifications\ActionCenterCache folder
App run count	Prefetch files
Storage analysis	
Credentials	Only username found using WinHex
Image sent	Mainfolder\AC\INetCache\Y6HVKNWO
Message	Not found manually
App info	Microsoft\install agent\checkpoints
Sender and receiver's DP	AppData\Local\Microsoft\Windows\Notifications\wpnidm folder
Job search	Main Folder\INetCache folder
Connections	Main folder\AC\INetCache\language.html
Member Id	Main folder\INetCache\li-evergreen-jobs-ad.html
Transferred/downloaded files	Mainfolder\Ac\INetCache\cacheID, AppData\Local\Microsoft\Windows\Explorer (checking images ADS ZoneID)
Deleted data	Unallocated folder
Uninstallation	ProgramFiles\WindowsApps\Deleted
Registry analysis	
App model ID	NTUSER\Software\Microsoft\Windows\Currentversion\authentication\LogonUI\Notifications\Background capability\s-1-15-2-1533305960\AppuserModelId
App info	Software\Microsoft\Windows\Currentversion\InstallAgent\Categorycache
Msg sender's DP	NTUSER\software\Microsoft\Windows\currentversion\pushnotifications\wpnidm\7821a69d
Contact photos	Computer\HKEY_ClASSES_ROOT\Localsettings\software\Microsoft\mangedbyApp\Specified container
App install time	Computer\HKEY_ClASSES_ROOT\Localsettings\software\Microsoft\Windows\CurrentVersion\AppModel\Repository\Families\7EE7776C.LinkedinforWindows_w1wdnht996qgy\installtime

Table 9.4 Comparison of techniques

Analysis covered	Asma [29]	Yang [32]	Chaudhry et al. [33]	Our technique
RAM	No	Yes	Yes	Yes
Storage	Yes	Partial	No	Yes
Registry	No	No	No	Yes
Manual	No	Partial	No	Yes
Deleted artefacts	Yes	Yes	No	Yes

AppData\local\Microsoft\Windows\Notifications, Microsoft\ install agent, ProgramFiles\WindowsApps and prefetch files, as tabulated in Table 9.3.

The results clearly show that artefacts of LinkedIn app were stored outside the Main folder as well. Hence if the criminal deletes the data from the main folder, many useful artefacts can still be gathered through analysis of RAM, storage and registry to reveal any illicit activity.

Table 9.4 shows the comparison of our proposed technique with some of the existing techniques that were relevant to our case. The comparison highlights that our proposed methodology is more effective than other techniques. The proposed technique can also be used to forensically analyze any application downloaded from Windows AppStore as well.

9.5 Conclusion

RAM, storage and registry analysis are equally important from application forensic analysis point of view. The literature review conducted regarding forensic analysis of messaging and VoIP apps on Windows 10 reveals that minimal work has been done regarding manual forensic analysis of applications of Windows Store. Hence, a comprehensive study is presented considering all scenarios to retrieve potential artefacts. Moreover, the paper has also validated the technique by identifying artefacts of LinkedIn found at various locations after user activity on Windows 10 OS. In future, other trendy Windows store apps like Outlook, Grammarly etc. can also be tested on Win 10 or a comprehensive tool be developed.

References

1. Adeyemi, I.R., Razak, S.A., Azhan, N.A.N.: A review of current research in network forensic analysis. Int. J. Digit. Crime Forensics. **5**(1), 1–26 (2013)
2. Carrier, B.: Defining digital forensic examination and analysis tools using abstraction layers. Int. J. Digit. Evid. **1**(4), 1–12 (2003)
3. Lancaster, D.T.: Windows 10 is now on more than 14 million devices just 24 hours after launch [online]. Available: http://www.windowscentral.com/windows-10-now-14-million-devices-just-24-hours-after-launch (July 2015). Accessed: 13 Sept 2015
4. W3schools.com: Web statistics: OS platform statistics. http://www.w3schools.com/browsers/browsersos.asp (2016)
5. Majeed, A., Saleem, S.: Forensic analysis of social media apps in windows 10. NUST J Eng Sci. **10**(1), 37–45 (2017)
6. Domingues, F.: Digital forensic artifacts of the Cortana device search cache on Windows10 desktop. In: 11th International Conference on Availability, Reliability and Security, ARES.2016.44 Salzburg, Austria, IEEE (2016)
7. Singh, B., Singh, U.: A forensic insight into windows 10 jump lists. Digit. Investig. **17**, 1–13 (2016)
8. Khatri, Y.: Forensic implications of system resource usage monitor (SRUM) data in windows 8. Digit. Investig. **12**, 53–65 (2015)
9. Boyd, J.: 35 Insightful and Valuable LinkedIn Statistics. Retrieved July 2, 2018., from https://www.brandwatch.com/blog/linkedin-statistics/
10. De Paula, A.M.G.: Security aspects and future trends of social networks. In Proceedings of the Fourth International Conference of Forensic Computer Science, Brazil (2009)
11. Iqbal, A., Alobaidli, H., Almarzooqi, A., Jones, A.: LINE IM app forensic analysis. In: 12th International Conference on High-Capacity Optical Networks and Enabling/Emerging Technologies (HONET-ICT 2015)
12. Fontein, D.: The ultimate list of LinkedIn statistics that matter to your business. Retrieved February 7, 2017., from https://www.linkedin.com/pulse/ultimate-list-linkedin-statistics-matter-your-business-dara-fontein
13. Poh, M.: 10 Most Bizarre crimes linked to Facebook. Retrieved June 21, 2015, from http://www.hongkiat.com/blog/bizarre-facebookcrimes/ (n.d.)
14. Weise, E.: Banker used LinkedIn to send photo to prospective hire. Retrieved June 15, 2017, from https://www.usatoday.com/story/tech/news/2017/06/15/recruiter-used-linkedin-send-sex-photo-prospective-hire/102882292/
15. Amber, U., Nanda, P., He, X.: Online social network information forensics. A survey on use of various tools and determining how cautious Facebook users are? In: IEEE Trustcom/BigDataSE/ICESS.2017.364 (2017)
16. Hay, B., Nance, K., Bishop, M.: Live analysis: progress and challenges. IEEE Secur. Priv. **7**(2), 30–37 (2009)
17. Hausknecht, K., Foit, D., Burić, J.: RAM data significance in digital forensics. In: 38th International Convention on Information and Communication Technology, Electronics and Microelectronics, MIPRO 2015 – Proceedings (May), pp. 1372–1375 (2015)
18. Thantilage, R., Jeyamohan, N.: A volatile memory analysis tool for retrieval of social media evidence in windows 10 OS based workstations. In: National Information Technology Conference (NITC), Sri Lanka (2017)
19. Ahmed, W., Aslam, B.: A comparison of windows physical memory acquisition tools. In: Milcom 2015 track 3 – cyber security and trusted computing, IEEE, FL, USA (2015)
20. Aljaedi, A., Lindskog, D., Zavarsky, P., Ruhl, R., Almari, F.: Comparative analysis of volatile memory forensics: live response Vs. memory imaging. In: Proceedings of 3rd IEEE International Conference on Privacy, Security, Risk and Trust, pp. 1253–1258 (2011)
21. Prem, T., Paul Selwin, V., Mohan, A.K.: Disk memory forensics analysis of memory forensics frameworks flow. In: International Conference on Innovations in Power and Advanced Computing Technologies [I-PACT2017]
22. Alazab, M., Venkatraman, S., Watters, P.: Effective digital forensic analysis of the NTFS disk image. Ubiquit. Comput. Commun. J. **4**(3), 1–8 (2009)
23. John, J.L.: Digital forensics and preservation. Digital Preservation Coalition. Digital preservation handbook, Denmark (2012)
24. Zhang, S., Wang, L., Zhang, L.: Extracting Windows Registry Info from Physical Memory. IEEE (2011)

25. Arshad, A., Iqbal, W., Abbas, H.: USB storage device forensics for windows 10. J. Forensic Sci. **63**(3), 856–867 (2017). https://doi.org/10.1111/1556-4029.13596

26. Al Mutawa, N., Al Awadhi, I., Baggili, I., Marrington, A.: Forensic artifacts of Facebook's instant messaging service. In: International Conference for Internet Technology & Secured Transactions (IC-ITST), IEEE (2011)

27. Zhang, S., Wang, L., Zhang, L.: Extracting windows registry information from physical memory. In: 3rd International Conference on Computer Research and Development (2011)

28. Saidi, R.M., Ahmad, S.A., Noor, N.M., Younas, R.: Window registry analysis for forensic investigation. In: Proceedings of the 2013 International Conference on Technological Advances in Electrical, Electronics and Computer Engineering (TAEECE), IEEE (2013)

29. Kumar, H., Majeed, P.G., Pundir, S.: Forensic analysis of windows server 2008 physical memory. IJSRD-Int. J. Sci. Res. Dev. **2**(01), 1–4 (2014)

30. Majeed, A., Zia, H., Saleem, S.: Forensics analysis of three social media apps in Window 10. In: 12th International Conference on High-capacity Optical Networks & Enabling/Emerging Technologies, IEEE (2015)

31. Lee, C., Chung, M.: Digital forensic analysis on Window8 style UI instant messenger applications. In: Park, J.J. (ed.) Computer Science & its Applications. Springer, Berlin (2015)

32. Dija, S., Suma, G.S., Gonsalvez, D.D., Pillai, A.T.: Forensic reconstruction of executables win 7 physical memory. In: International Conference on Computational Intelligence & Computing Research, IEEE (2016)

33. Yang, T.Y., Dehghantanha, A., Choo, K.-K.R., Muda, Z.: Windows messaging app forensics: Facebook and Skype as case studies. PLoS One. **11**(3), e0150300 (2016)

34. Choudhary, P., Singh, U., Bharadwaj, N.K., Singh, B.: Facebook forensics for Win 10. In: 11th Annual Symposium on Information Assurance, USA (2016)

35. Ababneh, A., Abu Awwad, M., Al-Saleh, M.I.: IMO forensics in android and windows systems. In: 8th International Conference on Information, Intelligence, Systems & Applications (2017)

36. Meyers, C., Ikuesan, A.R., Venter, H.S.: Automated RAM analysis mechanism for windows OS for digital investigation. In: IEEE Conference on Application, Information and Network Security (AINS) (2017)

37. Gaur, S., Chhikara, R.: Memory forensics: tools and techniques. Indian J. Sci. Technol. **9**(48), 1–12 (2016). https://doi.org/10.17485/ijst/2016/v9i48/105851

Analysis of Windows OS's Fragmented File Carving Techniques: A Systematic Literature Review

10

Noor Ul Ain Ali, Waseem Iqbal, and Narmeen Shafqat

10.1 Introduction

The primary task of digital forensics is to extract data from storage media which may include documents, files, folders, browsing history, etc. In digital devices, a complete file is often not stored at a same place, rather it is divided into fragments for easy placement in the memory. Most of the Operating Systems (OS) these days avoid fragmentation because it tends to make the process of writing and reading, to and from, the memory slower. However, there are still several situations in which the OS is bound to write the file in fragments of two or more:

- If there is not enough space on the disk on which the file can be written without being fragmented [1].
- If new data is added to a file that already exists on the disk and the disk is missing unallocated sectors at the end of this old file such that there are no sectors available for appending new data [2].
- If the file system doesn't support the writing of a specific type of file in adjacent blocks of memory [3].

Generally, it is possible to recover all the data from a storage device by using its metadata and the Application Program Interface (API) of a file system. However, in some cases, the API is damaged, and thus data needs to be extracted using the technique of file carving [4]. In digital forensics, file carving is referred to as the recovery of file from a storage media without using the file system's metadata.

Carving of a file from memory is comparatively easier if the file is stored on adjacent blocks of memory; however, in reality, a file is saved in multiple fragments that are scattered across the memory [3]. Hence, the existing carving tools miserably fail at recovering fragmented files from the memory.

Another dimension of file carving is its usefulness for data recovery in times of accidental loss. Amongst different file formats available, recovering data of Microsoft Word format is quite difficult, since it is a compound file format. As described by Microsoft, compound file formatis a binary file format that contains several virtual streams. These streams are just serial arrangements of the sectors but consist of both control streams and user data streams [5]. Since Microsoft Word format is the most used format of Microsoft Office [1] and is least researched upon with respect to file carving, we will therefore keep the focus of our research on Microsoft Word Format. Word Format is required for almost all the electronic documentation done using computers; from sending attachments via electronic emails to keeping textual records in the form of documents etc. Hence, this research will be highly beneficial for the forensic investigators.

The rest of the paper is structured as follows: Sect. 10.2 discusses the Systematic Literature Review, Sect. 10.3 includes Related Work being carried out, Sect. 10.4 covers the proposed technique and finally Sect. 10.5 contains the Discussion and Analysis. The paper is finally concluded with Sect. 10.6.

10.2 Systematic Literature Review

In order to review the current techniques available for data carving of Word documents, a systemic literature review has been carried out based on the guidelines available in [6].

N. U. A. Ali · W. Iqbal (✉) · N. Shafqat
Department of Information Security, National University of Sciences and Technology, Islamabad, Pakistan
e-mail: waseem.iqbal@mcs.edu.pk; narmeen_shafqat@mcs.edu.pk

© Springer Nature Switzerland AG 2019
S. Latifi (ed.), *16th International Conference on Information Technology-New Generations (ITNG 2019)*,
Advances in Intelligent Systems and Computing 800,
https://doi.org/10.1007/978-3-030-14070-0_10

The research is performed in order to answer the questions mentioned below:

- What type of file carvers are available for extraction of word document?
- How many types of file carvers are able to cater for fragmentation?
- What types of techniques are used for carving other than the conventional header and footer extraction techniques?
- How many file carvers perform carving on volatile memory?

The search was performed on many digital libraries and online databases including IEEEXplore, Science Direct, Digital Forensics Research Conference (DFRWS) and Springer Link. Other than these Google and Explorer search engines were also used. The focus was on the topics that were related to digital forensics or digital investigations, and thus other irrelevant topics were discarded. Table 10.1 shows the keywords that were used to perform the searches on the sources mentioned above.

Using the above keywords, the databases and digital libraries were searched. The selection of the study was a multi-phased process. Firstly, relevant studies were found using the keywords and then screening was done based on the title of the study. After that, the screening was done by reading the abstract of each publication. As a result, a large number of publications were excluded because of their irrelevance to the research questions. The publications that were left behind were completely studied and assessed based on their relevance. The results of the search are listed below in Table 10.2.

Table 10.1 Searched keywords

No	Keywords
1.	Fragmented data carving
2.	Word document carving techniques
3.	Data carving tools
4.	Carving on volatile memory
5.	Compound file format
6.	Data carving

Table 10.2 Summary of research papers consulted

Database/library	No. of papers	Filter based on title	Filter based on abstract
IEEEXplore	232	27	8
Science direct	392	17	9
DFRWS	112	12	8
Springer link	237	39	14
Scopus	42	24	6

As a result, a set of carefully chosen publications are reviewed whose relevance depends upon the research questions, clear objectives and approach.

10.3 Related Work

Initially, file carvers used the concept of "magic numbers," or more precisely, byte sequences at predefined offsets to identify and retrieve files. This concept has been slightly revised with time. Table 10.3 below shows the summary of the extensive literature review carried out for this research. The table also highlights the memory type and recovered artefacts in each of the consult research.

The conducted literature review shows that many researchers have developed various file carving tools, to help them automate the process of file carving. Also that a lot of techniques have been developed for carving file formats like bmp, png, jpeg, videos and other multimedia files. However the least worked on file format is the Microsoft Word File Format which became our motivation behind working on the Word File Format.

A summary of some of the open-source tools that can be used for file carving by digital investigators after complete validation, are tabulated below in "Table 10.4".

The literature review carried out for this research is a blend of carving techniques from both volatile and non-volatile memory. An effort was made to cover all types of artefact files i.e. image fragments (BPM, JPEG, PNG etc.), multimedia files, text documents (Word, PDF etc.), and even live responses.

It is pertinent to mention that one of the file types which has been studied and researched very less is the Microsoft file. This is primarily because like Zipped files, Microsoft Word is also a compound file format. It contains directories and hierarchal structure of streams in which the data is stored. It also contains images and textual element, which makes the extraction of word document impossible using the existing carving tools and techniques [19].

10.4 Proposed Technique for Carving Word Document from the Memory

As mentioned earlier, minimal research on carving Microsoft's word files on Windows OS has been carried out in the past. Past researches have either considered the extraction of images from word documents or the extraction of text, but no comprehensive research has been carried out in the extraction of both the components simultaneously.

Table 10.3 Summary of literature review

Paper	Memory type	Artefact	Summary
Cohen [2]	Non-volatile	File and byte images	Recovery of fragmented file is done by a generator that will map all possible functions. Weakness: The technique is processing intensive
Luigi et al. [7]	Non-volatile memory	BMP and JPEG files	Classification was done based on the contents of blocks of data. Weakness: Less support for pdf and document files
Lin et al. [8]	Non-volatile	Text document	Employs use of control streams to extract text documents from memory
Azzat et al. [9]	Non-volatile memory	Image fragments	Worked specifically on the extraction of image fragments from memory.
Pal et al. [4]	Non-volatile	Image and text	Provides summary of mistakes in existing recovery techniques used for carving. Suggestion: Points out need for customized software for formats like videos, exes and audio
Zha et al. [10]	Non-volatile	Raw disk images from memory	An improved algorithm for popular open source tool: Scalpel
Vassil et al. [11]	Non-volatile memory	JPEG, PNG	Suggested that technique of just using header and footer for carving is flawed because it cannot cater for complex and compound file structures like ZIP and DOC.
Garfinkel [12]	Non-volatile	All types of files	Covers carving of a file fragmented into two segments only.
Roux [13]	Non-volatile	ASCII text files	Recovery is done using support vector machines (SVM) classifiers Scalable for larger training sizes.
Rainer et al. [14]	Non-volatile	Multimedia files	Proposes an open source file carver for multimedia files Weakness: Doesn't cover fragmented multimedia files
Zaid et al. [15]	Volatile memory	PDF files	Highlights that extraction of pdf file from RAM has a greater chance of recovery when there is a large RAM.
Simson et al. [16]	Non-volatile	Files	Proposed a technique that uses hashes of files for carving
Wagner et al. [17]	Databases	All types of data	Presented a tool for extraction of data specifically for databases.
Aaron et al. [18]	Volatile memory	Live responses	Conducted research on importance of volatile memory. Then, proposed the need for advanced tools for live memory forensics.

Table 10.4 Tools available for carving

Name	Algorithm	Type of artefact
Foremost [4]	Performs sequential header to footer carving	Image files
Scalpel [19]	Defines signatures for beginning and end of the file and extracting data between these signatures	Non-fragmented files
Volatools	A command line open source tool that supports the analysis of memory pages	Extraction of data from a live resource

Moreover, whenever there is a presence of fragmented word document then the extraction process becomes more complex and recovery of image and text almost seems impossible. This section proposes a novel technique that can be used to carve Word documents from the memory easily.

When a word file is loaded into the memory, it is fragmented and saved as segments of data in memory. These segments are scattered randomly across the whole memory. Hence, it is recommended to commence by:

- Firstly, classify the segments of data and assign each of them a file-type. This is done by looking for known markers in the memory. One known marker in the compound file identifier is D0CF11E0A1B11AE1 which indicates the start of a file.
- After this, we will group together chunks of data having same file types. This is done using any of the four techniques mentioned below:

(a) Classification – Binary (yes or no)
(b) Clustering – Apply different clustering technique on each segment
(c) Regression – Numeric value
(d) Ranking – Rank based on value

- Next, all segments belonging to a particular file type are grouped together. To help us identify the class of each segment for each file type, we will design classifiers for each type of document i.e. separate classifiers for both image and text portions of a word file.

Based on these classifiers, it can be determined that which segment belongs to which file [18]. After each segment is identified and classified, it is put together and the original word document will be formed.

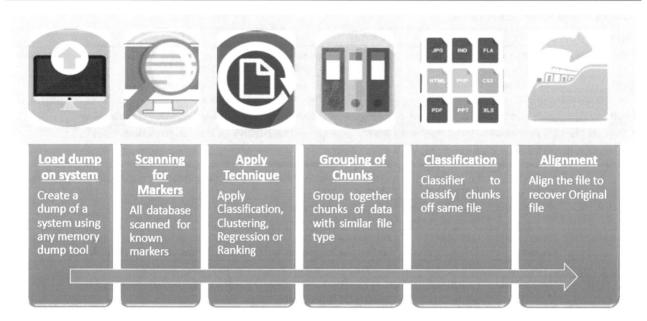

Fig. 10.1 Proposed methodology

10.5 Experimental Setup and Validation

This section briefly describes the manual testing of our proposed technique. The experimental setup included VMware workstation. The dump was created using the DumpIt tool, while WinHex was used for analysis of the memory.

We started off by searching the compound file identifier. After finding the identifier, we looked for the ending note. After going through many Hex values, we saw another marker showing the version number of the word format we were trying to cover. Using its address, we divided it by 512. Incase we obtain an even number, it means it is a starting of a sector. To recover the word document now, just copy the bytes between the identifier and save it with a.docx extension. Since we did the searching manually, we did not use any machine learning technique in the validation phase.

After each segment is identified and classified, it is put together and original word document will be formed. Our proposed method for the extraction of word document from memory is illustrated in "Fig. 10.1".

A simple comparison has been drawn between existing and our introduced technique in Table 10.5, in order to highlight the effectiveness of our approach.

10.6 Conclusion and Future Work

In this paper, we have analyzed various methods and techniques being deployed for carving of data from memory. The research basically focuses on the study of extraction of multiple formats from main memory using different tools

Table 10.5 Comparison between proposed and existing techniques

Features	Proposed technique	Existing techniques
Fixed size and known header	NO	YES
File recovered using missing header	YES	NO
Computationally intensive	NO	YES
Caters fragmentation	YES	NO
Machine learning classifiers	YES	NO

and methods and also lays emphasis on analyzing Microsoft Word document file format from carving point of view. But the proposed technique would not work in case the file is destroyed or corrupted. All the work above has been done by assuming that the file is not damaged or corrupted but is just fragmented across the memory.

From the above discussion, we have concluded that as future research, the recovery of fragmented Microsoft word document from volatile memory seems like a promising field to explore for forensic investigators and researchers.

References

1. Microsoft compound Document file format. Available at: https://msdn.microsoft.com/en-us/library/dd942138.aspx
2. Cohen, M.I.: Advanced carving techniques. Digit. Investig. **4**(3–4), 119–128 (2007)
3. Calhoun, W., Coles, D.: Predicting the type of file fragments, DFRWS USA S14-S20 (2008)
4. Pal, A., Memon, N.D.: The evolution of file carving. IEEE Signal Process. Mag. **26**(2), 59–71 (March 2009)
5. Foremost 1.53 [Online]. Available at: http://foremost.sourceforge.net

6. Nightingale, A.: A guide to systematic literature review. Surgery (Oxford). **27**(9), 381–384 (September 2009)
7. Sportiello, L., Zanero, S.: Context-based file block classification. i-Code: Real Time Malicious Code Identification; and by the EU Seventh Framework Program (FP7/2007–2013)
8. Lin, W., Xia, M.: A Microsoft word documents carving method base on interior virtual streams. Adv. Mater. Res. **433–440**, 3028–3032 (2012)
9. AI-Sadi, A., Yahya, M.B., Almulhem, A.: Identification of image fragments for file carving. In: 2013 World Congress on Internet Security (WorldCIS) (2014)
10. Zha, X., Sahni, S.: Fast in-place file carving for digital forensics. In: Part of International Conference of Forensics in Telecommunications, Information, and Multimedia, pp. 141–158 (2010)
11. Roussev, V., Garfinkel, S.L.: File fragment classification—the case for specialized approaches (10 Jul 2014)
12. Garfinkel, S.: Carving contiguous and fragmented files with fast object validation. In: Proceedings of 2007 Digital Forensics Research Workshop (DFRWS), Pittsburgh, PA, pp. 4S:2–12 (Aug 2007)
13. Roux, B.: Reconstructing textual file fragments using unsupervised machine learning technique. University of New Orleans Theses and Dissertations. 881 (Dec 2008)
14. Poisel, R., Tjoa, S., Tavolato, P.: Advanced file carving approaches for multimedia files. J. Wireless Mob. Networks, Ubiquitous Comput. Dependable Appl. **2**(4), 42–58 (2011)
15. Al-Sharif, Z.A., Odeh, D.N., Al-Saleh, M.I.: Towards carving PDF files in main memory. In: Proceedings of International Technology Management Conference (2015)
16. Garfinkel, S.L., McCarrin, M.: Hash based carving: searching media for complete files and file fragments with sector hashing and hash db. Digit. Investig. **14**(2015), S95–S105 (2015)
17. Wagner, J., Rasin, A., Grier, J.: Database forensic analysis through internal structure carving. Digit. Investig. **14**(2015), S106eS115 (2015)
18. Walters, A., Petroni, N.L.: Volatools: integrating volatile memory forensics into the digital investigation process. Digit. Investig., Elsevier (2007)
19. Richard, G.G. III, Roussev, V.: Scalpel: a frugal, high performance file carver. In: Proceedings of 2005 Digital Forensics Research Workshop (DFRWS), New Orleans, LA (Aug 2005)

Cybersecurity Certification: Certified Information Systems Security Professional (CISSP)

Ping Wang and Hubert D'Cruze

11.1 Introduction

There has been a large and fast growing workforce gap and demand for qualified cybersecurity professionals. According to the $(ISC)^2$ 2018 Cybersecurity Workforce Study, the shortage of cybersecurity professionals is close to three million globally and about half a million in North America and the majority of the companies surveyed reported concerns of moderate or extreme risk of cybersecurity attacks due to the shortage of dedicated cybersecurity staff [1]. An information security analyst is only one of the career titles in the cybersecurity profession. The latest career outlook published by the United States Labor Department Bureau of Labor Statistics shows that the employment of information security analysts, an example of cybersecurity jobs, is projected to grow 28% from 2016 to 2026, much faster and with better pay than the average for all occupations [2].

Education, training, and professional certifications are common solutions for alleviating shortage of professional staff. However, a recent study shows that top universities in the United States were failing at cybersecurity education with a lack of cybersecurity requirements for graduates and a slow change in curriculum and courses [3]. The national Centers of Academic Excellence in Cyber Defense Education (CAE-CDE) designation program jointly sponsored by the US National Security Agency (NSA) and Department of Homeland Security (DHS) has been a reputable standard for certifying and maintaining high quality of cybersecurity education with rigorous requirements for program evaluation and assessment of cybersecurity knowledge units. However, only about 200 (or 3%) colleges and universities in the U.S. have achieved the CAE-CDE designation status so far [4].

Recognizing the need to develop more and qualified cybersecurity professionals to meet the workforce demand, the National Initiative for Cybersecurity Education (NICE) recently published the NICE Cybersecurity Workforce Framework (NCWF SP800–181), which specifies cybersecurity professional categories, tasks, job roles as well as knowledge, skills, and abilities (KSAs) needed for cybersecurity jobs [5]. These KSAs are also mapped to the cybersecurity knowledge units (KUs) for college and university programs with CAE-CDE designations.

Professional certifications are an important supplemental credential system to help select talents and guide the training and development of cybersecurity workforce. In hiring information security analysts, for example, many employers prefer their candidates to have some relevant professional certification in the field, such as Certified Information Systems Security Professional (CISSP) in addition to a minimum of a bachelor's degree in order to validate the knowledge and best practices required for the job [2]. Ideally, the certification process used for developing and selecting qualified professionals in the cybersecurity field should incorporate the KSAs specified in the NCWF as well.

There are many different types of certifications for the cybersecurity field with various levels of requirements and rigor. The CISSP (Certified Information Systems Security Professional) certification stands out as a challenging but popular vendor neutral certification choice coveted by cybersecurity professionals and employers. Recent studies show CISSP as the top cybersecurity credential most valued by employers [6, 7].

This paper will discuss the rigorous requirements of the CISSP certification and explore the significant value and benchmark role of the CISSP certification in developing and maintaining cybersecurity workforce competencies. The

P. Wang (✉)
Robert Morris University, Moon, PA, USA
e-mail: wangp@rmu.edu

H. D'Cruze
University of Maryland, College Park, MD, USA

© Springer Nature Switzerland AG 2019
S. Latifi (ed.), *16th International Conference on Information Technology-New Generations (ITNG 2019)*,
Advances in Intelligent Systems and Computing 800,
https://doi.org/10.1007/978-3-030-14070-0_11

study will also map the CISSP knowledge domains and objectives to the model of competencies of the US cybersecurity industry and knowledge, skills, and abilities (KSAs) of the NICE cybersecurity workforce framework (NCWF). The goal of the study is to discover the value of the CISSP certification process to the cybersecurity workforce development and implications to cybersecurity education and training.

11.2 Background

A professional certification should be a process of independent verification of one's expertise of a certain level in a particular professional area and should require meaningful steps to examine one's knowledge, skills, and expertise before the certification designation is issued by the independent organization [8]. There are many organizations that provide certifications for professionals in the computing and information technology areas. These certifying organizations include industry vendors, such as Cisco Systems, Microsoft, Oracle, IBM, Amazon, and vendor neutral or vendor independent organizations, such as CompTIA, (ISC)^2, EC-Council, ISACA, GIAC. There should be three critical components included in a professional certification process: (1) An exam-based test for candidates to demonstrate mastery of a common body of knowledge in the area; (2) commitment and adherence to a code of ethical conduct for the professional community; and (3) Mandatory continuing education or professional development [8]. In addition, more rigorous and reputable professional certifications require certain minimum amount of relevant and verifiable professional experience for certification whereas some other and less rigorous certifications only recommend but not require such experience.

Professional certifications in computing and information technology (IT) areas that include Cybersecurity not only benefit employers in improving human resources, work productivity and employee performance but also give a competitive advantage to employees in financial compensation and future career and professional development and training [8–10]. Here are the important findings presented in the 2018 IT Skills and Salary Report published by Global Knowledge: (1) About 90% of IT professionals globally hold at least one certification, which is an increase of 3% over 2017; (2) The average salary of certified IT staff in the U.S. and Canada is $15, 913 or 22% more than non-certified peers, and in the Asia-Pacific region, certified IT professionals make 45% more than their non-certified peers; and (3) Professionals with cybersecurity certifications have significantly higher average salaries; In North America, for example, the average salary of security-certified professionals is $101, 083, or about 15% more than the average of all certified IT professionals [10].

Professional certifications help to alleviate the persistent shortage of supply of qualified cybersecurity professionals. Research shows that the shortage of cybersecurity professionals continues to leave companies and organizations vulnerable and is now the number one job concern among those working in the field [1]. 95% of IT industry leaders surveyed believe certifications added value of increased productivity and closing skillset gap, especially in Cybersecurity, which often requires further education and training; accordingly, cybersecurity certifications hold the top spots for IT salaries in the last 3 years [10]. For organizations to reach the maximum effectiveness of certifications in Cybersecurity, the certification domains and objectives should have substantial match with the competencies and knowledge, skills, and abilities (KSAs) for the cybersecurity industry, such as the Industry-wide Technical Competencies and the Workplace Competencies in the Cybersecurity Industry Model published by the U.S. Department of Labor and the specific KSAs in the NICE cybersecurity workforce framework (NCWF) [5, 11].

Professional certifications are usually intended to be a supplemental validation of formal education and training and a professional motivation or requirement for further professional development, including continued education and training. Therefore, the knowledge and skill domains and objectives of professional certifications should have substantial reflection or coverage of the education and training programs and learning outcomes in a particular field. Research shows that professional certifications can be used as a valuable guidance in designing and maintaining a vibrant cybersecurity curriculum as both professional certifications in Cybersecurity and cybersecurity curriculum and courses need to incorporate the important factors of cyber threat landscape, changing technology, workforce needs, industry standards, and government regulations [12]. In addition, the curriculum and course learning outcomes and activities should support the educational and professional and career goals of the students [13].

The next section of the paper will propose and explain a model of professional certifications in Cybersecurity in relation to the cybersecurity workforce demands and cybersecurity education and training curriculum design.

11.3 Proposed Model

Professional certifications are designed to meet the industry and workforce demands of a particular field. The industry needs for workforce development and validation are the main reason for a certain certification to exist. The certification domains and objectives should address the specific professional qualifications, technical and non-technical competencies, and specific knowledge, skills, and abilities (KSAs) required for successful performance of a certain profession. For a certification to be valuable and reputable, it should be mapped to industry standards at a national or international level. In addition, professional certifications should motivate and incorporate continued education and training for professionals to maintain their KSAs in the field. Based on these assumptions, this paper proposes a theoretical model of professional certifications as presented in Fig. 11.1 below.

Industry and workforce demands are the original driving force for professional certifications. Certifications help employers fill open positions and close skills gap as education and training are not sufficient [9, 10]. The industry and workforce demands, such as those in IT and Cybersecurity areas, are reflected in the specific professional competencies and KSAs. For example, the Cybersecurity Competency Model published by US Department of Labor includes not only the basic personal effectiveness competencies and common academic and workplace competencies often acquired through formal education and training but also more specialized

industry-wide technical competencies and industry-sector areas, such as cybersecurity technology, incident detection and response, and protection and defense against cyber threats [11]. In addition, the NICE Cybersecurity Workforce Framework (NCWF) presents similar and more comprehensive and detailed work roles and tasks as well as specific attributes of KSAs for each task in the cybersecurity industry [5].

In addition work experience and education and training, professional certifications are an important means of validating and updating employees' competencies and KSAs for needed for the industry. Accordingly, professional certification domains and objectives should and often do reflect the workforce skillset needs of the industry. Therefore, mapping the certification objectives to the industry competencies and KSAs is an important measure of the validity and value of the professional certification.

The certification domains and objectives in the proposed model also serve to inform and shape relevant education and training programs in terms program curriculum and course design. A recent case study shows that incorporating important content areas and objectives from top professional certifications is found to be valuable to the design and maintenance of a regular college degree curriculum in Cybersecurity [12].

Fig. 11.1 Proposed certification model

11.4   CISSP Case Study

This paper uses the case study methodology to discover and illustrate the value and benchmark role of professional certifications in developing and validating workforce qualifications for the cybersecurity industry. The case study focuses on exposing the features of the CISSP (Certified Information Systems Security Professional) certification. Relationships between cybersecurity industry competencies and workforce KSAs will be discussed in relation to CISSP certification requirements, domains and objectives.

CISSP is a cybersecurity professional certification credential issued by International Information Systems Security Certification Consortium, better known as (ISC)^2, which is a global non-profit membership association for information security and cybersecurity professionals and leaders. The CISSP certification program has been in existence since 1994. To qualify for the certification, a candidate must pass a comprehensive exam, have minimum relevant work experience, endorsement from a CISSP-certified professional, and agree to the (ISC)^2 Code of Ethics [14, 15].

The CISSP certification exam is a 6-hour traditional linear exam currently offered in eight different languages with the passing grade of 700 out of 100 points. The exam evaluates and validates the candidate's professional knowledge and expertise across the following eight cybersecurity domains in the CISSP Common Body of Knowledge (CBK):

Domain 1. Security and Risk Management
Domain 2. Asset Security
Domain 3. Security Architecture and Engineering
Domain 4. Communication and Network Security
Domain 5. Identity and Access Management (IAM)
Domain 6. Security Assessment and Testing
Domain 7. Security Operations
Domain 8. Software Development Security [15, 16].

Passing the exam is not sufficient for the CISSP certification. The candidate must have at least 5 years of verifiable paid work experience in at least two of the eight domains of the CISSP CBK. The candidate also needs a formal and signed endorsement from a CISSP-certified professional who can attest to the candidate's professional experience. In addition, the candidate must subscribe to and fully support the following (ISC)^2 Code of Ethics in order to qualify for the CISSP certification:

• Protect society, the common good, necessary public trust and confidence, and the infrastructure.
• Act honorably, honestly, justly, responsibly, and legally.
• Provide diligent and competent service to principles.
• Advance and protect the profession [15].

The CISSP certification program has been a positive response to help alleviate the workforce shortage and improve quality standard of professional skillsets for the cybersecurity industry. So far, over 140,000 professionals around the world have obtained the CISSP certification, and CISSP was ranked as the security credential most valued by employers by a margin of 3–1 in a recent cybersecurity trends report [6, 14].

The domains of expertise and objectives of the CISSP certification closely match the functional areas and technical competencies for the cybersecurity industry. The ideal target audience for CISSP are high-level experienced security practitioners, manager and executives, including the following positions with knowledge and expertise in a wide array of cybersecurity practices and principles [15]:

• Secure Acquisition
• Chief Information Security Officer
• Chief Information Officer
• IT Director/Manager
• Security Systems Engineer
• Security Analyst
• Security Manager
• Security Auditor
• Security Architect
• Security Consultant

These security positions are demanded for the industry-sector functional areas listed in the Cybersecurity Industry Model published by the US Department of Labor [11]. The specific industry-sector functional areas in the model that can be addressed by the CISSP target positions include the following [11]:

• Secure Acquisition
• Secure Software Engineering
• Systems Security Architecture
• Systems Security Analysis
• Enterprise Network Defense Analysis
• Network Defense Infrastructure Support
• Strategic Planning & Policy Development
• Security Program Management
• Security Risk Management

In addition, the domain areas and objectives in the CISSP exam include coverage of the following industry-wide technical competencies and corresponding critical work functions and associated knowledge, skills, and abilities (KSAs) listed in the US Labor Department Cybersecurity Industry Model [11, 16]:

• Cybersecurity Technology
• Information Assurance
• Risk Management
• Incident Detection
• Incident Response and Remediation

The CISSP knowledge domains and objectives also have substantial coverage of the specific cybersecurity knowledge, skills, and abilities (KSAs) published in the NICE Cybersecurity Workforce Framework (NCWF). Appendix 1 below shows the mapping between CISSP domains of expertise and exam objectives and NCWF KSAs [5, 16]. The mapping is a special contribution of this study that maps CISSP domain objectives to the specific KSAs. The content of each of the mapped KSAs is in the NCWF publication.

Maintenance of the CISSP certification requires continuous education, learning, and professional development. Each certification cycle is only valid for 3 years, during which the annual minimum number of Continuing Professional Education (CPE) credits must be earned, documented and submitted. To actively maintain and renew the CISSP certification, the certification holder must earn at least 40 CPE credits per year that are subject to auditing. The CPE credits can be earned through cybersecurity related education and training courses, research and publications, unique service contributions to the cybersecurity community, and general professional development activities [17].

11.5 Findings and Discussions

This study finds that industry and workforce demands for qualified and skilled talent are the driving force behind professional certifications that are expected to serve as independent validation of professional qualifications. The CISSP case study in this paper reveals the outstanding features of this certification in meeting the industry and workforce demands for the cybersecurity. The unique mappings between CISSP domain objectives and the industry-sector and technical competencies as well as specific knowledge, skills, and abilities (KSAs) for the cybersecurity industry indicate a strong correlation between CISSP domain objectives and the cybersecurity professional competencies and KSAs. The rigorous annual CPE requirement for maintaining the CISSP certification also positively motivates cybersecurity professionals towards further education and training in the field.

The market value of the CISSP certification further reinforces its significant benchmark role in cybersecurity workforce development. A recent independent study comparing CISSP certification holders and non-holders of the certification in Europe also finds that CISSP certification holders add more accreditation value to employers and are more attractive to industry recruiters [18]. In addition, the recent research reports on global IT and cybersecurity workforce skills and salaries have shown that CISSP certification is not only found to be "the most valued security credential" by the overwhelming majority of the employers but also holds the top spot in average global salary [6, 10, 14]. The rigorous exam, mandatory work experience and ethical compliance, as well as strict professional development requirement are all contributing factors to the strong value and credibility of the CISSP certification.

The rigor of the CISSP certification and its benchmark value to cybersecurity workforce development can be effectively incorporated into cybersecurity education and training programs and their curricula. There has been a gap between university programs and the cybersecurity workforce demands. "The increased demand for cybersecurity professionals is relatively new, and universities are still unable to respond to this demand by incorporating it in their curricula" [19]. Research shows that CISSP certification domains and objectives can be effectively integrated into a undergraduate cybersecurity curriculum with 100% coverage of the entire CISSP CBK domains in nine different cybersecurity courses [12].

The proposed model and study in this paper shows significant value and a benchmark role of the CISSP certification in cybersecurity workforce development. The adoption of the certification domain objectives can be useful to educational programs and curricula. However, it should be emphasized that even the most rigorous professional certifications such as CISSP are a valuable supplement to but not a replacement of formal long-term education in the field.

11.6 Conclusions

This study focuses on the important value and benchmark role of cybersecurity certifications in addressing workforce shortage and development of the cybersecurity industry. This research proposes a certification model that describes the close correlations between industry and workforce demands, professional competencies and KSAs, certification domains and objectives, and education and training curriculum and outcomes. The case study of CISSP certification for the cybersecurity industry is used to illustrate the model and the correlations. The study finds that CISSP certification is a rigorous, comprehensive and reputable credential highly valued by the cybersecurity industry and workforce.

In addition to the proposed certification model, this study has contributed valuable mappings between CISSP certification domain objectives and industry-sector and technical competencies and KSAs for specific work roles and tasks defined in the NICE cybersecurity workforce framework (NCWF). This study only presents the mapping between CISSP Domain 1 objectives and KSAs in the NCWF, which is a unique and pioneering contribution. Future studies may complete and present specific KSA mappings to other CISSP domains. Such detailed mappings will be very valuable to detailed credential evaluation for targeted industry recruiting and to the design of college and university cybersecurity program curriculum and course learning outcomes.

Appendix 1: Mapping Between CISSP Domain Objectives and NCWF KSAs

CISSP domain	NCWF KSA IDs		
Domain 1: Security and risk management	Knowledge (K) ID	Skill (S) ID	Ability (A) ID
1.1 Understand and apply concepts of confidentiality, integrity and availability	K0001, K0003, K004, K0005, K0019, K0037, K0038, K0044, K0168, K0203, K0260, K0262, K0295	S0006, S00367	A0094, A0119, A0123
1.2 Evaluate and apply security governance principles	K0002, K0003, K0005, K0026, K0044, K0048, K0070, K0168, K0203, K0262, K0267	S0034, S0367	A0033, A0094, A0111, A0119, A0123, A0170
1.3 Determine compliance requirements	K0003, K0004, K0027, K0028, K0037, K0038, K0040, K0048, K0049, K0054, K0168, K0169, K0260, K0261, K0262, K0267, K0624	S0034, S0367	A0033, A0094, A0111, A0123, A0170
1.4 Understand legal and regulatory issues that pertain to information security in a global context	K0003, K0004, K0019, K0037, K0038, K0040, K0048, K0049, K0054, K0059, K0126, K0146, K0169, K0199, K0267, K0322	S0034, S0367	A0033, A0077, A0090, A0094, A0111, A0117, A0118, A0119, A0123, A0170
1.5 Understand, adhere to, and promote professional ethics	K0003, K0206	S0367	A0123
1.6 Develop, document, and implement security policy, standards, procedures, and guidelines	K0002, K0003, K0004, K0005, K0006, K0013, K0019, K0027, K0048, K0059, K0070, K0146, K0168, K0179, K0199, K0203, K0267, K0264	S0034, S0367	A0033, A0094, A0111, A0117, A0118, A0119, A0123
1.7 Identify, analyze, and prioritize business continuity (BC) requirements	K0006, K0026, K0032, K0037, K0041, K0042	S0032	A0119
1.8 Contribute to and enforce personnel security policies and procedures	K0003, K0004, K0038, K0039, K0044, K0072, K0146, K0151, K0204, K0208, K0217, K0220, K0243, K0239, K0245, K0246, K0250, K0252, K0287, K0615, K0628	S0018, S0027, S0064, S0066, S0070, S0086, S0102, S0166, S0296, S0354, S0367	A0004, A0013, A0015, A0018, A0019, A0022, A0024, A0027, A0028, A0032, A0033, A0054, A0057, A0070, A0094, A0110, A0111, A0171
1.9 Understand and apply risk management concepts	K0005, K0006, K0013, K0019, K0027, K0028, K0037, K0038, K0040, K0044, K0048, K0049, K0054, K0059, K0070, K0084, K0089, K0101, K0126, K0146, K0168, K0169, K0170, K0179, K0199, K0203, K0260, K0261, K0262, K0267, K0295, K0322, K0342, K0622, K0624	S0001, S0006, S0027, S0034, S0038, S0078, S0097, S0100, S0111, S0112, S0115, S0120, S0124, S0128, S0134, S0136, S0137, S0171, S0238, S0367	A0028, A0033, A0077, A0090, A0094, A0111, A0117, A0118, A0119, A0123, A0170
1.10 Understand and apply threat modeling concepts and methodologies	K0005, K0006, K0021, K0033, K0034, K0041, K0042, K0046, K0058, K0062, K0070, K0106, K0157, K0161, K0162, K0167, K0177, K0179, K0180, K0198, K0199, K0203, K0221, K0230, K0259, K0287, K0288, K0332, K0565, K0624	S0003, S0025, S0044, S0047, S0052, S0077, S0078, S0079, S0080, S0081, S0120, S0136, S0137, S0139, S0156, S0167, S0173, S0236, S0365	A0010, A0015, A0066, A0121, A0123, A0128, A0159
1.11 Apply risk-based management concepts to the supply chain	K0001, K0002, K0005, K0013, K0019, K0038, K0048, K0049, K0054, K0057, K0065, K0103, K0122, K0126, K0147, K0148, K0149, K0154, K0165, K0169, K0179, K0195, K0214, K0263, K0264, K0296, K0297, K0298, K0322, K0506, K0527, K0530, K0621, K0623	S0022, S0170, S0171, S0331, S0368, S0373	A0009, A0077, A0090, A0111, A0120, A0132 A0133, A0134, A0135
1.12 Establish and maintain a security awareness, education, and training program	K0040, K0054, K0059, K0124, K0204, K0208, K0215, K0217, K0218, K0220, K0226, K0239, K0245, K0246, K0250, K0252, K0287, K0313, K0319, K0628	S0001, S0004, S0006, S0018, S0027, S0051, S0052, S0053, S0055, S0056, S0057, S0060, S0064, S0070, S0073, S0075, S0076, S0081, S0084, S0086, S0097, S0098, S0100, S0101, S0121, S0131, S0156, S0184, S0270, S0271, S0281, S0293, S0301, S0356, S0358	A0013, A0014, A0015, A0016, A0017, A0018 A0019, A0020 A0022, A0023 A0024, A0032 A0055, A0057 A0058, A0063 A0066, A0070 A0083, A0089 A0105, A0106 A0112, A0114 A0117, A0118 A0119, A0171

References

1. (ISC)^2: Cybersecurity Professionals Focus on Developing New Skills as Workforce Gap Widens: (ISC)^2 Cybersecurity Workforce Study 2018. Retrieved from https://www.isc2.org/research (2018)

2. US Labor Department BLS (Bureau of Labor Statistics).: Retrieved from https://www.bls.gov/ooh/computer-and-information-technology/information-security-analysts.htm (2018)

3. White, S.K.: Top U.S. Universities Failing at Cybersecurity Education. CIO. Retrieved from https://www.cio.com/article/3060813/it-skills-training/top-u-s-universities-failing-at-cybersecurity-education.html (25 Apr 2016)

4. Wang, P., Dawson, M., Williams, K.L.: Improving cyber defense education through national standard alignment: case studies. Int. J. Hyperconnectivity Internet Things. **2**(1), 12–28 (2018)

5. NICE (National Initiative for Cybersecurity Education): NICE Cybersecurity Workforce Framework (SP800-181). Retrieved from https://csrc.nist.gov/publications/detail/sp/800-181/final (2017)

6. (ISC)^2. Cybersecurity Trends: 2017 Spotlight Report. Retrieved from https://www.isc2.org (2017)

7. Information Security Careers Network (ISCN): What are the Best Cyber Security Certifications to have in 2019? (List of the Top 10). Retrieved from https://www.infosec-careers.com/2018/07/16/the-best-cyber-security-certifications-in-2019/ (2018)

8. Martinez, A.: Get Certified & Get Ahead, 3rd edn. Computing McGraw-Hill, New York (2000)

9. CompTIA: Reasons Why EmployersLook for IT Certifications. Retrieved from https://certification.comptia.org/why-certify/professionals/5-reasons-employers-look-for-it-certifications (2015)

10. Global Knowledge: 2018 IT Skills and Salary Report. Retrieved from https://www.globalknowledge.com/us-en/content/salary-report/it-skills-and-salary-report/ (2018)

11. US Department of Labor: Cybersecurity Industry Model. Retrieved from www.doleta.gov (2014)

12. Knapp, K.J., Maurer, C., Plachkinova, M.: Maintaining a cybersecurity curriculum: professional certifications as valuable guidance. J. Inf. Syst. Educ. **28**(2), 101–114 (2017)

13. Wang, P.: Designing a doctoral level cybersecurity course. Issues Inf. Syst. **19**(1), 192–202 (2018)

14. (ISC)^2: The Ultimate Guide to the CISSP. Retrieved from https://www.isc2.org/Certifications/Ultimate-Guides/CISSP (2018)

15. (ISC)^2: CISSP – The World's Premier Cybersecurity Certification. Retrieved from https://www.isc2.org/Certifications/CISSP (2018)

16. (ISC)^2: CISSP Certification Exam Outline. Retrieved from https://www.isc2.org/-/media/ISC2/Certifications/Exam-Outlines/CISSP-Exam-Outline-2018-v718 (2018)

17. (ISC)^2: (ISC)2 Continuing Professional Education (CPE) Handbook. Retrieved from https://www.isc2.org/-/media/ISC2/Certifications/CPE/CPE%2D%2D-Handbook-Digital-V2.ashx (2017)

18. Aijala, T.: CISSP Certification – Accreditation Value for Employees and Recruiters. Retrieved from http://www.theseus.fi/handle/10024/148953 (2018)

19. IEEE Cyber Security: The Institute: The Cybersecurity Talent Shortage Is Here, and It's a Big Threat to Companies. Retrieved from https://cybersecurity.ieee.org/blog/2017/04/13/the-institute-the-cybersecurity-talentshortage-is-here-and-its-a-big-threat-to-companies/ (2017)

Discovering Student Interest and Talent in Graduate Cybersecurity Education

12

Sherri Aufman and Ping Wang

12.1 Introduction

Cybersecurity is the process of protecting and defending critical or sensitive information systems and data assets. Cybersecurity has become a fast-growing career field and a significant workforce area with increasing demand and opportunities for higher education. In July 2018, H.R.3393, the New Collar Jobs Act, was introduced in congress [1]. The rationale behind the bill is to grow the cybersecurity workforce by providing debt relief, scholarships, increase training, and offer tax credits for companies advocating for academic degrees or certifications [2].

The workforce demand for cybersecurity talent is not temporary but long-term with continuous growth in the future. The latest career outlook published by US Labor Department Bureau of Labor Statistics shows that the employment of information security analysts, an example position title of cybersecurity jobs, is projected to grow 28% from 2016 to 2026, much faster and with better pay than the average for all occupations [3]. A recent CompTIA report shows that there are over 300,000 existing and unfilled cybersecurity job vacancies with the most popular positions listed for Cybersecurity Specialist, Cybersecurity Analyst, Penetration Tester, and Cybersecurity Engineer [4].

Influenced by the anticipated shortage of skilled professionals and world-wide demand, Burning Glass Technologies partnered with CompTIA to create Cyberseek [5]. The partnership's objective is to improve national security by assisting cyber defense in finding talent, and addressing the short supply of cybersecurity workers with online job opportunities and career resource planning [5]. However, such efforts by the industry are limited and not sufficient to keep up with the growth and demand for cybersecurity workforce, and the cybersecurity industry continues to outgrow the supply of talent.

To help mitigate the talent shortage, the US Department of Homeland Security and National Security Agency championed the Centers for Academic Excellence [7]. Centers for Academic Excellence in Cyber Defense Education (CAE-CDE) are 2 and 4 year colleges and universities that undergo a demanding certification process [7]. Approved CAE-CDE institutions teach a relevant and quality curriculum that encourages students to learn by doing [7]. CAE-CDE is considered the gold standard in providing quality assurance for cybersecurity education, but there are only about 3% of US colleges and universities have achieved the CAE-CDE designation so far [16].

Leadership talent with experience, technical expertise, and advanced or graduate education is the most valuable workforce component in any industry. Nearly 61% of cybersecurity job positions require a bachelor's degree, and 23% require a master's degree [3]. Over 83% of these job postings seek a minimum of 3 years' work experience, and depending upon the position, specific technical expertise is required as shown in Table 12.1 below [3]. In addition, Cybersecurity is a multidisciplinary and sophisticated field that involves information systems, computer science, information technology, business management, leadership and teamwork skills, communication skills, creativity, critical thinking, problem-solving and analytical skills, and advanced training and graduate education are important preparation for future cybersecurity leaders to acquire and develop these challenging and comprehensive skills to succeed [15].

Thus, desired qualifications for cybersecurity positions often expand beyond technical skills as shown in Fig. 12.1 below. Advanced college education from a relevant graduate program becomes valuable for future leaders in the

S. Aufman · P. Wang (✉)
Robert Morris University, Moon, PA, USA
e-mail: aufman@rmu.edu; wangp@rmu.edu

© Springer Nature Switzerland AG 2019
S. Latifi (ed.), *16th International Conference on Information Technology-New Generations (ITNG 2019)*,
Advances in Intelligent Systems and Computing 800,
https://doi.org/10.1007/978-3-030-14070-0_12

Table 12.1 Cybersecurity occupational skillset [3]

Cybersecurity occupations and skillset	
Information Security	Network Security, Cryptography, Information Assurance
Network Setup	Cisco Routers, Firewalls, Network Engineering, VPN
Database Coding	SQL, Oracle, Python, Java, Perl, C++
Auditing	Internal Audit, Risk Management, Risk Assessment, Legal Compliance
Network Protocols	Protocol (TCP/IP), SSL, DNS, LDAP, DHCP, Transmission Control Protocol
Systems Administration	System/Network Configuration, Disaster Recovery Planning, Windows Servers

Fig. 12.1 Ranking professional competencies for information security workforce [6]

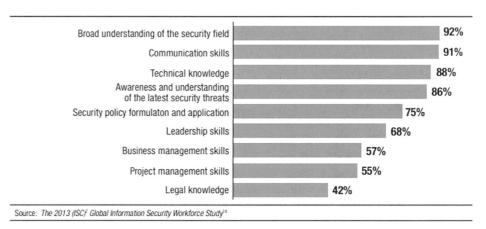

Source: *The 2013 (ISC)² Global Information Security Workforce Study*¹⁸

profession. Soft skills, written and oral communication, leadership, teamwork, and management skills are often sought after by employers. For example, unique challenges exist in hiring for cybersecurity positions in healthcare, accounting, and finance [3]. Employers need "hybrid" cybersecurity talent that understand the technical aspects as well as specific and advanced knowledge, analysis and judgment in the context of applicable information security regulations and compliance requirements from HIPPA and Sarbanes-Oxley Act [3]. A highly sought after cybersecurity professional will possess a combination of technical knowledge, business acumen, with advanced leadership and communication skills.

Among the large and increasing demand for cybersecurity workforce, leadership talent is the most challenging to discover and develop. Graduate education programs in the United States face the challenges of finding the best talent for student enrollment and producing qualified graduate students employable in the field of cybersecurity. The goal of this study is to propose and illustrate a new model for discovering and developing interest and talent in graduate cybersecurity education through effective career guidance and curriculum and course design. The following sections of the paper will review additional background for this study, present the proposed theoretical model, describe the case study method using the sample graduate admissions and curriculum data from a private US university, and discuss the findings.

12.2 Background

The motivation to continue one's education with post baccalaureate studies in a graduate program may vary by individual. Graduate education requires substantial investment of time, money and energy, and a common professional goal for graduate students is to enter and succeed in a satisfactory and rewarding full-time career with bright prospects for advancement and leadership [15]. For example, potential graduate students must commit minimum resources in the form of time and tuition. The average graduate program in the US will take 2 years to complete accompanied with an annual average tuition expense of $16,435 [8]. However, a graduate education may increase one's earning power substantially. The latest data from the US Department of Labor Bureau of Labor Statistics show that a young adult with a master's degree earned on average $68,000 per year as compared to a bachelor's degree holder earning an annual income of $56,000, which is a $12,000 or 21% annual increase in wage [8].

Students pursuing graduate education may do so for a variety of specific reasons, which include one or more of the following:

- Enhance their subject knowledge
- Prepare for graduate studies at the doctoral level

- Seeking career advancement
- Expanding their technical expertise in an evolving field of study

But a common general goal for master's and doctoral students is to succeed and lead in tomorrow's knowledge workforce and make important contributions in industry, government, non-profits, entrepreneurial venture, or in teaching and research, and the program curriculum and course design and activities should incorporate comprehensive technical and non-technical professional skills such as communication, teamwork, project management, and leadership to support students' professional development and career goals [15, 17].

The motivation for professional and career advancement may have contributed to increased completion rates in graduate programs. The National Center for Education Statistics reports that from 2000 through 2017, graduation rates, for those earning a master's degree or higher, increased from 5% to 9% [9]. The increase in workers with a master's degree in the job market adds to the list of reasons to continue with post baccalaureate education: to remain competitive in order to stay and succeed in the workforce.

Over the past 50 years, the perception of a master's degree has evolved. In the 1970s, a master's degree had the stigma of earning the "second place" for those not able to pursue doctoral studies in arts and humanities [10]. Today a graduate degree is a key professional credential, sought by employers, in areas of engineering, healthcare, computer science, and business [10]. Graduate curricula offer or aim to offer an environment of shared learning, problem solving, critical thinking, and technical skills valued by employers and not offered in undergraduate studies [10, 17].

For some professions in particular, a graduate degree has become non-negotiable. For example, data from the US Department of Labor Bureau of Labor Statistics show that thirty-three occupations require a master's degree in order to obtain an entry level position in fields ranging from anthropologist to psychology to urban planners [8]. Graduate education is increasingly important for the cybersecurity field as well. Research by Burning Glass Technologies reports the following findings [3]:

- Cybersecurity employers demand a highly educated, highly experience workforce.
- 84% of cybersecurity postings specify a minimum of a bachelor's degree
- Because of high education and experience requirements, a skills gap exists.
- Skills gap cannot be resolved through short-term solutions.
- Employers and training providers must work together to cultivate a talent pipeline.

In addition, Cybersecurity is essentially a people issue and people with better leadership talent are critical to the success of Cybersecurity [18]. Quality graduate programs will produce better leaders for the professional field. Accordingly, graduate programs offering advanced education in Cybersecurity should effectively advance students' knowledge, skills, and abilities in both technical and non-technical areas to prepare them for leadership in the cybersecurity field.

Partnership with business enterprises in the relevant industry to improve professional development opportunities for graduate students is one of the key recommendations in the recent study report by the Council of Graduate Schools [17]. Several enterprises have stepped up to coordinate market demands with education and training organizations and government agencies for a better educated workforce for the cybersecurity industry. For example, Burning Glass Technologies, CompTIA, and National Initiative for Cybersecurity Education (NICE) under the National Institute of Standards and Technology (NIST) in the U.S. Department of Commerce jointly created Cyberseek to promote cybersecurity education with useful tools and resources.

NICE also developed and published the NICE Cybersecurity Workforce Framework (NCWF) with specific work categories, job roles and tasks for the cybersecurity industry and corresponding expected knowledge, skills, and abilities (KSAs). The goal of NCWF is to provide a national standard for workers, academia, certification providers, and public and private industry sectors to define cybersecurity work and required skills [11]. NCWF begins by describing a broad grouping of functions, which fall into seven categories of cybersecurity work: analyze, collect and operate, investigate, operate and maintain, oversee and govern, protect and defend, and securely provision [11]. These categories drill down to "specialty areas", which can be described as concentrations specific to the role and job [11]. The framework delves deeper into the knowledge, skills, and abilities (KSAs) related to the respective position to nurture a competent cybersecurity talent pipeline [11].

In addition, the US Department of Labor published a Cybersecurity Competency Model that includes the basic personal effectiveness competencies, common academic and workplace competencies often acquired through formal education and training, and more specialized industry-wide technical competencies and industry-sector areas, such as cybersecurity technology, incident detection and response, and protection and defense against cyber threats [19].

The large and increasing demand for qualified professionals and competitive leadership talent for the cybersecurity industry brings abundant opportunities for colleges and universities to offer relevant graduate programs in Cybersecurity. The expected knowledge, skills, and abilities (KSAs) defined in the NCWF and the comprehensive competencies in the Cybersecurity Competency Model should be useful guidelines for quality graduate programs in Cybersecurity.

12.3 Proposed Model

Based on the background review, this paper proposes a new model for discovering student interest in pursuing graduate education in Cybersecurity. This model proposes that student interest in pursuing graduate education in Cybersecurity is motivated by promising and rewarding careers in Cybersecurity and by quality curriculum and courses in the graduate program. The model is illustrated in Fig. 12.2 below.

12.3.1 Careers in Cybersecurity

More and more graduate students from master's and doctoral programs have been pursuing careers in non-academic industry sectors [15, 17]. The cybersecurity industry offers abundant and growing career opportunities, high levels of job satisfaction, attractive and competitive rewards in compensation, as well as great potential for professional growth and leadership [20, 21]. This is an important factor for graduate students given their additional investment of time and money in advanced education beyond the undergraduate degree.

12.3.2 Curriculum and Courses

The curriculum and courses for the graduate program should match and support the program objectives as well as the educational and professional goals for students [15]. Students

are expected to reach the course learning outcomes and workplace competencies in the case of cybersecurity education. The course learning outcomes are course specific and should include knowledge and application skills on technical topics, research skills, professional skills, and abilities to define, describe, analyze problems, evaluate, and create solutions. The expected knowledge, skills, and abilities (KSAs) for relevant cybersecurity job roles defined in the NICE Cybersecurity Workforce Framework and the comprehensive technical and non-technical competencies in the Cybersecurity Competency Model from the US Department of Labor should be useful guidelines for designing quality curriculum and courses in Cybersecurity. In addition, the cybersecurity field is evolving with new technologies and developments. So the cybersecurity curriculum should be frequently reviewed and updated, and new courses addressing the industry needs and workforce demands should be included [22].

12.3.3 Student Interest

Student interest means that the student is willing and motivated to pursue advanced study and training in a certain field. This paper proposes that student interest in Cybersecurity is positively affected and motivated by promising and rewarding careers in the cybersecurity industry and reputable programs with good quality curriculum and courses for cybersecurity education. Student interest is measured by the student admissions to and enrollment in the graduate program of study.

12.4 Case Study

This research uses the case study of the graduate cybersecurity program at Robert Morris University (RMU), a national university located in the northeast of the United States. RMU is a private not-for-profit university comprised of five academic schools. Housed in the School of Communications and Information Systems (SCIS) is the MS in Cybersecurity and Information Assurance, a 30-credit graduate degree program. Course work prepares students to recognize and combat information systems threats and vulnerabilities [12]. As the leading program in the Computer Information Systems (CIS) department, it focuses on technology and management of information security and assurance [12]. Through course work, team projects, and extensive virtual hands-on practice in simulation labs, students are trained on how to recognize and defend against information threats, vulnerabilities, and security risks.

Since the rewarding careers and growing opportunities and prospects in the cybersecurity workforce have already been well documented in multiple study reports [3, 6, 20, 21],

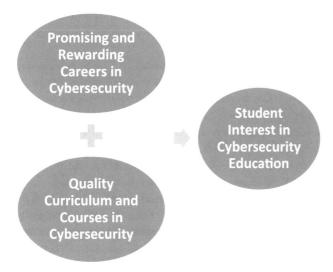

Fig. 12.2 Model for discovering student interest

this case study is to focus on the curriculum and outcome of student interest factors in the proposed model. In terms of curriculum and course design, the MS in Cybersecurity and Information Assurance program at RMU is of high quality and appealing to prospective student interest. The curriculum of the program consists of seven core courses and three elective courses. The core courses include the key subject areas of hardware and operating systems, network technology and management, database management systems, computer network security, IT security and assurance, and secure programming. Each of the courses has clear learning outcomes that support key cybersecurity job roles and tasks and knowledge, skills, abilities defined in the NICE Cybersecurity Workforce Framework. There are a variety of learning activities in the courses, including lectures, discussions, labs, team project, research, and presentations, to develop students' technical and non-technical competencies as described in the Cybersecurity Industry Model from the US Department of Labor. The curriculum design and course activities provide comprehensive learning and practice opportunities for students to develop their professional and leadership skills, such as teamwork collaboration, critical thinking and decision making in handling complex and evolving cybersecurity cases and situations.

Two of the core courses, INFS 6231 and INFS 6490, also help prepare students to sit for the CompTIA Network+ and Security+ certification exams. Additional certifications can be earned by strategically selecting elective courses that include certification as part of the course objective. Certifications not only validate one's technical background but also improve their competitiveness in the job market. Nearly 84% of government defense contractors require certifications to satisfy regulations, and 74% of private sector employers "view certification as an indicator of competency" [6]. For the elective courses, the student may incorporate a programming language(s) or an internship. Programming skills and internship experience add strong dimensions to the student's skillset, technical abilities, and marketability. Professional certifications and technical skills also give students a competitive edge for career advancement and leadership opportunities.

A special curriculum feature of RMU's graduate education including the MS in Cybersecurity and Information Assurance program is the Integrated 4 + 1 program available to students performing academically well in their undergraduate degree coursework. The integrated program offers a multitude of benefits. Students in the 4 + 1 program will earn both a bachelor's and master's degree in 5 years. The integrated students receive a financial incentive. They may continue to utilize their scholarship monies, which

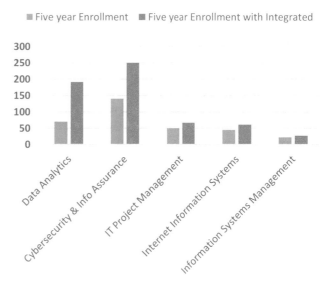

Fig. 12.3 Graduate CIS enrollments 2013–2018

will reduce the tuition expense for the graduate degree. Since the integrated applicants are current RMU undergraduate students, the application process is simplified, and students have a smoother transition to graduate study as they are already familiar with the academic environment at RMU.

Accordingly, student interest in the graduate and integrated program in Cybersecurity and Information Assurance at RMU has been strong and the highest among all CIS programs in the past 5 years as shown in Fig. 12.3 above.

In addition, post-graduation alumni surveys indicate strong professional success rates among RMU graduates [13]:

- 95% placement in a job or graduate school within a year of graduation
- 84% of graduates are employed in their field
- 77% of graduates work in a professional or managerial position

RMU graduate students have been employed by top corporations in Pittsburgh and beyond in positions such as security systems engineers, cybersecurity specialists, penetration testers, security analysts, information security analysts, forensic investigators, and consultants. The top employers of RMU graduates include a variety of sectors from banking, professional sports, healthcare, retail, and logistics [14]. They all need cybersecurity talent to protect their assets, information, and systems.

12.5 Conclusion

This paper proposes that student interest in the field of cybersecurity is positively driven by the promising and rewarding career opportunities in the field and quality curriculum and courses that prepare them for professional success. The demand for skilled cybersecurity workers and leaders illustrates the appeal of a rewarding career, in terms of financial compensation, intellectual challenges, and professional growth. The forecasted shortage of qualified workers and the never-ending need for cyber defense strategies in private, public, and military, lends itself to a long term positive prospect of the field. Advanced or graduate education adds competitive qualifications for future leaders in the fast growing cybersecurity field.

The case study in this research uses RMU's master's and integrated (4 + 1) graduate program in Cybersecurity and Information Assurance for illustration. The program has a quite well designed curriculum including core and elective courses and internships that develop student knowledge, skills, and abilities needed for the cybersecurity profession. The comprehensive course learning outcomes and activities also support the goal of developing technical and non-technical competencies as well as leadership skills expected for the cybersecurity industry. The student interest in the cybersecurity program, as evidenced in the program enrollment in the past 5 years, has been positive and strong and stands out among all graduate CIS programs.

However, the cybersecurity field is evolving and develops rapidly with emerging challenges, solutions, and technologies. Accordingly, there should be regular reviews and updates of the graduate curriculum and courses. Future studies in this area may include exploration of including new and relevant courses in the program, such as Python programming, cloud security, and Internet of Things (IoT) security. Regular program reviews and curriculum updates are important for maintaining the appeal of the program as qualified cybersecurity workers and leaders will need essential KSAs to identify risks and vulnerabilities and develop and evaluate solutions in the new technological frontiers. Future research may also address how to align graduate level curriculum and courses to the national standard, such as the CAE-CDE (Center of Academic Excellence in Cyber Defense Education) designation so as to maintain and strengthen the quality and appeal of graduate cybersecurity programs.

References

1. Barnes, D.: A Bipartisan Agenda to expand the new collar jobs, Aug. 1, 2017. [Online]. Available: https://thehill.com/blogs/pundits-blog/technology/344801-a-bipartisan-agenda-to-expand-new-collar-jobs (2017). Accessed 19 Oct 2018
2. Chadha, J.: Three ideas for solving cybersecurity skills gap, The Wall Street Journal, para. 11, Sept. 18, 2018. [Online]. Available: http://www.wsj.com/articles/three-ideas-for-solving-the-cybersecurity-skills-gap-1537322520. Accessed 15 Oct 2018
3. US Labor Department BLS (Bureau of Labor Statistics).: Retrieved from https://www.bls.gov/ooh/computer-and-information-technology/information-security-analysts.htm (2018)
4. Tauchman, E.R.: Cybersecurity jobs: everything you ever wanted to know, para. 2, Oct. 1, 2018. [Online]. Available: https://certification.comptia.org/it-career-news/post/view/2018/10/01/cybersecurity-jobs-everything-you-ever-wanted-to-know. Accessed 15 Oct 2018
5. Track the Cybersecurity job market with Cyberseek, Nov 9, 2017. [Online]. Available: https://www.burning-glass.com/blog/track-cybersecurity-job-market-cyberseek/. Accessed 15 Oct 2018
6. Suby, M.: ISC2-Global information security workforce study, [Online]. Available: https://iamcybersafe.org/wp-content/uploads/2017/01/2013-ISC2-Global-Information-Security-Workforce-Study.pdf (2013). Accessed 20 Oct 2018
7. National Centers of Academic Excellence, [Online]. Available: https://www.nsa.gov/resources/students-educators/center-academic-excellence. Accessed 8 Oct 2018
8. U.S. Department of Labor Bureau of Labor Statistics, Should I get a master's Degree. Sept. 2015. [Online]. Available: https//www.bls.gov/careeroutlook/2015/article/print/should-i-get-a-masters-degree.htm. Accessed 9 Oct 2018]
9. U.S. Department of Education, National Center for Education Statistics. The condition of education (NCES 2018-144). [Online]. Available: https://nces.ed.gov/fastfacts/display.asp?id=27 (2018). Accessed 9 Oct 2018
10. Gallagher, S.: In defense of the master's degree, [Online]. Available: https://www.forbes.com/site/realspin/2014/04/04/in-defense-of-the-masters-degree/#234d2631492e (2014, April 04). Accessed 10 Oct 2018
11. U.S. Department of Commerce, National Institute of Standards and Technology. NICE Cybersecurity Workforce Framework. [Online]. Available: https://www.nist.gov/itl/applied-cybersecurity/nice/resources/nice-cybersecurity-workforce-framework (2017). Accessed 11 Oct 2018
12. Cyber security and information assurance, rmu.edu, [Online]. Available: https://sentry.rmu.edu/OnTheMove/wpmajdegr.major_desc?iCalledBy=WPMAJDEGR&idegree=MS&imajor=CSIA&ischool=G. Accessed 17 Oct 2018
13. Reputation and outcome, rmu.edu, [Online]. Available: https://rmu.edu/why-rmu/reputation (2018, June 30). Accessed 31 Oct 2018
14. Master of Business Administration, rmu.edu, [Online]. Available: https://rmu.edu/academics/graduate/master-business-administration (2018, June 30). Accessed 31 Oct 2018
15. Wang, P.: Designing a doctoral level cybersecurity course. Issues Inf. Syst. 19(1), 192–202 (2018)

16. Wang, P., Dawson, M., Williams, K.L.: Improving cyber defense education through national standard alignment: case studies. Int. J. Hyperconnect. Internet Things. **2**(1), 12–28 (2018)
17. Denecke, D., Feaster, K., Stone, K.: Professional Development: Shaping Effective Programs for STEM Graduate Students. Council of Graduate Schools, Washington, DC (2017)
18. Hasib, M.: Cybersecurity Leadership, 3rd edn. Tomorrow's Strategy Today, LLC (2015)
19. US Department of Labor.: Cybersecurity industry model. Retrieved from www.doleta.gov (2014)
20. (ISC)^2.: Cybersecurity professionals focus on developing new skills as workforce Gap Widens: (ISC)^2 Cybersecurity Workforce Study 2018. Retrieved from https://www.isc2.org/research (2018)
21. Global Knowledge.: 2018 IT skills and salary report. Retrieved from https://www.globalknowledge.com/us-en/content/salary-report/it-skills-and-salary-report/ (2018)
22. Bicak, A., Liu, M., Murphy, D.: Cybersecurity curriculum development: introducing specialties in a graduate program. J. Inf. Syst. Educ. **13**(3), 99–110 (2015)

Alexander G. Eustis

13.1 Introduction

IoT device security and the Mirai Botnet are directly related due to the way Mirai operates. It is because of the proliferation of insecure IoT devices that the Mirai Botnet is so destructive. Unlike other typical pieces of malware, Mirai does not target personal computers or servers. Its targets are the various internet enabled appliances that have become so common throughout our everyday lives. Everything from security cameras, to DVRs to lightbulbs have become internet enabled [13]. With a projection of 25 Billion devices being connected to the internet by 2020, Mirai and its variants will have a plethora of potential targets [13].

Since it is these IoT devices that are the primary targets, security measures will need to be emphasized on the devices themselves. This is a known problem area in information technology as IoT devices tend to emphasize ease of use and simplicity over security. The average consumer, who may not have a high degree of technical knowledge, would rather have a Plug-and-Play device that requires little knowledge to set-up than having to go through an involved configuration process [11]. Many IoT products are just inherently insecure due to their design. They may only have stripped down operating systems installed and cannot support traditional security measures [3]. After Mirai wreaked havoc using IoT devices, the industry has become more aware of the need for better security on these devices. While typical cyber-security measures will still apply to IoT devices, there are several issues inherent in the devices themselves that need to be addressed. Industry professionals have proposed security measures, both basic and complex, to improve the security on IoT devices to prevent them from being ensnared in a botnet.

A. G. Eustis (✉)
Robert Morris University, Moon, PA, USA
e-mail: agest3@mail.rmu.edu

13.2 What is the Mirai Botnet?

The Mirai Botnet is a piece of Malware that was co-created by Paras Jha and Josiah White [15]. It is designed to target IoT devices and use them to launch Distributed Denial of Service Attacks (DDoS). It takes control of IoT devices by scanning for devices that are still protected by default administrator passwords. The source code for Mirai contains a list of 68 different user ID and password combinations for various manufacturers of IoT devices. The list includes default passwords for devices as varied as routers, security cameras, printers and DVRs [9]. Mirai ensnares these devices by connecting to them at the operating system level using Telnet and Secure Shell (SSH). This means that it can bypass any web-based interface and go straight for the built-in operating system. Changes on the web interfaces of the devices may not be able to circumvent an infection of Mirai because some passwords are hard-coded into the devices [9]. Once a device is infected, Mirai will run from the device's memory until the device is shut off or power is lost. When a device is disconnected or powered off, Mirai will be deleted from the device's memory, removing any trace of an infection. This makes it extremely difficult to tell if a device has even been compromised by Mirai. Rebooting a device will not get rid of Mirai entirely though. After a vulnerable device has been powered back on, it can be re-infected within minutes [8].

13.3 Mirai's Impact

At its peak, the Mirai Botnet had enslaved over 600,000 IoT devices to be used for DDoS attacks. With these devices, the developers would rent portions of their Botnet as a DDoS for hire service. If someone had the money, they could rent a portion of the network to launch a DDoS attack for them. The true scope of Mirai's potential was realized in 2016

© Springer Nature Switzerland AG 2019
S. Latifi (ed.), *16th International Conference on Information Technology-New Generations (ITNG 2019)*,
Advances in Intelligent Systems and Computing 800,
https://doi.org/10.1007/978-3-030-14070-0_13

with three major DDoS attacks; the attack against security researcher Brian Krebs's website, the attack against internet service provider OVH, and the attack against Dyn DNS a month later [11].

13.3.1 Krebs on Security

Security researcher Brian Krebs was the victim of a DDoS attack against his website, krebsonsecurity.com. A portion of the Mirai botnet was used to launch the attack, which took his site offline for 4 days. The attack against Brian's website was so incessant that his DDoS protection provider, *Akamai*, was forced to drop protection for his site. Krebsonsecurity.com became a financial liability for *Akamai* and was causing problems for their other customers. *Akamai* estimated that it would have cost them millions of dollars to devote resources to protecting Krebs's site [11]. Of the 600,000 devices ensnared by Mirai, only 24,000 of that total amount were used to attack Krebs's website with devastating effect [11]. These devices launched data blasts of up to 665Gbps against Krebs's site [6]. This was record breaking at the time.

13.3.2 OVH

OVH is an internet service provider that found itself falling victim to the Mirai Botnet. It was attacked the same day as Brian Krebs's website, September 21st, 2016 [5]. *OVH* held the distinction of being the victim of the largest amount of DDoS traffic being directed at it at the time. Up to 799Gbps of traffic was generated to attack *OVH* [6].

13.3.3 Dyn DNS

To borrow a phrase from popular culture, the Mirai Botnet quite literally "Broke the Internet" on October 21st, 2016. A DDoS attack was launched against *Dyn DNS* which slowed or outright blocked internet access for end users across the East Coast of the United States. Popular websites such as *Twitter, Reddit, GitHub* and *Soundcloud* were caught in this attack. *Dyn* was eventually able to mitigate the attack, but it shows how powerful DDoS botnets are becoming with attack strengths nearly reaching 1-Tbps [5].

13.4   IoT Device Vulnerabilities

It is the lack of security measures taken to protect IoT devices that makes them targets for Mirai and other hacking attacks. This is often by design as manufacturers often prioritize simplicity and ease of use over security. Most consumer IoT devices are low cost, low power, and have a limited amount of memory and processing power available [13]. Others are sold "as is" meaning that they are static and were not designed to be patched or upgraded [3]. Many consumer devices are only secured using a default administrator password which is easily searchable on the internet [11]. This same rule applies to the OS level passwords on these devices. Even if the vendors do not disclose them publicly to consumers, they are still easily deciphered [9]. The simplicity of the devices and the easily decipherable default passwords is what allowed Mirai to infect thousands of devices so easily.

13.4.1 Case Study: XiongMai Technologies

For example, Researchers from *FlashPoint Security* discovered that most of the compromised devices that were used in the attack against *Dyn DNS* were DVRs and IP cameras produced by the Chinese technology company, *XiongMai Technologies*. This company manufactures components for IoT devices which are then sold to venders who use them in their own products [10]. Components produced by *XiongMai Technologies* are incorporated into products from over 100 different manufacturers in a process known as "White Labeling" [7]. *SEC Consult,* a cybersecurity consulting company, discovered several glaring security vulnerabilities in these devices that allowed them to be easily compromised.

First, the *XiongMai* produced devices made use of a cloud service called XMEye P2P that bypasses firewalls and allows for remote connections to the devices. The Cloud IDs used by the devices connecting to this service were not sufficiently randomized which made it easier for attackers to determine that a vulnerable service was running on the device. Up to 9 million devices were found to be running this insecure service and no brute-force protection was enabled to detect suspicious activity against them [7].

A second major vulnerability in these devices was the use of weak default passwords. Mirai had such an impact because its code contained a list of default passwords for various devices. The devices produced by *XiongMai* were configured with a default account named "Admin" with a blank password. During setup, end users were not prompted or encouraged to change the default username and password for this account. These devices also contained an undocumented account that could be accessed by the XMEye Cloud service. This account was named "Default" with a password of "tluafed" (default spelled backwards) [7].

Third, *XiongMai* did not digitally sign their firmware updates for their devices. Using a digital signature protects the integrity of a piece of code by ensuring that it isn't tampered with. Ideally, a firmware update would not be installed on a device unless it had a digital signature that matched the one used by the manufacturer. By not validating their firmware with digital signatures, *XiongMai* created the potential for rogue, malicious firmware updates to be installed on the devices. These could easily be distributed using the company's own XMEye Cloud service. Unlike Mirai, which is deleted from memory after power is cut, firmware updates cannot be removed. Malicious, counterfeit firmware could be loaded on the devices and give an attacker permanent control of the device [7]. This combination of vulnerabilities led to the near entirety of *XiongMai's* product line being weaponized for DDoS attacks by Mirai [10].

13.4.2 Other Examples

Being ensnared by a Botnet isn't the only danger that exists from insecure IoT devices. It is possible to hack an IoT device for cyber-espionage and for financial gain. Even the most innocuous devices become a threat when they have internet connectivity. For example, a smart lightbulb has been used to discover the username and password for a Wi-Fi network [13]. There is even a report of a casino that had 10GB of data stolen through a hacked smart fish tank [14]. Aside from this, there are potentially deadly consequences to IoT device security flaws. For example, in 2015, *Chrysler* needed to release a security update for their vehicles' entertainment systems after a live demonstration showed that the vehicle's steering, engine and brakes could be hacked through it. This is a threat that could literally put people's lives in danger, and with the increasing presence of self-driving vehicles, a new attack surface is emerging [13].

13.5 Proposed IoT Security Solutions

With the exponential increase in potency of DDoS attacks and the growing proliferation of IoT devices, the Information Technology industry is becoming much more aware of the threats surrounding the insecurity of IoT devices. As such, industry professionals have developed best practices for securing IoT devices and are proposing new solutions in the forms of industry standard associations and innovative security tools.

13.5.1 IoT Security Best Practices

The same policies and methods that are used to secure personal computers and corporate networks can also be applied to securing IoT devices. *The Online Trust Alliance* recommends the following as a baseline for securing IoT devices within an enterprise network. These methods could also be applied to consumer devices [14].

- **Place IoT Devices on a separate, firewalled network.**
- **Do NOT use devices with hard-coded passwords.**
- **Govern permissions to the devices.**
- **Disable unnecessary functionality.**
- **Cover / Block Cameras and Microphones.**
- **Verify that physical access to a device does not allow for network intrusion.**
- **Disable automatic connections.**
- **Enable encryption where possible.**
- **Make sure that the controlling applications for IoT devices are secure.**
- **Keep firmware up to date.**
- **Follow proper device lifecycle procedures.**

13.5.2 Developing Industry Security Standards

The growing prevalence of IoT and associated cyber-attacks has prompted several organizations that control technology standards to take notice. Due to the number of manufacturers and different tiers of devices, there is a lack of standardized security measures within the IoT market. Organizations like *Institute of Electrical and Electronics Engineers (IEEE)* and The *National Institute of Standards and Technology (NIST)* control standards for various information technology products from personal computers to telecommunications systems. They, along with several other organizations, are working to develop security standards for the growing IoT device ecosystem. In February of 2018, the *National Institute of Standards and Technology (NIST)* drafted a document to introduce new standards in IoT device security. Titled "Interagency Report on Status of International Cybersecurity Standardization for the Internet of Things", this document aims to help policy makers and standards organizations develop and standardize IoT components, systems and services [12].

The full draft of the document can be found at: https://csrc.nist.gov/CSRC/media/Publications/nistir/8200/draft/documents/nistir8200-draft.pdf

13.5.3 Innovative Security Technologies

For decades, the username and password combination has been the primary means of authentication to a device or a network, but this authentication method has severe limitations when it comes to security. The implementation of multi-factor authentication has improved security for networks and end users, but IoT devices still mainly rely on a humble username and password combination for security. The inherent insecurities with the username & password authentication method have prompted the industry to develop other means to authenticate to and secure devices without having to use a password. There are some innovative technologies being developed that can be applied to IoT device security.

Context-Aware Authentication

Context-Aware Authentication is a means of security that uses machine learning to constantly evaluate risks without impacting a user's experience [2]. Context-Aware authentication works by constantly monitoring the resources that a user is accessing to determine a level of confidence in the user. If this authentication method detects potentially malicious activity, more disruptive authentication methods will be deployed to halt that activity. If the system detects that a user's behavior is within the norm, it will maintain the existing user experience [4]. This method of authentication will effectively halt any malicious activity and leave legitimate users unaffected. This can be applied to both network resources and IoT resources.

Physically Unclonable Functions

Physically Unclonable Functions (PUFs) are potentially useful as a highly secure hardware-based authentication method. A Physically Unclonable Function is defined as "A challenge-response mechanism in which the mapping between a challenge and the corresponding response is dependent on the complex and variable nature of physical material [1]." It works by taking advantage of the slight differences in the internal hardware of a device. The concept behind a PUF is that the slight variations in the construction of a computer chip will generate a different response to the same logical challenge, even if they are physically the same type of part. For example, two different chips used on a motherboard may have the same specifications and form factor, but they will still generate a different response due to how the variations in their physical construction process the challenge. Each chip will generate a unique response. This is accomplished by using variability-aware circuits which detect these slight differences between components [1]. Physically unclonable functions have multiple applications for device security at the hardware level. They can be used to generate cryptographic keys, authenticate to devices and protect intellectual property. PUFs are the machine equivalent to human biometrics [1].

13.6   Conclusions

In my research, I have come to two main conclusions. The first is that the threat surface posed by IoT devices is not going to go away any time soon. These devices are becoming more heavily integrated into consumer and enterprise prod-

ucts and potential attackers will make consistent attempts to take advantage of the vulnerabilities in these devices. Secondly, the information technology industry is well aware of the security threats inherent in these devices, but has been very slow to implement security solutions on these devices. If improved security measures on IoT devices are not implemented, more threats such as Mirai are likely to appear and launch increasingly larger DDoS attacks or more devastating information leaks.

Allison Nixon from *Flashpoint Security* has proposed that an Industry Security Association be created to publish security standards that all members of that organization must adhere to and are periodically audited against. Devices and manufacturers that meet these standards would be promoted with a seal of approval [10]. While no central body has yet been created to handle this task, multiple standards organizations are developing their own security policies to test devices against. Over time, these disparate organizations may work together to develop industry wide standards but it is currently too soon to tell. These standards organizations have only implemented these policies within the past 2 years. They are notably late in addressing the issue, but they are at least now getting to the point where they have a concrete plan to address this threat.

Another barrier that is affecting the security of IoT devices is the market for these devices. The majority of devices that were ensnared by the Mirai Botnet are consumer devices. Manufacturers of these consumer devices are much more inclined to make their devices easy to use and understand than secure. As stated before, consumers generally prefer plug-and-play devices over ones that would require significant effort on their part to set up and use. The market for IoT devices doesn't create much of an incentive for manufacturers to emphasize security on their devices. If a universal standards organization is created, they will be able to pressure manufactures into emphasizing the security on their devices. Until the nature of the IoT industry changes, following the general best practices of information security can provide a reasonable amount of protection against potential threats, but the industry will need to go further to address the growing threat of botnets and hacks against IoT devices. The creation of industry wide standards and pressuring manufacturers to emphasize security on their products is the most practical way to counter this threat. Hardening the security capabilities of consumer devices will make them less vulnerable to botnet malware such as Mirai or hacking and eavesdropping attacks. Manufacturers who produce IoT devices for Enterprise use should work to develop new security technology such as Adaptive Authentication and Physically Unclonable Functions. This will address the threats that weak IoT security poses to large enterprises while maintaining productivity and a positive user experience.

References

1. Casarona, J., McHale, L., McDougall, L., Gunreddy, V., Cantrell, M., Gora, M., Morozov, S., Maiti, A., Kim, I., Schaumont, P.: Research on Physical Unclonable Functions (PUFs) at SES Lab, VT, Retrieved Oct 2018 from Virginia Tech's Bradley Department of Electrical & Computer Engineering Website: http://rijndael.ece.vt.edu/puf/background.html (2011)
2. Hamilton, D.: Best practices for IoT security. Retrieved Sept 2018 from *Network World's* Website: https://www.networkworld.com/article/3266375/internet-of-things/best-practices-for-iot-security.html (2018, March 27)
3. Internet of Secure Things: what is really needed to secure the Internet of Things?. Retrieved Oct 2018 from *Icon Labs'* website: https://www.iconlabs.com/prod/internet-secure-things-%E2%80%93-what-really-needed-secure-internet-things (2018)
4. Kepes, B.: Forget two-factor authentication, here comes context-aware authentication. Retrieved 31 Oct 2018 from Computer World's Website: https://www.computerworld.com/article/3105866/application-security/forget-two-factor-authentication-here-comes-context-aware-authentication.html (2016, Aug 15)
5. Kerner, S.M.: DDoS Attack Snarls friday morning traffic. Retrieved 21 Oct 2018 from eWeek's Website: http://www.eweek.com/security/ddos-attack-snarls-friday-morning-internet-traffic (2016, Oct 21)
6. Kerner, S.M.: Mirai IoT Botnet creators plead guilty for roles in cyber-attacks. Retrieved 19 Sept 2018 from eWeek's Website: http://www.eweek.com/security/mirai-iot-botnet-creators-plead-guilty-for-roles-in-cyber-attacks (2017, Dec 13)
7. Kirk, J.: Review shows glaring flaws in Xiongmai IoT devices. Retrieved 17 Dec 2018 from Bank Info Security's website: https://www.bankinfosecurity.com/review-shows-glaring-flaws-in-xiongmai-iot-devices-a-11596 (2018, October 12)
8. Krebs, B.: Source code for IoT Botnet "Mirai" released. Retrieved 24 Oct 2018 from Brian Krebs's cyber-security blog, *Krebs on Security*. https://krebsonsecurity.com/2016/10/source-code-for-iot-botnet-mirai-released/ (2016, Oct 1)
9. Krebs, B.: Who makes the IoT things under attack?. Retrieved 11 Sept 2018 from Brian Krebs's cyber-security blog, *Krebs On Security*: https://krebsonsecurity.com/2016/10/who-makes-the-iot-things-under-attack/ (2016, Oct 3)
10. Krebs, B.: Hacked cameras, DVRS powered todays massive internet outage. Retrieved 14 Sept 2018 from Brian Krebs's cyber-security blog, *Krebs On Security*: https://krebsonsecurity.com/2016/10/hacked-cameras-dvrs-powered-todays-massive-internet-outage/ (2016 Oct 21)
11. Krebs, B.: Study, attack on Krebsonsecurity cost IoT device owners 323k. Retrieved 11 Sept, 2018 from Brian Krebs's cyber-security blog, *Krebs On Security*: https://krebsonsecurity.com/2018/05/study-attack-on-krebsonsecurity-cost-iot-device-owners-323k/ (2018, May 7)
12. Miller, S.: NIST maps Out IoT security standards. Retrieved 29 Oct 2018 from GCN's Website https://gcn.com/articles/2018/02/15/nist-iot-standards.aspx (2018, Feb 15)
13. O'Niell, M.: Insecurity by design: today's IoT Device security problem. Engineering. **2**(1), 48–49 (2016)
14. Solomon, H.: How to secure consumer IoT Devices in the enterprise. Retrieved Oct 2018 from *IT World Canada's* website: https://www.itworldcanada.com/article/how-to-secure-consumer-iot-devices-in-the-enterprise/404144 (2018, April 17)
15. Vanian, J.: Mirai Botnet: 3 Men Plead guilty to cybercrimes. Retrieved 19 Sept 2018 from Fortune.com: http://fortune.com/2017/12/13/%E2%80%AA%E2%80%AAmirai%E2%80%AC-%E2%80%AAbotnet%E2%80%AC-cybercrime-doj/ (2017, Dec 13)

The Study of the Effectiveness of the Secure Software Development Life-Cycle Models in IT Project Management

14

Saniora R. Duclervil and Jing-Chiou Liou

14.1 Introduction

Cyber-Security has become prevalent and important in today's society. According to Internet study, 1 in 3 Americans become victims of a cyber-attack. According to USA Today, last year, 15.4 million Americans were victims of Identity fraud. In the last year, Equifax underwent a data breach and 143 million Americans were victims. Information stolen included credit card numbers and social security numbers.

Information Technology Project Management is the process of developing an application by planning, designing, executing and monitoring that application in order to meet organizational needs. IT project managers use a process or model called the System Development Life-Cycle or Software Development Life-Cycle (SDLC).

There are a few different models of SDLC proposed and used in IT industry. The two most popular ones are: Waterfall and Agile.

The Waterfall model SDLC has a total of five to six phases which are: Requirement gathering and analysis, design, development/coding, testing, deployment, and maintenance. On the contrary, the agile model uses an iterative spiral SDLC and requires several iterations, called Sprints.

Regardless Waterfall or Agile, the SDLC is the process IT project managers and teams use to develop a software. However, conventionally security components are not incorporated into the SDLC process. It is when the software is completed and at its operation then security is taken into account. Hence, that leaves room for vulnerabilities and room for hackers to attack the system. That is when the Secure Software Development Life-Cycle comes in to fill the gap in recent development of Secure Software Development Life Cycle (SSDLC).

SSDLC is usually the same process as SDLC and it also has the same phases. But in this case, security is incorporated in each phase of the SSDLC. The only problem is that the SSDLC is not one size fits all. There are many new approaches/models of SSDLC that have been proposed or modified from existing SSDLC models and not all these models can work for all types of IT projects because of the different needs and specifications.

Most of the proposed SSDLC models are developed with additional activities or components inserted in the its corresponding SDLC. Some SSDLA were proposed primary based on the Waterfall modes [2–5, 7, 10]. And others are targeting on Agile model [1, 8] (Fig. 14.1).

It is best to find the most effective SSDLC that will work for most or all IT projects. A basic waterfall-based secure software Development Life-Cycle model is provided below for reference of discussion. These are the necessary components for an effective SSDLC.

14.2 Analysis of SSDLC Models

There are some comparisons of different SSDLC models done by other researchers [6, 13, 14]. But these articles did not perform actual comparison based any measurements or used any criteria derived from the characteristics embedded in the SSDLC models they compared. They simple listed the SSDLC models and described the security related activities/components inserted into the original SDLC.

To perform the comparison of SSDLC models in more technical depth, we need to look into the characteristics of those popular models and select the common criteria from the characteristics for comparison. To the end, we have to firstly determine the SSDLC models we would to compare.

S. R. Duclervil · J.-C. Liou (✉)
School of Computer Science, Kean University, Union, NJ, USA
e-mail: duclersa@kean.edu; jliou@kean.edu

© Springer Nature Switzerland AG 2019
S. Latifi (ed.), *16th International Conference on Information Technology-New Generations (ITNG 2019)*,
Advances in Intelligent Systems and Computing 800,
https://doi.org/10.1007/978-3-030-14070-0_14

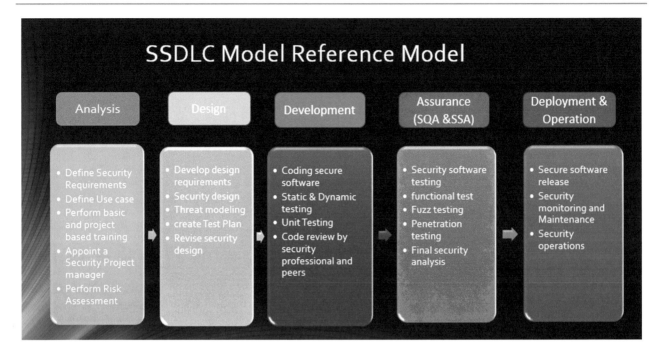

Fig. 14.1 SSDLC reference model

The first model that was selected was Microsoft's Trust-worthy Computing Security Development Life-Cycle model. The SDL (as Microsoft calls it) [8–10], was developed and adapted by Microsoft Corporation in 2004 for its personal use and as a way to prevent anymore vulnerabilities in their developed applications. This model has a total of six phases: Requirements, design, implementation, verification (which is the testing phase), the release phase, and the support and servicing phase. Most security components are the same as the basic SSDLC, but Microsoft has a security Kick-off meeting in the first phase, and the need to register with Microsoft's SWI (Secure Windows Initiative).

The second model was the Gary McGraw Touch-Point Model [7]. This model was developed by Gary McGraw in 2004 and the highlights software security touch points which is also known as best practices. It was later become "Building Security In" and was adopted by some organizations, such as the Software Engineering Institute (SEI). This model has 6 phases: Requirements and use case, Architecture and design, Test plans, code, Tests and Test results, and Feedback from the field. The security components are more or less the same but the difference is that the Touch Point model is a waterfall and an agile development model combined to make one model.

The third model that was identified was the Software Engineering Institute (SEI) Team Software Process (TSP) for Secure Software Development [12]. This SSDLC was developed by SEI. The software engineering Institute, like any other company or person, developed this model to decrease the likelihood of appearances of vulnerabilities in developed

applications. This model has only four phases which is not the same for the basic SSDLC model. It has a requirement, design, implementation, and testing phase but no operation and deployment phase. Organizational policies, management oversight, resources and training, project planning, project tracking, risk management, measurement and feedback are incorporated into all of the phases.

The last model that was identified was the SSDLC model developed by Abhinav Rastogi and Russell L. Jones [3]. This model has a total of five phases: Design, development, Testing, Operation and Maintenance, and the disposal phase. There are few differences in this model. This model has no requirement phase, but the requirement gathering and work-shops are done in the design phase. Training is not done until the operation and maintenance phase but it is ongoing even after the process. Certification is required and at the end there is a disposal phases that focuses on if or when the program is being disposed of, moved, remodeled, or archiving. This last phase is a plus because there are security protocols and components that need to be added when disposing of, the redeveloping of or moving the software.

14.3 Analysis of Characteristics and Assessment Criteria

To perform the comparison for the four models of SSDLC, we started with studying each model and observing sig-nificant characteristics from each individual one. Once all information was collected after studied all the models, we were able to identify some characteristics that all models or some of them have in common.

Based on those similar characteristics that concluded from the study of the four models, we have developed four sets of criteria: Focus area of Application, Implementation of model, security Implementations and Enhancement, and Security training and Staffing.

14.3.1 Focus Areas of Application

This group focuses on the area of the application or the software so whether it is agile or waterfall or if it is more specific to a sector, or if it contains all phases.

The first criterion is that the model has to be either Waterfall or Agile. When a model is a Waterfall model that is indication that the project is going to last 6 months to 2 years. When a SSDLC model is an agile model that is an indication that the project will take the most a month to complete. A SSDLC or an SDLC model for that matter would not be as effective if both Waterfall and Agile were combined. So as of now, the SSDLC model is more effective when it is either Waterfall or Agile. The measure of this criterion using a binary measurement which was yes or no. In this criterion, it determines if the SSDLC models would work well in a Waterfall approach, Agile approach, or both.

The second criterion in this group is that the model has to be universal for most organization. For this criterion an 1–5 scale measurement is applied: 1 being specific and 5 being general for public use. The purpose of this model is to find a model that is most effective for the use of most or all Information Technology projects.

The third criterion in this group is that the SSDLC has to contain all the phases that are required in the SSDLC model. The phases include: Analysis/requirement phase, design phase, development phase, assurance (SQA, SSA) phase, and deployment and operation phase. This criterion had a scaling measurement that was either yes or no. All phases must be included according to the reference model that was provided earlier in this study.

14.3.2 Implementation of Model

These set of criteria focuses on the implementation of the model and how the implementation will affect the application when it is being developed.

The first criterion is that the SSDLC model has to be compatible with the development of the software. It means that the model has to incorporate certain components that will enhance and benefit the development of the software. To that end, a percentage measurement for this criterion is chosen. 100% being that the model is completely compatible

with the development of the software. Which means that regular as well as security components are added into the model and the model is able to bring fourth the functionality and the security components in an effective manner. If the model is at 75% compatibility, that means that the model only satisfies only part of the regular function and security function requirements. This could happen if the model does not include all phases of the SSDLC or if one or more phases are lacking security components If the model is at 50% compatibility that means that only the regular components are added but not the security components. If the model is at 49% or below, that means that it does not [perform neither the regular or security requirements.

The second criterion is optional but necessary. A phase that focuses on new or continued development. This phase focuses on disposing the software, or developing a new software entirely or moving the software to a new location. The data involved with the software need to be protected to avoid data breaches or vulnerabilities while disposing of, changing the location of, or developing a new software from the existing software.

The third criterion for this group of criteria is that the preliminary/initiation phases have to be effective for the development of the software. In the preliminary phase which is mostly the requirement phase in SSDLC, the requirements analysis for the software is performed, project managers, security project managers and stakeholders are selected. Security training takes place for the team involved with the project as well. In a requirement phase, there can also be user stories which are usually in agile development. User stories are tasks that can be manipulated at a later time, even though it was already completed, that gives the team the opportunity to go back and change the user story or fix a problem that occurred. This is usually in agile development because agile development last a maximum of 1 month. This allows the team to go back and redo a sprint (task in agile), add to it, or correct it.

The measurements provided were highly effective, modernly effective, and not effective. A model's requirement phase is highly effective if it has both a requirement analysis and a user story in a waterfall model, if it has training (Security training, or a certification or coding lessons) for the project team or workshops, if there are only user stories in an agile model. Security requirements cannot be in an agile model. If a model's requirement phase is moderately effective, it has a requirement analysis and a user story but no training or workshop for the team involved with the project. If the model's requirement phase is ineffective, that means it has a security requirement phase in an agile model, and there is no training.

The last criterion requires a SSDLC model to have a deployment/operation phase. At the end of the development of the software, there needs to be security components and protocols in order release the software to users and monitor and maintain it. The measurement provided for this criterion was yes or no, yes if it has a deployment/operation phase, or no if it does not have a deployment/operation phase.

14.3.3 Security Implementation and Enhancements

This group focuses on the security components of the SDLC which is the ground for SSDLC.

The first criterion for this group is that security testing has to be done in at least two or more phase of the SSDLC. In the SSDLC there are security components in all of the SSDLC phases. Testing is usually done in the testing phase and during implementation/development phases. This makes the model more effective with its security components so that the software will be less vulnerable to attack. The measurements provided were: Yes, it it has two or more phases that does testing, or no if it does not.

The last criterion for this group is that all phases have to have a security component. That is the whole idea for the Secure Software Development Life-Cycle model. Security has to be emphasized throughput all phases to ensure in everything that is being done in SSDLC will ensure the safety of the software. The model is 100% effective if all phases have a security component. And it is 50% effective if it does not.

14.3.4 Security Training and Staffing

This group focuses on whether there is security staff and how difficult training is.

Security training is very essential for developing a model for the safest software. Security training involves teaching the team about vulnerabilities, secure coding, how to create

a threat model, security protocols, etc. This criterion is being measured by its difficulty in a 1–10 scale. 1–3 is easy, 4–6 is moderate, and 7–10 is difficult. Security training has to be easy and understandable. If a type of security training requires a team member to get security certified, they wouldn't know what area to get it in. Security training has to be specific and attainable.

The last criterion is that there needs to be a security staff that is on top of all security aspect of the SSDLC and make sure all security components are carried out successfully. Implementing a security staff is the same thing as implementing a staff/team for a regular System development life Cycle. In SSDLC, a security project manager, stakeholder and team (such as a tester and software engineer who is aware of the cyber security field) has to be present. A yes or no measurement is provided for this criterion.

14.4 Comparison of SSDLC Models

14.4.1 Comparison from Criteria

Based on information collected, As shown in the Table 14.1, for the Focus Area of Application group of criteria, no model as of now works for both waterfall and agile development. In the second criteria, the more universal model is the Rastogi and Jones model and Microsoft's. In the third criteria, the only mode that does not have all the phases required in the SSDLC is the SEI model. The more effective model was the Rastogi and Jones model, the second effective model was Microsoft's model.

For implementation and model group, SEI is at 75% compatibility because it does not have a deployment/operations phase that has security components and protocols to protect the software when it is released and being monitored. The Gary McGraw Touch Point model is at 0% compatibility

Table 14.1 Comparison Results from criteria 1–5

SSDLC models	Criterion 1: Agile, Waterfall Or both	Criterion 2: has to be Universal:	Criterion 3 Group Must have all phases	Criterion 4 Compatibility	Criterion 5: New or continued development
Microsoft	Waterfall: Yes	Scale of 2	Yes	100%	No
	Agile: No				
	Both: No				
Touch Point	Waterfall: Yes	Scale of 1	Yes	0%	Yes
	Agile: Yes				
	Both: No				
SEI Process	Waterfall: Yes	Scale of 2	No	75%	Yes
	Agile: Yes				
	Both No				
Jones & Rastogi	Waterfall: Yes	Scale of 3	Yes	100%	No
	Agile: No				
	Both: No				

Table 14.2 Comparison Results from criteria 6–11

SSDLC models	Criterion 6: Initiation phase effective	Criterion 7: Has operation & maintenance phase	Criterion 8 Is Security testing done in more than one phase of the SSDLC?	Criterion 9 Security is emphasized throughout phases of the SSDLC model	Criterion 10 Security training (Basic and specific) The difficulty of implementing training.	Criteron 11 Has to have a security project manager and staff
Microsoft	Highly effective	Yes	Yes	100% All phases	Scale of 2	Yes
Touch Point	Ineffective	Yes	Yes	100% All phases	Scale of 10	No
SEI Process	Highly effective	No	Yes	75% all Phases	Scale of 5	Yes
Jones & Rastogi	Highly effective	Yes	Yes	100% all phases	Scale of 8:	Yes

because as of now, agile and waterfall development cannot be combined. In the third criteria two models are the leading: The Rastogi and Jones model has a phase that has new or continued development, and the Touch Point model because of the fact that it can work with agile development which allows for continued development. SEI is the least effective in this criterion because it is the only one that does not have a deployment/operations phase. The more effective models in this group of criteria are the Jones and Rastogi model and the Microsoft model.

For the Security enhancements and implementation group, As depicted in the Table 14.2, the more effective models are the Jones and Rastogi model and Microsoft's model. Security testing is done in at least two of the phases for all models. All models have security components in each phase except the SEI model because it does not have a deployment/operations phase which has security components and protocols releasing and monitoring the software.

For the Security training and staff group of criteria, all models have a security staff except for the Touch Point model, the models that implemented an easy training were Microsoft's model and the SEI model. The Touch Point model has no security training. In the Rastogi and Jones model, what is involved in the security training is not specified and it is only implemented in the operations phase and continues on after that. The most effective model in this group is the Microsoft model and the SEI model.

14.4.2 Summary of the Comparison

Microsoft's model is an effective model to use, however, it was adapted by Microsoft Corporation and is not meant to be used by the general public. In other terms, this model is less effective for most or all IT projects.

The Gary McGraw is a model that has the potential to be a waterfall model but not both agile and waterfall combined. As of now we have not found a model that could work for both waterfall and agile development. If this model were to be used, considering the model leans more towards agile, there would not be enough time to complete all the phases

that is crucial to the successful development of the software because agile development only takes at most a month while waterfall could take 6 months up to 3 years. As of now, a model like this is not effective for SSDLC development yet alone being used by most organizations for software development.

The SEI team process for Secure Software Development has a great training process for the team members in the requirement phase. However, this model does not have the deployment and operation phase that contains components and protocols for releasing the software to the public and monitoring it to prevent any attacks. As of now, this model is not as effective.

The model that is most effective is the SSDLC model developed by Rastogi and Jones. It contains all of the phases and meets most of the important criteria. It has an operation and maintenance phase, it is universal enough for public use, and the process model is compatible with the development of the software. Although a requirements phase is not added, the requirement gathering and workshops are added into the design phase. Security training is not added until the Maintenance phase but it is ongoing even after the development of the software. There is a disposal phase which is not in the universal model but it does prove useful if the software needs to be moved, redeveloped, or disposed of.

14.5   Conclusion

There are many Secure Software Development Life-Cycle models available to the IT industry. Some of them are very similar to each other. However, since most IT projects are different in terms of sizes, operation objectives, as well as the sectors of industry, the SSDLC model has to be general enough to meet at least some of the criteria for an IT organization's software development.

This paper is only meant to compare existing popular models and to identify which one is most effective for being used for most or many IT projects and the one that would be most effective in protecting a software from various attacks and vulnerabilities.

Since cyber-attacks are becoming more common and more dangerous and unavoidable, it is important that our software is protected, yet alone the process used to develop our software.

14.6 Future Work

Considering the various software development models proposed in the IT industry, the four SSDLC models used for comparison in our study is not representing the whole spectrum of the research. Our plan is to research more models (such as Behavior Driven Development model) and rebuild our criteria to bring forth an effective model for future general public use.

Acknowledgment The authors would like to acknowledge Martha Salma and Wendy Alvarado of the McNair scholarship program at Kean University for the support of the research.

References

1. Ayalew, T., Kidane, T., Carlsson, B.: Identification and evaluation of security Activities I Agile projects. 2013 Nordic Conference on Secure IT Systems, pp. 139–153, Ilulissat, Oct 2013
2. Daud, M.I.: Secure software development model: a guide for secure software life cycle. In: Proc. The inetrnational MultiConference of Engineerings and Computer Scientist, vol. I, Hongkong (March 2010)
3. Jones, R.L., Rastogi, A.: Secure coding: building security into the software development life cycle. Inf. Syst. Secur. **13**(5), 29–39 (2004)
4. Keary, E., Manico, J.: Secure development lifecycle. Last retrieved 11/16/18, https://www.owasp.org/images/7/76/Jim_Manico_(Hamburg)_-_Securiing_the_SDLC.pdf (n.d.)
5. Morana, M.: Building security into the software life cycle, a business case. Last retrieved 11/16/18, https://www.blackhat.com/presentations/bh-usa-06/bh-us-06-Morana-R3.0.pdf (n.d.)
6. Manico, J.: Secure SSDLC. Retrieved 11/16/18, https://www.issala.org/wp-content/uploads/Jim-Manico-SDLC-Architecture-v14.pdf (n.d.)
7. McGraw, G.: Software Security. IEEE Secur. Priv. 32–35 (2004)
8. Microsoft.: Security development lifecycle for Agile development. Microsoft Security Development Lifecycle. Last retrieved 11/16/18, https://www.blackhat.com/presentations/bh-dc-10/Sullivan_Bryan/BlackHat-DC-2010-Sullivan-SDL-Agile-wp.pdf (2009, June 30)
9. Microsoft.: Microsoft Security Development Lifecycle (SDL) – process guidance. Retrieved June 2018, from msdn.microsoft.com: https://msdn.microsoft.com/en-us/library/windows/desktop/84aed186-1d75-4366-8e61-8d258746bopq.aspx (2012)
10. Microsoft Corporation Team.: The Trustworthy computing security development lifecycle. Last retrieved 11/16/18, https://msdn.microsoft.com/en-us/library/ms995349.aspx (2007)
11. Davis, N.: Developing secure software, in secure software engineering. The DoD software Tech News. **8**(2), 3–7. Last retrieved 11/16/18, http://www.sis.pitt.edu/jjoshi/devsec/securesoftware.pdf. (2005).
12. Over, J.W.: Team software software process for secure software development. Last retrived 11/16/2018. https://resources.sei.cmu.edu/asset_files/Presentation/2002_017_001_24393.pdf (2002)
13. Tiirik, K.: Comparison of SDLC and touch points. Last retrieved 11/16/18, https://courses.cs.ut.ee/MTAT.03.246/2013_spring/uploads/Main/essay09.pdf
14. Win, B.D.: Secure development lifecycles. Last retrieved 11/16/18, https://handouts.secappdev.org/handouts/2013/Bart%20De%20Win/SecAppDev2013%20-%20SDLC%20Session%20Bart%20De%20Win%20v1.0.pdf (2013)

Bruno Luiz Kreutz Barroso, Fábio Mangueira, and Methanias Colaço Júnior

15.1 Introduction

Money laundering (ML) usually refers to such activity or processes that deals with criminal proceeds to disguise their illicit origin and make them look legit [1]. ML is considered as a major crime in criminology, and is identified as one of the top group crimes in today's society [2], besides being, frequently, a transnational crime that occurs in close relation to other crimes, like illegal drug trading, terrorism, or arms trafficking [3].

Criminal elements in today's technology-driven society use every means available at their disposal to launder the proceeds from their illegal activities. In response, international community has made anti-money laundering (AML) efforts are being made [4]. Usually, financial institutions use semi-automated processes to flag suspicious ML transactions, based on medians and predetermined standard irregularities [5]. AML systems are pivotal and fundamentals to aid governments and institutions to fight against ML.

In this context, is necessary to identify the best practices to combat ML, the best techniques and opportunities to be explored. It is needed to disseminate a culture of repression to this kind of delict, which accompany, encourages and finances the apparatus and investment of many other daily delicts, presenting a dangerous threat to the society.

This article presents a Systematic Mapping that had as objective to identify and systematize the approaches, techniques and algorithms used to detect ML. With this purpose, articles from important databases of CS were mapped.

After answering the research questions it was identified that the main techniques explored were supervised classification techniques [1, 5–18] with 15 (28.3%) and clustering [2, 5, 7, 10, 15, 17, 19–25] with 14 techniques (26.42%).

As for the characterization of the publications, the amount oscillated a lot over the years, mostly because of the low amounts of publications. In relation to the countries, China was, by a large margin, the country with the most publications in the field. The peak of publications about the theme was in 2010, the IEEE International Conference on Machine Learning and Applications (ICMLA), the International Conference on Machine Learning and Cybernetics (ICMLC) and the IEEE International Conference on Data Mining Workshops (ICDMW) published the most papers. The conferences dominated the publications landscape. Finally, two similar papers from the same city, published by two different authors were identified.

This paper is organized as follows: in Sect. 15.2, the literature works related to the theme of this systematic mapping are presented; in Sect. 15.3, the method adopted in this mapping is presented; in Sect. 15.4, the results of the analysis are described; in Sect. 15.5, threats to validity are presented; Finally, in Sect. 15.6, the conclusion is presented.

15.2 Related Works

Secondary studies related to our research were found. Ngai et al. [26] presented a classification framework and a systematic review on the application of data mining techniques in the detection of financial fraud. D. Yue et al. [27], also presented a generic framework for understanding and classifying different combinations of financial fraud detection techniques and data mining algorithms. However, unlike the present study, the work referred [26] and [27] were not just about money laundering, this type of crime was only a subset of the

This work was conducted during a scholarship supported by FAPITEC/-CAPES.

B. L. Kreutz Barroso (✉) · F. Mangueira · M. C. Júnior
Postgraduate Program in Computer Science – PROCC, Federal University of Sergipe (UFS), São Cristóvão, Sergipe, Brazil

financial fraud classified. In the systematic review conducted by Ngai et al. [26], only one primary study dealing with money laundering was found.

This paper distinguishes itself by treating and focusing itself solemnly in money laundering, by not focusing solely on data mining and by emphasizing the techniques of outlier detection and time series. In addition, the present mapping contemplates more recent studies.

15.3 Method

Some researchers have been working to establish stable methods for applying the systematic review process in the literature [28–31]. One of these methods is the Systematic Mapping, which consists of a systematic protocol for searching and selecting relevant studies in the literature, with the objective of extracting information and mapping the results to a specific research problem [28,29]. The present study was based on the protocols proposed by Kitchenham et al. [28] and Petersen et al. [29].

The choice of performing Systematic Mapping was justified by allowing the analysis of primary studies in a broader way to answer the research questions, as well as collecting evidence to guide future research.

15.3.1 Research Questions

The objective of this paper was to perform a Systematic Mapping with the purpose of identifying and analyzing primary studies, to characterize the use of algorithms, methods and techniques to detect evidence of money laundering. For the research questions' elaboration, initially, it was decided to detail the approaches used for the detection of outliers and time series. This approach was based on control papers and the assumption that the problem of money laundering produces data that favors the discovery of anomalies, since the detection of suspicious activities can be seen as a outlier detection problem [22]. In addition, time series are used in the scope of several financial problems [32]. In this context, the study intended to highlight the researches that used these two techniques, while not failing to identify all the others. Furthermore, it was intended to identify the current panorama of the ML detection field, in order to guide future research. Thus, the following questions were elaborated:

- **Q1**: What are the most commonly used computational approaches and techniques as basis for money laundering detection?
- **Q2**: What specific outlier discovery methods or algorithms are used to identify transactions that may indicate money laundering?

- **Q3**: What specific time series analysis methods or algorithms are used to identify transactions that may indicate money laundering?
- **Q4**: Which countries have the highest number of research published on this context?
- **Q5**: Which years have had the most publications in this area?
- **Q6**: What are the main journals and conferences about the subject?
- **Q7**: Which is the most popular publication venue?

15.3.2 Search Strategy

The following bases were used to execute the Systematic Mapping: Scopus, IEEE and ACM. Download without restriction was granted through the Capes journals portal (https://www.periodicos.capes.gov.br). The Scopus base was chose due its comprehensively collection of articles from several databases: Science Direct, Springer and Elsevier are among them [33]. To supplement the Scopus results the ACM and IEEE bases were used. These databases are responsible for publishing the major journals and conferences in the area of CS.

Sources were selected through the keywords search according to their availability on the internet. Only English studies, works related to CS and articles published in conferences, periodicals or book chapters were selected.

The advanced search refinement option was used in the Scopus database to select only results within the field of CS whose language were English. Also, results referring to conference recapitulations and notes were excluded. In the other bases no refinement was made.

The search string used, generated with the keywords, was:

((“money laundering” OR “capital laundering”) AND (“data mining” OR “data analytics” OR “outlier” OR “forecasting” OR “time series” OR “big data” OR “business intelligence” OR “data science” OR “artificial intelligence” OR “machine learning”))

With the search conducted during August – November of 2017, 86 unique results were returned by the search strings. After this stage, the papers selection was started, which will be detailed below.

15.3.3 Selection Criteria

In order to filter the relevant papers to this Systematic Mapping, the inclusion and exclusion criteria were established. The study used the following inclusion criteria:

- The result should contain the theme of this study in the title, abstract or keywords;

- The result needs to explore an algorithm, technique, mechanism or approach for money laundering detection.

To confirm the inclusion criteria, the abstract and introduction of each paper were analyzed.

In parallel, the articles were analyzed according to the exclusion criteria. The exclusion criteria described below was also applied to them:

- Papers that do not belong to the field of CS;
- Secondary studies, as they deal with third-party approaches;
- Papers that were unavailable;
- Ongoing studies.

After the inclusion and exclusion criteria were applied, the relevance of the studies were evaluated. Among the 86 unique papers found, 35 were selected to compose the primary studies. As 2 of these 35 articles [34, 35] were considerably similar but had distinct authors, it was decided to count the two articles as one to avoid noise caused by the duplication of one of the studies, totaling, in the end, 34 primary studies. Therefore, when one of these articles is referred in this paper it means both are being referenced. Nevertheless, it is not the aim of this study to prove that there was any unethical behaviour or violation, further investigation, perhaps by IEEE and Springer, would be necessary, as they could contact the authors and give them the right to defend themselves if they judge there was any misconduct.

The chart in the Fig. 15.1 shows the amount of articles by scientific repository after the application of the selection criteria. Although, IEEE, ACM are under the Scopus umbrella, the papers counted as Scopus were the ones found in other bases.

15.4 Discussion

In this section, we present the analysis results of the primary studies, answering the research questions presented earlier.

In the chart displayed in Fig. 15.2, is presented the characterization of the main approaches found for **Q1**. The most relevant the techniques were supervised classification techniques [1, 5–18] with 15 instances (28.3%), followed by the ones based on clustering [2, 5, 7, 10, 15, 17, 19–25] with 14 (26.42%) of the main techniques from the papers. Outlier detection techniques [21, 22, 34, 36, 37] had 7 (13.21%), while association rules techniques [5, 24, 38] had 6 algorithms (11.32%). Time series analysis [2, 32, 39, 40] is the next with 4 (7.55%). Graph mining [41, 42], optimization algorithms [1, 36] and heuristics [7, 23] had 2 (3.77%) each. Finally, the paper that used a rule-based expert system did not disclose the specific algorithm used [10], representing 1 instance (1.89%).

Among the primary studies, 5 proposed the use of intelligent agents [4, 5, 13, 43, 44], but only 2 of them specified the techniques implemented: clusters, supervised classifiers and association rules [5], and supervised classifiers [13]. The others didn't go into the algorithm details.

Supervised classification builds up and utilizes a model to predict the categorical labels of unknown objects to distinguish between objects of different classes [26, 45]. Amid the supervised classifiers, decision trees had the most occurrences (7 out of 15, 46.66%), followed by neural networks (4 out of 15, 26.66%) and SVM (3 out of 15, 20%). The drawback of supervised classifiers is that they need labeled data to be trained and the process of labeling data when dealing with big data sets may be exhaustive.

Clustering is used to divide objects into conceptually meaningful groups (clusters), with the objects in a group be-

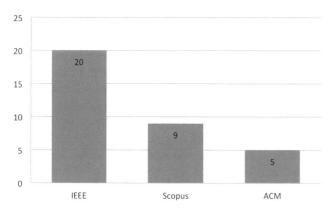

Fig. 15.1 Articles selected by base

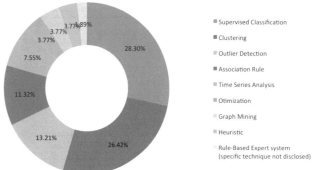

Fig. 15.2 Characterization

ing similar to one another but very dissimilar to the objects in other groups. Clustering is also known as data segmentation or partitioning and is regarded as a variant of unsupervised classification [26, 45, 46]. Clusters prominent use is due to their ability find meaningful structures in the data set. Some studies didn't specify the clustering algorithm implemented, using terms as: "centre-based clustering algorithm", "propri-etary clustering algorithm" and "modified algorithm", but in the ones that did, K-Means was the most popular with 3 instances in 14 (21.42%), followed by Improved Minimum Spanning Tree clustering Algorithm, DBSCAN, CLOPE, EM and CBLOF with 1 instance each (7.42%). The other clustering techniques were not specified.

The answer to **Q2** is presented in Table 15.1. Anomaly detection can identify unexpected activity in the regular data-flow [47]. Outlier detection is employed to measure the "distance" between data objects to detect those objects that are grossly different from or inconsistent with the remaining data set [26,45]. In this table, the outlier detection techniques were elucidated. The 7 outlier detection algorithms identified are: Cross Dataset Outlier Detection Model, Dissimilarity Metric, Isolation Forest, One class SVM, Gaussian Mixture Model, Hidden Markov Model and Local Outlier Factor. All algorithms had only one instance in the papers.

The response to **Q3** is presented in Table 15.2. Time series analysis comprises methods for analyzing a sequence of data points, measured typically at successive time spaced uniform intervals, in order to extract meaningful statistics and other characteristics of the data [48]. Time series analysis algorithms that were presented are further detailed in this table. All algorithms, methods or techniques for time series analysis were found only once, being: Time variant behav-ioral pattern, Sequence Matching Based Algorithm, Scan

Table 15.1 Algorithms and techniques (time series)

Time series technique	Reference
Time variant behavioral pattern	[39]
Sequence matching based algorithm	[32]
Scan statistics based method	[40]
Correlation analysis along timeline	[2]

Table 15.2 Algorithms and techniques

Outlier detection technique	Reference
Dissimilarity metric	[21]
Cross dataset outlier detection model	[34]
Isolation forest	[37]
One class SVM	[37]
Gaussian mixture model	[37]
Hidden Markov model	[36]
Local outlier factor	[22]

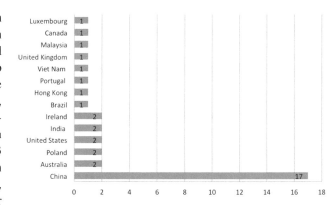

Fig. 15.3 Papers per country

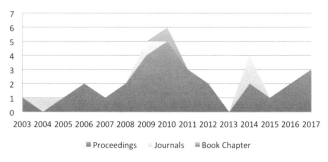

Fig. 15.4 Papers per year

Statistics Based Method and Correlation Analysis Along Timeline.

The answer to **Q4** is shown in Fig. 15.3. The amount of papers per country was counted using the affiliation of the authors as parameter, if the authors were affiliated to insti-tutes from different countries both countries were counted. China leads the number of publications, with a total of 17 papers. Australia, Poland, United States, India and Ireland appear next, with 2 publications each. The other countries, Brazil, Hong Kong, Portugal and Vietnam, United Kingdom, Malaysia, Canada and Luxembourg have 1 publication each.

China's leadership may be due to government policy changes in relation to the financial system, which began in 1976, when it was virtually non-existent and improved in the following decades with the increased role of independent financial activity [49], perhaps Chinese government may be more eager to fund researches on this topic. Another more obvious hypothesis, in this context, is that China may simply have a larger number of researchers working on data mining and artificial intelligence applications than other countries.

The answer to **Q5** is presented in Fig. 15.4, in which it can be observed that the year with the highest number of publications was 2010, with 6 papers. In 2009, 5 papers were published and in 2014, 4. The oldest publication on the subject dates back to 2003. There have been an year in which no paper was published: 2013. It is notorious the scarcity of publications on the field and the oscillations over the years.

In the above chart, Fig. 15.4, the distribution of publications per venue of publication can also be observed. The only publication whose primary source was a book chapter, occurred in the year in which the number of publications from conferences was higher, 2010. The conferences accounted for the largest number of publications in almost every year, except in 2014, year in which the peak of publications in journals occurred and tied the number of publications from conferences in that year.

The answer to **Q6** is that the ICMLA, ICMLC, ICDMW were the main conference identified with each one being the source of 2 publications. All other publications from conferences were originated from different sources. The journals were the source of 1 paper each: Expert Systems with Applications, IEEE Intelligent Systems, International Journal of Security and Applications and Journal of Theoretical and Applied Information Technology.

The low absolute number of papers published in the main conferences identified indicates that others can catch up to them in the near future. It also indicates that there may be a lack of conferences dedicated exclusively to fraud detection techniques in general, probably because the highly specificity of the field.

Finally, the answer to **Q7** is shown in Fig. 15.5, showing that the most popular venue to publish papers on this topic are conferences. This pattern is not surprising, since the most accessible medium for scientific publications are known as conferences. For example, over 100,000 conference events worldwide are indexed in the Scopus database, whilst only nearly 22,000 journals are indexed [33]. Surprising is the low absolute number of publications found in journals, which may denote that papers published at conferences have not been sufficiently worthy for their extension in journals, that is, they may not have been of sufficient quality. In addition, one possibility is that the results were not instigating enough for the deepening and continuity of the studies.

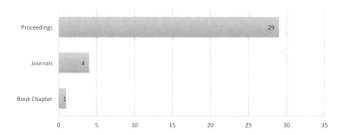

Fig. 15.5 Publication distribution

15.5 Threats to Validity

Construction Validity The search string may not cover the whole money laundering detection area. To mitigate this threat, we sought to construct the most comprehensive string possible utilizing a control paper and the opinion of one of the three researchers, a member of government staff that investigates Money Laundering.

Internal validity: (Data extraction) Researchers were responsible for extracting and classifying the main algorithms of each publication, biases or data extraction problems can threaten the validity of data characterization; **(Selection Bias):** Some papers may have been categorized incorrectly as the articles were included or excluded in the systematic mapping according to the researchers' judgment.

To mitigate these threats, selection and extraction reviews were made by all three researchers involved, with a final vote on disagreements.

External Validity Although Scopus is the largest database of scientific literature, with over 60 million records and 21,500 journals [33], it's impossible to state that the results of this systematic mapping covered all of CS. Nevertheless, this study presented evidence of the main techniques used and gaps to be explored, serving as a guide for future works in this line.

15.6 Conclusion

In this work, a systematic mapping was carried out, aiming to identify scientific papers related to the analysis and evaluation of algorithms, methods and techniques in the field of CS, to detect and combat ML. Since no other similar work of Mapping or Systematic Review specifically about ML has been found, we assumed that this is the first work of this type in this specific area of academic scientific knowledge.

This mapping was conducted following the research protocol and selection of studies presented in Sect. 15.2. With this method, data from 34 primary studies was extracted and analyzed, identifying trends in this area.

As results, it was identified that the most relevant techniques identified were supervised classifiers, firstly, and clusters, secondly. Between the former, decision trees (first), neural network and SVM were the most used algorithms and, amid the clusters, K-Means, first of all, followed by Improved Minimum Spanning Tree clustering Algorithm, DBSCAN, CLOPE, EM and CBLOF were the main ones ((**Q1**).

There was no repetition between the algorithms used for Outlier detection and Time series analysis, which indicates that there is no consolidate approach for both. Thus, all algorithms classified as Outlier detection and Time series analysis in the studies were highlighted (**Q2, Q3**).

In the global scenario, China stands out firstly, followed by Poland, Australia, Ireland, India and the United States, all in second place, as the countries that have published the most papers, presenting, respectively, 17, 2, 2, 2, 2 and 2 publications each (**Q4**).

Over the years, the number of publications fluctuates a lot, the oldest publication dates back to 2003. The year with the most publications was 2010, when 4 papers were published. It is notorious the scarcity of publications in this field of research (**Q5**).

The ICMLA, ICMLC and ICDMW were the main conferences identified, each one publishing 2 of the primary studies, while the main journals had one publication each. In this case, the low absolute number of papers published indicates that other journals and conferences can challenge their spot in the near future (**Q6**).

Finally, the most popular venue for publications on the topic are conferences, as expected, as the number of conferences indexed in the databases used is greater than the number of scientific journals (**Q7**)

Besides the apparent need to deepen the researches in the discussed area, the results found in this work map the state of the art of detecting transactions suspicious of money laundering, making it clear that it is an area of interest for researchers around the world and it has great growth potential.

We believe that this work is relevant to the academy, governments and the community at large, presenting them with trends in the detection of money laundering. In addition, it can offer yet another approach in the search for the best solutions to the current scenario and to combat the threat of organized crime against society.

References

1. Lv, L.-T., Ji, N., Zhang, J.-L.: A RBF neural network model for anti-money laundering. In: International Conference on Wavelet Analysis and Pattern Recognition, ICWAPR'08, vol. 1, pp. 209–215. IEEE (2008)
2. Zhang, Z.M., Salerno, J.J., Yu, P.S.: Applying data mining in investigating money laundering crimes. In: Proceedings of the Ninth ACM SIGKDD International Conference on Knowledge Discovery and Data Mining, pp. 747–752. ACM (2003)
3. Schott, P.A.: Reference Guide to Anti-money Laundering and Combating the Financing of Terrorism. World Bank Publications, Washington, DC (2006)
4. Gao, S., Xu, D.: Conceptual modeling and development of an intelligent agent-assisted decision support system for anti-money laundering. Exp Syst Appl **36**(2), 1493–1504 (2009)
5. Alexandre, C., Balsa, J.: Integrating client profiling in an anti-money laundering multi-agent based system. In: Rocha, A., Correia, A.M., Adeli, H., Reis, L.P., Teixeira, M.M. (eds.) New Advances in Information Systems and Technologies, pp. 931–941. Springer, Cham (2016)
6. Luo, X.: Suspicious transaction detection for anti-money laundering. Int. J. Secur. Its Appl. **8**, 157–166 (2014)
7. Le Khac, N.A., Markos, S., Kechadi, M.-T.: A data mining-based solution for detecting suspicious money laundering cases in an investment bank. In: 2010 Second International Conference on Advances in Databases Knowledge and Data Applications (DBKDA), pp. 235–240. IEEE (2010)
8. Ju, C., Zheng, L.: Research on suspicious financial transactions recognition based on privacy-preserving of classification algorithm. In: First International Workshop on Education Technology and Computer Science, ETCS'09, vol. 2, pp. 525–528. IEEE (2009)
9. George, I., Kavakli, M.: Data mining in the investigation of money laundering and terrorist financing. In: Surveillance Technologies and Early Warning Systems: Data Mining Applications for Risk Detection, p. 228 (2010)
10. Freedman, R.S., Sobkowski, I.: Surveillance of parimutuel wagering integrity using expert systems and machine learning. In: IAAI (2010)
11. Schmidhuber, J.: Deep learning in neural networks: an overview. Neural Netw. **61**, 85–117 (2015)
12. Wang, S.-N., Yang, J.-G.: A money laundering risk evaluation method based on decision tree. In: 2007 International Conference on Machine Learning and Cybernetics, vol. 1, pp. 283–286. IEEE (2007)
13. Kingdon, J.: Ai fights money laundering. IEEE Intell. Syst. **19**(3), 87–89 (2004)
14. Keyan, L., Tingting, Y.: An improved support-vector network model for anti-money laundering. In: 2011 Fifth International Conference on Management of e-Commerce and e-Government (ICMeCG), pp. 193–196. IEEE (2011)
15. Le Khac, N.A., Kechadi, M.-T.: Application of data mining for anti-money laundering detection: a case study. In: 2010 IEEE International Conference on Data Mining Workshops (ICDMW), pp. 577–584. IEEE (2010)
16. Tang, J., Yin, J.: Developing an intelligent data discriminating system of anti-money laundering based on SVM. In: Proceedings of 2005 International Conference on Machine Learning and Cybernetics, 2005, vol. 6, pp. 3453–3457. IEEE (2005)
17. Liu, R., Qian, X.-L., Mao, S., Zhu, S.-Z.: Research on anti-money laundering based on core decision tree algorithm. In: Control and Decision Conference (CCDC), 2011 Chinese, pp. 4322–4325. IEEE (2011)
18. Paula, E.L., Ladeira, M., Carvalho, R.N., Marzagão, T.: Deep learning anomaly detection as support fraud investigation in Brazilian exports and anti-money laundering. In: 2016 15th IEEE International Conference on Machine Learning and Applications (ICMLA), pp. 954–960. IEEE (2016)
19. Cao, D.K., Do, P.: Applying data mining in money laundering detection for the Vietnamese banking industry. In: Asian Conference on Intelligent Information and Database Systems, pp. 207–216. Springer (2012)
20. Yang, Y., Lian, B., Li, L., Chen, C., Li, P.: DBSCAN clustering algorithm applied to identify suspicious financial transactions.

In: 2014 International Conference on Cyber-Enabled Distributed Computing and Knowledge Discovery (CyberC), pp. 60–65. IEEE (2014)

21. Wang, X., Dong, G.: Research on money laundering detection based on improved minimum spanning tree clustering and its application. In: Second International Symposium on Knowledge Acquisition and Modeling, KAM'09, vol. 2, pp. 62–64. IEEE (2009)

22. Gao, Z.: Application of cluster-based local outlier factor algorithm in anti-money laundering. In: International Conference on Management and Service Science, MASS'09, pp. 1–4. IEEE (2009)

23. Cheong, T.-M., Si, Y.-W.: Event-based approach to money laundering data analysis and visualization. In: Proceedings of the 3rd International Symposium on Visual Information Communication, p. 21. ACM (2010)

24. Umadevi, P., Divya, E.: Money laundering detection using TFA system (2012)

25. Chen, Z., Nazir, A., Teoh, E.N., Karupiah, E.K., et al.: Exploration of the effectiveness of expectation maximization algorithm for suspicious transaction detection in anti-money laundering. In: 2014 IEEE Conference on Open Systems (ICOS), pp. 145–149. IEEE (2014)

26. Ngai, E., Hu, Y., Wong, Y., Chen, Y., Sun, X.: The application of data mining techniques in financial fraud detection: a classification framework and an academic review of literature. Decis. Support Syst. **50**(3), 559–569 (2011)

27. Yue, D., Wu, X., Wang, Y., Li, Y., Chu, C.-H.: A review of data mining-based financial fraud detection research. In: International Conference on Wireless Communications, Networking and Mobile Computing, WiCom 2007, pp. 5519–5522. IEEE (2007)

28. Kitchenham, B.: Procedures for performing systematic reviews. Keele, UK, Keele University, vol. 33, no. 2004, pp. 1–26 (2004)

29. Petersen, K., Feldt, R., Mujtaba, S., Mattsson, M.: Systematic mapping studies in software engineering. In: EASE, vol. 8, pp. 68–77 (2008)

30. Brereton, P., Kitchenham, B.A., Budgen, D., Turner, M., Khalil, M.: Lessons from applying the systematic literature review process within the software engineering domain. J. Syst. Softw. **80**(4), 571–583 (2007)

31. Wohlin, C., Runeson, P., Neto, P.A.d.M.S., Engström, E., do Carmo Machado, I., De Almeida, E.S.: On the reliability of mapping studies in software engineering. J. Syst. Soft. **86**(10), 2594–2610 (2013)

32. Liu, X., Zhang, P., Zeng, D.: Sequence matching for suspicious activity detection in anti-money laundering. In: Intelligence and Security Informatics, pp. 50–61 (2008)

33. Elsevier, B.: Scopus Content Coverage Guide. Elsevier. Available at: https://www.elsevier.com/solutions/scopus/how-scopus-works/content (2017). Accessed Feb 2018

34. Zhu, T.: An outlier detection model based on cross datasets comparison for financial surveillance. In: IEEE Asia-Pacific Conference on Services Computing, APSCC'06, pp. 601–604. IEEE (2006)

35. Jun, T.: A cross datasets referring outlier detection model applied to suspicious financial transaction discrimination. In: Intelligence and Security Informatics, pp. 58–65. Springer, Berlin (2006)

36. Li, Y., Duan, D., Hu, G., Lu, Z.: Discovering hidden group in financial transaction network using hidden markov model and genetic algorithm. In: Sixth International Conference on Fuzzy Systems and Knowledge Discovery, FSKD'09, vol. 5, pp. 253–258. IEEE (2009)

37. Camino, R.D., State, R., Montero, L., Valtchev, P.: Finding suspicious activities in financial transactions and distributed ledgers. In: 2017 IEEE International Conference on Data Mining Workshops (ICDMW), pp. 787–796. IEEE (2017)

38. Dreżewski, R., Dziuban, G., Hernik, Ł., Pączek, M.: Comparison of data mining techniques for money laundering detection system. In: 2015 International Conference on Science in Information Technology (ICSITech), pp. 5–10. IEEE (2015)

39. Krishnapriya, G., Prabakaran, M.: Money laundering analysis based on time variant behavioral transaction patterns using data mining. J. Theor. Appl. Inf. Technol. **67**(1), 12–17 (2014)

40. Liu, X., Zhang, P.: A scan statistics based suspicious transactions detection model for anti-money laundering (AML) in financial institutions. In: 2010 International Conference on Multimedia Communications (Mediacom), pp. 210–213. IEEE (2010)

41. Michalak, K., Korczak, J.: Graph mining approach to suspicious transaction detection. In: 2011 Federated Conference on Computer Science and Information Systems (FedCSIS), pp. 69–75. IEEE (2011)

42. Li, X., Cao, X., Qiu, X., Zhao, J., Zheng, J.: Intelligent anti-money laundering solution based upon novel community detection in massive transaction networks on spark. In: 2017 Fifth International Conference on Advanced Cloud and Big Data (CBD), pp. 176–181. IEEE (2017)

43. Gao, S., Xu, D., Wang, H., Wang, Y.: Intelligent anti-money laundering system. In: IEEE International Conference on Service Operations and Logistics, and Informatics, SOLI'06, pp. 851–856. IEEE (2006)

44. Alexandre, C., Balsa, J.: A multiagent based approach to money laundering detection and prevention. In: ICAART (1), pp. 230–235 (2015)

45. Tan, P.-N., Steinbach, M., Kumar, V.: Introduction to Data Mining, 1st edn. Pearson Addison Wesley, Boston (2005)

46. Han, J., Pei, J., Kamber, M.: Data Mining: Concepts and Techniques. Elsevier, San Diego (2011)

47. Reddy, Y.B.: Event-based anomalies in big data. In: Information Technology-New Generations, pp. 33–42. Springer (2018)

48. Lin, R.A.K.-l., Shim, H.S.S.K.: Fast similarity search in the presence of noise, scaling, and translation in time-series databases. In: Proceedings of the 21st International Conference on Very Large Data Bases, pp. 490–501. Citeseer (1995)

49. Keidel, A.: China's financial sector: contributions to growth and downside risks. In: Barth, J.R., Tatom, J.A. (eds.) China's Emerging Financial Markets, pp. 111–125. Springer, New York (2009)

Mauricio Xavier Zaparoli, Adler Diniz de Souza,
and Andre Henrique de Oliveira Monteiro

16.1 Introduction

Travelling or renting a property through applications like Airbnb or other similar services might bring some discomfort to the parties involved. The key exchanging process between host and guest is problematic because generally, they need to meet for that. The meeting time might be a problem. The property owner might be working or the guest arrival might be in the early morning. The host must be present in the rented property's city or depend on a trusted third party to deliver the keys. Besides that, the guest might lose the keys or not give them back in worst cases [1, 2].

These problems bring risks for both hosts and future guests. They both can have their belongings stolen or even their lives put at risk, especially in areas with big touristic activities [3]. There are many reports of guests that were assaulted inside the property [4, 5], or even hosts whose belongings were stolen by guests [6] who copied the property's keys. Another problem is the need to control the entrance of employees, cleaning service for example, in the property. Eventually, these service providers might be ill intended and steal the guests [7–9]. In such specific cases, controlling who enters, spent time in the property e leaving time is essential to minimize these presented risks.

Therefore, the research problem addressed in this article: How to guarantee the safety for host and guest in accessing shared physical environments?

Many electronic and computational solutions are already available to address the problems mentioned earlier. However, a great part of them can be easily hacked by computer specialists. Therefore, there is a need for searching for a solution that increases the security of this kind of access systems.

For that matter, the last years were marked by the popularization of systems based on blockchain technology. Blockchain is a decentralized managing technology, the base of any cryptocurrency, responsible for issuing and transferring of money between its users [10]. It's a continuously growing record of transactions between users confirmed by the participating nodes of the blockchain network, available to all, and controlled and owned by no one [10, 11]. The advantage of blockchain is that this record can't be modified neither deleted once the data of the transactions have been confirmed by all nodes in the network [12]. By this means the blockchain is known for its data integrity and security, which extend its use for other services and applications [10] other than money transaction between users.

Thus, this work proposes the development of an electronic lock controlled by smart contracts registered in the Ethereum blockchain platform.

This work aims at presenting the developed solution. The following characteristics were evaluated: (i) convenience and (ii) security for both guest and host brought by using smartphones for unlocking electronic locks aggregating the security offered by the Ethereum blockchain platform.

16.2 Blockchain Technology

16.2.1 Blockchain

Blockchain, a term that has been gaining much more attention since it first appeared in 2008 with Bitcoin [13]. It's a technology that allows the public, distributed, encrypted and unalterable record of the transactions of any cryptocurrency [14]. It uses the proof-of-work method to maintain the system in sync [15].

M. X. Zaparoli (✉) · A. D. de Souza · A. H. de Oliveira Monteiro
Institute of Mathematics and Computation, Federal University of
Itajubá, Itajubá, MG, Brazil

© Springer Nature Switzerland AG 2019
S. Latifi (ed.), *16th International Conference on Information Technology-New Generations (ITNG 2019)*,
Advances in Intelligent Systems and Computing 800,
https://doi.org/10.1007/978-3-030-14070-0_16

It's public because any computer can access the blockchain network [15, 16]. It's decentralized because all the participating nodes of the network, the miners, have a copy of that record [10, 11, 15, 16]. No trusted intermediary organization is needed to manage it [10, 11, 15, 16] which allows the users to transact money directly between each other. So there isn't a central database to be hacked and compromise the data integrity [16].

Cryptographic because of two reasons. First, the transactions must be verified to guarantee that the user sending money owns it [15]. Finally, in the way that transactions are added to blocks that are chained with each other by the proof-of-work process [10, 15].

Proof-of-work is the mechanism in which the miners compete with each other to solve a cryptographic puzzle in exchange of payment in cryptocurrency tokens. This process involves the combination of four variables: timestamp, the hash of the transactions, the identity of the previous block and a nonce. This nonce is an arbitrary number than when combined with the other three by a hash function dictates the process difficulty. Through this mechanism, the participating nodes get in the consensus of which new block is going to be chained [17]. As information of past blocks are used none of them can be altered without altering the information of those that follow [10, 15].

The strong aspects of blockchain are: (i) the data integrity and (ii) security. When added together they become attractive to other services and applications that go beyond money transaction [10].

Thereby, there are blockchains alternatives other than Bitcoin where the focus isn't in the currency. They implement a platform for smart contracts, records and other applications [18]. They can be used for public and private records, digital identity, and physical and intangible asset registration [11]. For example, to protect an idea, instead of patenting it, it could be coded in blockchain for future proof.

Zhao et al. [19] indicates many other applications to the blockchain technology, among them, are: (i) private data protection systems, (ii) more transparent voting systems, (iii) product tracking systems in the supply chain, and much more.

The blockchain functionality can revolutionize many areas like finances, accounting, managing, law, politics, and government [11, 18, 19] and many more.

16.2.2 Ethereum

Considering the existence of other cryptocurrencies, with more dynamic properties compared with Bitcoin, it was chosen to work with the Ethereum token. It was selected because it allows not only the electronic transfer of assets, from one person to another, but does this in an automatic manner through smart contract execution.

Ethereum is the second largest blockchain network and the fastest growing one [16]. Each node runs the Ethereum Virtual Machine (EVM), where applications, called smart contracts, are built and can be accessed globally [15]. These applications are executed exactly as they were programmed, with no censorship, fraud or third party intervention [16].

It has its own cryptocurrency called ether which is necessary to pay for the contract's execution [18]. These contracts are developed using Solidity programming language [20]. The platform has a big developer community working in decentralized applications development and in the system scalability [21].

16.2.3 Smart Contract

In its traditional form, a contract is an agreement between two people or organizations, or a legal document that explains the details of this agreement [22]. Each of the parties involved acquiring rights and duties relative to the other party [23].

Smart contracts also settle agreements between two or more parties the same way that traditional contracts do but do so in an automated and decentralized manner, forcing the execution of the contract's terms without fraud and third-party intervention [11, 16]. All of this eliminating the need for trust between parties [11].

According to Nick Szabo [24], the smart contract objective is to satisfy the contractual terms, minimize exceptions both accidental and malicious and eliminate the need of a trusted third party, thus minimizing the costs.

Essentially a smart contract is an algorithm that is executed on all the participating nodes of a blockchain platform [11,20,25]. They are capable of storing data, receive and send payments with cryptocurrencies and store them [11].

16.2.4 Smart Property

Smart property is a term first introduced by Nick Szabo [25], which he defines as it's an extension on the smart contract to property. Smart property can be created embedding smart contracts in physical objects.

The general concept is to transact any kind of property in blockchain models [11]. A property coded in blockchain may have its ownership or access controlled by smart contracts subjective to existing law [11, 13, 26], in an efficient, automatic and decentralized way [26]. This is applicable for any kind of asset: physical (homes, vehicles, bicycles, and others) or intangible (ideas, votes, health information, and others) [11].

16.3 Methodology

In the proposed solution the guest choose a property, makes a reservation and access the property using an application on his smartphone. The lock is controlled by a Raspberry Pi 3 B+ that communicates with the Android application using a data transfer protocol through sound called Chirp.io operating in ultrasonic mode.

The guest approaches his smartphone's speakers to the microphone connected to the single-board computer and transmits his credentials. With this information, the Raspberry Pi access the project's server database and identifies which smart contract it should communicate with. Verifying the data exchanged in this communication the solenoid lock is activated and unlocks the door.

16.3.1 Literature Review

A research based on systematic review [27] was conducted in this work to identify researches and applications related to blockchain and smart contracts employment. The objective was to evaluate the current market and academic contexts.

Among the identified researches it's worth mentioning [28] that studied blockchain usage in vehicle insurance. To achieve this, smart contracts were used to record vehicle historical information, paths taken by it and driver behavior to create insurance quotes. Alexander Masluk and Mikhail Gofman [29] also proposed an accounting mechanism for personal data collection, retention, and exchange.

Analyzing other works and applications, one proposed a smart property implementation. In it, a physical asset access control system was implemented using a Raspberry Pi [30]. This control was achieved through a Bitcoin cryptocurrency token tracking. The access control and property ownership could be sold transferring this specific token from seller to buyer.

The development of an electronic lock controlled by smart contracts would allow its usage in many sectors involving asset rental, from physical locations to vehicles using the smart property characteristics of the currency.

16.3.2 Project Proposal

Application Proposal/Business Model

The purpose of this work was to develop an electronic lock controlled by smart contracts recorded in the Ethereum blockchain platform. The objective was to provide convenience and security for the AirBnB users or the hotel sector.

The business model proposal is described as follows: The guest has to book the physical space he wants and pay it through a web application. After that, the server records information about the reservation on the database and deploys a smart contract on the Ethereum blockchain.

The Raspberry Pi, the controller of every lock, queries the smart contract to validate the guest's credentials that he presented using an application on his smartphone. Thus, the property owner guarantees that his keys won't be copied, the guest guarantees that he will be the one with access to the property during his reservation and eliminates the need for a meeting to exchange keys for both of them or, in the case of the hotel sector, eliminates the need for a check-in.

Prototype

The initial prototype was implemented using an Android application, an electronic circuit to control the solenoid lock, a Python script and a smart contract manually recorded in the Ethereum blockchain platform. The application transmits a hardcoded credential using a data transfer protocol through sound called Chirp.io. This sound is captured by the microphone connected in the Raspberry Pi. The Python script decodes this credential and query the smart contract and verifies if it matches the one recorded there. After the credential verification, the lock's electronic circuit is activated by the Raspberry Pi and unlocks the door.

Architecture

The website was built using the Laravel PHP framework to speed up the project's entities CRUDs (Create-Read-Update-Delete) using the MVC (Model-View-Con- troller) model, bootstrap for the frontend and MySQL for the database. The architecture overview is shown in Fig. 16.1.

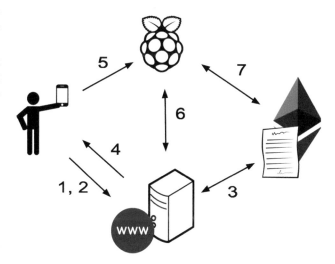

Fig. 16.1 Project's architecture

1. **Registration** The project proposed in this work starts with the guest's registration in the project's website with his personal data. This registration is validated via email.

   After the guest's registration, the login is available in the smartphone application. This application was developed natively on Android and uses the Chirp.io data transfer protocol.

2. **Reservation** On the Reservation Menu, the guest chooses a property and specifies the reservation period based on the property availability and confirms the booking. The reservation is then pending approval of the hotel manager or the property owner.

3. **Deployment** After the reservation's approval, the server is responsible for the smart contract deployment and records its address on the database. The smart contract, developed with the Solidity programming language, also records data related to the reservation.

4. **Profile and reservations access** Accessing the smartphone or the website the guest will have access about his profile and reservations information. Choosing a reservation gives him the option of transmitting his credentials.

5. **Identification** In the booked hotel room or property, the guest chooses to transmit his credentials using his smartphone. These credentials are transmitted through the smartphone speakers, captured by the microphone connected to the Raspberry Pi and decoded by the Chirp.io protocol.

6. **Database query** The Raspberry Pi queries the database for the reservation and receives the smart contract address.

7. **Smart Contract query** With that address the smart contract can be accessed and the credentials and date can be verified. After this validation, the electronic circuit of the lock is activated and the door is unlocked.

16.3.3 Implementation

Raspberry Pi

The communication of the application on the user's smartphone with the Raspberry Pi was made using the Chirp.io data transfer protocol through sound. Through this protocol, the user transfers his credentials using his smartphone speakers and the Raspberry Pi captures them through the connected microphone.

In the communication with the smart contract registered on the Ethereum blockchain, the web3.py Python library was used. Its API is derived from the Javascript web3.js API. To configure this communication the smart contract address on the blockchain, an HTTPProvider that connects the web3 with Ethereum and the smart contract Application Binary Interface (ABI).

The electronic circuit that controls the lock is activated using the general-purpose input/output (GPIO) pins of the Raspberry Pi.

Server

The server was designed to support some independent applications them being: hosting the online booking system, database and a communication middleware between Raspberry Pi and the Ethereum blockchain.

The booking system is an application designed to record users, properties and bookings data, being developed on the Laravel framework. The communication with the blockchain is achieved using the web3.js library. The objective of this layer is to record data from the user's bookings in order to deploy the smart contracts on the blockchain so they can be later used by the Raspberry Pi for access validation.

The Raspberry Pi uses the middleware supplied by the server as a communication layer provided to centralize and to better control the information that flows inside the system. The middleware provides secure communication to exchange smart contract's information. It's worth highlighting that the access information aren't recorded in the server but are registered on the block- chain in a cryptographic way.

Smartphone App

The smartphone application was developed natively in Android. It also uses the Chirp.io protocol to transfer the user's credentials and a MySQL connector to connect to the server's database.

The application is able to communicate with the database and access data such as user's profile, properties, and the user's bookings. He is also provided with the option of submitting his credentials to be able to access the property. The Chirp.io protocol is used to encode credential's information and send it via sound through the smartphone's speakers.

16.4   Conclusion

Using the blockchain in the development of an electronic lock to control hotel rooms and properties made possible to notice that smart contracts carry along the security and data integrity characteristics of the block- chain.

When a smart contract is recorded on the blockchain its data will stay there permanently due to its data integrity. So the guest has the guarantee that he is going to be the only one that can access the room or property during his reservation.

The hotel manager guarantees that the keys won't be copied ensuring that whoever was staying in that room won't be able to access it after the reservation period. This guarantees that ill minded past guests won't access it again and steal their belongings. With the lock, it's also possible to control the employee's entrance and exits on rooms.

The applications of the system presented in this article go beyond the ones described, with its rapid adaptation to the implementation of smart property concept being possible (definitive real estate transfer, vehicles, and other assets), which access is made through some kind of electronic lock. The advantage of this type of operation using the blockchain technology is that they are carried out in a safe and fast way without any banks, registry office, lawyers and other middlemen, reducing the costs and time taken to the contract implementation.

Using blockchain technology and smart contracts has great potential to revolutionize nowadays existing business models in many different information systems, due to its core characteristics of security and data integrity.

References

1. Dealing with AirBnB for a lost key is ridiculous. Airbnb Hell, 28-July-2016. [Online]. Available: https://www.airbnbhell.com/dealing-airbnb-lost-key-ridiculous/. Accessed 02 Oct 2018
2. Elderly AirBnB guest in Germany kicks cat, steals keys. Airbnb Hell, 15-Sept-2016. [Online]. Available: https://www.airbnbhell.com/elderly-airbnb-guest-germany-kicks-cat-steals-keys/. Accessed 02 Oct 2018
3. XU, Y.-H., Kim, J.-W., Pennington-Gray, L.: Explore the spatial relationship between AirBnB rental and crime. Tourism Trabel and Research Association: Advancing Tourism Research Globally, 2017 ttra International Conference
4. White, B.: Short-term rental guests robbed, pistol whipped in hollywood. NBC Southern California, 11-July-2018. [Online]. Available: https://www.nbclosangeles.com/news/local/AirBnB-guests-robbed-beaten-in-Hollywood-487738771.html. Accessed 02 Oct 2018
5. Robin, N.: British tourists flee New Orleans after AirBnB home invasion. http://www.fox8live.com, 06-June-2018. [Online]. Available: http://www.fox8live.com/story/38355584/british-tourists-flee-new-orleans-after-airbnb-home-invasion/. Accessed 02 Oct 2018
6. Thomas, E.: EXCLUSIVE: discovery bay AirBnB host says her home was brutally ransacked by renters. ABC7 San Francisco, 11-May-2018. [Online]. Available: https://www.airbnbhell.com/dishonest-host-refuses-to-admit-shoes-have-been-stolen/. Accessed 02 Oct 2018
7. Clothes stolen by host, AirBnB does nothing. Airbnb Hell, 26-June-2018. [Online]. Available: https://www.airbnbhell.com/clothes-stolen-by-host-airbnb-does-nothing/. Accessed 02 Oct 2018
8. Dishonest host refuses to admit shoes have been stolen. Airbnb Hell, 26-Sept-2018. [Online]. Available: https://www.airbnbhell.com/dishonest-host-refuses-to-admit-shoes-have-been-stolen/. Accessed 02 Oct 2018
9. Guests robbed in salo AirBnB, host possessions untouched. Airbnb Hell, 16-Nov-2017. [Online]. Available: https://www.airbnbhell.com/guests-robbed-salo-airbnb-host-possessions-untouched/. Accessed 02 Oct 2018
10. Yli-Huumo, J., Ko, D., Choi, S., Park, S., Smolander, K.: Where is current research on blockchain technology? – a systematic review.

11. Swan, M.: Blockchain: Blueprint for a New Economy. O'Reilly Media, Sebastopol (2015)
12. Gatteschi, V., Lamberti, F., Demartini, C., Pranteda, C., Santa-maría, V.: To blockchain or not to blockchain: that is the question. IT Prof. 20(2), 62–74 (2018)
13. Nakamoto: Bitcoin: a peer-to-peer electronic cash system (2008). [Online]. Available: https://bitcoin.org/bitcoin.pdf. Accessed 03 Oct 2018
14. Beck, R., Avital, M., Rossi, M., e Thatcher, J.B.: Blockchain technology in business and information systems research. Bus. Inf. Syst. Eng. 59(6), 381–384 (2017)
15. Burniske, C., Tatar, J.: Cryptoassets the Innovative Investor's Guide to Bitcoin and Beyond, 1st edn. McGraw-Hill Education, New York (2018)
16. Tapscott, D., Taspcott, A.: Blockchain Revolution: How the Technology Behind Bitcoin Is Changing Money, Business, and the World. 1st edn. Penguin Random House LLC, New York (2016)
17. Nofer, M., Gomber, P., Hinz, O., e Schiereck, D.: Blockchain. Bus. Inf. Syst. Eng. (BISE) 59(3), 183–187 (2017)
18. Antonopoulos, A.M.: Mastering Bitcoin: Unlocking Digital Cryptocurrencies, 1st edn. O'Reilly Media, Sebastopol (2015)
19. Zhao, J.L., Fan, S., Yan, J.: Overview of business innovations and research opportunities in blockchain and introduction to the special issue. Financ. Innov. 2, 1–7 (2016)
20. Ethereum: Solidity. Solidity – solidity 0.4.24 documentation. [Online]. Available: https://solidity.readthedocs.io/en/v0.4.25/. Accessed 04 Oct 2018
21. D'Aliessi, M.: How does ethereum work? – Michele D'Aliessi – medium. Medium, 11-Feb-2018. [Online]. Available: https://medium.com/@micheledaliessi/how-does-ethereum-work-8244b6f55297. Accessed 04 Oct 2018
22. Cambridge Dictionary: Contract meaning in the Cambridge English dictionary. Cambridge dictionary. [Online]. Available: https://dictionary.cambridge.org/dictionary/english/contract#dataset-cbed. Accessed 03 Oct 2018
23. What is contract? definition and meaning. BusinessDictionary.com. [Online]. Available: http://www.businessdictionary.com/definition/contract.html. Accessed 03 Oct 2018
24. Szabo, N.: Smart Contracts: Phonetic Sciences, Amsterdam (1994). [Online]. Available: http://www.fon.hum.uva.nl/rob/Courses/InformationInSpeech/CDROM/Literature/LOTwinterschool2006/szabo.best.vwh.net/smart.contracts.html. Accessed 04 Oct 2018
25. Szabo, N.: Nick Szabo – smart contracts: building blocks for digital markets. Literature Review on Reaction Time (1996). [Online]. Available: http://www.fon.hum.uva.nl/rob/Courses/Information-InSpeech/CDROM/Literature/LOTwinterschool2006/szabo.best.vwh.net/smart_contracts_2.html. Accessed 04 Oct 2018
26. Acheson, N.: What is smart property, and what can it be used for? fintechblue, 08-May-2016. [Online]. Available: http://www.fintechblue.com/2015/12/smart-property-what-does-that-mean-for-the-blockchain/. Accessed 04 Oct 2018
27. Kitchenham, B., Charters, S.: Guidelines for performing systematic literature reviews in software engineering. Software Engineering Group School of Computer Science and Mathematics Keele University, Keele, UK, 9 July 2007
28. Vo, H.T., Mehedy, L., Mohania, M., Abebe, E.: Blockchain-Based Data Management and Analytics for Micro-insurance Applications. IBM Research, Australia (2017)
29. Masluk, A., Gofman, M.: Protecting personal data with blockchain technology. In: Latifi, S. (ed.) Information Technology – New Generations. Advances in Intelligent Systems and Computing, vol. 738. Springer, Cham, Princeton, NJ. USA. (2018)
30. Dhore, K., Stallworth, B., Xu, K.: BitTrade: A Pure Implementation of Smart Property. Princeton, NJ, USA (2015)

PLoS ONE 11(10), e0163477 (2016). [Online]. Available: https://doi.org/10.1371/journal.pone.0163477. Accessed 03 Oct 2018

Biometric System: Security Challenges and Solutions

Bayan Alzahrani and Fahad Alsolami

17.1 Introduction

In the recent years, biometrics data have achieved a rapid growth as an identity authentication tool. Biometric authentication considered a promising technology used across multiple systems such as government, police station and commercial application. Biometric-based authentication system provides a reliable verification while overcoming some concerns related to the traditional password-based system. Among many biometric traits, a fingerprint is the most popular traits that extensively studied for recognition purpose [1]. The reason for the popularity is the strong uniqueness since not likely to have two people share the same fingerprint pattern. Another reason that the fingerprints have several matching algorithms and many features can be used as the authentication tool, which improves the accuracy and enhances the performance of the recognition system [2]. However, fingerprint-based identification systems can be very extensive in term of scalability and need many resources for processing and storage [3].

Modern cloud-computing systems are the best solution to provide scalability for big data such as fingerprint data. Cloud computing has gained wide acceptance for individuals as well as organizations in terms of the computation, performance and storage. In spite of the advantages, there are some concerns related to the privacy and security of biometrics data in the cloud computing since it is not immune to threats [4]. Due to the sensitivity of biometrics data, these concerns have been raised especially when the biometrics data compromised, they cannot be revoked and reissued like PINs and passwords [5].

In cloud computing, the sensitive data (i.e., Biometrics data) will be stored on remote servers. Thus, the data will be owned and operated by others and more than that will be accessed through the Internet easily. In other words, there is a high chance that the data will be compromised. For this reason, confidentiality is one of the main concerns in the cloud storage. Confidentiality in the cloud can be compromised by insider attack (i.e., cloud service providers) [6]. Since the biometric data stored in their plaintext form, the owner of the cloud could sell or share the data for unauthorized purposes [4]. On the other hand, there are some systems use the encryption methods in order to achieved confidentiality. Using encryption techniques may introduce the risk of a brute-force attack [7]. Obviously, encryption alone may not provide sufficient security.

In addition to the confidentiality issue in the cloud storage, there is also the issue of availability [8]. The availability of the cloud is important to allow the authorized user access and use the system and the stored data from different locations at any time. The availability issue occurs in cloud computing system when the attacker uses all the available resources, to make it impossible or difficult for legitimate users to use them, which cause the denial of service attack [9]. Also, cloud maintenance could increase the downtime and then affect the availability [10].

On the other hand, biometric data itself can be vulnerable to different attacks either doppelganger attack or biometric dilemma [11]. These attacks destroy the value of biometric, reduce the privacy and could lead to more security threats. Another security concern occurs during the biometric matching process, when the biometric data vulnerable to be hacked by the man-in-the-middle attack to gain illegal access to the data [12]. According to that, several schemes have been proposed to overcome security issues and protect biometric data.

B. Alzahrani · F. Alsolami (✉)
Department of Information Technology, King Abdulaziz University, Jeddah, KSA
e-mail: balzhrani0043@stu.kau.edu.sa; fjalsolami@kau.edu.sa

© Springer Nature Switzerland AG 2019
S. Latifi (ed.), *16th International Conference on Information Technology-New Generations (ITNG 2019)*,
Advances in Intelligent Systems and Computing 800,
https://doi.org/10.1007/978-3-030-14070-0_17

Template protection schemes [5, 13] was proposed to study the problem of achieving the security in biometrics system. Template protection method provides a variety of schemes that address the security of biometric but still suffering from some issues. First, there are some schemes that store the biometrics templates in the plaintext format, which expose the templates to different attacks. Second, some of the other schemes use encryption technique to encode the biometrics data in order to achieve the privacy [14]. However, at the matching phase, the encoded template need to be decrypted in order to match the data, which makes the system vulnerable to the unauthorized attempts [14]. Finally, some of the template protection approaches suffering from the accuracy and performance issues [15, 16], which lead to prevent many systems from using these approaches.

Regarding biometric applications, security is the key issue that has a lot of remaining challenges. In this study, several types of biometrics identification approach are reviewed. In each biometric system, two phases process data: the enrollment and matching phases. During enrollment, the system takes the biological trait of each user and saves it in the database as a gallery image to be used later for identification purpose, along with the user ID. Later in the matching phase, a new copy of the biological trait from the user is captured, (i.e., a probe image), and compared with the previous gallery images saved in the database to determine if the two biological traits relate to the same person or not. The main objective of this paper is to discuss security issues with the biometric system. The existing reviewed solutions are examined and analyzed.

The paper is organized as follows: Sect. 17.2 presents a brief overview of the fingerprint matching algorithm steps. Section 17.3 summarizes the common attacks and threats that affect security of the biometric system. The proposed security schemes are presented in Sect. 17.4. The finding and future direction are discussed in Sect. 17.5. Finally, the conclusion is presented in Sect. 17.6.

17.2   Background on the Fingerprint Identification

Fingerprint features used in the recognition system can be categorized into three types: orientation ridge flow, minutiae point, and ridge contour [17]. A minutiae-based structure is a commonly used feature since the minutiae point has a unique structure that makes each fingerprint image distinct from others in the recognition system. In fact, the minutiae point can be indicated by the ridge bifurcation or ridge ending [18]. To demonstrate, Fig. 17.1 depicts a bifurcation is the point or area in which the ridge divides into two branches or parts while a ridge ending is the point in where the ridge line is terminated, the second image highlights the ridge ending with a

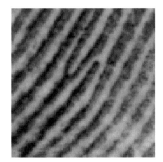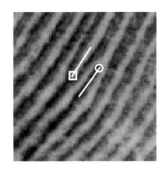

Fig. 17.1 Minutiae point: ridge ending (circle); bifurcation (square) [18]

circle and the bifurcation with a square. The orientation and coordinate location of the ridge for all minutiae points need to be extracted to match the fingerprints. To understand the details of our proposed system, we must know the detailed description of the NIST Bozorth matcher [18].

17.2.1 NIST Bozorth Matcher

The NIST Bozorth matcher [18] is a minutiae-based fingerprint matching algorithm. The natural form of the Bozorth algorithm generates the minutiae point by a Minutiae Detection, or MINDTCT, algorithm [18]. A MINDTCT algorithm locates the minutiae points and other details in the fingerprint images. A MINDTCT algorithm takes the fingerprint image from the sensor and extracts all minutiae in that image. After that, a minutiae file is created for all the minutia points of the fingerprint. For each minutia point, the algorithm assigns the location of minutiae on the fingerprint image (x coordinate, y coordinate), the orientation angle (θ), and the quality (q). An example of the minutia file is illustrated in Table 17.1.

After creating the minutia file for each fingerprint image, the matching algorithm passes through three major steps:

Table 17.1 Example of a minutiae file that contains all the minutia points of a particular fingerprint. This minutiae file from FV C2002Db2a [19]

Minutiae points	X coordinate	Y coordinate	Orientation angle	Quality
MP 1	48	86	5	19
MP 2	63	104	159	81
MP 3	85	37	56	89
MP 4	56	48	67	88
MP 5	35	178	180	39
MP 6	84	209	40	29
MP 7	71	72	135	21
…….	….	….	….	….
MP n	198	132	86	79

1. The algorithm constructs a pair table with the values (d_{k_j}, β_1, β_2, k, j, θ_{k_j}), as illustrated in Fig. 17.2. The system takes each pair of minutia and generates an entry in the pair table for them. Each pair table entry stores seven elements of information: where the distance between the minutiae pair (k, j) is d_{k_j}, β_1 and β_2 are the angles of each minutiae with respect to the line between them, and θ_{k_j} is the orientation of the line between the two minutiae points [18]. The pair table stores the entries of the minutiae pair according to the distance between the minutia pair (k, j) from small to the large distance. The algorithm constructs a pair table of the probe fingerprint (fingerprint image being tested) and one table of each gallery fingerprint (fingerprint images stored in a database) to be matched.

2. The algorithm constructs a match table, wherein the algorithm compares each entry in the pair table of probe against each entry in pair tables of all gallery. The algorithm generates the match table for compatible entries between the probe and the gallery in which the distances and the angles between them are within a given threshold, as shown in Fig. 17.3. Each row in the match table includes one pair from a gallery pair table and the corresponding pair from a probe pair table with the difference in the angle between them. An example of entries in the match table is illustrated in Table 17.2.

3. The algorithm creates links between nodes in the match table by traversing the rows of the table to form clusters.

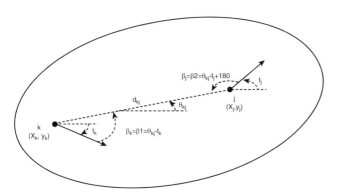

Fig. 17.2 The distance between two minutia points (k, j) with (β_1, β_2, θ_{k_j}) angles on a fingerprint image [18]

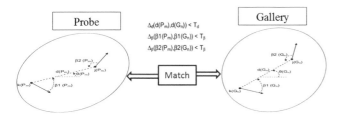

Fig. 17.3 The match between two fingerprints in the NIST algorithm [18]

Table 17.2 Example of the entries in the match table. This match table from FV C2002Db2a dataset [19]

Rows	$\Delta(\theta(Probe),\ \theta(Gallery))$	Probe minutiae indices k	Probe minutiae indices j	Gallery minutiae indices k	Gallery minutiae indices j
1	59	1	2	4	10
2	−25	1	4	5	9
3	−1	2	8	8	11
4	20	3	5	11	5
5	116	4	7	11	9
....	....	....	....	....	....
n	−153	8	9	7	14

The algorithm uses these clusters to compute the final score of matching to determine if both the gallery fingerprint image and the probe fingerprint image relate to the same person [18].

17.3 Biometric Security Issues

Biometric systems have become a universal method since use of the biometric traits increased, particularly for authentication and security goals. This results in an increased motivation for attacking and abusing the biometric data. Boult et al. [11] reviews the threats that affect the value of biometrics over a long-term in security applications. The researchers discuss two type of threats: biometric dilemma and doppelganger attack. In the biometric dilemma, the attacker obtains the features set of the fingerprint from a low security level of biometrics system to gain access to the high security level of biometrics system by using the stolen features. In a doppelganger attack, the attacker tries to detect the closer match of the fingerprint in the biometric system database to impersonate the people.

There are several reasons for biometric system vulnerability, which can be classified as intrinsic failure and adversary attack [5]. In intrinsic failures, there are two mistakes, known as false reject and false accept, that occur when the biometrics system makes an improper decision.

The falsely rejected decision happens to a legitimate user and occurs when there is a large variation between requested biometric feature and the stored template of the user. The false accept occurs when there is an insufficiency of distinction in the biometrics data, which causes substantial similarity between many users in the biometric feature sets. An adversary attack attempts to take advantage of the intrinsic failures of the system, trying to trick the biometric system for personal interests [5].

A possible attack on the biometric system is presented by Jain et al. [17]. This research categorizes the biometric system attacks into either insider or external attacks. The insider attack occurs when the attack comes from the database system owner for enrollment fraud, collusion, and coercion the biometric data. External attacks are those such as an obfuscation, Trojan horse, man-in-the-middle, or template attack. The obfuscation attack occurs during the enrollment operation by using fake biometric data. Trojan horse and man-in-the-middle attack commonly occur through verification and authentication in the matching process. The template attack targets the template stored in the database system for personal interests.

17.4 Secure Biometrics Data

Many approaches have been proposed to protect biometrics data in both local storages and in the cloud. We review a template protection scheme, encrypted biometrics data scheme, and biometrics data in the cloud.

17.4.1 Biometric Template Protection

A biometrics template is generated from the biometric sample; the template should not reveal any biometric information [13]. The security of the stored template is important since the template contains very sensitive information that is vulnerable to be stolen, which compromises user privacy as well as system security. Therefore, many secure biometrics schemes have been proposed to protect the template to prevent any leakages. Most of these schemes categorized into cancelable biometrics and biometrics cryptosystem [13]. In cancelable biometrics, the biometric templates can be renewed and revoked, and are unique to every application, much like a password. The cancelable approach further categorizes into biometric salting and irreversible transformations [13]. The salting approach combines the biometric features along with the user-specific password. Jin et al. [20] proposes a biometric salting approach called Bio-Hashing. Bio-Hashing combines biometric features along with random-number (token) to produce Bio-Codes. The generated Bio-Code is stored in the database rather than the original biometric. In the Bio-Hashing authentication, the real users give the token and their original biometric data. Moreover, a new Bio-Code can be generated by changing the token for every new application or when the stored Bio-Code intercepted. However, if the attacker can gain access to the transformed template and steal the token, the attacker is then able to generate a similar match of the original template. This renders the security of the data susceptible to threats.

The irreversible transformation function is a one-way function modulate the data into a new shape. A realization of the irreversible transform function is presented by

Tulyakov et al. [21]. In their proposed system, the biometric information is corrupted by using symmetric hash functions. The hash functions provide better security than the salting approach due to their one-way attribute as irreversible functions. In addition, a new template can be generated by changing the hash function in a case of disclosure. However, the main drawback of the proposed system is the difficulty of preserving and satisfying both the irreversibility and accuracy of the system, which can be attributed to the fact that, by using the hash functions, some information from the original biometric possess is lost.

In the biometric cryptosystem, a key is associated with the biometric data. The key is either bounded (key binding schemes) or extracted (key generation schemes). The template is constructed in such way that not much information about the key and the biometric data is exposed [22]. Juels et al. [23] proposes a key-binding biometric cryptosystem called fuzzy vault. The fuzzy vault has been widely used to protect stored templates in fingerprint matching system. This system hides a secret key vault inside a large amount of data. The secret vault is unlocked only when there is another set of data that closely overlaps with the original one. This approach provides adequate protection for stored template, but, at the same time, the recognition accuracy is very low with a GAR of 89%. Another key-binding scheme approach is the Bipartite biotokens [24]. In this system, the biometric data is transformed using some transform parameters depending on the biometric data of the user. In addition, the transformed data is folded with user password. The system then embeds the secret key into generated biotokens. This approach allows the secret key to release only when the data matches. The method significantly enhances the template security as compared to other systems. However, in the current form, if the user has already forgotten the password, the system requires the user to provide a new biometric sample to revoke and reissue a new fingerprint template. Additionally, the main limitations of the Bipartite biotokens system are the scalability and the speed of the matching algorithm since the system stores all the data at a local storage. Moreover, the availability of the system is poor and there is no guarantee of continuity in service to the user if the local storage is compromised.

A widely known example of the key generation schemes is a fuzzy extractor proposed by Juels et al. [15]. The researchers propose a key generation scheme by introducing two components: the fuzzy extractor and secure sketch. In the enrollment process, fuzzy extractor uses the features of biometric data to produce cryptographic key and public parameters, which are stored in the database rather than the template itself. While in the verification process, when a query comes from the user is similar enough to the references parameters stored in the database, the secure sketch authenticates the user then retrieves the secret key stored in the database. The fuzzy extractor approach is very useful in cryptographic applications since the secret key can be generated with the user's biometric template. On the other side, the approach makes it easier to accept an attacker as a legitimate user, as demonstrated with the obtained ratio of FAR is 4.42% with GAR of 90.33%. Moreover, a recent key generation technique based on the cloud computing is proposed by Wu et al. [25] to improve the level of the matching accuracy. The researchers develop a new bio key generation algorithm called FVHS. A FVHS directly generates bio key sequences from finger vein samples using the machine-learning system. In the authentication framework, a unique sequence is obtained by combining the bio keys of users and specific characteristics of cloud computing services to provide identity authentication. The proposed system can extract a finger vein bio-key with a GAR of more than 99.9%, and the FAR is less than 0.8%. However, FVHS's speed is extremely slow.

17.4.2 Biometric in the Encrypted Domain

Biometric is sensitive data that must be protected regularly; one common technique used to secure biometric systems is encryption. Some approaches use the symmetric encryption techniques, such as the Advanced Encryption Standard algorithm, to store the encrypted form of biometric data. However, in the matching process of this approach, the biometric data must be decrypted first, which make it susceptible to attacks during the authentication process [14]. On the other hand, other approaches use homomorphic encryption [16, 26]. Since the implementation of homomorphic encryption is recent, only a few biometric systems that apply homomorphic encryption have been proposed so far.

Barni et al. [26] presents a new fingerprint verification system based on a Finger-Code. The Finger-Code is fixed length representation of the fingerprints data. The encrypted representation of the Finger-Code is constructed to produce the related template of the client. The generated template coming from the client is in encrypted form, while the biometric template stored in the database server is in clear form. To make it impossible for the server to misuse the resulting matching values, the homomorphic cryptosystems processes the matching operation in encrypted domain. However, the low accuracy of the recognition method based on the Finger-code template is the main problem of the proposed scheme. Therefore, the few security applications using the Finger-Code system consider the system more suitable for applications where the privacy of the data is more significant than the accuracy of the system.

To satisfy both security and accuracy rate, a privacy-preserving biometrics authentication scheme is proposed by Torres et al. [16]. The proposed approach protects the privacy

of biometric system using fully homomorphic encryption method (FHE) [27]. In the enrollment process of FHE, the image of the iris biometric from the user is processed as a biometric template, then encrypted using FHE and stored in the database with a corresponding user ID. In the authentication process, the encrypted biometric template, which comes from the user, is examined against the encrypted biometrics templates in the database to decide if the user is allowed or rejected. The FHE approach can give a remarkable level of privacy, since all the data going into the channels are encrypted. The main limitations of FHE scheme are the size of the ciphertext and the performance of the system.

Boult et al. [11] proposes cryptographic secure biotokens to secure fingerprints system. This proposed scheme is a non-invertible approach. In this approach, the system transforms each biometric feature (v) by scaling and translation operation as in the equation:

$$v' = (v - t) * s \tag{17.1}$$

After that, the result from Eq. (17.1) divides into integer stable part (q) and residual part (r). The stable part (q) encrypts and transforms by applying the public key technique with the hash function. Then the encrypted part (w) and residual part (r) are stored as the secure biometric template. The results show that the proposed approach has the ability to perform matching in the encrypted domain without affecting the accuracy of the system. However, since there are many parameters to be stored for each identity, this approach requires a scalable storage. Otherwise, the performance of the system may be affected.

17.4.3 Biometric in the Cloud Storage

According to the growth of biometric technology in recent years, storing the biometrics data is rapidly rising in the cloud rather than the local storage. Kohlwey et al. [28] presents a cloud-based biometric system. In this system, the researchers design a prototype for matching a combination of human iris images. They also discuss the design considerations for the next generation of the biometric system. Finally, they present some underlying component that can operate in the cloud environment, such as Apache Hadoop [29, 30], Apache ZooKeeper [31] and Apache HBase [32].

Another proposed framework for storing the biometrics data in the cloud-computing is presented by Raghava et al. [33]. The researchers use Apache Hadoop for an iris recognition application. Hadoop can easily handle a large amount of data on the cloud in parallel. The achieved results show that the use of Hadoop could enhance the efficiency in the cloud-computing. Cloud-ID-Screen is another approach proposed by Alsolami et al. [34, 35] in which the fingerprint features are split into smaller subsets, then spread over multiple clouds at the same time using Hadoop. The system provides privacy and security in the cloud by ensuring that no individual cloud can store all the subsets of the fingerprint. At the end, all these approaches can positively affect the biometric data computation processes. However, since the cloud store the biometrics data in a plaintext form, the two approaches have some shortcomings in both privacy and security.

17.5 Findings and Direction for Further Research

Several research studies have contributed to secure biometric systems and address different issues in the previous section. However, there are some challenges and constraints that need to be addressed further. In this section, we present our findings with respect to the biometric system needs.

1. There is a need for finding a novel solution that simultaneously improves both the privacy and security of biometric data.
2. There is a need for further research to examine the conflict of interest between the owner of the biometric system and user privacy. The organization focuses on interoperability and the ability to share data across all biometrics databases, which can potentially violate user privacy.
3. There is a need for developing a biometric identification system that covers all the security standards of smart devices, especially with the growing number of smart devices connected to the Internet.

   The future direction of research is expected to integrate the new security tools (i.e. blockchain) with the biometric system to enhance the security of the system.

17.6 Conclusion

Security presents a significant challenge for biometric implementations according to the increased use of biometric technology in recent years. This paper presents the detailed description of the fingerprint identification process. Moreover, the paper explains some security issues in the biometric data. Therefore, this paper presents several mechanisms used to protect biometric data with the different security techniques that should take place at the different layers such as template protection scheme, encryption mechanism, and cloud approaches. Finally, several findings and potential directions of research in biometric security are presented.

References

1. Hahn, C., Hur, J.: Efficient and privacy-preserving biometric identification in cloud. ICT Exp. **2**(3), 135–139 (2016). Special Issue on ICT Convergence in the Internet of Things (IoT)
2. Gupta, P., Gupta, P.: An accurate slap fingerprint based verification system. Neurocomputing **188**, 178–189 (2016). Advanced Intelligent Computing Methodologies and Applications
3. Asha, S., Chellappan, C.: Biometrics: an overview of the technology, issues and applications. Int. J. Comput. Appl. **39**, 35–52 (2012). Full text available
4. Tian, Y., Gofman, M., Villa, M.: Biometrics in cloud computing and big data (Chapter 8). In: Mitra, S., Gofman, M. (eds.) Biometrics in a Data Driven World: Trends, Technologies, and Challenges, pp. 245–262. CRC Press, Boca Raton (2016)
5. Jain, A.K., Nandakumar, K., Nagar, A.: Biometric template security. EURASIP J. Adv. Signal Process **2008**, 113:1–113:17 (2008)
6. Duncan, A., Creese, S., Goldsmith, M., Quinton, J.S.: Cloud computing: insider attacks on virtual machines during migration. In: 2013 12th IEEE International Conference on Trust, Security and Privacy in Computing and Communications, pp. 493–500, July 2013
7. Kumar, P.R., Raj, P.H., Jelciana, P.: Exploring data security issues and solutions in cloud computing. Proc. Comput. Sci. **125**, 691–697 (2018). The 6th International Conference on Smart Computing and Communications
8. Armbrust, M., Fox, A., Griffith, R., Joseph, A.D., Katz, R., Konwinski, A., Lee, G., Patterson, D., Rabkin, A., Stoica, I., Zaharia, M.: A view of cloud computing. Commun. ACM **53**, 50–58 (2010)
9. Thangavel, M., Varalakshmi, P., Sridhar, S.: An analysis of privacy preservation schemes in cloud computing. In: 2016 IEEE International Conference on Engineering and Technology (ICETECH), pp. 146–151, Mar 2016
10. AlZain, M.A., Pardede, E., Soh, B., Thom, J.A.: Cloud computing security: from single to multi-clouds. In: 2012 45th Hawaii International Conference on System Sciences, pp. 5490–5499, Jan 2012
11. Boult, T.E., Scheirer, W.J., Woodworth, R.: Revocable fingerprint biotokens: accuracy and security analysis. In: 2007 IEEE Conference on Computer Vision and Pattern Recognition, pp. 1–8, June 2007
12. Li, S.Z., Jain, A.K. (eds.) Encyclopedia of Biometrics. Springer US, Boston (2015)
13. Kaur, H., Khanna, P.: Biometric template protection using cancelable biometrics and visual cryptography techniques. Multimed. Tools Appl. **75**, 16333–16361 (2016)
14. Scheirer, W.J., Boult, T.E.: Cracking fuzzy vaults and biometric encryption. In; 2007 Biometrics Symposium, pp. 1–6, Sept 2007
15. Juels, A., Wattenberg, M.: A fuzzy commitment scheme. In; Proceedings of the 6th ACM Conference on Computer and Communications Security, CCS '99, New York, NY, USA, pp. 28–36. ACM (1999)
16. Alberto Torres, W.A., Bhattacharjee, N., Srinivasan, B.: Privacy-preserving biometrics authentication systems using fully homomorphic encryption. Int. J. Pervasive Comput. Commun. **11**, 151–168 (2015)
17. Jain, A.K., Nandakumar, K., Ross, A.: 50 years of biometric research: accomplishments, challenges, and opportunities. Pattern Recognit. Lett. **79**, 80–105 (20160
18. Watson, C., Garris, M., Tabassi, E., Wilson, C., McCabe, R., Janet, S., Ko, K.: User's Guide to NIST Biometric Image Software (NBIS). Department of Commerce, National Institute of Standards and Technology, Gaithersburg (2007)
19. Maltoni, D., Maio, D., Jain, A.K., Prabhakar, S.: Handbook of Fingerprint Recognition. Springer, London (2009)

20. Jin, A.T.B., Ling, D.N.C., Goh, A.: Biohashing: two factor authentication featuring fingerprint data and tokenised random number. Pattern Recognit. **37**(11), 2245–2255 (2004)
21. Tulyakov, S., Farooq, F., Govindaraju, V.: Symmetric hash functions for fingerprint minutiae. In: Singh, S., Singh, M., Apte, C., Perner, P. (eds.) Pattern Recognition and Image Analysis, pp. 30–38. Springer, Berlin/Heidelberg (2005)
22. Álvarez Mariño, R., Álvarez, F.H., Encinas, L.H.: A crypto-biometric scheme based on iris-templates with fuzzy extractors. Inf. Sci. **195**, 91–102 (2012)
23. Juels, A., Sudan, M.: A fuzzy vault scheme. Des. Codes Cryptogr. **38**, 237–257 (2006)
24. Scheirer, W.J., Boult, T.E.: Bipartite biotokens: definition, implementation, and analysis. In: Tistarelli, M., Nixon, M.S. (eds.) Advances in Biometrics, pp. 775–785. Springer, Berlin/Heidelberg (2009)
25. Wu, Z., Tian, L., Li, P., Wu, T., Jiang, M., Wu, C.: Generating stable biometric keys for flexible cloud computing authentication using finger vein. Inf. Sci. **433–434**, 431–447 (2018)
26. Barni, M., Bianchi, T., Catalano, D., Di Raimondo, M., Donida Labati, R., Failla, P., Fiore, D., Lazzeretti, R., Piuri, V., Scotti, F., Piva, A.: Privacy-preserving fingercode authentication. In: Proceedings of the 12th ACM Workshop on Multimedia and Security, MM&Sec '10, New York, NY, USA, pp. 231–240. ACM (2010)
27. Gentry, C.: A fully homomorphic encryption scheme. Ph.D. thesis, Stanford, CA, USA (2009). AAI3382729.
28. Sussman, A., Trost, J., Maurer, A., Kohlwey, E.: Leveraging the cloud for big data biometrics: meeting the performance requirements of the next generation biometric systems. In: 2011 IEEE World Congress on Services (SERVICES 2011), vol. 00, pp. 597–601 (2011)
29. Apache hadoop. [Online]. Available: http://hadoop.apache.org/
30. Dean, J., Ghemawat, S.: Mapreduce: simplified data processing on large clusters. In: Proceedings of the 6th Conference on Symposium on Opearting Systems Design & Implementation – Volume 6, OSDI'04, Berkeley, CA, USA, pp. 10–10. USENIX Association (2004)
31. Apache zookeeper. [Online]. Available: http://zookeeper.apache.org/
32. Apache hbase. [Online]. Available: http://hbase.apache.org/
33. Shelly, Raghava, N.S.: Iris recognition on hadoop: a biometrics system implementation on cloud computing. In: 2011 IEEE International Conference on Cloud Computing and Intelligence Systems, pp. 482–485, Sept 2011
34. Alsolami, F.J.: Toward secure sensitive data in the cloud. Ph.D. thesis, University of Colorado at Colorado Springs (2015)
35. Alsolami, F., Alzahrani, B., Boult, T.: Cloud-id-screen: secure fingerprint data in the cloud. In: 2018 IEEE 4th International Conference on Identity, Security, and Behavior Analysis (ISBA), pp. 1–8, Jan 2018

Access Controls in Internet of Things to Avoid Malicious Activity

Yenumula B. Reddy

18.1 Introduction

The IoT consists of physical objects embedded with sensors, software, cloud, and network connectivity of these objects to store, exchange, process, and make decision of complex data. In the smart city, the data generates from activities, events, and other sources of making data (mostly from sensors) for decision-making on a real-time basis. The other sources of data may be traffic signals, smart devices, embedded systems, and traffic. The data generated is in large volume, continuous, and in high velocity. This type of data is called big data in current technology.

The word big data was coined recently and became famous due to its storage (beyond the size of the standard database), retrieval, analysis, and production of useful results. It is an unstructured, vast volume, continuously growing on a real-time basis, and challenging to process. The technology changes influence the classification changes of such data over a period. Therefore, its definition changes time to time and organization to organization (challenging to have a perfect description). Every organization has vested interest in the classification of such unstructured data. As the technology changed from computers to hand devices, the processing became a big problem. Due to this reason, cloud requirement exists. Cloud helps to take most of the storage of large volume of data, complex computations, and generation of customer output.

Cloud requirement for big data is due to its size, a different type of data from multiple sources, the velocity of its flow (incoming and outgoing), potential value if adequately classified and processed, and confidentiality. Cloud computing is a paradigm with unlimited on-demand services. It can virtualize hardware and software resources, high processing power, storage, and pay-per-usage. Moreover, it transfers cost calculation responsibilities to the provider and minimizes the great setup of computing facilities at small enterprises. It has negotiable natural resources and gets computing power as required. The cloud services provide infrastructure as a service, software as a service, and platform as a service.

Cloud computing delivers storing and processing of sizeable unstructured volume of continuously generated data, resource availability, and fault tolerance through its various hardware and software facilities. Many companies including Nokia, redBus, Google, IBM, Amazon, and Microsoft provide consumers to consume service on-demand. The big business decided to migrate to the Hadoop Distributed File System (HDFS) that integrates data into the same domain and uses supplicated algorithms to get proper results for its customers. The advantage of using Hadoop is cheaper storage compared to traditional databases. Currently, HDFS helps Nokia, redBus, Google, and other companies to fulfill their needs. The facility helps these companies to concentrate on their businesses rather than on technical details and requirements.

Big data technology solves many problems irrespective of volume, velocity, and source of generation. It is a constantly changing technology, and many industries, customers, and government agencies are involved in usage and managing. Further, the data is in a cloud environment. Due to this reason, we need to create security policies, access rights, and secure storage and retrieval. Even though data is continuously growing, controlling is required (means valuable data need to be stored). Therefore, we need to enforce the data governance policies like organizational practices, operational practices, and relational practices.

Disaster recovery (in the case of dangerous accidents including floods, earthquakes, fire, and accidental loss of data) for valuable data is a requirement. The big corporations

Y. B. Reddy (✉)
Department of Computer Science, Grambling State University, Grambling, LA, USA
e-mail: ybreddy@gram.edu

© Springer Nature Switzerland AG 2019
S. Latifi (ed.), *16th International Conference on Information Technology-New Generations (ITNG 2019)*,
Advances in Intelligent Systems and Computing 800,
https://doi.org/10.1007/978-3-030-14070-0_18

define a set of procedures for a disaster recovery plan to restore the data. In addition to security policies, disaster recovery is strongly recommended (fault-tolerant depends on disaster recovery). The other problems include the secure transfer of data to the cloud, incorporating high-performance computing, and data management. Big data in the cloud has many research and practical challenges. Storing the data using an encryption technique takes extra time. Standardization of procedures to minimize the impact of heterogeneous data. Data governance, recovery plans, quality of services for secure transfer of data, and petaflop computing are some of the problems for implementation.

IoT generates a significant amount of complex data from multiple sources and requires to process real-time basis with defined policies and privacy requirement. There is a need to represent high-level aggregating requests which often involves inference techniques. This data needs various controls, human intervention, and feedback. Also, there is a need to identify the devices, customers, and their limits (control level). The restrictions include the data access and retrieval (inside the system model and hierarchical control). Besides, the customer or connecting device must be trusted to eliminate the malicious activity.

The successful deployment of complex data activity on cloud requires building a business case with an appropriate strategic plan to use the cloud. The project must develop productivity, extract more significant value, continuous improvement, customer acquisition, satisfaction, loyalty, and security. Assess suitable cloud environment (private or public) and develop a technical approach. Next, address the governance, privacy, security, risk, and accountability requirements. Finally, deploy the operational environment. The provider meets many challenges depending on the cloud data environment. Security, cost factor, customer satisfaction, and service reliability are central issues.

18.2   Literature Review

Storage, processing, and retrieval of big data in the cloud are significant problems in current research. Pedro et al. [1] studied the overview of present and future issues. The document discusses scalability and fault tolerance of various vendors including Google, IBM, Nokia, and redBus. The authors further considered the security, privacy, integrity, disaster recovery, and fault-tolerant issues. The authors [2] discussed the review of current service models, import concepts of cloud computing, and processing of big data. Elmustafa and Rashid [3] presented the survey issues of big data security in cloud computing.

Linda et al. [4] showed the environmental examples of big data use in government that includes Environmental Protection Agency, Department of the Interior, Department of Energy, and Postal Services. The study consists of the government open-access initiatives, the federal data center consolidation initiative, and the enforcement of compliance online. James [5] presented a roadmap to the success of big data analytics and applications. The report discusses the definition and description of unstructured data, relevant use cases in the cloud, potential benefits, and challenges associated with deploying in the cloud.

The impact of cloud computing on healthcare was studied in [6]. The study includes on-demand access to computing and large storage, supporting big data sets for electronic health records, and ability to analyze and track the health records. Alan [7] presented the environmental sustainability of big data, barriers, and opportunities. It also includes new opportunities for partnership-based collaboration, sustainability to organizations to big data efforts, and emerging business models.

Yan et al. [8] discussed the access control in cloud computing. The paper contains the temporal access control in cloud computing using encryption techniques. Yuhong et al. [9] data confidentiality in cloud computing. The article uses the trust-based evaluation encryption model. In this model, the trust factor decides the access control of user status. Young et al. [10] discussed the security issues in cloud computing. The paper describes the access control requirements, authentication, and ID management in the cloud.

Ali and Erwin [11] reviewed security and privacy issues on big data and cloud aspects. They concluded that cloud data privacy and safety is based on the cloud provider. They also discussed big data security challenges and cloud security challenges. Their paper examines the security policy management and big data infrastructure and programming models. They did not suggest any particular model but discussed all possible solutions for the security of big data in the cloud. Marcos et al. [12] presented approaches and environments to carry out big data computing in the cloud. The paper discusses the visualization and user interaction, model building, and data management. Venkata et al. [13] examined issues in a cloud environment for big data. The primary focus is security problems and possible solutions. Further, they discussed MapReduce and Apache environments in the cloud and needed for the security.

Saranya and Kumar [14] addressed the security issues associated with big data in a cloud environment. They suggested few approaches for the complicated business environment. The paper discusses unstructured big data characteristics, analytics, Hadoop architecture, and real-time big data analytics. The authors did not present any particular model in the article. They explained a few concepts related to security in a cloud environment. Avodele et al. [15] presented issues and challenges for deployments of big data in the cloud. They suggested solutions that are relevant to organizations to deploy the data in the cloud. The authors indicated the importance of authentication controls and access controls.

Security in IoT was discussed in [19–26]. The authors conclude that IoT networks are hugely needed to ensure confidentiality, authentication, access control, and integrity, among others. The reason for immediate attention to security is a dramatic increase in the number of connected devices. These devices create technical problems such as attacks with a broader scope of influence and attacks that last longer. Therefore, immediate attention and procedures are required to detect the hacker to avoid the damage to sensitive information.

The remaining paper discusses the problem formulation that leads to the authentication model in the cloud in Sect. 18.3, simulations in Sect. 18.4, and conclusions and future work in Sect. 18.5.

18.3 Problem Formation

The security model involves the cloud customer data security at storage, retrieval, transfer, processing, and updates (insert, modify, delete). The security needs to set at the log entry at user and cloud level. It also requires the automatic validation of stored data status and verifies the trust level. The framework of the proposed model includes the data encryption, correctness, and processing. These three modes depend upon the access rights of the user as discussed at the beginning of the current section. For storage and retrieval of data, the basic encryption techniques AES and RSA and steganography model are sufficient. If the data requires storage and processing, the recommendations in [16] may be useful. The paper discussed the various techniques to search cipher text and query isolation (avoid the untrusted server). Controlled searching, dealing with variable word lengths, searching encrypted index, and supporting hidden search are part of the research. In this paper, the proposed access control model with encrypted processing data is useful to avoid untrusted provider and malicious users in the cloud.

The access control model for IoT of cloud storage incorporates the authentication of customer and its current access level. The token identification (TID) is attached as soon as the user log in into the system. To maintain the security of data and its trust level, we have to define many control parameters to the user access in the cloud. The current TID model in Eq. (18.1) explains with seven parameters:

$$TID = UID, IID, MDT, TA, PA, LGE, SA \qquad (18.1)$$

where

UID   User identification and access rights
IID   Issue date
MDT   Maximum date (expiration date)
TA   Time of access
PA   Place of access (current place and node ID)

LGE   Log entry (UID, IID, MDT, TA, PA, LGE)
SA   Security alarm

The customer is an owner of the data or another customer. In either case, the customer is a client with different access rights. Once the customer log in into cloud network, the authentication access token connects to the user account. The token verifies the customer/user access limits and allows or denies the appropriate file access. Further, the system does the entries in customer/user and cloud log table for each attempt of a user to a particular file with all details. The various validation and verification check the modifications help to find unauthorized access. The trustworthiness of provider or customer/user can be calculated using the log values.

The trustworthiness of a customer/user can be calculated using trust function in Eq. (18.2). For each entry of the user, the weight "W" is assigned. The entry $W_{i,j}$ means, ith user and jth entry. Let Ng be the number of times the user has right behavior, and Nb is the number of times of bad behavior of the ith user. Multiply the user entry value with weight "W" with right or wrong actions, and calculate the trustworthiness of a customer/user. The trustworthiness T_i of ith user/customer is calculated as follows:

$$T_i = \frac{\sum\limits_{i,j} W_{i,j}{}^{*}Ng_j + x}{\left(\sum\limits_{i,j} W_{i,j}{}^{*}Ng_j + x\right) + \left(\sum\limits_{i,k} W_{i,k}Nb_k + y\right)} \qquad (18.2)$$

If T_i the trust value is above the threshold, the customer/user is considered as good; otherwise false alarm alerts the owner. The user may be a customer, provider, or owner. The weight varies between 0 and 1, and the number of times the user accesses the data (or data files) will be 0–10. If weight $= 0$, then $x = 1$; else $x = 0$. Similarly, if the number of times $= 0$, then y value is 1; otherwise $y = 0$.

18.4 Simulations

The simulations on Eq. (18.2) were performed and provided in Fig. 18.1a–c. The threshold value for legal (trusted) user was fixed at 0.85 and above in the current situation. We decide the login user is a hacker or trusted using the average trust level. For example, if the threshold value is greater than 0.85 for an average of ten (10) attempts, then the user is legal (assumption). The program was developed in MATLAB language to create a graph for Eq. (18.2). Figure 18.1a–c presents the sample results. The random data generated for each user and plotted the average of ten access values. Figure 18.1c shows that the user is malicious and messaged to security manager as a hacker (trust level of user accesses are below. 8). The user access values depend upon the random values and the corresponding weights selected. If the

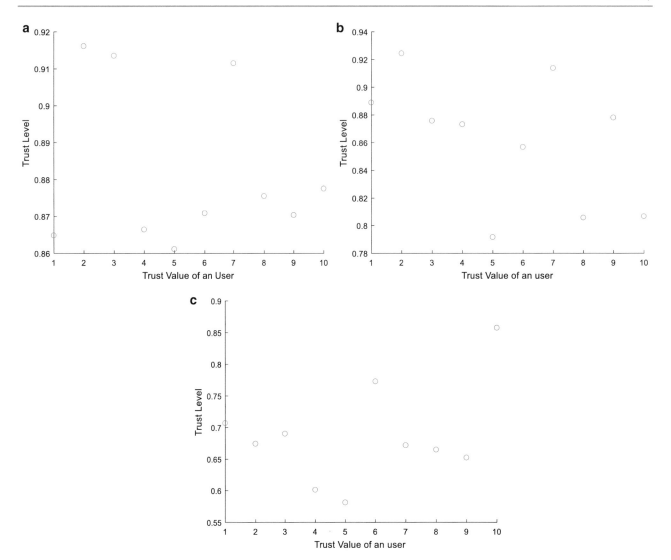

Fig. 18.1 (a) Experiment1: Average of ten random accesses of a user to data by each user. (b) Experiment 2: Average of ten random accesses of a user to data by each user (sample graphs). (c) Experiment 3: Average of ten random accesses of a user to data by each user (sample graphs)

malicious user attempts the data, then the alarm signals to the security manager using threshold values. The proposed data is a random selected sample for the test calculation. The complex calculation requires a real data (not provided) since it has various parameters for each user logging and processing the data.

18.5   Access Controls on Sensitive Data

Access to sensitive data cannot satisfy pure trustworthiness. Along with trustworthiness, the procedure requires the user access limits, day, time of the day, and log entry for validation. The UID contains access rights and user ID issue date, expiration date, time of access, and location of access (depends upon sensitiveness of data). Once the user logs in into the system, the cloud log and owner logs are entries that are automatically registered. For hackers, only cloud log

entry appears. The various validation and verification checks reveal the hacking. The token ID parameters in Eq. (18.1) are used in objective function G.

$$G = \{N, A, D, U\} \tag{18.3}$$

The objective function G replaces TID, N replaces UID (contains IID, MDT, TA, and PA), D is a data file (or database), and U replaces LGE. The security alarm will be activated depending upon the hacker identification or trust failure. Therefore the parameters are explained further as below:

N the set of users $(n_1, n_2, \ldots \ldots n_m)$
A set of access rights $(a_1, a_2, \ldots \ldots a_p)$
D set of allowed resources in file or database $(d_1, d_2, \ldots \ldots d_q)$
U the result of the query and log entries for verification and validation

Once the authenticated user $n_i (n_i \in N)$ logs in into cloud environment, the CCCRN service attaches a service token to a resource within its domain with a set of access types a_i. The limitation helps to control the user for resource access. For every service requested by the user, the system generates a set of access permissions to the resources. The services required should not exceed the user access limits. If the resource requirements are outside the user boundaries, then the system alarms the security and denies the request. Hacker is a user that does not have any role in the system. An authorized user will be treated as a hacker if the user tries to access unauthorized information. For example, the healthcare staff member will be considered as an intruder if the user accesses unauthorized data or misuses (for instance, printing and forwarding) the authorized information

In the proposed CCCRN environment, the user with complete authorization access is called a super user (S). The super user "S" possesses access rights of all users $S \supseteq \underset{i=1,n}{Y} a_i$ where $\supseteq$ means contains. All accesses of the super user on the database must be recorded. The user that does not have authorization to resource(s) is called hacker (h_i) and represented as H $(h_i \in H)$ and $\forall H$ (hackers); the access right $a_{ih} \mapsto d_i \equiv \phi$ is true; a_{ih} is access rights of the hackers ($\mapsto$ implication to and $\equiv$ is equivalent to). Using this information, we design two algorithms.

Algorithm 18.1

If the query $Q(n_i, d_i)$ matches the n_i as owner for token identification (TokenID), then the corresponding utility function u_i will be generated; else the query reflects as $Q(n_i, hd_i)$, where h is a hacker.
If the hacker is an internal user, then

$hu_i \supseteq u_i + h' d_i$ (u_i internal user) alarms security manager about internal hacker.

If $Q(n_i, d_i) \subset u_i$, then exit;

else
if $Q(n_i, d_i) \not\subset u_i$ & & $Q(n_i, d_i) \cong hu_i$, then
convert $Q(n_i, d_i)$ as $Q(n_i, hd_i)$, and generate $hu_i \supseteq u_i + h' d_i$.

Store the user utility hu_i that contains $u_i + h' d_i$, inform security, and keep the counter (log) in alert for further attempts.

Algorithm 18.1 helps to detect the hacker if the user tries to gain the information with unauthorized access from the database. The following query and Table 18.1 explain the unauthorized access to information.

If $Q(n_i, d_i) \equiv Q(hn_i, d_i) \not\subset u_i$ or $Q(hn_i, d_i) \approx hu_i$, then
$Q(hn_i, d_i) = hu_i$, retrieving hu_i (utility from the Hacker alarm to database) and alerting the security alarm

Table 18.1 Hacker log and action

Hacker	Status	Result	Action
A	New	hu_i	New hacker, alarm
A	Repeat	hu_i	Alarm and freeze

where hu_i is available in log or identified as a new hacker and logged as new entry. The log is provided in Table 18.1.

In general, if the hacker attempts to gain access to the database at different trimmings, the time attribute plays an important role to detect the hacker. Algorithm 18.1 is modified as Algorithm 18.2.

Algorithm 18.2

If $Q(n_i, t_i, d_i)$ is genuine and attempted during duty times, then corresponding utility function u_i will be generated; else

the query reflects as $Q(n_i, t_j, hd_i)$, and then user
will get $hu_i \supseteq u_i + h' d_i$ (where u_i is internal user information and $h' d_i$ is the hacker alarm at time t_j).

If $Q(n_i, t_i, d_i) \subset u_i$, then exit (user access accepted);

else

if $(Q(n_i, t_j, d_i) \not\subset u_i)$ & & $(Q(n_i, t_j, d_i) \cong hu_i)$.
Convert $Q(n_i, t_j, d_i)$ as $Q(n_i, t_j, hd_i)$, and generate $hu_i \supseteq u_i + h' d_i$ (alarm alert to security manager).

Note Store the user utility hu_i that contains $u_i + h' d_i$, alert security, and keep the counter for further attempts. If the hacker is external, then divert to the KDS. If the user hacks with authentication, then the time stamp will help to detect the hacker. For example,

if $Q(n_i, t_j, d_i) \equiv Q(hn_i, t_j, d_i) \not\subset u_i$ or $\subseteq hu_i$, then
$Q(hn_i, t_j, d_i) = hu_i$, retrieving hu_i and alarming the security,

where hu_i is available in log or identified as a new hacker and logged as a new entry. Table 18.2 provides the log entries.

Depending upon the security level, Algorithm 18.2 will be modified by adding the terminal type and log-on timings. Terminal type and time of access attributes along with access type attributes will protect the secret and top secret information.

Let us assume the hospital environment in the healthcare system. A doctor and nurse have the same access rights to individual patient data (doctor prescribes the medicine which was implemented by the nurse). Then the attributes patient ID, type of medication, and scheduled time dose to be given to a patient are accessible by the nurse. The same attributes are also available by the doctor. Therefore, the system security depends upon the merge and decomposition of two or more users.

Table 18.2 Hacker log and detection

Hacker	Status	Time	Result	Action
A	New, internal	Outside-bounds	hu_i	Detect as internal hacker and alarm
A	Repeated, internal	Within-bounds	hu_i	Check for presence of real user and alarm and find real user

18.6 Conclusions

The issues and challenges in IoT [19–26], processing of complex data in cloud, and security issues were discussed in [8–15, 17–18]. We found that it is required to develop a trust and access control methodology in IoT and cloud environment for real-time access to data and processing for decision-making. Therefore, in the current research, an objective function was proposed with a set of users, associated access rights, resources, and return result verification. The proposed model is appropriate for the big data in a cloud environment where IoT devices are involved. Further, two algorithms were presented where the model can be extended to Hadoop Distributed File Systems to detect the external and internal hackers in a cloud environment. The tables were presented for hacker detection through algorithms. The user entry logs, authentication, and access rights have a significant role in providing the hacker information to the security administrator.

The future work improves the algorithms to protect sensitive data in business and government through access rights in a cloud environment. Further, incorporating honeypots will mislead the malicious users and hackers in sensitive data places.

Acknowledgments The research work was supported by the CMAST program funded by NSF through award number HRD 1719523. The author wishes to express appreciation to Dr. Connie Walton, Director of Sponsored Programs at Grambling State University and PI of NSF Big Data grant for her continuous support. The author also wishes to appreciate Dr. Stacy Duhon, Dean, College of Arts and Sciences, for continuous encouragement.

References

1. Neves, P.C., Schmerl, B., Bernardino, J., Camara, J.: Big data in cloud computing: features and issues. In: International Conference on Internet of Things and Big Data, January 2016
2. Branch, R., Tjeerdsma, H., Wilson, C., Hurley, R., McConnell, S.: Cloud computing and big data: a review of current service models and hardware perspectives. J. Softw. Eng. Appl. **7**, 686–693 (2014)
3. Ahmed, E.S.A., Saeed, R.A.: A survey of big data cloud computing security. Int. J. Comput. Sci. Softw. Eng. (IJCSSE). **3**(1), 78–85 (2014)
4. Breggin, L.K., et al. (eds.): Big Data, Big Challenges in Evidence-Based Policy Making. West Academic Press, St. Paul (2014)
5. Kobielus, J.: Deployment in big data analytics applications to the cloud: road map for success. Cloud Standards Customer Council, Technical report (2014)
6. Impact of cloud computing on Healthcare. Cloud Standards Customer Service, Technical report (2012)
7. Keeso, A.: Big data and environmental sustainability: a conversation starter. Smith School Working Paper Series, Working Paper 14–04 (2014)
8. Zhu, Y., Hu, H., Ahn, G.J., Huang, D., Wang, S.: Towards temporal access control in cloud computing. In: Proceedings of IEEE INFOCOM (2012)
9. Liu, Y., Ryoo, J., Rizvi, S.: Ensuring data confidentiality in cloud computing: an encryption and trust-based solution. In: Proceedings of IEEE 23rd Wireless and Optimal Communication Conference (WOCC) (2014)
10. Min, Y.-G., Shin, H.-J., Bang, Y.H.: Cloud computing security issues and access control solutions. J. Eng. **9**(2), 135–140 (2012)
11. Gholami, A., Laure, E.: Big data security and privacy issues in the cloud. Int. J. Netw. Secur. Appl. (IJNSA). **8**(1), 59–79 (2016)
12. Asuncao, M.D., Calheiros, R.N., Bianchi, S., Netto, M.A.S., Buyya, R.: Big data computing and clouds: trends and future directions. J. Parallel Distrib. Comput. **79–80**, 3–15 (2015)
13. Inukollu, V.N., Arsi, S., Ravuri, S.R.: Security issues associated with big data in cloud computing. Int. J. Netw. Secur. Appl. (IJNSA). **6**(3), 45–56 (2014)
14. Saranya, R., Muthukumar, V.P.: Security issues associated with big data in cloud computing. Int. J. Multidiscip. Dev. **2**(4), 580–585 (2015)
15. Avodele, O., Izang, A.A., Kuyoro, S.O., Osisanwo, F.Y.: Big data and cloud computing issues. Int. J. Comput. Appl. (0975–8887). **133**(12), 14–19 (2016)
16. Song, D.X., Wagner, D., Perrig, A.: Practical techniques for searches on encrypted data. IEEE Symposium on Security and Privacy, pp. 44–55 (2000)
17. Kiran, J.S., Sravanthi, M., Preethi, K., Anusha, M.: Recent issues and challenges on big data in cloud computing. IJCCCSSST. **6**(2), 98–102 (2015)
18. Zardari, S., Khan, N.A., Memon, M.A.: Systematic analysis of risks in cloud architecture. Int. J. Comput. Sci. Inf. Secur. (IJCSIS). **14**(11), 1184–1190 (2016)
19. Mann, J.: The internet of things: opportunities and applications across industries. International Institute for Analytics, December 2015
20. Ali, Z.H., Ali, H.A., Badawy, M.M.: Internet of things (IoT): definitions, challenges and recent research direction. Int. J Comput. Appl. (0975–8887). **128**(1), 37–47 (2015)
21. Razzaq, M.A., Qureshi, M.A., Gill, S.H., Ullah, S.: Security issues in the internet of things (IoT): a comprehensive study. Int. J. Adv. Comput. Sci. Appl. (IJACSA). **8**(6), 383–388 (2017)
22. Zhou, W., Zhang, Y., Liu, P.: The effect of iot new features on security and privacy: new threats, existing solutions, and challenges yet to be solved, pp. 7–17 (2018). pp. 1–11
23. Tanaka, S., Fujishima, K., Mimura, N., Ohashi, E.T., Tanaka, M.: IoT security: IoT system security issues and solution approaches. Hitachi Rev. **65**(8), 59–73 (2016)
24. Barcena, M.B., Wueest, C.: Insecurity in the internet of things, semantic, 12 March 2015
25. O'Donnell, L.: IoT security concerns peaking – with no end in sight. In: HackerOne CEO Talks Bug Bounty Programs at RSA Conference, April 2018
26. Simon, M.: Internet of Things security issues bleed into 2018, January 2018. https://www.helpnetsecurity.com/2018/01/16/internet-of-things-security-issues-2018/

Hesham H. Alsaadi, Monther Aldwairi, and Eva-Marie Muller-Stuler

19.1 Introduction

Quantum computers have been the ideal dream for weather forecasting [1], cryptanalysis [2], particle physics [3] and financial modeling [4]. As quantum computers move out of research labs into real commercial use, security becomes a main concern. Classical computing encodes all of their information into a series of binary digits (bits), 0 and 1, corresponding to either on or off states. Bits can only be in one state or the other at a particular moment in time. If there are several different values to be examined, each must take its turn separately [5]. Quantum computers on the other hand, work on an entirely different abstraction. They encode information using "Quantum Binary Digits" or "Qubits". Information in qubits is typically represented by a quantum state and described in 0 or 1, or both at the same time [6]. An ordinary bit can compute or store 0 or 1, but a Qubit can work and operate in between the values of 0 and 1 [7]. Quantum computing requires complex numbers to characterize the quantum state. It takes an exponential number of bits of memory on a classical computer to store

the quantum state, which puts conventional computers in a contrary position to simulate a quantum system [8].

D-Wave computer system architecture does quantum computation using quantum physics rule known as "superposition", which allows quantum bits "Qubits" to exist as both 0 and 1 at the same time [9]. D-Wave can efficiently perform simple calculations in parallel using a logic-fate model called quantum annealing, also known as adiabatic quantum computing, allowing it to solve simple programming model for physical annealers such as boolean satisfiability problems [10].

Conventional solutions been developed to translate conventional program and interface them to D-Wave systems such as Solver Application Program Interface (SAPI) [11], Qbsolv [12] and QSage [13]. The measured approach does not implement the use of D-Wave programming model directly, but assist the effectiveness of a particular computational algorithm to D-Wave in a conventional manner. These tools use certain functions programmed in C++ language, which ease the translation process into D-Wave search space for solutions [14].

Recently Scott Pakin's of Los Alamos National Laboratory released new open source Python tool called QMASM [15]. The tool allows programmers to investigate how to map arbitrary computation onto a quadratic unconstrained binary optimization (QUBO) for execution on a quantum annealing in D-Wave 2X system [16].

One of the biggest questions in quantum computing is the breadth of tools developed to program computations onto quadratic execution in the D-wave machine. The D-Wave 2X system has a Web API with client libraries available for C, C++, Python, and MATLAB. The tools contain GUI interface that allows interaction with the machine for access to cloud service over a network. By submitting problems to the system, users can use a high-level programming in C, C++, Python or MATLAB to create and execute a quantum

H. H. Alsaadi
College of Technological Innovation, Zayed University, Abu Dhabi, UAE
e-mail: M80005998@zu.ac.ae

M. Aldwairi (✉)
College of Technological Innovation, Zayed University, Abu Dhabi, UAE

Department of Network Engineering and Security, Jordan University of Science and Technology, Irbid, Jordan
e-mail: monther.aldwairi@zu.ac.ae; munzer@just.edu.jo

E.-M. Muller-Stuler
Laniakea Labs, London, UK
e-mail: emms@gmx.de

© Springer Nature Switzerland AG 2019
S. Latifi (ed.), *16th International Conference on Information Technology-New Generations (ITNG 2019)*,
Advances in Intelligent Systems and Computing 800,
https://doi.org/10.1007/978-3-030-14070-0_19

machine instruction. It is also possible to directly program the system using Quantum Machine Language to issue the quantum machine instruction [17].

Because D-Wave programming requires specialized knowledge of what qubits are coupled with others, which ones are missing and which ones represent a certain variable. QMASM computes a simple heuristic computation algorithm, present symbolic variable names and assembly like code blocks that might be reusable. Developers may integrate the solution into D-wave system directly or by using local simulator, which raises questions about the reliability security of such tool [18].

In this paper, we intend to investigate the security threats found in QMASM by using the latest static analysis tools at hand. The main contributions of this work are:

- Execute a comprehensive static analysis on QASM Python source code.
- Investigate the threats found in QMASM that could lead to D-Wave system been left vulnerable to multiple security breaches.

The rest of this paper is structured as follows. Section 19.2 presents literature review. Section 19.3 describes the tools, and experimental setup needed. In Sect. 19.4 we present the results of the experimental evaluation and the possible threats found in QMASM. Section 19.5 concludes this work with recommendation for future directions.

19.2 Literature Review

Securing Python open source tools is still relatively new field due to the extensive in-depth knowledge and expertise required. Therefore, research efforts are limited to the best of the authors' knowledge. There have been, however, some of the significant pioneering attempts that paved the way for best practices in analyzing Python flaws as summarized below.

In 2013 the National Vulnerability Database found that Python 2.6 through 3.2 does create race conditions that allow local users to obtain credentials after they access */.pypirc* under certain circumstances [19]. It is considered an access control vulnerability. Python seems to have fewer vulnerabilities than some of the other languages [20].

Highlighting the challenges in the field of securing Python and providing a comprehensive analysis of bugs found in open source tools programmed in Python. Brett Cannon from the University of British Columbia provided detailed history about the security flaws in Python and the weak mechanisms enforced by Python policies. The research provided a solution model "*interpreter*" that applied to Python greater safety enhancement called "*Pythonic*". The study pointed

out providing security to Python tools could be improved by using data modules functionality such as *Built-In Functions, Built-In Types* and *Import statement*.

Python is considered an open source project developed by volunteers, indicating that no security expert on the development team has fulfilled the gap in securing Python or provide tools that enable developers to analyze their development in Python [21]. Having said that, researchers may potentially improve the security of Python tools by using data flow assertions. Alexander Yip and Xi Wang demonstrated a new programming language that helped prevent security vulnerabilities. The study introduced a new programming language called "*RESIN*", which operated within a language runtime, such as the Python or PHP interpreter [22]. It allowed developers to specify application-level data flow assertions with application data stream. The programming language worked as a translator to check defined objects along with data that move throw the application. The research demonstrated how "RESIN" can prevent a range of cyber attacks such as SQL injection and cross-site scripting. The researchers presented their solution analysis by explaining real-time scenario implementation for programmers to explicitly specify data flow vulnerabilities in existing PHP and Python applications.

Up to our knowledge there have not been any serious attempts at studying the security of quantum assemblers and similar software.

19.3 Tools and Experimental Setup

The setup for the experiments required the arrangements of a computer with Python 3.5.2 (Anaconda3 version 4.2.0 64-bit) [23] and installation of QMASM tool. To complete the setup for the experiments, we have downloaded the latest update of QAMSM Python tool. Listing 1 shows a sample of installed QMASM v1.0 files and folders as installed on Windows 7 64bit operating system.

We select several source code analysis tools to investigate the various resulting threats found in QMASM. The experimentation is targeted to detect taint style vulnerabilities by performing static source code analysis on QMASM open source Python project hosted by GitHub [24]. *Tainted* data denotes data that originates from possibly malicious users, and that can potentially cause security problems at vulnerable points in the program (called *sensitive sinks*) [25]. Tainted data may enter the program at specific places, and can spread across the program via assignments and similar constructs [26].

The experimental setup required the arrangement of directories that contain Python scripts, any folder that contains Python script will be examined separately and independently alongside with global testing.

Listing 19.1 QMASM v1.0 directory files and folders

```
C:\Users\User\Anaconda3\Lib>python listpythondirectory.py
qmasm-1.0\.gitignore
qmasm-1.0\LICENSE.md
qmasm-1.0\MANIFEST.in
qmasm-1.0\qasm.py
qmasm-1.0\README.md
qmasm-1.0\setup.py
qmasm-1.0\build\lib\qasm\cmdline.py
qmasm-1.0\build\lib\qasm\dwave.py
qmasm-1.0\build\lib\qasm\globals.py
qmasm-1.0\build\lib\qasm\output.py
qmasm-1.0\build\lib\qasm\parse.py
qmasm-1.0\build\lib\qasm\problem.py
qmasm-1.0\build\lib\qasm\utils.py
qmasm-1.0\build\lib\qasm\__init__.py
qmasm-1.0\build\scripts-3.5\qasm.py
qmasm-1.0\examples\1of5.qasm
qmasm-1.0\examples\circsat.qasm
qmasm-1.0\examples\comparator.qasm
qmasm-1.0\examples\gates.qasm
qmasm-1.0\examples\README.md
qmasm-1.0\examples\sort4.qasm
qmasm-1.0\extras\qasm.ssh
qmasm-1.0\extras\README.md
qmasm-1.0\qasm\cmdline.py
qmasm-1.0\qasm\dwave.py
qmasm-1.0\qasm\globals.py
qmasm-1.0\qasm\output.py
qmasm-1.0\qasm\parse.py
qmasm-1.0\qasm\problem.py
qmasm-1.0\qasm\utils.py
qmasm-1.0\qasm\__init__.py
qmasm-1.0\QASM.egg-info\dependency_links.txt
qmasm-1.0\QASM.egg-info\PKG-INFO
qmasm-1.0\QASM.egg-info\SOURCES.txt
qmasm-1.0\QASM.egg-info\top_level.txt
```

Because they provide a comprehensive security debugging analysis of Python scripts, the tools described below were chosen for the vulnerability assessment of QMASM.

Bandit Python AST-based static analyzer from OpenStack Security Group. Bandit is a tool designed to find common security issues in Python code. Bandit processes each file, builds an Abstract Syntax Tree (AST) from it and runs appropriate plugins against the AST nodes. Once Bandit has finished scanning all the files, it generates a report. Bandit is currently a stand-alone tool, which can be downloaded by end-users and run against arbitrary Python source code modules [27].

Mypy is an experimental optional static type checker for Python that aims to combine the benefits of dynamic (or "duck") typing and static typing. Mypy combines the expressive power and convenience of Python with a powerful type system and compile-time type checking. Mypy type checks standard Python programs developed in Python 2 and 3 (PEP484); run them using any Python VM with basically no runtime overhead [28].

Py-find-injection Python tool that uses various heuristics look up to perform SQL injection vulnerabilities in Python source code scripts [29].

Pyflakes Python tool that is similar to PyChecker in scope [30], the difference is that it only executes the modules faster and safer. Pyflakes checks only logical errors in Python source code and does not perform any style check [31].

vulture Python static analysis tool that finds unused classes, functions, and variables in Python programs. Vulture is compatible with Python 2.6, 2.7 and 3.x and it helps to clean up errors in high-level Python scripts. Vulture can be used together with pyflakes, however, in this research they will be examined independently [32].

xenon Python code complexity monitoring tool based on Radon. Xenon is programmed using given command with various sets of thresholds for code complexity analysis. This tool is utilized in the research to accurately investigate the threat level complexity found in Python blocks and modules [33].

The use of different tools provides different analysis and execution techniques, which is necessary to segregate the files for precise analysis.

19.4 Experimental Results and Discussion

QMASM tool has several versions released regularly by the developers [34]. The researchers needed to install all versions to provide precise, comprehensive analysis results. Table 19.1 summarizes the main description of all QMASM releases. The aforementioned open source tools were used for testing and evaluating the latest release of QMASM v1.2 for vulnerabilities.

The results provided by Bandit analysis tool on all versions of QMASM releases resulted in the same infected files with the low severity threat level. We used recursive scan type to reverse engineer the code for analysis. Moreover, the confidence of the code blocks issues rated from medium to high. In all releases, Python files that contain moderate severity threats and medium confidence provide the culmination of hard coded password strings. The affected files *dwave.py* are Python files that were found in all releases in following directories:

- qmasm-1.0\build\lib\qasm\dwave.py
- qmasm-1.0\qasm\dwave.py
- qmasm-1.1\build\lib\qmasm\dwave.py
- qmasm-1.1\qmasm\dwave.py
- qmasm-1.2\build\lib\qmasm\dwave.py
- qmasm-1.2\qmasm\dwave.py

Bandit also discovered Python files that contain moderate severity threats and high confidence, mostly issues of standard pseudo-random generators functions. Below list shows Python files generated with random acceleration features that are not suitable for security/cryptographic purposes in high integration environment. The affected files *problem.py* are Python files found in each release in the following directories:

Table 19.1 Comparison of QMASM releases

Name	Version	Date	No. Python files	Description	Size
QASM	v1.0	August 26,2016	19	The first formal release of QASM. Tested using Python 2.7.12, SAPI 2.4, and qOp 2.2, all on Ubuntu Linux 16.04 LTS (Xenial Xerus)	216 KB
QMASM	v1.1	October 27,2016	19	Renamed QMASM, the code contains miscellaneous, sources, disconnected variables, and both proxied and non-proxied network access	232 KB
QMASM	v1.2	November 12,2016	19	Contains new output format: the MiniZinc constraint-modeling language by using –format=minizinc on the qmasm command line	380 KB

Table 19.2 Mypy analysis of code checking problems of modules and annotation variables

Python file name	Type of error	LOC	Description
__init__.py	Module error	1	cmdline
__init__.py	Module error	2	dwave
__init__.py	Module error	3	globals
__init__.py	Module error	4	output
__init__.py	Module error	5	parse
__init__.py	Module error	6	problem
__init__.py	Module error	7	utils
dwave.py	Module error	7	dwave_sapi2.core
dwave.py	Module error	8	dwave_sapi2.embedding
dwave.py	Module error	9	dwave_sapi2.local
dwave.py	Module error	10	dwave_sapi2.remote
dwave.py	Module error	11	dwave_sapi2.util
output.py	Module error	9	dwave_sapi2.util
problem.py	Module error	7	dwave_sapi2.util
setup.py	Module error	10	setuptools.command.install
globals.py	Module error	10	progname
globals.py	Module error	13	sym2num
globals.py	Module error	16	next_sym_num
globals.py	Module error	19	program
globals.py	Module error	22	chain_strength
globals.py	Module error	23	pin_strength
parse.py	Annotation variable	145	macros = { }
parse.py	Annotation variable	146	current_macro = (None, [])
parse.py	Annotation variable	147	aliases = { }
parse.py	Module error	148	program

- qmasm-1.0\build\lib\qasm\problem.py
- qmasm-1.0\qasm\problem.py
- qmasm-1.1\build\lib\qmasm\problem.py
- qmasm-1.1\qmasm\problem.py
- qmasm-1.2\build\lib\qmasm\problem.py
- qmasm-1.2\qmasm\problem.py

Although Py-find-injection was able to read the code blocks of all versions, no data representing any threats was retrieved. That might be because the tool needs to be configured for particular analysis scenarios.

Mypy resulted in high severity threats and problems in all QMASM releases. There are 22 errors of missing modules and unable to be located during debugging the code, as well as three variables that needed annotations. The examples of errors related to modules and annotation variables are shown in Table 19.2.

Debugging all QMASM releases resulted in high impact code mistakes that need patches. By performing a logical scan on QMASM source codes, Pyflakes flagged six infected Python files that need to be corrected to avoid serious threats. The errors commonly are invalid syntax, unused import variables, and undefined object names (Table 19.3). There are 18 unused import variables, two invalid syntax errors related to printing solution energy example and 16 undefined object names as seen in Fig. 19.1. The Python

Table 19.3 Pyflakes analysis results of QMASM v1.2

Python file name	Block type	LOC	Rank	Risk	Description
qmasm\output.py\	Function	147	C	Moderate – slightly complex block	output_minizinc
	Function	97	C	Moderate – slightly complex block	output_qbsolv
	Function	22	C	Moderate – slightly complex block	coupler_number
	Function	41	C	Moderate – slightly complex block	output_qubist
build\lib\qmasm\output.py\	Function	147	C	Moderate- slightly complex block	output_minizinc
	Function	97	C	Moderate – slightly complex block	output_qbsolv
	Function	22	C	Moderate – slightly complex block	coupler_number
	Function	41	C	Moderate – slightly complex block	output_qubist
qmasm\parse.py\	Function	149	E	High – complex block, alarming	parse_file
	Function	311	C	Moderate – slightly complex block	parse_lhs
build\lib\qmasm\problem.py\	Function	117	D	More than moderate – more complex block	convert_chains_to_aliases
build\lib\qmasm\parse.py\	Function	149	E	High – complex block, alarming	parse_file
	Function	311	C	Moderate – slightly complex block	parse_lhs
qmasm\problem.py\	Function	117	D	More than moderate – more complex block	convert_chains_to_aliases
build\lib\qmasm\dwave.py\	Function	49	D	More than moderate – more complex block	find_dwave_embedding
qmasm\dwave.py\	Function	49	D	More than moderate – more complex block	find_dwave_embedding
qmasm\output.py\	Module	NAN	B	Low – well structured and stable block	NAN
build\lib\qmasm\output.py\	Module	NAN	B	Low – well structured and stable block	NAN
build\lib\qmasm\problem.py\	Module	NAN	B	Low – well structured and stable block	NAN
qmasm\problem.py\	Module	NAN	B	Low – well structured and stable block	NAN
build\lib\qmasm\dwave.py\	Module	NAN	B	Low – well structured and stable block	NAN
qmasm\dwave.py\	Module	NAN	B	Low – well structured and stable block	NAN
qmasm.py	Invalid syntax	343	B	Low – well structured and stable block	print
build\scripts-3.5\qmasm.py\	Invalid syntax	343	B	Low – well structured and stable block	print

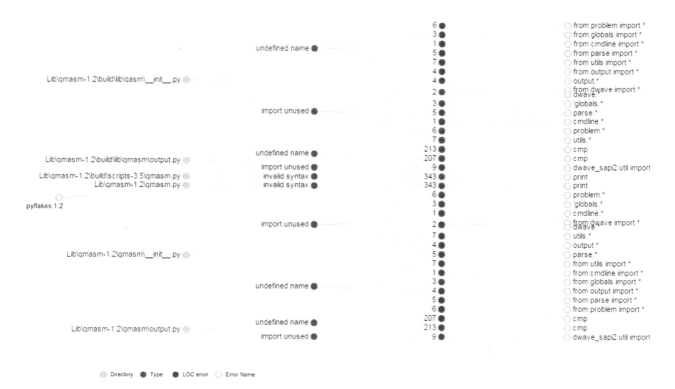

Fig. 19.1 Sample analysis visualization of pyflakes debugging tool for code errors found in QMASM latest version 1.2

files are found in all releases within the following directories:

- qmasm-1.2\qmasm.py
- qmasm-1.2\build\lib\qmasm\output.py
- qmasm-1.2\build\lib\qasm__init__.py
- qmasm-1.2\build\scripts-3.5\qmasm.py
- qmasm-1.2\qmasm\output.py
- qmasm-1.2\qmasm__init__.py

Furthermore, Vulture resulted in same errors in all QMASM versions in the following directory qmasm-1.2\build\lib\qmasm. The infected Python files are *cmdline.py*, *dwave.py*, *output.py*, *parse.py* and *problem.py*. These Python files contain 14 unused function names and one unused class name. Moreover, the difference in the latest releases of QMASM v1.2, is that there are two specific unused features that has not been found in other versions. One resides in *dwave.py* with unused function name *solution_is_intact*, and the other in *problem.py* with unused function name *find_disconnected_variables*. Note that the latest release contains unused class name in *problem.py* with the name *Problem*. An example of the detailed errors is specified in Table 19.4.

Xenon's command line resulted in different threshold complexity across all the code base of QMASM versions.

The actual threshold values described as the following options:

- -a, --max-average: Threshold for the average complexity across all the codebase.
- -m, --max-modules: Threshold for modules complexity.
- -b, --max-absolute: Absolute threshold for block complexity.

All of these options are inclusive, and can be combined into single command condition as follow:

```
C:\Users\User>$ xenon -b B -m A -a A C:\Users\User\qmasm-1
```

The above command analyzes Python blocks and computes Cyclomatic Complexity based on Randon. Xenons uses Radon to analyzes the AST tree of QMASM tool and compute the Cyclomatic Complexity. Every block ranked from A (best complexity score) to F (worst one) [35]. All ranks correspond to complexity scores, see Table 19.5 for complexity scores.

Xenon's results indicated that the highest threshold complexity error rate was found in QMASM version 1.2. The results show that there are eight risks of modules failure of rank B, and ten threats moderated as complex functions of rank C. Additionally, there are four threats in functions of rank D and two highly complex functions with rank E.

Table 19.4 Vulture results of QMASM v1.2 infected Python files using logical error scan

Infeacted Python File Name	LOC	Type of Error	Description
cmdline.py	10	Unused function	parse_command_line
dwave.py	19	Unused function	connect_to_dwave
dwave.py	137	Unused function	embed_problem_on_dwave
dwave.py	171	Unused function	update_strengths_from_chains
dwave.py	180	Unused function	scale_weights_strengths
dwave.py	200	Unused function	solution_is_intact
dwave.py	215	Unused function	submit_dwave_problem
output.py	113	Unused function	write_output
parse.py	259	Unused function	parse_files
problem.py	22	Unused class	Problem
problem.py	32	Unused function	assign_chain_strength
problem.py	56	Unused function	assign_pin_strength
problem.py	69	Unused function	pin_qubits
problem.py	113	Unused function	convert_chains_to_aliases
problem.py	199	Unused function	find_disconnected_variables

Table 19.5 Xenon ranks corresponding to complexity scores

CC score	Rank	Risk
1–5	A	Low – simple block
6–10	B	Low – well structured and stable block
11–20	C	Moderate – slightly complex block
21–30	D	More than moderate – more complex block
31–40	E	High – complex block, alarming
41+	F	Very high – error-prone, unstable block

We concluded that the branch of "high" and "more than moderate" is harder to understand and change reliably than code in which the branching is low. Moreover, code blocks contain unstructured stable modules and functions that need to be less moderated to increase stability and reduce the complexity of the tool. Analysis results illustrated by Table 19.2.

19.5 Conclusions and Future Work

D-wave computer system architecture allows small programs to be computed using quantum physics rules. Python tool QMASM allows programmers to investigate arbitrary computations by integrating the tool into D-Wave systems or by using a local simulator. Moreover, the QMASM contains very high technical scripts, which add on the ability and flexibility of executing computations and commands on robust systems such as D-Wave 2X. Currently, over-the-counter tools are built to aid programmers and examiners to analyze and debug the source codes for any potential errors that may cause systems and hardware to crash. Ignoring the fact that some vital errors are often left out while programming, white-boxing source-codes are essential to run the software in a high confidence environments.

In this research, most of the identified functions and modules, though well structured, are either undefined or unused. It is hard to apply fixes, because QMASM is highly dependent on D-Wave's proprietary SAPI library (programmed in C++) and will not work properly without it. Some of the vulnerability analysis tools used in this paper's did yield serious and important issues in the Python scripts that prompt fast fixes. In addition, they identified errors with line of code (LOC), and further secure coding is needed to ensure the reliability of the tool and fix the security flaws found.

The paper also presented technical aspects of analyzing Python scripts and provides knowledge of strength and limitations of tools. The work gives developers in-depth insight and expertises onto the right tool to use during debugging. We're currently working on vulnerability assessment of recently released Quantum computing software tools including IBM Qiskit [36] and Google Cirq [37].

Acknowledgements This work was supported by Zayed University Research Office, Research Incentive Fund #R18054.

References

1. Spencer, B., Al-Obeidat, F.: Temperature forecasts with stable accuracy in a smart home. Proc. Comput. Sci. **83**, 726–733 (2016), The 7th International Conference on Ambient Systems, Networks and Technologies (ANT 2016)
2. Aldwairi, M., Aldhanhani, S.: Multi-factor authentication system. In: The 2017 International Conference on Research and Innovation in Computer Engineering and Computer Sciences (RICCES'2017). Malaysia Technical Scientist Association, Aug 2017
3. Yaseen, Q., Aldwairi, M., Jararweh, Y., Al-Ayyoub, M., Gupta, B.: Collusion attacks mitigation in internet of things: a fog based model. Multimed. Tools Appl. **77**(14), 18249–18268 (2018)
4. Al-Ayyoub, M., Jararweh, Y., Rabab'ah, A., Aldwairi, M.: Feature extraction and selection for arabic tweets authorship authentication. J. Ambient. Intell. Humaniz. Comput. **8**(3), 383–393 (2017)
5. Nielsen, M.A., Chuang, I.: Quantum Computation and Quantum Information. Cambridge University Press, Cambridge (2002)
6. Bernhardt, C.: Quantum Computing for Everyone. The MIT Press, Cambridge (2019). ISBN-10: 0262039257, ISBN-13: 978-0262039253
7. Sharma, K.: Understanding quantum computing. IJSEAS **1**(6), 370–388 (2015)
8. Nagy, M., Akl, S.G.: Quantum computation and quantum information. Int. J. Parallel Emergent Distrib. Syst. **21**(1), 1–59 (2006)
9. Hsu, J.: D-wave's year of computing dangerously [news]. IEEE Spect. **50**(12), 11–13 (2013)
10. Mohammad, M., Lin, E.B., Darweesh, A., Howari, F.: Special b-spline tight framelet and it's applications. J. Adv. Math. Comput. Sci. **29**(5), 1–186 (2018). http://www.journaljamcs.com/index.php/JAMCS/article/view/24142
11. Karimi, S., Ronagh, P.: A subgradient approach for constrained binary programming via quantum adiabatic evolution. arXiv preprint arXiv:1605.09462 (2016)
12. Booth, M., Dahl, E., Furtney, M., Reinhardt, S.P.: Abstractions considered helpful: a tools architecture for quantum annealers. In: High Performance Extreme Computing Conference (HPEC), pp. 1–2. IEEE (2016)
13. Glover, F.: Future paths for integer programming and links to artificial intelligence. Comput. Oper. Res. **13**(5), 533–549 (1986)
14. Mohammad, M., Lin, E.-B.: Gibbs effects using daubechies and coiflet tight framelet systems. Contemp. Math. **01**, 271–282 (2018)
15. Pakin, S.: A quantum macro assembler. In: High Performance Extreme Computing Conference (HPEC), pp. 1–8. IEEE, (2016)
16. Rieffel, E.G., Venturelli, D., O'Gorman, B., Do, M.B., Prystay, E.M., Smelyanskiy, V.N.: A case study in programming a quantum annealer for hard operational planning problems. Quantum Inf. Process **14**(1), 1–36 (2015)
17. D-Wave-Systems: The d-wave 2x™ quantum computer technology overview. Available at https://www.dwavesys.com/sites/default/files/D-Wave. Accessed 29 Apr 2017
18. AlRoum, K., Alolama, A., Kamel, R., Barachi, M., Aldwairi, M.: Detecting malware domains: a cyber-threat alarm system. In: Proceedings of the First International EAI Conference on Emerging Technologies for Developing Countries. AFRICATEK, Marrakech, vol. 10, 27–28 Mar 2017
19. NIST: Vulnerabilities detail cve-2011-4944. [Online]. Available: https://nvd.nist.gov/vuln/detail/CVE-2011-4944 (2013)
20. Aldwairi, M., Noman, H.: A zero-day attach exploiting a yahoo messenger vulnerability. Int. J. Sci. Eng. Res. **3**(8), 1–4 (2012)
21. Cannon, B., Wohlstadter, E.: Controlling access to resources within the python interpreter. Proc. Second EECE **512**, 1–8 (2010)
22. Yip, A., Wang, X., Zeldovich, N., Kaashoek, M.F.: Improving application security with data flow assertions. In: Proceedings of the ACM SIGOPS 22nd Symposium on Operating systems principles, pp. 291–304. ACM (2009)
23. Anaconda: Anaconda for windows. [Online]. Available: https://www.continuum.io/downloads (2016)
24. Pakin, S.: Quantum macro assembler. [Online]. Available: https://github.com/lanl/qmasm (2016)
25. Jovanovic, N., Kruegel, C., Kirda, E.: Static analysis for detecting taint-style vulnerabilities in web applications. J. Comput. Secur. **18**(5), 861–907 (2010)

26. Jovanovic, N., Kruegel, C., Kirda, E.: Pixy: a static analysis tool for detecting web application vulnerabilities. In: 2006 IEEE Symposium on Security and Privacy (SP'06), pp. 263–268, May 2006

27. Hinds, L., Jaeger, A.: Bandit: Python ast-based static analyzer. [Online]. Available: https://github.com/openstack/bandit (2015)

28. Fisher, D.: Mypy: static type checker for python source codes. [Online]. Available: https://github.com/python/mypy (2017)

29. Brown, J.: Py-find-injection: Sql injection python source code. [Online]. Available: https://github.com/uber/py-find-injection (2013)

30. Newton, E., Shue, J., Norwitz, N.: Thomasvs. Pychecker: static analysis tool for finding bugs in python source code. [Online]. Available: https://sourceforge.net/projects/pychecker/ (2010)

31. pyflakes: Pyflakes: python static analysis errors debugging tool. [Online]. Available: https://github.com/PyCQA/pyflakes (2016)

32. Seipp, J.: Vulture: python static analysis tool that find dead classes and mdules. [Online]. Available: https://github.com/jendrikseipp/vulture (2016)

33. Lacchia, M.: Xenon: python static analysis monitoring tool based on radon. [Online]. Available: https://github.com/rubik/xenon (2014)

34. Pakin, S.: Qmasm releases. [Online]. Available: https://github.com/lanl/qmasm/releases (2016)

35. Lacchia, M.: Radon documentation. [Online]. Available: https://media.readthedocs.org/pdf/radon/latest/radon.pdf (2017)

36. Bello, L., Challenger, J., Cross, A., Faro, I., Gambetta, J., Gomez, J., Javadi-Abhari, A., Martin, P., Moreda, D., Perez, J., Winston, E., Wood, C.: Ibm qiskit terra v0.6. [Online]. Available: https://qiskit.org/ (2018)

37. Gidney, C., Bacon, D., Babbush, R., Omole, I., Mruczkiewicz, W., Arya, K., Duckering, C.: Cirq. [Online]. Available: https://github.com/quantumlib/Cirq/blob/master/README.rst (2018)

Part II

Software Testing

Christopher de Souza Lima Francisco and Adler Diniz de Souza

20.1   Introduction

The Earned Value Management (EVM) technique has been applied in several projects over the last 40 years [1]. This technique had a positive influence on several aspects related to project results, such as: improved planning, risk assessment, monitoring, reporting, control, among others [2]. However, few studies have been conducted with the purpose of analyzing and evaluating the stability of cost and time performance indicators, taking into account the quality of the project being produced. Although EVM has been used by several companies for more than 35 years to predict results of time and cost, many studies such as [3] found vulnerabilities.

Amongst the vulnerabilities pointed out by these studies is the lack of integration of quality data to the EVM technique. In order to solve this problem, articles were published trying to contribute to the effort to include quality data in the EVM technique, such as [4–6] and [7]. Solomon [8] indicates the use of Capability Maturity Model Integration to strengthen the adhesion of EVM, especially related to quality assurance. [6] proposed the integration of critical quality metrics to EVM. [4] shows a set of principles and guidelines that specify effective technical performance measures to use in conjunction with EVM. The lack of quality measures in the EVM technique can cause erroneous projections and contribute to the delivery of projects out of time, out of the cost and without conformity to the needs of the clients. An indicator, for example Cost Performance Index (CPI) or Schedule Performance Index (SPI), in isolation may not convey the actual state of the project if the number of bugs or nonconformities is higher than expected and the project man-

ager is unable to obtain this measure causing erroneous projections. Wrong projections can pass optimistic results and postpone the execution of corrective or preventive actions that would improve the final performance of the projects, avoiding delays and costs higher than the ones estimated [9]. There are many elements to a project: (i) requirements, (ii) schedule, (iii) cost, (iv) quality, (v) human resources. Projects can be complex and difficult to manage in order to achieve positive results, even those considered "small." Most of the life cycle of a project occurs during the development/execution phase. Training, professional meetings and conferences do not deposit enough energy to the methods and techniques to prepare the project manager to perform follow-up and performance reports. There is insufficient focus to address performance measures and indicators, or use them to control the project [3]. The real business environment, with few resources and high competition, requires efficient management tools to deliver projects on time, on budget, and with quality. In this scenario EVM is recognized as an efficient tool to measure performance and provide feedback to the project in progress. However, the methodology does not include the quality component in its method [6, 10].

In order to develop or evolve new EVM techniques that take into account the quality component, it is necessary to characterize the main articles that study in some way the EVM technique integrated with quality. For this reason, a study based on a systematic review was carried out to reduce the bias of an informal review and also to allow such bibliographic research to be updated with new publications made available over time. Thus, a study was carried out based on the integration of the quality component into EVM. Different data sets used in several researches were analyzed to find the main contributions and quality measures added to EVM. Therefore, techniques that relate quality to EVM have been identified and have been applied, through case studies or simulations, to prove their effectiveness. Based on

C. de Souza Lima Francisco (✉) · A. D. de Souza
Federal University of Itajubá, Itajubá, Minas Gerais, Brazil
e-mail: christopher@inatel.br

© Springer Nature Switzerland AG 2019
S. Latifi (ed.), *16th International Conference on Information Technology-New Generations (ITNG 2019)*,
Advances in Intelligent Systems and Computing 800,
https://doi.org/10.1007/978-3-030-14070-0_20

this identification, it was possible to perform an analysis of the main quality measures used by researchers and software developers that integrates the quality component into the EVM method.

It is of interest to identify studies that point to problems in the traditional EVM technique, or to compare it with other proposals, and to identify evolutions of it in order to increase predictability and reliability of the generated performance indicators. Therefore, all the studies that proposes evolutions of the technique must have been validated, through case studies or simulation that prove the efficiency of the new proposal. The problem that will be addressed in this research is the lack of the quality component in the cost and schedule results of the software projects that use the EVM technique.

The success of a project is one of the most discussed topics in the field of project management [11], but it is the least agreed problem [12, 13]. The success of the project refers to the achievement of the objectives of cost, time, scope [14] and quality [4, 6, 8, 10, 13, 15]. According to management literature, what can not be measured can not be managed [16]. Project success can be measured using Key Performance Indicators (KPIs) – which include, but are not limited to, cost, time, quality, safety and stakeholder satisfaction [11].

This article aims to analyze reports of experience and scientific publications through a study based on systematic review. As well as identify problems and proposals for evolution or improvement in the EVM technique taking into account the quality component. The final objective is to answer the primary question: "Are there any quality measures that relate to the EVM technique that may be used in the future to improve the performance of its traditional indicators (CPI – Cost Performance Index and SPI – Schedule Performance Index) in software projects?". Where Schedule Performance Index (SPI) indicates how efficiently you are actually progressing compared to the planned project schedule and Cost Performance Index (CPI) helps to analyze the efficiency of the cost utilized by the project. It measures the value of the work completed compared to the actual cost spent on the project.

20.2   Selection Procedures and Criteria

Digital sources will be accessed via the Web, through pre-established search expressions. If it is not possible to obtain the complete article through the search sites, the authors of the articles will be contacted via e-mail. Publications of the non-digital sources will be analyzed manually, when available, considering the defined search expression. For articles in English, the search expression below will be used: ("Earned Value Management" AND Quality). When the search is manual, the keywords present in the search expression should be in the titles and abstracts of the articles. Only publications that describes proof of concept and/or experience reports will be accepted.

Publications should be excluded if the Exclusion Criteria (EC) matches:

- EC1-01 – Publications will not be selected where the keywords are not present in the publication and there are no variations of these keywords (except plural).
- EC1-02 – Publications will not be selected in which the keywords of the search do not appear in the title, abstract and/or text of the publication.
- EC1-03 – Publications that describe and/or present keynote speeches, tutorials, courses, workshops will not be selected.
- EC1-04 – Publications will not be selected in which the acronym EVM does not mean "earned value management".
- EC1-05 – Publications will not be selected in which the acronym CPI does not mean "cost performance index".
- EC1-06 – Publications in which the SPI acronym does not mean "schedule performance index" will not be selected.
- EC1-07 – Publications in which the acronym EAC does not mean "estimated at completion" will not be selected.
- EC1-08 – Publications in which the TAC acronym does not mean "time at completion" will not be selected.
- EC1-09 – Publications in which TCPI does not mean "to cost performance index" will not be selected.
- EC1-10 – Publications that present tools to support EVM will not be selected. Except if the tool presents quality data integrated into EVM.
- EC1-11 – Publications that show the adherence of a particular project management approach to EVM will not be selected.
- EC1-12 – Publications that describe improvements in EVM by inserting the quality component but do not present subsidies that allows the identification of whether if it was applied or simulated.
- EC1-13 – Publications that describe the use of EVM, such as project control, or sub-processes of specific domains will not be selected.
- EC1-14 – Publications that simply cite EVM as a monitoring and control technique or that explain how to use it as a monitoring and control technique will not be selected.
- EC1-15 – Publications that EVM has not been implemented integrated with quality measures will not be selected.

Publications can only be included if the Inclusion Criteria (IC) matches:

- IC1-01 – Publications that mention EVM and quality can be selected.

- IC1-02 – Publications that describe new cost and schedule control techniques, using or integrated with quality measures and indicators can be selected.
- IC1-03 – Publications that discuss the stability of the schedule and cost performance indicators with the quality component can be selected.
- IC1-04 – Publications describing extensions of the EVM technique (not improving the predictability of the Estimate at Completion (EAC) and Time at Completion (TAC) indicators) aligned with the quality component can be selected.
- IC1-05 – Publications describing proposals for extensions of the EVM technique can be selected, aiming to improve the predictability of Estimate at Completion EAC and Time at Completion (TAC) considering the quality component.
- IC1-06 – Publications highlighting the most common quality-related EVM problems can be selected.
- IC1-07 – Publications that present statistical methods or statistical process control, applied to EVM, may be selected.
- IC1-08 – Publications comparing traditional EVM with new quality control techniques can be selected.
- IC1-09 – Publications that add new quality variables to EVM can be selected to improve the predictability of Estimate at Completion (EAC) and Time at Completion (TAC).

20.3 Data Extraction Procedures

Data extracted from the selected publications should be stored in a database and should contain:

- Title,
- Author(s),
- Publication Date,
- Conference or Journal,
- Data from the characteristics of interest stated in the study objective:
  - General: Represents a category of general items about the article.
    * Objective: Article's objective description.
  - Quality measures integrated into EVM, how they were integrated and if results were obtained in real or simulated projects:
    * Conclusion: Conclusion on the integration of quality measures into EVM (measures integrated, not integrated or if it can not be observed).
    * Justification: Justification for the conclusion found.

* Context:
  · Project area: Information Technology, Aerospace, Aviation, Civil Construction, Petrochemical, among others.
- Researcher's additional comments:
- Publication's contribution (from 01 (very bad) to 05 (excellent)).

20.4 Scope Definition and Preliminary Studies

The first stage of planning was the prospection on the topic of interest to the study. The objective of the study was to identify and characterize possible problems when using the EVM technique related to quality and the solutions indicated to solve these problems. Keyword options related to EVM and "earned value and quality" and their synonyms (such as earned value management and quality, earned value and quality control, earned value management and quality control, amongst others) were tested. Key words and synonyms were tested in singular and plural. The results obtained with the preliminary searches showed good results on SCOPUS library (http://www.scopus.com/home.url) for the subject under study, returning 48 publications, of which only 8 were false positives (publications not related to the research topic). Preliminary results on Compendex library (http://www.engineeringvillage.com/) and IEEE (http://www.ieee.org/portal/site), brought a list of 20 articles by Compendex and 34 by IEEE, being that 13 articles found on Compendex and 10 found on IEEE search engines were also present on SCOPUS and only 2 publications were false positives.

As the objectives in this first step were: (i) to define the scope of the topic of interest and (ii) to carry out preliminary studies on the subject, this activity met its purpose.

20.4.1 Identification of Control Publications and Keywords

At first, the database of articles was populated with about 55 papers from SCOPUS, COMPENDEX and IEEE that appeared as a result of the searches conducted in the tests of the preliminary studies (described in the previous section). Eight out of the 55 articles returned were classified as within the control group [10, 17–22] and [23]. The control group consists of articles that, in any way, reference the main point of study of this systematic review. The 8 articles included in the control group referred to EVM and quality, as well as other related topics such as cost and schedule.

At this point the tests began with the search expression:

(("Earned Value Management" AND Quality) OR ("Earned Value Management Technique" AND Quality) OR ("Earned Value Management Method" AND Quality) OR ("Earned Value Management Methodology" AND Quality) OR ("Earned Value Management" AND "Quality Control") OR ("Earned Value Management Technique" AND "Quality Control") OR ("Earned Value Management Method" AND "Quality Control") OR ("Earned Value Management Methodology" AND "Quality Control")).

But this search expression did not fit SCOPUS's advanced search and was broken in two in order to continue with the tests, resulting in the search expression below:

(("Earned Value Management" AND Quality) OR ("Earned Value Management Technique" AND Quality) OR ("Earned Value Management Method" AND Quality) OR ("Earned Value Management Methodology" AND Quality)).

This simplified search string worked well on SCOPUS, Compendex and IEEE search engines, resulting in 48, 20 and 34 publications respectively. An even simpler form of the previous string was obtained with exactly the same results on SCOPUS, Compendex and IEEE. The resulting search expression was:

("Earned Value Management" AND Quality).

In the second round of tests, the articles listed in the control group were reviewed and two more publications [4] and [15] related to EVM integrated with quality were added for a total of 10 publications in the control group from a universe of 57 publications. Both papers found are indexed by the IEEE search engine.

For manual search, NDIA Systems Engineering Conference and Systems and Software Technology Conference were considered, which are indexed by the search engine of SCOPUS and IEEE. The selection of this journal for manual review was due to two important articles added to the control group being editions present in these journals.

20.5   Study Results

The objective of this systematic review was to answer the primary question: "Are there any quality measures that relate to the EVM technique that may be used in the future to improve the performance of its traditional indicators (CPI – Cost Performance Index and SPI – Schedule Performance Index) in software projects?".

A total of 10 articles met the selection criteria. Therefore, these were all articles that, to a certain extent, referred to EVM integrated with the quality component, as well as related topics such as cost and schedule control. Figure 20.1 shows the proportion of articles sources in relation to articles published by journals or conferences and Fig. 20.2 shows the number of publications per year.

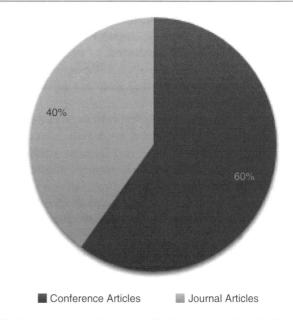

Fig. 20.1 Proportion between publications in Journals or Conference

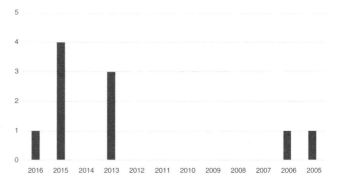

Fig. 20.2 Number of publications per year

The EVM technique uses two basic indicators – Cost Performance Index (CPI) and Schedule Performance Index (SPI) – to measure the past performance of projects [24]. To improve the conventional EVM technique and expand its focus and functionality, it is necessary to involve other indicators and take into account the uncertainty that the project may face in the future. These indicators may provide a better insight into the performance of other key aspects of the project.

In general, the studies carried out and presented in this systematic review suggests that the adoption of practices that integrate quality to EVM in the execution of projects, reach their goal in a more realistic way. This is in agreement with [4], which says that organizations that use this technique tend to achieve greater performance.

The results obtained after the studies showed that there are some ways to integrate the quality component into the EVM method effectively. The articles that contributed the most, proposed new equations capable of measuring quality with input data easily obtained at the beginning, middle and end of the execution of a project, such as the quantity of quality requirements – required by customers and also the quality performance indicator (QPI) – [10], as well as fuzzy quality indicators [17]. Other important and missing measures in the conventional method are also discussed: (i) Quality Variance, Quality Earned Value and Quality Index Number [10], (ii) Rework, Cost of Development [17]. The traditional EVM has a variation of 6.1% of the initial cost. After performing the rework, this variation rises to 27.7% of the real cost at the end of the project [17]. The new models presented by [17] and [10] present a more realistic variation of 7.9% and 8.5% of the real cost including rework. The present result in [17] was obtained by means of simulated projects in the area of computer science, while the results obtained by [10] were through tests in real projects in the agricultural area.

Performance-Based Earned Value (PBEV) and its extensions is introduced by [4], explaining that this is an improvement of the EVM system. PBEV is based on standards and models for Systems Engineering, Software Engineering and project management, so it can overcome the problems presented in the traditional EVM system related to technical performance and quality. The main difference of PBEV is the focus on customer requirements. An explanation of the patterns presented by [4, 8, 15] and [23].

An extension of the EVM technique integrating the history of quality performance as a means of improving cost predictability is proposed by [18]. The proposal is evaluated and compared with the traditional EVM technique through different hypothesis tests. This technique was validated through simulation in [22] and later tested with 23 real software projects in [18].

The QPI – Quality Performance Index – introduced by [18] is an indicator that shows how efficient the quality of a specific process is. Given date, the indicator shows whether the number of nonconformities (NC) is higher or lower than expected, allowing projections on future quality performance through the NC Estimate to Complete (NCEC). QPI is given by the following equation:

$$QPI = \frac{ENC(d)}{INC(d)} \qquad (20.1)$$

where:

- ENC (d): represents the total expected NC for a given date;
- INC (d) represents the total NC identified for a given date.

Values below 1 for the indicator means that more than the expected number of nonconformities are being found. Values above 1 indicate that fewer nonconformities are being en-countered. The objective of the quality performance indicator (QPI) is to predict the amount of future nonconformities, given current performance, and to assess the impact of quality on project costs. This article presents a proposal to extend the EVM technique that adds the quality component through several new equations that take into account the history of quality performance in projects. Several tests were conducted to evaluate the proposed technique and the results were satisfactory, showing that the new technique has more accuracy than the traditional EVM technique at the beginning (25% execution), during (50% execution) and end (75% execution) of a project [22] and [18].

20.5.1 Exploring Relationships Between Studies

The main articles studied by this systematic review introduce new variables that can be integrated into EVM in order to enhance its results related to cost and schedule forecast. Part of the study findings shows that these variable have a relationship that can be established between them, though this is not directly pointed out, it is possible to identify factors that proves this relationship. This is done by exploring the characteristics of the studies and their reported findings.

Quality Requirements, Compliance Cost and Noncompliance Cost

The Quality Requirements (QR) variable is introduced by [10]. QR is the quality requirements for a given task, the unit of QR may vary according to the project needs. This variable can be further explored by changing its unit to symbolize number of QRs or the cost associated with the number of QRs.

Part of the proposal from [22] includes quality costs (QC), which is introduced by [25] and consists of:

- Compliance costs (CC): costs that are allocated throughout the project for activities to prevent failures such as: (i) training, (ii) documentation of processes, (iii) tests and (iv) inspections.
- Noncompliance costs (NCC): costs that are allocated in the course and after the project execution, attributed to failures. They can be divided into two categories, internal failure costs, e.g.: (i) re-work, (ii) wastes and external failure costs, such as: (i) loss of reliability, (ii) product warranty and (iii) loss of market.

QR introduced by [10] can be related to Compliance Cost (CC) in the sense that each activity to prevent failure related to CC is one QR, by associating CC directly with QR the final result is QR expressed in monetary terms, that is, the cost of Quality requirements in one task or in a entire project.

Noncompliance Cost (NCC) [22, 25] can also relate do QR [10] when calculating total Quality Costs (QC):

$$QC = CC + NCC \quad (20.2)$$

$$QC = QR(cost) + NCC \quad (20.3)$$

Equations 20.2 and 20.3 shows that it is possible to merge the variables presented by [10, 22] and [25] to establish a relationship between them and further enhance their usage.

Quality Performance Index

Quality Performance Index (QPI) is used to indicate how efficiently the project, task or process is conducted. As stated by [10], QPI equals to 1 when all QR are met, and 0 if none are. For [22], given a certain date, the QPI shows if the number of noncompliance is higher or lower than expected. As explained in Sect. 20.5.1, Quality Requirements (QR), Compliance costs (CC) and Noncompliance costs (NCC) relate to each other, thus QPI presented by [10] and [22] have a relationship as well. The number of QR met and total QR can be used to calculate QPI.

$$QPI = QR(met)/QR \quad (20.4)$$

This relates to Eq. 20.1 in the sense that ENC(d) is the number of QR met given a certain date and INC(d) is the number of QR established on a certain date as well.

Quality Earned Value (QEV)

As defined by [10], Quality Earned Value (QEV) is used to measure the project ability to deliver the quality requirements defined by the project's stakeholder, throughout the project execution. It focuses on providing a snapshot of the project efficiency to deliver the project quality requirements based on the used time and spent money – Actual Cost (AC). This QEV variable reflects the earned value of the work that met the QR of the performed work. It can be calculated by multiplying the QPI by the Actual Cost (AC) expressed in monetary terms, as explained by [10]:

$$QEV = QPI * AC \quad (20.5)$$

As discussed in Sects. 20.5.1 and 20.5.1, QPI introduced by [10] and QPI introduced by [18] and [22] can be related, thus QEV can also be related to [18] and [22].

Quality Variance

Quality Variance (QV) [10] is used to describe cumulative quality efficiency of the project. The formula to estimate this value is:

$$QV = QEV - AC \quad (20.6)$$

QV is dependent on QEV and can be related to CC and NCC introduced by [18] and [22] as well.

Exploring Relationships Between Studies Conclusion

By exploring the most significant data of the articles present in the control group, it is possible to see that new variables that relate quality and EVM are introduced, such as the previously discussed Quality requirements (QR), Quality costs (QC), Compliance costs (CC), Noncompliance costs (NCC), Quality Performance Index (QPI), Quality Earned Value (QEV) and Quality Variance (QV) and also that Different authors are adding their contribution on how to improve the traditional EVM technique by incorporating quality measures.

By identifying patterns and exploring relationships emerging from the data during the preliminary studies it is possible to confirm a strong relationship between articles of the control group and their presented variables, providing relevant information on how to add new quality measures to EVM in order to achieve better overall project performance. Also it is possible to verify that these studies are, in different ways, aiming for the same goal, which is to add quality variables to EVM. All the variables presented in previous section can be related to one another and are introduced in different articles by different authors.

20.6   Conclusion

This systematic review aimed to contribute to the studies that are interested in enhancing the EVM method by including the quality component, and to answer the primary question: "Are there any quality measures that relate to the EVM technique that may be used in the future to improve the performance of its traditional indicators (CPI – Cost Performance Index and SPI – Schedule Performance Index) in software projects?". This research shows that incorporating the quality component is supposed to enhance the ability of the EVM in estimating the performance indicators and the future cost and schedule in a more reliable manner. The contributions to this study shown in this systematic review provides a better understanding on how to create new measures and apply them in real life or simulated projects. The primary question was answered and examples of new quality measures that relate to the EVM technique were demonstrated.

The quality variables introduced by articles in the control group and further discussed in Sect. 20.5, are going to be the the underlying set of facts and assumptions for the development of a proposal that will aim to integrate quality measures into the EVM technique and to improve the performance of its traditional indicators (CPI – Cost Performance Index and SPI – Schedule Performance Index) in software projects.

References

1. Lipke, W.: Project duration forecasting...a comparison of earned value management methods to earned schedule. Meas. News **21**(01), 24–31 (2008)
2. Lipke, W.: Statistical methods applied to EVM: the next frontier. CrossTalk **19**, 20–23 (2006)
3. Lipke, W., Zwikael, O., Henderson, K., Anbari, F.: Prediction of project outcome. Int. J. Proj. Manag. **27**(4), 400–407 (2009)
4. Solomon, P.J.: Performance-based earned value. In: INCOSE International Symposium, vol. 15 (2007)
5. Leu, S.-S., Lin, Y.-C., Chen, T.-A., Ho, Y.-Y.: Improving traditional earned value management by incorporating statistical process charts. In: 23rd International Symposium on Automation and Robotics in Construction (2006)
6. Yerabolu, R., Institute, P.M.: Framework for Integrating Project Quality, Risk Management, and Integration Management Disciplines Into Earned Value Management (EVM) for Deriving Performance Based Earned Value (PBEV) (2010), pp. 275–279. Tokyo, Japan (2006)
7. Ma, X., Yang, B.: Optimization study of earned value method in construction project management. In: 2012 International Conference on Information Management, Innovation Management and Industrial Engineering, vol. 2, pp. 201–204 (2012)
8. Solomon, P.: Using cmmi to improve earned value management. Technical Report CMU/SEI-2002-TN-016, Software Engineering Institute, Carnegie Mellon University, Pittsburgh (2002)
9. Lipke, W.: Schedule is different. PMI CPM J. Meas. News. **1**, 31–34 (2003).
10. Dodson, M., Defavari, G., de Carvalho, V.: Quality: the third element of earned value management. Proc. Comput. Sci. **64**, 932–939 (2015), Conference on ENTERprise Information Systems/International Conference on Project MANagement/Conference on Health and Social Care Information Systems and Technologies, CENTERIS/ProjMAN/HCist 7–9 Oct (2015)
11. Siddiqui, S., Ullah, F., Thaheem, M.J, Gabriel, H.: Six sigma in construction: a review of critical success factors. **7**, 06 (2016)
12. Thomas, G., Fernández, W.: Success in it projects: a matter of definition? Int. J. Proj. Manag. **26**(7), 733–742 (2008) Special Issue: Achieving IT Project Success
13. Baccarini, D.: The Logical Framework Method for Defining Project Success. Project Management Institute, Newtown Square (1999)
14. Ika, L.A.: Project success as a topic in project management journals. Proj. Manag. J. **40**(4), 6–19 (2009)
15. Solomon, P.J.: Pratical perfomance-based earned value. In: Systems and Software Technology Conference (2006)
16. DeMarco, T.: Controlling Software Projects: Management, Measurement, and Estimates. Yourdon Press, New York (1982)
17. Khalid, T.A.: Controlling software cost using fuzzy quality based EVM. In: International Conference on Computing, Control, Networking, Electronics and Embedded Systems Engineering (2015)
18. de Souza, A.D., Rocha, A.R.C., Cristina, D., Constantino, B.A.: A proposal for the improvement of project's cost predictability using earned value management and quality data – an empirical study, pp. 170–181. Springer, Berlin/Heidelberg (2014)
19. Salari, M., Bagherpour, M., Reihani, M.: A time -cost trade-off model by incorporating fuzzy earned value management: a statistical based approach. J. Intell. Fuzzy Syst. **28**, 11 (2014)
20. Iranmanesh, S.H., Hojati, Z.T.: Intelligent systems in project performance measurement and evaluation, pp. 581–619. Springer, Cham (2015)
21. Lipke, W.: Is something missing from project management? CrossTalk **26**, 16–20 (2013)
22. de Souza, A.D., Rocha, A.R.C.: A Proposal for the Improvement Predictability of Cost Using Earned Value Management and Quality Data, pp. 190–201. Springer, Berlin/Heidelberg (2013)
23. Solomon1, P.J.: Basing earned value on technical performance. CrossTalk **26**, 25–28 (2013)
24. Naeni, L.M, Shadrokh, S., Salehipour, A.: A fuzzy approach for the earned value management. Int. J. Proj. Manag. **29**(6), 764–772 (2011)
25. Putnam, L.H., Myers, W.: Five Core Metrics: Intelligence Behind Successful Software Management. Dorset House Publishing Co., Inc., New York (2003)

Urgent and Emergency Care: An Academic Application System Case Study

Daniela America da Silva, Fabio Kfouri, Samara Cardoso dos Santos,
Luiz Henrique Coura, Wilson Cristoni, Gildarcio Sousa Goncalves,
Leonardo Guimaraes dos Santos, Jose Crisostomo Ozorio Junior,
Breslei Max Reis da Fonseca, Jean Carlos Lourenco Costa, Juliana Pasquini,
Alexandre Nascimento, Johnny Marques, Luiz Alberto Vieira Dias,
Adilson Marques da Cunha, Paulo Marcelo Tasinaffo, Beatriz Perondi,
Anna Miethke-Morais, Amanda Cardoso Montal,
Solange Regina Giglioli Fusco, and Thiago Sakamoto

21.1 Introduction

This paper tackles the development of an academic project using the Collaborative Interdisciplinary Problem-Based Learning (Co-IPBL).

Two issues are emerging in health care as clinicians face the complexities of current patient care: the need for applying new technologies in health care management; and the need for these professionals to collaborate with professionals from engineering and computer science background.

Interdisciplinary health care teams with members from many professions usually answer calls, by working together, collaborating, and communicating closely to digitize patients' care [13, 15, 16].

This research work provides an integration of 3 different courses taught at the Brazilian Aeronautics Institute of Technology (Instituto Tecnológico de Aeronáutica – ITA): CE-240 Database Systems Projects, CE-245 Information Technologies, and CE-229 Software Testing. It involved some cooperative work with technical Physicians from Urgent and Emergency Health Care at the Hospital of Clinics from the Faculty of Medicine at the University of Sao Paulo, Brazil.

It also describes a practical application of collaborative and interdisciplinary concepts, by using the Scrum Agile Method [4, 5, 17].

This project was named in Portuguese as *"Soluções Tecnológicas Aplicáveis ao Gerenciamento de Informações Hospitalares Ostensivas com Big Data – STAGIHO-BD"*, meaning in English, "Technological Solutions Applicable for Managing Ostensive Hospital Information with Big Data – TSA4MOHIBD".

It has considered the development of a Software System for decision making support, involving Patients, Doctors, Hospitals, and Suppliers as actors. It encompassed the usage of: Scrum Agile Method; Value Engineering; Big Data; Blockchain Hyperledger, Software Quality, Reliability, Safety, and Testability [7–10, 12].

The best practices on Scrum Agile Method and Value Engineering were used, in order to assure computer system adherence to the project requirements in a time frame of just 17 academic weeks [2, 3].

This TSA4MOHIBD Project [18] was divided into two groups of application: External Regulation and Internal Regulation, by sharing its development among four student teams, which were responsible for developing different functional requirements involving the verification of quality, reliability, safety, and testability.

D. A. da Silva (✉) · F. Kfouri · S. C. dos Santos · L. H. Coura
W. Cristoni · G. S. Goncalves · L. G. dos Santos · J. C. O. Junior
B. M. R. da Fonseca · J. C. L. Costa · J. Pasquini · A. Nascimento
J. Marques · L. A. V. Dias · A. M. da Cunha · P. M. Tasinaffo
Brazilian Aeronautics Institute of Technology, Sao Jose dos Campos,
Sao Paulo, Brazil

B. Perondi · A. Miethke-Morais · A. C. Montal · S. R. G. Fusco
T. Sakamoto
Hospital of Clinics at the Faculty of Medicine, University of Sao
Paulo, Sao Jose dos Campos Sao Paulo, Brazil

21.2 The Urgent and Emergency Health Care in the HCFMUSP

Before starting the academic year, some students from the Brazilian Aeronautics Institute of Technology (ITA) had a

© Springer Nature Switzerland AG 2019
S. Latifi (ed.), *16th International Conference on Information Technology-New Generations (ITNG 2019)*,
Advances in Intelligent Systems and Computing 800,
https://doi.org/10.1007/978-3-030-14070-0_21

planning meeting with some members of the Emergency Care in the Hospital of Clinics from the Faculty of Medicine at the University of São Paulo (in Portuguese, *Hospital da Clínicas da Faculdade de Medicina de São Paulo –* HCFMUSP).

On that opportunity, some internal members of the hospital have presented some details of the internal and external regulation and there was also an on-site visit to some emergency areas and hospital sectors.

Selected members of the HCFMUSP technical team for the meeting were those who participate in the internal regulation of the hospital for patient care authorization, such as the director of internal regulation, and two physicians who coordinate the internal regulation.

Selected ITA members for the meeting were one PhD Candidate student and one Masters' degree student for the preparation of the requirements specification.

As a result it was possible to understand the characteristics of urgent and emergency service and to assign degrees of importance to each service attribute.

On top of most valuable attributes, to simplify the Proof of Concept and to reduce the scope and complexity of the system to be developed, an assigned mission was developed based on the rescue of motorcycle accident victims scenario.

The main reason of this choice of the mentioned scenario was the its high frequent of occurrence in the city of São Paulo [1, 11].

21.2.1 Regulations

Regulations of Hospitals and Health Units attending Urgencies and Emergencies are usually carried out by the Health Care Supply Regulatory Center (CROSS).

It is through the CROSS that the HCFMUSP use to be contacted, in order to respond to urgencies and emergencies.

An emergency patient may also arrive by the regular entrance of the HCFMUSP and after contacting the CROSS (usually via email or through the CROSS Web Portal). The HCFMUSP will screen through an Internal Hospital Regulation to filter relevant cases for hospital care.

21.2.2 The External Regulation

The Health Care Secretary of Sao Paulo understood the regulation as an important tool for the management of public health systems.

This regulation has among its objectives the equity of the access implemented through dynamic actions executed in an equitable, orderly, timely, and rational way, creating the CROSS.

The CROSS brings together actions aimed at regulating access in hospitals and outpatient areas, contributing to the integrality of the assistance and providing the adjustment of the available health care supply to the immediate needs of the citizen as described at http://www.cross.saude.sp.gov.br/ [6].

21.2.3 The Internal Regulation

The Internal Regulation is the area in the Hospital responsible for verifying availability to receive a Patient. Also, It makes the contact with the appropriate hospital institute to evaluate whether or not a Patient could be transferred to the Hospital.

21.3 The Agile Development

21.3.1 The Business Model Canvas

The Business Model Canvas is a visual management tool developed by Alex Osterwalder and Yves Pigneur in their book Business Model Generation, becoming it widely used for mapping the business model of a new project in a practical and clear way.

The main idea of Canvas is to allow project developers to elaborate a model in a free and creative form. It is said that the framework is dynamic and should be altered whenever necessary, however making evident the feasibility of such a model.

This Canvas tool consists of a framework that approach the four aspects of a new business project: (1) for whom the business is being developed, the segment of customers that may impact; (2) how it will be developed, an aspect that addresses issues such as available resources, main activities and partnerships; (3) how much investment will be required for the business to take place and the source of such investments; and, finally, (4) what value proposition the business will offer to its customers.

Regarding it is a project for a Software instead of a business enterprise, the use of the Canvas frame in STAGIHO-BD was adapted according to the need of developers to map essential aspects of feasibility and value delivery.

It was prepared by the Project Product Owner and Scrum Masters. Then, it was validated with project clients, as representatives of the Hospital das Clinicas, as shown in Fig. 21.1.

21.3.2 Sprints

The Scrum agile management method was adopted for the product development. The product was developed over 3

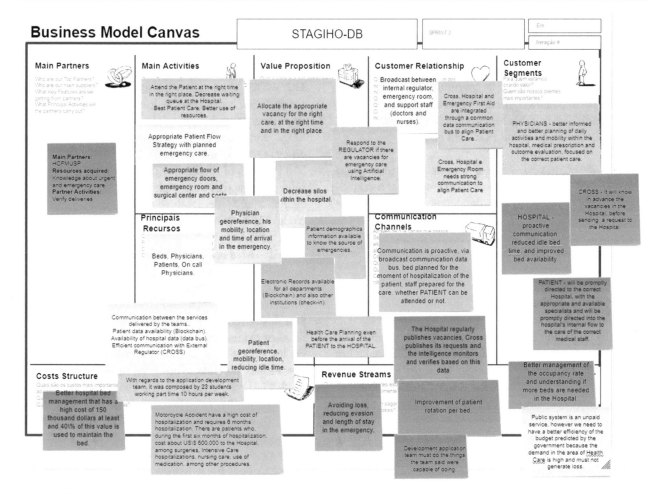

Fig. 21.1 The business model canvas

Sprints of 4 weeks each. It was developed by 4 Scrum teams supported by the Scrum Masters and one Product Owner.

Each the Scrum team met virtually once a week to check on each team the member's task development, as well as, the Scrum Master information sharing and the main developer's difficulties.

The Product Owner was responsible for directing teams to develop value solutions, according to the customer's main needs. Such needs were divided into Users Stories and then distributed over the teams.

Scrum Masters were responsible for keeping the synchronism and integration among the teams. Also, they ensured the proper teamwork and managed the activities developed by each member of each discipline involved in the project.

In the first Sprint, Users Stories were developed that aimed to communicate the involved institutions to identify the best places available for a motorcycle accident victim considering factors such as the distance of the hospital to the accident site, the severity of the injuries and the availability of specialized medical teams to perform care.

During the second Sprint, the development was focused on the patient flow already inside the hospital and on the share of relevant information to optimize this flow aiming a faster and more effective care.

The third Sprint was dedicated to the improvement of some micro-services, as well as the development of the Blockchain network and its integration with the API.

At the end of each Sprint, a Sprint Retrospective ceremony was held to encourage the sharing of opinions and suggestions from all project participants and to debrief on what worked well during Sprint and on improvements for the execution of the following Sprint.

21.4 The POC Demo as an Assigned Mission

To reduce the scope and complexity of the system to be developed, an assigned mission was developed to the rescue of motorcycle accident victims, mainly because it is a frequent occurrence in the city of Sao Paulo.

On a regular working day, every 15 min, the Emergency mobile care service (SAMU) assists a motorcyclist in the city of São Paulo.

Most of victims are not made up of couriers known as "motoboys". Usually, they are motorbikes driver using the motorcycles to go to work or are weekend drivers.

The rehabilitation of those who are injured is usually time-consuming and laborious, as described below:

- 40% of motorcycle accident victims need to undergo complex surgeries and long physiotherapy treatments;
- The most serious injuries caused by motorcycle accidents are usually in the skull and spine;
- Most motorcycle riders use the vehicle as transport, only for 2 h a day, usually to move between home and work place;
- Most of them have been injured before;
- Of the total number of accidents initially investigated, 2% resulted in life losses;
- Of the total number of injuries: 48% had serious injuries, 17% in the legs and feet; 12% in arms, and 23% had other types of trauma;
- Most of them were discharged immediately after care and only 18% had to be hospitalized; and
- Considering annual costs of about R$ 100 million are invested by the Orthopedics Institute of the HC, exclusively for the recovery of motorcyclists: "These are patients who, in the first six months of hospitalization, cost about R$ 300 thousands to the Hospital, on surgeries, ICU hospitalizations, occupations of wards, use of medications, among other procedures."

Based on previously reported motorcycle accident data, the development of the TSA4MOHIBD project should be able to provide an adequate management for the control of victim assistance only considering motorcycle accidents, in such a way that:

- Those motorcycle accident victims (PATIENTS) may be appropriately diagnosed, promptly identified, and/or attended to;
- PHYSICIANS may have computers and/or computer tools capable of providing preventive and appropriate planning, scheduling, and controlling of motorcycle accident emergency services, for example, by identifying the needs of: hospitalization period, procedures, medical teams, procedures for surgery preparations, as well as the availability of surgical rooms and Intensive Therapy Units;
- The HOSPITAL and/or the INTERNAL CONTROLLER shall have computers and/or computer tools capable of using appropriate technologies to provide efficient screening methods, prioritization of care, patient, and physician

locations within the HOSPITAL, in order to locate, if necessary, other HOSPITALS to attend emergencies of motorcycle accident victims, according to the severity and needs of specialized treatments for the PATIENTS;
- HOSPITAL must have adequate computers and/or computer means to collaborate in the process of managing large data flows, involving efficient care for injured motorcycle patients, by using appropriate technologies and efficient screening methods for this type of emergency;
- SUPPLIERS of medicines, devices, and/or technologies must have appropriate tools to participate in logistics and/or supply process which are used in care of motorcycle accident victims attended by urgency and emergency of HOSPITALS;
- PHYSICIANS and NURSES must have appropriate tools to provide care to motorcycle accident victims attended by the HOSPITAL emergency care, for example, to identify them from their entrance hall, by verifying and controlling also the movement of the stretchers, according to internal flows of required services;
- Each confirmed case of motorcycle-injured PATIENT can and should generate, within the inventory control of a HOSPITAL, the release of care kits containing materials and medications, according to the size of events. In these cases, in addition to medicines, other supplies can and should be considered important, such as blood bags for transfusions, as accident victims need quick rescue and the high risk of life losses because of blood losses;
- The population/society should have access to computers and/or computer tools of the System Project involving the TSA4MOHIBD and/or the Real-time TSA4MOHIBD capable of providing appropriate records, management, controls, and governance of resources used in the area of Health Care; and
- Public administration must have reliable data to provide a comprehensive situational awareness to support decision making for accidents and/or crises, involving motorcycle accidents.

21.5 The Final Product

In order to provide the proper speed of the communication to the Regulator and also to allow the Regulator to know in advance the number of bed vacancy in the Hospital, it was suggested a communication architecture between the Regulator and the Hospital similar to the AirBnB model.

Then, the hospital would publish the number of vacancy, physicians, and specialists available in a data bus communication. Also, the Regulator would publish an attendance request for a patient and its initial conditions in the data bus communication.

After that, an Artificial Intelligence algorithm would perform some analysis to identify what kind of treatment is necessary for the Patient, and to search among the Hospitals which one is the closest one, has vacancy and suitable resources to receive this Patient.

The Artificial Intelligence algorithm, after finding the appropriate Hospital, would place in the same data bus a message to the Hospital.

In the Hospital, the Internal Regulator operator would analyze the request and place a message in the Data Bus to inform the Artificial Intelligence algorithm about the Patient acceptance.

It would also inform straight the Urgency Care Unit that a Patient would arrive and the estimated date and time of arrival.

The patient information would be available in the Blockchain Hyperledger, and then data from the CROSS and data about the Patient will be internally matched in the Hospital. From that point onward, the planning for Patient Care would have been started prior to the Patient arrival.

When the Patient arrives, the technical team, the bed, the required examination, and health care steps would proceed as the suggested planning. After the Internal Controller acceptance, the intelligence of the system would place a message to the Regulator to inform if the Patient was accepted by the HCFMUSP.

After the Patient arrival in the Hospital, his Electronic Health Record could be read by any Institute or professional inside the HCFMUSP. The information about patient care steps as well as all the current bed management in the Hospital would be presented through a dashboard that could be available for the governance team, by physicians and nurses looking at the next steps of the Patient Care.

For the Urgency and Emergency Care at the HCFMUSP, our proposal is to use two separate Blockchain networks: one for Patient and other for Attendance.

The Blockchain Patient network has overall patient information and is a private network with information that can be accessed by the CROSS and any other Institution that would like in the future to access Patient Identification, e.g. Single Health System, in Portuguese, *Sistema Unico de Saude – SUS*.

The Blockchain Attendance network has detailed patient information and all the health care details performed by the Hospital. It is a private network with information that can be accessed only by the Hospital and internal institutes. For future work, examination details could be also stored in this Blockchain network.

The main benefit of isolating patient's data from patient's own care is to provide the data safety, because confidential information will be managed only by the hospital.

Another benefit was to allow that two different groups of students learning Blockchain could also learn independently. The integration has worked smoothly during the 3rd Sprint of the TSA4MOHIBD project, which has focused in the Integration.

The Blockchain Hyperledger implementation also makes use of encapsulated micro-services allowing smoother integration of funcionalities, which had been developed during the 1st Sprint (focusing in the External Flow implementation), and also during the 2nd Sprint (focusing in the Internal Flow Implementation).

21.6 Deliverables

The final product was created by 4 teams. Each team based on team members skills has focused on specific USs and all 4 teams have been working together since the 2nd Sprint to integrate the development.

21.6.1 The Agile Team 1

The Agile Team 1 was responsible for the creation of hospital micro services focused on vacancies, attendees, and specialists. It has developed micro services in NodeJS for the creation of reservation and publication of beds, attendees, and specialists available, as shown in Fig. 21.2.

For this, the following methods were implemented: POST, GET, PUT, and DELETE to insert, retrieve, edit, and delete hospital data. A communication integration interface was developed using Blockchain technology in hospital data transfer.

21.6.2 The Agile Team 2

The Agile Team 2 was responsible for bed management for patient accommodation upon request of the reservation service, as shown in Fig. 21.3. After the reservation, it's possible to recover, edit, and remove reservations.

This team was also responsible for creating the Patient Blockchain Business Network in the Hyperledger Composer Fabric Blockchain Architecture for: registering patient information; building the API in NodeJS for integration with the Hyperledger Composer Fabric Blockchain stagi-hobd.paciente through the GET, POST, DELETE, and PUT methods developed in Hyperledger; and integrating with a graphical web interface for viewing patient's Blockchain data and hash.

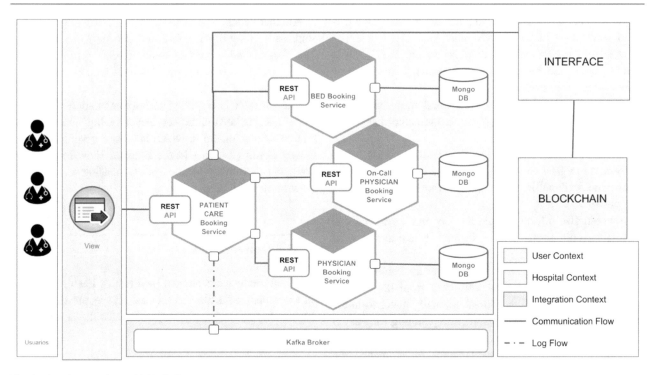

Fig. 21.2 Micro services with NodeJS

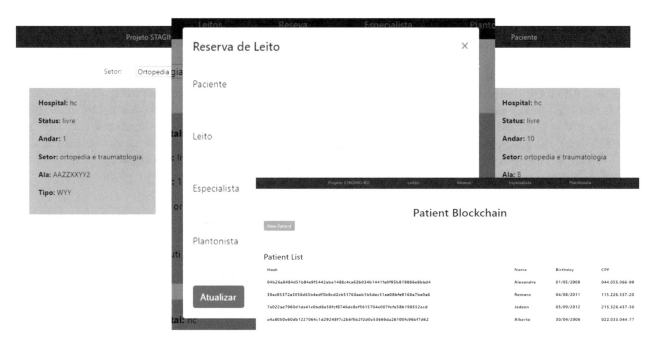

Fig. 21.3 The bed management

21.6.3 The Agile Team 3

The Agile Team 3 was responsible for the development of the Intelligence algorithm for the resource planning to the medical care inside the hospital and also for the conception of a dashboard.

To support all these applications it was constructed a streaming channel to link all stakeholders and to transform the Hospital in an active agent able to inform its available resources, instead of just be waiting for care needs.

The Intelligence algorithm was based in three variables: Urgency based on Manchester Protocol; body region with injure; and the type of injure.

Each of these variables were scored based in our experience, but could be better adjusted by Intelligence Artificial and also by medical board. The following Tables show how each variable was scored.

Id	Body region	Weight
1	Lower members	1
2	Upper limbs	1
3	Pelvis	2
4	Abdomen	3
5	Neck	3
6	Spine	4
7	Thorax	4
8	Head	5

Id	Damage type	Weight
1	Excoriation	1
2	Injury	1
3	Bruise	1
4	Dislocation	1
5	Contusion	2
6	Amputation	4
7	Trauma	5
8	Poly-trauma	5

Id	Protocol manchester	Weight
1	Not urgent	1
2	Little urgent	2
3	Urgent	3
4	Very urgent	4
5	Emergency	5

It was assumed as a premise that the body of emergency rescue could classify and provide these three variables to the system to perform a previous medical screening and so on, to search among the vacancies offered by several hospitals, which one is the nearest and if it contains resources available to receive the patient to a fast medical care.

The combination of these variables have allowed the building of an oracle that gave us an overview of medical diagnosis and the kind of tie breaker to set levels of priority for each patients, considering the hospital environment receiving different cases all the time. Once was defined the best choice, the system makes a reservation.

The resource planning is a solution applied inside the hospital, especially in the ICU (Intensive Care Unit) through an overview Panel.

After receiving a reservation, the system schedules resources like physicians, nurses, rooms, and others assigning how to attend the new patient, as shown in Fig. 21.4.

This schedule consider also other patients that are already hospitalized and who are under medical supervision. All patients information are enriched by data from the blockchain bringing enough data to provide agility and fluidity in the emergency care.

The building of the dashboard artifact made possible to follow planned versus realized occupancy rate of each resource, service time, waiting time, and other resources.

21.6.4 The Agile Team 4

The Agile Team 4 was responsible for creating technical solutions to work with geolocation information and creating a relational database about occurrences, patients, doctors and their specialties. The main developed requirements of team 4 were:

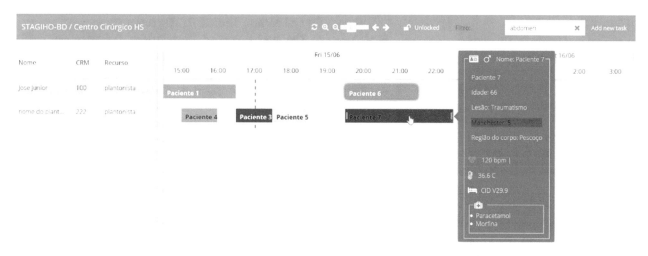

Fig. 21.4 The planning of patient care

Fig. 21.5 Geolocation capabilities

- To provide patient demographics to identify: the transport type used to take the patient into the hospital and the esteem time of his locomotion based on latitude and longitude of accidents and also the nearest hospitals, where patients could be moved to as shown in Fig. 21.5; and
- To provide the estimation time for patient stay at the hospital helping to verify quality of services to be provided. Besides, team 4 has developed an application for dealing with patient's and doctor's information, calculating distance and time between both in hours, minutes, or even seconds, and displaying graphic representations of their positions in time. This geolocation tool can be used also for showing the room where the patient can make an examination within a hospital.

21.7 Conclusion

The use of interdisciplinarity in 3 courses of Computer Science has worked as expected, aggregating data and integrating sectors such as External and Internal Regulations through its PATIENTS, HOSPITALS, PHYSICIANS, and HEALTH SUPPLIERS, for the decision making process related to Urgency and Emergency Care, involving motorcycle accidents.

The TSA4MOHIBD project was developed by students from three different Computer Science courses taught at the Brazilian Aeronautics Institute of Technology (*Instituto Tecnologico de Aeronautica* – ITA), on the 1st Semester of 2018.

This paper has described the development of a software system based on Big Data, Blockchain Hyperledger, Micro services, and other emerging technologies for governmental organizations and private sectors [14–16].

21.7.1 Specific Conclusions

The Scrum framework has been adapted to the reality of the interdisciplinary academic environment of ITA, helping 4 teams of more than 20 students to deliver value to stakeholders at the end of each sprint and also just one product at the end of the TSA4MOHIBD project.

The application of Test Driven Development (TDD) and Acceptance Testing Driven Development (ATDD) techniques in the project was closely related to the interdisciplinarity approach adopted, since acceptance tests were created and implemented by CE-229 Software Testing course students. Blockchain and NodeJS applications were implemented by CE-240 Database System Project and CE-245 Information Technologies course students.

The Blockchain Hyperledger, NodeJs, MongoDB, and MySQL Databases were hosted on the AWS cloud services and have represented the main tools applied for integrating services from External and Internal Regulations.

21.7.2 General Conclusions

The authors believe that is possible to integrate products generated with Blockchain Hyperlegder, from different teams and functional segments such as the Internal and External Regulation, with different technologies like Micro services, NodeJs, MongoDB and MySQL Databases.

This becomes possible considering a minimal organization effort between teams, by defining separated Blockchain networks with medical ostensive data to be exchanged between each blockchain network, allowing collective planning among all involved entities.

21.7.3 Recommendations

It is strongly recommended to align expectations and results that may be compromised and quickly adjusted to review deliverables, when working with emerging technologies [19, 20].

It is fundamental to have team members that could identify in advance where the team could fail using new technologies, proposing improvements, and allowing repositioning the team itself to deliver what is agreed in consensus on time.

It is also recommended the building of some tutorials to facilitate the Blockchain Hyperledger learning process, which can be used to quickly demonstrate new technologies' acquisition.

21.7.4 Future Works

For future works, it is suggested that the process used in this TSA4MOHIBD academic project prototype could be extended to other Blockchain Hyperledger projects and predictive systems, in order to improve estimation of efforts and resources to attend Urgency and Emergency Care.

It is also suggested to expand some cooperation among ITA, hospitals, innovation foundations, medical suppliers, industries, and public and private enterprises, in order to get a selection of academic projects aligned with updated market needs.

Acknowledgements The authors would like to thank: Hospital of Clinics from the Faculty of Medicine at The University of Sao Paulo; Brazilian Aeronautics Institute of Technology; Ecossistema Negocios Digitais Ltda; and Casimiro Montenegro Filho Foundation (FCMF), for their general and finantial support, during the development of this TSA4MOHIBD Project.

References

1. Bonacim, C., Araujo, A.: Cost management applied to public university hospitals, Public Administration Journal (2010). Accessed from www.scielo.br/pdf/rap/v44n4/v44n4a07.pdf in 10 June 2018
2. Ci&T: Presentation Ci&T Value Engineering Framework, Ci&T (2018). Accessed from https://www.ciandt.com/card/business-value-engineering-framework in 10 June 2018
3. Conceicao, A., Silva, F., Rocha, V., Locoro, A., Barguil, J.: Eletronic Health Records Using Blockchain, XXXVI Brazilian Symposium on Computer Networks and Distributed Systems (SBRC) (2018). Accessed from http://www.sbrc2018.ufscar.br/wp-content/uploads/2018/04/07-181717-1.pdf 10 June 2018
4. Cohen, D., et. al.: An introduction to agile methods, Fraunhofer center for experimental software engineering. Adv. Compu. **62**, 1–66 (2004)
5. Cohn, M.: Succeeding with Agile: Software Development Using Scrum, 1st edn. Pearson Education, Inc., Upper Saddle River (2010)
6. CROSS: Health care supply regulatory center (2018). Accessed from http://www.cross.saude.sp.gov.br/ 10 June 2018
7. FMUSP: Medical School of São Paulo (2018). Accessed from http://www.fm.usp.br/fmusp/portal/ in 10 June 2018
8. Forbes: This is why blockchain will transform healthCare (2017). Accessed from https://www.forbes.com/sites/bernardmarr/2017/11/29/this-is-why-blockchains-will-transform-healthcare/#13be90781ebe in 10 June 2018
9. Forbes: Business model canvas: a simple tool for designing innovative business models, Forbes (2012). Accessed from https://www.forbes.com/sites/tedgreenwald/2012/01/31/business-model-canvas-a-simple-tool-for-designing-innovative-business-models/ in 10 June 2018
10. GitHub: Open-source landscape map for healthcare-related blockchain (2018). Accessed from https://github.com/acoravos/healthcare-blockchains in 10 June 2018
11. Lima, M.A.B.: Target costing in public hospital services, Accounting Journal (2013). Accessed from https://periodicos.ufpe.br/revistas/ricontabeis/article/download/7966/8040 in 10 June 2018
12. Linux Foundation Projects: Blockchain Hyperledger (2018). Accessed from https://www.hyperledger.org/ in 10 June 2018
13. Martins, J.C., Mancilha, A.F.P., Basseto, E.E.J., Goncalves, G.S., Louro, H.D.B., Gomes, J.M., Filho, L.A.L., Coura, L.H.R.S., Rodrigues, R.A., Neto, W.C., da Cunha, A.M., Dias, L.A.V.: Using Big Data, Internet of Things, and Agile for Crises Management. In: 14th International Conference on Information Technology: New Generations (ITNG 2017), Las Vegas (2017)
14. Pugh, K.: Lean-Agile Acceptance Test-Driven Development: Better Software Through Collaboration. Addison-Wesley (2011). ISBN 978-0321714084
15. Silva, D.A., Goncalves, G., Santos, S., Pugliese, V., Navas, J., Santana, R., Queiroz, F., Vieira L.A., Cunha, A., Tasinaffo, M.: Health care information systems: a crisis approach. In: 15th International Conference on Information Technology: New Generations (ITNG), Las Vegas (2018)
16. Silva, D.A., Santana, R., Navas, J., Goncalves, G., Vieira, L.A., Cunha, A., Tasinaffo, M.: Health care transformation: an academic application system case study. In: 10TH IFAC Symposium on Biological and Medical Systems – International Federation of Automatic Control (IFAC BMS 2018), São Paulo (2018)

17. Sutherland, J.: SCRUM Handbook. Scrum Training Institute Press (2010). Available online at https://www.researchgate.net/publication/301685699_Jeff_Sutherland's_Scrum_Handbook

18. TSA4MOHIBD Project: Technological solutions applicable for the management of ostensive hospital information with big data (2018). Accessed from https://sites.google.com/site/projetointerdisciplinar2018/ in 10 June 2018

19. Vieira, P.: Engineering and value analysis: a question of efficiency and survival, e-Disciplinas University of São Paulo (2018). Accessed from https://edisciplinas.usp.br/mod/resource/view.php?id=891829 in 10 June 2018

20. WHO: Emergency Care System Framework Infographic, WHO (2018). Accessed from http://www.who.int/emergencycare/emergencycare_infographic/en/ in 10 June 2018

Decentralizing Rehabilitation: Using Blockchain to Store Exoskeletons' Movement

Daniela America da Silva, Claudio Augusto Silveira Lelis,
Luiz Henrique Coura, Samara Cardoso dos Santos, Leticia Yanaguya,
Jose Crisostomo Ozorio Junior, Isaias da Silva Tiburcio,
Gildarcio Sousa Goncalves, Breslei Max Reis da Fonseca,
Alexandre Nascimento, Johnny Cardoso Marques, Luiz Alberto Vieira Dias,
Adilson Marques da Cunha, Paulo Marcelo Tasinaffo,
Thais Tavares Terranova, Marcel Simis, Pedro Claudio Gonsales de Castro,
and Linamara Rizzo Battistella

22.1 Introduction

This paper tackles the development of an academic project using the Collaborative Interdisciplinary Problem-Based Learning (Co-IPBL).

Clinicians face many issues when dealing with the complexities of current patient care. Among them, two can be listed: (i) the growing demand for new technologies in health care management; and (ii) the need of collaborative efforts of professionals from Engineering and Computer Science background.

As a result, interdisciplinary health care teams with members from many professions have been working together, collaborating, and communicating closely to digitize patients' care [1–3].

This research was motivated by two driving forces. The first one was the promising results reported on a research using exoskeletons for rehabilitation [4–7]. The second one was related to the opportunity brought by the transition from paper-based medical records to electronic records initiated over the past few years and still happening in different areas in Health Care [8–11].

In this context, the purpose of this research is to analyze the opportunity of improving rehabilitation data storage and analysis as well as to improve patient information availability, privacy, and inviolability. As a result, it can also help to decentralize data analysis, during the rehabilitation process, better understanding the clinical recovery.

This paper presents the preliminary results of an ongoing research effort involving the integration of 3 distinct on going courses at the Brazilian Aeronautics Institute of Technology (Instituto Tecnologico de Aeronautica – ITA): CE-230 Software Quality, Reliability, and Safety; CE-235 Real-Time Embedded Systems Project; and CE-237 Advanced Topics in Software Testing.

This effort involved a cooperative work with rehabilitation professionals from the HC-FMUSP in Brazil [12]. It also involved a practical application of collaborative and interdisciplinary concepts supported by the Scrum Agile Method [13–15].

The adopted project code-name in English was TSA4MOHITR (Technological Solutions Applied for Managing Ostensive Hospital Information in Real Time), in Portuguese, "Soluções Tecnológicas Aplicáveis ao Gerenciamento de Informações Hospitalares Ostensivas em Tempo Real – STAGIHO-TR".

It encompasses the development of a Proof of Concept (PoC) software system to validate the viability of innovative combination of emerging technologies to support the main needs of distinct actors involved in the rehabilitation of patients with lower limb impairment after motorcycle accidents [1, 2, 16–18].

In this context, this PoC explores aspects of rehabilitation phases and presents an integrative architecture by using

D. A. da Silva (✉) · C. A. S. Lelis · L. H. Coura · S. C. dos Santos
L. Yanaguya · J. C. O. Junior · I. da Silva Tiburcio · G. S. Goncalves
B. M. R. da Fonseca · A. Nascimento · J. C. Marques · L. A. V. Dias
A. M. da Cunha · P. M. Tasinaffo
Brazilian Aeronautics Institute of Technology, Sao Jose dos Campos, Sao Paulo, Brazil

T. T. Terranova · M. Simis · P. C. G. de Castro · L. R. Battistella
Rehabilitation Hospital Lucy Montoro, Hospital of Clinics at the Faculty of Medicine, University of Sao Paulo, Sao Paulo, Brazil

S. Latifi (ed.), *16th International Conference on Information Technology-New Generations (ITNG 2019)*,
Advances in Intelligent Systems and Computing 800,
https://doi.org/10.1007/978-3-030-14070-0_22

disruptive technologies such as Blockchain to analyze data from patients in Rehabilitation Centers.

22.2 Background

22.2.1 Blockchain

Blockchain is a database architecture that maintains continuously growing set of data records into blocks. Those blocks are linked by using cryptography, ensuring information security [19–21].

This technology was initially created to address the double spending problem of crypto-currencies. Most recently, a variety of applications started to consider Blockchain mainly due to its properties.

The main advantage of using blockchain are its distributed nature and also its security. The distributed nature means that there is no single master computer that holds the entire dataset exclusively.

In fact, all the computer nodes in the network have a copy of data. As a consequence, a Blockchain is resistant to data modification by design. Also, it is public, clear and auditable [19–21].

22.2.2 Rehabilitation Using Exoskeletons

In this work, the same definition of exoskeleton adopted by [6] is used: "an active mechanical device that is essentially anthropomorphic in nature, is worn by an operator and fits closely to his or her body, and works in concert with the operator's movements".

The terminology "exoskeleton" is also employed for devices that augments the performance of an able-bodied wearer [6]. Norms and Standards are under analysis for exoskeletons and for this work it is under the IEC 80601-2-78 ED1 (Medical Electrical Equipment – Part 2-78: Particular requirements for the basic safety and essential performance of medical robots for rehabilitation, assessment, compensation, or alleviation) [22, 23].

In the context of rehabilitation, the application of exoskeletons has been reporting encouraging results. After more than two years of research, testing exoskeletons with eight volunteers in Sao Paulo, scientists came across with the following result: continued use of devices has helped to recover some patients.

The most dramatic case was about a patient who took steps with a walker relying only on his own legs strength, without the help of the nurse [4, 5].

The Rehabilitation process for functions recovery consists usually of four steps:

- The process begins already in the wheelchair, with the patient immersed into a virtual reality environment wearing

a tactile suit, which transmits sensations of movement to the body;
- Then he/she moves on to the Lokomat [24], a machine to move his paralyzed legs on a treadmill, with the patient body held by suspended handles;
- Right after, with the neural activity sensor re-installed, the patient transmits commands to move the legs at the same time as they are moved by the Lokomat; and
- Finally, the exoskeleton walks when the patient sends the same trained brain commands in the other phases. In addition to locomotion, physiological improvements occur in the muscles for example.

The complexity of performing simple movements such as lifting and sitting, due to injuries to the spine, implies that the intention to move requires actuation and performance of muscles such as the anterior tibial, quadriceps, gluteus maximus, abdominals, trapezius, among others.

It outlines an ideological conception that an exoskeleton application tends to interpret motor performance and provide the necessary force to perform the movements without the aid of crutches [25].

It is possible to consider them as the first stage of the process until the individual go to walk, with the definition of the primordial parameters that it will grant the capacity of balance and static stability.

All the complexity of those movements are orchestrated by a hierarchical Command and Control strategy. In the Command and Control strategy there are three related levels of hierarchical controllers (Fig. 22.1), each one with a specific function:

- High-level Controller – Understands Movement Intention;

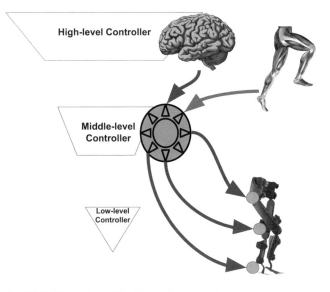

Fig. 22.1 Illustration of the hierarchical control strategy

- Middle-level Controller – Translation Layer maps (states) the intent of movement for tracking on low level devices and centralizes the control of multiple joints; and
- Low-level Controller – Articulating level controller drives the actuator as force, torque, and position or angle of the exoskeleton joint.

22.2.3 Motorcycle Accidents

According to a research from the Institute of Orthopedics and Traumatology of the Hospital of Clinics from the Medicine School at the University of São Paulo (Hospital das Clinicas da Faculdade de Medicina da Universidade de Sao Paulo – HCFMUSP), accidents involving motorcycle are one of the main source of serious injured patients that require an intensive rehabilitation process.

In fact, in a 3-month study on a 24-hour regimen, on alternate days, the following outcome was identified in the care of the victims by motorcycle accidents:

- 44% had injuries considered serious;
- 17% had fractures of lower limbs and 12% had fractures of upper limbs, 9% had poly-trauma and 5% had crane-encephalitic trauma;
- 67% of the unqualified victims had serious injuries and 43% of the disabled ones were severe; and
- 28% of the victims were hospitalized, 2% died, and 56% were discharged.

22.3   Methods and Materials

22.3.1 Key Concepts and Tools

In order to develop a computer system to satisfy both the project and learning requirements in an aggressive time frame of just 17 academic weeks, a unique combination of key concepts and tools was adopted.

The concepts employed in the project were Scrum Agile Method, Value Engineering, Model-Driven Development (MDD), Integrated Computer Aided Software Engineering Environment (I-CASE-E), and Software Quality dimensions (reliability, safety, and testability).

The used technological tools were the SCADE Suite®(Safety Critical Application Development Environment, ANSYS Inc./Esterel Technologies), Blockchain Hyperledger, Kafka Services, Grafana, and Lego MindStorms (to emulate an exoskeleton).

Hyperledger is an open-source project hosted by The Linux Foundation aiming to advance the adoption of Blockchain technologies cross-industry. Grafana is an open-source and general purpose dashboard web application framework. Apache Kafka is an open-source stream-processing software

platform to provide a unified, high-throughput, low-latency, and massively scalable platform for handling real-time data feeds.

22.3.2 The Agile Development Best Practices

The TSA4MOHITR Academic Project has involved the following three graduate courses:

- CE-230 Software Quality, Reliability, and Safety, involving the understanding of the development of software as process, product, and/or service with quality, reliability, and safety;
- CE-235 Real-Time Embedded Systems Project, involving the understanding of agile development of embedded and real-time software systems, and also the understanding of software as process, product, and/or service; and
- CE-237 Advanced Topics in Software Testing, tackling the main advanced techniques used for software testing.

During this TSA4MOHITR academic project development, all students participated as members of Team Developers (TDs).

The CE-235 course students were responsible for modeling and programming, by using SCADE. Also, they were responsible for evaluating the compatibility with other software tools, such as Lego for Scrum, Docker, Kafka, and Python libraries.

The CE-237 course students were responsible for the definition and execution of the acceptance testings and its oracles.

At the same time, the CE-230 course students were responsible for validating and delivering intermediate product values with quality, reliability, safety, and testability.

During this 2nd Semester of 2018, students from these three disciplines were involved in acquiring knowledge about Blockchain, implementing it as a repository for commands to be sent to the exoskeleton.

Product Owners (POs) and Scrum Masters (SMs) offered guidance to teams and tried to correct: the development progress in the event of any delay; the prioritization of the Backlog Product; and also the creation of some Scrum quick guides and videos to facilitate the understanding of the underlying process. Scrum Masters helped also teams to define and adhere to the process and to make sure the work would be performed to the best of students' ability.

22.3.3 The Project Organization

The TSA4MOHITR Project [26] was divided into three groups of hierarchical controllers: high, middle, and low level controllers. Its development was shared among four student teams, which were responsible for developing different functional requirements, involving the verification of distinct dimensions of quality, reliability, safety, and testability [27, 28].

22.3.4 The Acceptance Tests

The CE-237 course students were responsible for the definition and execution of acceptance tests and their oracles. Based on the content of this subject the students were able to prepare tests at the beginning of each sprint for each one of the user stories (USs) chosen by each team.

As a result, there was a better performance during the development of USs and a faster development helping to ensure better quality in a Proof of Concept (PoC).

22.4   Analysis and Major Findings

22.4.1 The Rehabilitation Process at the Lucy Montoro Hospital

As required by the Agile Scrum Method, during this project there was an involvement of final users – the rehabilitation team.

Therefore, a cooperation effort was established among students and professionals from the Rehabilitation sector of Lucy Montoro Hospital [12] at the HCFMUSP.

On this context, visits to the rehabilitation facilities of the Lucy Montoro Hospital and interviews were performed for information gathering.

This session presents the main highlights of the collected information supporting the main considered needs used to help to delineate a chosen PoC of User Stories.

The process is customized to each patient's specific needs. This requires preparation of a distinct program for each patient as well as its exclusive execution.

Each patient presents different responses to treatments, requiring specific adjustments into patient's program. This requires an intensive effort to collect, organize, store, and analyze data from each patient for treatments progress and for several ongoing researches on patient's evolution rehabilitation.

Also, the rehabilitation requires a well-coordinated multidisciplinary and distributed team involved to analyze encompassing Physiatrists and Physiotherapists, among others.

The first one is responsible for treating a wide variety of diseases that cause some degree of disability. The second is responsible for the development, maintenance, and rehabilitation of patients' capacities.

There are many apparatuses involved to support the different portions of the rehabilitation process, such as Motor Rehabilitation, NeuroPhysiological Study, Magnetic Stimulation, and Electroencephalogram.

However, despite the needs of a holistic approach involving all the data combined to be analyzed by multidisciplinary and decentralized team of professionals, each used equipment has its own panel for data analysis and stores its own data into its internal storage.

Although they could be connected to a network, they are not, so there is no communication among the machines or to a central repository. For back-up purposes, an operator performs a copy of the data of each equipment and stores them into a local server. However, there are back-up files into distinct file formats and data-types.

Therefore, professionals need to manually collect the data from machines and insert then into spreadsheets for later analysis of training progress. This manual process imposes a challenge to the required decentralization of patients information as well as threatens security and privacy aspects of patients personal data.

As a result, it was identified an opportunity to enhance data collection, storage, distribution and security management straight from the distinct equipment used for rehabilitation (Fig. 22.2).

This opportunity can be summarized as a set of features to support health-care professionals involved in the rehabilitation efforts to:

- access data collected from different equipment independent of the physical location (decentralized) and agnostic of data types and formats;
- analyze data collected by the rehabilitation apparatuses, according to each distinct patient's training program independent of the physical location (decentralized);
- correlate the applied training data with improvements data acquired, as the clinical portrait improves from any physical location (decentralized); and

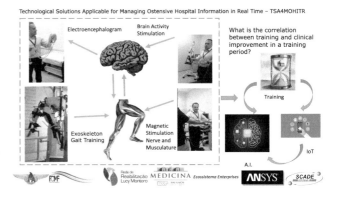

Fig. 22.2 Opportunities to read different equipment data

- support health-care actors in decision making along the evolution of rehabilitation training from any physical location (decentralized)

.

22.4.2 Assigned Mission

A scope reduction effort was performed to adjust the scope to enable the main user needs support facing available resources and academic goals. As a result, the scope was delineated as a focus on the rehabilitation of patients with impairment of the lower limbs, victims of motorcycle accidents, because of the reported high frequent of occurrence in Urgent and Emergency Care in Brazil.

Therefore, the assigned mission for this project was the development of PoC that should be capable of providing appropriate management of the rehabilitation of patients with impairment of the lower limbs, a frequent occurrence in emergencies from motorcycle accidents, in such a way that:

- PHYSICIANS may have computational tools capable of identifying the eligibility of the PATIENT for rehabilitation treatment (ASSUMPTION);
- PHYSICIANS may have computational tools capable of providing preventive planning, scheduling, and control of rehabilitation services for motorcycle accident victims, for example, to predict the type of assisted technology required during treatment;
- The HOSPITAL must have computational capabilities to store recovery movements of each PATIENT in database that allows PATIENT data privacy and its inviolability;
- The HOSPITAL must have computational capabilities to store and retrieve movement patterns (stand, sit, and walk) of each patient to analyze the evolution of the rehabilitation and fine tune the training as measured by the recovery of muscle stimuli;
- The population/society must have a computational tool of the System Project involving STAGIHO-DB and/or STAGIHO-TR (Real Time) capable of providing records, management, controls, and governance of resources used in the rehabilitation area of health (ASSUMPTION); and
- Public administration must have reliable data to provide a comprehensive situational awareness, aiming to support decision making for patient rehabilitation, involving motorcycle accidents, and lower limb involvement (ASSUMPTION).

22.4.3 Proposed Product

The developed product was a PoC oriented prototype with some fictional elements to enhance students' involvement and help to contextualize the product.

Considering the summarized set of features listed, the automation of data collection from the used equipment, its proper management, and decentralized access assuring the its security are the key elements considered by the developed product.

Therefore, the suggested product must store each patient's movement in a database that allows patient's data privacy and inviolability. This requirement motivated the use of Blockchain.

Blockchain is suitable for storing and retrieving movement patterns of each patient to get up and sit-down only and analyze the evolution of rehabilitation.

Figure 22.3 illustrates the main components of the adopted product architecture and its inter-relations.

Figure 22.4 illustrates the proposed control strategy by keeping the 3 hierarchical layers and showing how the main architectural components were distributed over them, described as follows:

Display CDS – The proposed solution uses a CDS Display for user interaction, developed through SCADE. This display implements the first level of control (high-level) associated with motion intent. The user, for example, thinks

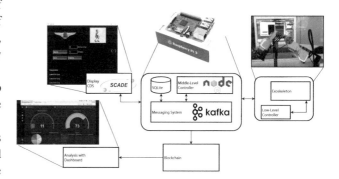

Fig. 22.3 The proposed product

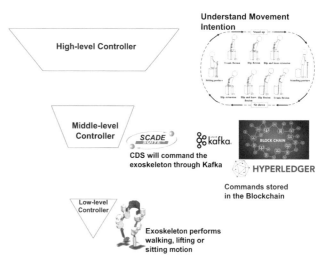

Fig. 22.4 The proposed control strategy components

about raising, and by using the display he triggers the lift command;

Messaging System and Middle-Level Controller – The command received from the display triggers a message in the messaging system. The implemented Messaging Hub is supported by Kafka. There are multiple channels to avoid cross-over and loss of messages. The implemented mid-level controller follows the logic presented by the technical literature [29]. It is responsible for receiving the CDS message, interpreting it and verifying its validity and integrity. After its validation, the command is sent to the low-level controller to control the robot;

Exoskeleton and Low-level controller – For the PoC purposes, a Lego MindStorms robot was used to test the control output, emulating an exoskeleton. It receives parameters from the low-level controller to effectively execute the movement. This controller directs the specific commands to the 4 distinct actuators (motors) controlling the robot structure. In addition, it verifies the movement and provides a feedback about the movement confirmation and the current state of the robot current and of its engines current using the message system;

Return and Situational Awareness – This message is captured by the mid-level controller that updates an internal control table with the states of the motors and the availability of each motor to perform movements. The command, the confirmation, and the patient data are stored in a SQLite database, for later submission to the blockchain network. At the same time, a status update message is returned to the CDS for displaying the system current status. This is fundamental to the situational awareness of both the healthcare team supporting the patient and the patient himself about all steps involved from the CDS to the robot, including the mid-control. The system was deployed into a Raspberry Pi 3, which allows connection Wi-Fi, Bluetooth, and also the connection with a physical screen for the CDS;

Blockchain – The sending of data to the Blockchain network occurs whenever the user has availability of connection to the Internet. The data is downloaded via Wi-Fi, respecting the protocol specified by the Blockchain network and ensuring the integrity and inviolability of such sensitive patient data and their treatment. The proposed solution used for Blockchain was the Hyperledger; and

Analysis – Once data is in Blockchain, it is possible to analyze the evolution of the patient and its treatment. It is also possible to realize predictions for the change in the treatment stage. Some dashboards for health-care professionals were implemented by using Grafana.

22.4.4 Preliminary Results

From the academic standpoint, the use of Co-IPBL has demonstrated to be a successful experience, considering a

PoC prototype could be built by using emerging technologies under a Scrum Agile Method paradigm in only few weeks with limited available resources.

Also, some preliminary tests performed in the PoC prototype has shown very encouraging results. Some simulations have also demonstrated that data with distinct data-types and formats could be collected from multiple data-sources, with low latency and potential high-scalability.

The proposed and adopted solution based on Blockchain proved to be viable and to support some requirements of decentralized and secure access to information.

Moreover, tests used to validate the PoC prototype could support users to analyze data independently collected from physical locations, by using dashboards to facilitate multidimensional analyses and data correlations.

As a result, the tests using the PoC prototype could demonstrate it is possible to use a technology that stores the results achieved by patients in movements to get up, walk, and sit in a decentralized manner.

Also, it has allowed the creation of dashboards according to the needs of a medical team without being restricted to what is presented on the console of an equipment. It made possible data to be available to specialists from different areas.

In addition, it was shown that it is possible to store exoskeleton by using Blockchain. Therefore, the assigned mission was accomplished.

Finally, the PoC prototype helped to shed a light into some additional needs such as using voice command to control the exoskeleton and another interface for the patient to interact with the device.

22.5 Conclusion

This paper aimed to describe the development of an academic interdisciplinary project by using Scrum Agile Method and its best practices, in order to develop a prototype for a PoC, with cloud-computing resources, Blockchain, and a suite for embedded real-time systems applicable to rehabilitation with exoskeletons.

Software quality, reliability, safety, testability, integrity, and availability are important issues for protecting health data and services for rehabilitation too.

Physiatrists need to know more about how exoskeletons are helping in the clinical recover of patients and when physiological improvements occur.

The Technological Solutions Applicable for Managing Ostensive Hospital Information in Real Time – TSA4MOHITR project was developed by students from three different courses taught at the Brazilian Aeronautics Institute of Technology, on the 2nd semester of 2018.

The application of Test Driven Development (TDD) in this project was closely related to the adopted interdisciplinary approach, since the test and the development were created by students from all three courses.

22.5.1 Specific Conclusions

The project described in this paper was successful, considering its academic and PoC goals. Academically, it enabled students to experiment a close to real-life software development project, by using the Agile Scrum Method and emerging technologies to develop a real PoC.

In this context, the Scrum framework has been adapted to the reality of the interdisciplinary academic environment of ITA, helping the entire team of more than 20 students to offer value to stakeholders at the end of each sprint and also at the end of this project.

Also, the application of Test Driven Development (TDD) and Acceptance Testing Driven Development (ATDD) techniques in the project was closely related to the interdisciplinary approach adopted, since acceptance tests were created by CE-237 Software Testing course students, while the embedded software was implemented by CE-230 and CE-235 course students.

The students were able to learn new emerging technologies from scratch and help each other to move along the learning curves efficiently.

As a result, they were able to deliver and test a PoC integrating multiple components and technologies such as Kafka, Lego MindStorms, Blockchain, Hyperledger, and Grafana.

The SCADE was used to build an exoskeleton embedded system. It was possible to see it by a screen of commands where an exoskeleton movement is selected and then the commands are sent by the Apache Kafka to the exoskeleton to move, walk, and sit down.

Finally, the PoC results were very promising, showing the viability of using emerging technologies to automate the data collection directly from rehabilitation equipment, enabling its usage by a multidisciplinary distributed, and ensuring security and privacy of patients' rehabilitation data.

22.5.2 General Conclusions

The authors believe that is possible to quickly develop still more sophisticated products based on integration of multiple emerging technologies, by using a collaborative approach.

It requires only some minimal organization among teams such as the definition of separate functionalities to be developed by each team and a collective planning effort shared among all the personnel involved.

In addition to that, it is highly desirable the usage of real time integration programming suites such as SCADE.

22.5.3 Recommendations

It is strongly recommended to keep track of the potentially affected results to quickly adjust to stakeholders' expectations and review deliverables, when working with emerging technologies. Also, it is recommended to have participants in the team able to identify in advance potential failures risks with the new technology to allow the team to pivot in order to reach its goals in the proper time-frame.

Finally, it is also recommended to rely on previously defined standard, because it can help with particular non-functional requirements, such as safety and performance, when dealing with mission critical applications like those from medical devices.

22.5.4 Future Work

It is suggested that the process used in this TSA4MOHITR academic project prototype can be extended to other Exoskeleton in Rehabilitation projects and predictive systems, in order to improve decision making in the rehabilitation and to better understand clinical recover of patients. It is also suggested for future work to expand some cooperation among the ITA, hospitals, innovation foundations, medical suppliers, industries, and public and private enterprises, in order to get a selection of academic projects aligned to updated needs from the market.

Acknowledgements The authors thank: Esterel Technologies/ANSYS Inc.; Hospital of Clinics from Faculty of Medicine of University of São Paulo; Rehabilitation Hospital Lucy Montoro; Ecossistema Negocios Digitais Ltda; ITA; and FCMF, for all their general and finantial support, during the development of this TSA4MOHIBD Project.

References

1. Silva, D.A., Goncalves, G., Santos, S., Pugliese, V., Navas, J., Santana, R., et al.: Health care information systems: a crisis approach. In: 15th International Conference on Information Technology: New Generations (ITNG 2018), Las Vegas (2018)
2. Silva, D.A., Santana, R., Navas, J., Goncalves, G., Vieira, L.A., Cunha, A., et al.: Health care transformation: an academic application system case study. In: 10TH IFAC Symposium on Biological and Medical Systems – International Federation of Automatic Control (IFAC BMS 2018), São Paulo (2018)
3. de Castro Martins, J., Pinto, A.F.M., Junior, E.E.B., Goncalves, G.S., Louro, H.D.B., Gomes, J.M., et al.: Using big data, internet of things, and agile for crises management. In: Information Technology-New Generations, pp. 373–382. Springer, Cham (2018)

4. SuperInteressante: The exoskeleton changes the game (2017). Accessed in Sept 2018. Available from: https://super.abril.com.br/tecnologia/o-exoesqueleto-vira-o-jogo/

5. Donati, A.R., Shokur, S., Morya, E., Campos, D.S., Moioli, R.C., Gitti, C.M., et al.: Long-term training with a brain-machine interface-based gait protocol induces partial neurological recovery in paraplegic patients. Sci Rep. **6**, 30383 (2016)

6. Dollar, A.M., Herr, H.: Lower extremity exoskeletons and active orthoses: challenges and state-of-the-art. IEEE Trans. Robot. **24**(1), 144–158 (2008)

7. Moreno, J.C., Figueiredo, J., Pons, J.L.: Exoskeletons for lower-limb rehabilitation. In: Colombo, R., Sanguineti, V. (eds.) Rehabilitation Robotics, pp. 89–99 (2018)

8. Gates, C.: Electronic Medical Record Reminder to Improve Human Papillomavirus Vaccination Rates among Adolescents (2018). http://thescholarship.ecu.edu/handle/10342/6887

9. Hillestad, R., Bigelow, J., Bower, A., Girosi, F., Meili, R., Scoville, R., et al.: Can electronic medical record systems transform health care? Potential health benefits savings, and costs. Health Aff. **24**(5), 1103–1117 (2005)

10. Scott, J.T., Rundall, T.G., Vogt, T.M., Hsu, J.: Kaiser Permanente's experience of implementing an electronic medical record: a qualitative study. BMJ. **331**(7528), 1313–1316 (2005)

11. Fraser, H., Biondich, P., Moodley, D., Choi, S., Mamlin, B., Szolovits, P.: Implementing electronic medical record systems in developing countries. J. Innov. Health Inform. **13**(2), 83–95 (2005)

12. Montoro, L.: Rede de Reabilitação Lucy Montoro (2018). Accessed in Sept 2018. Available from: http://www.redelucymontoro.org.br/

13. Sutherland, J.: SCRUM Handbook. Scrum Training Institute Press (2010). https://www.researchgate.net/publication/301685699_Jeff_Sutherland's_Scrum_Handbook

14. Dea, C.: An Introduction to Agile Methods. Fraunhofer Center for Experimental Software Engineering. Advances in Computers, vol. 62 (2004). http://www.cse.chalmers.se/~feldt/courses/agile/cohen_2004_intro_to_agile_methods.pdf

15. Cohn, M.: Succeeding with Agile: Software Development Using Scrum. Addison-Wesley, Upper Saddle River (2010)

16. Forbes: This is why blockchain will transform healthCare (2017). Accessed in June 2018. Available from: https://www.forbes.com/sites/bernardmarr/2017/11/29/this-is-why-blockchains-will-transform-healthcare/

17. Projects, L.F.: Blockchain Hyperledger. Accessed from in June 2018. Available from: https://www.hyperledger.org/

18. Cross, H.: Care supply regulatory center. Accessed from in 10 June 2018. Available from: http://www.cross.saude.sp.gov.br/

19. Nakamoto, S.: Bitcoin: a peer-to-peer electronic cash system (2008). Available from: https://bitcoin.org/bitcoin.pdf

20. Banafa, A.: IoT and blockchain convergence: benefits and challenges. IEEE Internet Things (2017). https://iot.ieee.org/newsletter/january-2017/iot-and-blockchain-convergence-benefits-and-challenges.html

21. Dorri, A., Kanhere, S.S., Jurdak, R., Gauravaram, P.: Blockchain for IoT security and privacy: the case study of a smart home. In: 2017 IEEE International Conference on Pervasive Computing and Communications Workshops (PerCom Workshops). IEEE, pp. 618–623 (2017)

22. IEC/DIS: Medical electrical equipment – Part 2–78: Particular requirements for the basic safety and essential performance of medical robots for rehabilitation, assessment, compensation or alleviation (2018). Accessed in Sept 2018. Available from: https://www.iso.org/standard/68474.html

23. Jacobs, T., Veneman, J., Virk, G.S., Haidegger, T.: The flourishing landscape of robot standardization [Industrial Activities]. IEEE Robot. Autom. Mag. **25**(1), 8–15 (2018)

24. Jezernik, S., et al.: Robotic orthosis lokomat: a rehabilitation and research tool. Neuromodulation Technol. Neural Interface **6**(2), 108–115 (2003)

25. Correia, P.: Aparelho locomotor: Função neuromuscular e adaptações a atividade física (2012). Available from: http://www.fmh.utl.pt/indices/aparelocvol2.pdf

26. Project, T.: Technological solutions applicable for the management of ostensive hospital information with big data (2018). Accessed in June 2018. Available from: https://sites.google.com/site/projetointerdisciplinar2018/.

27. Pugh, K.: Lean-Agile Acceptance Test-Driven Development: Better Software Through Collaboration. Addison-Wesley (2011). ISBN 978-0321714084

28. GitHub: Open-source landscape map for healthcare-related blockchain (2018). Accessed in June 2018. Available from: https://www.forbes.com/sites/bernardmarr/2017/11/29/this-is-why-blockchains-will-transform-healthcare/

29. Aliman, N., Ramli, R., Haris, S.M.: Design and development of lower limb exoskeletons: a survey. Robot. Auton. Syst. **95**, 102–116 (2017)

Analysis and Comparison of Frameworks Supporting Formal System Development based on Models of Computation

Augusto Y. Horita, Ricardo Bonna, and Denis S. Loubach

23.1 Introduction

Design techniques for embedded systems range from traditional development to the use of formal models. According to [1], systems should be modeled at a high level of abstraction, using formal models, without any reference to implementation code or platforms, thus targeting a correct-by-construction development. In this sense, the use of *models of computation* (MoC) is considered a key approach to formally describe the behavior of a system as a set of laws that govern the interaction of components in a design.

Researchers have been implementing frameworks to aid developers in modeling and simulating applications using a formal base. Among the main available frameworks, one can mention Ptolemy II [2], ForSyDe [3], SDF3 [4], and Simulink [5].

In view of this, our paper introduces a comparison of development frameworks. To select from a range of the available ones, we defined a set of criteria that allow a more comprehensive analysis. The frameworks have to be free and *open-source*, which allows a deeper analysis of how each framework implements MoCs, besides the shared knowledge and collaboration environment of the developer's community [6]. Moreover, the compared frameworks have to be based on different programming paradigms, so that one can observe the benefits and drawbacks of each paradigm. Based on that,

the selected ones were Ptolemy II and ForSyDe. Both of them supports formal modeling based on MoC.

Related to MoCs, there is also a wide range. In this case, we considered MoCs that are widely used and have different *timing abstractions*. One is the *synchronous reactive* (SR), representing the *timed* MoCs, and the other is the *synchronous dataflow* (SDF), representing the *untimed* MoCs.

The following *aspects* are analyzed in the comparison carried out in this paper: framework maturity, programming paradigm, scalability, model execution, programming entry method, and design methodology.

We show frameworks comparison considerations with respect to these aspects, pointing out the main benefits and drawbacks of each framework, highlighting their modeling approaches, through a case study developed and implemented in the frameworks under comparison.

23.2 Background

23.2.1 Models of Computation (MoC)

A *model* is an abstraction or simplification of an entity, which can be a physical system or even another model. It includes the tasks' relevant characteristics and properties for that particular model. In this context, a MoC is an abstraction of a real computing device [7], which can serve different objectives and thus, different MoCs can be used for modeling different systems depending on its behavior. Essentially, MoCs are collections of abstract rules that dictate the semantics of execution and concurrency in heterogeneous computational systems.

A meta-model, or framework, for reasoning about properties of MoCs, named *tagged signal model* (TSM) is introduced in [8]. Within that framework, systems are regarded as compositions of processes acting on signals.

A. Y. Horita (✉) · R. Bonna
Advanced Computing, Control & Embedded Systems Laboratory/FEM, University of Campinas–UNICAMP, Campinas, SP, Brazil
e-mail: ahorita@fem.unicamp.br; rbonna@fem.unicamp.br

D. S. Loubach
Department of Computer Systems, Computer Science Division, Aeronautics Institute of Technology – ITA, São José dos Campos, SP, Brazil
e-mail: dloubach@ita.br

© Springer Nature Switzerland AG 2019
S. Latifi (ed.), *16th International Conference on Information Technology-New Generations (ITNG 2019)*,
Advances in Intelligent Systems and Computing 800,
https://doi.org/10.1007/978-3-030-14070-0_23

Definition 1 (Signal) A signal s, belonging to the set of signals S, is a set of events $e_i = (t_i, v_i)$, which are elementary units of information composed by a tag $t_i \in T$ and a value $v_i \in V$. A signal can be viewed as a subset of $T \times V$.

Definition 2 (Process) A process P is a set of possible behaviors, and can be viewed as relations between input signals S^I and output signals S^O. The set of output signals is given by the intersection between the input signals and the process $S^O = S^I \cap P$. A process is functional when there is a single value mapping $F : S^I \rightarrow S^O$ which describes it. Therefore, a functional process has either one behavior or no behavior at all.

Models within the TSM are classified as *timed* or *untimed*. In a *timed* MoC, the set of tags T is *totally ordered*, meaning that one can order every event in every signal of the MoC based on its tag. In an *untimed* MoC, the set of tags T is *partially ordered*, meaning that not all, but only local groups of events can be ordered based on its tag, e.g. the ones belonging to the same signal.

23.2.2 Synchronous Reactive (SR) MoC

In a SR MoC, every signal is synchronized, meaning that for any event in one signal, there is an event with the same tag in every other signal. Because of this, the SR MoC is classified as a *timed* MoC within the TSM.

The SR MoC is based on the *perfect synchrony hypothesis*, which states that neither computation nor communication takes time [9]. It means that, according to this MoC, the processes receives an event, computes the output at zero time and waits for the next event. To make this hypothesis possible, this MoC splits the time axis into slots, where the evaluation cycle and process occur. For this model to work faithfully as the system physical behavior, the time slot may be chosen in a way that the system is able to respond *fast enough*, which means to compute the output event of an input within an evaluation cycle, and before the next input arrives, leading to a separation of functional and temporal behavior of the system.

23.2.3 Synchronous Dataflow (SDF) MoC

SDF belongs to the family of *untimed* MoCs, named *dataflow*. According to [10], dataflows are directed graphs where each node represents a process and each arc represents a signal path. Each process, when activated, i.e. fired, consumes a certain amount of data, denominated tokens, from its input ports and generate a certain amount of data for its output ports.

In the SDF case, each input and output port is associated with a fixed natural number. Such numbers, i.e. *token rates*, define the token consumption, for input ports, or the token production, for output ports. An actor can fire only if the signal paths have enough tokens to supply the amount needed by all input ports of the actor. As consequence, no signal path can have a negative amount of tokens.

The fact that an SDF actor always consumes and generates the same amount of tokens, i.e. fixed token rates, allows efficient solutions for problems like finding a static schedule for single and multi-processor implementations and buffer size definitions [10].

23.3 Framework Analysis and Comparison

23.3.1 General Frameworks Overview

Ptolemy II is a framework based on *actor-oriented* design targeting formal modeling of heterogeneous cyber-physical systems (CPS). It provides a graphical user interface (GUI) named *Vergil*. In Ptolemy II, a model is implemented as a hierarchical interconnection of actors, concurrently executed and that communicate through messages via interconnected ports. A specific component, named *Director*, drives the semantics of each hierarchical level. Ptolemy II is open-source. Its code is free for academic and commercial use under the Berkeley Standard Distribution (BSD) license style, which means that the user needs to include the copyright notice into products. The application stable versions, compatible with multiple operating systems, can be downloaded from [11].

Formal System Design (ForSyDe), on the other hand, is composed by a set of different libraries that aims the formal and correct-by-construction system design, comprehending different phases of the design process, including modeling and simulation, design space exploration, synthesis and compilation. *ForSyDe-Shallow* is one of the main tools for system modeling and simulation. It is based mainly on *Haskell functional programming language*, and uses the concept of *process constructors*, which are higher-order functions that take side effect-free functions and values as arguments to create processes. Each MoC implemented in ForSyDe is essentially a collection of process constructors that enforce the semantics of each specific model, avoiding the usage of a Director. ForSyDe has the same license base as Ptolemy II, BSD style, but most of its libraries include an extra clause that prohibits the use of project or contributors names without written consent. Its libraries source code are under version control and can be found at [12].

23.3.2 Framework Maturity

The first steps in Ptolemy II development were taken in the 1980s in the University of California, Berkeley, as a software framework called *Blosim*, made only for simulating signal processing systems. After that, the framework kept its evolution, implementing new MoCs and, even switching its base language to leverage the framework logic, coming from C, passing through Lisp and C++ (Ptolemy Classic), and currently Java (Ptolemy II) [2]. Nowadays, Ptolemy II has subsets and branches which are sponsored by commercial partners (e.g., Microsoft, Denso, Ford, National Instruments, Siemens, Toyota), like *CyPhySim*, which simulates CPS, and *CapeCode*, a modeling tool focused on internet of things systems. These branches are part of *iCyPhy*, a project that aims to conduct a research on architectures design, modeling, and analysis techniques for CPS.

The development of ForSyDe started in the Royal Institute of Technology (KTH) in 1999 [13]. Although it had not this specific name back there, the authors presented the principles and methodologies involved in the framework, combining the synchrony hypothesis with the functional language paradigm. Since then, an increasing number of researchers and other people from academia have been working on this framework to add new MoCs implementation and additional libraries. Currently, the main tool for modeling and simulation of heterogeneous systems using MoCs in the ForSyDe project is the ForSyDe-Shallow, implemented as a shallow-embedded domain specific language (EDSL) in the functional programming language Haskell. Apart from ForSyDe-Shallow, the project has other tools such as ForSyDe-Atom and ForSyDe-SystemC. ForSyDe-Atom, also implemented as an EDSL in Haskell, is a spin-off from the ForSyDe-Shallow that aims to push the concept of *orthogonalization of concerns* to its limits using the concept of *layers* [14]. ForSyDe-SystemC is also an EDSL, but implemented in the IEEE standard language SystemC.

23.3.3 Programming Paradigm

Ptolemy II is based on the imperative programming paradigm, being *Java* its base implementation language. Java is a concurrent and strongly typed language, which facilitates the modeling of concurrent process of a system [15]. It is also object-oriented, allowing new MoCs to inherit attributes and methods from existing ones.

ForSyDe relies on the functional programming paradigm, developed mainly in Haskell. The main characteristic of *Haskell* [16] is that it is a purely functional programming language, based on *Lambda Calculus* system, which means that functions does not depends on system's state, i.e. the output of a function is always the same, provided that the same parameters are inputted. Alone, this feature facilitates the application of formal methods for transformation, synthesis, verification [3]. Besides that, it is a lazy language, its functions only evaluate arguments when the results are needed. It also includes variable type inference, which improves coding performance. These specifications, together with others, makes it possible to have a higher abstraction level when modeling systems with this language.

23.3.4 Scalability

Ptolemy II includes *Ontologies package* [17] as an extension to manage scalability, based on application's semantic concepts annotations and lattice-based ontologies. The scalability modeling is described in [2]. We argue that implementing models in *Vergil*, which is a GUI, may become a problem for very large systems, such as state machines with many states and state transitions. In such systems, the visual representation provided by the GUI may become incomprehensible and may be very costly to perform small modifications on the model.

Implemented in Haskell, ForSyDe inherits all the features of the host language, such as lazy evaluation, allowing to model infinite input streams in a pure functional, side-effect-free way. Although it seems costly to implement a model in ForSyDe due to the lack of a GUI, for large systems it is actually an advantage because modifications can be performed in a few lines of code, instead of dragging blocks, connecting many transitions or signal paths and configuring each new object added using a GUI. Besides, Haskell is a compiled language, which means it has a better performance when executing than interpreted languages such as Java.

23.3.5 Model Execution

The main way to simulate a model in Ptolemy II is using Vergil. After modeling the system, the user can execute and run the model by pressing a specific command button. The output is presented depending on the output actor that was included in the model, which can be graphical, text or files. It is also possible to include more than one output actor.

All the actors, including the Director and Output, have their implementation source code accessible and they can be customized in the model, just by "right click" on the actor and selecting the desired option.

In ForSyDe, one way to execute a model is to use the *ForSyDe-Shallow* library. After implementing the model, the user must load that library into a Haskell compiler iterative environment, e.g. GHCi, and call the processes defined in the model. If applicable, the output will be printed as defined in the model, on the same interface.

23.3.6 Programming Entry Method

Ptolemy II uses Vergil GUI as its main programming entry method. The developer can drag-and-drop components from an available list and define parameters included in each actor. The *Director* is also taken from the same list. It is possible to customize components as needed by the application.

In ForSyDe, the developer can model the application by directly writing a source code in Haskell. For each MoC, there is a list of processes constructors that may be used to model each one of the application processes.

23.3.7 Design Methodology

Ptolemy II is based on an imperative paradigm, implemented basically in Java, leading to a well defined package structure.

The *kernel package* defines the structure of MoCs, specifying the relations between components, domains and their hierarchy. *Data package* refers to the classes responsible for data transfer between models, the main class in this package is the *Token*, the base for all units of data exchanged between components. *Math package* basically treats operations with matrices and vectors. *graph package* provides support to analyze and manipulate mathematical graphs. The *actor package* leads with I/O ports and actors, which are executable entities that exchange data through I/O ports, this is also the package that includes the *Director* class, which are customized to drive each domain, i.e. MoC, in an application. *Gui package* offers user interface methods to edit parameters of model components and managing windows. Also providing user interface components, *vergil package* implements *Vergil*, the GUI for Ptolemy II. The *moml package* provides a parser for modeling markup language (MoML) files, which is the XML schema used to store models.

ForSyDe aims to push system design to a higher level of abstraction using the functional programming paradigm [3], implemented using Haskell language. This leads to well defined and structured models, and also gives a solid base for formal analysis. In ForSyDe, the first step when implementing a target platform is to develop a formal and purely functional *specification model*, allowing stepwise design refinement through formal design transformations, reaching the *implementation model*. The processes are constructed by higher-order functions, called process constructors, which have software or hardware semantics [18]. This first phase is called *Functional Domain* and it is all expressed using ForSyDe semantics. Having the Implementation Model, the development reaches the *Implementation Domain*, in which the allocation of resources and code generation are performed. Finally, the model is described in hardware, e.g. VHDL, and software, e.g. C language.

23.4 Case Study

Two different systems are modeled in Ptolemy II and ForSyDe, as case study. The first is an *encoder-decoder system* (EDS) modeled based on the SR MoC, and the second is a *frequency spectrum system* (FSS) based on the SDF MoC.

We used the following tools versions: Ptolemy II v11.0.1_20180619, and Java development kit v10.0.0; ForSyDe v3.3.2.0, and Glasgow Haskell compiler (GHC) v8.0.1. The complete case study source codes are available in [19].

23.4.1 Encoder-Decoder System (EDS)

The EDS, Fig. 23.1, is a reconfigurable system introduced in [20] modeled with the SR MoC. That system has 2 input signals: s_{key} and s_{input}. s_{key} is a signal containing the *keys* to the encoder/decoder process, and s_{input} is the input signal to be encoded.

The process *genEnc* generates an encoding function based on each *key* value, inputted via the s_{key} signal. The chosen encoding function is the basic addition function, therefore the output of *genEnc* is the function $f_{Enc}(x) = x + key$. The process *genDec* behaves in the opposite way *genEnc* does, it outputs the decoding function $f_{Dec}(x) = x - key$. The processes ap_{Enc} and ap_{Dec} apply the functions f_{Enc} and f_{Dec} to its data inputs to encode and decode data.

To model the EDS in Ptolemy II, we first define the semantics as the SR MoC using the *SR Director*. To include it, the user needs to select the Director and drag it into the graphic editor contained in Vergil. In this model, the *Sequence* actor was used to generate the input signals s_{key} and s_{input}.

We used the *Expression* actor to model the processes *genEnc*, *genDec*, ap_{Enc} and ap_{Dec}. Finally, to create an output as a plain text, it was used the *NonStrictDisplay*

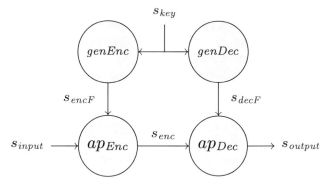

Fig. 23.1 EDS modeled according to the SR MoC. (Adapted from [20])

actor, which shows the output as illustrated in *OutputDisplay* window, Fig. 23.2.

To model the EDS in ForSyDe, we first create each process using the appropriated process constructor. To model the *genEnc* and *genDec* processes, we use the `combSY` process constructor, which takes a function as argument and generates a process that takes a single input signal and applies the function, generating the output signal, i.e. *currying* technique. To model ap_{Enc} and ap_{Dec}, we use the `zipWithSY` process constructor with the function application operator, $ in Haskell, to generate processes that takes one input signal of functions and one input signal of values and outputs a signal resulted from the application of the functions to the values. With all process defined, we finally connect the processes to each other forming the desired process network.

Listing 23.1 shows the EDS implemented in ForSyDe. The functional paradigm of Haskell makes this particular case study straightforward to implement due to the fact that functional languages handles functions as first-class citizens, allowing them to be used as any other value, including being exchanged via signal paths.

Listing 23.1 EDS SR MoC in ForSyDe/Haskell code

```
1   module EDS where
2   import ForSyDe.Shallow
3   -- Process Definitions
4   genEnc = combSY (+)
5   genDec = combSY (\x y -> y-x)
6   ap_enc = zipWithSY ($)
7   ap_dec = zipWithSY ($)
8   -- EDS process network
9   eds s_keys s_input = (s_enc, s_output)
10    where s_enc = ap_enc s_encF s_input
11          s_output = ap_dec s_decF s_enc
12          s_encF = genEnc s_keys
13          s_decF = genDec s_keys
14   -- Input Example
15   s_keys_example = signal [1, 4, 6, 1, 1]
16   s_input_example = signal [1, 2, 3, 4, 5]
17   -- to execute the model:
18   -- eds s_keys_example s_input_example
19   -- result: ({2,6,9,5,6},{1,2,3,4,5})
```

23.4.2 Frequency Spectrum System (FSS)

The FSS model, Fig. 23.3a, is based on a multi-rate SDF example [2], which models a system that generates the frequency spectrum from a noisy sine wave.

There are two input signals in the FSS model, a sine wave s_{sine} and a random noise signal s_{noise}. The $\sum$ actor adds events of each input signal and outputs the noisy sine wave s_{noisy}. The *Spectrum* actor consumes 256 events of s_{noisy} to generate 256 events representing the frequency spectrum of the input signal.

To model a system using SDF in Ptolemy II, the *SDF Director* block must be chosen to enforce its semantics into the model. In this implementation based on [2], Fig. 23.3b, we used a *Sinewave* actor to generate the sine wave and a *Noise* source, which outputs pre-generated random values stored in a file, to generate the noise signal. The *Spectrum*, pre-implemented in Ptolemy II, actor then generates the frequency spectrum of the noisy signal using 256 input tokens.

FSS implementation in ForSyDe follows the same procedure used to implement the EDS. We create each process with the appropriate process constructor and connect each one to form the process network. The main challenge is to implement the *discrete Fourier transform* (DFT) function in Haskell to use it in the Spectrum actor, since there is no pre-implemented Spectrum actor such as in Ptolemy II.

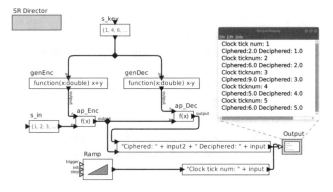

Fig. 23.2 EDS SR MoC modeled in Ptolemy II

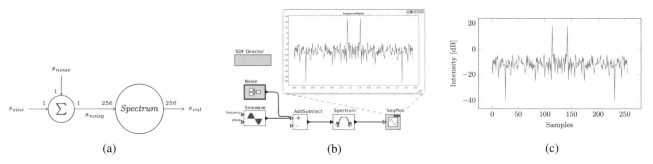

(a)   (b)   (c)

Fig. 23.3 FSS SDF models. (**a**) Frequency spectrum system (FSS) modeled according to the SDF MoC. (**b**) FSS SDF MoC modeled in Ptolemy II. (Adapted from [2]) (**c**) FSS SDF MoC output using ForSyDe

Listing 23.2 shows the FSS implemented in ForSyDe. The actor *Spectrum*, defined as `spectrum`, was created using the process constructor `actor11SDF`, which takes, as arguments, the token consumption rate 2^8, the token production rate 2^8, and the function, i.e. the DFT implemented as the function `spec`. The result is an actor that, whenever fired, consumes 256 input tokens and produces 256 tokens corresponding to the input's frequency spectrum. The $\sum$ actor was created using the process constructor `actor21SDF`, 21 indicates it has 2 inputs and 1 output, which takes, as arguments, the token consumption rates of each input port in a tuple, the token production rate and the function, a basic add function.

Listing 23.2 FSS SDF MoC in ForSyDe/Haskell code

```
1   module FSS where
2   import ForSyDe.Shallow
3   import Data.List.Split
4   import System.IO
5   -- Sine wave parameters example
6   sampleFreq = 8000
7   sineFreq = 440
8   sinePhase = 0
9   -- Spectrum process parameter
10  order = 8 :: Int
11  -- Discrete Fourrier Transform Calculation (answer in
        dB)
12  spec :: (RealFloat a, Enum a) => [a] -> [a]
13  spec x = zipWith (\re im -> 20 * logBase 10 (sqrt $ re
        ^2 + im^2) - 10 * logBase 10 m)
14          [sum $ zipWith (*) x2 [cos (2*pi*k*n/m) | n
            <- [0 .. (m-1)]] | k <- [0 .. (m-1)]]
15          [sum $ zipWith (*) x2 [sin (2*pi*k*n/m) | n
            <- [0 .. (m-1)]] | k <- [0 .. (m-1)]]
16    where m = fromIntegral $ length x
17          x2 = zipWith (*) (cycle [1,-1]) x
18  -- Spectrum actor
19  spectrum :: (RealFloat a, Enum a) => Signal a ->
        Signal a
20  spectrum = actor11SDF (2^order) (2^order) spec
21  -- Adder actor
22  adder :: Num a => Signal a -> Signal a -> Signal a
23  adder = actor21SDF (1,1) 1 (\[x1] [x2] -> [x1+x2])
24  -- Adder in series with spectrum
25  freqSpec :: (RealFloat a, Enum a) => Signal a ->
        Signal a -> Signal a
26  freqSpec s1 s2 = spectrum $ adder s1 s2
27  -- Sine wave generator
28  sineGenerator :: RealFloat a => (a, a, a) -> [a] ->
        Signal a
29  sineGenerator (amp, freq, phase) time = signal $ map
        (\t -> amp * sin (2*pi*freq*t + phase)) time
30  -- Sine wave source actor
31  s_sine :: Signal Double
32  s_sine = sineGenerator (1, sineFreq, sinePhase) [0, 1/
        sampleFreq .. (2^order - 1)/sampleFreq]
33  -- Noise source actor
34  s_noise :: String -> Signal Double
35  s_noise noise_inp = signal $ map (\x -> read x ::
        Double) (splitOn "\n" noise_inp)
36
37  -- Format the output to save in a file
38  outputFormat :: (RealFloat a, Show a) => Signal a ->
        String
39  outputFormat NullS = []
40  outputFormat (x:-xs) = show x ++ "\n" ++ outputFormat
        xs
41
42  main = do
43    noise <- readFile "NoiseSignal.txt"
44    let s_out = freqSpec s_sine (s_noise noise)
45    writeFile "OutputSignal.txt" $ outputFormat s_out
```

Figure 23.3c shows the output generated by the FSS ForSyDe implementation. Since the same random signal was used in both Ptolemy II and ForSyDe implementations, the result displayed by Fig. 23.3c matches the result of Fig. 23.3b.

23.5   Related Work

A comparison of object-oriented (OO) model-based specification languages was performed in [21]. They argued that OO concept leads to profits including complexity management, secure data, reuse and polymorphic expressions. They presented differences between the languages OCL, Z and VDM.

Regarding formal verification tools, [22] presented a semantics comparison of two untimed SystemC/Transaction-Level Modeling, the non-preemptive semantics and the concurrent one. They used Lotos and CADP SystemC toolboxes as a framework to implement comparison experiments, concluding that, with an encoding contribution in Lotos, the concurrent semantics generalize the non-preemptive one, with similar semantics although using global lock.

Edwards et al. [1] presented a study on the design of embedded systems, analyzing principles that should be the base for formal modeling, specification, validation and synthesis. To describe modeling, they went through MoCs definition, also discussing about modeling languages and their interactions with the implementation languages. When discussing validation, they mentioned a range of software and hardware co-simulation techniques, and formal verification methods.

23.6   Conclusion

We presented in this paper analysis and comparison of two open-source frameworks that support formal modeling and system simulation, namely Ptolemy II and ForSyDe. We highlighted their main differences considering aspects such as framework maturity, project size, programming paradigm, availability, scalability, model execution, model simulation, programming entry method, and underlying language. To support the comparison, we modeled two systems using synchronous reactive (SR) and synchronous dataflow (SDF) MoCs.

By analyzing the compared aspects, the following can be stated. When compared to ForSyDe, Ptolemy II has been developed for a long time and counts on a wider development team. Besides, it has a more intuitive interface and the advantage of object-oriented programming when creating new actors or MoCs. On the other hand, ForSyDe elevates the abstraction level and has better performance when executing/simulating models. It also has better scalability, as it does not depend on any GUI.

As future work, one could compare these frameworks describing their tools and libraries for modeling other MoCs or heterogeneous systems.

Acknowledgements The authors thank *Ecossistema Negocios Digitais Ltda*, FCMF and ITA for their financial support in this paper.

References

1. Edwards, S., Lavagno, L., Lee, E., Sangiovanni-Vincentelli, A.: Design of embedded systems: formal models, validation, and synthesis. Proc. IEEE **85**(3), 366–390 (1997)
2. Ptolemaeus, C. (ed.): System Design, Modeling, and Simulation Using Ptolemy II. Ptolemy.org, Berkeley (2014)
3. Sander, I., Jantsch, A., Attarzadeh-Niaki, S.-H.: ForSyDe: system design using a functional language and models of computation. In: Handbook of Hardware/Software Codesign, pp. 1–42. Springer, Netherlands (2016)
4. Stuijk, S., Geilen, M., Basten, T.: SDF3: SDF for free. In: Sixth International Conference on Application of Concurrency to System Design (ACSD'06), pp. 276–278 (2006)
5. Simulink Documentation. [Online]. Available: https://www.mathworks.com/help/simulink/index.html
6. Davis, D., Jabeen, I.: Learning in the gnu/linux community. In: Proceedings of the 2011 Conference on Information Technology Education, ser. SIGITE '11, pp. 21–26. ACM, New York (2011). [Online]. Available: https://doi.org/10.1145/2047594.2047600
7. Jantsch, A.: Models of embedded computation. (2005) Embedded Systems Handbook, Chapter Models of Embedded Computation. CRC Press, Boca Raton
8. Lee, E., Sangiovanni-Vincentelli, A.: A framework for comparing models of computation. IEEE Trans. Comput. Aided Des. Integr. Circuits Syst. **17**(12), 1217–1229 (1998)
9. Benveniste, A., Berry, G.: The synchronous approach to reactive and real-time systems. Proc. IEEE **79**(9), 1270–1282 (1991)
10. Lee, E., Messerschmitt, D.: Synchronous data flow. Proc. IEEE **75**(9), 1235–1245 (1987)
11. B.-O.: University of California. Ptolemyii download page. [Online]. Available: http://ptolemy.berkeley.edu/ptolemyII/ptII11.0/index.htm (2018)
12. S. KTH University.: Forsyde tools page. [Online]. Available: https://forsyde.github.io/tools.html (2018)
13. Sander, I., Jantsch, A.: Formal system design based on the synchrony hypothesis, functional models, and skeletons. In: Proceedings Twelfth International Conference on VLSI Design. (Cat. No.PR00013). IEEE (1999)
14. Ungureanu, G., Sander, I.: A layered formal framework for modeling of cyber-physical systems. In: Design, Automation & Test in Europe Conference & Exhibition (DATE). IEEE (2017)
15. Gosling, J., Joy, B., Steele, G.: The Java Language Specification. Addison-Wesley Professional, Boston (2000)
16. The Haskell Purely Functional Programming Language Home Page (2018). [Online]. Available: https://www.haskell.org
17. Leung, M.-K., Mandl, T., Lee, E.A., Latronico, E., Shelton, C., Tripakis, S., Lickly, B.: Scalable semantic annotation using lattice-based ontologies. In: Model Driven Engineering Languages and Systems, pp. 393–407. Springer, Berlin/Heidelberg (2009)
18. Sander, I.: The forsyde methodology. In: Swedish System-on-Chip Conference (2002)
19. Horita, A.Y.: Ptolemy II and ForSyDe Source Code Examples (2018). [Online]. Available: https://github.com/AugustoHorita/ptolemy-forsyde-examples
20. Sander, I., Jantsch, A.: Modelling adaptive systems in forsyde. Electron. Notes Theor. Comput. Sci. **200**(2), 39–54 (2008). [Online]. Available: https://doi.org/10.1016/j.entcs.2008.02.011
21. Jalila, A., Mala, D.J.: Object-oriented model-based specification languages: a comparison. SIGSOFT Softw. Eng. Notes **39**(5), 1–4 (2014) [Online]. Available: https://doi.org/10.1145/2659118.2659132
22. Helmstetter, C., Ponsini, O.: A comparison of two SystemC/TLM semantics for formal verification. In: 2008 6th ACM/IEEE International Conference on Formal Methods and Models for Co-Design. IEEE (2008)

Maria Cristina Tenório, Roberta Vilhena Vieira Lopes, Joseana Fechine, Tarsis Marinho, and Evandro Costa

24.1 Introduction

The software development is a non-trivial task, which may involve a complex framework composed of different artifacts of Software Engineering, whose main objective is to deliver a quality product that meets the requirements specified by the business [4]. In order to guarantee this quality for the software in development, the software engineering established a set of activities, separated in three stages, that must compose the process of construction of the systems: Verification, Validation and Testing.

The Mutation Test, or Mutation Analysis, emerged in the late 1970s as a defect-based technique useful for guiding the test process [5, 9]. It is a technique based on defects, which assumes that competent programmers write correct programs, or very close to the correct one (hypothesis of the competent programmer). Based on this hypothesis, it can be stated that errors are introduced in the programs through small syntactic deviations that, although they do not cause syntactic errors, alter the semantics of the program and, consequently, lead the program to an incorrect behavior.

Thus, the mutation test identifies the most common syntactic deviations and, based on changes made by mutation operators, on the program to be tested, assists the tester in writing test cases that reveal the transformations that would lead to the construction of an incorrect program [4]. In addition, this technique also explores the hypothesis of the coupling effect, in which the set of tests capable of revealing the small changes in the code, are also able to differentiate programs with more complex defects, that is, real defects (Table 24.1).

Mutation analysis is a strategy that, based on coverage, defines the test criteria [14]. To fully meet the coverage criteria it is necessary that all the useful mutants be killed [6]. Thus, in mutation tests, a suitable test set is considered when the test suite reaches 100% score on its mutation score [16].

Although it has proven effective in the selection of test data, mutation tests have a high computational cost. One of the main factors that make the cost so high is the generation, during the mutation process, of duplicate mutants and the equivalent (useless mutants) [6]. Duplicate mutants are those that, although they have distinct syntax, are semantically the same. Already the equivalent mutants are semantically similar to the original code. Another problem related to useless mutants is that they eventually inflate the mutation score, since the number of dead increases. Thus, one of the major challenges of software engineering for mutation testing is to reduce the set of generated mutants while maintaining quality of the corresponding tests (Fig. 24.1).

The term subsumption was initially used in coverage tests by [3], establishing that subsumption occurs when a criterion C1 subsumes, or dominates, a criterion C2 if, and only if, each set of execution paths P satisfies C1 also satisfies C2 (Table 24.2).

M. C. Tenório (✉) · E. Costa
Federal University of Campina Grande, Campina Grande, PB, Brazil

Federal University of Alagoas, Maceió, AL, Brazil
e-mail: cristinaescarpini@copin.ufcg.edu.br; evandro@ic.ufal.br

R. V. V. Lopes
Federal University of Campina Grande, Campina Grande, PB, Brazil
e-mail: roberta@ic.ufal.br

J. Fechine
Federal University of Alagoas, Maceió, AL, Brazil
e-mail: joseana@computacao.ufcg.edu.br

T. Marinho
Federal Institute of Alagoas, Arapiraca, AL, Brazil
e-mail: tarsis.souza@ifal.edu.br

© Springer Nature Switzerland AG 2019
S. Latifi (ed.), *16th International Conference on Information Technology-New Generations (ITNG 2019)*,
Advances in Intelligent Systems and Computing 800,
https://doi.org/10.1007/978-3-030-14070-0_24

Table 24.1 Matrix of mutation

	t_1	t_2	t_3	t_4
m_1	1	1	0	0
m_2	1	0	0	1
m_3	0	1	0	0
m_4	1	1	1	1

$$MS(P,T) = \frac{K}{(M-E)} \text{, where:}$$

P: program under test
T: test suite
K: number of killed mutants
M: number of generated mutants
E: number of equivalent mutants

Fig. 24.1 Mutation score

Table 24.2 Initial population – GAADT

Chromosome	Adapt
$\{< m_1, m_2, m_0 >\}$	0
$\{< m_1, m_3, m_0 >\}$	2
$\{< m_1, m_4, m_0 >\}$	0
$\{< m_2, m_1, m_0 >\}$	0
$\{< m_2, m_3, m_0 >\}$	0
$\{< m_2, m_4, m_0 >\}$	0
$\{< m_3, m_1, m_0 >\}$	0
$\{< m_3, m_2, m_0 >\}$	0
$\{< m_3, m_4, m_0 >\}$	0
$\{< m_4, m_1, m_0 >\}$	2
$\{< m_4, m_2, m_0 >\}$	2
$\{< m_4, m_3, m_0 >\}$	2

Mutation testing has proven to be an effective way to test programs. It is able to simulate the effect of other white box testing techniques, while providing improved fault detection. Several researchers have invested in strategies to reduce the cost of mutation analysis, with the main focus being the non-generation of these useless mutants [1, 12, 13, 15]. One of the proposed solutions was the selective mutation, where the number of mutation operators is limited to a small set, chosen carefully, causing a small number of mutants to be created. One of the problems with this technique is that although the number of mutants generated is smaller than if we used all the available operators, it does not guarantee that redundant mutants are not generated, causing considerable noise to remain in the mutation score. For example, some mutants are killed by almost all tests. Thus, deleting these mutants from the set to be used does not affect which tests are chosen, but results in a different mutation score [11].

Another proposed solution for mutant redundancy is the Minimal Mutation, a recent research that accurately defines redundancy among mutants, identifying mutants called Mu-

tators. The notion of dominance has traditionally been used to compare test criteria: a C1 criterion subjects C2 if each set of tests satisfying C1 also satisfy C2 [10]. Thus, it can be said that given a set of M mutants any is said to be dominant D a minimal subset of M, so that any test set fitting for D will be suitable for M. The dominence can be roughly classified as static or dynamic decrease.

Thus, contributing to the creation of tools the present work presents a model for automatic identification of subsumption in mutation tests, using an instance of the GAADT – genetic algorithm based on abstract data types. We compared our model to the dynamic subsumption method proposed by [2] and then the visualization model from the generation of graphs presented by [10] are discussed.

24.2 Genetic Algorithms

In the 1970s, a University of Michigan doctoral student in computer science called John H. Holland attempted to develop a computational method that would be able to address phenomena generated by complex adaptive systems. In the course of his work, Holland realized that there was a clear similarity between the phenomena he studied and the process of evolution of the species, for just as the interaction between the adaptive agents determined the result of the phenomena investigated by him, the interaction between environmental factors determined the next population of a given region. Based on this finding, Holland proposed a computational method to simulate the process of evolution of the species, called Genetic algorithm [8].

In Holland's perception, the process of biological evolution is summarized in the transformation of one population into another, by the action of selection, crossing, mutation, inversion and substitution operators. In this perception, the population is a fixed-length vector of chromosomes, which are also fixed-length vectors of elements belonging to the set 0,1. Holland denominates locu the positions of the chromosome, allele the value that occupies a locu in the chromosome and gene to the set of alleles that can occupy a locu in the chromosome (Table 24.3).

The operators of the genetic algorithm of Holland behave as follows. First, the selection operation chooses the population chromosomes that can generate offspring for the next population. Next, the crossing, mutation and inversion operations are applied to the selected chromosomes in order to generate new chromosomes. Finally, the substitution constructs the new population with some of the chromosomes of the current population and the population of chromosomes generated.

More recently, the name genetic algorithm has been used to name a field of evolutionary computation [7] that comprises both the algorithms inspired by the genetic algorithm

Table 24.3 Population – 1st generation

Origin	Chromosome	Adapt
P_0	$\{< m_1, m_3, m_0 >\}$	2
P_0	$\{< m_4, m_1, m_0 >\}$	2
P_0	$\{< m_4, m_2, m_0 >\}$	2
P_0	$\{< m_4, m_3, m_0 >\}$	2
cross	$\{< m_4, m_1, m_0 >, < m_1, m_3, m_0 >\}$	6
cross	$\{< m_4, m_1, m_0 >, < m_4, m_2, m_0 >\}$	6
cross	$\{< m_4, m_1, m_0 >, < m_4, m_3, m_0 >\}$	6
cross	$\{< m_4, m_2, m_0 >, < m_4, m_3, m_0 >\}$	6
mut	$\{< m_1, m_3, m_4 >\}$	3
mut	$\{< m_4, m_1, m_3 >\}$	6
mut	$\{< m_4, m_1, m_2 >\}$	6
mut	$\{< m_1, m_3, m_4 >\}$	3

Table 24.4 Population – 2nd generation

Origin	Chromosome	Adapt
P_1	$\{< m_4, m_1, m_0 >, < m_4, m_2, m_0 >\}$	6
P_1	$\{< m_4, m_1, m_0 >, < m_4, m_3, m_0 >\}$	6
P_1	$\{< m_4, m_2, m_0 >, < m_4, m_3, m_0 >\}$	6
P_1	$\{< m_4, m_1, m_3 >\}$	6
P_1	$\{< m_4, m_1, m_2 >\}$	6
cross	$\{< m_4, m_1, m_0 >, < m_1, m_3, m_0 >, < m_4, m_2, m_0 >\}$	24
cross	$\{< m_4, m_2, m_0 >, < m_4, m_1, m_0 >, < m_4, m_3, m_0 >\}$	12

of Holland to work the limitations of this algorithm when applied to problems of different natures, as the "limitation" of the fixed size of the chromosome and also of the fixed size of the population, as well as the artificiality of the selection operation and the renewal of the generated populations. A proposal to work on these limitations was proposed by [17] and called the genetic algorithm model based on Abstract Data Types (GAADT) (Table 24.4).

In GAADT the adoption of the idea of fixed representation for the chromosome or population, as well as the definition of different genetic operations for each class of problem and type of representation adopted, will be treated with abstract idea types and data associated with a more real modeling of the Genetic procedures:

- for problems with chromosome and population representation, GAADT adopted a representation of the stratified chromosome at three levels of perception (chromosome, gene and base) to model a representation of the chromosome independent of the solution adopted and the problem to be treated. Depending on the interpretation given to the types (chromosome, gene and base) it is possible to reproduce any of the representations of existing chromosomes;

- for problems concerning the quality of chromosome adaptation to the environment, GAADT exists an environment modeling defined as a structure in which one of the components is the population. According to this model, environmental changes are seen as the beginning of a new period of evolution during which the chromosomes of the current population will suffer the action of the genetic operators in order to construct a new population formed only by chromosomes that satisfy the requirements of the current environment. After the evolution period comes the period of stagnation, during which the population does not evolve. The period of stagnation is finalized when a new environmental change occurs, initiating a new cycle of a period of evolution followed by a period of stagnation. The result of the problem for the current environment is the most adapted chromosome in the population of stagnation reached, if it is compatible with a stopping condition, which should be defined during the design period of the GAADT;

- for the problems arising from the need to define a different genetic operator every time the chromosome representation changes, was solved by specifying operators based on the abstract data types. Soon the instantiation of the types genes, chromosomes and population can be applied to all operators without problems;

- for problems arising from the generation of the new population, in GAADT a chromosome will die only if its adaptation to the current environment is less than a stipulated value, but before dying, in an attempt to survive this chromosome will generate all possible mutations, to if it is certain that it has no chance of adapting to the environment, to the problems of premature convergence, the GAADT avoids this fact with the representation of the population by a set, which can have any size and the generation of a population of all possible chromosomes whenever a genetic operator of crossing or mutation is called.

24.3 GAADT Specification for Dynamic Dominance Identification

In this section we will present an instantiation of the base, gene, chromosome and population types for the proposed problem, taking into account that the only information provided to GAADT is the *mutation* matrix, which relates the submitted mutant programs to the test. In GAADT the chromosomes will be represented by their genetic material, which has on the bases their elementary units of formation. The set $B = X \cup \{b_\lambda\}$, where X is the set of *mutants* considered in the analysis and b_λ is base-innocuous.

The base elements are grouped into sequences to form the characteristics (genes) of the chromosomes. But not every base sequence represents a characteristic for the chromosome. Therefore, there must be a law of formation to indicate how the bases must be grouped to form a given characteristic.

In this work, the law of formation of characteristics will be represented by the set of Axioms of Genes Formation (AFG), which should be defined for each case according to the semantics attributed to the gene. The set $G = \{g =< b_0, b_1, \ldots, b_n > |b_1, \ldots, b_n \in B \wedge g \in AFG\}$, where AFG is formed by the following axioms:

- $afg_1 = (g =< b_0, \ldots, b_n >\in G(b_0 = b_\lambda \implies (\forall i \in \{1, 2, .., n\}b_i = b_\lambda)))$, says that if the root of the dominance tree is base-innocuous, then all leaves must also contain an innocuous mutation;
- $afg_2 = (g =< b_0, \ldots, b_n >\in G(b_0 \neq b\lambda \implies (\forall i \in \{1, 2, .., n\}(b_0 \neq b_i)) \wedge (\forall j \in \{1, 2, .., n\} brokentests(b_0) \subseteq brokentests(b_j)))))$, says that if the root of the dominance tree is different from the base-innocuous then the information contained in the root must be different from the information contained in all the leaves, and that the root-broken tests should contain the tests broken by the leaves;
- $afg_3 = (\forall g =< b_0, \ldots, b_n \in G((\forall i \in \{0, .., n\}(b_i \neq b_\lambda)) \implies (\forall i, j \in \{0, .., n\}(i \neq j \implies b_i \neq b_j)) \cap (\forall j \in \{1, .., n\}(linha(b_0) < linha(b_j)))))))\}$, says that if the root and leaf information of the simple dominance tree are all different from base-innocuous then they must also be different from each other.

The $g_\lambda =< b_\lambda, b_1, \ldots, b_n >$ sequence will be called the gene-innocuous and the $g =< b_0, b_1, b_2, \ldots, b_n >$ sequence will be called the single dominance tree of the b_0 mutation on the $b_1, b_2, \ldots, b_n$ mutations, which can be graphically represented as a non-human tree of a only depth level with root b_1, with leaves $b_1 \ldots, b_n\}$. The $b_0, b_1, \ldots, b_n$ elements of g will be denoted from now on by $root(g)$ and $leaf(g)$.

Genes are grouped into clusters to form the population's chromosomes. The set of genes $c = \{g_1, g_2, \ldots, g_n\}$ that makes up a given chromosome serves to identify this chromosome within the population. But not every set of genes can be considered chromosomes, to inform when a set of genes will have formed a chromosome will be defined the set of Axioms of Formation of Chromosomes (AFC), which should be defined for each situation according to the adopted semantics for the chromosome. The set chromosome $C = \{c|c \subseteq G \wedge c \in AFC\}$, where $AFC = \{afc_1 = (\forall c \in C(\exists_1 g_i \in c(/\exists g_j \in c - \{g_i\}(root(g_i) = root(g_j)) \wedge (root(g_i) \in leaf(g_j)))\}$. The set $c_\lambda = \{g_\lambda\}$ will be called the chromosome-innocuous and the set $c = \{g_1, g_2, \ldots, g_w\}$ will be called a compound dynamic dominance tree. The chromosome formation axiom states that each chromosome has only one root, so it is a tree formed by the combination of the simple trees represented by the genes (Table 24.5).

The population is a set of all chromosomes constructed according to the axioms of gene and chromosome formation. Soon the population in this modeling will be precisely the result of the problem.

GAADT works with two types of genetic operators: reproduction and mutation. The genetic reproduction operator is characterized by combining the genes of two parent chromosomes to form their child chromosomes, whereas the genetic operator of mutation is characterized by altering the genes that make up one chromosome to form another mutant chromosome (Table 24.6).

The gene of the parent chromosomes for a given characteristic that will be part of the child chromosomes is the one that best satisfies the constraints of the problem on the characteristic expressed by this gene, which will be called the dominant gene. Given two genes g_1 and g_2, they are said to express the same feature if there is a relevant attribute to the problem in focus that is satisfied by the given genes. The dominant gene is an $domi : G \times G \to G$ as can be seen in (24.1).

$$domi(g_i, g_j) = \begin{cases} g_\lambda, & if \quad (\neg same(g_i), g_j)), \\ g_i, & if \quad (same(g_i, g_j) \wedge (degree(g_i) \leq degree(g_j)), \\ g_j, & if \quad (same(g_i, g_j) \wedge (degree(g_i) > degree(g_j)). \end{cases} \quad (24.1)$$

It is said that two genes express a same characteristic, $same : G \times G \to Boolean$, if they share a property relevant to the problem in question, in the case of the construction of dominance trees the relevant characteristic will be the root of the simple trees that compose it.

$$same(g_i, g_j) = \begin{cases} true, if & (root(g_i) = root(g_j)), \\ false, cc & . \end{cases} \quad (24.2)$$

In this work, the degree of adaptation of a gene is given by an $grau : G \to \Re, degree(g) = card\{x|\forall f \in leaf(g)(x \in (broken(root(g))broken(f)))\} * card\{x|\forall x \in B((x \subset leaf(g) \vee (x = root(g))\}$ function, where $card(W)$ is the size of the set W. The degree of adaptation of the innocuous gene g_λ is less than the degree of adaptation of any other element of the gene type.

The production of new chromosomes during the evolutionary process of a population serves to direct the search

Table 24.5 Population – 3rd generation

Origin	Chromosome	Adapt
P_2	$\{< m_4, m_1, m_0 >, < m_1, m_3, m_0 >,$ $< m_4, m_2, m_0 >\}$	24
P_2	$\{< m_4, m_2, m_0 >, < m_4, m_1, m_0 >,$ $< m_4, m_3, m_0 >\}$	12

Table 24.6 Population – 4th generation

Origin	Chromosome	Adapt
P_3	$\{< m_4, m_1, m_0 >, < m_1, m_3, m_0 >,$ $< m_4, m_2, m_0 >\}$	24

for more adapted chromosomes through the transmission of the characteristics of greater degree of adaptation present in the chromosomes of the current population. The adaptation of a chromosome is given by the function $adapt : C \to \mathbb{R}$, $adapt(c) = \sum_{g \in c} \theta_{g,c} \times grau(g)$, where $\theta_{g,c}$ is the weight with which the g gene contributes to the adaptation of the c chromosome.

The weight parameter in the adaptation calculation of a chromosome will be used to represent epistasis among the genes that make up the chromosome. In other words the presence of one gene may inhibit or potentiate the characteristic of another gene. In the instantiation of GAADT this parameter will be treated as: $\theta_{g,c} = 1 + card\{x | \forall y \in c - \{g\}(root(y) \in (leaves(g))\}$.

The crossing operation receives two parent chromosomes, able to cross, and returns a population whose chromosomes are formed by the fertilization of the dominant genes present in the genetic material of the given chromosomes. In this instantiation it will be allowed that all the chromosomes of the population cross each other. Fertilization is an $fec : C \times C \to \mathcal{P}(G)$, $fec(c_i, c_j) = \{g | \forall g_1 \in c_1 \forall g_2 \in c_2(g = domi(g_1, g_2))\}$ function. And crossover is a $cross : C \times C \to \mathcal{P}(P)$, $cross(c_i, c_j) = \{c | c \subseteq fec(c_1, c_2)\}$ function.

The genetic operator of mutation is composed of the functions of exchange, such that the chromosomes resulting from the action of these operators will present part of the genes contained in the chromosome that gave rise to it. The swap operation is an $swap : C \to \mathcal{P}(C)$ function.

$$swap(c, I, O) = \begin{cases} (c \cup I) - O, if(((c \cup I) - O) \in AFG), \\ (c \cup I), if((c \cup I) \in AFG), \\ c - O, if((c - O) \in AFG), \\ c, se \quad cc. \end{cases}$$

(24.3)

The mutation is a function $mut : C \to \mathcal{P}(C)$, $mut(c) = \{x | \exists I, O \in \mathcal{P}(G)((card(I) \leq \frac{card(c)}{2}) \land (card(O) \leq \frac{card(c)}{2}) \land x = swap(c, I, O) \land (degree(x \geq degree(c))\}$.

A genetic algorithm operates on populations of chromosomes that evolve according to the characteristics of an $\mathcal{A}$ environment. A $\mathcal{A}$ environment is an 8-tuple $P, \mathcal{P}(P), Rq, AFG, AGC, Tx, Ic, P_0$, where:

- P is the population,
- $\mathcal{P}(P)$ is the power set of P,
- Rq is the set of requirements (characteristics expressed through formulas in a first order language) of the problem that influence the genealogy of the population P,
- AFG is the set of axioms of formation of the genes of the chromosomes of the population P,
- AFC is the set of axioms of formation of the chromosomes of the population P e
- Tx is the set of pairs of chromosomes (x, y), where x is a chromosome constructed from the chromosome y, by the action of the crossing or mutation, thus registering the genealogy of the chromosomes belonging to the populations generated by the GAADT during its execution,
- $\hat{I}c$ is the set of genealogical operators that act on the population P,
- P_0 is a sub-population belonging to $mathcal P$, called the initial population, with at least one chromosome.

GAADT is an $GAADT : \mathcal{A} \to \mathcal{A}$ function, the population being only one element of this environment. The chromosomes of population P_t are the chromosomes of populations $P_0, P_1, \ldots, P_{t-1}$ that still satisfy the requirements of problem Rq. The function $GAADT$, working with three important predicates the criterion of preservation of the chromosomes $p_{corte} \in Rq$, the criteria of stop maximum number of iterations $T \in Rq$ and value of the adaptation of the chromosomes considered satisfactory for the result of the problem in analysis $K \in Rq$.

$$GAADT(\mathcal{A}_t) = \begin{cases} \mathcal{A}_t, if \quad (P_t = \{c | \forall c \in P_t(adapt(c) \geq K)\}, \\ \mathcal{A}_t, if \quad (t = T), \\ GAADT(\mathcal{A}_{t+1}), cc. \end{cases}$$

(24.4)

where $P_{t+1} = cross(a, b) \cup mut(c) \cup p_{cut}(P_t)$ com e $a, b, c \in P_t$.

24.4 Results and Discussion

In order to develop a graphical model to describe redundancy among mutations, [11] probed the dynamic subsumption relations for a specific program. His work demonstrates the generation of a dynamic subsumption graph from a simple example, where the initial dominance relationship is established through a mutation matrix, correlating tests that break with the mutants used to perform the tests.

The dynamic subsumption states that a mutant mx subsumes dynamically my in relation to the test set T if, and only if, there is at least one test that matches mx, where every test that kills m_x also kills m_y. f mutant x was killed by the test and $S\ (i,\ x)\ S\ (i,\ y),\ i = 1 .. \mid T \mid$, we say that m_x dynamically subsumes my with respect to T. Taking into account only the context of the tests, the work in question seeks to establish how many and which mutants are necessary to evaluate a certain set of tests, where the minimum set of Mutants M is useful when it does not have redundant mutants. This set is considerably smaller than those provided by current approaches regarded as good practice in mutation testing.

In addition, the dynamic subsumption differs from the notion used in white box mutation analysis in one crucial respect: it does not allow "empty" subsumption. Another interesting aspect is its transitive nature. The dynamic subsumption is modeled as a logical and thus transitive implication. Thus, if a dynamically subsumed mutant is removed from a set of mutants, such deletion does not affect the subsuming relation in the other dynamically subsumed mutants. Thus, dynamically subsumed mutants can be removed arbitrarily.

Our experiment was based on the mutation matrix used by [11], since our main objective is to demonstrate the possibility of automatic identification of the dynamic subsumption relation in mutation test. Since the term "mutation" is common in the two areas studied, it is necessary, for the purpose of understanding this work, to distinguish them. Thus, for evolutionary computation, *mutation* refers to the genetic operation that arbitrarily changes one or more components of a chosen structure, generating new individuals in the population. In the test context, **mutation** is the process of generating artificial defects of a given unit of code. For this, the code is duplicated and suffers small syntactic changes.

This matrix shows a relationship between the artificial defects (mutants) and the tests used during the mutation analysis. The numbers 0 and 1 represent "test did not break" and "test broke", respectively. Thus, the initial population was formed by all genes with two or more base elements, because to find the dominance will require at least two **mutations**. Below is the initial population modeled for the GAADT genetic algorithm and its degree of adaptation.

Note in this population the presence of the base m_0 which is considered the innocuous base of this instantiation of ours. Chromosomes with zero matching occur whenever the tests broken by leaf knots are not contained in the root nodes of their gene. The chromosomes of this population that have an adaptation above the average 0.66 will be kept in the next population and along with the descending chromosomes of the current population.

The next iteration of the GAADT will consider as cutoff the value of the arithmetic mean of the chromosomes of this

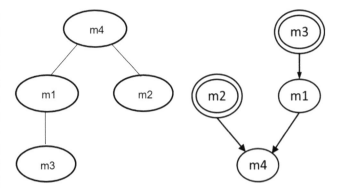

Fig. 24.2 GAADT Chromosome vs. solution tree [11]

population which is 4.27. Generating new population shown below.

In this iteration the genetic operators of GAADT began the construction of chromosomes that already existed in the population under construction, and since this algorithm does not work with several copies of the same chromosome, they do not appear in the table, as is the case of some chromosomes generated by crossing and of all chromosomes generated by mutation, the cutoff value for this population is 9.42.

At this moment the GAADT will present a population formed by only one chromosome, that for the current environment has no more how to improve. The moment when it is said that a configuration of evolutionary stagnation was reached, this population being the solution of the problem.

In GAADT, the chromosome is composed of elemental branches – genes. Each gene is represented by a simple n-ary tree (only one level) where the first element is the root and the other elements are the leaves. Considering that a chromosome has a set of genes, its representation is given by the combination of several n-ary trees, where the child node of one tree is the leaf of another. The Fig. 24.2 shows the solution tree generated by the GAADT (left) and the solution tree presented by [11] (right). It can be observed that, although the n-ary tree presented by GAADT is inverted, they are similar to each other, demonstrating that from the proposed model it is possible to automatically identify dynamic subsumption relations in mutation tests.

24.5   Conclusion

In this paper we present an automated model based on a genetic algorithm approach, aiming the identification of dynamic subsumption in mutation tests. Concerning the used Genetic Algorithm, it is necessary to consider some theoretical and operational aspects, which are a little different of the classical approach and makes it satisfactory to reach the solution of the addressed problem, as proposed by Holland [8]:

- **Representation Restrictions:** Holland approach adopts the binary alphabet to represent a chromosome, thus limiting the scope of the possible domains to be treated, as representations by means of graphs. On the other hand, GAADT approach is based on TAD and thus allows the user to instantiate their representation in up to three levels of perception (base, gene and chromosome);
- **Restrictions on operator behavior:** The GA proposed by Holland works with three genetic operators: crossing a cut-off point, mutation by complement and inversion. All these operators for the problem we are dealing with are likely to generate a kind of individuals "monsters", i.e., chromosomes that do not belong to the solution space of the problem. Using GAADT, we do not have that negative effect, that is guaranteed by the use of axioms in the formation of genes and chromosomes, leading to every chromosome generated belongs to the solution space of the proble;
- **Convergence:** With respect to the GA of Holland, it conveys but does not guarantee that its convergence is directed towards a globally optimal solution, since by the schema theory there is a probability that the algorithm converges to a great location. The theory of evolutionary processes assurances that GAADT as being a monotonically growing function converges in the direction of global optimum, guaranteed;
- **Time of convergence:** To the best of our knowledge, none of the genetic algorithms in the literature are concerned with securing a small convergence time since they apply to solve NP-complete problems or NP-hard, but technologically this has been addressed by the community through hardware or microprocessor implementation.

Thus, to evidence the importance of our approach, we used a baseline example with subsumption graph already established in the literature. Such example offers a parameter for comparison with the output of our genetic algorithm, ensuring that for the same problem, the result of our approach was similar to result found from a manual evaluation. One of the central contributions of our proposal is that the automated identification of the dynamic subsumption reduces the effort in generating the subsumption graphs, which represents an important step towards the success in the mutation test area. As future work, we intend to apply our approach to complex systems, and incorporate it into mutation tools for testing.

References

1. Adamopoulos, K., Harman, M., Hierons, R.M.: How to Overcome the Equivalent Mutant Problem and Achieve Tailored Selective Mutation Using Co-evolution. Springer, Berlin/Heidelberg (2004)
2. Ammann, P., Delamaro, M.E., Offutt, J.: Establishing theoretical minimal sets of mutants. In: Proceedings of the 2014 IEEE International Conference on Software Testing, Verification, and Validation, ICST'14, Washington, DC, pp. 21–30. IEEE Computer Society (2014)
3. Clarke, L.A., Podgurski, A., Richardson, D.J., Zeil, S.J.: A comparison of data flow path selection criteria. In: Proceedings of the 8th International Conference on Software Engineering, ICSE'85, Los Alamitos, pp. 244–251. IEEE Computer Society Press (1985)
4. Delamaro, M., Jino, M., Maldonado, J.: Introdução ao teste de software. In: Introdução ao Teste de Software. Elsevier Editora Ltda
5. DeMillo, R.A., Lipton, R.J., Sayward, F.G.: Hints on test data selection: help for the practicing programmer. Computer **11**(4), 34–41 (1978)
6. Fernandes, L., Ribeiro, M., Carvalho, L., Gheyi, R., Mongiovi, M., Santos, A., Cavalcanti, A., Ferrari, F., Maldonado, J.C.: Avoiding useless mutants. In: Proceedings of the 16th ACM SIGPLAN International Conference on Generative Programming: Concepts and Experiences, GPCE 2017, New York, pp. 187–198. ACM (2017)
7. Goldberg, D.E.: Genetic Algorithms in Search, Optimization and Machine Learning, 1st edn. Addison-Wesley Longman Publishing Co., Inc., Boston (1989)
8. Holland, J.H.: Adaptation in Natural and Artificial Systems: An Introductory Analysis with Applications to Biology, Control and Artificial Intelligence. MIT Press, Cambridge (1992)
9. Jia, Y., Harman, M.: An analysis and survey of the development of mutation testing. IEEE Trans. Softw. Eng. **37**(5), 649–678 (2011)
10. Kurtz, B., Ammann, P., Delamaro, M.E., Offutt, J., Deng, L.: Mutant subsumption graphs. In: Proceedings of the 2014 IEEE International Conference on Software Testing, Verification, and Validation Workshops, ICSTW'14, Washington, DC, pp. 176–185. IEEE Computer Society (2014)
11. Kurtz, B., Ammann, P., Offutt, J., Delamaro, M.E., Kurtz, M., Gökçe, N.: Analyzing the validity of selective mutation with dominator mutants. In: Proceedings of the 2016 24th ACM SIGSOFT International Symposium on Foundations of Software Engineering, FSE 2016, New York, pp. 571–582. ACM (2016)
12. Mathur, A.P., Wong, W.E.: Reducing the cost of mutation testing: an empirical study. J. Syst. Softw. **31**,185–196 (1995)
13. Mresa, E.S., Bottaci, L.: Efficiency of mutation operators and selective mutation strategies: an empirical study. Softw. Test. Verif. Reliab. **9**, 205–232 (1999)
14. Offutt, A.J., Untch, R.H.: Mutation testing for the new century. Chapter Mutation 2000: Uniting the Orthogonal, pp. 34–44. Kluwer Academic Publishers, Norwell (2001)
15. Offutt, A.J., Rothermel, G., Jefferson, A., Rothermel, O.G., Zapf, C.: An experimental evaluation of selective mutation. In: Proceedings of 1993 15th International Conference on Software Engineering, Baltimore, pp. 100–107 (1993)
16. Offutt, A.J., Lee, A., Rothermel, G., Untch, R.H., Zapf, C.: An experimental determination of sufficient mutant operators. ACM Trans. Softw. Eng. Methodol. **5**(2), 99–118 (1996)
17. Vilhena Vieira Lopes, R.: Um algoritmo genético baseado em tipos abstratos de dados e sua especificação em Z. PhD thesis, Universidade Federal de Pernambuco, Recife

Mateus de Oliveira, Adriana Prest Mattedi, and Rodrigo Duarte Seabra

25.1 Introduction

Communication and knowledge have always been the basis of human civilization, but the difference now is that current communication is largely formed by technological networks, which give a new meaning to an old form of social organization, that is, as social networks [1]. Egler [2] states that techno-social networks allow constructing new possibilities of action for changes in public policies, in which individuals act directly on their objects of interest. Hence, relations with the State are redefined in relation to public policies, their implementation and effectiveness; to the democratic participation of municipal management; and to the implications of daily life, given by urban and behavioral reconfiguration. Examples of the networks contribution to mass mobilization are movements such as the Arab Revolution, the 15M Spanish, Occupy Wall Street, demonstrations against the government of Erdogan in Turkey in 2013, among others [3]. Similarly, collaborative self-organization occurs in communities to address violence issues and to cause changes in programs and policies regarding relevant community issues [4]. Examples of mobile applications used as tools for collaborative self-organization for violence of communities are Commisur, Life360, Agentto, ProtectMe, CityCop e Haus.

The social network formation guarantees the maintenance of cooperativeness, which is the central point for the success of human societies [5]. These networks organize human relationships, which are dynamic and change each time an individual breaks or creates connections with others. Virtual and real social networks currently merge through broad access to Internet-connected devices, which use software and websites that promote interactions among individuals, groups, and networks of dynamic relationships [6]. This scenario motivates researchers to seek an understanding the society's relations with technology. Thus, while it is possible to evaluate the use of these technological resources to mitigate the effects of unwanted events, such as robberies and kidnappings, it is also possible to evaluate them as a support tool for culture, socioeconomic reality, training and people attitude.

In this sense, the goal of this research is to understand the perception of users regarding the adoption of the Life360 app, focused on personal and family safety, based on the diffusion theory of innovation. Perception is here understood as the way in which individuals organize and interpret their sensory impressions in relation to facts, objects and people in order to understand their surroundings [7].

25.2 Theoretical Foundation

In the process of adopting a new technology, besides functional requirements, social aspects must also be observed. In Brazil, problems such as cellular operators' difficulty in offering their services to the whole Brazilian territory at an affordable price, poor distribution of income, and the deficit in digital inclusion reflect barriers to access Internet technology [8, 9]. The goal of digital inclusion is to increase the participation of the population in the use of technologies. However, for inclusion to actually occur, available physical access is not enough. This access should also consider the social, cultural and psychological context that occurs in stages, as shown in Fig. 25.1 [10].

M. de Oliveira
Institute of Production and Management Engineering,
Federal University of Itajubá, Itajubá, MG, Brazil

A. P. Mattedi · R. D. Seabra (✉)
Institute of Mathematics and Computing, Federal University of Itajubá,
Itajubá, MG, Brazil
e-mail: rodrigo@unifei.edu.br

© Springer Nature Switzerland AG 2019
S. Latifi (ed.), *16th International Conference on Information Technology-New Generations (ITNG 2019)*,
Advances in Intelligent Systems and Computing 800,
https://doi.org/10.1007/978-3-030-14070-0_25

Fig. 25.1 Four successive kinds of access in the appropriation of digital technology [10]

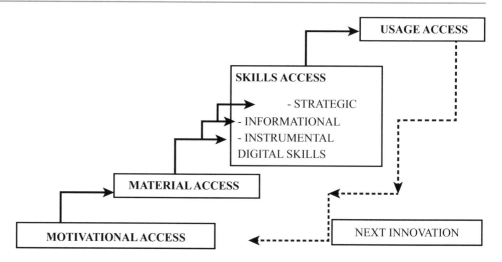

The first stage of the model refers to the motivation to use technology. The lack of this motivation may be related to the absence of utility and need of perception, or of financial resources or even for not feeling comfortable with the technology. The second stage (material access) refers to the physical means of access to technology. In the third stage of the model, the utilization degree of the technology is evaluated in three levels: instrumental ability – ability to work with hardware and software; informational ability – ability to search, select and process information in networks; and strategic ability – the ability to use technologies to achieve personal goals and to enhance one's social position. At the last level of the model is the actual or productive use, which can be measured by the time, application and diversity, the bandwidth of the connection and the intensity of the active and/or creative use of the Internet.

The motivational factor of access can still be observed from the perspective of the Rogers Innovation Theory [11]. In this case, diffusion is equivalent to the information dissemination process of an innovation. Access to this information can reduce an individual's degree of uncertainty or perception of risk in relation to innovation. Thus, diffusion is the process by which an innovation is communicated through certain channels over time, between certain elements of a social system. The social system and the communication channels are very important since the social and communication structures interfere with the diffusion of an innovation. The author argues that there is a temporal sequence for adopting a new technology that diffuses through five categories of adopters, starting with the innovators, followed by the first followers, then by the initial majority, followed by the later majority, and lastly by the latecomers.

In addition to these elements, the process of adopting a new technology depends on five attributes: Relative Advantage (degree to which an innovation is perceived to be better than its precursor); Compatibility (degree to which an innovation is perceived to be consistent with the existing values, needs, and past experiences of potential adopters); Complexity (relative degree in which innovation is perceived as difficult to understand and to use); Experimentation (degree to which an innovation can be experienced on a limited basis); and Observability (degree to which the results of an innovation are observable by others) [11].

Similarly, a new application will be accepted as the adoption process progresses among the categories of adopters, being disseminated by opinion makers in a particular social context. This progression is directly influenced by the perceptions of the adopters on each of the five attributes mentioned above.

Regarding the user's attitude towards sharing information in the networks, most people are apprehensive about privacy in relation to their location; privacy preferences are complex and vary based on a number of contextual factors [12]. Toch et al. [13] found evidence that people are more comfortable sharing their location when it refers to a place visited by a large and diverse group of people. However, Duh et al. [14] showed that the occurrence of errors in experiments performed in the field is higher than those performed in the laboratory motivated, among other factors, by the noise, movement and physical and mental conditions of the users in field environments. In the field of technological adoption, Pavlou [15] presents an evaluation model of e-commerce user acceptance that has been adapted from the TAM model. Regarding the relationship between smartphones and personal security, Save et al. [16] present a smartphone app based on IHC concepts to make it easier to activate the application's emergency functions. Regarding users' perceptions of security applications, Czeskis et al. [17] conducted a study of the perception of parents and adolescents about the use of security applications for smartphones.

25.3 Method

In order to evaluate the security and collaborative surveillance app, Commisur, Life360, CityCop, ProtectMe and Haus apps were pre-selected and tested for defining which would be used in the study. The selection criteria took into account the scenario "protection network based on communication and monitoring in a social group" and the expected functionalities were: (i) Group creation and control: for information security reasons, group moderators should have full control of who composed the network; (ii) Messages and alerts exchange: network members communicate through them to inform risk situations or ask for help; (iii) Geolocation incorporated into mobile devices: in case of need, it makes possible for aid to go directly to the applicant's place.

Life360 was chosen for having a creation and a control group, message exchange, alerts and geolocation. Other applications were discarded due to incomplete translation to the Portuguese language (regarding accessibility, consideration should be given to the heterogeneity of users' profiles [18]), the impossibility of group management (excluding unwanted members); and for not having all the expected functionality. Life360 is an application available for Android and iOS platforms used to alert groups of people, called circles, about the arrival or departure of the user in a region determined by the group. The features are based on the combination of location composite, via Global Positioning System (GPS) and Internet network, by Wi-Fi or by telephone operator data network. Other functions are available, such as individual message rooms and/or group and sending requests for help [19].

A field study [20, 21] was carried out in Itajubá city, MG, Brazil, with the participation of two distinct groups. The first one consisted of five people, four of whom had a technology higher education, aged between 18 and 39 years old and income between three and five Brazilian minimum wages. The second group consisted of four people with varied profiles regarding income, age, education and profession.

Since it is a security app test, the fear of sharing private information in an unknown application requires a kind of trust between participant and researcher. Given these circumstances, the participant's selection occurred by the non-probabilistic method of sample for convenience. Thinking aloud was conducted during the tests phase and voice records were utilized to transcript the user's perceptions over the app. The materials used were: participating user's smartphone, activity list and demographic and usage questionnaire. The activities list (Table 25.1) contains 21 tasks that guided users to use the Life360 app. With the demographic questionnaire, we sought information on sex, age, education, income, professional activity, with any type of device use. For the

Table 25.1 Activities list

Activities
A1 - Download and install the Life360 app
A2 - Create an account and sign in
A3 - Create a "Family" circle
A4 - Add a relative to the circle "Family"
A5 - Check your location on the map
A6 - Create a place with arrival and departure alert
A7 - Activate a family member to trigger alerts in the location created in the previous step
A10 - Check-in to your location
A11- Send the message "Hello friend!" to a member of the circle
A12 - Send message "Hello group!" to all the circle members
A13 - Send a picture message to a group member
A14 - Send Help alert (no need to add other phones)
A17 - Change settings to disable your location
A18 - Enable all the alert options in settings
A19 - Enable all the map elements in settings
A20 - Change the profile picture
A21 - Check notifications received by the application (messages and alerts)

last characteristic (use), five alternatives were available to classify participants according to their degree of ability to use smartphones: (1) only for calls and text messages; (2) for calls, sending messages and some applications such as Whatsapp, Facebook and Instagram; (3) to play and for some applications such as WhatsApp, Facebook and Instagram; (4) for everything, but there was some difficulty in setting up and in installing applications; (5) for everything, with good ability to configure and install applications.

According to the theory of the diffusion of innovation, the use questionnaire has six questions that seek to evaluate aspects that influence the process of adhesion to the technology. These questions were extracted from the PSSUQ (Post-Study System Usability Questionnaire), which aims to evaluate computer systems users' satisfaction [22], and is described below. Relative Advantage (Q1): *I believe the application contributes to improving my family's safety.* Compatibility (Q2): *I am satisfied with the application and intend to continue to use it.* Observability (Q3): *I want to encourage others to use this app*; Complexity (Q4): *It was easy to learn how to use the application* and (Q5): *I understood how the application works the first time I used it*; Experimentation (Q6): *I was able to complete tasks quickly using the application.*

In the beginning, initial instructions to perform the activities were given to participants and after 2 weeks of use, the results were collected.

25.4   Data Analysis

The respondent's profile is shown in Table 25.2. Participants are identified by the letter P followed by an identification number and family income is shown within the minimum wage range (US$287,35). Study level is shortened by: Pos (Postgraduate degree), Grad (Graduate degree), HS (High school) and Prim (Primary school) and professional activity areas are shortened: Tecn (technology), Educ (education), Stud (student), Adm (administration) and W (worker). The devices operating system for all the participants was Android, described with their trade names; and the last column (Use) shows the degree (from 1 to 5) to which respondents perceive their intensity of smartphone use.

Concerning the questionnaire on use, the answers obtained in question Q1 brought ethical and information security considerations. Respondent P1 understood that Life360 should be used to monitor children and that, among adults, some privacy issues may occur, leading to conflicts. The P2 perception was fear of having her data stolen and used for evil purposes; however, similarly to P5, she believes that in specific cases, Life360 can help. For P8, Q1 was seen as neutral, probably because he was an adolescent and did not understand the complexity of the factors involved in security and privacy issues. In contrast, P3 and P4 understood that the application has no risks and that its use may be beneficial. Another important point, indicated by P7, was the difficulty in monitoring her son transfer between home and school with the app since her child cannot take the device with him due to the risk of being robbed on the way, and still being able to disrupt a lesson.

Regarding question Q2, respondents expressed overall satisfaction. However, P1, P2, P5 and P8 did not understand the need to use Life360 since WhatsApp already works as a versatile application. P3 and P4 have seen various utilities for Life360, such as finding lost or stolen devices, meeting friends in an unfamiliar city and even notifying parents when they arrive at/from any destination in travel cases. P6, P7 and P9 understood that although Life360 may meet the needs

of family surveillance, they did not perceive a substantial sense of utility at the moment they used it. According to the technology access model [23], the motivational aspect is a barrier for users with profiles similar to the last four respondents (P6 to P9).

The objective in Q3 is to investigate the diffusion potential of Life360 as defined by Rogers [11, 24]. As a result, participants P1 to P5 would indicate Life360 if they realized the need for relatives or friends to monitor children or elderly parents. For P6 to P9, answers were vague, which shows their doubts about not being able to imagine a situation in which the indication would occur. In this case, income and education influence the participant's perception, as pointed out by Van Dijk [10].

The goal of Q4 was to analyze facility of use of Life360 and the degree of compliance with users. Higher skill users (level 5) pointed to several inconsistencies in function names, navigability, and lack of application responses about the results of their actions. On the other hand, participants with lower levels of use ability (levels 2 and 3) complained about connection failures, slow image sending or slow execution of the application. We realized that the experiences of using other applications might have evoked some expectation regarding usability. In this sense, lower skill users are less critical regarding the usability of the system, while those with greater familiarity with technology present a more accurate analysis. We must mention that, because it is a security application; Life360 presents a different proposal from Whatsapp, which would justify a logic of differentiated use.

Another point to note is the convergence between Compatibility and Complexity. For P6 and P9, the device used slowed down the operation app, and for P7 and P8, the Internet connection was unstable. At this point, the social aspect is latent and reveals two critical items. First, the lack of physical access of lower-income people to the resources necessary for the full and satisfactory use of features offered by the application; and second, interference of skill level and technical knowledge about the technology used in understanding factors inherent to the use and performance of functionalities.

Question Q5 evaluates the immediate understanding of application usage. In this case, respondents P1 to P5 gave more favorable responses than respondents P6 to P9. Probably, the social condition interferes with the understanding and the ability to judge usability results from this understanding, and not just from experiences.

Finally, the Q6 goal was to understand the respondents' perception of efficiency in basic procedures execution using Life360. Three points were brought up in the analysis of this question. The first one was pointed out by P1 when he says that the term "alert" is used for two distinct functions, which generated confusion and delayed task accomplishment. The

Table 25.2 Respondents' profile

	Sex	Age Range	Income	Schol.	Prof.	Device	Use
P1	F	18–39	05–10	Pos	Tecn.	Galaxy S6	5
P2	F	18–39	03–05	Pos	Tecn.	Moto G4	5
P3	M	18–39	03–06	Grad	Tecn.	Zenphone 6	5
P4	M	18–39	03–07	Pos	Tecn.	Moto G Pl	5
P5	F	18–39	03–08	Pos	Educ.	Moto G5 Pl	5
P6	F	18–39	01–02	HS	Stud	Galaxy Duos	3
P7	F	18–39	02–03	HS	Adm.	Moto Gt	2
P8	M	<18	01–02	Prim	Stud.	Gran Prime	2
P9	F	40–59	02–03	Prim	W.	Galaxy Duos	2

second point was raised by P3, who found it difficult to disable the location function. The third point was highlighted by participant P5, who considered the navigation a bit confusing. For P7 and P9, experimentation was not satisfactory given the low performance of Life360 in their smartphones, whereas, for P6 and P8, experimentation was partially satisfactory, because the connection instability made it difficult to send images.

25.5 Final Considerations

The descriptive analysis model proposed in this qualitative case study captured participants' use perceptions about Life360. Also, the access model to digital technologies allows understanding users' profiles regarding their position in access level to technology. Therefore, the model can be said that it is to be low-cost and easy-to-apply, which helps to understand the perception of new users of collaborative apps. For technology adherence to be successful, the social conditions and physical access of the target public have to be taken into account. This model thus allows identifying the limitations faced by the use of technology in different social contexts. In addition, critical usability factors could be observed during the process. We concluded that Life360 is pertinent to be used by people with better levels of education and income since this kind of user is closer to the technology strategic use level. This solution showed not to be convenient for use between couples. This fact limits Life360 to "parental control" since reports point to its use for this purpose. Some limitations of this research were detected. First, it is necessary to develop research that incorporates insecurity perception aspects caused by the use of Life360 over time in a given social context. Also, quantitative studies on the proposed model should be conducted, with statistical analyses of confidence and redundancy of the constructs used. Lastly, it is necessary to increase the sample size to allow quantitative and probabilistic analyses.

References

1. Castells, M., Cardoso, G.: A Sociedade em Rede: Do Conhecimento à Acção Política. Imprensa Nacional - Casa da Moeda, Belém (2005)
2. Egler, T.T.C.: Redes tecnossociais e democratização das políticas públicas. Sociologias. **12**(23), 208–236 (2010)
3. Antoun, H., Malini, F.: A internet e a rua: ciberativismo e mobilização nas redes sociais. Editora Sulina, Porto Alegre (2013)
4. Fawcett, S.B., et al.: Using empowerment theory in collaborative partnerships for community health and development. Am. J. Community Psychol. **23**(5), 677–697 (1995)
5. Rand, D.G., Arbesman, S., Christakis, N.A.: Dynamic social networks promote cooperation in experiments with humans. Proc. Natl. Acad. Sci. **108**(48), 19193–19198 (2011)
6. McLoughlin, C., Lee, M.J.W.: Social software and participatory learning: Pedagogical choices with technology affordances in the Web 2.0 era. ICT: Providing choices for learners and learning. Proceedings Singapore. pp. 664–675 (2007)
7. Robbins, S.P., Judge, T.A., Sobral, F.: Comportamento organizacional: teoria e prática no contexto brasileiro. Pearson Prentice Hall (2010)
8. The World Bank.: DataBank – World Development Indicators. Available in: http://databank.worldbank.org/data/reports.aspx?source=world-development-indicators (2017). Accessed 29 Nov 2017
9. Cetic.: TIC Domicílios. Available in: http://cetic.br/pesquisa/domicilios/indicadores (2016). Accessed 12 Oct 2017
10. Van Dijk, J.A.G.M.: Digital divide research, achievements and shortcomings. Poetics. **34**(4-5), 221–235 (2006)
11. Rogers, E.M.: Diffusion of Innovations. Simon and Schuster (2010)
12. Sadeh, N., et al.: Understanding and capturing people's privacy policies in a mobile social networking application. Pers. Ubiquit. Comput. **13**(6), 401–412 (2009)
13. Toch, E., et al.: Empirical models of privacy in location sharing. Proceedings of the 12th ACM international conference on Ubiquitous Computing, pp 129–138 (2010)
14. Duh, H.B., et al.: Usability evaluation for mobile device: a comparison of laboratory and field tests. Proceedings of the 8th conference on human-computer interaction with mobile devices and services, pp. 181–186, (2006)
15. Pavlou, P.A.: Consumer acceptance of electronic commerce: Integrating trust and risk with the technology acceptance model. Int. J. Electron. Commer. **7**(3), 101–134 (2003)
16. Save, S., et al.: Applying human computer interaction to individual security using mobile application. In: Communication, Information & Computing Technology (ICCICT), pp. 1–6 (2015)
17. Czeskis, A., et al.: Parenting from the pocket: Value tensions and technical directions for secure and private parent-teen mobile safety. In: Proceedings of the Sixth Symposium on Usable Privacy and Security, p. 15. ACM (2010)
18. Rodrigues, G.J., et al.: Métodos, Técnicas e Ferramentas de Processos de Usabilidade Alinhado com as Diretrizes de Acessibilidade: Uma Revisão Sistemática da Literatura. Anais do XIII Brazilian symposium on information systems, pp. 182–189 (2017)
19. Life360.: Welcome to your new family circle. Available in: https://www.life360.com/. Accessed 31 Jan 2018
20. Gil, A.C.: Métodos e técnicas de pesquisa social, p. 216. Atlas, São Paulo (2008)
21. Günther, H.: Pesquisa qualitativa versus pesquisa quantitativa: esta é a questão. Psicologia: teoria e pesquisa. **22**(2), 201–210 (2006)
22. Lewis, J.R.: Psychometric evaluation of the PSSUQ using data from five years of usability studies. Int. J. Hum. Comput. Interact. **14**(3-4), 463–488 (2002)
23. Perez, G.: Adoção de inovações tecnológicas: Um estudo sobre o uso de sistemas de informação na área de saúde. Tese de Doutorado. Universidade de São Paulo (2006)
24. Rossetto, A.M., et al.: Implicações de variáveis organizacionais na adoção de inovações tecnológicas em organizações públicas: estudo de caso de implantação de sistema de informações geográficas em prefeitura de médio porte. Revista de Administração Pública. **38**(1), 109–136 (2004)

An Experience Report from the Migration of Legacy Software Systems to Microservice Based Architecture

Hugo Henrique S. da Silva, Glauco de F. Carneiro, and Miguel P. Monteiro

26.1 Introduction

Microservices relate to an architectural style inspired by service-oriented computing [1] and comprise a promising solution to efficiently build and manage complex software systems [2]. Adoption of a microservices-based architecture promises to obtain cost reduction, quality improvement, agility, and decreased time to market. Microservices can be approached as the software equivalent of *Lego* bricks: after they are proven to work they fit together appropriately. They are an option to construct complex solutions in less time than with traditional architectures [2].

Many legacy software systems moved to the cloud without prior adjustments in their architecture for the new infrastructure. Many of them have been originally placed in virtual machines and deployed in the cloud, assuming the characteristics of resources and services of a traditional data center. This approach fails to reduce costs, improve performance and maintainability [3].

A open question concerns the steps that should be followed to migrate a monolithic legacy system to a microservices-based architecture. To the best of our knowledge, just a few works discussed this issue [4–6]. To fill this gap, we present the lessons learned when migrating two legacy systems, which were acquired in a two-phase study. It addresses the following Research Question (RQ): *Which steps should be performed to support the migration of legacy software systems to microservices-based architecture?*

H. H. S. da Silva · G. de F. Carneiro (✉)
Universidade Salvador (UNIFACS), Salvador-Bahia, Brazil
e-mail: contato@hugohenrique.com.br; glauco.carneiro@unifacs.br

M. P. Monteiro
Universidade Nova de Lisboa (UNL), Lisboa, Portugal
e-mail: mtpm@fct.unl.pt

Lessons learned helping practitioners from industry and academia in migrating legacy systems to microservices can contribute in encourage the embracing of this challenge.

The rest of this paper is organized as follows. Section 26.2 discusses the main shortcomings of a monolithic legacy system and map them to possible solutions provided by the microservice-based architecture. In Sect. 26.3, we report a two-phase study in which the first phase is a *pilot study* aimed at identifying key steps of the migration process as well as improvement opportunities. The second phase is presented in Sect. 26.4 and comprises a *case study* to apply a reviewed and improved version of the steps performed in the first study. Section 26.5 reports lessons learned and discusses tasks related to the migration based on experience acquired. Section 26.6 discusses opportunities for future research and provides concluding remarks.

26.2 Monolithic vs Microservices

The use of single executable artefacts or *monoliths* and the modularization of their abstractions, rely on the sharing of resources of the same machine (memory, databases and files) [1]. Since the parts of a monolithic system depend on shared resources, they are not independently executable [1, 7, 8]. Large monolithic systems are difficult to maintain and evolve due to their complexity [1]. Tracking down bugs in these conditions requires much effort and is thus likely to diminish team productivity [1]. Moreover, adding or updating libraries risks producing inconsistent systems that either do not compile/run or worse, *misbehave* [1]. A change performed on a monolithic system entails the re-building of the whole application. As the system evolves, it becomes ever more difficult to maintain it and keep track of its original architecture. This can result in recurring downtimes, specially for large projects, hindering development, testing,

© Springer Nature Switzerland AG 2019
S. Latifi (ed.), *16th International Conference on Information Technology-New Generations (ITNG 2019)*,
Advances in Intelligent Systems and Computing 800,
https://doi.org/10.1007/978-3-030-14070-0_26

and maintenance activities [1]. Monolithic systems under these conditions are prone to stop working and become unable to provide part or all of their functionality. They also suffer from scalability issues. To deal with the shortcomings of this type of applications and to handle an unbound number of requests, developers create new instances of them and split the load among these instances. Unfortunately, this approach is not effective, since the increased traffic will be targeted only to a subset of the modules, causing difficulties for the allocation of new resources for other components [1]. Microservices should be small and independent enough to allow the rapid development, (un)pluggability, independent evolution and harmonious coexistence. Microservices have been referred as a solution to most of the shortcomings of monolithic architecture. They use small services to remove and deploy parts of the system, enable the use of different frameworks and tookits and to attain increased scalability and better overall system resilience. A microservice architecture can make use of the elasticity and better pricing model of cloud environments [9].

Next follows a non-exhaustive list of advantages that stand out when using microservices: cohesive and loosely coupled services [10]; independent implementation of each microservice and thus adaptability [11]; independence of multifunctional, autonomous and organized teams to provide commercial value, not just technical characteristics [11]; independence of domain concepts [10]; freedom from potential side effects (SPoF) in services; encourage of the *DevOps* culture [9], which basically represents the idea of decentralizing skills concentration into multifunctional teams, emphasizing collaboration between developers and teams, ensuring reduced lead time and greater agility during software development.

26.3　A Two-Phase Study

This section describes the design and settings of a two-phase exploratory study with the goal of identifying relevant and effective steps of a migration of legacy systems to a microservices-based architecture. Exploratory studies are intended to lay the groundwork for further empirical work [12]. The study aims to address the following RQ: *What would be the set of effective steps to migrate legacy systems to a microservices-based architecture?* The specific research questions (SRQ's) derived from the base RQ are as follows: *SRQ1: How to find features in a legacy application so that they can be subsequently modularized and become a candidate to a microservice-based architecture?* and *SRQ2: How to migrate the best candidate features to a microservice-based architecture?*
The Study Protocol. The first author of this paper carried out the tasks of the reported study, after discussing the strategies,

experiences and impressions with the other two authors. To answer the research questions (primary and secondaries), all steps registered by the first author in manuscripts were analyzed. *The Legacy Systems*. Candidate applications for this study should match the following characteristics: (1) be a legacy application, (2) have a monolithic architecture that does not have its functionalities modularized, (3) show symptoms of scattering and tangling, and (4) structurally correspond to the *Big Ball of Mud* anti-pattern [13]. *Expected Outcome.* In contrast, we expect the evolved version of the application to be more coherent, loosely coupled, showing a modular decomposition more aligned to the services it provides [14]. We also expect to witness an increase in the autonomy of developing teams within the organization, as new functionalities can be localized within specific services [14].

DDD Key Concepts. To accomplish the tasks of this study, we used key Domain-Driven Design (DDD) concepts to translate functionalities into domain and subdomain and thereby support the migration. A *Bounded Context* is a subsystem in the solution space with clear boundaries that distinguish it from other subsystems [15]. *Bounded Context* aids in the separation of contexts to understand and address complexities based on business intentions. The *Domain* in the broad sense is all the knowledge around the problem one is trying to solve. Therefore, it can refer to either the entire *Business Domain*, or just a basic or support area. In a *Domain*, we try to turn a technical concept with a model (*Domain Model*) into something understandable. The *Domain Model* is the organized and structured knowledge of the problem. This model should represent the vocabulary and main concepts of the domain problem and identify the relationships between all entities. It should act as a communication tool for all involved, creating a very important concept in DDD, which is *Ubiquitous Language*. This model could be a diagram, code examples or even written documentation of the problem. The important thing is that the *Domain Model* must be accessible and understandable by all involved in the project.

26.3.1　The Pilot Study

The *ePromo* system was selected as the subject of the Pilot study. It comprises a typical example of a corporate/business coupon web system implemented in the PHP programming language for the management of outreach campaigns. The web server is *Nginx* and its features include: creation of personalized offers and issuance of tickets made by the customer. All functionalities are implemented in a large artifact, connected to a single relational database (*MySQL*), whereas *Memcached* is used as a memory cache system, including data related to the sessions – signs of a mono-

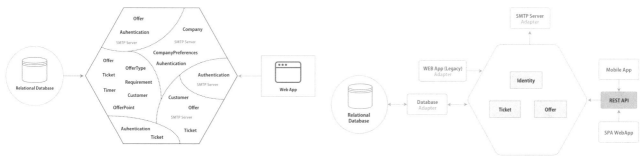

Fig. 26.1 Entities and associated features scattered and tangled in ePromo

Fig. 26.2 ePromo modularized version (End of the pilot study)

lithic application. Due to the sudden growth of demand for coupons, the application started to face problems in this specific component, which led to interruptions in the system operation.

In order to answer *SRQ1 (find features to be subsequently modularized and turned into microservice candidates)*, the participant applied a manual identification of candidate features and their respective relationships, by navigating among the directories and files and identifying the purposes of each class. Figure 26.1 illustrates the identified entities were: `Offer`, `OfferPoint`, `Ticket`, `Requirement`, `Timer`, `User`, `Company` in the beginning of the pilot study. By analysing the features associated to these entities, we acquired an initial perception of how they are tangled and scattered in the code. In fact, the functionalities are the reference to build the context map. It is worth mentioning that during the elaboration of the context map based on information retrieved from the source code, it was possible to recognize the entities and the candidates for value object's and aggregate roots. At this time, we had the opportunity to spot code tightly coupled to the web framework, right at the initial browsing stage.

We decided to deal with one feature at a time, based on the list of features. We started with the functionality that would have lowest impact when compared to the others. This would enable the validation of boundaries between features with the least risk of side effects. Considering that the business rules were scattered throughout the controllers with significant duplication, additional effort to identify the various functionalities involved was required. This scenario also indicated a symptom of tangling.

When analyzing the `TicketsController` artifact it was noticeable that it has many responsibilities and that its business rules were scattered. It needed extensive refactoring, including extraction of clear layers for different levels of abstraction. Each layer would be represented by a folder, which entails structural changes at that level, within the repository's source root.

New directories have been created: `Application`, `Domain` and `Infrastructure`. Folder `Application`

is to be devoid of business logic and be responsible for connecting the user interface to the lower layers, i.e., the application layer will be able to communicate with the domain layer, which will act as a sort of public API for the application. It will accept requests from the outside world and return answers appropriately. Folder `Domain` is to harbour all concepts, rules and business logic of the application, such as the user entity or the user repository. These files will be stored according to the context identified in previous steps. Folder `Infrastructure` is to host the implementations concerning technical features, which provide support to the layers above, namely persistence, and communication over networks.

At this point, we applied the *Command* pattern [16] to minimize coupling and deal with the tangled code with scattered business rules and identified in the controllers of the application. *Command* encapsulates a request as an object, thereby parametrizing clients with different requests, queue or log requests, and support undoable operations [16]. Based on `TicketController`, *Command* was used to uncouple the controller from the user interface logic. When looking at the commands, we should be able to spot the goal of that code snippet. The controller is to pass just the information needed by the command – `CreatingTicket` in this case – to forward to the handler, which will handle the acceptance of the command and will complete its task. This approach brings several advantages, namely: (1) the functionality can run in any part of the application; (2) the controller will no longer have business rules, doing just what is proposed above; (3) as a result of decoupling, the tests can be made easier. The new version of the modularized system is presented in Fig. 26.2.

26.3.2 Lessons Learned from the Pilot Study

Based on the experience gained in the pilot study, we can answer *SRQ1* as explained in the following. In Sect. 26.3.1, we described that the identification of functionalities faced difficulties due to the existence of lots of classes with repeated business rules and scattered throughout. This situation is typified as the *Anemic Model* anti-pattern.[1] Therefore, identifying business resources requires much effort. During the identification and mapping of the business contexts, we noticed that despite the sudden growth of demand for coupons, the number of features candidates for microservices may not be indicative of the use of a microservice architecture. There is not a positive trade-off between the advantages of microservices and the corresponding costs and effort required to manage it [2]. Although microservices approaches offer substantial benefits, the corresponding architecture requires extra machinery, which may also impose substantial costs [2]. This would will give rise to greater complexity, which is incompatible with the relative simple scenario now perceived through the map of contexts. Therefore, the decision for the migration should bear in mind the extra effort required to work on automated deployment, monitoring, failure, eventual consistency, and other factors introduced by a microservice architecture. For these reasons, we decided not to opt for the migration, and keep *ePromo* in its new modularized version.

At this point, we reached a preliminary list of lessons learned comprised of two main parts: *part 1* related to the restructuring of the legacy system to a modularized version and *part 2* related to migration of the modularized version to microservices. *Part 1* of the lessons learned are related to the (a) identification of candidate functionalities that can be modularized in legacy applications; (2) analysis of relationships and organizational dependencies in the legacy system; (3) identification of each domain and sub-domain. In the sequence, *part 2* of the lessons learned is related to the (4) selection of the candidates according to their importance to the domain and the application itself; (5) conversion of the candidate functionalities to microservices.

26.4 The Case Study

The goal of this case study is to analyze an effective way to find candidate functionalities to be modularized in legacy applications to be later converted into microservices. *Target System for the Case Study*. Figure 26.3 illustrates a typical scenario of the *eShop* system. It is an online store in which users can browse a product catalog. The system provides functionalites such as user authentication, catalogue of products, special offers, and payments. All functionalities are implemented in the PHP programming language in a "big module", connected to a single relational database (MySQL). The system runs as a single artifact on a Nginx web server. The size of the source code increased dramatically over the years, as stakeholders asks for ever more changes and

Fig. 26.3 A traditional monolithic legacy software system (Case study)

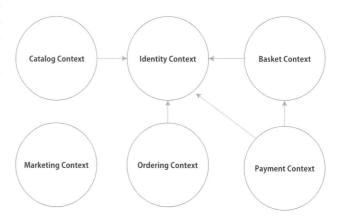

Fig. 26.4 A context map for the monolithic legacy software system (Case study)

new functionalities. To deal with such requests, developers struggled to deliver new releases, which demanded ever more effort.

Part I – Migrating the Legacy System to a Modularized Version. We manually identified the candidate functionalities by navigating among the directories and files to find out the purpose of each artifact as was done for the pilot study. Figure 26.3 illustrates the entities Identity, Basket, Marketing, Catalog, Ordering and Payment related to the identified functionalities. This is the result of the first step aimed at identifying main functionalities and responsabilities in view of a tentative establishment of boundaries between them. Next, we planned to break down the main module into units. The key to this task was the use of bounded contexts ans their respective relationships, as represented in Fig. 26.4. We applied in each bounded context the following DDD key concepts: *aggregate root*, *value objects* and *domain services*. These concepts help to manage domain complexity and ensures clarity of behavior within the domain model. After identifying contexts, we sorted them by level of complexity, starting with the simpler ones to validate the context mapping. We also placed the contexts into well-defined layers, expressing the domain model and business logic, eliminating dependencies on infrastructure, user interfaces and application logic, which often get mixed with it.

We should concentrate all the code related to domain model in one layer, isolating it from the user interface, application and infrastructure parts [15]. In some cases, we can apply the *Strangler* pattern [6] to deal with the complexity of the module to be refactored.

A folder should be created for each of the bounded contexts and within each folder, three new folders should be added, one for each layer: `Domain`, `Application`, `Infrastructure`. They contain the source code necessary for this bounded context to work. It is crucial to consider the domain models and their invariants and to recognize entities, value objects and also aggregate roots. We should maintain the source code in these folders as described in the sequence. Folder `Application` contains all application services, command and command handlers. Folder `Domain` contains the classes with existing tatical patterns in the DDD, such as: `Entities`, `Value Objects`, `Domain Events`, `Repositories`, `Factories`. Folder `Infrastructure` provides technical capabilities to other parts of the application, isolating all domain logic from the details of the infrastructure layer. The latter contains, in more detail, the code for sending emails, post messages, store information in the database, process `HTTP` requests, make requests to other servers. Any structure and library related to "the outside world", such as network and file systems, should be used or called by the infrastructure layer.

Part II – Migrating from the Modularized version to Microservices At this point, our focus is the analysis of the previously developed context map and the assessment of the feasibility of decomposing each identified context into microservice candidates. In this case, during the analysis of the context map, it is required understanding and identifying the organizational relationships and dependencies. This is analogous to domain modeling, which can start relatively superficially and gradually increase levels of detail.

The most commonly used way to decompose an application into smaller parts is based on layered segmentation based on user interface, business logic and database responsibilities. However, this is prone to give rise to coupling between modules, causing the replication of business logic in the application layers [1] – *coupling* defines the degree of dependency between components or modules of an application. The microservice proposal to circumvent this problem entails segmenting the system into smaller parts with fewer responsibilities. In addition, it also considers domain, focus and application contexts, yielding a set of autonomous services, with reduced coupling.

In order to answer *SRQ2*, the bounded contexts from DDD are used to organize and identify microservices [17]. Many proponents of the microservice architecture use Eric Evans's DDD approach, as it offers a set of concepts and techniques that support the modularization in software systems. Among these tools, *Bounded Context* is used to identify and organize

the microservices. Evans made the case for bounded contexts as facilitating the creation of smaller and more coherent components (models), which should not be shared across contexts. In the context map shown in Fig. 26.4, the arrow is used to facilitate identification of upstream/downstream relationships between contexts. When a limited context has influence over another (due to factors of a less technical nature), provision of some service or information this relationship is considered upstream. However, the limited contexts that consume it comprise a downstream relationship [15].

Correct identification of bounded contexts using DDD and breaking a large system across them is an effective way of defining microservice boundaries. Newman points out that bounded contexts represent autonomous business domains (i.e., distinct business capabilities) and therefore are the appropriate starting point for identifying boundaries for microservices. Using DDD and bounded contexts lowers the chances of two microservices needing to share a model and corresponding data space, risking a tight coupling. Avoiding data sharing facilitates treating each microservice as an independent deployment unit. Independent deployment increases speed while still maintaining security within the overall system. DDD and bounded contexts seems to make a good process for designing components [14]. Note however, that it is still possible to use DDD and still end up with quite large components, which go against the principles of the microservice architecture. In sum, *smaller is better*. An important service feature is its low number of responsibilities, which is reinforced by the definition of the *Single Responsibility Principle* (SRP) [18]. Each service must have a well-defined boundary between the modules, which should be independently created and published, through an automated deployment process. A team can work on one or several *Bounded Context*'s, with each serving as a foundation for one or several microservices. Changes and new features are supposed to related to just one *Bounded Context* and thus just one team [10].

Keeping all data on a single basis is contrary to the decentralized data management feature of microservices. The strategy is to move resources vertically by decoupling the primary feature along its data and redirect all front-end applications to the new APIs. Having multiple applications using the data from a centralized database is the primary lock to decouple the data along with the service (Fig. 26.5).

Migrating data from an existing application is a complex process. It requires special care, which depends on the specific situation. During the migration of the *eShop* database, we decided to perform it in small chunks. We selected the tables related to each service and create a new database schema (MySQL) for the respective service. We then migrated them one by one. The database was not particularly large and this approach was applied without side effects. However, this

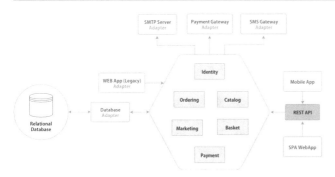

Fig. 26.5 An evolved monolithic legacy system (Case study)

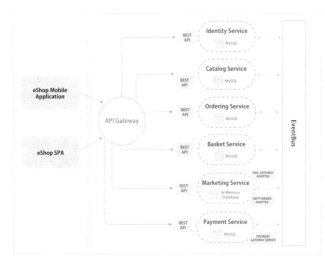

Fig. 26.6 A new based microservices software system (Case study)

approach may not be the most efficient, depending on the size of the database to be migrated. Each specific scenario must be analyzed in its terms. To perform the migration, we used *Doctrine Migrations*.[2] Figure 26.6 shows the architecture of the new system based on microservices.

26.5 Lessons Learned

As a result of the experience gained in the two-phase study reported before, we identified four key challenges in the migration process. The first, is related to the identification of functionalities. This is not trivial, especially in cases of large modules through which functionalities are scattered and tangled among themselves. This is in fact a recurring issue already discussed in the literature [19]. The second challenge is the definition of optimal boundaries among candidate features for microservices. Once these limits are established, there follows the third challenge, to decide which will be

converted to microservices. After this decision, we should face the fourth challenge, related to carefully analyze these candidate microservices regarding their respective granularity and respective cohesion.

The literature already addressed the *decomposition problem* for identifying modules, packages, components, and "traditional" services, mainly by means of clustering techniques upon design artifacts or source code. However, boundaries between modules defined using these approaches were too flexible and allowed software to evolve into instances of *Big Ball of Mud* [13]. Although much was written on the value of cohesive services and the power of bounded contexts, there appears to be a void in the guidance on how to identify these in practice [20]. The main issue is that those people trying to determine service boundaries are technologists looking for a technological solution but defining cohesive, capability-aligned service boundaries instead requires domain expertise. To accomplish this, a modelling exercise should be carried out independently of the specific technology used.

Applying the aforementioned strategies yielded multiple autonomous microservices, each with its own database. For communication between the microservices, we use `HTTP` communication mechanisms as API *Restful* and also asynchronous communication with an `EventBus` implementation, running *RabbitMQ*.[3] As shown in Fig. 26.6, each of the microservices now work with an independent relational database, except the `Marketing` service because it is an auxiliary service. For this one, we chose to use an in-memory database.

26.6 Conclusions

Migrating a legacy application is often a hard and complex work. It rarely can be performed without significant effort. To the best of our knowledge, there are frameworks that can be used to support practitioners during the development (forward engineering) of microservice-based systems, such as *Spring Cloud* [4] and Hystrix,[5] just to name a few. However, none of them provides full support to the three migration phases. To fill this gap, this paper present the lessons learned to support the stated migration. We believe that the availability these lessons learned can support and encourage practitioners from the industry and academia to perform this type of migration. Considering that these lessons learned is based on our experience on a specific two-phase study. We also plan to conduct a survey with practitioners from the industry to characterize their perception regarding challenges faced

---

[2]https://www.doctrine-project.org/projects/migrations.html

[3]https://www.rabbitmq.com

[4]http://projects.spring.io/spring-cloud/

[5]https://github.com/Netflix/Hystrix

during this type of migration, characteristics of possible processes they may have used for this purpose and opinion about the lessons learned reported.

References

1. Dragoni, N., Giallorenzo, S., Lafuente, A.L., Mazzara, M., Montesi, F., Mustafin, R., Safina, L.: Microservices: yesterday, today, and tomorrow. In: Mazzara, M., Meyer B. (eds.) Present and Ulterior Software Engineering, pp. 195–216. Springer, Cham (2017)
2. Singleton, A.: The economics of microservices. IEEE Cloud Comput. **3**(5), 16–20 (2016)
3. Toffetti, G., Brunner, S., Blöchlinger, M., Spillner, J., Bohnert, T.M.: Self-managing cloud-native applications: design, implementation, and experience. Futur. Gener. Comput. Syst. **72**, 165–179 (2017)
4. Kalske, M., Mkitalo, N., Mikkonen, T.: Challenges when moving from monolith to microservice architecture. In: Current Trends in Web Engineering, pp. 32–47. Springer, Cham (2017)
5. Leymann, F., Breitenbcher, U., Wagner, S., Wettinger, J.: Native cloud applications: why monolithic virtualization is not their foundation. In: Cloud Computing and Services Science, pp. 16–40. Springer, Cham (2016)
6. Taibi, D., Lenarduzzi, V., Pahl, C.: Processes, motivations, and issues for migrating to microservices architectures: an empirical investigation. IEEE Cloud Comput. **4**(5), 22–32 (2017)
7. Richardson, C.: Microservices: decomposing applications for deployability and scalability. InfoQ 25: 15–16 (2014). Available at https://www.infoq.com/articles/microservices-intro. (Accessed 21 Aug 2018)
8. Richardson, C.: Pattern: monolithic architecture. Posjećeno **15**, 2016 (2014)
9. Balalaie, A., Heydarnoori, A., Jamshidi, P.: Microservices architecture enables devops: migration to a cloud-native architecture. IEEE Softw. **33**(3), 42–52 (2016)
10. Wolff, E.: Microservices: Flexible Software Architecture. Addison-Wesley Professional, Boston (2016)
11. Millett, S.: Patterns, Principles and Practices of Domain-Driven Design. Wiley, Indianapolis (2015)
12. Seaman, C.B.: Qualitative methods in empirical studies of software engineering. IEEE Trans. Softw. Eng. **25**(4), 557–572 (1999)
13. Coplien, J.O., Schmidt, D.C.: Pattern Languages of Program Design. ACM Press/Addison-Wesley Publishing Co., Reading (1995)
14. Newman, S.: Building Microservices: Designing Fine-Grained Systems. O'Reilly Media, Inc., Sebastopol (2015)
15. Evans, E.: Domain-Driven Design: Tackling Complexity in the Heart of Software. Boston, Addison-Wesley Professional (2004)
16. Gamma, E.: Design Patterns: Elements of Reusable Object-Oriented Software. Pearson Education India, Reading (1995)
17. Nadareishvili, I., Mitra, R., McLarty, M., Amundsen, M.: Microservice Architecture: Aligning Principles, Practices, and Culture. O'Reilly Media, Inc., Beijing/Boston (2016)
18. Martin, R.C.: The single responsibility principle. The principles, patterns, and practices of Agile Software Development, vol. 149, p. 154. Pearson Education, Upper Saddle River (2002)
19. Ossher, H., Tarr, P.: Multi-dimensional separation of concerns and the hyperspace approach. In: Software Architectures and Component Technology, pp. 293–323. Springer, Boston (2002)
20. McLarty, M.: Designing a microservice system. [Online]. Available: http://www.apiacademy.co/designing-a-system-of-microservices

Kênia Santos de Oliveira and Stéphane Julia

27.1 Introduction

In a business environment, transformation of business requirements in a system specification is a crucial task and for any Software Engineering project it is fundamental that the architectural model possess the requirements specified in the analysis model.

According to [1], the validation of non-functional and functional requirements throughout the design process is an economical mechanism when compared to a test validation process of a compiled product. Besides other advantages, this verification maximizes the guarantee of software quality and minimizes rework costs [2]. Moreover, performance requirements are also important to the guarantee of software quality because it specifies the speed or operational effectiveness of a capability that must be delivered by a system architecture. According to [3], performance is an important quality attribute that is usually affected negatively in Service-Oriented Architecture (SOA). As an example of performance requirement, the response time can be analyzed. A longer response time can cause users to think the system is down, for example. According to [4], there is an insufficiency of appropriate methods and tools to support such performance of service-oriented systems. The approaches presented in the literature regarding requirement verification in an architectural model deal with a variety of architectures, most of which present some kind of informal traceability of requirements.

As for any type of software architecture project, it is crucial to ensure that a SOA project adequately meets the requirement specifications, and therefore needs to be verified. In this article, after verifying the equivalence between the scenarios specified in a public requirement model and the deadlock-free scenarios that exist in the corresponding SOA model through a kind of bisimulation equivalence (functional requirement equivalence for SOA models [5]), it is possible, in particular, to certify if the main business relationship between different organizations sharing a same common goal are also equivalent in terms of performance considering a public requirement model.

Such an analysis will be possible considering the symbolic dates computed when the sequent of Linear Logic that correspond to the existing deadlock-free scenarios of a SOA model is proved to be correct. The requirement of a collaborative system is then specified through a kind of public WorkFlow net and the architectural model is represented by Interorganizational WorkFlow net, which may not be sound. In both models time intervals are then associate to each activity of the corresponding business process.

The remainder of this article is organized as follows. In Sect. 27.2 is presented the definition of an Interorganizational WorkFlow net. An overview of t-time WorkFlow net is presented in Sect. 27.3. In Sect. 27.4 the main concepts about Linear Logic is presented. The approach to formally verify performance requirements in SOA models is defined in Sect. 27.5. Finally, the conclusion of the work is presented in Sect. 27.6.

K. S. de Oliveira (✉)
Computing Faculty, Federal University of Uberlândia, Uberlândia, Brazil

Federal Institute of Brasília, Brasília, Brazil

S. Julia
Computing Faculty, Federal University of Uberlândia, Uberlândia, Brazil
e-mail: stephane@ufu.br

27.2 Interorganizational WorkFlow Net

Basically, an interorganizational workflow process is a set of 'local' workflow processes connected in a same 'global' workflow process, which are loosely coupled [6]. The formal

© Springer Nature Switzerland AG 2019
S. Latifi (ed.), *16th International Conference on Information Technology-New Generations (ITNG 2019),*
Advances in Intelligent Systems and Computing 800,
https://doi.org/10.1007/978-3-030-14070-0_27

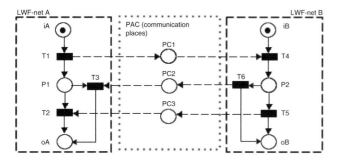

Fig. 27.1 An IOWF-net example

definition of an Interorganizational WorkFlow net (IOWF-net) that can be used to represent an interorganizational workflow processes is the following [7]:

$$IOWF - net = \{PN_1, PN_2, \ldots, PN_n, P_{AC}, AC\}, \text{ such that}$$

- For each $k \in \{1, \ldots, n\} : PN_k$ is a Local WorkFlow net (LWF-net) with source place i_k and sink place o_k.
- P_{AC} is a set of elements (places) communicating asynchronously.
- AC consist to the asynchronous communication relation. For each asynchronous communication element, it specifies a set of input and output transitions.

Figure 27.1 show an IOWF-net example with LWF-nets A and B. The source place for LWF-net A is 'iA' and for LWF-net B is 'iB' while the sink place are 'oA' and 'oB' respectively. The communication places is represented by PC1, PC2 and PC3. In PC1, for example, the input transition is T1 and the output transition is T4.

To transform an IOWF-net into a simple WF-net, in [7] was defined the Unfolded Interorganizational WorkFlow net U(IOWF-net). For the purpose of respecting the basic structure of a simple WF-net, must be added to the IOWF-net a global source place 'i' and a global sink place 'o' and the asynchronous communication elements existing between the different LWF-nets are mapped into ordinary places.

The Soundness property is used to verify the correctness of a workflow process. Despite the importance of Soundness correctness criteria, in practice, most of the time, workflow processes do not meet this criterion [8]. Therefore, in the case of non-sound interorganizational workflows the variants of the Soundness criterion, such as Relaxed Soundness and Weak Soundness, may be considered [9]. In this work, the proposed approach accepts in particular relaxed sound processes. In relaxed sound processes all the system activities must appear in at least one process that ended correctly.

27.3 T-Time WorkFlow Nets

To evaluate the performance of a process modeled by a Petri net, it is necessary to consider the modeling of explicit time constraints. To deal with time in Petri net models two approaches are detach: Merlin's Time Petri nets [10] and Ranchamdani's Timed Petri nets [11]. In these two temporal Petri net models, time inscriptions are always associated to transitions (t-time).

The time constraints in a t-time Petri net is represented by an time interval $[\theta_{min}, \theta_{max}]$ associated with each transition, which corresponds to an imprecise enabling duration [10]. For example, the time interval [5, 11] associated with a specific transition denote that, after the transition has been enabled, it will be fired at least five time units and at the most eleven time units.

Considering the definitions about t-time Petri nets, a t-time WorkFlow net is a WorkFlow net extended with time constraints. The definition of a t-time WorkFlow net is [12]:

$$N = (P, T, F, I), \text{ such that}$$

- (P, T, F) is a WorkFlow net, where: P is a finite set of places, T is a finite set of transitions and F is a set of arcs.
- I relates with each transition $t \in T$ an enabling time interval $I(t) = [\theta_{min(t)}, \theta_{max(t)}]$ where $\theta_{min(t)}$ represents the minimum firing time and $\theta_{max(t)}$ represents the maximum firing time of the transition t.

Figure 27.2 show an example of a t-time WorkFlow net. As can be observed in Fig. 27.2, transitions $T1$, $T2$ and $T3$ have associated an enabling interval.

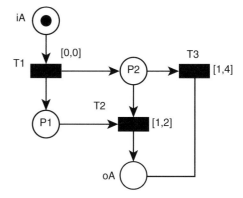

Fig. 27.2 A t-time WorkFlow net example

27.4 Linear Logic

Introduced by Jean-Yves Girard [13], Linear Logic emphasizes the role of formulas as resources instead of emphasizing proof, as in intuitionistic logic, or truth, as in classical logic.

Several connectives were defined in Linear Logic; however, in this article only the *times* connective (denoted by '$\otimes$') and the *linear implies* connective (denoted by '$\multimap$') will be used.

The *times* connective express simultaneous availability of resources. For example, '$C \otimes D$' express that the resources 'C' and 'D' are simultaneously available.

The *linear implies* connective express a state change. For example, '$C \multimap D$' denotes that consuming 'C', 'D' is produced and 'C' will not be available anymore after the production of 'D'.

A Petri net model can be translated into Linear Logic formulas since there is an almost direct translation of the structure of a Petri net into formulas of Linear Logic [14, 15]. The following definition were presented in [15] to translate a Petri net model into Linear Logic formulas:

- A marking M of a Petri net is represented by $M = A_1 \otimes \ldots \otimes A_k$ where A_i are place names. For example, in Fig. 27.2 the marking M of the WF-net is $M = iA$.
- A transition of a Petri net is represented by $M_1 \multimap M_2$ where M_1 and M_2 are markings. In the WF-net shown in Fig. 27.2, for example, the transition $T1$ is represented by $iA \multimap P1 \otimes P2$.
- $M, t_i \vdash M'$ is a Linear Logic sequent and represents a scenario of a Petri net. M is the initial marking, M' is the final marking and t_i is a list of non-ordered transitions. For example, the sequent $iA, T1, T2 \vdash o$ represents one possible scenario of the WF-net shown in Fig. 27.2.

A sequent can be proven building a proof tree by applying the rules of the sequent calculus. Only three rules will be used in this article:

- $\multimap_L$ rule – indicate a transition firing and generates a right sequent, which represents the subsequent remaining to be proved, and a left sequent, which represents the consumed tokens by this firing;
- $\otimes_L$ rule – transforms a marking in a list of atoms;
- $\otimes_R$ rule – transforms a sequent of type $A, B \vdash A \otimes B$ into two identity ones $A \vdash A$ and $B \vdash B$.

More detail about rules of the sequent calculus can be find in [15].

The proof tree of the Linear Logic is read from the bottom-up. The proof of a WorkFlow net represent by Linear Logic formulas stops when the identity sequent of the atom that finalize the process is produced ($o \vdash o$, where 'o' correspond to a sink place), when all the leaves of the proof tree are identity sequents or when there is not any rule that can be applied.

In a proof tree of the Linear Logic, each transition of a t-time Petri net can produce a symbolic date associated to each atoms consumed or produced [15]. Consider D_i a date and d_i a duration associated to a transition firing (t_i). A pair (D_p, D_c), representing respectively the production and the consumption date of atoms, will be associated with each atom of the proof tree.

The date calculus in the canonical proof tree is obtained by the following steps [15]:

(1) attach a production date D_i to all initial markings;
(2) for each $\multimap_L$ rule occurrence, calculate the firing date of the transition (this corresponds to the maximum of the production dates of the consumed atoms, incremented by the enabling duration d_i of the considered transition);
(3) update all the temporal stamps of the atoms which have been consumed and produced.

As an example, the Linear Logic proof tree with symbolic dates for the sequent $iA, T1, T2 \vdash o$ of the WF-net shown in Fig. 27.2 is the following:

$$
\cfrac{
\cfrac{
\cfrac{P1(D_1+d_1, D_1+d_1+d_2) \vdash P1 \quad P2(D_1+d_1, D_1+d_1+d_2) \vdash P2}{P1(D_1+d_1,..), P2(D_1+d_1,..) \vdash P1 \otimes P2} \otimes_R \quad oA \vdash oA
}{P1(D_1+d_1,..), P2(D_1+d_1,..), P1 \otimes P2 \multimap oA \vdash oA} \multimap_L
}{
\cfrac{iA(D_1, D_1+d_1) \vdash iA \quad P1(D_1+d_1,..) \otimes P2(D_1+d_1,..), T2 \vdash oA}{iA(D_1,..), iA \multimap P1 \otimes P2, T2 \vdash oA} \multimap_L
} \otimes_L
$$

As can be observed in the previous example, the Linear Logic proof tree is correctly finalized because the identity sequent $oA \vdash oA$ (corresponding to the sink place) is produced and there are no more sequents to be proven.

For better viewing, the production and consumption dates associated with each atom of the proof tree can be described in a table. Table 27.1 show the production and consumption dates for the Linear Logic proof tree shown above. As can be observed in Table 27.1, the production date of the atom iA is D_1 because it is an initial marking token. The atom oA does not has consumption date because it is a final marking token.

Table 27.1 Example of production and consumption dates

Atoms	Production dates	Consumption dates
iA	D_1	$D_1 + d_1$
$P1$	$D_1 + d_1$	$D_1 + d_1 + d_2$
$P2$	$D_1 + d_1$	$D_1 + d_1 + d_2$
oA	$D_1 + d_1 + d_2$	Unknow

The production and consumption dates also depends on time intervals, that is, an enabling duration d_i takes its values within a time interval $\Delta_i = [\delta_i min, \delta_i max]$ [15]. For example, considering the production dates presented in Table 27.1, the production date interval of the atom $P1$ is $[D_1 + d_{1min}, D_1 + d_{1max}]$.

27.5 Performance Verification of Requirement Scenarios in SOA Models

The approach presented in this article is an extension of the approach presented in [5]. The requirement model consist of a public WF-net, that is, a WF-net that containing the activities of all parties involved in the process. The architectural model of SOA-type is a set of private WF-nets that communicate through asynchronous mechanisms with the purpose of produce the services specified by the requirement model. To allow the analyzes in the architecture model is considered the unfolded model; therefore, the architectural model consist of an U(IOWF-net). The services of the SOA model are scenarios of the corresponding U(IOWF-net). A scenario consist of a well defined route traced into the corresponding WF-net and more than one scenario has to be considered if the WF-net has more than one route, that is, places with two or more output arcs.

In the approach presented in [5] the objective was to show that, in the context of SOA, all scenarios in terms of behaviour present in a requirement model (public WF-net) are also present in the corresponding architectural model (U(IOWF-net)). The approach used the Linear Logic proof trees and the precedence graphs, which show the operational semantic of distinct models, to verify the equivalence between the models.

In this present article, the purpose is to verify if a non-function requirement, in relation to performance of SOA, defined in a temporal model of requirement analysis are present in the corresponding temporal architectural model. The time intervals associated with the transitions are used to verify if the response time of a scenario of the architectural model is within the time interval of the corresponding scenario in the requirement model. Therefore, the scenarios identified in the architectural model that satisfy the behaviour of the scenarios specified in the requirement model are compared in relation to performance. For this, the proposed method use Linear Logic proof trees with symbolic date associated to each atom of the proof tree. Once calculated, the symbolic dates can be reused when the numeric values associated with the activities of the processes are changed. According to [15], symbolic values decreases imprecision on dates and the causality information is not lost.

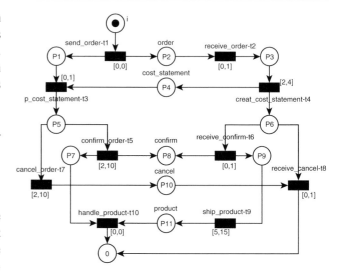

Fig. 27.3 Public WF-net (requirement model)

Therefore, the steps of the proposed method in this article are:

(1) identify the scenarios of the requirement and architectural models that are equivalents in terms of behaviour using the approach presented in [5];
(2) for each scenario identified in step 1 calculate for each atom the corresponding production and consumption dates and their respective symbolic date intervals, as explained in Sect. 27.4;
(3) calculate numerical date intervals considering the time interval defined in the temporal requirement and architectural models and replacing these values in the symbolic date intervals found in step 2;
(4) the scenarios of the architectural model that satisfy the behaviour of the scenarios specified in the requirement model will be equivalents in terms of performance if the numerical date interval that end a scenario of the architectural model belongs to the numerical date interval that end the equivalent scenario of the requirement model.

The proposed approach is illustrated using the examples presented in Fig. 27.3 (public WF-net) and Fig. 27.4 (U(IOWF-net)). These examples represent a simplified situation of the real world where a customer makes an order for a particular company and from a cost statement sent by the company to the customer, it decides whether or not to continue with the order. If the customer decides to continue with the order, the company will prepare the order to send to the customer. Two business partners are involved in the process, which are a contractor and a subcontractor represented respectively by the LWF-net contractor and LWF-net subcontractor.

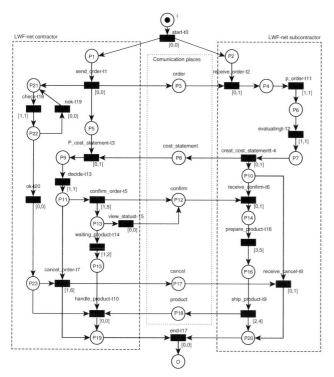

Fig. 27.4 U(IOWF-net) (architectural model)

In the U(IOWF-net) of Fig. 27.4 there is an iterative route (*check*-18 and *nok-t*19) that in this work will be replaced by a simple global tasks (*check*-18′), as defined in [16].

The transitions (activities) are represented only by their corresponding numbering; for example, the transition *send-order-t_1* is represented only by t_1. Translating the WF-net of Fig. 27.3 (requirement model) into formulas of the Linear Logic, the transitions are represented as following:

$$t_1 = i \multimap P_1 \otimes P_2, \quad t_2 = P_2 \multimap P_3,$$
$$t_3 = P_1 \otimes P_4 \multimap P_5, \quad t_4 = P_3 \multimap P_4 \otimes P_6,$$
$$t_5 = P_5 \multimap P_7 \otimes P_8, \quad t_6 = P_6 \otimes P_8 \multimap P_9,$$
$$t_7 = P_5 \multimap P_{10}, \quad t_8 = P_6 \otimes P_{10} \multimap o,$$
$$t_9 = P_9 \multimap P_{11}, \quad t_{10} = P_7 \otimes P_{11} \multimap o.$$

The scenarios of the requirement model (public Work-Flow net) are called respectively of Sr1 and Sr2, and their corresponding sequents are:

- $Sr1 = i, t_1, t_2, t_3, t_4, t_7, t_8 \vdash o.$
- $Sr2 = i, t_1, t_2, t_3, t_4, t_5, t_6, t_9, t_{10} \vdash o.$

Translating the U(IOWF-net) of Fig. 27.4 (architectural model) into formulas of the Linear Logic, the transitions are represented as following:

$$t_0 = i \multimap P_1 \otimes P_2, \quad t_1 = P_1 \multimap P_3 \otimes P_5 \otimes P'_{21},$$
$$t_2 = P_2 \otimes P_3 \multimap P_4, \quad t_3 = P_5 \otimes P_8 \multimap P_9,$$
$$t_{11} = P_4 \multimap P_6, \quad t'_{18} = P'_{21} \multimap P'_{22},$$

$$t_{12} = P_6 \multimap P_7, \quad t_4 = P_7 \multimap P_8 \otimes P_{10},$$
$$t_{13} = P_9 \multimap P_{11}, \quad t_{20} = P'_{22} \multimap P_{23},$$
$$t_7 = P_{11} \otimes P_{23} \multimap P_{17} \otimes P_{19}, \quad t_8 = P_{10} \otimes P_{17} \multimap P_{20},$$
$$t_5 = P_{11} \multimap P_{12} \otimes P_{13}, \quad t_{15} = P_{13} \multimap P_{12},$$
$$t_6 = P_{10} \otimes P_{12} \multimap P_{14}, \quad t_{14} = P_{13} \multimap P_{15},$$
$$t_{16} = P_{14} \multimap P_{16}, \quad t_9 = P_{16} \multimap P_{18} \otimes P_{20},$$
$$t_{10} = P_{15} \otimes P_{18} \otimes P_{23} \multimap P_{19}, \quad t_{17} = P_{19} \otimes P_{20} \multimap 0.$$

The scenarios of the architectural model (U(IOWF-net)) are called respectively of Sa1, Sa2 and Sa3, and their corresponding sequents are:

- scenario $Sa1 = i, t_0, t_1, t_2, t_{11}, t_{12}, t_4, t'_{18}, t_3, t_{13}, t_{20}, t_7, t_8, t_{17} \vdash o.$
- scenario $Sa2 = i, t_0, t_1, t_2, t_{11}, t_{12}, t_4, t'_{18}, t_3, t_{13}, t_{20}, t_5, t_6, t_{14}, t_{16}, t_9, t_{10}, t_{17} \vdash o.$
- scenario $Sa3 = i, t_0, t_1, t_2, t'_{18}, t_{11}, t_{12}, t_4, t_3, t_{13}, t_5, t_{20}, t_{15}, t_6, t_{16}, t_9, t_{10}, t_{17} \vdash o.$

In the first step of the approach presented in this article, the scenarios of the architectural model that satisfy the behaviour of the scenarios described in the requirement model need to be identified. In this step, the approach present in [5] is used. Therefore, the precedence graphs of the public WF-net and the precedence graphs of the U(IOWF-net) (obtained from the Linear Logic proof trees) are compared to verify the equivalences in terms of behavior. Following the approach present in [5], it was identified that scenario Sa1 produce the same behaviour as scenario Sr1 and scenario Sa2 produce the same behavior as scenario Sr2. To simplify the demonstration of the approach, is not shown here the process of comparison to find the equivalent scenarios in terms of behavior; however, for more details, the article [5] can be consulted.

For the equivalent scenarios in terms of behavior identified in step 1, it can be verified the performance comparing date intervals produced when the execution of the scenarios end. Therefore, in the second step of the approach it is necessary to calculate for each atom the corresponding production and consumption dates and their respective symbolic date intervals during the proof tree construction of the valid sequents that correspond to the scenario of both models (requirement and architectural models).

The proof trees build in the first step can be reused to calculate for each atom the corresponding production and consumption dates. To simplify the demonstration of the proposed approach, the proof trees with symbolic date for scenarios Sr2, Sa1 and Sa2 are not shown is this article. Considering $Seq_1 = D_1 + max(d_1, d_4) + d_3$, the corresponding proof tree with symbolic date for scenario Sr1 is shown on the top of the next page.

$$\cfrac{\cfrac{P_6(D_1+d_1+d_2+d_4, D_1+d_1+d_2+max(d_4,d_7)+d_8)\vdash P_6 \quad P_{10}(Seq_1+d_2+d_7, Seq_1+d_2+max(d_4,d_7)+d_8)\vdash P_{10}}{P_6(D_1+d_1+d_2+d_4, D_1+d_1+d_2+max(d_4,d_7)+d_8), P_{10}(Seq_1+d_2+d_7, Seq_1+d_2+max(d_4,d_7)+d_8)\vdash P_6\otimes P_{10}}\ \otimes R \qquad o\vdash o}{P_5(max(Seq_1,Seq_1+d_2))\vdash P_5 \qquad P_6(D_1+d_1+d_2+d_4), P_{10}(Seq_1+d_2+d_7,.), P_6\otimes P_{10}-\!\!\circ o\vdash o}\ \multimap L}$$

$$\cfrac{\cfrac{P_1(D_1+d_1,Seq_1)\vdash P_1 \quad P_4(D_1+d_1+d_2+d_4,Seq_1+d_2)\vdash P_4}{P_1(D_1+d_1,Seq_1), P_4(D_1+d_1+d_2+d_4,Seq_1+d_2)\vdash P_1\otimes P_4}\ \otimes R \qquad P_6(D_1+d_1+d_2+d_4,.), P_5(max(Seq_1,Seq_1+d_2)), P_5-\!\!\circ P_{10},t_8\vdash o}{P_1(D_1+d_1,.), P_4(D_1+d_1+d_2+d_4,.), P_6(D_1+d_1+d_2+d_4,.), P_1\otimes P_4-\!\!\circ P_5,t_7,t_8\vdash o}\ \multimap L}$$

$$\cfrac{P_3(D_1+d_1+d_2, D_1+d_1+d_2+d_4)\vdash P_3 \qquad P_1(D_1+d_1,.), P_4(D_1+d_1+d_2+d_4,.)\otimes P_6(D_1+d_1+d_2+d_4), P_1\otimes P_4-\!\!\circ P_5,t_7,t_8\vdash o}{P_1(D_1+d_1,.), P_3(D_1+d_1+d_2,.), P_1\otimes P_4-\!\!\circ P_5, P_3-\!\!\circ P_4\otimes P_6,t_7,t_8\vdash o}\ \otimes L$$

$$\cfrac{P_2(D_1+d_1, D_1+d_1+d_2)\vdash P_2 \qquad P_1(D_1+d_1,.), P_3(D_1+d_1+d_2,.), P_1\otimes P_4-\!\!\circ P_5,t_4,t_7,t_8\vdash o}{P_1(D_1+d_1,.), P_2(D_1+d_1,.), P_2-\!\!\circ P_3,t_3,t_4,t_7,t_8\vdash o}\ \otimes L$$

$$\cfrac{i(D_1, D_1+d_1)\vdash i \qquad P_1(D_1+d_1,.)\otimes P_2(D_1+d_1,.),t_2,t_3,t_4,t_7,t_8\vdash o}{i(D_1,.),i-\!\!\circ P_1\otimes P_2,t_2,t_3,t_4,t_7,t_8\vdash o}\ \multimap L$$

For better viewing, the production and consumption dates associated with each atom of the proof trees can be shown in tables. However, for reasons of space the tables were omitted in this article. The information most important in the tables is about the dates that correspond to the conclusion of the scenarios. Such dates will be used to compare the performance between the scenarios of the requirement model and the scenarios of the architectural model. In the examples used in this article, the atom 'o' represent the end of the scenarios; therefore, for scenarios Sr1, Sr2, Sa1 and Sa2 only the symbolic date intervals of production of the atoms 'o' are shown in Table 27.2.

In the third step of the approach, it is necessary to calculate numerical date intervals considering the symbolic information produced in Table 27.2. Taking into account that the process starts at the date 0 (that is, $D_1 = 0$) and the time intervals of the models shown in Figs. 27.3 and 27.4, the numerical date intervals of production of the atom 'o' for scenarios Sr1, Sr2, Sa1 and Sa2 are shown in Table 27.3.

The final step of the approach is to compare the performance of the requirement model scenarios (public WorkFlow net) that are present, in term of behavior, in the architectural model (U(IOWF-net)). As shown in Table 27.3, the numerical date interval [5, 18] of scenario Sa1 does not belong to the numerical date interval [4, 17] of scenario Sr1, then it is possible to conclude that scenarios Sa1 and Sr1 are not also equivalent in terms of performance. However, scenarios Sa2 and Sr2 are equivalent in terms of performance, because the numerical date interval [9, 21] of scenario Sa2 belongs to the numerical date interval [9, 32] of scenario Sr2.

With the proposed approach it was possible to identify that, although the architectural model (private models) respects the functional requirements (behaviour) of the analysis model (public model), the non-functional requirements (performance) are not fully respected.

Table 27.2 Symbolic date intervals of production for scenarios Sr1, Sr2, Sa1 and Sa2

Atom o	Production Symbolic Date Interval
Sr1	$[D_1 + d_{2min} + max(d_{1min}, d_{4min}) + d_{3min} + max(d_{4min}, d_{7min}) + d_{8min},$ $D_1 + d_{2max} + max(d_{1max}, d_{4max}) + d_{3max} + max(d_{4max}, d_{7max}) + d_{8max}]$
Sr2	$[D_1 + d_{2min} + max(d_{1min}, d_{4min}) + d_{3min} + max(d_{4min}, d_{5min}) + d_{6min} + max(d_{5min}, d_{9min}) + d_{10min},$ $D_1 + d_{2max} + max(d_{1max}, d_{4max}) + d_{3max} + max(d_{4max}, d_{5max}) + d_{6max} + max(d_{5max}, d_{9max}) + d_{10max}]$
Sa1	$[max(d_{4min}, d_{7min}) + max(d_{13min}, d_{20min}) + max(D_1 + max(d_{0min}, d_{1min}) + d_{2min} + d_{11min} + d_{12min} +$ $max(d_{1min}, d_{4min}) + d_{3min}, D_1 + d_{0min} + d_{1min} + d_{18min}) + max(d_{7min}, d_{8min}) + d_{17min},$ $max(d_{4max}, d_{7max}) + max(d_{13max}, d_{20max}) + max(D_1 + max(d_{0max}, d_{1max}) + d_{2max} + d_{11max} + d_{12max} +$ $max(d_{1max}, d_{4max}) + d_{3max}, D_1 + d_{0max} + d_{1max} + d_{18max}) + max(d_{7max}, d_{8max}) + d_{17max}]$
Sa2	$[max(d_{9min}, d_{10min}) + d_{17min} + max(d_{9min}, d_{14min}, d_{20min}) + max(D_{1min} + d_{0min} + d_{1min} + d_{18min},$ $D_1 + max(d_{0min}, d_{1min}) + d_{2min} + d_{11min} + d_{12min} + max(d_{1min}, d_{4min}) + d_{3min} + d_{13min} +$ $max(d_{4min}, d_{5min}) + d_{6min} + d_{16min}),$ $max(d_{9max}, d_{10max}) + d_{17max} + max(d_{9max}, d_{14max}, d_{20max}) + max(D_{1max} + d_{0max} + d_{1max} + d_{18max},$ $D_1 + max(d_{0max}, d_{1max}) + d_{2max} + d_{11max} + d_{12max} + max(d_{1max}, d_{4max}) + d_{3max} + d_{13max} +$ $max(d_{4max}, d_{5max}) + d_{6max} + d_{16max})]$

Table 27.3 Comparison of the numerical date intervals

	Scenarios	Production numerical date intervals of the atom o
Requirement model scenarios	Sr1	[4, 17]
	Sr2	[9, 32]
Architectural model scenarios	Sa1	[5, 18]
	Sa2	[11, 24]
Scenarios equivalents in terms of behaviour	Sr1 and Sa1	[4, 17] and [5, 18]
	Sr2 and Sa2	[9, 32] and [11, 24]
Scenarios equivalents in terms of performance	Sr2 and Sa2	[9, 32] and [11, 24]

27.6 Conclusion

This article introduced an approach for verify performance requirements in SOA models represented by Interorganizational WorkFlow nets and Linear Logic. The objective was to show that, in context of SOA, all scenarios of a requirement model (public model) and architectural model (private models), which are equivalents in terms of behavior, are also equivalents in terms of performance. The method was based specially on the construction of Linear Logic proof trees with symbolic dates; thereby, instead of numerical dates, the execution dates of the activities can be given by symbolic dates.

The main advantage of working with symbolic dates is that, once calculated, they can be reused when the numeric values associated with the activities of the processes are changed.

The architectural models (type SOA) considered in the approach presented in this article is not necessarily sound. However, when a WorkFlow net is not sound, only the deadlock-free scenarios may be considered in order to prove formally the performance requirements within the context of SOA.

An advantage of working with the acyclic structure of a WorkFlow net through the proof of sequents of Linear Logic is that all the deadlock-free scenarios are encountered.

An important observation is that the proof trees can be used to verify behaviour requirements, as shown in the approach presented in [5], and performance requirements, as shown in this work. Therefore, the same proof tree can be reused for qualitative analysis (functional requirement) as well as quantitative analysis (performance requirement).

As a future work proposal, we will define a procedure that should allow the detection and eventual removal of negative functional requirements and the detection of new scenarios that may appear in the architectural model and which can be documented in the public model (new scenarios in the requirement model).

References

1. Hoyos, H., Casallas, R., Jiménez, F.: Hiles-t: an ADL for early requirement verification of embedded systems. In: Proceedings of the 5th International Workshop on Model Based Architecting and Construction of Embedded Systems, pp. 7–12. ACM (2012)
2. Goknil, A., Kurtev, I., Berg, K.V.D.: Generation and validation of traces between requirements and architecture based on formal trace semantics. J. Syst. Softw. **88**, 112–137 (2014)
3. O'Brien, L., Merson, P., Bass, L.: Quality attributes for service-oriented architectures. In: Proceedings of the International Workshop on Systems Development in SOA Environments, SDSOA '07, Washington, DC, p. 3. IEEE Computer Society (2007)
4. Grundy, J., Hosking, J., Li, L., Liu, N.: Performance engineering of service compositions. In: Proceedings of the 2006 International Workshop on Service-oriented Software Engineering, SOSE '06, pp. 26–32. ACM, New York (2006)
5. Oliveira, K.S., de Oliveira, V.F., Julia, S.: Using linear logic to verify requirement scenarios in SOA models based on interorganizational workflow nets relaxed sound. In: 19th International Conference on Enterprise Information Systems (2017)
6. van der Aalst, W.M.P.: Loosely coupled interorganizational workflows: modeling and analyzing workflows crossing organizational boundaries. Inf. Manag. **37**, 67–75 (2000)
7. van der Aalst, W.M.P.: Modeling and analyzing interorganizational workflows. In: International Conference on Application of Concurrency to System Design, pp. 262–272. IEEE Computer Society Press, 1998
8. Fahland, D., Favre, C., Koehler, J., Lohmann, N., Volzer, H., Wolf, K.: Analysis on demand: instantaneous soundness checking of industrial business process models. Data Knowl. Eng. **70**, 448–466 (2011)
9. Passos, L.M.S.: A metodology based on linear logic for interorganizational workflow processes analysis. Ph.D. dissertation, Federal University of Uberlândia (2016)
10. Merlin, P.M.: A study of the recoverability of computing systems. Ph.D. dissertation (1974) aAI7511026
11. Ramchandani, C.: Analysis of asynchronous concurrent systems by timed petri nets. Ph.D. dissertation, Massachusetts Institute of Technology, Cambridge (1973)
12. Ling, S., Schmidt, H.: Time petri nets for workflow modelling and analysis. In: 2000 IEEE International Conference on Systems, Man, and Cybernetics, vol. 4, pp. 3039–3044. IEEE (2000)
13. Girard, J.-Y.: Linear logic. Theor. Comput. Sci. **50**, 1–102 (1987)
14. Girault, F., Pradin-Chezalviel, B., Valette, R.: A logic for petri nets. Journal Européen des Systèmes Automatisés **31**, 525–542 (1997)
15. Riviere, N., Pradin-Chezalviel, B., Valette, R.: Reachability and temporal conflicts in t-time petri nets. In: 9th International Workshop on Petri Nets and Performance Models (2001)
16. Valette, R.: Analysis of petri nets by stepwise refinements. J. Comput. Syst. Sci. **18**, 35–46 (1979)

Part III

Data Mining and Big Data Analytics

A Big Data Experiment to Assess the Effectiveness of Deep Learning Neural Networks in the Mining of Sustainable Aspects of the Hotels Clients Opinions

Thiago de Oliveira Lima, Methanias Colaço Júnior, Kleber H. de J. Prado, and Adalberto dos S. Júnior

28.1 Introduction

The applications of natural language processing (NLP), traditionally use the words in a sentence as a characteristic vector that is considered an input for the machine learning algorithms. In the last years, the concept of embeddings has been used with efficacy in the development of more modern natural language processing systems [1]. Therefore, different algorithms with such purpose were developed for the generation of embeddings, among them the *Word2Vect* [2]. Such technique considers that given one or several words in the context, it tries to predict the following word. Another way for the generation of embeddings is related with the approaches that use word co-occurrence matrices, such as *Glove* [3].

Besides the approaches that map the words of the text into in numeric vectors (embeddings), the learning machine based on deep learning neural networks, also named Deep Learning, is being used for natural language processing tasks nowadays. Programs like *Siri*, *Alexa* and *Cortana* are able to interact as humans through questions and answers. However, in the field of sentiment analysis, the *Long Short-Term Memory Units algorithm (LSTMs)* has been used with success in the sequential processing of words in a sentence [4].

In an increasingly globalized world, some factors are unavoidable in the process of business management, such as: increased use of technology information and communication, energy crisis and scarcity of natural resources, climate change, increasing global competitiveness, new needs and desires of customers, growth and development of cities, development of new markets, strengthening of the economy of the countries development, innovation, increase in global consumption, growth of the world population, sustainability and increased international tourism flow. In this context of major transformations, companies develop strategies to provide their stakeholders with a superior value, not only through the excellence of products and services, but also through the creation of a sustainable management system [44].

Thus, the work in question proposes, through a controlled experiment, to assess the efficiency and efficacy of the deep learning algorithms based on the *LSTM* neural networks in the task of mining hotels clients opinions with focus on sustainable aspects. With this in mind, the proposal is to investigate how identified sustainable aspects can influence positively or negatively on the hotels clients opinions.

The content of this article is organized in the following way. Section 28.2, we will approach the related works. Section 28.3, we will report the adopted methodology. Section 28.4, we will present the necessary conceptual basis for the understanding of this work. Section 28.5, we will describe the definition and planning of our controlled experiment. Section 28.6, we will detail the operation of the experiment, since the preparation up to the data collection. Section 28.7, we will analyze and interpret the obtained results. Lastly, Sect. 28.8, we will present the conclusion and provide guidelines for future works.

28.2 Related Works

Some researchers performed sentiment classification through neural networks. Mikolov [5], proposed a method named Skip-gram model, that makes the analysis of each sentence

T. de Oliveira Lima (✉) · M. C. Júnior
Postgraduate Program in Computer Science – PROCC, Federal University of Sergipe (UFS), Aracaju, Brazil

K. H. de J. Prado · A. dos S. Júnior
Postgraduate Program in Computer Science – PROCC, Federal University of Sergipe (UFS), Aracaju, Brazil

Universidade Federal de Pelotas – Rio Grande do Sul, Pelotas, Brazil
e-mail: adalberto.santos@ufpel.edu.br

© Springer Nature Switzerland AG 2019
S. Latifi (ed.), *16th International Conference on Information Technology-New Generations (ITNG 2019)*,
Advances in Intelligent Systems and Computing 800,
https://doi.org/10.1007/978-3-030-14070-0_28

(sentence embedding), thus substituting the *BOW – Bag of Words* approach, which analysis the sentiment for each word. Such approach was developed by the Google engineers and achieved an accuracy of 72%.

In 2015 [6] proposed a unidirectional *LSTM*. The work achieved an accuracy of 88% in the sentiment polarity classification (positive or negative), by using Glove vectors [3].

Tang et al. [7] proposed an LSTM neural network based on targets of interest. In the sentence processing, the *Target-Connection LSTM (TC-LSTM)* algorithm has as input three distinct characteristics: The embedding vectors of the targets of interest, embeddings of the preceding words and embeddings of the subsequent words. Wang et al. [12] extended the *TC-LSTM* network and additionally proposed to learn an embedding vector for each aspect.

Lastly, in the work of Hartmann et al. [1], several embedding models were trained in a large corpus of the Portuguese language, and were made available in the NILC Word Embeddings repository. For the construction of such repository, 17 different corpus were used, totaling 1.395.926.282 tokens. The training of the vectors occurred in algorithms like Word2vec [2], *FastText* [8], *Wang2vec* [9] and *Glove* [3].

As a differential, we propose to conduct an experimental process to evaluate the application of deep learning neural networks, based on sustainable aspects in Portuguese opinions, with the purpose of evaluating how sustainable practices influence positively or negatively of the hotels clients opinions.

28.3 Methodology

The methodology used in the work in question consists of an exploratory research [10] in classification terms, because a systematic literature mapping was performed [11], with systematic approaches, in which the relevance algorithms were collected in the context of this research.

After the definition and collection of the algorithms, two classifiers based on LSTM neural networks were chosen: *TC-LSTM* [7] and *AT-LSTM* [12]. These algorithms were chosen because Deep learning methods employ multiple processing layers to learn hierarchical representations of data, and have produced state-of-the-art results in many domains [13].

At last, to achieve the main objective of this research and consequently collect data, the execution of a controlled experiment (in vitro) was proposed, which involved the hotel review database. According to [14], an experimentation is not a simple task, because it involves preparing, performing and analyzing data correctly. The authors highlight the control of subjects, objects and instrumentation as one of the main advantages of the experimentation, which makes it possible to extract more general conclusions on the investigated subject.

Other advantages include the ability to perform statistical analyses, by using hypothesis test methods, and opportunities for replication. The authors of [15] also assert that the scientific research cannot be based on opinions or commercial interests. Scientific investigations are represented by studies based on observation and/or experimentation in the real world and its measurable behaviors, as in this research. Such aspects must also be taken into consideration in the construction and assessment of algorithms and software.

During the execution of the experiment, with the definition of the algorithms, a training dataset was submitted to the free data mining API and open-source Tensor-Flow [16], in which knowledge models were generated aiming to perform tests on the algorithms and compare the resulting metrics. In such context, this work was also classified as a laboratorial and experimental one, due to the planning and execution of a controlled experiment. Aiming to assist the calculations and verifying the existence of significant differences in the accuracy, recall, precision, average training time and average execution time of the algorithms, the SPSS – IBM data analysis tool [42] was used, applying basic and advanced statistical techniques. Summarizing, the experiment can be divided into three main stages: planning; data cleaning operation, dataset generation and data collection; and analysis of results. The experiment in question is detailed in Sects. 28.5 and 28.6.

28.4 Conceptual Basis

28.4.1 Opinion Mining

One of the most important factors for the success of a company is to know its target public. With the great advance of technology, mainly in the social network area, it is possible to access recently disclosed information at a fascinating speed and reach a broader target public. Web users have to opportunity to register and publish their ideas and opinions in several communication means, like discussion forums, blogs, Twitter and social networks. Thus, many organizations have given a special attention to such way of researching opinions because it is faster, more economic and dynamical.

Opinion mining can be applied to parts of the text independent of its size and format, and they can be posts, tweets, product reviews or comments. Any opinion is formed by at least two key elements: the target and the sentiment. The target can be represented by an entity or topic, like a product, a person, a product, an organization, a brand or even an event. However, a sentiment represents an attitude, opinion or emotion that the author showed regarding such target [17], and its polarity corresponds to a value in some scale that represents a positive, neutral or negative assessment of the meaning of such sentiment [18].

Lately, a convenient approach for the advertising and promotion of the tourist industry that has grown, is the use of web pages. Thus, many researchers noticed that the reviews and comments gathered by the sites are useful for guests and hotel managers [19]. However, the analysis of the large array of evaluative texts is a tough task, not only for the guests, but also for the stakeholders in hotel management. On the other side, the analysis of the reviews based on few items can generate biased situations [20]. In the past years, some researches have proposed opinion extraction systems, mainly independent of domain, to automatically extract structured opinion representations contained in the texts [21–23].

28.4.2 Evaluate Metrics

Based on the confusion matrix presented [24–26], we can use the main quality metrics: accuracy (28.1), precision (28.2), recall (28.3) and F-measure (28.4) [27, 28]. In addition, we will consider the average training times and classification models (28.5) as performance metric.

$$acurracy = \frac{TP + TN}{TP + TN + FP + FN} \qquad (28.1)$$

$$precision = \frac{TP}{TP + FP} \qquad (28.2)$$

$$recall = \frac{TP}{TP + FN} \qquad (28.3)$$

$$fmeasure = \frac{2 \times precision \times recall}{precision + recall} \qquad (28.4)$$

$$T = \frac{\sum_{k=1}^{n} t_{ik} - t_{fk}}{n} \qquad (28.5)$$

TP-True Positive, TN-True Negative, FP-False Positive and FN-False Negative

28.5 Definition and Experiment Planning

In this and in the two next sections, this work is presented as an experimental process. The same follows the Wohlin's guidelines [14] and experimental processes with recent publications [29, 30]. This section will focus on the definition of the objective and planning of the experiment.

28.5.1 Goal Objective

The objective of this study is to assess, through a controlled experiment, the efficiency and efficacy of the *TC-LSTM* and *AT-LSTM* algorithms in the context of opinion mining oriented toward sustainable aspects, aiming to predict the polarity degree (positive or negative) of the opinions stated in text assessments. The target of the experiment will be a corpus with sustainable aspects of reviews of the *Booking* and *Tripadvisor* social media regarding the hotels in the state of Sergipe (Brazil). Such corpus was built with the use of expert knowledge in the ambit of sustainable hotel management [44]. The objective was formalized using the GQM model proposed by Basili et al. [31]:

Analyze the Target-Connection LSTM (TC-LSTM) [7] and Attention-based LSTM (AT-LSTM) algorithms [12], applied to the context of natural language processing, **with the purpose of** evaluating them, **with respect to** accuracy, recall, precision, F-measure, and average training time and execution, **from the viewpoint of** the Data Analytics researchers and professionals, **in the context of** the clients opinions polarity about sustainable aspects, contained in the reviews on the Booking and Tripadvisor social medias about Sergipe hotels (Brazil).

28.5.2 Planning

Context Selection

Initially, we captured the reviews on the *Booking* and *Tripadvisor* social media of the hotels of Sergipe (Brazil) to compose the experimentation dataset. According to a survey conducted in 2015 [32], the state of Sergipe (Brazil) presents an index of tourism competitiveness below the national index. Therefore, it is fundamental to contribute and improve these indicators in the national scenario. With the use of expert knowledge [44] and information contained in the standard NBR 15.401 [33], we elaborated a data dictionary of sustainable aspects, which guide indicators for a good sustainable management of hotels. Besides, a work for the creation of labels on the reviews with the sentence, sustainable aspects and the phrase polarity (positive or negative) was performed.

After the pre-processing, which consisted of the removal of stopwords, slang, orthographic accentuation and correction, 2,200 reviews were selected, 50% positive and 50% negative.

During the experiment, the assessed algorithms used the 10-Fold Cross-validation approach *10-Fold Cross-validation*

[34]. In such process, the data are divided into 10 parts, keeping their proportions. In each test, one part of the data is separated for the initial phase, it means, the training one. The other, for the posterior one, the test phase.

Hypothesis Formulation

The following research question, which answer aims to accomplish the objective of the work, was elaborated to guide the study:

- What is the best algorithm in terms of accuracy, recall, precision, F-measure, average training time and average classification time?

Six metrics will be used for assessing the question: (1) Accuracy; (2) Precision; (3) Recall; (4) F-measure, (5) Average training time and (6) Average classification time.

With the objectives and metrics defined, the following hypothesis will be considered (for each metric):

H_0: The algorithms have the same metric mean.

$$\mu 1(\text{metric}) = \mu 2(\text{metric})$$

H_1: The algorithms have distinct metric means.

$$\mu 1(\text{metric}) \neq \mu 2(\text{metric})$$

Note that H_0 is the hypothesis that we want to refute.

Selection of Participants

According to [35], the participants are the selected individuals from the population with the purpose of carrying out the experiment. Such participants are responsible for informing parameters to the experiment, like the value of the variables.

The data were divided in the training and test sets. The first one composed by 64% of the reviews and the second one by 36%. The decision on the distribution was made by convenience, knowing that the basis was available with such distribution. Table 28.1 hows the amount of reviews by data set versus polarity.

Experiment Project

The experiment project refers to the following stages: Preparation of the development environment, it means the downloading and installation of all the items described in Sect. 28.5.2. Subsequently, the implementation of the

Table 28.1 Distribution of datasets vs. polarity

DataSet	Polatiry	
	Negative	Positive
Train dataset	700	700
Test dataset	400	400

TC-LSTM and AT-LSTM algorithms, making use of the hotel datasets already labeled according to the sustainable aspects and finally, the execution of the statistical tests for the assessment of the defined hypotheses.

Independent Variables

The TC-LSTM [7] e AT-LSTM Algorithms[12] were used for the classification task.

Dependent Variables

Accuracy (ACU), precision (PRE), recall (REC), F-measure (F1), average training time (ATT) and average classification time (ACT).

Instrumentation

The instrumentation process started with the environment configuration for the achievement of the controlled experiment; data collection planning; and the development and execution of the assessed algorithms. The used materials/resources were:

- The Jupyter Notebook tool for the development [36].
- *TensorFlow* [37]
- *Scikit-learn* [38]
- Word vectors (Word Embeddings) [45].
- A computer with Intel(R) Core(TM) i-5500U CPU @ 2.40 GHz, 8 GB RAM – 64 bits.

28.6 Experiment Operation

28.6.1 Preparation

The adequate environment, aiming to store the data used in the controlled experiment, was prepared for the achievement of this work. Thus, the following stages to achieve such objective were performed:

- **Collecting of reviews:** collecting reviews from the *Booking* and *Tripadvisor* social media of the hotels in the state of Sergipe (Brazil).
- **Creation of the data dictionary of sustainable aspects:** Creating a data dictionary related to the sustainable aspects built by experts in the sustainable tourism area [44]. The standard NBR 15.401 was also considered.
- **Definition of labels and reviews:** Labeling the database according to the sustainable aspects collected in the item above, using the expert knowledge [44].
- **Configuration of the development environment**: Installation of the development environment Jupyter Notebook to perform the implementation, interaction and execution of the algorithms assessed in this article.

- **Python libraries:** downloading and configuration of the TensorFlow, NLTK, SkLearn, Numpy and Pandas library in the Jupyter Notebook environment.
- **Word vectors:** Downloading of the pre-trained Word Embeddings model Word2Vect [2] containing 100 dimensions.

28.6.2 Execution

In synthesis, after the preparation of the environment and selection of the participants, topics approached in Sects. 28.6.1 and 28.5.2, respectively, the pre-processing stages were initiated, by applying the tokenization, substitution of the abbreviated words, removal of stopwords, and orthographic correction stages. Such stages aim to improve data quality, as well as to transform the texts into a more structured and adequate representation for the classificatory processes.

At the end of the pre-processing stages, the new structured set was subjected to the algorithms approached in this work, in order to accomplish the classification.

It is important to emphasize that the average classification time was calculated by using 100 random reviews of the test dataset. However, the average training time was calculated based on five executions (training ones) in the training dataset.

28.6.3 Data Validation

As assistance for the analysis, interpretation and validation, it was planned to used three types of statistical tests: the Shapiro-Wilk Test [39], the Levene's Test and the Anova Test. The Anova Test aims to compare more than two value groups. The test requires that the sample have a normal distribution, besides the existence of homoscedasticity among the treatments (homogeneous variances) [40]. Therefore, the Shapiro-Wilk and the Levene's Tets were used [41]. The first one aims to validate the normality of the samples, while the second one verifies the homoscedasticity.

In the cases, which the distribution of the samples do not observe the requirements demanded by the Anova Test, it means, normal distribution and homoscedasticity, the Wilcoxon Test was used. The Wilcoxon Test compares the media of the paired samples, by verifying the difference magnitude. All the statistical tests were conducted by using the SPSS – IBM Tool [42].

28.7 Results

The following dependent variables were analyzed to answer the questions: Accuracy, precision, recall, F-measure, average training time and average classification time.

28.7.1 Analysis and Data Interpretation

After the execution of the algorithms, the classifications of the results were obtained by using the 10-cross-validation approach. The averages for each metric and algorithm are presented in Table 28.2.

As we can observe in Table 28.2, the TC-LSTM algorithm did not obtain a very satisfactory development in the verified metrics, being inclusive overcome by the AT-LSTM method in all aspects. In relation to the Accuracy, Precision and F-measure metrics; the AT-LSTM classifier presented a far superior performance than the other method. On the other side, the percentage difference of the recall metric among the approached methods was not as significant in comparison with the other analyzed variants. In relation to the training and test times, the AT-LSTM algorithm presented less elevated average times, thus characterizing a better performance. However, it is not possible to make such statements without sufficiently conclusive statistical evidence.

So, firstly the Shapiro-Wilk Test was conducted. Then, a 0,05 significance level was defined for all the experiment. When applying the Normality Test (Shapiro-Wilk) and obtaining the significance levels (p-values) of the approached methods, it was observed that the values are below the adopted level, concluding that the distributions are not normal. Thus, based on such results, it was not possible to use the Anova as planned. Then, we adopted a nonparametric statistical test, in this case the Wilcoxon Test [43], to accomplish

Table 28.2 Comparative of the metrics of the algorithms

Averages (μ)	Algorithm	
	TC-LSTM	*AT-LSTM*
Accuracy (ACU)	42.84%	74.58%
Precision (PRE)	44.46%	95.54%
Recall (REC)	71.40%	77.20%
F-measure (F1)	54.48%	85.37%
Average training time (seconds) (ATT)	14.05s	7.35s
Average classification time (seconds) (ACT)	3.90s	1.12s

Table 28.3 Result of the wilcoxon statistical test

Hypotheses	Action	
	p-value	Result
HO^{ACU}	0.000	Refute H0
HO^{PRE}	0,000	Refute H0
HO^{REC}	**0,517**	**Retain H0**
HO^{F1}	0,000	Refute H0
HO^{ATT}	0,000	Refute H0
HO^{ACT}	0,000	Refute H0

the hypothesis tests of this work. The results of the tests using Wilcoxon can be observed in Table 28.3.

When analyzing the result of the listed significance levels (p-value), we found indicators for the rejection of the new Wilcoxon null hypotheses related to accuracy, precision, F-measure, average training time and average classification time, because the found values (0.00) are below the stipulated significance level for the experiment (0.05). However, regarding the recall, we did not have statistical significance that demonstrates differentiation between the algorithms, it means, the null hypothesis will be kept.

The AT-LSTM algorithm presented the best results, mainly in terms of accuracy, F-measure, average training time and average classification time. The first with 74,58%, the second with 85,37%, the third with 7,35 s and the last one with 1,12 s. However, as mentioned before, although AT-LSTM has also been higher, with 77,20%, recall was the only metric whose difference for the TC-LSTM (71.40%) did not reach statistical significance. In other words, regarding Wilcoxon Test, both ones have the same distribution of recalls.

28.7.2 Threats to Validity

One of the main questions of an experiment is how valid the results obtained by it are. Threats to validity may limit the ability to interpret and/or describe the data results obtained in an experiment. Thus, there is no way to ignore the following threats found during the experiment:

Threats to the Construct Validity
- The implementation performed to reproduce the algorithms may have suffered some changes related to its original function.
- Even though using the expert knowledge [44] to construct the data dictionary, it still becomes an activity with a certain level of subjectivity. After the classification of sentences, the four sustainable tourism experts conducted the inter-pair review to mitigate such risk.

Threats to the External Validity
- The experiment was only applied to the reviews made in the hotels in the state of Sergipe (Brazil) and with texts in Portuguese, thus the result may undergo alterations when it is applied to other languages or even if applied to diverse Brazilian regions.

28.8 Conclusion and Future Works

The objective of this work was to assess, through a controlled experiment, the performance and quality of the TC-LSTM and AT-LSTM algorithms, aiming to mine the opinions in the context of hotel reviews. After the selection of the participants, a dataset containing 2,200 reviews was obtained and used in the experiment.

At the end of the classification and result analysis stages, it was evident that the AT-LSTM model presented the best results in face of the metrics proposed and assessed by this article, mainly in terms of accuracy, precision, F-measure, average training time and average classification time. The first with 74,58%, the second with 95,54%, the third with 85,37%, the fourth with 7,3 s and the last one with 1,12 s. Recall was the only metric whose difference for the TC-LSTM did not reach statistical significance. This algorithm can be considered for mining opinions based on specific aspects of tourism and others peculiar market niches.

For future works, we intend to use other embedding models (for example FastText, Wang2Vec and Glove) and its dimensional variants will be analyzed aiming to achieve better results, making use of the NILC (Interinstitutional Center for Computational Linguistics) Word Embeddings repository [45]. Moreover, we intend to extend the work by applying a bigger data volume, considering other Brazilian states.

References

1. Hartmann, N., et al.: Portuguese word embeddings: evaluating on word analogies and natural language tasks. arXiv preprint arXiv:1708.06025 (2017)
2. Mikolov, T., Chen, K., Corrado, G., Dean, J.: Efficient estimation of word representations in vector space. In: Proceedings of International Conference on Learning Representations Workshop (ICLR-2013) (2013)
3. Pennington, J., Socher, R., Manning, C.D.: Glove: global vectors for word representation. In: Proceedings of the 2014 Conference on Empiricial Methods in Natural Language Processing (EMNLP-2014), vol. 12, pp. 1532–1543 (2014)
4. Greff, K., et al.: LSTM: a search space odyssey. IEEE Trans. Neural Netw. Learn. Syst. **28**(10), 2222–2232 (2017)
5. Mikolov, T., et al.: Efficient estimation of word representations in vector space. arXiv preprint arXiv:1301.3781 (2013)
6. Tai, K.S., Socher, R., Manning, C.D.: Improved semantic representations from tree-structured long short-term memory networks. arXiv preprint arXiv:1503.00075 (2015)

7. Tang, D., et al.: Effective LSTMs for target-dependent sentiment classification. arXiv preprint arXiv:1512.01100 (2015)
8. Bojanowski, P., Grave, E., Joulin, A., Mikolov, T.: Enriching word vectors with subword information. arXiv preprint arXiv:1607.04606 (2016)
9. Ling, W., Dyer, C., Black, A., Trancoso, I.: Two/too simple adaptations of word2vec for syntax problems. In: Proceedings of the 2015 Conference of the North American Chapter of the Association for Computational Linguistics: Human Language Technologies. Association for Computational Linguistics (2015)
10. Severino, A.J.: Metodologia do trabalho científico. Cortez editora (2017)
11. Lima, T., Junior, M.C., Augusta, M., Nunes, S.N.: Mining on line general opinions about sustainability of hotels: a systematic literature mapping. In: International Conference on Computational Science and Its Applications. Springer, Cham (2018)
12. Wang, Y., Huang, M., Zhao, L.: Attention-based LSTM for aspect-level sentiment classification. In: Proceedings of the 2016 Conference on Empirical Methods in Natural Language Processing (2016)
13. Young, T., et al.: Recent trends in deep learning based natural language processing. arXiv preprint arXiv:1708.02709 (2017) 18
14. Wohlin, C., et al.: Experimentation in Software Engineering. Springer Science & Business Media, Berlin/New York (2012)
15. Juristo, N., Moreno, A.M.: Basics of Software Engineering Experimentation. Springer Science & Business Media, New York (2013)
16. Abadi, M., et al.: Tensorflow: a system for large-scale machine learning. OSDI, vol. 16 (2016)
17. Liu, B.: Sentiment analysis and opinion mining. Synth. Lect. Hum. Lang. Technol. 5(1), 1–167 (2012)
18. Tsytsarau, M., Palpanas, T.: Survey on mining subjective data on the web. Data Min. Knowl. Disc. 24(3), 478–514 (2012)
19. Srisuan, J., Hanskunatai, A.: The ensemble of Naïve Bayes classifiers for hotel searching. In: 2014 International Computer Science and Engineering Conference (ICSEC). IEEE (2014)
20. Songpan, W.: The analysis and prediction of customer review rating using opinion mining. In: 2017 IEEE 15th International Conference on Software Engineering Research, Management and Applications (SERA). IEEE (2017)
21. Chaabani, Y., Toujani, R., Akaichi, J.: Sentiment analysis method for tracking touristics reviews in social media network. In: International Conference on Intelligent Interactive Multimedia Systems and Services. Springer, Cham (2017)
22. Hung, C.: Word of mouth quality classification based on contextual sentiment lexicons. Inf. Process. Manag. 53(4), 751–763 (2017)
23. Hu, Y.-H., Chen, Y.-L., Chou, H.-L.: Opinion mining from online hotel reviews–a text summarization approach. Inf. Process. Manag. 53(2), 436–449 (2017)
24. James, G., et al.: An Introduction to Statistical Learning, vol. 112. Springer, New York (2013)
25. Patro, V.M., Ranjan Patra, M.: Augmenting weighted average with confusion matrix to enhance classification accuracy. Transactions on Machine Learning and Artificial Intelligence 2(4), 77–91 (2014)
26. Makhtar, M., Neagu, D.C., Ridley, M.J.: Binary classification models comparison: on the similarity of datasets and confusion matrix for predictive toxicology applications. In: International Conference on Information Technology in Bio-and Medical Informatics. Springer, Berlin/Heidelberg (2011)
27. Caelen, O.: A Bayesian interpretation of the confusion matrix. Ann. Math. Artif. Intell. 81(3–4), 429–450 (2017)
28. Dwivedi, A.K.: Performance evaluation of different machine learning techniques for prediction of heart disease. Neural Comput. Appl. 1–9, 29:685–693 (2016)
29. de Oliveira, N., Robert A., Junior, M.C.: Experimental analysis of stemming on jurisprudential documents retrieval. Information 9(2), 28 (2018)
30. Santos, B.S., Junior, M.C., de Souza, J.G.: A initial experimental evaluation of the neuromessenger: a collaborative tool to improve the empathy of text interactions. In: Latifi, S. (ed.) Information Technology-New Generations, pp. 411–419. Springer, Cham (2018)
31. Basili, V.R., Weiss, D.M.: A methodology for collecting valid software engineering data. IEEE Trans. Softw. Eng. 6, 728–738 (1984)
32. do Turismo, M.: Índice de Competividade do Turismo Nacional. Available in: https://goo.gl/c3agrm. Access in: 01 out (2018)
33. de Matos, J.K.E., Costa, M.A.N.: Sustentabilidade nos meios de hospedagem no Brasil: a norma NBR 15401: 2006. Congresso de Arquitetura, Turismo e Sustentabilidade, vol. 1 (2012)
34. John Lu, Z.Q.: The elements of statistical learning: data mining, inference, and prediction. J. R. Stat. Soc. A. Stat. Soc. 173(3), 693–694 (2010)
35. Travassos, G.H., Gurov, D., Amaral, E.A.G.G.: Introdução àã Engenharia de Software Experimental. UFRJ, Rio de Janeiro (Brazil) (2002)
36. Kluyver, T., et al.: Jupyter Notebooks-a publishing format for reproducible computational workflows. ELPUB (2016)
37. Abadi, M., et al.: Tensorflow: a system for large-scale machine learning. OSDI, vol. 16 (2016)
38. Pedregosa, F., et al.: Scikit-learn: machine learning in Python. J. Mach. Learn. Res. 12, 2825–2830 (2011)
39. Shapiro, S.S., Wilk, M.B.: An analysis of variance test for normality (complete samples). Biometrika 52(3/4), 591–611 (1965)
40. Field, A.: Descobrindo a Estatística Usando o SPSS-2. Bookman Editora, São Paulo (Brazil) (2009)
41. Levene, H.: Robust tests for equality of variances. In: Olkin, I., et al. (eds.) Contributions to Probability and Statistics: Essays in Honor of Harold Hotelling, pp. 279–292. Stanford University Press, Stanford (1961)
42. SPSS IBM.: IBM SPSS Statistics for Windows, Version 20.0. IBM Corp, New York (2011)
43. Litchfield, J.T. Jr., Wilcoxon, F.: A simplified method of evaluating dose-effect experiments. J. Pharmacol. Exp. Ther. 96(2), 99–113 (1949)
44. del Pino Medina Brito, M., dos Santos Júnior, A.: La gestión sostenible de los establecimientos de hospedaje de la provincia Sur de Brasil. I Foro Internacional de Turismo Maspalomas Costa Canaria. Universidad de Las Palmas de Gran Canaria (2013)
45. Hartmann, N., et al.: Portuguese word embeddings: evaluating on word analogies and natural language tasks. arXiv preprint arXiv:1708.06025 (2017)

Sérgio Luisir Díscola Junior, José Roberto Cecatto,
Márcio Merino Fernandes, and Marcela Xavier Ribeiro

29.1 Introduction

Solar Flares (SF) are sudden releases of large amounts of energy from the solar atmosphere [1]. They are categorized into 5 classes, namely, *A, B, C, M and X*, respectively in order of their strength, where SF of class *A* are the least harmful, while *X* flares are the most powerful and dangerous ones. They are categorized according to the level of X-ray emitted by Sun during the event. These phenomena impact satellite communications [2], Global Positioning System (GPS) and may also produce electricity power blackouts. So, it is imperative to develop robust solar flare forecasting systems.

The features that influence the forecasting process are not known. We can find papers that use features derived from magnetogram vector [3–5], sunspot area [6], radio flux or X-ray flux [7] and [8] the X-rays time series.

Solar Flare datasets are extremely imbalanced. Most work in literature using traditional classifiers that deal with imbalanced datasets have the drawback of producing biased results [9–11]. An alternative to handle the poor results of the learning in imbalanced data is the usage of an Ensemble of Classifiers (EC) [12–14]. The EC main goal is to improve "weak" classification methods by applying many of weak classifiers (also called base inducers), so that the final classification may produce more accurate results. In this sense,

we propose an EC tuned-up for the domain of the Solar Flare forecasting.

Most of the previous works of Solar Flare forecasting perform binary forecasting, classifying solar flare only as "Positive" or "Negative". Few works predict individual classes. In the latter case, they usually use purely statistical methods in the forecasting process. Some methods consider "Positive" results for classes greater than or equal to "C" [15], others consider "Positive" for forecasts greater than or equal to a class M [3–5, 7]. Furthermore, a recurrent practice in the previous literature works is the forecasting of just the maximum Solar Flare of a given day.

Accordingly, we propose a method called ECID (Ensemble of classifiers for imbalanced datasets) that tackles some important open issues:

(1) Perform individual class forecasting producing a multi-class result for a given day, so that the method provides to the astrophysicist a tool that shows possible Solar Flare categories that may happen in a given day;
(2) Treat the imbalanced dataset issue using Ensemble with a stratified random sampling for the training of the inducers;
(3) Perform also a multi-label solution, giving the possibility to the astrophysicist to decide (when adjacent classes are indiscernible).

S. L. Díscola Junior (✉)
Department of Computer Science, Federal University of São Carlos, São Carlos, SP, Brazil

Federal Institute of São Paulo, São Carlos, SP, Brazil
e-mail: sergio.discola@ufscar.br

J. R. Cecatto
National Institute for Space Research, São José dos Campos, SP, Brazil

M. X. Ribeiro · M. M. Fernandes
Department of Computer Science, Federal University of São Carlos, São Carlos, SP, Brazil

29.2 Method Description

In this section, we describe ECID (Ensemble of Classifiers for Imbalanced Datasets) and its pre-processing steps. Figure 29.1 presents its overview.

Steps 1, 2 and 3 are the preprocessing steps responsible for obtaining, cleaning and transforming the data for the learning task. Traditional classification methods produce models that

© Springer Nature Switzerland AG 2019
S. Latifi (ed.), *16th International Conference on Information Technology-New Generations (ITNG 2019)*,
Advances in Intelligent Systems and Computing 800,
https://doi.org/10.1007/978-3-030-14070-0_29

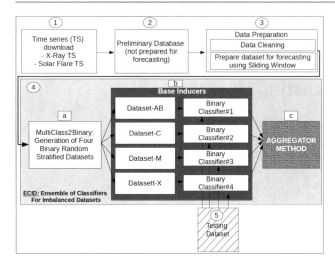

Fig. 29.1 Overview of the proposal for solar flare forecasting: the method ECID and its pre-processing

classify instances with **"current" events**. For the forecasting purpose, it is necessary to map current values with future events to turn such classifiers in forecasting methods. Thus, our method prepares the original dataset to the forecasting by using the "Sliding Window" algorithm proposed in [8].

The prepared data are submitted to the proposed ensemble method ECID. The issue of imbalanced data is tackled by employing a stratified random sampling, which produce the datasets employed to train the base inducers. This is done by splitting the original dataset into four balanced subsamples, one for each forecasting class, using the new "MultiClass2Binary Balancing" schema (see Fig. 29.1 – Step 4(a)).

The base inducers are weak binary classifiers (see Fig. 29.1 – Step 4(b)): the "Binary Classifier#1" generates a "Positive" result if a class "A or B" is predicted by the model and "Negative", otherwise; "Binary Classifier#2" generates a "Positive" result if a class "C" is predicted and "Negative", otherwise; and the same logic is applied for the classes "M" and "X". The output of the base inducers are submitted to the aggregator method (see Fig. 29.1 – Step 4(b)), which combines the individual votes of each inducer, producing the final multi-class forecasting. The forecasting is also multi-label when the aggregator function returns more than a class.

Step 5 provides the testing dataset to validate the model, which comprises the time series of the current day. Hence, ECID provides forecasting for the next day. A detailed description of the steps of ECID is given in following.

29.2.1 The Preprocessing Steps

As shown in Fig. 29.1, in Steps 1 and 2, the time series of X-ray intensity and the solar flare report are collected from their sources. These time series have different sample rates: the X-ray time series have a sample rate of 12 min, and the

solar flare report has a "varying" sample rate depending on the duration of the Solar Flare events. So, for one entire day, the X-ray time series produces 120 instances, and the solar flare report produces a varying amount of instances. The next task is to map the X-ray intensity and the solar flare report, as formally explained next:

- the time series of X-ray intensity is defined as $X = \{instant\,of\,X\text{-}ray\,obs,$
  $X\text{-}ray\,intensity\}$, where:
  - $instant\,of\,X\text{-}ray\,obs$ is the observation instant of time;
  - $X\text{-}ray\,intensity$ is the intensity of X-ray measured given by W/m^2.
- the solar flare report contains summarized information about solar events and it is defined as $E = \{instant\,of\,Solar\,Flare,$
  $Solar\,Flare\,Class\}$, where:
  - $instant\,of\,Solar\,Flare$ is the instant of time that a solar flare was observed;
  - $Solar\,Flare\,Class$ is the class of a solar flare.

The mapping between these time series produces the "Preliminary Database" defined as:

- a tuple $T = \{instant\,of\,XRay\,observation,$
  $X\text{-}ray\,intensity, Solar\,Flare\,Class\}$;
- $instant\,of\,XRay\,observation$ is the instant when the X-ray was emitted by the Solar Flare;
- $X\text{-}ray\,intensity$ is the X-ray emitted by the Sun in $instant\,of\,XRay\,observation$ and,
- $Solar\,Flare\,Class$ is the class of the Solar Flare occurred in that instant.
- the "Preliminary Database" is $M^{XE} = \cup T$.

In Step-3, the data is cleaned by discarding tuples with troubled values and, then, the "Sliding Window" approach [8] is applied to map current instances to future events (classes of solar flares), enabling the learning model to forecast. The Sliding Window approach builds a set formally defined as:

- $|M^{XE}|$ is the number of instances of the "Preliminary Dataset", M^{XE};
- $S^{SFD} = \{x_i(t), x_i(t+1), \ldots,$
  $x_i(t + current\,Window\,Size),$
  $maximum\,Class\,Of\,Future\,Window,$
  $1 \leq i \leq |M^{XE}|\}$, where S^{SFD} is the "Slided Solar Flare Dataset", $x_i(t)$ is the $X - ray_i\,intensity$ of the current Window, $curren\,Window\,Size$ is the size of the current window, and $maximum\,Class\,Of\,Future\,Window$ is the maximum solar flare class occurred in the future window (for a detailed explanation of current and future window, see [8]);

29.2.2 The Proposed Method ECID (Ensemble of Classifiers for Imbalanced Datasets)

The preprocessed data is submitted to ECID that employs a modified bootstrap strategy: ECID builds multiple learning models of the same type from different subsamples of the training dataset. Specifically, it splits the training dataset in subsamples using a strategy that we named *MultiClass2Binary*.

"MultiClass2Binary" strategy is the key to provide a strong multi-class forecasting for 4 classes of solar flares (AB, C, M and X). This strategy builds 4 distinct balanced datasets using stratified random undersampling. The sampling strata are the solar flares classes, and the sampling schema is detailed next. The first dataset is composed of 50% of the tuples classified as classes *A* or *B* and 50% of remaining; the second one is composed of 50% of tuples class C and 50%, the remaining; the third is balanced in the same way for class M, and the last one for class X.

In Step-4(b) Fig. 29.1, each dataset is applied to a binary inducer from the same type (eg. a decision tree algorithm). Accordingly, the **Binary Classifier#1** provides the forecasting for class *A* or *B*, **Binary Classifier#2**, for class C, **Binary Classifier#3**, for class M, and, **Binary Classifier#4**, for class X. We propose to employ all inducers of the same type because preliminary experiments did not demonstrated improvements by combining different types of inducer. However, the combination of different inducers types could also be performed.

In Step-4(c), a testing dataset is applied to each model to produce specific results for each solar flare class. For solar flare prediction, the sample rate of the input is 12 min. Different from traditional approaches, which the input of testing is a single tuple, the testing input is a set of 120 tuples (collected in a day to forecast the most high class for the next day). The Aggregator method combines the 120 results obtained in the last task, providing unified multi-class forecasting for a given day.

The Aggregator method is based on a voting count schema. Table 29.1 shows a simplified example of the output produced by the base inducers, which are binary classifiers. The "instant" column corresponds to the daily 12 min sample vote by each inducer. Columns z_Y (where $z \in \{AB, C, M, X\}$) is set to 1 if the correspondent inducer produced positive forecasting for the given class. Consider the tuple with "Day_1_00:00", for example, if **Binary Classifier#1** produced a "Positive" forecasting, the AB_Y column is set to 1; if **Binary Classifier#2** produced a "Positive" forecasting, the C_Y column is set to 1; and so on. Note that, the forecasting is originally given in a 12 min rate, but the goal of this work is to produce a daily forecasting. So, the Aggregator combines the votes of the base inducers producing a daily result.

Table 29.1 Example of the base inducers individual forecasting

Instant	AB_Y	C_Y	M_Y	X_Y
Day_1_Instant_00:00	1	1	0	0
Day_1_Instant_00:12	...	...	...	...
Day_1_Instant_00:24	...	...	...	...

After calculating the daily sum of the z_Y columns of Table 29.1, the Aggregator decides among the classes, considering the three most voted ones as formally defined: Let *firstYes, secondYes, thirdYes* be the number of votes obtained, respectively, by the first three classes most daily voted in the aggregated table.

The parameters p_1, p_2 and p_3 are the least vote frequency acceptable to the class to be considered for *firstYes, secondYes*, and *thirdYes*, respectively, i.e. parameters p_1, p_2 and p_3 denote the daily minimum percentage of votes required for the class to be chosen. We found empirically that the best results are obtained using $p1 = 20\%$, $p2 = 5\%$ and $p3 = 2\%$. These parameters values show that as the class voting becomes smaller, the class becomes rarer, so it should be kept in the results. In fact, as the class frequency becomes smaller, it will probably be less foreseen, then the importance of the minority class should be increased.

29.3 Experiments

Experiments 1–6 were performed to validate the proposed method. All the experiments:

- used the same solar dataset composed of the time series of X-ray intensity emitted by Sun and the solar flare report from the period 2010 to 2017;
- employed the data pre-processing described in Sect. 29.2.1;
- used 70% of the dataset for training and 30% for testing to validate the method. More specifically, the training dataset have 67678 instances (50% of class AB, 35% of class C, 13% of class M, and 2% of class X). The dataset is extremely imbalanced, and according to astrophysicist, the most relevant class are M and X, which are the most infrequent ones.

For experiments 1 to 3, we employed a traditional classification method alone to produce the forecasting model: Experiment 1 used IBK; Experiment 2 employed SVM; and Experiment 3 employed J48. The implementation of these methods was obtained from Weka [16].

For experiments 4 to 6, we employed the proposed ensemble method ECID (see Sect. 29.2.2). The stratified undersampling performed in the Step 4(a) of ECID (see Fig. 29.1) is shown in Fig. 29.2. The training dataset contains 33839

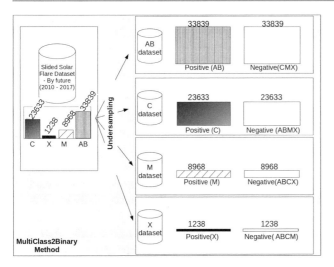

Fig. 29.2 ECID sampling (see Fig. 29.1)

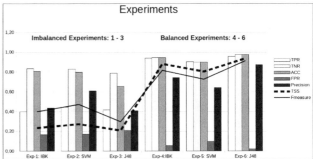

Fig. 29.3 Experiments – arithmetic mean

tuples labeled as class "A or B" (called "AB"), 23633 labeled as "C", 8969 as "M" and 1238 as class "X". The method randomly undersamples the original dataset according to each class. For example, the balanced dataset-AB is composed of 33839 tuples labeled "A" or "B" of the original dataset, these tuples will be re-labeled as "Positive(AB)", the other 33839 tuples are composed by a subsample of the original tuples labeled as classes "C", "M" and "X", which will also be re-labeled as "Negative(CMX)". The experiments 4–6 used this strategy for undersampling, but they vary the base inducers in each experiment: in Experiment 4 the base inducers were IBK; in Experiment 5 were SVM; and in Experiment 6 the base inducers were J48.

The metrics used to analyze the results must be carefully interpreted. Analyzing only accuracy (ACC) may incur in error due to the majority classes have higher probability to be predicted then the minority ones. In that case, the overall ACC may perform well, but the most relevant classes (the minority) should be poorly predicted. Then, for a better analysis, the True Positive Rate (TPR) and the Precision for each class may be balanced. In this case, it means that either the model truly predicted most classes. Some metrics aim to give numeric values for this "balance", such as **TSS** which is the difference between recall and False Positive Rate (FPR) [4], and **F-Measure**, that relates Precision and Recall. Our method was validated using the joint analysis of accuracy (ACC), True Positive Rate (TPR), True Negative Rate (TNR), False Positive Rate (FPR), TSS and F-Measure. As the results obtained are multi-class, and sometimes multi-label, we took some cautions:

(1) First, it was calculated individual metrics for each class (AB, C, M and X) by considering "Positive" for a specific class, and "Negative" as all the remaining classes;

(2) Second, the arithmetic mean of each metric for all classes was obtained to compare the experiments performed (shown in Fig. 29.3).

As observed in Table 29.2, the classifiers alone produced relatively good results for predictions to classes AB and C (which are the majority classes), and very poor results for classes M and X (the minority and most important classes). The highest results for classes AB and C were 72.8% of TPR and 64.1% of Precision in Experiment-2 (SVM), and 76% of TPR and 70.8% of Precision for class C in Experiment-1 (IBK). In the other hand, it reached 34.2% of TPR and 12.4% of Precision for class M, and 73.3% of TPR and 10.8% of Precision for class X. Although a good TPR was obtained for class X, a very poor Precision drastically decreased the level of reliability of the forecasting model. TSS and F-measure metrics corroborate these results, which were reasonable for classes AB and C, and very poor for classes M and X.

Table 29.3 shows the results of Experiments 1–6, which used our proposed method ECID. The results obtained for the minority, but most important, classes were higher than the previous experiment using the classifiers alone. The **best results for class X** were achieved in Experiment 6 using J48 as the base inducers of ECID: 86.6% of TPR and 76.4% of Precision, TSS of 0.85 and an F-Measure of 0.812. Also, the predictions of class M resulted in 100% of TPR and 80% of Precision. Predictions of classes AB and C achieved results of more than 95%.

Figure 29.3 shows the increase of the individual metrics from the first three experiments (not using ECID) and the last three ones (using ECID). TPR had an increase of more than 50%, TNR, an increase of 12.5%, FPR, a decrease of 33%, and the Precision an increase of 56%. As shown in Fig. 29.3, TSS and Precision increased with the proposed method.

29.4   Related Works

In literature, Solar flare forecasting methods usually aggregate more than one class as the "Positive" and "Negative",

Table 29.2 Experiments 1–3 – Metric Details

	Experiment-1: IBK				Experiment-2: SVM				Experiment-3: J48			
	AB	*C*	*M*	*X*	*AB*	*C*	*M*	*X*	*AB*	*C*	*M*	*X*
TPR	0,635	0,760	0,197	0	0,728	0,649	0,131	0,266	0,309	0,286	0,342	0,733
TNR	0,833	0,542	0,962	1	0,741	0,606	0,971	0,998	0,978	0,757	0,597	0,825
ACC	0,757	0,638	0,853	0,971	0,736	0,625	0,851	0,977	0,693	0,549	0,561	0,822
TSS	0,469	0,302	0,160	0	0,469	0,255	0,103	0,264	0,288	0,043	−0,06	0,558
FPR	0,166	0,457	0,037	0	0,258	0,393	0,028	0,001	0,021	0,242	0,402	0,174
Precision	0,708	0,566	0,468	0	0,641	0,565	0,434	0,8	0,914	0,482	0,124	0,108
F-Measure	0,670	0,649	0,277	0	0,681	0,604	0,202	0,4	0,462	0,359	0,182	0,189

Table 29.3 Experiments 4–6 – Metric Details

	Experiment-4: IBK				Experiment-5: SVM				Experiment-6: J48			
	AB	*C*	*M*	*X*	*AB*	*C*	*M*	*X*	*AB*	*C*	*M*	*X*
TPR	0,900	1	0,916	0,933	0,892	0,910	0,875	0,933	0,966	1	1	0,866
TNR	0,978	0,906	0,945	0,943	0,907	0,893	0,902	0,898	0,985	0,962	0,974	0,983
ACC	0,942	0,942	0,942	0,942	0,900	0,900	0,900	0,900	0,977	0,977	0,977	0,977
TSS	0,879	0,906	0,861	0,876	0,799	0,804	0,777	0,831	0,952	0,962	0,974	0,850
FPR	0,021	0,093	0,054	0,056	0,092	0,106	0,097	0,101	0,014	0,037	0,025	0,016
Precision	0,973	0,870	0,628	0,5	0,892	0,844	0,477	0,358	0,983	0,943	0,8	0,764
F-Measure	0,935	0,930	0,745	0,651	0,892	0,876	0,617	0,518	0,975	0,971	0,888	0,812

Table 29.4 Comparison among literature works and ECID

	TPR	TNR	ACC	FPR	Precision	TSS	F-Measure
ECID	0,96	0,98	0,98	0,02	0,87	0,94	0,91
Nishikawa (=X)	0,90	0,99	0,99	0,0003	0,89	0,91	0,89
Nishikawa (>=M)	0,91	0,99	0,99	0,002	0,92	0,91	0,92
Bobra (>=M)	0,71	0,98	0,97	–	0,80	0,70	0,75
Li (>= M)	0,73	0,78	0,77	–	–	–	–

turning it a binary classification model. Instead of this, our method provides the contribution to give multi-class (and multi-label) forecasting, because, in a single day, it is important to identify the correct class of solar flare that occurs in a day. Additionally, as the astrophysicists do not fully understand these phenomena, each work considers a different set of solar features as input. Taking those facts into account, we can cite some relevant works that we named:

- *Nishizuka*: In [3], it was developed a solar flare forecasting method that labels: (1) *classes = X* as "Positive", in a first experiment, and (2) *classes ≥ M* as "Positive", in the last set of experiment;
- *Bobra*: In [4], it was developed a solar flare forecasting method that labels *classes ≥ M* as "Positive";
- *Li*: In [5], it was developed a solar flare forecasting method that labels *classes ≥ M* as "Positive" according to an equation of flare importance;

Table 29.4 presents the metrics obtained by the *Nishizuka, Bobra and Li* forecasting methods and our proposed method ECID:

As shown in Table 29.4, ECID achieved TSS and Precision higher than *Nishikawa* and the other studies. If we consider just the forecasting of class "X", our method obtained a TSS of 0,85 and an F-measure of 0,81. These results are slightly lower than *Nishikawa* ones, but our method also predicted flares of classes "AB", "C" and "M" consistently, instead of *Nishikawa*, which performed a binary classification. Thus, we believe that our proposed approach is an important contribution for the task of solar flare forecasting.

29.5 Conclusions

Most works for solar flare forecasting performs a binary prediction. Therefore, in that scenario the problems of imbalanced data and high similarity between adjacent classes are minimized. Our contribution was to deal with these problems by proposing the new ensemble method ECID, producing a multi-class forecasting. For each solar flare class, ECID employs a stratified random sampling for the training of base one-class inducers, strengthen their sensitivity to one-class. Using a modified bootstrap approach, the aggregator method combines the inducers results enabling a strong multi-class forecasting, which can also be multi-label in case of indiscernible classes. The results obtained showed that our proposal is well-suited to solar flare forecasting, achieving 86.6% of TPR and 76.4% of Precision for class X, 100% of TPR and 80% of Precision for class M, 100% of TPR and 94.3% of Precision for class C, and 96.6% of TPR and 98.3% of Precision for class AB. For future work, we intend to add more solar features in the process.

Acknowledgements The authors thank INPE, IFSP, FAPESP, CAPES and CNPq.

References

1. Holman, G.D.: The mysterious origins of solar flares. Sci. Am. **294**(4), 38–45 (2006)
2. Basu, S., Basu, S., MacKenzie, E., Bridgwood, C., Valladares, C.E., Groves, K.M., Carrano, C.: Specification of the occurrence of equatorial ionospheric scintillations during the main phase of large magnetic storms within solar cycle 23. Radio Sci. **45**(5), 1–15 (2010)
3. Nishizuka, N., Sugiura, K., Kubo, Y., Den, M., Watari, S., Ishii, M.: Solar flare prediction model with three machine-learning algorithms using ultraviolet brightening and vector magnetograms. Astrophys. J. **835**(2), 156 (2017)
4. Bobra, M.G., Couvidat, S.: Solar flare prediction using SDO/HMI vector magnetic field data with a ML algorithm. Astrophys. J. **798**(2), 135 (2015)
5. Yu, D., Huang, X., Hu, Q., Zhou, R., Wang, H., Cui, Y.: Short-term solar flare prediction using multiresolution predictors. Astrophys. J. **709**(1), 321–326 (2010)
6. Gallagher, P.T., Moon, Y.-J., Wang, H.: Active-region monitoring and flare forecasting I. Data processing and first results. Sol. Phys. **209**(1), 171–183 (2002)
7. Li, R., Zhu, J.: Solar flare forecasting based on sequential sunspot data. Res. Astron. Astrophys. **13**(9), 1118–1126 (2013)
8. Discola Jr, S., Cecatto, J., Fernandes, M., Ribeiro, M.: SeMiner: a flexible sequence miner method to forecast solar time series. Information **9**(1), 8 (2018) [Online]. Available: http://www.mdpi.com/2078-2489/9/1/8
9. García, S., Herrera, F.: Evolutionary undersampling for classification with imbalanced datasets: proposals and taxonomy. Evol. Comput. **17**(3), 275–306 (2009)
10. Wallace, B.C., Small, K., Brodley, C.E., Trikalinos, T.A.: Class imbalance, redux. In: 2011 IEEE 11th International Conference on Data Mining, pp. 754–763. IEEE (2011)
11. Han, H., Wang, W.-Y., Mao, B.H.: Borderline-smote: a new over-sampling method in imbalanced data sets learning. In: Advances in Intelligent Computing, pp. 878–887. Springer, Berlin/Heidelberg (2005)
12. Sagi, O., Rokach, L.: Ensemble learning: a survey. Wiley Interdisciplinary Rev. Data Min. Knowl. Discov. **8**(4), e1249 (2018)
13. Galar, M., Fernandez, A., Barrenechea, E., Bustince, H., Herrera, F.: A review on ensembles for the class imbalance problem: bagging-, boosting-, and hybrid-based approaches. IEEE Trans. Syst. Man Cybern. Part C Appl. Rev. **42**(4), 463–484 (2012)
14. Rätsch, G., Onoda, T., Müller, K.-R.: Soft margins for adaboost. Mach. Learn. **42**(3), 287–320 (2001)
15. Ahmed, W., Qahwaji, R., Colak, T., Higgins, P.A., Gallagher, P.T., Bloomfield, D.S.: Solar flare prediction using advanced feature extraction, machine learning, and feature selection. Sol. Phys. **283**(1), 157–175 (2013)
16. Hall, M., Frank, E., Holmes, G., Pfahringer, B., Reutemann, P., Witten, I.: The WEKA data mining software: an update. SIGKDD Explor. **11**(1), 10–18 (2009)

Fault Diagnostics for Multivariate Non-normal Processes

Denwick Munjeri, Mali Abdollahain, and Nadeera Gunaratne

30.1 Introduction

Several multivariate process capability indices (MPCIs) were developed to measure multivariate process capability and are mainly applicable to multivariate normal processes [1, 2]. However, in practice, some multivariate quality characteristics do not follow the normal distribution. Furthermore, to improve performance of multivariate processes, fault diagnostics should be carried out to identify and rank quality characteristics responsible for process poor performance. Reference [3] has developed a fault diagnostic approach that identifies and ranks GD variables using Burr XII Distribution. However, there is a shortcoming with this approach as the identified GD variable may be made up of multiple original variables, therefore identifying a cluster of variables may not help the quality practitioner that much when it comes to prioritizing individual quality characteristics for process performance improvement.

Machine learning (ML), using artificial neural net input gain measurement approximation (*ANNIGMA*), has also been used as a fault diagnostic tool to identify and rank responsible characteristics [4, 5]. Though, this approach does the identification and ranking of responsible characteristics, it does not measure capability or yield, of the multivariate non-normal process, which is vital in guiding the practitioner to the desired process performance level.

This paper proposes to address these challenges by developing a fault diagnostic hybrid for multivariate non-normal processes. The next section discusses multivariate process capability analysis. In Sect. 30.3, the proposed hybrid is developed. Application of the proposed hybrid on real and simulated data are carried out in Sect. 30.4 Results are discussed in Sect. 30.5 and conclusion is in Sect. 30.6.

30.2 Multivariate Capability Analysis

A full multivariate capability analysis does include, among other things, a measure of multivariate process capability, yield and fault diagnostics [4–8]. However, most of the current procedures are applicable to multivariate normal processes. Some approaches have been developed to measure performance of multivariate non-normal processes [3, 7].

This paper aims to integrate these two approaches with machine learning to perform full multivariate non-normal process capability analysis.

30.2.1 Geometric Distance Approach to Reduce Dimensionality

Geometric Distance approach reduces the dimension of the multivariate process data and render them more tractable for a statistical analysis [9]. The GD approach utilizes the Euclidean distance which is defined as follows: Let $\mathbf{X}$ represent the n quality characteristics $(X_1, X_2, . ., X_n)$ and $\mathbf{T}$ be the corresponding target vector with $(T_1, T_2, . ., T_n)$. The covariance matrix of $\mathbf{X}$ identifies and groups correlated variables into GD variables. The GD variables are defined as,

$$GD = \sqrt{(X_1 - T_1)^2 + \cdots + (X_n - T_n)^2} \qquad (30.1)$$

D. Munjeri (✉) · M. Abdollahain · N. Gunaratne
School of Sciences, RMIT University, Melbourne, VIC, Australia
e-mail: denwick.munjeri@rmit.edu.au

© Springer Nature Switzerland AG 2019
S. Latifi (ed.), *16th International Conference on Information Technology-New Generations (ITNG 2019)*,
Advances in Intelligent Systems and Computing 800,
https://doi.org/10.1007/978-3-030-14070-0_30

When the underlying distribution is often unknown. Reference [7] combined correlated quality characteristics to form G and determined the distribution that best fit G. The maximum radial distance (MRD) is the upper limit of the GD variables and, in this case, zero is the lower specification limit [9]. MRD is defines by:

$$MRD = \sqrt{\left(Tol_{X_1}\right)^2 + \cdots + \left(Tol_{X_n}\right)^2} \qquad (30.2)$$

where $\left(Tol_{X_1}\right)^2$ is the specification tolerance of the ith quality characteristic. The GD variables are independent and often follows different distributions. In this paper, we will deploy Burr XII distribution to model individual GD variables.

30.2.2 Fitting Burr XII Distribution to GD Variables

The two-parameter Burr XII distribution can be used to describe the data in the real world [10, 11]. Underlying distribution of the data can be fitted using Burr distribution. We define the Burr XII distribution according to the identified parameters k and c, measuring skewness and kurtosis, respectively. The probability density function is defined as,

$$f(x) = \frac{ckx^c}{(1+x^c)^{k+1}} \qquad (30.3)$$

The cumulative distribution function of the Burr XII distribution is defined as,

$$F(x) = 1 - \frac{1}{(1+x^c)^k} \qquad (30.4)$$

where c, $k \geq 1$ and $x \geq 0$. We then fit Burr XII distribution to individual GD variables. The parameters c and k, for the individual GD variables, are estimated using the simulated annealing search algorithm [3]. When the appropriate Burr XII distributions has been obtained, we then estimate proportion of nonconformance (PNC) and yield [3, 10, 11]. The PNC is defined by,

$$PNC = 1 - \int_0^{MRD} f(x)dx \qquad (30.5)$$

where f(x) is the density function of the GD variable and $\int_0^{MRD} f(x)dx$ is the probability of a product conforming. Since GD variables are independent, therefore, the total PNC for the process is estimated by,

$$PNC_{Total} = 1 - \prod_{i=1}^{k} \int_0^{MRD_i} f_i(y)dy \qquad (30.6)$$

where $f_i(y)$ is the density function of the ith GD variable. Reference [12] introduced two MPCIs, MC_p and MC_{pk}, which are defined as:

$$MC_p = -\frac{1}{3}\Phi^{-1}\left(\frac{p}{2}\right) \qquad (30.7)$$

And

$$MC_{pk} = -\frac{1}{3}\Phi^{-1}\left(2^{n-1}p_{max}\right) \qquad (30.8)$$

respectively, where Φ is the cumulative distribution function, p is the total proportion of non-conforming products and p_{max} is the maximum proportion of non-conforming products.

30.2.3 Machine Learning Approach for Source Identification

For this study, we have chosen the artificial neural net input gain measurement approximation (ANNIGMA) which is an improved neural network based on feature ranking approach [5, 13]. ANNIGMA reduces the impact on output from noisy or irrelevant input features by way of reducing weights associated with those irrelevant features. ANNIGMA uses relevance based on the weights associated with each quality characteristic for ranking, consequently, speeding up the fault diagnostic task [5, 13].

For a two-layer neural network, if i, j and k are the input, hidden and output layer and F is a logistic activation linear function $F(x) = 1/(1+ exp(-x))$, then the output of the network is given by Eq. 30.9.

$$O_k = \sum_j F\left(\sum_i A_i \times W_{ij}\right) \times W_{jk} \qquad (30.9)$$

where W_{ij} and W_{jk} are the network weights. The local gain can be defined as:

$$LG_{ik} = \left|\frac{\Delta O_k}{\Delta A_i}\right| \qquad (30.10)$$

According to [5], the local gain LG_{ik} can be defined in terms of network weights by

$$LG_{ik} = \sum_j |W_{ij} \times W_{jk}| \qquad (30.11)$$

The ANNIGMA score for i th input and k th node [5] is defined as:

$$ANNIGMA_{ik} = \frac{LG_{ik}}{\max(LG_k)} \qquad (30.12)$$

ANNIGMA estimates the relative importance of input features. It is worth mentioning that this machine learning approach is an effective fault diagnostic tool but does not measure multivariate process performance, which is a vital reference component in process improvement tasks.

30.3 Proposed Fault Diagnostic Hybrid

This paper proposes a fault diagnostic hybrid that uses Geometric Distance approach, Burr XII distribution and machine.

learning approaches to reduce data dimension, estimate yield, process capability and carry out the task of identifying and ranking original variables responsible for poor performance of high dimension non-normal processes. Table 30.1 outlines, in detail, the proposed fault diagnostic approach for multivariate non-normal processes.

30.4 Application of the Roposed Hybrid

In this section, the proposed fault diagnostic hybrid has been deployed to perform a multivariate non-normal capability analysis and fault diagnostics on a process that produces a computer component with 7 correlated quality characteristics. The data [3] consists of 100 samples of size 1 with key quality characteristics; X_1 (contact gap X), X_2 (contact loop T_p), X_3 (LLCR), X_4 (contact x T_p), X_5 (contact loop diameter), X_6 (LTGAPY) and X_7 (RTGAPY), respectively. The specification limits are: 0.10 ± 0.04, 0 ± 0.50, 11 ± 5, 0 ± 0.2, 0.55 ± 0.06, 0.07 ± 0.05, and 0.07 ± 0.05 respectively. Using GD approach, the variables were clustered

into 4 independent variables: GD_1 (X_1, X_2, X_3), $GD_2(X_4)$, $GD_3(X_5)$ and $GD_4(X_6, X_7)$.

30.4.1 Simulation Study

To assess the robustness of the proposed hybrid in fully assessing the capability of the multivariate non-normal processes, we have used the original data [3] to simulate 4 different scenarios using the fast fifth-order polynomial transforms method developed by Headrick [14]. In each scenario, 600 samples are randomly generated such that the sample size, n, is n = 1.

Scenario A: Is the original (or current) state for comparison.

Scenario B: The variable X_7 is identified from **Scenario A** and has been improved by shifting the process mean μ_7 from 0.08 to 0.07 (the target mean).

Scenario C: Variable X_7 and X_5 are both improved: μ_7 from 0.08 to 0.07 and μ_5 from 0.54 to 0.55.

Scenario D: Variable X_1, X_2 and X_3 are shifted away from the target by 25%, 15% and 35% increase to $\mu_1 = 0.138$, $\mu_2 = 0.092$ and $\mu_3 = 15.11$, respectively.

30.5 Discussion

In this section, results of the simulation study (Table 30.2) will be analysed using *Scenario A* as reference for the other simulation scenarios.

Scenario A: The original (or current) state of the process before adjustments.

From the results shown, in Table 30.2, the estimated process yield is 0.9550 and multivariate process capability is 0.9878. The poor performance is caused by variables GD_3 and GD_4 which have yield below the threshold, that is less than 0.9973. GD_4 has the lowest yield of .09714. Therefore,

Table 30.1 Flowchart for the proposed methodology

Step 1:	Read the multivariate data.
Step 2:	Construct the covariance matrix and cluster correlated variables.
Step 3:	Deploy Geometric Distance approach to reduce dimensionality of multivariate data by fitting GD variables to correlated clusters and use MRD to estimate the specification limits for the individual GD variables (Eqs. 30.1 and 30.2).
Step 4:	Fit the appropriate Burr XII distribution to the GD variables using simulated annealing algorithm [3] to estimate c and k parameters.
Step 5:	Use the fitted Burr XII distribution to estimate the yield. (Eqs. 30.5 and 30.6). If yield is acceptable, go to **Step 9**.
Step 6:	Rank GD variables, in terms of yield, starting with the lowest.
Step 7:	Estimate the multivariate process capability index for the process the yield (Eq. 30.7). If multivariate process capability index is acceptable, go to **Step 9**.
Step 8:	Use machine learning fault diagnostic approach, that is ANNIGMA, to identify and rank characteristics responsible for poor performance, within the GD variable with the lowest yield or MPCI value. Go to **Step 2**.
Step 9:	STOP.

Table 30.2 Simulation results for proposed hybrid

Attributes		Scenario A	Scenario B	Scenario C	Scenario D
Yield for GD_i	GD_1	0.9999	0.9999	0.9999	**0.9601**
	GD_2	0.9999	0.9999	0.9999	0.9999
	GD_3	**0.9833**	**0.9833**	0.9998	**0.9833**
	GD_4	**0.9714**	0.9980	**0.9995**	**0.9714**
Process Yield		0.9550	0.9811	0.9976	0.9170
Process Capability		0.9878	0.9912	**1.02**	0.9717
Ranked GD variables		**GD_4**, GD_3, GD_1, GD_2	**GD_3**, GD_4, GD_1, GD_2	**GD_4**, GD_3, GD_1, GD_2	**GD_1**, GD_3, GD_4, GD_2
Identified GD_i		GD_4	GD_3	GD_4	GD_1
ML Ranked X_i variables		X_7, X_6	X_5	X_6, X_7	X_3, X1, X2,
Identified X_i		X_7	X_5	X_6	X_3

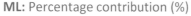

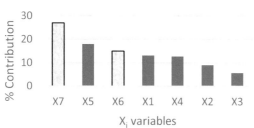

Fig. 30.1 The Machine Learning ranking of the contributions of quality characteristics

GD_4 is the most significant GD variable responsible for poor performance. GD_4 is formed by X_6 and X_7.

Machine learning results show that X_6 and X_7 contribute 36% and 64%, respectively, to GD_4's poor performance, making X_7 the first variable to be targeted for process improvement.

To confirm this result, we run fault diagnostics on all the 7 variables to rank variables responsible for process poor performance. Figure 30.1 results prove that the most significant variables are X_7, X_5, X_6 corresponding to GD_4 and GD_3. Since resources are usually limited, process improvement must be guided by these rankings, in this case, we address X_7 then X_5, then X_6, in that order, until we achieve the desired overall process performance.

Scenario B: In this scenario, the variable X_7, identified from **Scenario A,** is improved. Deploying the proposed algorithm results in GD_3 being ranked as the most significant, with a yield of 0.9833, and subsequently, X_5, since it is the only variable in GD_3. This is confirmed by Fig. 30.1.

Scenario C: Variable X_7 and X_5 were both improved, and all GD variables have acceptable yield, that is yield above 0.9973, though GD_4 has the least yield. The multivariate process is also capable with a multivariate process capability

index equal to 1.02, that is above the threshold of 1.0. Applying machine learning source identification approach to GD_4 results in X_6 being ranked the most significant, since X_7 has already been improved to acceptable levels. So, as original variables are being improved, the simulation study proves that the proposed hybrid can pick the next significant variable.

Scenario D: Variable X_1, X_2 and X_3 are shifted away from the target to $1.25X_1$, $1.15X_2$ and $1.35X_3$, respectively. The results show that these three variables affect the yield of GD_1 and hence it was picked by the proposed hybrid that GD_1 has the lowest yield, of 0.9170, hence contributes the most to poor process performance. Amongst the GD_1 variables, X_3 was ranked as the most significant variable.

It can be shown from these results, that the proposed hybrid is robust in carrying out multivariate process capability analysis and fault diagnostics, up to the original variable responsible for poor performance by only targeting quality characteristics within the GD variable with the least yield.

30.6 Conclusion

The paper identified the challenges of performing multivariate process capability analysis and fault diagnostics on a multivariate non-normal process with correlated quality characteristics. A hybrid was proposed to address the shortcomings of the existing approaches. The proposed hybrid is based on Geometric Distance approach, fitting Burr XII distributions and machine learning approach to carry out fault diagnostics, given that the responsible GD variable may represent multiple original variables. The novelty of the proposed hybrid is that it affords the practitioner an opportunity to measure process performance using either process yield or multivariate process capability, while identifying and ranking the quality characteristic responsible for poor performance. The results confirm that, instead of investigating all the variables, the proposed hybrid can identify and rank the most significant variable responsible for process poor performance by concentrating on the GD variable with the lowest yield, consequently, reducing the cost and time of the fault diagnostic task.

References

1. Shahriari, H., Hubele, N.F., Lawrence, F.P.: A multivariate process capability vector. Proceedings of The 4th industrial engineering research conference, Institute of Industrial Engineers. pp. 304–309, (1995)
2. Taam, W., Subbaiah, P., Liddy, J.W.: A note on multivariate capability indices. J. Appl. Stat. **20**, 339–351 (1993)
3. Ahmad, S., Abdollahian, M., Zeephongsekul, P., Abbasi, B.: Multivariate non-normal process capability analysis. Int. J. Adv. Manuf. Technol. **44**, 757–765 (2009)
4. Gunaratne, N.G.T., Abdollahian, M.A., Huda, S., Yearwood, J.: Exponentially weighted control charts to monitor multivariate process variability for high dimensions. Int. J. Prod. Res. **55**(17), 4948–4962 (2017)
5. Hsu, C., Huang, H., Schuschel, D.: The ANNIGMA-Wrapper approach to fast feature selection for neural nets. IEEE Trans. Syst. Man Cybern. B Cybern. **32**, 207–212 (2002)
6. Dharmasena, L., Zeephongsekul, P.: A new process capability index for multiple quality characteristics based on principal components. Int. J. Prod. Res. **54**, 4617–4633 (2016)
7. Wang, F.K.: Quality evaluation of a manufactured product with multiple characteristics. Qual. Reliab. Eng. Int. **22**, 225–236 (2005)
8. de-Felipe, D., Benedito, E.: A review of univariate and multivariate process capability indices. Int. J. Adv. Manuf. Technol. **92**, 1687–1705 (2017)
9. Wang, F.K., Hubele, N.F.: Quality evaluation using Geometric Distance approach. Int. J. Reliab. Qual. Saf. Eng. **6**, 139–153 (1999)
10. Burr, I.W.: Cumulative frequencu distribution. Ann. Math. Stat. **13**, 215–232 (1942)
11. Liu, P.H., Chen, F.L.: Process capability analysis of non-normal process data using Burr XII distribution. Int. J. Adv. Manuf. Technol. **27**, 975–984 (2009)
12. Shiau, J.H., Yen, C.L., Pearn, W.L., Lee, W.T.: Yield-related process capability indices for processes of multiple quality characteristics. Qual. Reliab. Eng. Int. **29**, 487–507 (2013)
13. Huda, S., Abdollahian, M., Mammadov, M., Yearwood, J., Shafiq Ahmed, S., Sultan, I.: A hybrid wrapper–filter approach to detect the source(s) of out-of-control signals in multivariate manufacturing process. Eur. J. Oper. Res. **237**, 857–870 (2014)
14. Headrick, T.C.: Fast fifth-order polynomial transforms for generating univariate and multivariate nonnormal distributions. Comput. Stat. Data Anal. **40**, 685–711 (2002)

Business Intelligence: Determination of Customers Satisfaction with the Detection of Facial Expression

Fariba Rezaei Arefi, Fatemeh Saghafi, and Masoud Rezaei

31.1 Introduction

Face reading means discovering individuals' Moral and personality traits just by looking at their faces. Face reading has been done by people in different ages and it is an interesting subject to know people based on only their appearances. Elegant and detailed points and many complexities exist in the science of facial reading and throughout the centuries mankind have been attempting to understand others thoughts with making a link between their facial expressions and their personality traits. Over the year's facial science has also expanded and has divided into different branches. Examples of facial science includes: Metastopscopy which is the science of finding out in human beings from the lines on the forehead; Physiognomy which teaches us how to get to people personality based on their faces.

Today Scientists call this pseudoscience and pay not much attention to them and look at them like sciences like astrology, that are, unproven, but innocuous sciences. However, many of people consider face science as a sure way to judge others. In this study, we use face science and data mining to capture data from individuals' images to final conclusions about customer satisfaction with the performance of an organization. For this, in the first step, we will explain how to perform facial recognition using data mining and then,

in the second step, with data mining, on the face mode of customers when entering and leaving the organization, we will come to the final results on how satisfied or dissatisfied our customers are. All results and outputs in this research are carefully analyzed and documented by the clementine software, and the steps involved in the research are presented.

31.2 Face Detection and Its Methods

Nowadays, with the increasing spread of tools for working with digital images in various fields such as acquisition, storage, processing and image retrieval, there is an emerging need to introduce new algorithms and improve the methods of processing and analyzing digital images. The use of digital images has become commonplace in various fields. From public applications such as digital photography cameras, cell phones to specialized applications such as medical, geographic and satellite decision making tools. Our demand for image processing is to make a series of corrections to an image, to make it clearer and more practical for a person, or to extract the necessary information to guide an automated system. In recent years, one of the active areas of image processing has been the recognition of facial expressions. The six main modes of the face that are often considered for detection are happiness, sadness, anger, surprise, fear, hatred. Face shapes are created with changes in different points of the face, and each of these parts is called an active unit (AU). We can't describe a facial mode with just a single AU. Nowadays the detection of facial expressions has been widely used; including the human-computer relationship. Robots that have feelings and get happy with your happiness and get sad with your sadness are an example of this field. In designing animation and cartoon films, designing different characters in games and in computer graphics, face detection can be used greatly. If a cartoon character is inspired by a person's face to

F. Rezaei Arefi
Faculty of New science and Technology, University of Tehran, Tehran, Iran

F. Saghafi (✉)
Faculty of Management, University of Tehran, Tehran, Iran
e-mail: fsaghafi@ut.ac.ir

M. Rezaei
Alborz Campus, Department of Management, University of Tehran, Tehran, Iran
e-mail: Rezaei2017@ut.ac.ir

© Springer Nature Switzerland AG 2019
S. Latifi (ed.), *16th International Conference on Information Technology-New Generations (ITNG 2019)*,
Advances in Intelligent Systems and Computing 800,
https://doi.org/10.1007/978-3-030-14070-0_31

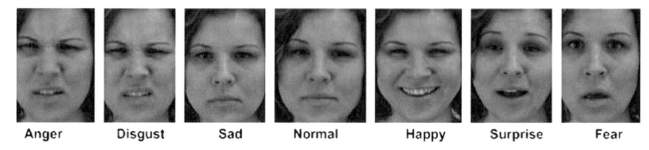

Anger   Disgust   Sad   Normal   Happy   Surprise   Fear

Fig. 31.1 Six main facial expressions of the face

Fig. 31.2 The changes of the AUs of different facial areas to represent the six main face modes

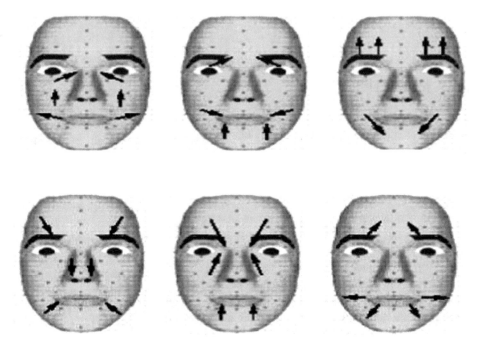

display a mode, the presentation of the desired state is much more realistic and the design is also simpler. Detecting facial expressions is based on facial features, which most important of them are eyes, mouth and eyebrows. Using the rules, it is possible to divide the movements of these parts, determined by different AUs, and describe each face-specific state with a number of AUs in different individuals. But the variation in AUs in different individuals is different for displaying a particular state. Figure 31.1 depicts the six main facial expressions of the face. Figure 31.2 shows the changes of the AUs of different facial areas to represent the six main face modes.

In the field of analyzing and synthesizing facial expressions, many things have been done. Considering the importance of the analysis stage, most of the work and research carried out on the stage of detection of concentrated states. In general, three steps should be taken for an automatic recognition system of facial expressions.

- Detection of the face and its characteristics
- Extraction of information about facial expressions from a fixed image or a sequence of images
- Classification of images [1]

31.2.1 Glossary of Face Detection

So far, different ways of detecting faces have been provided by scholars using various sensitive points. Some of these methods use less sensitive points of faces and others count more points in the faces. Some methods have more success rates than other methods. Cohn and colleagues have presented a method for the face detection with incomplete AUs [2].

Kobayshi and Hara, by manually identifying 30 Facial Feature Points (FFP Points) and extracting several features from the points and using fuzzy neural network, provided a method to detect states of fear, surprise, hatred, sadness, pleasure and anger with the ability to recognize them about 78% with training 5 faces, 83% with training 10 faces and 88% with training 15 faces [3].

Ushida and his colleagues using CFS and taking into account 27 FFP points (12 for eyebrows, 7 for the right eye

Fig. 31.3 Facial Feature Points

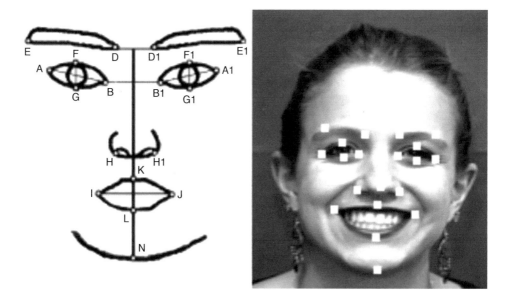

and 8 for the mouth) provided a method for detecting anger, pleasure and sadness. By examining 56 images depicted in one of the three modes mentioned, their proposed system presented detection power of 78.7% [4].

Nakamura and Ebine with using fuzzy logic provided a way to distinguish four pleasure, anger, sadness and surprise modes. Their method detected correctly in 86.7% of cases experimenting 59 images. The method presented by them was carried out in several stages [5].

- Preprocessing involving extracting the processed area of face.
- Removing head movement and normalizing the surface processed on the face.
- Assigning surfaces related to different moving areas on the face and extraction of movements related to them.

Using optical flux, Yacoob and his colleagues obtained different AUs in sequential frames that represented a particular state. These frames started from a normal mode and ended up in one of the main face modes. With this method for each face mode, a feature vector was obtained based on the movement of AUs and was used to train a fuzzy RBF neural network. The system described for division of frames to one of the six main face modes has of detection power of about 80% [6].

Tsapatsoulis and Piat, provided an automated follow-up method based on fuzzy inference for these features by comparing the relative distances defined by the FFPs in the first frame, which began from normal mode and followed up from one frame to another. The membership functions used for each of the features were of the triangular type, which its value was limited by the maximum and minimum, and its head centered on the average value of that feature. Using 8

features, they obtained diagnostic power from 0% to 90% by FIS system for sequential frames [7]. In this research, we use 20 sign points of the face, eyebrows, mouth, nose, and cheeks and extract these points from images received from the cameras of a hypothetical organization to examine the facial expressions of individuals. We specify the points according to [8] presented in this field base on this reason that this has been introduced as one of the most reliable an simplest ways in face detection in many resources based on our studies and panel expert suggestion. We extract the data from the received images and store them in the final file to analyze these data. Figure 31.3 shows the 20 FFP points that have been considered in this study.

8 points are for the eyes, 4 points are for the eyebrows, 3 points for the tip of the nose, 4 points are for mouth and 1 point is for the chin. All 20 determined points of 400 images were extracted from the customers of an organization and the specified coordinates were finalized for clementine software and analyzes were performed on these data.

31.2.2 Image Mining

In 2007, Sanjay introduced an Image processing technique with using microwave conversion. This method utilizes replicating common patterns, identifying patterns, and data mining models with the assumption that a real world image or feel can be related to a specific group. This method uses a three-step process including collecting images, learning and categorizing. Since the conversion of the microwave uses a time-frequency relation, it can be used to reproduce as an alternative to the Fourier series. The microwave conversion decomposes an image into non-spatial frequency ranges and uses a small frequency range to analyze fundamental components (PCAs). Categorizing helps to determine which group each image is associated with [9].

Image mining was introduced by Pattnaik in 2008 using clustering and data compression techniques. Satellite images of clouds rebuild a major role in weather forecasting. The frequency of receiving images from one image per minute to an image per hour varies according to the weather conditions. The results will lead up to a large stock of images. In this method, data clustering with grading vector was implemented to cluster and compact static images [10].

In 2002, Penner argued about the use of a framework and a standard tool for analyzing images. A tool and technique for data mining in archiving systems for images were provided by him. These tools and techniques are standard and can be used for other image mining tasks. They used this method for medical activities, such as X-ray image mining and MRI [10].

Image mining based on decision trees was introduced by Kum. In this method, a general frame was given based on the decision tree for processing and exploring visual data. The features of image pixels are extracted and converted to tables similar to database tables, in order to execute various types of data mining algorithms for searching and exploring them. Each row of table contains a descriptor that is generated in relation to the characteristics of a given pixel. Thus, they presented a decision tree in order to express the relationship between the features and pixels of the image [11].

In 2007, Sheela and Shanti provided image mining techniques for categorizing and dividing MRI images. Images categorizing plays an essential role in most medical applications. They developed a system in which image mining techniques were used to categorize images into normal and abnormal groups, and then abnormal brain tissue fragments were identified to diagnose brain-related diseases [10].

Another technique of image mining was described in 2003 by Jing and Ngo. For a given number of images, they first used the described relational graph to characterize the images. They applied an image segmentation algorithm that was presented by Felzenswlab and Huttenlocher in 1998 on incoming images so each part was encoded as an apex of the ARG graph. They used ARG graphs to express input images. The next step was to discover the latent patterns in these ARG graphs. An inaccurate algorithm for the conventional maximal graph was proposed by them in order to find these patterns [10].

In 2004, Srivastava and Oza proposed a knowledge-based image mining technique. The authors discuss the automatic exploration of multidimensional images using the theory of centers. A central function is defined as a product of the data mapped in the specification space. The authors proposed a new technique for exploring an automated knowledge based image based on the theory of central points, which provides a completely non-linear symmetric positive mapping of the main images to a high-dimensional, potentially uncertain, multidimensional specification space [10].

In 2003, Qin explained the method of exploring remote imagery on a conceptual network. The author proposed a conceptual network theory. Then these techniques are used to explore sentimental images, test and display spectral features, explore texture specifications, explore shapes, and locate spatial distribution rules. Applications for exploring remote sense imagery, such as automatic sorting, smart image retrieval a sense of distance were analyzed by him [10].

Victor and Peter presented a new minimal spanning tree in 2010 based on a clustering algorithm for image mining. The minimum spanning tree clustering algorithm was used to detect clusters with irregular ranges. The authors presented a minimal spanning tree based on clustering techniques, weighing the edges based on the Euclidean distance, which is very important in extracting the graph from the images [10].

In 2003, a conventional way of categorizing images with a model of knowledge acquisition was automatically presented by the web site by Yanai. This method uses three phases for processing. The first phase is collecting images that are related to the keywords of a specific category from the web. The second phase is learning period which extracts the characteristics of the collected images and link them to the corresponding category. The third phase categorizes an unknown image in the corresponding class according to the image properties [10].

31.3 Reaearch Methodology

Considering the fact that the subject of the article is image mining on the received images and data mining to measure customer satisfaction, we needed to use the software for data mining and image processing. In this method, we extracted points from the images and stored in a database file. We used image mining system presented in [8]. This system runs image mining process and we can get data to analyze using SPSS Clementine software. The number of our saved images in this study was 400 images of 200 people who were logged in and out during their arrival and departure to an organization. Regarding the constant recourse to organizations, unfortunately, the organizations concerned were not willing to cooperate with us for security reasons. Therefore, it was difficult to provide real photos of customers due to data defects in an organization, which is why we used the database of images available on the web for this research. It is hoped that this study will be considered as the starting point for using and applying this method to organizations by using real photos of customers.

Finally, according to the determination of the faces, the satisfaction or dissatisfaction of the customers is examined,

200 records of the 200 people whose images were stored in the database of images were investigated and a good prediction of customer satisfaction or dissatisfaction has shown.

31.4 How to Extract Information from Images

The five steps of image mining in this research are as follows:

Step 1: Pictures of the organization's customers from the CCTV cameras during the arrival and exit are extracted by using a software module to search and find each customer's photos at the time of arrival and exit and then these photos are stored at a constant scale and size in our database.

Step 2: At this point, 20 sensitive points described (these sensitive points can be based on our selective method or the amount of sensitivity employed and the number of different sensitive points used) are extracted from each image and stored in a document. In the coordinates of 20 points, 8 points are for the eyes, 4 points are for the eyebrows, 3 points for the tip of the nose, 4 points are for mouth and 1 point is for the chin.

Step 3: The document containing the coordinates of all the images received from the cameras forms the input of image mining system presented by [8] and get the data for data mining software for face detection and image analysis. Since the ultimate goal is to determine customer satisfaction, using a panel of experts consisting of 6 image processing professionals with more than 5 years of experience in this field, it has been found that some facial expressions do not play a role in determining the degree of satisfaction. Similar examples can be considered for more face modes, but because, for example, the face of amazed person does not clearly describe the satisfaction or dissatisfaction of a customer properly, so in this paper, according to the results of the Panel of Experts We have just outlined four sad, happy, normal and angry modes from seven face modes.

Step 4: This file is provided as an input to the clementine software and forms our source for image mining operations. Then we define the type of fields and all fields are of the input type and the result field or final detection of the face is specified as the output field. Next, we set the training and testing set in this data mining. In this section, we select 75% of the images as Training and the remaining 25% as the Testing Set.

The concept of partitioning in this sample is that the clementine data mining software employs 75% of the images and coordinates entered into the tutorial and tests its knowledge on 25% of the images, to measure the validity

of this study, we defined True Positive, False Positive, True Negative and False Negative criteria as follow:

- True Positive: normal and happy face modes correctly identified as normal and happy.
- False Positive: normal and happy face modes incorrectly identified as angry and sad.
- True Negative: angry and sad face modes correctly identified as angry and sad.
- False Negative: angry and sad face modes incorrectly identified as happy and normal.

Sensitivity and specificity were calculated with following equations: Sensitivity = true positive/(true positive + false negative), Specificity = true negative/(true negative + false positive) [12].

Step 5: In this step, which is the last step in face detection using the data mining method with the help of the clementine software, the method with the least error and the highest accuracy of prediction and diagnosis is correct and always use this method.

31.5 How to Define the Most Accurate Method for Determinig the Facial Expressions

At this point, depending on the input of the data mining software, which is the coordinates of different faces modes, we will receive different outputs. A method that has the least error and the highest success rate will be the basis for our action. In the following, some of the different outputs for image mining are presented on existing data and the results of these methods are specified.

31.5.1 Decision Tree Method C5.0

In the clementine software, there are different types of decision trees [13], here, according to the data, an output from the decision tree C5.0 and the existing data is displayed and also the results of the output analysis is shown in Table 31.1.

In this method, considering the 75% of the training set, 286 records are considered as training set, which detected

Table 31.1 Final result of decision tree C5.0

Results for output field RESULT
Comparing $C-RESULT with RESULT

'Partition'	1_Training		2_Testing	
Correct	258	90.21%	105	92.92%
Wrong	28	9.79%	8	7.08%
Total	286		113	

258 records with regular logic and did not log 28 records. Finally, the Test set has also considered 113 records, which correctly performed 105 facial recognition cases and did not correctly predict 8 cases in terms of logic. In general, this method has a success rate of over 90%.

31.5.2 Bayesian Network Method

The Bayesian method is based on one of the laws of the statistics and the probability of exploration operations, which is governed by a conditional probability or based on a bias relationship [14].

In this section, the possibility of the occurrence of facial expressions in Bayesian network method is done by Clementine. According to the results shown in Table 31.2, the output of the Bayesian network method is significant in software and is more than 95% accurate.

The results in more detail include: False Positive: 8, False Negative: 11, True Positive: 214, True Negative: 166, which cause Sensitivity rate of 95.11% and Specificity rate of 95.40%.

31.5.3 SVM Method

In the SVM method, information is separated using a data vector and find significant results [15]. As the third method used in our image mining, we used this method to differentiate and categorize facial expressions. As shown in Table 31.3, in general, this technique has a success rate of over 90%.

The results in more detail include: False Positive: 11, False Negative: 18, True Positive: 201, True Negative: 169, which cause Sensitivity rate of 91.78% and Specificity rate of 93.88%.

31.6 Data Mining and Methods Used to Determine Customer Satisfaction

At this stage, according to the data in the data mining software, it is attempted to determine customer satisfaction.

31.6.1 Neural Network Method

In this way, which suggests the natural neural network, there are neurons that can acquire knowledge and predict some of the cases using the knowledge gained [16]. Table 31.4 shows the results for the neural network to predict customer satisfaction or dissatisfaction. This table uses face mode of the person's login and face mode while leaving the organization for forecasting by the software.

The results in more detail include: False Positive: 37, False Negative: 21, True Positive: 64, True Negative: 78, which cause Sensitivity rate of 75.29% and Specificity rate of 67.82%.

31.6.2 CRT Decision Tree

In this research, CRT decision tree as a predictor of customer satisfaction is a very interesting situation and has made a good prediction of 200 records in the input file. The output is plotted in Table 31.5.

The results in more detail include: False Positive: 3, False Negative: 0, True Positive: 103, True Negative: 94, which cause Sensitivity rate of 100% and Specificity rate of 96.90%.

31.6.3 The Decision Tree of the GUEST

In Table 31.6 the results are shown [17–18].

Table 31.2 Final result of Bayesian network method

Results for output field RESULT
  Comparing $B-RESULT with RESULT

'Partition'	1_Training		2_Testing	
Correct	272	95.1%	108	95.58%
Wrong	14	4.9%	5	4.42%
Total	286		113	

Table 31.3 Final result of SVM

Results for output field RESULT
  Comparing $S-RESULT with RESULT

'Partition'	1_Training		2_Testing	
Correct	262	91.61%	108	95.58%
Wrong	24	8.39%	5	4.42%
Total	286		113	

Table 31.4 Final result of Neural Network Method

Results for output field Satisfaction
  Comparing $N-Satisfaction with Satisfaction

'Partition'	1_Training		2_Testing	
Correct	99	70.71%	43	71.67%
Wrong	41	29.29%	17	28.33%
Total	140		60	

Table 31.5 Final result of CRT decision tree

Results for output field Satisfaction
  Comparing $R-Satisfaction with Satisfaction

'Partition'	1_Training		2_Testing	
Correct	138	98.57%	59	98.33%
Wrong	2	1.43%	1	1.67%
Total	140		60	

Table 31.6 Final result of GUEST decision tree

⊟ Results for output field Satisfaction
  ⊟ Comparing $R-Satisfaction with Satisfaction

'Partition'	1_Training		2_Testing	
Correct	136	97.14%	56	93.33%
Wrong	4	2.86%	4	6.67%
Total	140		60	

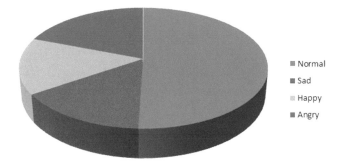

■ Normal

■ Sad

■ Happy

■ Angry

Fig. 31.4 Final result of customers' satisfaction

The results in more detail include: False Positive: 6, False Negative: 2, True Positive: 123, True Negative: 69, which cause Sensitivity rate of 98.40% and Specificity rate of 92.00%.

31.7   Conclusion

In this research, using the image mining method, customers' faces are identified at the time of arrival and departure of the organization, and in the next step, we used an image mining system and then their satisfaction is determined using data mining. Data mining software SPSS Clementine 12 has been used for data mining. In this research, the image mining was performed on 400 images and the customers' faced modes were determined from among the four considered states. For image capture, usually seven face modes are used. A review of related works showed that some facial expressions, such as surprise do not contain information about customer satisfaction. Four normal, sad, happy and angry modes were identified to determine customer satisfaction. In the data mining, results from three decision tree trees C5.0 and Baysian network and SVM method were investigated and all of them showed accuracy more than 90% in face recognition. In data mining, three decision-making CRT algorithms, neural network and decision tree were reviewed. Among the tested image mining methods considered in this study, the Bayesian method has the least error method. Among the three methods of data mining, CRT decision tree method has been selected as the least error and most successful method. As a final result Fig. 31.4 shows the percentage of satisfied customers is revealed and in what condition customers will be happy to leave the organization and in what state they will leave the organization dissatisfied. Our method for determining

customers' satisfaction is more appropriate to fill in the questionnaire because the customers may not complete the questionnaire for various reasons accurately or honestly, or the questionnaire has not been given to the customers at the right time. We were having difficulty obtaining real photos from customers. Hope this research is the starting point for researching and reviewing real photos of customers.

References

1. Khanum, A., Mufti, M., Javed, M.Y., Shafiq, M.Z.: Fuzzy case-based reasoning for facial expression recognition. Fuzzy Sets Syst. 231–250 (2009)
2. Shang, Z., Joshi, J.: Hoey, J.: Continuous facial expression recognition for affective interaction with virtual avatar (2017)
3. Sabri, M., Kurita, T.: Facial Expression Intensity Estimation Using Siamese and Triplet Networks. Neurocomputing. **313**, 143–154(3) (2018)
4. Pandey, K.K., Mohanty, P.K., Parhi, D.R.: Real time navigation strategies for webots using fuzzy controller. IEEE 8th International Conference on Intelligent Systems and Control (ISCO) (2014)
5. Jamshidnezhad, A., Nordin, M.J.: A modified genetic model based on the queen bee algorithm for facial expression classification. J. Comput. Theor. Nanosci. **9**, 1109–1114(6) (2012)
6. Danilov, D.I., Lakhtin, A.S.: Optimization of the algorithm for determining the hausdorff distance for convex polygons. Ural Math. J. **4**, 14 (2018)
7. Piat, F., Tsapatsoulis, N.: Exploring the time course of facial expressions with a fuzzy system. International conference on. Vol. 2. IEEE (2000)
8. Valstar, M., Pantic, M.: Fully automatic facial action unit detection and temporal analysis. IEEE Comput. Soc. (2006)
9. Tumula, S., Fathima, S.S.: Probabilistic graphical models for medical image mining challenges of new generation. In: Knowledge Computing and its Applications. Springer, Singapore (2018)
10. Sudhir, R.: A survey on image mining techniques: theory and applications. Comp. Eng. Intell. Syst. **2**(6), 44–53 (2011)
11. Kun-Che, L., Don-Lin, Y.: Image Processing and image mining using decision trees. J. Inf. Sci. Eng. **25**, 989–1003 (2009)
12. Parikh, R., Mathai, A., Parikh, S., et al.: Understanding and using sensitivity, specificity and predictive values. Indian J. Ophthalmol. **56**(1), 45–50 (2008)
13. Delen, D., Cemil, K., Ali, U.: Measuring firm performance using financial ratios: a decision tree approach. Expert Syst. Appl. **40**(10), 3970–3983 (2013)
14. Dai, Q., Li, J., Wang, J., Chen, Y., Jiang, Y.: A Bayesian hashing approach and its application to face recognition. Neurocomputing. **213**, 5–13 (2016)
15. Kh.Lekdioui, , R. Messoussi, Y. Ruichek, Y. Chaabi, R. Touahni, "Facial decomposition for expression recognition using texture/shape descriptors and SVM classifier." Signal Process. Image Commun. 58 (2017): 300–312
16. Guo, Y., Li, G., Wang, J., et al.: Optimized neural network-based fault diagnosis strategy for VRF system in heating mode using data mining. Appl. Therm. Eng. **125**, 1402–1413 (2017)
17. Kellehe, J.D., Namee, B.M., Aoife, D.A.: Fundamentals of Machine Learning for Predictive Data Analytics: Algorithms, Worked Examples, and Case Studies. MIT Press, Cambridge, MA, USA (2015)
18. Barros, R.C., et al.: A survey of evolutionary algorithms for decision-tree induction. IEEE Trans. Syst. Man Cybern. Part C Appl. Rev. **42**(3), 291–312 (2012)

Data Integrity Model for Environmental Sensing

32

Daniel T. Siegel, Leanne D. Keeley, Poorva P. Shelke, Andreea Cotoranu, and Matt Ganis

32.1 Introduction

Since the Spring of 2017, a real-time water quality monitoring device has been continuously developed by students of Pace University's Seidenberg School of Computer Science and Information Systems. The device dubbed ADA, has been utilized for environmental research, as well as studies pertaining to Internet-of-Things technologies. With four different sensors floating a few feet beneath the surface of the Choate Pond, ADA collects several types of environmental data, and sends a set of data to a central computer at Pace University over a Wi-Fi connection. This process occurs in real-time at 15-minute intervals. Initially, this real-time data was presented in a simple chart with limited scalability [1]. By the Fall of 2017, a web-based application was developed to present the data with more data visualization options for the user [2]. Earlier in 2018, a study was conducted to simulate management and analysis of "Big Data" by expanding the data set to include an instance of data taken from a sensor over the course of several years. With this newly implemented mock-data, a team proceeded to incorporate Charts.js into the application to extend the visualization of such a large data set over an extended period of time [3].

With this in mind, there are several challenges to face with this newly incorporated set. While big data has emerged with many opportunities for businesses and organizations to interpret vast amounts of data in real-time and make more informed data-driven decisions than ever before, it has also presented a troubling issue commonly referred to as drift. Data drift is defined as "The unpredictable, unannounced and unending mutation of data characteristics caused by the operation, maintenance and modernization of the systems that produce the data" [4]. Hence, any changes made to the systems on which ADA operates, combined with the jump from traditional data to big data, affect the data.

Several reasons may be listed for how big data contributes to drift, but they can all be summarized by one word: Decentralization [5]. Under traditional data processing methods, data sources were managed by one or more entities within a single organization. The data technologies were usually static, managed from within, and changed through internal IT governance. The data sources had a defined schema that rarely changed, and when they did, those changes were agreed upon by the organization. Finally, under one roof, processing and analysis could be performed in single batches. The shift to the cloud architecture made drastic changes to the ease of traditional data handling. Changes to the schema occur more often and abruptly due to data often arriving with less structured data. The data sources are now managed by third parties and are often done so poorly. The data processing is done by a collection of separate software components with changes implemented outside of an organization's IT governance. Processing and analysis have been made more complicated, requiring combinations of batch, event-driven, and streaming operations [4]. Basically, any processing technology that relies on a stable, unchanging environment under centralized control, is set up for failure in a big data-driven world.

Drift leads to loss of data fidelity and deterioration of data operations, both having an adverse impact on an or-

D. T. Siegel (✉) · L. D. Keeley · P. P. Shelke · A. Cotoranu · M. Ganis
Seidenberg School of CSIS, New York, NY, USA
e-mail: ds68328p@pace.edu; lkeeley@pace.edu; ps54279n@pace.edu; acotoranu@pace.edu; mganis@pace.edu

© Springer Nature Switzerland AG 2019
S. Latifi (ed.), *16th International Conference on Information Technology-New Generations (ITNG 2019)*,
Advances in Intelligent Systems and Computing 800,
https://doi.org/10.1007/978-3-030-14070-0_32

ganization. Corrupt, missing, or misinterpreted data can lead to poor and overlooked insights. Any degradation of an organization's data will harm its reputation. Currently, deviations from normal data readings are being displayed on ADA's web application. Incorporating a data integrity model is the ideal way to counter drift as well as chance errors. Before the model is designed, this study needs to determine the potential type of drift affecting the sensors. From there, various drift compensation methods and algorithms will be tested to try to reduce the error rate on the data readings.

32.2 Literature Review

32.2.1 The Jefferson Project

The research surrounding this large-scale environmental research project has been very influential to similar-scale projects such as ADA. A joint effort of IBM Research, Rensselaer Polytechnic Institute, and the FUND for Lake George, this project utilizes a network of 42 sensor platforms, all connected by "Internet of Things" technology [6]. These platforms gather measurements of data relating to water temperature, air temperature, pH, oxygen content, salinity, and dissolved organic matter at regular intervals. This data is then shipped to several supercomputers over a cellular Internet connection.

In 2014, a report by the FUND revealed that rising salinity due to road-salt runoff was a threat to the state of Lake George, and recommended further research. While road-salt runoff from tributaries had long been thought to be a non-issue, the report found that the salt level in the 32-mile lake had tripled since 1980. This was a cause for alarm [7]. Knowing that the application of road-salt during and after storms leads to increases in salt runoff, the researchers reacted. Their response, was the development of sensors with the added function of predicting and adjusting to changing weather and lake conditions. These sensors have been applied to a four-pronged model which includes a weather model, a runoff mode, a salt model, and a circulation model. Each of these are designed to make predictions for how compounds, especially road salt, are affecting the water in the lake, and the precipitation that transports these compounds [8].

With this newfound ability to predict data drift, The Jefferson Project has been able to convince surrounding municipalities to be more tactical regarding road-salt application. The data integrity model in this instance has played a significant role in maintaining the quality of the water, essential to the regional economy.

32.2.2 The River and Estuary Observatory Network (REON)

Concerned with the amount of human impact on areas where rivers and estuaries meet the shoreline, The Beacon Institute launched its own collaboration with IBM and Clarkson University. The goal of REON was to create a water monitoring and forecasting network of its own in 2007. Their first monitoring platform, B1, was deployed on the Hudson River. This, along with two additional "B" units, laid the foundation for a second generation of monitoring technology. The web of interconnected sensors was dubbed the Real Time Hydrologic Stations (RTHS). REON's array of sensors gathered massive amounts of data for temperature, pressure, salinity, and turbidity by the minute [9]. For rapid analysis of the high volume of data, REON makes use of IBM's "System S" to stream from the various sources [10].

As the operations expanded to the capital region, New York City, and Cornwall, REON began to offer its own web application providing data visualization on its website. A user may choose one or more variables to observe over the date range of their choice. It is in a similar vein to ADA, although there are multiple rivers or estuaries to choose from on their application. Furthermore, REON's application displays charts of single variables before combining them into a single chart [11].

32.3 Research Requirements

32.3.1 Data Integrity

Barbara Martin of The American Water Works Association defines data integrity from a business standpoint stating that the concept "refers to the accuracy and consistency of stored data. Data integrity is imposed . . . through the use of standard rules and procedures, and is maintained through the use of error checking and validation routines" [12]. Martin claims that it is equally applicable to data analysis of water quality. When applying this definition to the current study, the two points that resonate are "error checking" and "validation routines". Establishing a baseline for our data integrity model to deal with outliers, or an error mode in software setup as Martin suggests, are potential solutions to explore.

32.3.2 Data Cleaning

Dirty data needs to be cleaned. Cleaning entails checking for outliers, normalizing raw data, and deciding how to fill in missing values. Data for Project ADA has been received in a CSV file, and outliers have been observed as well as values not consistent with standards. For instance, several cells show a turbidity of 404,404, a number far out of the normal range, and represents the HTTP response code for error. We also observe an occasional value of 999.9 for dissolved oxygen levels. Even outliers that represent legitimate data points can deliver undesirable results to data models. While simply deleting records with bad data seems easy enough, if done too often it will reduce the accuracy of our data integrity model. Data cleaning alone takes up to 60% of the time in most data mining processes [13]. Cleaning can be accomplished when working with the data using a script that locates error codes and outliers and replaces them with flag values such as "NaN" or "?", indicating that a proper data point is missing.

In addition, sanity checks need to be implemented in order to ensure that the values returned by the sensors are valid according to their attribute. For example, pH values below 0 or above 14 are not possible, and water temperatures of over 100 °C would represent water vapor. Such values outside of a specified range are best removed prior to analysis (D. Fischer, V. Kelly, personal communication, 18 October 2018).

32.3.3 Drift Type

Data drift has an adverse impact on the reliability of the data, the operations pertaining to it, and ultimately the productivity of the end-users of the data. The transition from traditional data processing to "Big Data" introduced this challenge, with its changes to processing architecture typically based on the assumption of stability. There are three types of data drift; Schematic Drift, Semantic Drift, and Infrastructure Drift.

1. *Schematic Drift:* Also referred to as structural drift, schematic drift occurs when the data schema changes at the source. Such changes may include the addition, deletion, or reordering of attributes. In addition, changes to the structure and incompatible changes to existing attributes are a significant source of schematic drift [14].
2. *Semantic Drift*: This type of drift refers specifically to the interpretation of the data. In the business world, a common example occurred during the transition from IPv4 to IPv6 protocol. This transition led to misinterpretations of data, leading to false positives originally thought to be revenue spikes [14].
3. *Infrastructure Drift*: As the transition to Big Data continues to occur, control of data repositories becomes more decentralized. Since each governing body of data source systems has their own standards, any change to the software components can create an incompatibility for existing operations [14].

In the context of this study, there is reason to believe that the type of drift that the team is faced with is Semantic Drift. The reason lies in the interpretation of the ADA data. When looking beyond chance errors, and more into systemic errors relating to values outside an appropriate range, the question becomes "What is the source of these abnormal values being displayed?". Have the sensors, or any part of the infrastructure for that matter, been contaminated, thus corrupting the data? Or are these data values legitimate, and the users are witnessing an entirely different phenomenon, such as the effects of climate change?

Identifying drift in the data can be done by comparing it to similar data collected independently. One way is through the deployment of separate sensors to audit the data against. Another is by testing the sensors in a known environment pre- and post-deployment. For example, placing a water temperature sensor in a solution for which the temperature is known in a lab setting (G. Gastil-Buhl, personal communication, 18 October 2018). Comparing data against data which has been collected elsewhere, but is comparable, is another method. A final way is by finding baseline numeric relationships between the different environmental metrics collected. If these relationships are accepted as representing normal data, if new data fails to maintain these relationships, the data can be said to have drifted.

32.3.4 Drift Compensation Methods

1. *Orthogonal Signal Correction:* This technique corrects for variance in an array or matrix of sensor data by finding a vector which is orthogonal to the observed concentration vector [15]. This allows it to counter drift without altering the existing relationships between the individual data points.
2. *Component Correction:* The process for Component Correction is as follows; First, find a score vector for a reference metric. Then, calculate the drift component correction using that score vector. Finally, apply the resulting component correction to the non-reference metrics to remove the effects of drift [16].
3. *Component Deflation:* A method that uses Canonical Correlation Analysis to find linear dependencies between sets of data and uses that to find a regression model that describe the drift that is affecting the data. It then deflates the data by applying the inverse of the regression expression to counter the drift [17].

32.3.5 Linear Regression Function

The linear regression function is a common knowledge discovery method that contributes to prediction as well as other data mining tasks [13]. This classifier takes numeric inputs and learns a linear regression model. Some implementations find one attribute that provides the lowest square error, others create a function that incorporates multiple attributes. The resulting function can be used to define the relationship between variables [13] and detect changes in the relationships.

Linear regression is a powerful machine learning algorithm that generates a simple function that can be easily implemented and comprehended, even by those not familiar with machine learning techniques. Additionally, the limited number of variables in a linear regression equation allows it to be easily rebuilt on a regular basis as more data is incorporated into the model.

32.4   Analyses

For this study, air temperature data collected by NOAA's NCDC at the Westchester County Airport and air and water temperature data collected by the HRECOS project at Piermont Pier in the Hudson River was used to audit data collected by ADA [18, 19]. This is done by merging the data on time stamps and plotting the data in a line graph to visualize trends in the data (Fig. 32.1).

An increasing difference between the value of the ADA sensor data and the other temperature data is visualized by plotting the delta value between the ADA data and the other data against time to see if the difference is growing over time, indicating that ADA's data has drifted (Fig. 32.2).

Another method of detecting data drift is by auditing the various sensors against one other. We know that temperature affects dissolved oxygen [21]. If a function can be found that adequately expresses the relationship between these data points, we can use it to tell if one or both of the sensors is drifting (Fig. 32.3).

This can be deployed on a live server and used to compare new data with expected data. If new data does not match the expected data to within a certain confidence, we can issue an alert that the data from one of the sensors is drifting.

32.5   Integrity Model and Visualizations

32.5.1   Static Graph

A static line graph for the data has been implemented using Plotly for Javascript as well as Chart.js. Plotly allows more flexibility in terms of zoom options, as well as interactivity. Charts.js offers a smoother, easier to read image. A dropdown menu gives the user the ability to choose which sensor data to observe. The start date and end date inputs have remained, allowing the user to choose a desired time frame to view the data. An additional dropdown menu allows the user to choose from a variety of analyses. The default option selected is "Raw", while other options include "Daily Temperature Range", and "Growing Degree Days". A checkbox input allows the user to view a table of statistics describing the data they are looking at.

The data to be visualized is pulled from the database in which the sensor data is stored. While the raw data stored is not altered by any data processing, any records requested via user input will undergo a process before being presented on the static graph. Even when the "Raw" analysis is chosen by the user, the data is still run through the cleaning algorithm. An error checking code is applied. Any records consisting of "NaN" values or impossible values are removed. After the selected data has been cleaned, Plotly and Chart.js plot the sensor data on the Y-axis and time on the X-axis. A user may gain a more detailed view of

Fig. 32.1 Comparison of data trends between four sources [20]

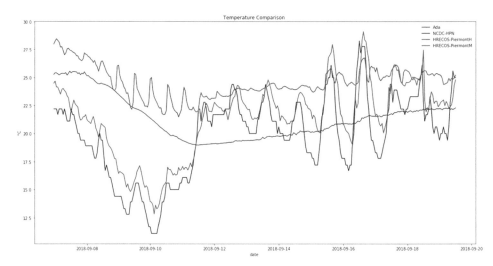

Fig. 32.2 Temperature delta between ADA and three other data sources [20]

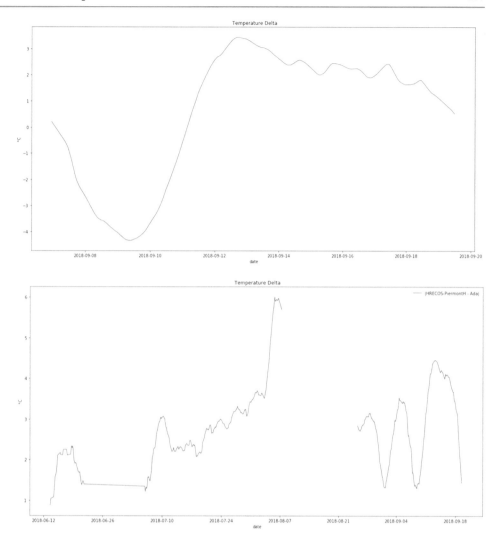

```
Linear regression on Temp
DOpct = 9.3 * Temp - 150.09
Predicting 87.52 if attribute value is missing.
```

Fig. 32.3 Linear regression for DOpct found using WEKA

data points by hovering their cursor over the line, a feature provided by Plotly. When the "Daily Temperature Range" analysis is selected, the difference between the highest value and the lowest value recorded for each date is plotted. Upon selection of the "Growing Degree Days" analysis, the cumulative growing degree days (represented by the function $DD = (Tmax - Tmin)/2) - BaseTemp$) for each month within the selected range is plotted. The statistics that can be calculated and displayed at the request of the user include the range, the standard deviation, the slope calculated for a linear regression line fitted to the data, and Pearson's Correlation Coefficient, which measures the linear relationship between the data and timestamp (Figs. 32.4 and 32.5).

32.5.2 Animated Graph

In addition to a static graph representing data in a specified range, users have the option of viewing an animated graph showcasing a week's worth of data returned by the sensors graphed together. No input is required on the user's part other than to select the animated graph option on ADA's web application.

Preprocessing of a week's worth of data is as follows. The application pulls the latest 672 data records. Recall that a new record is created every 15 minutes. Four records multiplied by 24 hours multiplied by 7 days yields 672 records. The same data cleaning function used for the static graph is applied to the animated graph before the data is presented. Then, the application applies a linear scale to the datasets returned by each sensor, in order to make the ranges presented by each sensor equal, so as to graph them together. Since each sensor returns data with different units,

Fig. 32.4 Graph of daily average conductivity with statistics table [22]

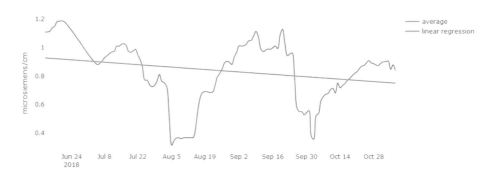

Range   0.316 - 1.187
Standard Deviation   0.216
Slope   -0.001
Pearson's Correlation Coefficient -0.228

Fig. 32.5 Graph of growing degree days [22]

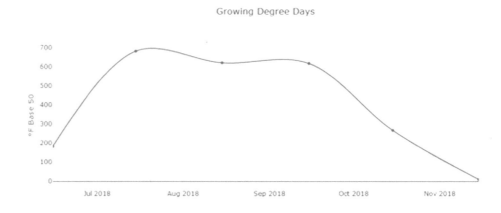

the data is presented as percentages of the observed range so that they will occupy the same vertical space on the graph. Thus, the relationships between the different lines can be more precisely observed. This is achieved with the function $f(x) = z' + ((y' - z')/(y - z)) * (x - z)$ where x is the variable; y is the maximum observed and y' is a given range, in this case 100; z is the minimum observed and z' is 0.

The data is displayed smoothly across as much of the screen as possible. Each sensor line is of a different color and drawn simultaneously to highlight how the data changes overtime and to visualize the relationship between different types of data. To achieve this, the domain and range of the graph is set before drawing the lines, so the graph is not zooming out and panning left to right as the graph is drawn. The data is then graphed two points at a time for a smooth progression through time. Animation time is minimized and

the graph is extended as opposed to being redrawn to make the animation proceed quickly.

32.5.3 Admin Dashboard

The health of the ADA sensor platform is presented to the administrator in chart format. The past 24 hours of data from each sensor are presented. If some problem is detected within this time period, the sensor will be flagged, or marked as bad. In those cases, the first problematic data point is presented with its timestamp. Health is determined by doing a validation check and a drift detection check. For validation, the data is checked against a variation of the data cleaning function. Instead of replacing a value with "NaN", the sensor is simply marked as returning problematic data, represented by a red X mark beside the sensor's title and printing the offending data point and its timestamp.

Drift is determined by auditing the sensors against one another. This is achieved by running the full set of available data at the time when the application is built as a training set from the sensor through WEKA's Simple Linear Regression classifier with one sensor selected as the value to be predicted. This results in a simple equation that describes the relationship between the sensor being predicted and the sensor that best predicts the observed value. This linear equation (in addition to the value to be predicted in the case of a missing value) is used to build a function which can be implemented by a webserver to predict a value. The mean absolute error observed by WEKA in the training dataset is then used to determine if the observed value falls within an acceptable range of the predicted value. If the absolute value of the difference between the predicted and observed values is greater than the mean absolute error, the value is marked as possibly representing data drift (Fig. 32.6).

This should then prompt the administrator to take further action, such as auditing the sensor in question in a controlled environment to double-check its calibration. For instance, water temperature can be used to audit the DOpct sensor as the latter has a documented inverse relationship with the former [21] and is confirmed by the WEKA machine learner which picked out temperature as the best predictor

of DOpct, as alluded to in Fig. 32.3. It is important to note that no reciprocal drift detection checks should be allowed. The reason for this is, if these two forms of data drift in parallel, this could go undetected. A drift detection algorithm has been implemented for each sensor with the exceptions of the DO gain sensor, which is not currently functioning; and the Depth sensor, which cannot logically be predicted using any of the available other sensors.

32.6 Conclusion

The next major steps for future teams of the ADA project involve refining the models developed by this study. The ADA platform has not been in the pond for a full year. Hence, higher-quality data collected in the future will allow researchers to enhance the model with data solely from the actual source, and not an analogous source. For now, the current study proves that it is indeed possible to implement various techniques into "Big Data" technologies to correct errors and detect data drift. Across various industries, data integrity models such as these may prove to be a valuable safeguard against consumer distrust, as well as a key to more reliable insights of data.

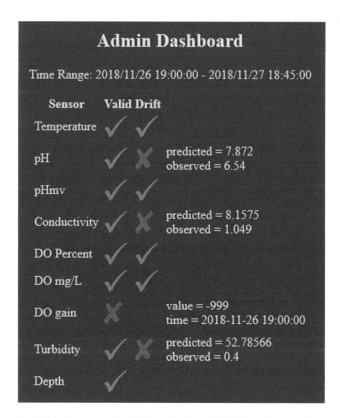

Fig. 32.6 Example of validation check and drift detection on admin console [22]

References

1. Adelman, J., Lamaute, N., Reicher, D., Van Norden, D., Ganis, M.: Remote Sensing in a Body of Water Using an Adafruit Feather. Seidenberg School of Computer Science & Information Systems, Pace University Pleasantville, NY 10570 (May 2017)
2. Andari, S., Caruso, M., Ganis, M., Robbins, C.B., Whit, C., Zada, A.: Web Application for Environmental Sensing. Seidenberg School of Computer Science & Information Systems, Pace University Pleasantville, NY 10570 (Dec 2017)
3. Caruso, M., Hassan, J., Keeley, L., Nikam, S., Zada, A.J.: Web Application for Environmental Sensing: Monitoring and Analyzing Water Temperatures. Seidenberg School of Computer Science & Information Systems, Pace University Pleasantville, NY 10570 (Feb 2018)
4. Taming data drift – the silent killer of data integrity. StreamSets Inc. https://19ttqs47cfw33zkecq3dz58m-wpengine.netdna-ssl.com/wp-content/uploads/2016/07/Taming-Data-Drift-White-Paper.pdf
5. Prabhakar, A.: Continuous ingest in the face of data drift (Part 1). Cloudera. http://vision.cloudera.com/continuous-ingest-in-the-face-of-data-drift/ (1 Feb 2016)
6. Picard, K.: The Jefferson project turns lake george into the world's smartest lake. Seven Days. https://www.sevendaysvt.com/vermont/the-jefferson-project-turns-lake-george-into-the-worlds-smartest-lake/Content?oid=18412829 (25 Jul 2018)
7. Johnson, S.K.: Science by Robot: outfitting the world's "smartest" lake. Ars Technica. https://arstechnica.com/science/2015/04/science-by-robot-outfitting-the-worlds-smartest-lake/ (18 Apr 2015)
8. In the Lab. The Jefferson project at lake George. Department of Biological Sciences, BT2149 Rensselaer Polytechnic Institute Troy, NY 12180 http://jeffersonproject.rpi.edu/lab. Accessed 1 Oct 2018

9. River and Estuary Observatory Network. Beacon Institute for Rivers and Estuaries. https://www.bire.org/river-and-estuary-observatory-network/. Accessed 1 Oct 2018

10. Why REON? Beacon Institute.https://www.thebeaconinstitute.org/approach/whyreon.php. Accessed 1 Oct 2018

11. Real-Time Hydrologic Sensing. REON. http://rths.us. Accessed 1 Oct 2018

12. Martin, B.: Tech-Tip – ensuring water quality data integrity. American Water Works Association. https://www.awwa.org/resources-tools/water-and-wastewater-utility-management/partnership-for-safe-water/partnership-resources/partnership-resources-details/articleid/4134/tech-tip-ensuring-water-quality-data-integrity.aspx (12 Apr 2016)

13. Larose, D.T.: An Introduction to Data Mining. Wiley-Interscience, John Wiley & Sons Inc, Hoboken, NJ (2005)

14. Pancha, G.: Big data's hidden scourge: data drift. CMS Wire. https://www.cmswire.com/big-data/big-datas-hidden-scourge-data-drift/ (8 Apr 2016)

15. Wold, S., Antti, H., Lindgren, F., Öhman, J.: Orthogonal signal correction of near-infrared spectra. Chemom. Intell. Lab. Syst. **44**, 1–2 (1998)

16. Artursson, T., Eklov, T., Lundström, I., Mårtensson, P., Sjöström, M., Holmberg, M.: Drift correction for gas sensors using multivariate methods. J. Chemom., Special Issue: Proceedings of the SSC6. **14**(5–6), 711–723 (2000)

17. Gutierrez-Osuna, R.: Signal processing methods for drift compensation. In: PRISM 2nd NOSE II Workshop, Department of Computer Science, Texas A&M University College Station, TX 77843 (May 2003)

18. Hourly Data. NOAA National Centers for Environmental Information, Westchester Co Airport. Station WBAN:94745 Hourly Data. https://www.ncdc.noaa.gov/cdo-web/datatools/lcd. Accessed 17 Oct 2018

19. Piermont Pier Hydrologic Station Data. Hudson River Environmental Conditions Observing System. www.hrecos.org/ (2018). Accessed 17 Oct 2018

20. Keeley, L.: Sonetteira/ADA-data-viz: ADA Data Visualizations – Release 1 (Version v1.0.1). Zenodo. https://doi.org/10.5281/zenodo.2533272 (7 Jan 2019)

21. Beal, B.: Examining the relationship between dissolved oxygen and water temperature. Project Watershed. http://projectwatershed.org/sites/projectwatershed.org/files/Relat_dissolved_oxygen_temperature.pdf. Accessed 2 Nov 2018

22. Keeley, L.: Sonetteira/ADA: ADA Website – Release 1 (Version v1.0.0). Zenodo. https://doi.org/10.5281/zenodo.2533186 (7 Jan 2019)

Analyzing the Similarity-Based Clusterability of the Vertices in a Complex Network

33

Md Atiqur Rahman and Natarajan Meghanathan

33.1 Introduction

Complex network analysis is fundamental to the understanding and modeling of relationships in several domains such as social networks [1], communication networks [2], biological networks [3], citation networks [4], cyber-attack networks [5] etc. Centrality metrics are the most common node-level metrics and quantify the topological importance of a node (vertex) with respect to one or more aspects [6]. There are two major categories of centrality metrics: neighborhood-based and shortest path-based. While the degree (DEG) and eigenvector centrality (EVC) metrics are considered representative metrics for the neighborhood-based category, the betweenness (BWC) and closeness (CLC) centrality metrics are considered representative metrics for the shortest path-based category. The DEG of a vertex is a measure of the number of neighbors of the vertex. The EVC of a vertex [7] is a measure of the degree of the vertex as well as the degrees of its neighbors. The BWC of a vertex [8] is a measure of the fraction of the shortest paths between any two vertices that go through the vertex. The CLC of a vertex [9] is a measure of the sum of the shortest path distances from the vertex to every other vertex. For more information about these centrality metrics, the interested reader is referred to [6–9]. Throughout the paper, the terms 'node' and 'vertex', 'link' and 'edge', 'network' and 'graph', 'cluster' and 'community' are used interchangeably. They mean the same.

Community detection is a classical problem in complex network analysis wherein one or more clusters of vertices are identified based on the topological distribution of the vertices and the edges connecting the vertices [6]. The primary focus of the community detection algorithms (e.g., [10]) has been to determine highly modular communities such that the intra-cluster density is as high as possible and the inter-cluster density is as low as possible [11]. With the above approach for clustering in complex networks, two vertices are likely to belong to the same cluster only if they have an edge between them or have a multi-hop shortest path of fewer intermediate edges. Even within a cluster of high density, it is possible that vertices significantly differ with respect to the values for one or more node-level metrics (see Sect. 33.2 for a motivating example) if the extent of similarity of the vertices (with respect to one or more node-level metrics) is not the primary criteria for cluster formation.

Our research objective is to quantify the extent to which vertices in a complex network are logically clusterable on the basis of the similarity among the values incurred for one or more node-level metrics. The nodes in such a logical cluster need not be connected, but would have very similar values for the node-level metrics. Identification of such logical clusters would be very informative for several network domains. For example: in an organizational network, a logical cluster of employees with similar profiles (but in different departments) would help in forming an integration team that could coordinate across departments without any ego differences.

Md. A. Rahman
Computational and Data-Enabled Science and Engineering, Jackson State University, Jackson, MS, USA
e-mail: md.a.rahman@students.jsums.edu

N. Meghanathan (✉)
Department of Computer Science & Engineering, Jackson State University, Jackson, MS, USA
e-mail: natarajan.meghanathan@jsums.edu

© Springer Nature Switzerland AG 2019
S. Latifi (ed.), *16th International Conference on Information Technology-New Generations (ITNG 2019)*,
Advances in Intelligent Systems and Computing 800,
https://doi.org/10.1007/978-3-030-14070-0_33

In a health information network, a quantitative measure of clusterability of patients with similar health parameters is critical to decide whether treatment plans need to be individualized or can be generalized at the cluster-level. The Hopkins Statistic measure (ranges from 0 to 1) has been traditionally used to assess the clusterability of a dataset [12]. We use the Hopkins Statistic to assess the similarity-based clusterability of the vertices in a complex network.

The rest of the paper is organized as follows: Section 33.2 motivates the proposed research for similarity-based logical clustering with an illustrative example. Section 33.3 illustrates the computation of the Hopkins Statistic on a coordinate system of the normalized centrality values for the example graph of Sect. 33.2. Section 33.4 presents the Hopkins Statistic values computed for 47 real-world networks (of diverse degree distributions) and evaluates the logical clusterability of the vertices on the basis of the similarity with respect to the neighborhood-based centrality metrics vs. the shortest path-based centrality metrics. Section 33.5 reviews related work on similarity assessment and clustering in complex networks, and highlights our contribution. Section 33.6 concludes the paper and outlines plans for future work.

33.2 Proposed Approach and Example

Our proposed approach to quantitatively assess similarity-based clusterability is as follows: Let K be the number of centrality metrics considered for the analysis. We determine the centrality values of the vertices in the complex network with respect to each of these metrics. We then individually normalize the centrality values of the vertices for each metric. The normalization of the centrality values with respect to a metric is done using the square root of the sum of the squares of the centrality values. We distribute the vertices in a K-dimensional coordinate system (wherein the range for each dimension is from 0 to 1) such that the coordinate for a vertex is a K-tuple, with an entry for each centrality metric. We refer to the above distribution as the logical topology of the vertices. If two vertices have similar values for the centrality metrics, then they should be closer to each other in the logical topology. There could exist one or more logical clusters of such similar vertices so that the Euclidean distance between vertices within a cluster is much smaller than the Euclidean distance between vertices in two different clusters. If the centrality values of the vertices are not similar to each other, then the vertices could be further away from each other in the logical topology and such a topology would not be effectively clusterable. The Hopkins Statistic is an effective measure to quantify the extent to which vertices could be clustered based on the similarity of their centrality values. The larger the Hopkins Statistic, the more effective is the clusterability of the vertices on the basis of similarity in their centrality values.

Figure 33.1a presents the communities determined by the classical Girvan-Newman algorithm [10] on a toy example graph. We observe the nodes in the individual clusters of Fig. 33.1a to be more connected to each other than to the nodes in the other clusters (i.e., a larger intra-cluster density and a lower inter-cluster density). However, from a centrality point of view, we observe the nodes in the individual clusters to be much different from each other (see Fig. 33.1b and c). Figure 33.1b and c respectively present a logical clustering of the vertices based on the similarity of the normalized values of the DEG and EVC metrics. The vertices within such logical clusters need not be physically connected to each other (Fig. 33.1b) or even if connected, the logical clusters need not have a high modularity (Fig. 33.1c), but would have similar (either identical or very close) values for the centrality metric considered.

If two or more centrality metrics are to be considered together for similarity assessment, then we distribute the vertices in a coordinate system of the normalized centrality values (ranging from 0 to 1 for each dimension) and assess the clusterability of the vertices based on their proximity to each other. Figure 33.1d presents one such distribution of the vertices based on a coordinate system of the normalized DEG and EVC values. Though the exact clusters depend on the clustering algorithm used, we observe vertex 5 to be far away from the rest of the vertices in the normalized (DEG, EVC) co-ordinate system and such a conclusion cannot be arrived at in Fig. 33.1a, b or c. Likewise, vertex 1 gets grouped with vertices 0, 2 and 3 under the traditional approach (see Fig. 33.1a), but is reasonably far away from them in the (DEG, EVC) co-ordinate system.

33.3 Hopkins Statistic Measure

We use the Hopkins Statistic measure [12] as the basis to quantify the similarity-based clusterability of the vertices in a complex network. The Hopkins Statistic (ranges from 0 to 1) is traditionally used to assess the clustering tendency of any dataset. The idea behind the Hopkins Statistic is as follows: Given a dataset of size n, we pick m samples (constituting the set X such that $m \ll n$) from this dataset as well as generate m uniformly random samples (constituting the set Y) exhibiting the same variation as that of the given dataset. We determine the sums of the distances of the samples in the sets X and Y to the nearest data points (i.e., one nearest data point for each sample) in the given dataset. The Hopkins Statistic is the ratio of the sum of the nearest neighbor distances of the samples in the set Y to the sum of the nearest neighbor distances of the samples in the sets X and Y. If the data points in the given dataset are randomly distributed, then the sums of the nearest neighbor distances for the samples in the sets X and Y would be approximately the same, and the Hopkins Statistic is expected to be 0.5. If the data points in the given

Fig. 33.1 Example to illustrate
the clustering of the vertices
based on the traditional approach
vs. the normalized
centrality-based coordinate
system. (**a**) Physical clusters
identified by the traditional
community detection approach.
(**b**) Logical clustering of the
vertices based on the similarity
with respect to degree centrality.
(**c**) Logical clustering of the
vertices based on the similarity
with respect to eigenvector
centrality. (**d**) Distribution of
vertices in the logical topology
respect to degree and eigenvector
centrality metrics

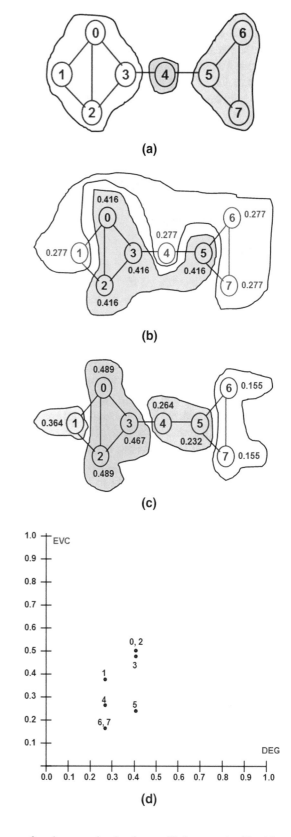

dataset are highly clusterable, then the sum of the nearest
neighbor distances for the samples in the set X would be
very negligible compared to the sum of the nearest neighbor

distances for the samples in the set Y; hence, the Hopkins
Statistic is expected to be close to 1.0 for a highly clusterable
dataset. If the data points in the original dataset are uniformly

distributed, then the sum of the nearest neighbor distances for the samples in the set X would be significantly larger than the sum of the nearest neighbor distances for the samples in the set Y; the Hopkins Statistic in such a case would be closer to 0. A value of 0.75 or higher for the Hopkins Statistic indicates a clustering tendency at the 90% confidence level [5] [13–15].

We now illustrate the procedure to compute the Hopkins Statistic measure for the logical topology of the vertices illustrated in Fig. 33.1d. The distance between two vertices in the logical topology is computed as the Euclidean distance between their coordinates (represented by the normalized centrality values with respect to degree and eigenvector centrality). There are a total of eight vertices ($n = 8$) in the toy example graph of Fig. 33.1. Let the parameter 'm' be 5. That is, we randomly pick five of the eight vertices from the graph and constitute the set X. Let the five vertices picked be {0, 3, 4, 6, 7}, and the set X would comprise of their corresponding (DEG, EVC) values as coordinates: {(0.416, 0.489), (0.416, 0.467), (0.277, 0.264), (0.277, 0.155), (0.277, 0.155)}. From Fig. 33.1b and c, we could identify the range of the DEG and EVC values to be [0.277, …, 0.416] and [0.155, …, 0.489] respectively. While the normalized DEG values for the vertices in the original dataset are either 0.277 or 0.416, the normalized EVC values do not appear to follow any particular pattern. Hence, to generate the set Y with five data points, the DEG values need to be either 0.277 or 0.416 (we randomly choose one of these two values for each data point) and the EVC values need to be uniform randomly distributed in the range [0.155, …, 0.489]. Accordingly, the five samples constituting the set Y are generated to be: {(0.416, 0.168), (0.277, 0.328), (0.416, 0.479), (0.416, 0.183), (0.277, 0.217)}.

Figure 33.2 summarizes all the calculations. The sums of the nearest distances to a vertex for the samples in the sets X and Y are respectively 0.122 and 0.276; the Hopkins Statistic for the (DEG, EVC) coordinate system is $0.276/(0.276 + 0.122) = 0.69$. Since it is relatively closer to

0.5, we could conclude that the vertices in the given complex network graph are not very much clusterable on the basis of the similarity with respect to the DEG and EVC values.

33.4 Similarity-Based Clusterability of Real-World Networks

We applied the proposed approach to a suite of 47 real-world networks of diverse degree distributions and computed the values for the Hopkins Statistic for the logical topologies of the vertices with respect to the neighborhood-based DEG and EVC metrics and the shortest path-based BWC and CLC metrics. The real-world networks (for more details, refer to [16]) considered fall under one of these domains (the number of networks under each category is indicated within the parentheses): I. Acquaintance network (12), II. Friendship network (9), III. Co-appearance network (6), Employment network (4), Citation network (3), Collaboration network (3), Literature network (3), Biological network (3), Political network (2), Game network (2), Transportation network (1), Geographical network (1) and Trade network (1). Table 33.1 presents the Hopkins Statistic values for each network with respect to the (DEG, EVC) and (BWC, CLC) measures along with the number of nodes and edges as well as the average degree (k_{avg}) and spectral radius ratio for node degree [17] for each of these networks. The spectral radius ratio for node degree (λ_{sp}) is a measure of the variation in node degree.

Figure 33.3 plots the Hopkins Statistic values for the two coordinate systems/logical topologies: if a data point is below the diagonal line, then the Hopkins Statistic value for the (DEG, EVC) coordinate system is relatively larger; if a data point is above the diagonal line, then the Hopkins Statistic value for the (BWC, CLC) coordinate system is relatively larger. Table 33.1 highlights the Hopkins Statistic values that are (overall) relatively larger for a real-world network with respect to the two coordinate systems. From both Fig. 33.3 and Table 33.1, we could come to the following conclusions:

Fig. 33.2 Calculations for the Hopkins Statistic for (DEG, EVC)-Similarity based Clusterability for the graph of Fig. 33.1

Vertex ID	DEG	EVC
0	0.416	0.489
1	0.277	0.364
2	0.416	0.489
3	0.416	0.467
4	0.277	0.264
5	0.416	0.232
6	0.277	0.155
7	0.277	0.155

Hopkins Statistic

$$= \frac{0.276}{(0.276 + 0.122)}$$

$$= 0.693$$

Vertex in Set X	DEG	EVC	Nearest Distance	Nearest Vertex
0	0.416	0.489	0.000	2
3	0.416	0.467	0.022	2
4	0.277	0.264	0.100	1
6	0.277	0.155	0.000	7
7	0.277	0.155	0.000	6
			Sum = 0.122	

Data Point in Set Y	DEG	EVC	Nearest Distance	Nearest Vertex
1	0.416	0.168	0.064	5
2	0.277	0.328	0.036	1
3	0.416	0.479	0.080	0
4	0.416	0.183	0.049	5
5	0.277	0.217	0.047	4
			Sum = 0.276	

Table 33.1 Hopkins statistic values with respect to the neighborhood-based (DEG, EVC) metrics vs. shortest path-based (BWC, CLC) metrics

#	Net.	Network Name	# nodes	# edges	k_{av}	λ_{sp}	(DEG, EVC)	(BWC, CLC)
1	ADJ	Word Adjacency Net	112	425	7.6	1.7	0.92	0.84
2	AKN	Anna Karenina Net	140	494	7.1	2.5	0.93	0.92
3	JBN	Jazz Band Net	198	2742	27.7	1.4	0.92	0.96
4	CEN	C. Elegans Neural Net	297	2148	14.4	1.7	0.96	0.98
5	CLN	Centrality Literature Net	118	613	10.4	2.0	0.93	0.85
6	CGD	Citation Graph Dr. Net	259	640	4.1	2.2	0.84	0.95
7	CFN	Copperfield Net	89	407	9.1	1.8	0.96	1.00
8	DON	Dolphin Net	62	159	5.1	1.4	0.79	0.73
9	DRN	Drug Net	212	284	1.9	2.8	0.89	0.96
10	DLN	Dutch Lit 1976 Net	37	81	4.4	1.5	0.78	0.64
11	ERD	Erdos Collab. Net	433	1314	6.1	3.0	0.94	0.99
12	FMH	Faux Mesa High Sc. Net	147	202	1.9	2.8	0.86	0.96
13	FHT	Friendship Hi-Tec Firm	33	91	5.1	1.6	0.81	0.77
14	FTC	Flying Teams Cad. Net	48	170	7.1	1.2	0.77	0.76
15	FON	US Football Net	115	613	10.7	1.01	0.94	0.71
16	CDF	College Dorm Fraternity	58	967	33.3	1.1	0.91	0.79
17	GD96	Graph Drawing '96 Net	180	228	2.5	2.4	0.96	0.95
18	MUN	Marvel Universe Net	167	301	3.6	2.5	0.93	0.98
19	GLN	Graph Glossary Net	67	118	3.3	2.0	0.87	0.93
20	HTN	Hypertext 2009 Net	115	2164	38.3	1.2	0.93	0.84
21	HCN	Huckleberry Co. Net	76	302	7.9	1.7	0.93	0.99
22	ISP	Infec. Socio-Patt. Net	309	1924	12.4	1.7	0.91	0.92
23	KCN	Karate Club Net	34	78	4.6	1.5	0.87	0.80
24	KFP	Korea Family Plan. Net	37	85	4.3	1.7	0.85	0.73
25	LMN	Les Miserables Net	77	254	6.6	1.8	0.87	0.96
26	MDN	Macaque Dom. Net	62	1167	37.6	1.04	0.90	0.75
27	MTB	Madrid Train Bomb. Net	64	295	9.2	1.9	0.89	0.86
28	MCE	Manufact. Comp. Empl	77	1549	40.2	1.1	0.89	0.84
29	MSJ	Social Net. Journal Net	475	625	2.6	3.5	0.99	0.93
30	AFB	Author Facebook Net	171	940	10.0	2.3	0.94	0.96
31	MPN	Mexican Pol. Elite Net	35	117	6.7	1.2	0.74	0.65
32	MMN	ModMath Net	30	61	4.1	1.6	0.74	0.62
33	PBN	US Politics Books Net	105	441	8.4	1.4	0.84	0.85
34	PSN	Primary Sc. Contact Net	238	5539	46.5	1.2	0.93	0.85
35	PFN	Prison Friendship Net	67	142	4.2	1.3	0.82	0.80
36	SJN	San Juan Sur Family Net	75	155	4.1	1.3	0.78	0.80
37	SDI	Scotland Cor. Interlock	230	359	2.9	1.9	0.96	0.98
38	SPR	Senator Press Rel. Net	92	477	10.4	1.6	0.91	0.78
39	SWC	Soccer World Cup Net	35	118	6.7	1.4	0.85	0.70
40	SSM	Sawmill Strike Com. Net	24	38	3.2	1.2	0.59	0.75
41	TEN	Taro Exchange Net	22	39	3.5	1.06	0.67	0.56
42	TWF	Teenage Fem. Fr. Net	47	77	3.3	1.5	0.82	0.70
43	UKF	UK Faculty Friend Net	83	578	14.2	1.3	0.86	0.88
44	APN	US Airports 1997 Net	332	2126	14.3	3.2	0.96	0.94
45	RHF	Residence Hall Fr. Net	217	1839	16.9	1.3	0.92	0.91
46	WSB	Windsurfers Beach Net	43	336	15.6	1.2	0.84	0.78
47	WTN	World Trade Met. Net	80	875	21.8	1.5	0.93	0.88

The median of the Hopkins Statistic values for the (DEG, EVC) and (BWC, CLC)-based logical topologies are 0.89 and 0.85 respectively. Hence, majority of the real-world networks are effectively clusterable with respect to both the neighborhood and shortest path-based centrality metrics. On a relative basis: out of the 47 real-world networks, for 29 networks (i.e., for about 60% of the real-world networks): the (DEG, EVC)-based Hopkins Statistic values are larger. Also, the Hopkins Statistic values for the (DEG, EVC)-based logical topologies are relatively closer to 1.0 compared to the Hopkins Statistic values for the (BWC, CLC)-based logical topologies. We could thus conclude that real-world networks are more effectively clusterable with respect to the neighborhood-based centrality metrics compared to shortest path-based metrics.

Figure 33.4 plots the distribution of the Spectral radius ratio for node degree vs. the Hopkins Statistic measure values for the (DEG, EVC) and (BWC, CLC)-based logical topologies. We observe for both the coordinate systems, the logical topologies are more effectively clusterable with increase in the spectral radius ratio for node degree. This implies: scale-free networks (larger values for the spectral radius ratio for node degree) are more effectively clusterable with respect to both categories of centrality metrics. Whereas, the random networks (that have lower values for the spectral radius ratio for node degree) are (relatively) less effectively clusterable with respect to both categories of centrality metrics, especially the shortest path-based centrality metrics.

33.5 Related Work and Our Contribution

Clustering in complex networks has been traditionally on the notion of distance between the nodes [13]. Nodes that are

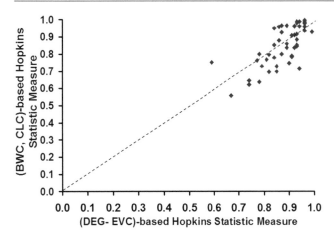

Fig. 33.3 Comparison of the Hopkins Statistic values for the (DEG, EVC)-based logical topologies vs. (BWC, CLC)-based logical topologies of the real-world networks

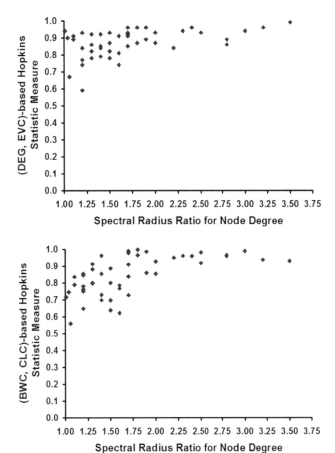

Fig. 33.4 Spectral radius ratio for node degree vs. the Hopkins statistic measure values

closer to each other are preferred for being grouped within a cluster. Various forms of distance measures (Euclidean, Manhattan, Minkowski, Chebyshev, Mahalanobis, etc) have been used in the literature [14]. Similarity-based clustering has been typically considered only for non-network data [15, 18], broadly as a problem in data mining. Given a distribution

of data points, data points that are similar with respect to one or more parameters tend to be clustered together [19, 20]. The typical approach [19] is to associate a multi-dimensional vector to each data point, measure the pair-wise similarity (using functions such as cosine similarity [6], Jaccard Index [19], SimRank [21], PathSim [22]) or the distances between the data points (using some distance function), and then cluster the data points that are similar and closer to each other. Our research proposes a combination of the above two approaches. We first determine the values for the centrality metrics (node-level parameters) of the nodes in a complex network, then treat the nodes as data points with an associated multi-dimensional vector of the normalized values of the centrality metrics, and finally analyze the clusterability of these data points.

The proposed Hopkins Statistic-based clusterability assessment of node similarity could be the first step in pursuit of developing a network-level similarity index for complex networks. If the Hopkins Statistic measure is high, there could exist one or more clusters of similar nodes (with respect to the centrality metrics) in the network. Though there could exist two or more such clusters in the network (with the nodes in one cluster significantly different from the nodes in the other clusters with respect to the centrality metrics), a lower value for the Hopkins Statistic measure is definitely an indication that the nodes in the logical topology of centrality-based coordinates are more randomly distributed and not very similar to each other (i.e., not clusterable). In other words, the computation of the Hopkins Statistic measure could be used as a prerequisite step for any algorithm to quantify/assess the network-wide similarity of all the nodes in the network; there would not be any need for the algorithm to proceed further with assessing the network-wide similarity of the nodes if the value for the Hopkins Statistic measure is low.

33.6 Conclusions and Future Work

Our high-level contribution in this paper is a proposal to assess the logical clusterability of the nodes in a network on the basis of node-level metrics (such as centrality metrics). We propose to distribute the nodes in a K-dimensional coordinate system (referred to as the logical topology) of the normalized centrality values of the vertices (where 'K' is the number of centrality metrics considered and we assign a dimension for each centrality metric). Two vertices that need not be physically connected in the complex network could be very close to each other in the logical topology if they incur similar values for the centrality metrics. The Hopkins Statistic has been identified as the quantitative measure to assess the similarity-based clusterability of nodes in such a logical topology. A lower value for the Hopkins Statistic is definitely an indication that the nodes do not exhibit similar values for the centrality metrics considered. On the other hand, a higher value for the Hopkins Statistic is an indication that there could exist one or more clusters of similar vertices in the network with respect to the centrality metrics considered. We analyzed a suite of 47 real-world networks of diverse degree distributions and determined their Hopkins Statistic values with respect to the neighborhood-based degree and eigenvector centrality metrics and the shortest path-based betweenness and closeness centrality metrics. We observed about 60% of the real-world networks to exhibit relatively larger Hopkins Statistic values with respect to the neighborhood-based centrality metrics. Also, the Hopkins Statistic measure values for the neighborhood-based centrality metrics are relatively closer to 1 compared to the values for the shortest path-based centrality metrics. As part of future work, we plan to develop a network-wide node similarity index (NSI) measure for complex network analysis and test our hypothesis that a higher Hopkins Statistic is a prerequisite for a network to exhibit a higher value for the NSI.

Acknowledgements This research was funded by the NASA EPSCoR subaward #: NNX14AN38A from the University of Mississippi; the NSF MRI Grant 13-38192 and NSF CNS Grant 14-56638.

References

1. Grando, F., Noble, D., Lamb, L.C.: An analysis of centrality measures for complex and social networks. In: Proceedings of IEEE Global Communications Conference, pp. 1–6, Washington, DC, USA (Dec 2016)
2. Meghanathan, N.: On the use of centrality measures to determine connected dominating sets for mobile ad hoc networks. Int. J. Ad Hoc Ubiquitous Comput. **26**(4), 205–221 (2017)
3. Ozgur, A., Vu, T., Erkan, G., Radev, D.R.: Identifying gene-disease associations using centrality on a literature mined gene-interaction network. Bioinformatics. **24**(13), 277–285 (2008)
4. Ding, Y.: Scientific collaboration and endorsement: network analysis of coauthorship and citation networks. J. Informet. **5**(1), 187–203 (2011)
5. Eslami, M., Zheng, G., Eramian, H., Levchuk, G.: Anomaly detection on bipartite graphs for cyber situational awareness and threat detection. In: Proceedings of the 2017 IEEE International Conference on Big Data, pp. 4741–4743. Boston, MA, USA (Dec 2017)
6. Newman, M.E.J.: Networks: An Introduction. Oxford University Press, Oxford (2010)
7. Bonacich, P.: Power and centrality: a family of measures. Am. J. Sociol. **92**(5), 1170–1182 (1987)
8. Freeman, L.: A set of measures of centrality based on betweenness. Sociometry. **40**(1), 35–41 (1977)
9. Freeman, L.: Centrality in social networks: conceptual clarification. Soc. Networks. **1**(3), 215–239 (1979)
10. Huang, J., Sun, H., Han, J., Deng, H., Sun, Y., Liu, Y.: SHRINK: a structural clustering algorithm for detecting hierarchical communities in networks. In: Proceedings of the 19th ACM International Conference on Information and Knowledge Management, pp. 219–228, Toronto, Canada (Oct 2010)
11. Girvan, M., Newman, M.E.J.: Community structure in social and biological networks. Proc. Natl. Acad. Sci. U. S. A. **99**(12), 7821–7826 (June 2002)
12. Banerjee, A., Dave, R.: Validating clusters using the hopkins statistic. In: Proceedings of the IEEE International Conference on Fuzzy Systems, pp. 149–153, Budapest, Hungary (July 2004)
13. Szabo, G., Alava, M., Kertesz, J.: Clustering in complex networks. Complex Networks, Lect. Notes Phys. **650**(1), 139–162 (2004)
14. Soler, J., Tence, F., Gaubert, L., Buche, C.: Data clustering and similarity. In: Proceedings of the 26th International Florida Artificial Intelligence Research Society Conference, pp. 492–495, St. Pete Beach, FL, USA (May 2013)
15. Aalam, P., Siddique, T.: Comparative study of data mining tools used for clustering. In: Proceedings of the 3rd International Conference on Computing for Sustainable Global Development, pp. 3971–3975, New Delhi, India (Mar 2016)
16. Meghanathan, N.: Randomness index for complex network analysis. Soc. Netw. Anal. Min. **7**(25), 1–15 (2017)
17. Meghanathan, N.: Spectral radius as a measure of variation in node degree for complex network graphs. In: Proceedings of the 3rd International Conference on Digital Contents and Applications, pp. 30–33, Hainan, China (Dec 2014)
18. Taghva, K., Veni, R.: Effects of similarity metrics on document clustering. In: Proceedings of the 7th International Conference on Information Technology: New Generations, pp. 222–226, Las Vegas, NV (2010)
19. Han, J., Kamber, M., Pei, J.: Data Mining: Concepts and Techniques, 3rd edn. Morgan and Kaufmann Publishers, Burlington, MA (July 2011)
20. Blooma, M.J., Chua, A.Y.K., Goh, D.H.: Quadripartite graph-based clustering of questions. In: Proceedings of the 8th International Conference on Information Technology: New Generations, pp. 591–596, Las Vegas, NV (2011)
21. Jeh, G., Widom, J.: SimRank: a measure of structural-context similarity. In: Proceedings of the 8th ACM SIGKDD International Conference on Knowledge Discovery and Data Mining, pp. 538–543, Edmonton, Alberta, Canada (July 2002)
22. Yu, W., Lin, X., Zhang, W., Chang, L., Pei, J.: More is simpler: effectively and efficiently assessing node-pair similarities based on hyperlinks. Proc. VLDB Endowment. **7**(1), 13–24 (Sept 2013)

Helton Franco de Sousa, Leandro Guarino de Vasconcelos, and Laercio A. Baldochi

34.1 Introduction

The pervasiveness of the Web has changed several aspects of modern life. As business, government and banking services became online, people have no other option but using the Internet. However, it takes more than an Internet connection and a hardware device to benefit from online services. Using a web application is far from trivial for a large amount of users, especially those who were not born in the digital age. On the other hand, those who were born after 1995 are usually very comfortable with the web. It is common that they help their parents and grandparents to perform tasks online [11].

As people grow older, motor control and visual acuity decreases. Moreover, cognitive abilities such as learning, memory retrieval and attention are impacted negatively. The first attempts to support users affected by age-related issues were based on heuristics, for example, "if the user presents visual acuity problems, make the UI elements bigger". The problem with this approach is that it causes trade-offs in the design, as changes in the application's interface may compromise other aspects of its design – navigation, for instance.

Another problem with this approach is that a "one size fits all solution" is not appropriate. Studies show that age,

alone, has little or no impact on web literacy [5, 7]. Recent studies show that senior citizens that got old in contact with technology usually perform better than those who did not have this contact over the years [15]. Thus, low technological experience is also an issue that prevents users to perform well when using web applications.

A key point in the design of a web application is that it is, in general, targeted for a large audience of users. A study performed by Mbipom and Harper [14] shows that accessibility is correlated to good aesthetics only in web pages that present a clean design, with few interaction elements. However, most e-commerce applications present pages with tens of interaction elements [16]. Therefore, one cannot compromise the user experience of the majority of the users in order to make the application accessible.

An approach that seems promising to tackle this problem is the usage of multi-layered interfaces [12], which are designed to support users with different abilities, offering simplified interfaces designed for seniors and people with low web literacy, as well as full fledged interfaces designed for advanced users. A relevant issue in order to make the multi-layered interface approach feasible for web applications is transparently identifying the user, so as to provide the correct interface layer. To this end, it is paramount to track the user's actions in order to detect usage patterns.

In previous work, we proposed the RUM approach [19] – Real-time Usage Mining – which allows analyzing users' logs during navigation. Towards identifying behavior patterns, RUM exploits Web Usage Mining (WUM) techniques over client logs, i.e., logs collected on the user's browser. These fine-grained logs report details regarding each interaction, such as mouse and keyboard events. Moreover, it reveals which HTML element was the target of an event [1].

The RUM approach has proven to be effective in order to analyze the user behavior in web applications. However, an application specialist was required for defining usage

H. F. de Sousa
POSCOMP, Federal University of Itajubá, Itajubá, Brazil
e-mail: helton.franco@unifei.edu.br

L. G. de Vasconcelos
Brazilian Institute for Space Research, São José dos Campos, Brazil
e-mail: leandro.guarino@lit.inpe.br

L. A. Baldochi (✉)
Institute of Mathematics and Computing, Federal University of Itajubá, Itajubá, Brazil
e-mail: baldochi@unifei.edu.br

© Springer Nature Switzerland AG 2019
S. Latifi (ed.), *16th International Conference on Information Technology-New Generations (ITNG 2019)*,
Advances in Intelligent Systems and Computing 800,
https://doi.org/10.1007/978-3-030-14070-0_34

patterns and user profiles, and matching patterns to profiles. Therefore, our mining approach was application dependent.

This work reports an extension in our previous work called RUM++, which allows the detection of usage patterns associated to the elderly and to users that present low web literacy – the target users of our work – in any web application.

In order to achieve our goal, we reviewed the literature in order to understand the interaction problems presented by our target users. Then, we investigated which attributes of the client logs are related to these problems. Following, we performed an experiment to find out how these attributes vary among target users and regular users. Based on the results of this experiment, we defined usage patterns that characterize our target users. Finally, we tested RUM++ using real logs from a gamification application. Results showed that the found patterns are effective in order to identify our target users by analyzing their interactions in web applications.

This paper is organized as follow. Section 34.2 presents a literature review on aging and technology. Section 34.3 presents our previous work, which aims at understanding the behavior of users in order to support the construction of adaptive web applications. Section 34.4 presents an extension to our previous work that aims at supporting aging users. In Sect. 34.5 we present a case study in which RUM++ is used to classify 44 users of a gamification learning application. Following, in Sect. 34.6, we discuss the results of our study. Finally, Sect. 34.7 presents our concluding remarks.

34.2 Literature Review

The impact of aging in the usage of computer applications dates back to the 1980s [6,21]. This seminal research pointed out important findings regarding older users, such as they are more likely to commit mistakes [6], they take more time to complete tasks and to learn how to use new applications [21].

As expected, these problems increased with the advent of the Web [7, 17]. Experiments performed with old and young users using web applications reported differences regarding the usage of input devices, application browsing and to the amount of time needed to perform tasks.

Chaparro et al. [4] proposed an experiment where young and older adults had to perform click and click-drag tasks. The results showed that older users performed the tasks more slowly. The work by Carvalho et al. [2] reports that older users frequently lost the pointer of the mouse and spend time trying to find it, which negatively impacts their performance. When typing is concerned, older users present a low performance both in terms of speed and accuracy when compared to young adults [3].

More recently, researches have investigated the performance differences in touch interaction. Findlater et al. [9] showed that, when using touchscreen devices, older users were still slower than young users, however their performance were better than when using the mouse.

The research by Hwang et al. [10] suggests that low visual acuity and motor control negatively affects the performance of users. In their experiment, after magnifying the interaction targets, the time to complete a task has decreased 14% and the error rate dropped 50%.

The behavior of older users also present differences in web browsing. A study by Liao et al. [13] noted that, before an action, older users take into consideration all the possible options, thus taking more time, while younger users chose an interaction target more quickly. As a result, older users browse a small group of pages and present page re-visitation rate significantly smaller.

Finally, it is worth to notice that interaction issues related to age are aggravated when the user has low technological experience. Moreover, even younger users that present low technological experience may experience the difficulties faced by older users when using web applications [5, 7]. Therefore, both age and technological experience may be associated to low web literacy.

As a result of our literature review, we found out that older users and people with low technological experience present interaction issues associated to point and click [2,4], and that these issues may be associated to low visual acuity and motor control [10]. Moreover, issues associated to the usage of the keyboard [3] and to web browsing [13] was also reported.

34.3 The RUM Approach

In order to enhance the usability and the user experience in web applications, we proposed RUM [19], an approach that allows mining client logs in real-time with the aim of understanding the behavior of users and profiling them. As a result, application developers may profit from our approach to code adaptive web applications. Figure 34.1 depicts RUM's architecture, which is organized in five modules:

(1) **Log collection:** collects and stores the user's actions performed in the application's interface;
(2) **Task analysis:** provides remote and automatic usability evaluation during navigation;
(3) **Automated KDD:** detects behavior or usage patterns exploring the navigation history of past users. This module exploits KDD techniques to uncover patterns from logs.
(4) **Knowledge repository:** stores and processes the detected behavior patterns, using parameters provided by the application specialist;
(5) **Service:** listens to requests from the web application, providing information regarding the user's behavior during navigation.

In Fig. 34.1, the arrows depict the data flow among modules and the numbers on arrows indicate the flow sequence. Initially, as illustrated by arrow 1, the Logging module detects the user's actions on the application's interface, considering the specificities of the input device (desktop, tablet, smartphone). Following, as depicted by arrows 2 and 3, the detected actions are converted to logs in order to be processed by the Task Analysis and Automated KDD modules. The detected behavior patterns feed the Knowledge Repository module (arrow 4), which is responsible for defining the relevance of each pattern.

While the user is browsing, the web application may interact with the Service module to request information regarding the user's actions (arrow 8). Based on this information, preprogrammed interface adaptations may be triggered. Arrow 9 depicts the response for a given request. Possible responses are the last actions of the current user (arrow 5), the result of the usability evaluation during navigation (arrow 6), and the behavior patterns performed by the active user (arrow 7).

Therefore, RUM provides two main services to the application developer: (i) task analysis and (ii) usage patterns detection. The first one exploits previous work on usability evaluation [18]. The Task Analysis module evaluates the execution of tasks by calculating the similarity among the sequence of events produced by users and those previously defined by the application specialist. This service is relevant to detect users who are struggling to execute tasks.

The usage patterns detection, on the other hand, relies on a KDD process to uncover patterns from logs. The KDD process implemented in RUM finds sequential patterns, i.e., sequences of actions performed by users. The approach then

relies on an application specialist to match sequential patterns to user profiles. Moreover, the specialist may associate a given action to be triggered when a user performs an action that is associated to her profile. It is important to notice that these actions are implemented by the application developer. Thus, the role of RUM is to inform the web application about the occurrence of an action.

The original KDD process implemented in RUM is not able to effectively detected the performance difference between two users that execute the same task, as its pattern detection mechanism only considers the difference in the sequence of actions performed by each user. In order to understand how well a user performs a task, it is paramount to analyze other information hidden in logs, such as the time to execute each task, the movement of the mouse, the amount of clicks, scrolling, and so far. In order to classify users according to performance profiles, we developed an extension to RUM's KDD module. The resulting approach was renamed RUM++.

34.4 RUM++

In order to find patterns that may reveal the user ability when using a web application, we developed an approach based on the classical KDD process proposed by Fayyad et al. [8]. Our approach, depicted in Fig. 34.2, is composed of four steps, as follows:

(A) **Selection and Preprocessing:** Initially logs are cleaned up from irrelevant data and noise. Following, we extract attributes from logs and harness them to profile data collected from users.

(B) **Transformation:** The goal of this step is to reduce the dimensionality of the data to be analyzed. In order to

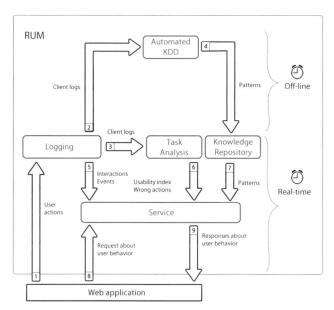

Fig. 34.1 Architecture of the RUM approach [19]

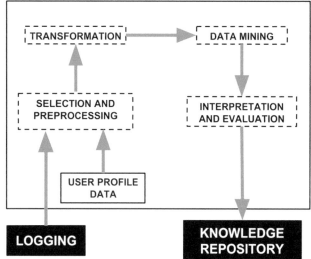

Fig. 34.2 Architecture of the RUM++ approach

achieve this goal, we select attributes that are correlated to the behavior of our target users.

(C) **Data Mining:** In this step, the selected attributes are used as input to data mining algorithms in order to classify users according to patterns presented in their logs.

(D) **Interpretation and Evaluation:** Finally, an specialist evaluates the mining results. In case the classification is correct, the used patterns are considered valid and, thus, are inserted in the knowledge repository.

As happens in the classical KDD process, our approach is iterative and incremental, possibly discovering new patterns as new data is processed. Following, we present the steps of our approach in detail.

34.4.1 Selection and Preprocessing

There are two input sources to our approach. The most important source are the collected logs, which present noisy data, such as requests to CSS files, javascript code and references to images. Thus, it is required to clean up logs from data that is not relevant for the mining process. The other data source is composed of user profiles, which need to be fed in the system in order to assure that the classification result is correct. Therefore, a task that needs to be done in this step is to harness logs to user profiles.

The log preprocessing is performed exploiting a tool for managing logs called Logstash. We use Logstash together with two other tools: a search engine called ElasticSearch and a tool for data visualization called Kibana. These tools form the so called ELK stack.[1] This way, after being preprocessed, collected logs are fed to Logstash, from where they can be searched using ElasticSearch. Finally, using Kibana one may visualize the logs and perform user friendly searches. We also provide a Python application in order to facilitate the execution of queries in the log data.

By the end of this step, the 20 following attributes are extracted from logs and made available.

(1) Amount of clicks;
(2) Amount of double clicks;
(3) Page elements associated to mouse interactions;
(4) Use of shortcut commands;
(5) Amount of zoom;
(6) Path within the application during navigation;
(7) Amount of key pressed;
(8) Sequence of events performed by the user;
(9) Amount of errors when filling forms;
(10) Orientation changes (for mobile devices);

(11) Mouse over amount for each page element;
(12) Total amount of events;
(13) Browsing duration;
(14) Amount of time spent per page;
(15) Scrolling speed;
(16) Amount of scroll;
(17) Typing speed;
(18) Amount of visited pages;
(19) Number of focus events;
(20) Number of visits per page.

34.4.2 Transformation

In transformation, the goal is to reduce the dimensionality of our data. In order to do so, we need to select the log attributes that can reveal behavioral differences among our target users and regular users. In order to achieve this goal, we analyzed the work presented on Sect. 34.2 in order to understand which log attributes are related to the interaction problems discussed in the literature. As a result, a group of 12 attributes have been selected as candidates for mining. In order to make sure the relevance of these attributes, we used a correlation technique to ensure that our selection was correct.

As discussed in Sect. 34.2, older users and people with low technological experience present interaction issues associated to point and click [2, 4, 10], use of the keyboard [3] and browsing [13]. Moreover, the literature also report that interaction with small targets tend to be more difficult. Therefore, log attributes associated to these issues are important candidates for investigation. Based on these findings, we selected 12 attributes – *1, 2, 5, 7, 9, 12, 13, 15, 16, 17, 18 and 19* – from the total of 20 attributes generated in the preprocessing step.

The number of relevant attributes may be narrowed using the Pearson Correlation. The approach for this is to calculate the correlation coefficient among the 12 selected attributes and data features that represent our target users. Section 34.5 explains this procedure.

At the end of this step, log data is transformed into data which is ready to be used as input to the mining algorithms.

34.4.3 Data Mining

In order to facilitate our mining process, we resort to Weka, a machine learning workbench for data mining. Among the several state-of-the-art algorithms provided by Weka, we selected *RepTree* (Reduces Error Pruning Tree Classifier), as we needed to classify users according to patterns found in their logs. Moreover, *RepTree* is considered fast [20] and may be used with numeric data. This algorithm builds a

---

[1] https://www.elastic.co/elk-stack

decision tree using information gain/variance and prunes this tree using reduced-error pruning.

From the available training options, we secleted *Cross-validation*. This technique allows evaluating the generalization ability of the model. The result of the classification provided by Weka is forwarded to the next step of our approach.

34.4.4 Interpretation and Evaluation

In the final step, the classification rules found by RepTree must be analyzed in order to verify if they are effective in order to classify users accordingly. In this step we need an application specialist to check if the classification provided by a given rule is effective. If so, the provided rule is added to the knowledge repository. Otherwise, it is discarded.

In order to evaluate RUM++, we performed an experiment with a real web application, involving 44 different users. The results of this experiment is discussed in the following section.

34.5 Case Study

In order to refine our approach, we performed an experiment using data from a real web application. As our aim was to evaluate how well RUM++ classifies users according to age and technological experience, we recruited users from two different group ages: the first group of users with age ranging from 18 to 39 years-old (group A), and second with ages ranging from 40 to 59 years-old (group B). In total, 44 subjects participated in the experiment, 16 from group A and 28 from group B.

Besides age, it was important to classify users according to their technological experience. In order to acquire this information, we developed a questionnaire in which users report the frequency (per week) that they use computers and smartphones, as well as the number of years using these devices. We also collected information regarding the level of education and the gender of each participant.

We used the number of years using computer devices, plus the frequency of use to compute a value between 0 and 1, which we called *Technological Experience Coefficient (TEC)*. A TEC value close to 0 means a subject with low technological experience and close to 1, a subject with high technological experience.

After classifying the recruited users according to their age and technological experience, we asked them to perform tasks in a learning gamification system called Level Learn (www.levellearn.com.br). The aim of this system is to support learning by providing an environment based on challenges and rewards. A challenge consists of a task to be performed in the system, and when the student performs the task correctly, she receives a reward.

In order to evaluate our users, we planned a challenge that required users to browse several pages, fill forms and interact with specific buttons in the user interface, sometimes requiring scrolling the page to find targets.

All 44 users performed the task using a desktop computer, with standard mouse and keyboard. Firstly, users were briefed about the Level Learn application, learning how to complete an example challenge. Following, the test challenge was explained and the user was left by herself to perform the required tasks. At the end, users were instructed to close the browser to quit the collection of logs.

34.6 Results and Discussion

All 44 participants succeeded in completing the task. Table 34.1 summarizes the data related to Young and Older users, in terms of mean age, amount of time using computer devices, frequency of use per week and the TEC calculated from this data.

As discussed in Sect. 34.4, we exploited the data on Table 34.1 to calculate the correlation among this data and the 12 attributes extracted from logs in the preprocessing step. Table 34.2 presents the results.

It is worth to notice that correlation values above 0.5 indicate significant positive correlation, while values bellow −0.5 indicate significant negative correlation. Interestingly,

Table 34.1 Collected data for young and older adults

	Young	Older
Age (years)	23.59	54.27
Usage time (years)	8.21	5.93
Usage frequency (per week)	6.62	2.4
TEC	0.88	0.44

Table 34.2 Correlation among log attributes and user data

	Description	Age	Frequency	Time	TEC
1	Amount of clicks	−0,213	0,199	0,100	0,166
2	Amount of double clicks	0,054	0,046	−0,165	−0,063
5	Amount of zoom	0,0	0,0	0,0	0,0
7	Amount of key pressed	0,597	−0,756	−0,689	−0,784
9	Amount of errors	0,178	−0,104	−0,046	−0,084
12	Total amount of events	0,569	−0,703	−0,654	−0,736
13	Browsing duration	0,753	−0,855	−0,641	−0,815
15	Scrolling speed *Scroll's*	−0,156	0,107	−0,066	0,026
16	Amount of scroll	0,648	−0,616	−0,601	−0,659
17	Typing speed	0,700	−0,750	−0,605	−0,737
18	Amount of visited pages	0,127	−0,101	0,011	−0,063
19	Number of focus event	−0,173	0,190	0,037	0,127

Table 34.3 Classification efficiency of RUM++ using REPTree

	Efficiency
Age	68.18%
Technological experience	84.09%

as shown in highlighted lines on Table 34.2, attributes 7, 12, 13, 16 and 17 correlates positively to age and negatively to usage time, usage frequency and technological experience. Hence, these attributes are more likely to reveal usage differences associated to young and older users. As a result, we were able to reduce the dimensionality of our data from 12 to 5 attributes, which makes the mining process much more efficient.

After concluding the transformation step, the resulting log data is used as input to the algorithm *REPTree* in Weka workbench using *Cross-Validation* with fold value 43 (dataset size - 1), with the aim of classifying users according to their age and technological experience. The efficiency of REPTree to classify users is shown in Table 34.3. As it can be noticed from the presented results, technological experience is more related to low web literacy than age. However, the experiment confirms that both of them impact the performance of users in web applications.

An interesting feature provided by Weka is the possibility to analyze the decision tree created by the REPTree algorithm. The tree created for classifying users according to age shows that the attribute 17 alone is enough to perform the classification. On the other hand, when the goal is to classify users according to their technological experience, two attributes are needed to obtain a good classification: 7 and 16.

Upon analyzing the classification results we concluded that the generated rules were sound and, therefore, we included them in the knowledge repository.

34.7 Conclusion

It is well known that the benefits brought about by web applications does not reach everyone. Aging users, for instance, usually struggle to perform tasks online. This work proposed an approach to support these users by providing a way to detect them transparently as they browse the web.

Our approach leverages the traditional KDD method proposed by Fayyad et al. [8]. Our main contribution lies in the preprocessing and transformation steps of the method, as we use both knowledge from the literature and a correlation technique to reduce the dimensionality of the data. The transformed data provided for the mining algorithm allows good results in the classification of users.

As future work, we plan to implement the analysis in real time, so as to support the construction of adaptive web applications.

References

1. Atterer, R., Wnuk, M., Schmidt, A.: Knowing the user's every move: user activity tracking for website usability evaluation and implicit interaction. In: Proceedings of the 15th International Conference on World Wide Web, WWW'06, New York, pp. 203–212. ACM (2006). ISBN:1-59593-323-9
2. Carvalho, D., Bessa, M., Magalhaes, L.: Different interaction paradigms for different user groups: an evaluation regarding content selection. In: Proceedings of the XV International Conference on Human Computer Interaction, New York, pp. 40:1–40:6. ACM (2014). ISBN:978-1-4503-2880-7
3. Carvalho, D., Bessa, M., Magalhães, L., Carrapatoso, E.: Age group differences in performance using diverse input modalities: insertion task evaluation. In: Proceedings of the XVII International Conference on Human Computer Interaction, Interacción'16, New York, pp. 12:1–12:8. ACM (2016). ISBN:978-1-4503-4119-6
4. Chaparro, A., Bohan, M., Fernandez, J., Choi, S.D., Kattel, B.: The impact of age on computer input device use:: psychophysical and physiological measures. Int. J. Ind. Ergon. **24** (5), 503–513 (1999)
5. Crabb, M., Hanson, V.L.: Age, technology usage, and cognitive characteristics in relation to perceived disorientation and reported website ease of use. In: Proceedings of the 16th International ACM SIGACCESS Conference on Computers & Accessibility, ASSETS'14, New York, pp. 193–200. ACM (2014). ISBN:978-1-4503-2720-6
6. Czaja, S.J., Hammond, K., Blascovich, J.J., Swede, H.: Age related differences in learning to use a text-editing system. Behav. Inf. Technol. **8** (4), 309–319 (1989)
7. Fairweather, P.G.: How older and younger adults differ in their approach to problem solving on a complex website. In: Proceedings of the 10th International ACM SIGACCESS Conference on Computers and Accessibility, Assets'08, New York, pp. 67–72. ACM (2008). ISBN:978-1-59593-976-0
8. Fayyad, U., Piatetsky-Shapiro, G., Smyth, P.: The KDD process for extracting useful knowledge from volumes of data. Commun. ACM **39** (11), 27–34 (1996). ISSN:0001-0782
9. Findlater, L., Froehlich, J.E., Fattal, K., Wobbrock, J.O., Dastyar, T.: Age-related differences in performance with touchscreens compared to traditional mouse input. In: Proceedings of the SIGCHI Conference on Human Factors in Computing Systems, CHI'13, New York, pp. 343–346. ACM (2013). ISBN:978-1-4503-1899-0
10. Hwang, F., Hollinworth, N., Williams, N.: Effects of target expansion on selection performance in older computer users. ACM Trans. Access. Comput. **5** (1), 1:1–1:26 (2013). ISSN:1936-7228
11. Lara, S.M., Fortes, R.P., Russo, C.M., Freire, A.P.: A study on the acceptance of website interaction aids by older adults. Univers. Access Inf. Soc. **15** (3), 445–460 (2016). ISSN:1615-5289
12. Leung, R., Findlater, L., McGrenere, J., Graf, P., Yang, J.: Multilayered interfaces to improve older adults initial learnability of mobile applications. ACM Trans. Access. Comput. **3** (1), 1:1–1:30 (2010). ISSN:1936-7228
13. Liao, C., Groff, L., Chaparro, A., Chaparro, B., Stumpfhauser, L.: A comparison of website usage between young adults and the elderly. In: Proceedings of the Human Factors and Ergonomics Society Annual Meeting. **44** (24), 4–101–4–101 (2000)

14. Mbipom, G., Harper, S.: The interplay between web aesthetics and accessibility. In: The Proceedings of the 13th International ACM SIGACCESS Conference on Computers and Accessibility, ASSETS'11, New York, pp. 147–154. ACM (2011). ISBN:978-1-4503-0920-2

15. O'brien, M.A., Rogers, W.A., Fisk, A.D.: Understanding age and technology experience differences in use of prior knowledge for everyday technology interactions. ACM Trans. Access. Comput. **4** (2), 9:1–9:27 (2012). ISSN:1936-7228

16. Paz, F., Paz, F.A., Pow-Sang, J.A.: Evaluation of usability heuristics for transactional web sites: a comparative study. In: Latifi, S. (ed.) Information Technology: New Generations, Cham, pp. 1063–1073. Springer International Publishing (2016). ISBN:978-3-319-32467-8

17. Priest, L., Nayak, L., Stuart-Hamilton, I.: Website task performance by older adults. Behav. Inf. Technol. **26** (3), 189–195 (2007) ISSN 0144-929X.

18. Vasconcelos, L.G., Baldochi, L.A., Jr.: Towards an automatic evaluation of web applications. In: SAC'12: Proceedings of the 27th Annual ACM Symposium on Applied Computing, New York, pp. 709–716. ACM (2012). ISBN:978-1-4503-0857-1

19. Vasconcelos, L.G., Baldochi, L.A., Santos, R.D.C.: Rum: an approach to support web applications adaptation during user browsing. In: Gervasi, O., Murgante, B., Misra, S., Stankova, E., Torre, C.M., Rocha, A.M.A., Taniar, D., Apduhan, B.O., Tarantino, E., Ryu, Y. (eds.) Computational Science and Its Applications – ICCSA 2018, Cham, pp. 76–91. Springer International Publishing (2018). ISBN:978-3-319-95165-2

20. Witten, I.H., Frank, E., Hall, M.A.: Data Mining: Practical Machine Learning Tools and Techniques. Morgan Kaufmann Series in Data Management Systems, 3rd edn. Morgan Kaufmann, Amsterdam (2011). ISBN:978-0-12-374856-0

21. Zandri, E., Charness, N.: Training older and younger adults to use software. Educ. Gerontol. Int. Q. **15** (6), 615–631 (1989)

Smart Food Security System Using IoT and Big Data Analytics

35

Sazia Parvin, Sitalakshmi Venkatraman, Tony de Souza-Daw, Kiran Fahd, Joanna Jackson, Samuel Kaspi, Nicola Cooley, Kashif Saleem, and Amjad Gawanmeh

35.1 Introduction

Considering one of the fasted growing areas of technology for both large and small-scale farming operations, precision farming is now a major participant in the quest to expand world food production. While the concept of crop yield monitoring has been presented for almost two decades, the development and implementation of smarter farm machines, crop sensors, and the software to analyse data collected by these devices has recently become a game-changer. The advent and adoption of smart intelligent farming systems in agriculture, such as collaborative spraying, humidity detector, smart irrigation techniques and other intelligent infrastructure provide the capacity to enable unmanned situation awareness, on a 24/7 basis. According to the Beecham Research report [1], smart farming will allow farmers and growers to improve productivity and reduce waste, ranging from the quantity of fertiliser used to the number of journeys made by farm vehicles. Food wastage is a major threat to food security, with waste estimated to be as high as 50% of all food produced due to lack of proper management from production to retail. In the Report by Savvas [1], an expert Romeo added: "While the Mobile to Mobile (M2M) agricultural sector is still emerging, M2M and Internet of Things (IoT) technologies will be key enablers for transforming the agricultural sector and creating the smart farming vision".

It is critical to realise that over the years, the agriculture sector has generated and accumulated big data and information resources, whose economic value would be much more than the information housed in social media networks such as Facebook. In order to achieve food security however, recognition must first be made of the multifaceted relationship between the elements of agricultural food production; the preservation, processing and marketing right through to the distribution and marketing of the foods. However, little or no effort has been made in the past few years to unlock the intelligence encapsulated within the large amount of information accumulated and stored with the agriculture Sector. Therefore, in addressing food security challenges, there is a need to take a holistic approach to the food supply chain to ensure innovation in food processing through the use of Information and Communication Technology (ICT) tools.

S. Parvin (✉)
Melbourne Polytechnic, Melbourne, VIC, Australia

Australia and University of New South Wales, Canberra, ACT, Australia
e-mail: SaziaParvin@melbournepolytechnic.edu.au

S. Venkatraman · T. de Souza-Daw · K. Fahd · J. Jackson
S. Kaspi · N. Cooley
Melbourne Polytechnic, Melbourne, VIC, Australia
e-mail: SitaVenkat@melbournepolytechnic.edu.au;
TonydesouzaDaw@melbournepolytechnic.edu.au;
KiranFahad@melbournepolytechnic.edu.au;
JoannaJackson@melbournepolytechnic.edu.au;
SamkKaspi@melbournepolytechnic.edu.au;
NicolaCooley@melbournepolytechnic.edu.au

K. Saleem
Center of Excellence in Information Assurance (CoEIA), King Saud University, Riyadh, Kingdom of Saudi Arabia
e-mail: ksaleem@ksu.edu.sa

A. Gawanmeh
Department of Electrical and Computer Engineering, Khalifa University of Science and Technology, Abu Dhabi, UAE
e-mail: amjad.gawanmeh@ku.ac.ae

35.2 Literature Review

Agriculturists around the world now recognise the pressing need for Data Science capabilities to unlock and action up the intelligence stored in the accumulated big data and information in Agribusiness in order to ensure food security.

© Springer Nature Switzerland AG 2019
S. Latifi (ed.), *16th International Conference on Information Technology-New Generations (ITNG 2019)*,
Advances in Intelligent Systems and Computing 800,
https://doi.org/10.1007/978-3-030-14070-0_35

The concepts of IoT have been widely used in various fields such as logistics, food security, health care, transportation, robotics and smart grid and in other fields where a number of significant and extensive applications have been revealed. Recent development of IoT has widely been adopted in the physical world in a number of sectors, including agriculture. For example, the IoT provides a critical component of the ICT infrastructure to manage the data in agriculture. The farming industry must embrace IoT if it is to feed the 9.6 billion global population expected by 2050, according to research [1]. Ashford [2] mentioned in the report that "The internet of things could be key to the farming industry meeting the challenge of increasing food production by 70% by 2050". The existing technical solutions employed in the control, monitoring and documentation of agricultural farm processes; transportation, logistics and environment, are just selected examples of how this "new wave" of services made accessible by Farm Management Information Systems (FMISs) [3]. IoT is generally defined as a network which unites all relevant elements with the internet through the utilisation of radio frequency identification (RFID), global positioning systems and sensors among other information detection devices in accordance with the agreed protocol. This enables the monitoring, management, tracking and intelligent identification of essential information [5]. The network can realize the automatic identification of objects and location, track, monitor and trigger the corresponding event [6]. It makes use of RFID technology for scanning and reading EPC tags on the items and achieves automatic identification of goods and information sharing.

Current farming technologies sustain a number of communication standards to exchange information through sensors, devices and actuators ("things"). A number of communication standards such as Zigbee, WiFi, and ISOBUS are the supported technology in existing farming to exchange information [4]. Using IoT framework, sensors/actuators can be connected, invoked, de-activated or disabled over the internet. Using the IoT, the system provides for two way communication between the sensor/actuator and other devices. Enabling agriculture domain with IoT would provide real-time data input on various assets, however, this raises the issue of being able to analyse, make sense of and derive intelligence from large amounts of data streams coming from agriculture sector through the IoT framework. There is a critical need to develop a set of system utilities and methodologies to help Agriculture to manage, analyse, make sense of, derive intelligence, and visualise big data being input through the IoT. This would enable a functional architecture of a farm management system (FMS) utilizing Future Internet (FI) capabilities to focus on those functionalities with scope for improvement the Farm Management System (FMS) that would enable farms with capabilities to develop resources not currently possible, such as advertisement, discovering

trustable stakeholders and most prominently, combining the functionality of different management systems without the risk of unstable data network and links etc).

35.3   Research Objectives

The overall objective of this research work is to propose a smart food security framework by using ICT and IoT to improve the supply chain for agribusinesses. The specific objectives of the research are to:

1. Develop an IoT enabled monitoring framework and information infrastructure to provide real-time track and trace of agricultural goods. Additionally, this would integrate existing silo-based logistics technology including GPS, GPRS, Smart Phones, Container and Agricultural Goods tracking systems to provide mechanisms to track and trace of goods movement, asset management and partnership, both with the manufacturer network. The information can be queried through product traceability including the basic information of agricultural products, product name, manufacturer, production date, inspection conclusion, the supply chain process list and sales records, etc.
2. Develop software component systems (big data mining utilities) that can analyse, make sense of, and derive intelligence from the incoming big data stream in agriculture. This objective will use: (1) ontologies for providing uniform knowledge representation about constituent elements, and sharing it among different logistics systems; and (2) lightweight semantic annotations using RDF that discovers data pollution (including non-compliant or incorrect input data), aberrant behaviour from the incoming big data stream, and identifying and correcting missing mandatory information from the incoming data stream in agriculture.
3. Develop a lightweight security mechanism to store the incoming data streams in a reliable and secure manner. This objective will be achieved through developing intelligent Intrusion Detection System (IDS) and to protect both the incoming and stored data from predators, loss and destruction and to provide a traceable RFID carrier to confirm the authenticity of the agricultural products.
4. Develop advanced Big Data analytics and intelligent techniques for representing and interpreting the stored big data, so as to enable informed decision making. As a result of this, decision makers will be able to visualize information, on a real time basis, and make informed and reliable decisions depending on the gathered data.
5. Build a smart prototype system to evaluate the feasibility of the approach, techniques and methodologies developed in objectives (1)–(4) in Agricultural supply chain to protect from food loss.

6. Use of an adaptation cost-effective farm technology and apply the system on a project area (e.g., South Asia) to improve efficiency of agribusiness supply chain by reducing the amount of food wasted due to poor storage and inadequate grain logistics and supply chain infrastructure during food transportation, storage, handling, packaging, distribution, processing.

35.4 Proposed Smart Food Security System Architecture

The proposed Food Security through Smart Agribusiness Supply Chain Management System incorporates multiple sensing devices to capture the real data from the agriculture sector. The proposed food supply chain requirements model based on the IoT technology is able to develop the food safety information platform and carry out the information on the monitor and analyse the data which can help to realize monitoring from food source to the final consumer, provide all-round, multi-angle services to the enterprise, government and the public. Remembering these objectives, the large amount of real time data streams from different sectors in agriculture are being input into the logistics network (Big Data). The sources of Big Data are RFID and Wireless Sensors data, auto and semi-automated asset tracking data; warehousing and transportation communication data; GPS, GPRS and position location systems for vehicle and shipment tracking; Surveillance Systems for situation awareness; public communications related to goods and assets' local and international movement; containers and contents tracking systems; and vehicle and engineering tracking systems such as black-boxes on the heavy vehicles. The proposed system will be able to capture the incoming data in various formats, and from various devices; store them in a secure platform, and analyse the incoming Big Data and derive intelligence from it in order to improve the operational efficiency of the supply chain of agricultural products and promote the development of agriculture sector.

The objective of the research is achieved by developing improved techniques for Big Data capturing, gathering, and applying advanced analytics to integrate Big Data. The data representation in the agriculture sector is designed to avoid the food loss in order to ensure food security. Figure 35.1 depicts the overall framework for Big Data integration for managing the work flow in agricultural supply chain systems. The realization of the proposed framework is explored through five research issues as follows:

Step 1: Analyse data from Agricultural Supply Chain through Semantic annotation and XML profiling.

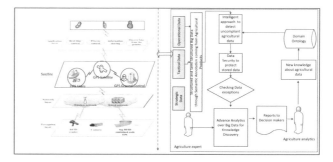

Fig. 35.1 Internet of things and big data analytics in agricultural supply chain

In the traditional agricultural product supply chain, the data about the information of agricultural products is mainly collected by means of manual and bar code. This can easily result in delays, errors and lack of information for agricultural supply chain. And it will make logistics and information flow distorted in the transmission process. The application of the internet of things on agricultural supply chain can build a system that can control and trace the quality of agricultural products by combining agricultural supply chain with farmers' purchase. The system architecture of the Internet of Things for agricultural supply chain is shown in Fig. 35.2. The technology of RFID and cloud computing integrate the information of production, distribution and safety of quality in the agricultural supply chain effectively. It will ultimately combine the farmers with the IoT and make transparency of the entire agricultural supply chain process. It is useful to establish a system that monitors and traces the quality of agricultural materials. The agricultural supply chain management based on the IoT can process the logistics information of every aspect including the production, procurement, storage, transportation and sale. The system can send exact number and right quality of agricultural products such as pesticides, fertilizers and seeds to appropriate places for meeting the needs of farmers in right price at the right time.

Data received from Agricultural materials through IoT need to be cleaned and filtered from noisy and unstructured data with respect to structured data. Some important considerations such as extraction, identification, associations of data should be taken into account during text annotation to enrich the unstructured data semantically. We will use two different approaches to address this: (i) by adopting techniques from automatic spelling and grammar checkers [7]; and (ii) semantic annotation methods which use structured data from agricultural databases. Once the semantic annotation is complete, an XML profile is generated that contains the annotated Big Data and agricultural logistics and supply chain systems profile.

Fig. 35.2 Proposed internet of things architecture for smart agricultural supply chain system

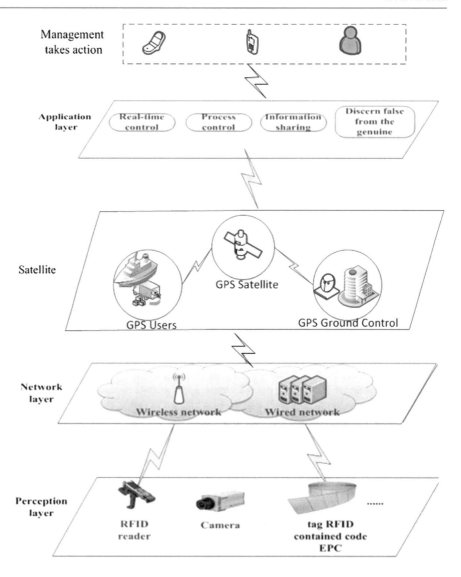

Step 2: Develop intelligent approaches to detect uncompliant agricultural data

Agricultural products are sent to the farmers mainly through production, transportation, storage, sales and other stages. In the production process of agricultural products, the entire items in product line including raw materials, products, semi-finished products and finished products should be identified and tracked to achieve a balanced and steady production. Each agricultural product is labeled with RFID tag encoded with the EPC. The EPC code contains the information of product such as product name, manufacturer, grade (classification), and place of origin, net weight, batch number, production date, and shelf life and so on. Agricultural management business process is shown in Fig. 35.2.

When the processing plant products are shipped after storage, RFID/EPC tag information in fixed locations RFID can be read into the wagon inside the product variety and quantity information. People can order this information, once found error, return the working personnel and processing factory communication; if shipment is correct, then classify products. Each batch of agricultural products will be stacked together and it is convenient for storing and linking goods.

In the transportation stage of agricultural products, installing GPS positioning system enables managers to know the accurate location of vehicles transporting product and the installation of wireless data acquisition system on the vehicles would allow adjustments to be made in the case of an emergency while simultaneously facilitating tracking and prevention of lost/stolen goods during transportation. Database about agricultural products should be established and set product code as its key information. It will not only help the enterprises to develop storage utilization, minimize inventory and save costs, but also the systems also have the potential to enable enterprises with an awareness of business inventory, allowing for precise decisions to be made during production and procurement. Agricultural production management system infrastructure encompasses elements

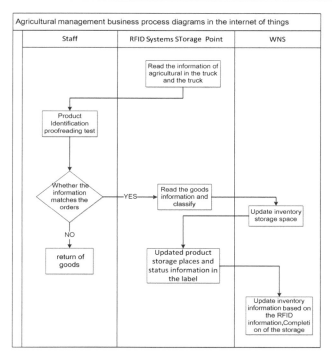

Fig. 35.3 Agricultural management business process diagram using the IoT

such as greenhouse facilities as well as an environmental and digital management/monitoring subsystem which is demonstrated in Fig. 35.3.

To obtain valuable insights into the working of agricultural supply chain system, it is important to find useful associations among concepts extracted from the different components and parts of the supply chain. In order to achieve this, we need to develop domain ontology [9] for information interoperability, Big Data integration (between information coming from different parts of the supply chain) and knowledge assimilation. One of the ways by which useful associations achieved is by singling out an auto-alarm about anomalous state to identify objects whose behavior is aberrant with respect to a codified and standard set of rules. The main aim of this abnormality detection is to identify such errors or indication of the events to reduce the impact of the outlying values in the process of knowledge discovery in databases. These are usually referred to as outliers in the literature [7]. Many data mining algorithms [7–9] in the literature are used to handle the RFID dataset to find the outliers. However, our proposed methodology treats outliers as intriguing new information which may give a clue about important occurrences or new information about agricultural products in agricultural transport and logistics systems. Once the new information has been identified or existing information needs to be updated about the agricultural product, the changes will be incorporated into the domain ontology and then the integrated Big Data is analyzed for further knowledge discovery.

Step 3: Lightweight secure mechanism to protect the incoming and stored data in Agricultural Supply Systems

The agricultural supply chain based on the technology of the IoT is a chain of setting production, storage, distribution and retail in one to provide a traceable RFID carrier to confirm the authenticity of the agricultural products. Relevant information about agricultural products can be found from tag RFID. This will face the producers directly and eliminate the fake products completely. It will strengthen quality control and purify agricultural market by controlling the import and export channels. In the IoT, resource-constrained things are connected to the unreliable and untrusted Internet via IPv6 and 6LoWPAN networks. Therefore, ensuring the security issues during the information retrieval of agricultural products is one of the critical things. Of the functionalities the system offers, security is of most importance in order to convey a sense of trust and reliability in the usage of the system. The farms must be assured that while the data is stored in the "cloud", it remains private. This can be ensured through the use of robust fraud and intrusion detection mechanisms. While some of these security elements have already been resolved, the following still requires attention:

- Authentication and authorization: A single sign on the cloud, allowing access to several functions but still maintaining effective authentication and authorization protection
- Privacy management: Mechanisms ensuring that external applications cannot subscribe to, or access data without the consent of the data's owner.
- Service registry and repository: Identities are managed in a service registry and repository and able to be accessed within the repository for publication and management when applicable.
- Confidentiality and integrity is maintained during the payment process through the application of an effective Revenue settlement and sharing system.

Therefore, it is important to not only protect communication and networks but to also safeguard the stored sensitive data in an IoT device, particularly for agricultural logistics applications. IoT requires multi-faceted security solutions where the communication is secured with confidentiality, integrity, and authentication services. However, there is no mechanism or intrusion detection system, developed for securing IoT in agricultural supply chain system. Such a mechanism is needed to ensure that the delivery and usage of services is trustworthy and meets security and privacy requirements. Therefore, there is a need to develop (i) a lightweight IDS for 6L0WPAN networks that use RPL as routing protocol in the IoT for agricultural supply chain system; and (ii) develop a lightweight yet highly reliable data security mechanism in the context of IoT in agribusiness.

Step 4: Advanced analytics applied to Big Data for knowledge discovery using agricultural product information

As a part of the integration and mining of Big Data, this research develops a methodology to resolve, aggregate and integrate information automatically utilizing experts' knowledge system via the methodologies of Knowledge Representation and Reasoning. The Local farm management systems (FMS) mainly aggregates sensor values collected through its interface of the configuration and communication module, it can send commands through the same interface and also it can take control of the overall management if the link to the Internet is not operational. In this project, all the arguments received by different agriculture experts will be analyzed to resolves conflicts and build argumentation trees separating and linking the arguments in favor and against the situation under consideration, and generate a report for decision maker to support the decision making process depending on the agricultural product information.

35.5   Conclusions and Future Work

Food losses in industrialized countries are as high as in developing countries, but in developing countries more than 40% of the food losses occur at post-harvest and processing levels, while in industrialized countries, more than 40% of the food losses occur at retail and consumer levels. Food waste at consumer level in industrialized countries (222 million ton) is almost as high as the total net food production in sub-Saharan Africa (230 million ton) [10]. This paper proposed an architecture for smart food security systems. Future work will involve validating the applicability of the proposed food security architecture by creating an intelligent agricultural data monitoring system. The validation process will be two-phased. During the first step, the developed system will be tested based on synthetic generated information. The system

will be further refined according to the results obtained. The actual system will be deployed and applied in agricultural setting. The results obtained from the field tests will be used to further refine the developed system. It is envisaged that such a two-phase comprehensive testing would ensure the validity and practicality of the developed methods as future work.

References

1. Savvas, A.: Farming industry must embrace the Internet of Things to "grow enough food". http://www.techworld.com/news/big-data/farming-industry-must-embrace-internet-of-things-3596905/ (2015)
2. Ashford, W.: IoT could be key to farming, says Beecham Research. http://www.computerweekly.com/news/2240239484/IoT-could-be-key-to-farming-says-Beecham-Research (2015)
3. Sørensen, C.G., Fountas, S., Nash, E., Pesonen, L., Bochtis, D., Pedersen, S.M., Basso, B., Blackmore, S.B.: Conceptual model of a future farm management information system. Comput. Electron. Agric. **72**(1), 37–47 (2010)
4. Thessler, S., Kooistra, L., Teye, F., Huitu, H., Bregt, A.: Geosensors to support crop production: current applications and user requirements. Sensors. **11**, 6656–6684 (2011)
5. Grosicki, E., Abed-Meraim, K., Hua, K.Y.: A weighted linear prediction method for near-field source localization. IEEE Trans. Signal Process. **53**(10), 3651–3660 (2005)
6. Lavate, T.B., Kokate, V.K., Sapkal, A.M.: Performance analysis of MUSIC and ESPRIT DOA estimation algorithms for adaptive array smart antenna in mobile communication. Int. J. Comput. Netw. **2**(3), 152–158 (2010)
7. Laurikkala, J., Juhola, M., Kentala, E.: Informal identification of outliers in medical data. In: The Fifth International Workshop on Intelligent Data Analysis in Medicine and Pharmacolog (2000)
8. Guha, S., Rastogi, R., Shim, K.: CURE: an efficient clustering algorithm for large databases. In: Proceedings of the 1998 ACM SIGMOD International Conference on Management of Data, vol. 27, issue: 2, pp. 73–84 (1998)
9. Öhgren, A.: Ontology development and evolution: selected approaches for small-scale _application contexts. Tech. Rep. **2004**, 7 (2004). School of Engineering, Jönköping University, _JTH, Computer and Electrical Engineering
10. Jenny, G., Christel, C., Sonesson, U., van Robert, O., Alexandre, M.: Global food losses and food waste. Food and Agriculture Organization of the United Nations Rome (2011)

Making Music Composing Easier for Amateurs: A Hybrid Machine Learning Approach

36

Jiaming Xu

36.1 Introduction

Recent advances in computing technologies have led to major developments in musical technology. Many machine learning algorithms and human-computer interfaces have been developed to promise new perspectives into computer-assisted music composition. However, the substantive use of music technologies have been limited to music professionals. The full potential of music technology has yet to be unlocked to make playing and writing music accessible to musically untrained users, so music entertainment and education has yet to be promoted among the general population. Existing machines learning methods for composing music are not suitable for taking the input by amateur users into account while maintaining musical appropriateness. Supporting musical creativity of untrained people has important social implications and requires multi-disciplinary research in the interface of music technology, machine learning, and human-computer interactions.

Our goal is to make music entertainment and education accessible to musically untrained people. More specifically, we propose a model that takes users' original musical ideas in an intuitive way, automatically modifies the original musical scores, and outputs musically appropriate melodies.

To achieve this, we propose a model that unites the Hidden Markov model (HMM) with the Recurrent neural networks (RNN) that assists in the composition of music for untrained users. The HMM allows us to take users' original music score as input and generate a melody that conforms to music harmony and retains user-specified preferences. Convolutional neural networks (CNN) preserve the music semantics and Long Short-Term Memory neural networks capture the time dependency or consonance in the context of a whole measure. Thus, the hybrid model leverages the power of deep learning and builds a holistic model that learns both the local and global correlations of music notes. Trained on a large existing digital library of different genres of music, the hybrid deep learning algorithm is further extended to allow users to specify magnitude of revision, duration of music segment to be revised, choice of music genres, popularity of songs, and co-creation of songs in social settings. These extensions enhance users' music knowledge, enrich their experience of self music learning, and enable the social aspects of making music.

We make the following contributions to the literature and practice of computer-aided music making. First, unlike the majority of music composition literature that focuses on automatic music composing by computers only, we propose a deep learning algorithm that takes human creation as its input and modifies the input according to musical rules and user preference. Thus, the model focuses on the human side instead of the machine side. Second, we are the first to apply a hybrid model combining an HMM, a CNN, and an LSTM neural network to music composition. The hybrid approach performs well in improving the speed of parameter estimation and increasing music variety. Third, the model is further extended to allow users to specify the magnitude of revision, duration of music segment to be revised, choice of genres, and popularity of songs. Finally, we also explore the technology and business aspects of this research. Our model could be applied to developing an app that allows for the co-creation of songs in social gatherings and concert halls. As presented in the Appendix, we develop the technical specifications of a mobile application and a plan for launching the application. The tools and user interface we

J. Xu (✉)
Columbia Business School, New York, NY, USA
e-mail: jx2355@columbia.edu

© Springer Nature Switzerland AG 2019
S. Latifi (ed.), *16th International Conference on Information Technology-New Generations (ITNG 2019)*,
Advances in Intelligent Systems and Computing 800,
https://doi.org/10.1007/978-3-030-14070-0_36

develop are intuitive, interactive, and flexible, suitable for the elderly and young children.

36.2 Review of Machine Learning Approaches in Computer-Aided Music Composing

36.2.1 Maintaining the Integrity of the Specifications

Several promising approaches to music composition have been developed in recent years, and algorithmic music composition has made remarkable progress. HMMs and RNNs are the most popular ways to make algorithmic music. Because of its ability to efficiently capture the transitions between musical notes, Hidden Markov chains have been widely adopted to generate new musical compositions. Power, Falk, and Chau use the HMM to compose German and American folk music involving human perception[1]. Gillick, Tang and Keller cluster similar abstract melodies together as states and use the HMM to generate jazz music [2]. Kitahara and Tsychiya use Fourier transformation to transform a user-generated melodic outline into discrete notes, and the output melody is a balance between user input and transition probabilities trained from the HMM [3]. However, there are several problems that HMM might encounter. First, trained on a set of compositions, HMMs can only produce subsequences that also exist in the original data. Second, the transition probabilities to the next state only depend on the last state. However, in music composition, notes are correlated in both time and pitch dimensions, and typically contains longer history information.

The Long Short-Term Memory network is a special kind of RNN models that is capable of retaining longer term memories than regular RNNs [4, 5]. With recurrent connections, LSTM neural networks can handle a situation where the information lapse is large and thus can exhibit temporal dynamics. Doug Eck first updated this approach by switching from standard RNN cells to LSTM cells and applied his architecture to improvise blues [6]. Bob Sturm used a character-based model and token-based model with LSTM neural networks to generate a textual representation of songs [7]. Liang et al. developed a sequential encoding scheme to compose Bach music with a deep LSTM sequential prediction model without prior knowledge or explicit supervision [8]. Coca, Correa and Zhao combined an LSTM neural network and chaotic composition algorithm [9].

Widely used in image compression and recognition, CNNs can compress an image and keep its features because of their shared-weights architecture and translation invariance characteristics [10]. However, this model has not been formally introduced to computer-aided music composition.

Recently, there has emerged sparse literature on the combination of HMMs and RNNs for increasing the interpretability of deep learning algorithms. The majority focuses on training RNNs to predict HMM states [11]. Krakovna and Doshivelez developed a hybrid HMM-RNN algorithm to increase the interpretability of RNNs and jointly train the hybrid model [12]. Tran et al. used the hybrid model in text mining and discovered that it greatly improves the prediction speed and accuracy [13]. Deng and Kwok proposed a hybrid Gaussian-HMM-Deep-Learning approach to estimate automatic chords [14]. In our work, we propose a model combining an HMM, a CNN, and an LSTM neural network to improve music composition consonance. We make the first attempt to apply the unified model to human-aided music composition.

36.3 Hidden Markov Model for Melody Generation

The easiest way for musically untrained users to express their music creations is to draw a melodic outline on two lines. The top of the line represents high-pitched notes while the bottom represents low-pitched notes. In our work, we want to develop a model that can take the user's music creation as input and then maximize the consonance of generated music while maintaining the user's original music ideas.

Hidden Markovian Model has been proven to be successful in a variety of tasks. It models a sequence of stochastic processes with probability distributions performs well in various tasks, including speech recognition and music composition [2, 15, 16]. In this work, we use HMM as the basic music composition framework.

36.3.1 The Hidden Markov Model

An HMM models a stochastic process where states are assumed to be unobservable. It represents a system with a sequence of states using conditional probabilities to model transitions between states [17, 18]. In our proposed HMM, we assume that every note state is generated by a latent state, and the current state at time t is conditioned on the $t-1$ state (this assumption will be relaxed in section 36.4.3). The transition between note states is a Markov process. An HMM can find a sequence of notes that balances the trade-off between the closeness to the user's melodic outline and the musical appropriateness while maximizing the consonance of a piece of music based on the observed sequence of notes generated by the user. The closeness to the outline is captured by the emission probabilities, and the musical appropriateness is captured by the transition probabilities.

Specifically, an HMM consists of five components:

(1) An initial state probability distribution.
(2) A space of NO observed state features:
$$S_O = O_1, O_2, O_3, \ldots, O_M.$$
(3) A space of NH discrete and hidden stochastic states representing the characteristic embeddings of note characteristic:
$$S_H = H_1, H_2, H_3, \ldots, H_N.$$
(4) A set of transition probabilities among stochastic states:
$$P(H_i \text{ at time t}| H_j \text{ at time t}-1), H_i, H_j \in S_H.$$
(5) A set of emission probabilities between hidden states and observed features:
$$P(O_i \text{ at time t}| H_j \text{ at time t}), O_i \in S_O, H_j \in S_H.$$

We use a vector h_t to represent the latent note states at time t and a vector o_t to represent the observed states. The observed states are obtained from user input. Each vector includes note characteristics. The sequence of notes chosen by the user during T time periods will become the observed note trajectory $O = O_1 O_2 \ldots O_T$, and we want to find the optimal sequence of latent classes, $H = H_1 H_2 \ldots H_T$, given an observed note sequence. The observed notes are connected by the underlying path of hidden states. The goal is to maximize the likelihood function of observed sequence of notes multiplying transitions and emissions across all time steps:

$$p(O_1 = o_1, \ldots, O_T = o_T)$$

$$= \sum_{H_1=1}^{NH} \sum_{H_2=1}^{NH} \cdots \sum_{H_T=1}^{NH} P(H_1 = h_1)$$

$$\prod_{t=2}^{T} p(H_t = h_t | H_{t-1} = h_{t-1}) \prod_{t=1}^{T} p(O_t = o_t | H_t = h_t)$$

$$(36.1)$$

In Fig. 36.1, the empty circles denote the notes submitted by the users and the shaded ones denote latent states.

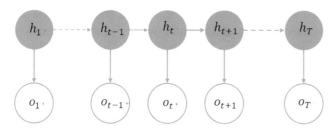

Fig. 36.1 Hidden Markov model

36.3.2 Transition Matrix for Musical Appropriateness

The transition probability $P(h_t|h_{t-1})$ governs the musical appropriateness at time t given the note sequence at time $t - 1$. More specifically, the $NH \times NH$ transition matrix at time t is defined as:

$$\begin{bmatrix} p_{t11} & p_{t21} & \cdots & p_{tNH1} \\ p_{t12} & p_{t22} & \cdots & p_{tNH2} \\ p_{t13} & p_{t23} & \cdots & p_{tNH3} \\ \vdots & \vdots & \vdots & \vdots \\ p_{t1NH-1} & p_{t2NH-1} & \cdots & p_{tNHNH-1} \\ p_{t1NH} & p_{t2NH} & \cdots & p_{tNHNH} \end{bmatrix}$$

Each element p_{tij} in the transition matrix represents the probability of transiting from state s at $t - 1$ to state s at time t. Hence, $0 \leq p_{tij} \leq 1$ and the row sum is one. Thus, the transition among states is represented by a first-order discrete-time and discrete-state HMM.

36.3.3 Emission Probability for Closeness

The emission probability $P(O_t = o_t | H_t = h_t)$ captures the closeness between user input and music appropriateness at time t. In user music composition, O_t is the note vector generated from a Fourier transformation based on user input. The emission probability is conditional on a hidden state k, the probability that the user input notes are observed. Traditionally, generalized EM can be used to estimate the HMM parameters when posteriors are tractable [19]:

$$p(O) = \sum_{H} P(O, H)$$

$$= E_{q(z)}[\ln p(O, H|\theta)] + H[q(H)] \qquad (36.2)$$

$$+ KL(q(H)||p(H|O, \theta))$$

Since $H[q(H)]$ is a constant, and $q(H)$ is chosen to set $KL(q(H)||p(H|O, \theta))$ to zero. Thus, updating θ only requires maximizing $E_{q(z)}[\ln p(O, H|\theta)]$. The gradient of joint probability is defined as:

$$J(\theta) = \sum_{H} p(H|O) \frac{\partial \ln p(O, H|\theta)}{\theta} \qquad (36.3)$$

Previous works focus on applying neural networks or HMMs separately to music composition [20–23]. In the next section, we incorporate deep learning neural network approaches, an LTSM neural network with a CNN into the HMM.

36.4   A Hybrid of the HMM and Neural Network Algorithms

HMMs are subject to several limitations. First, the Markov Process is memoryless. Each timestep is only affected by the information from one timestep earlier. Music consists of melodies, and in the composing process, composers regard each melody as an indivisible whole. If the melody is separated, harmony, which is the most important element of music, will not exist because the notes are not correlated. Not only are the adjacent notes related, but each note in the melody is related. As a result, HMMs suffer from a lack of global structure. Second, training an HMM is complex, and the training process can easily get stuck at local optimum[24]. Furthermore, the input vectors have high dimensions and they are usually sparse. Sparse vectors may take up a large computation space and lower computation speed and accuracy.

We combine HMMs with RNNs and propose a hybrid Neural Hidden Markov Model to train transition probabilities and emission probabilities.

36.4.1 LSTM Neural Networks

For regular RNNs, they have unsatisfying performance on memorizing long period of information [25, 26]. An LSTM neural network is a special case of RNNs and it can handle the situation where information time lap is quite large. Using LSTM neural networks to calculate the HMM state posteriors makes LSTM more interpretable [6, 27]. We combine the LSTM outputs and HMM state probabilities to obtain a final prediction. In previous research, there are two ways to combine HMM and LSTM neural networks. The first method is to put all latent and observed variables into the neural network, and its final layer outputs provide transition and emission probabilities, which are the inputs of the HMM. The second method is to estimate the parameters of neural networks and HMMs jointly by combining the HMM factorization of joint probability $p(O_t, H_t)$ with gradient J [13]. The output of the LSTM neural network are re-scales gradients. In our work, we choose the second method as joint estimation introduces uncertainty in gradients and increases converge speed.

36.4.2 Convolutional Neural Network

In reality, multiple notes can be played at each timestamp. To preserve the richness of musical effects, we allow the music input at each timestep to be a high dimensional vector that contains music features such as pitch, beat, and context. For music consonance, the number of notes that can be played is limited to a few choices. Therefore, the input vectors might be quite sparse. The sparse vectors often take up large computation space and suffer from Curse of Dimensionality. CNNs are effective in both allowing for the high dimensional input vectors and overcoming the sparseness of these inputs.

Thus, we further extend the hybrid model to a CNN framework that can compress the vector of note characteristic. Using a convolutional kernel with a pre-specified width allows the model to automatically compress vectors and keep the main features at the same time.

The first layer of CNN is the original characteristic embeddings. Then a convolutional layer with multiple filter matrices of different widths and a max-over-time pooling operation is applied to obtain a compressed and fixed dimensional expression of original input vector. Different filters capture different kinds of characteristics. In this way, the original input matrix is compressed into a dense matrix, and each element is represented by continuous variables. Note, we compress the vector of note characteristics and use the note number to represent the index of clusters described.

36.4.3 Clustering Algorithms

In previous research, a hidden state in an HMM is assumed to correspond to a single note. If there is only one note at each state, then the music generated would be quite boring. To make the music output more interesting, we relax this assumption by allowing each state to be a melody that is clustered manually. Since there are various music genres and musicians have their own composing style, manually clustering melody would increase "surprising" factors of composition style.

To increase the variety of the melody generated, we cluster similar compressed vectors with an unsupervised clustering algorithm. Each cluster has a unique index indicating the note number that appears most frequently as well as the genre of this cluster. Clustering algorithms is based on the Euclidean distance measure between vectors. We calculate the statistics that represent each cluster and use clusters as hidden states of the HMM. The transitions are between clusters and the ultimate output at each timestep is randomly chosen from the given cluster[1]. In order to add "surprising" variation of music, we allow the ultimate output at each timestep to be randomly chosen from a cluster. Thus, while the HMM and LSTM hybrid model ensure music consonance, and the random extraction algorithm enriches music diversity. The clustering algorithms are repeated for several iterations until few vectors switch clusters between each iteration.

36.4.4 Producing Transition Probabilities with the LSTM Neural Network

Let $c_i \in R^D$ be the vector embedding of cluster i and $n_k \in R^D$ be the note vector embedding processed by the convolutional neural network. There are P elements in cluster i. The emission probabilities are given by:

$$P(\mathbf{n}_k|\mathbf{c}_i) = \frac{\exp(\mathbf{c}_i^T)\mathbf{n}_k + b_k}{\sum_{p=1}^{P}\exp(\mathbf{c}_i^T)\mathbf{n}_p + b_k} \quad (36.4)$$

In our work, we use LSTM neural networks to calculate the emission probability. n_k is the output layer of the LSTM network.

When training of the transition probabilities, the LSTM neural network receives the output of CNN as input. It performs better at solving the learning long-term dependencies than regular RNNs. The output is denoted as conditional probability distribution $P(s_t, s_{1:t-1})$, where $s_{1:t-1} = [s_1, s_2, , s_{t-1}]$.

36.4.5 Training the LSTM Neural Network

According to Ke et al., the gradient can be written as [13]:

$$J(\theta) = \sum_{s}[p(\mathbf{s}|\mathbf{o})\frac{\partial p(\mathbf{s}, \mathbf{o}|\theta)}{\partial \theta}$$
$$= \sum_{t}\sum_{h_t}[p(\mathbf{s}_t|\mathbf{o})\frac{\ln p(\mathbf{o}_t|\mathbf{s}_t, \theta)}{\theta} \quad (36.5)$$
$$+ p(\mathbf{s}_t, \mathbf{s}_{1:t-1}|\mathbf{o})\frac{\partial \ln p(\mathbf{s}_t|\mathbf{s}_{1:t-1}, \theta)}{\partial \theta}$$

The posterior $p(\mathbf{s}_t|o)$ and $p(\mathbf{s}_t, \mathbf{s}_{1:t-1}|\mathbf{o})$ are obtained by the Baum-Welch algorithm, and $p(\mathbf{o}_t|\mathbf{s}_t, \theta)$ and $p(\mathbf{s}_t|\mathbf{s}_{1:t-1}, \theta)$ are trained by the LSTM neural network. Thus, the gradient is repeatedly recomputed until a convergence threshold is reached [28].

In summary, we propose a hybrid of an HMM, an LSTM neural network and CNN. The model has the following characteristics: (1) it takes the user's original music idea as input and involves users' creativity; (2) the prediction is based on a longer time series of the original score; (3) it allows multiple notes to be planned at the same time and allows the selection of a coherent chord. The whole structure of our model is shown in Fig. 36.2.

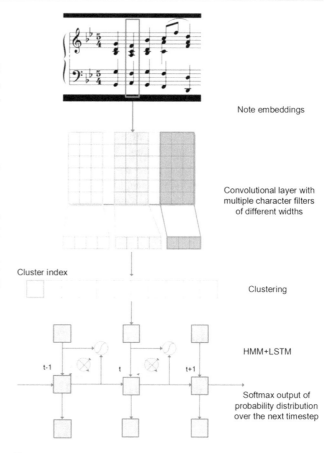

Note embeddings

Convolutional layer with multiple character filters of different widths

Cluster index

Clustering

HMM+LSTM

Softmax output of probability distribution over the next timestep

Fig. 36.2 Structure of the proposed hybrid model

36.5 Training NHMM from Existing Music Data

In training the model, a network's adaptable weights are initialized with random values drawn according to some distribution. Using numerical optimization methods such as gradient descent techniques and simulated annealing, the network is trained on the MIDI files of existing songs to perform a certain task until some training criterion is met.

36.5.1 LSTM Neural Networks Training

The timestep of input and output is a 1/16th beat. The input vector at time t is v_{it}, containing the following characteristics:

(1) Notes played simultaneously: The MIDI note value of each note played at time t. In our work, we consider 36 note numbers in 3 octaves.

(2) Last timestep notes: The MIDI note value of each note played at time $t-1$. We take the same 36 note numbers at time t.

(3) Consonance: Whether it is a single note at time t, a major chord, or a minor note.

(4) Previous context: The value of index k is the number of times a note with pitch index k was played during the last timestep at time $t-1$. There are 12 notes in an octave. Pitch names repeat every 12 notes. The index of C's is 1 and that of B flat's is 12. If there are 3 C's during the last time step, the value at index 1 would be 3.

(5) Beat: The position in the bar. Each row is one of the beat inputs and each column is a time step that repeats following the pattern:

$$0101010101010101$$

$$0011001100110011$$

$$0000111100001111$$

$$0000000011111111$$

36.5.2 Generating Music with the NHMM and User Input

After a user draws a melody outline, the Fourier transform is applied to the outline, and the inversion Fourier transformation is applied to obtain the note sequence in MIDI format. The note vector at timestep t is projected as a cluster index and is denoted as $O_t = \mathbf{o}_t$.

At each timestep, the model selects the hidden state that maximizes the joint probability $p(O_t = \mathbf{o}_t | H_t = \mathbf{h}_t) p(H_t = \mathbf{h}_t | H_{t-1} = \mathbf{h}_{t-1})$ with parameters trained by NHMM and then randomly chooses a vector from the hidden state as the output at time t.

In addition to imposing music regulation, we can also train the switching probabilities on different music genres. With different training data sets, it can recognize different music patterns such as jazz and classical.

36.6 Conclusion

Musical neurophysiologists have conducted extensive experiments and have demonstrated the profound effects of making music on the functioning of the human brain. They have found that the music not only improves participants' well-being and decreases depression, but also stimulates thinking, memory, combination skills and mental control. More interestingly, they illustrate that making music is far more incisive and formative in improving quality of life than listening to music.

Technology has made the learning of music theory and composing accessible for every user. We propose a promising algorithm that makes music composition available to musically untrained users and helps them better understand music theory. The framework we propose unifies probabilistic model (HMM) and a deep learning algorithm (RNN), incorporates long-term memory and semantics of music (LSTM and CNN). To the best of our knowledge, there is no related work that applies both the hybrid model and joint estimation method to human-aided music composition. The proposed solution fully takes user original creation into consideration, modifies the raw scores with machine learning technologies, and outputs musically appropriate melodies that reflect users' original creation. We also design an intuitive mobile user interface with senior citizens and young children as the initial users.

This research is a multi-disciplinary study in the interaction of statistics, machine learning, human-computer interfaces, and artificial intelligence. Different from most machine learning and computer-aided music making algorithms, our model recognizes human input and improves the original music creation of musically untrained users. We believe that our approach has a promising future in a real application.

Appendix

For users untrained in music, the design of the user interface should be intuitive, interactive, and visual. In our research, we mainly focus on resolving two issues. First, we design a human interface is suitable for musical untrained people to implement their musical ideas. Second, the music composed by our model can maintain user intuition and keep music consonance as much as possible. Machine learning technologies are expected to play a significant role for the second issue.

A fully automatic music composer is a system that generates a musical piece when the user presses the RUN button (often with a few parameter inputs). Typically, such software reads the inputs from the user, such as a sequence of the pitches, onset times, and offset times of notes to be played in a time-pitch plane displayed on the computer screen. After the user hits the RUN button, the sound of each computer-generated note will be played by the chosen instruments such as piano, ocarina, and harmonica. Given the advantage of the HMM in taking user input, our model offers users the flexibility of imposing a user's preferences on the computer algorithm.

(1) Users' Choice of Magnitude of Revision

The balance between the closeness to the outline and musical appropriateness can be controlled by changing σ^2.

(2) Users' Choice of Music Genre

We train the HMM with music data of different types of music such as jazz, classical, pop, and rock and roll. The difference among music genres is reflected by the transition matrix $P(h_t|h_{t-1})$. This allows the users to select styles according to their preferences.

(3) Incorporate Popularity of the Song

Our model can incorporate information about the popularity of a song and allow users to choose to incorporate the characteristics of the music that are deemed popular. This can be done by training the model on selected songs. Additional information from the Hits Board or ranking on social media can be incorporated into the model.

(4) Allow for Crowd-Making of Songs

Our model recognizes multiple notes played at the same time as input. This permits our model to be expanded to allow several users to input their co-creation, revise as an integrated piece, and then play the result as one melody.

(5) Rate the Music Creation

To help the users gain more knowledge, we measure how well the initial music creation approximates the target style chosen by the users as well as the incidences where musical regulations are violated. This can be done using the intermediate output from our model. Furthermore, we can also build a community and invite other users to comment and rate the musical pieces created by untrained users.

References

1. Power, S.D., Falk, T.H., Chau, T.: Classification of prefrontal activity due to mental arithmetic and music imagery using hidden Markov models and frequency domain near-infrared spectroscopy. J. Neural Eng. **7**(7), 26002 (2010)
2. Gillick, J., Tang, K., Keller, R.M.: Machine learning of Jazz grammars. Comput. Music J. **34**(3), 56–66 (2010)
3. Kitahara, T., Tsuchiya, Y.: A Machine Learning Approach to Support Music Creation by Musically Untrained People
4. Mozer, M.C., Soukup, T.: Connectionist music composition based on melodic and stylistic constraints. In: Advances in Neural Information Processing Systems (1990)
5. Hochreiter, S., Schmidhuber, J.: Long short-term memory. Neural Comput. **9**(8), 1735–1780 (1997)
6. Eck, D., Schmidhuber, J.: A first look at music composition using LSTM recurrent neural networks. Istituto Dalle Molle Di Studi Sull Intelligenza Artificiale (2007)
7. Sturm, B.L., et al.: Music transcription modelling and composition using deep learning (2016)
8. Liang, F., et al.: Automatic stylistic composition of Bach Chorales with deep LSTM (2017)
9. Coca, A.E., Correa, D.C., Zhao, L.: Computer-aided music composition with LSTM neural network and chaotic inspiration. In: International Joint Conference on Neural Networks (2014)
10. Zhang, W., et al.: Parallel distributed processing model with local space-invariant interconnections and its optical architecture. Appl. Opt. **29**(32), 4790–4797 (1990)
11. Mirghafori, N., Morgan, N., Bourlard, H.: Parallel training of MLP probability estimators for speech recognition: a gender-based approach. In: Neural Networks for Signal Processing (1994)
12. Krakovna, V., Doshivelez, F.: Increasing the Interpretability of Recurrent Neural Networks Using Hidden Markov Models (2016)
13. Ke, T., et al.: Unsupervised Neural Hidden Markov Models (2016)
14. Deng, J., Kwok, Y.K.: Automatic Chord estimation on sevenths-bass Chord vocabulary using deep neural network. In: IEEE International Conference on Acoustics, Speech and Signal Processing (2016)
15. Verbeurgt, K., Dinolfo, M., Fayer, M.: Extracting Patterns in Music for Composition via Markov Chains, pp. 1123–1132. Springer, Berlin/Heidelberg (2004)
16. Trentin, E., Gori, M.: A survey of hybrid ANN/HMM models for automatic speech recognition. Neurocomputing **37**(1), 91–126 (2001)
17. Rabiner, L., Juang, B.: An introduction to hidden Markov models. Current protocols in bioinformatics/editoral board, Andreas D. Baxevanis ... [et al.], 2007. Appendix 3(Appendix 3): p. Appendix 3A
18. Netzer, O., Lattin, J.M., Srinivasan, V.: A hidden Markov model of customer relationship dynamics. Mark. Sci. **27**(2), 185–204 (2008)
19. Dempster, A.P., Laird, N.M., Rubin, D.B.: Maximum likelihood from incomplete data via the EM algorithm. J. R. Stat. Soc. **39**(1), 1–38 (1977)
20. Mozer, M.C.: Induction of multiscale temporal structure. In: International Conference on Neural Information Processing Systems (1991)
21. Todd, P.M., Loy, G.: Creation by Refinement and the Problem of Algorithmic Music Composition (1991)
22. Hild, H., Feulner, J., Menzel, W.: HARMONET: a neural net for harmonizing chorales in the style of J. S. Bach. In: Advances in Neural Information Processing Systems (1991)
23. Bellgard, M.I., Tsang, C.P.: Harmonising music using a network of Boltzmann machines (1992)
24. Johnson, M.: Why doesn't EM find good HMM POS-taggers? In: EMNLP-CoNLL 2007, Proceedings of the 2007 Joint Conference on Empirical Methods in Natural Language Processing and Computational Natural Language Learning, Prague, 28–30 June 2007
25. Sutskever, I.: Training Recurrent Neural Networks. Doctoral (2013)
26. Liu, I., Ramakrishnan, B.: Bach in 2014: Music Composition with Recurrent Neural Network. Eprint Arxiv (2014)
27. Bengio, Y., et al.: Global optimization of a neural network-hidden Markov model hybrid. IEEE Trans. Neural Netw. **3**(2), 252–259 (1992)
28. Welch, L.R.: Hidden Markov models and the Baum-Welch algorithm. IEEE Inf. Theory Soc. Newsl. **53**(2), 194–211 (2003)

## Part IV

High Performance Computing Architectures

A Generator and Corrector of Parametric Questions in Hard Copy

Francisco de Assis Zampirolli, Fernando Teubl, and Valério Ramos Batista

37.1 Introduction

Test generators have been easing the arduous tasks of elaborating and correcting numerous questions for exams given by teachers and panels. Usually these generators make tests automatically according to a databank of questions and predefined criteria, so that each student sits the exam with the same content but in a different version.

A test generator can include at least one of the following steps: (a) performing a random choice of questions from a databank, (b) making different versions of a question either in its statement (*parametric*) or in the order of answers (*multiple-choice*), (c) formatting and consolidating individual tests, and (d) following automatic methods to correct the tests, these either online or in hard copy.

Steps (a) and (c) are relatively simple, whereas (b) and (d) are pretty challenging. For instance, in (b) the user is frequently a teacher who must establish well-defined criteria upon which the question can be changed, so that the different answers are achieved with the same difficulty. Such parametrizations may require technical skills the users do not have unless they undergo some training to elaborate questions, mainly of the sort that fits a specific course. This is a typical requirement that risks turning inviable the popularisation of parametric question generators.

Regarding (d), on the one hand it is simple for tests carried out electronically. On the other hand, for handwritten tests it turns out to be a laborious step. The automatic corrector has to interpret and recognize the answers through computer vision and compare them with the answer keys. Usually each version of the test corresponds to a different answer key.

This paper introduces one of our contributions to the research area of Education. Herewith we present an automatic generator and corrector of random tests with parametric questions, which is devoted to exams in hard copy. Our programs were implemented in Python, together with open-source libraries, and they are applicable to large classes that consist of students in hundreds. Of course, for such a class any exam requires efficient processes of massive test generation and correction. Our present work contributes to meet the increasing demand for massive test generators that also ensure the integrity of the students' marks. There are several other similar generators in the literature but herewith we propose an accessible and free-of-charge program that also includes a new method to encode parametric questions.

37.2 Related Works

There are several test generators in the present day. Some of them include built-in tools for parametric questions and automatic correction of tests. As mentioned at the Introduction these features are already implemented in our proposal but we not only strive for a user-friendly platform with graphical interface: our program can be used for free and the encoding of the answer keys is inviolable.

Now we briefly describe some of these generators.

The *QuizPACK*[1] is a system developed in C programming language for both production and correction of questions. Questions must be parametrized by the teachers themselves directly in C language. Afterwards they have to upload

F. de Assis Zampirolli (✉) · F. Teubl · V. Ramos Batista
Centro de Matemática, Computação e Cognição, Universidade Federal do ABC (UFABC), Santo André, Brazil
e-mail: fzampirolli@ufabc.edu.br; fernando.teubl@ufabc.edu.br; valerio.batista@ufabc.edu.br

[1] *Quizzes for Parameterized Assessment of C Knowledge.*

S. Latifi (ed.), *16th International Conference on Information Technology-New Generations (ITNG 2019)*,
Advances in Intelligent Systems and Computing 800,
https://doi.org/10.1007/978-3-030-14070-0_37

their program files into the system. Parameters are identified through a pseudo-variable Z in the code, and then steadily replaced with random integers while generating the tests. These integers range in an interval given by the user [1, 2].

In [3] the author describes his *Matlab Implementation of the Automatic Generator of the Parameterized Tasks*. He developed a self-sufficient structuring in XML to write parametrized questions. There one can generate questions whose answers may be numerical, phrasal, multiple-choice or cloze. The generating process begins with the uploading of a text whose parameters are marked with *tags*, namely <input> containing both name and type of the variables [3]. After uploading the questions in XML the generator will produce the tests with their parametrized questions by attributing random values to these variables within a certain interval. The final output is also written in XML but can be re-formatted, for example into *Moodle XML* or LATEX [3].

SmartQuestion (sQ) is a software that automatically generates questions with multiple-choice answers. It follows the concept of "a way to store all static versions of a question in the same parametric question". This tool offers three tabs: sQd (*sQ design*), sQp (*sQ parameterize*) and sQg (*sQ generate*) [4]. The software sQ starts with the *bitmap* or *JPEG* image of the static question given by the users. With sQd they can define a place for each parameter, which may be either in the statement of the question (regular) or in the answer (for the alternatives). With sQp the users select parameters and insert new values in a textbox compatible with LATEX. Then sQg makes sub-images of the changed parameters and incorporates them into the respective places. The final result is a complete image of the new question. With sQg one can also shuffle the alternatives so that versions vary in both statement and order of answers [4].

MEGUA (*Mathematics Exercise Generator, Universidade de Aveiro*)[2] is an open-source software that enables one to make databanks of parametrized questions with their respective answers all in LATEX. It works with the mathematical software *SageMath*, which uses Python programming language [5]. Its question databanks are called "*Books*" and built through either PDFLatex (for hard copy) or HTML and MathJAX (for web publishing) [6]. The elaboration of a question essentially occurs inside the *Notebook* of *SageMathCloud*, whose name has changed to *CoCalc* since May 2017. This one consists of three steps: Firstly, on a new *worksheet* we make a cell to import the whole MEGUA library and open/create a databank in which the questions will be stored. Secondly, the code of the question is introduced into another cell that consists of a text in LATEX and the programming in Python. The LATEX part is divided in sections (cataloguing and description of the

exercise), "%problem" (name and statement of the question) and "%answer" (its solution). Finally *CoCalc* is concluded by the programming part, which contains two functions: "make_random" (generates random values for the statement) and "solve" (computes the right solution and gives others for the multiple-choice). The output of this cell execution is two files, one in PDF and other in TEX [7]. There is also a resource to add parametrized graphs to the exercises. However, MEGUA is not endowed with automatic correction of parametric questions in hard copy, a desirable feature at evaluating hundreds of candidates.

TestMakPro3 is a commercial software that generates parametric questions. It was implemented in *JavaScript* [8] and enables the user to elaborate the statement of a question with at most three variables, together with their range of values. For each variable we set its margin of error and its number of decimals to be given in each answer.

AMC is a free software to produce and manage multiple-choice quizzes, which also performs their automatic correction. Tests are written in LATEX but any user unfamiliar with this document preparation system can give them written in a special format called *AMC-TXT*, whose syntax is quite simple. One of the facilities of *AMC* is the option to shuffle questions so that each student's version turns out to be unique. Another is an automatic correction of the scanned tests that already brings their marks [9, 10]. But *AMC* does not work with parametric questions. Moreover, though *AMC* includes a database of students the tests are first associated to them only in the correction process, hence each student must fill out their identification data. Corrections are carried out inside the very statements of the questions, so that one has to digitize all exam pages for the correction process.

MakeTests[3] is an open-source program written in Python. Its main purpose is to generate tests out of a totally random choice of questions. For this purpose *MakeTests* enables the user to write questions programmed in Python and to classify them through directories. These correspond to subjects and levels of difficulty. *MakeTests* writes the exam headers and the random questions in PDF through specifications described in a configuration JSON file. The most important feature of *MakeTests* is the production of questions in a totally independent way, since one writes each question directly in Python (interfaced with LATEX). The user can set any type of answers: multiple-choice, true/false, numerical, etc. The greatest difficulty in *MakeTests* is that the user must have advanced knowledge of Python to programme questions, and this is impracticable in several areas of Education. Moreover, *MakeTests* does not include automatic correction: it generates answer keys but the correction is performed manually.

In general all approaches in the literature are somehow efficient at tackling production and/or correction of

---

[2]The name "MEGUA" stands for a trademark of the University of Aveiro since 2012 [5].

[3]Available at https://github.com/fernandoteubl/MakeTests

Table 37.1 Comparison between test generators

	QuizPack	[3]	SmartQuestion	MEGUA	TestMakPro3	AMC	MakeTests	MCTest
Language	C	Matlab		Python	JavaScript	Perl	Python	Python
Commercial					×			
Open-Source						×	×	×
Parametric Questions	×	×		×	×		×	×
Automatic Correction of Hard Copies						×		×
User Has to Programme	×	×		×			×	†

†: Only if you make use of parametric questions.

tests. However, their main common point is the inaccessibility either because the users have to master programming languages (e.g. [1, 6]) and/or to pay for the software (e.g. [3, 8]). Table 37.1 summarizes this section, together with our proposal MCTest,[4] whose previous versions could only circumvent the absence of parametric questions, as explained in Sect. 37.3.1, and also a mechanism for them that produces alternatives without repetition, detailed in Sect. 37.3.2. Our software present day's version is called MCTest 4.1.

37.3 Methods

Our method to produce parametric questions is now presented here. Firstly we describe the MCTest platform in the standard configuration. Afterwards we explain further implementations that now bolster the new functionalities.

37.3.1 MCTest

History

Nowadays the version 4.0 of our free software MCTest is finally available. MCTest has started in 2010 to automate an exam for openings in a specialization course at a Brazilian university. More than thousand candidates enrolled for that exam.

In the present day we apply MCTest to produce and correct over ten thousand tests every year. Version 1 was implemented in Matlab with automatic correction performed by snapping the exams with a computer webcam. Until version 3 the exams were written in traditional text editors. Version 2 was implemented in Java for Android. The 3rd version used Python, still only for correction of tests, which were digitized as input. Finally in the version 4.0 we added production of tests in Python through LaTeX.

In the present version multiple-choice tests are automatically corrected by uploading the answer cards digitized in PDF. This must be done via the ftp server vision.ufabc.edu.br by following the steps explained in a previous publication.

---

[4]Available at https://github.com/fzampirolli/MCTest4

```
QE::topic 1:: Question Q1-example of equation:
$\sin A \cos B =
\frac{1}{2}\left[ \sin(A-B)+\sin(A+B) \right]$
A: answer 1a
A: answer 1b
A: answer 1c
A: answer 1d
A: answer 1e

QE::topic 2::a:: Question Q2
A: answer 2a
A: answer 2b
A: answer 2c
A: answer 2d
A: answer 2e

QE::topic 2::a:: Question Q3
A: answer 3a
A: answer 3b
A: answer 3c
A: answer 3d
A: answer 3e

QM::topic 3::   Question Q4
A: answer 4a
A: answer 4b
A: answer 4c
A: answer 4d
A: answer 4e

QH::topic 4:: Question Q5
A: answer 5a
A: answer 5b
A: answer 5c
A: answer 5d
A: answer 5e
```

Fig. 37.1 Example of questions written for MCTest

Present Day's Version of MCTest

MCTest can produce exams corresponding to students of given classes. The students' data must be given by CSV spreadsheet, and questions are taken from a databank in which they must be written according to a specific template in TXT format (see Fig. 37.1).

In Fig. 37.1 we see five multiple-choice questions written in a TXT file. Their levels of difficulty are assigned as QE (easy), QM (medium) and QH (high). For instance, in

Federal University of ABC
Course: Introduction of Programming - BCM0505 - First Exam
Teacher(s): Name of Teacher
Period: 3/2018   **Modality:** Face-To-Face   **Date:** 25/10/2018

Student: Name of Student 1   **Registration:** 11000123   **Room:** 2018_BC0505_q3_A1

Sig.:_____

```
        A   B   C   D   E
    1   O   O   O   O   O
    2   O   O   O   O   O
    3   O   O   O   O   O
    4   O   O   O   O   O
```

1. Question Q3   A. answer 3a   B. answer 3c   C. answer 3d   D. answer 3b   E. answer 3e

2. Question Q1 - example of equation: $\sin A \cos B = \frac{1}{2}\left[\sin(A-B)+\sin(A+B)\right]$ A. answer 1c   B. answer 1b   C. answer 1e
   D. answer 1d   E. answer 1a

3. Question Q4   A. answer 4e   B. answer 4d   C. answer 4b   D. answer 4c   E. answer 4a

4. Question Q5   A. answer 5a   B. answer 5c   C. answer 5d   D. answer 5b   E. answer 5e

Fig. 37.2 A test generated by MCTest

Fig. 37.1 there are three easy questions, one medium and one difficult. For multiple-choice questions each alternative must begin with "A:", and MCTest will always take the 1st one to compose the whole answer key. Of course, MCTest performs a nested shuffling: of questions and of their *attached* alternatives. The user can also include dissertation questions by toggling them with "QT". MCTest produces all exams written in LaTeX with a custom header. A resulting PDF can be seen in Fig. 37.2, already with the nested shuffling. In Fig. 37.1, when the same character appears after the phrase `topic #::`, for instance "a::", this means that for each student only one of the two questions in Fig. 37.1 will be drawn.

When we just want to change some values manually, either in the statement or in the alternatives of a question, and then perform a random choice of them for each student, the model presented in Fig. 37.1 is inefficient. This is because duplicating and then changing a text manually can lead to errors. Another resource not available in MCTest 4.0 is the automatic computation of the right answer, which should then be included as one of the alternatives.

These are the main reasons for us to carry on version 4.1 of MCTest, which includes methods to generate parametric questions and to compute the right answer, together with the automatic production of false alternatives, as we are going to see in the next subsection.

37.3.2 MCTest with Parametric Questions

In order to implement parametric questions in MCTest we had to include two special delimiters `[[code: ...]]` and `[[def: ...]]`. The former must come in the statement, either of the question or of the alternatives. With `[[code: ...]]` we define the parameters (or variables) to be used at generating each student's test. The latter must be defined at the end of the TXT file. In this case each parametric question must be in a separate file. Moreover, `[[def: ...]]` has to contain the possible values applied beforehand, together with a method called `algorithm`. In the next section we show some examples of parametric questions included in MCTest 4.1 by means of these delimiters.

37.4 Results

By the methods just presented in the previous section, now we illustrate how they solve two practical problems that involve parametric questions.

37.4.1 Parametric Question for ULM

Uniform Linear Motion (ULM) is a typical subject of parametric questions. In Fig. 37.3 we see a complete example for a Physics exam written in TXT.

```
QE::ulm:: % question text
A car moves on a road with an hourly function
$s=[[code:a0]] + [[code:a1]]t$, where $s$ is
given in miles and $t$ in hours. The car passes
the mile [[code:a2]] exactly at:

% answers
A: [[code:correctAnswer]]
A: [[code:correctAnswer-1]]
A: [[code:correctAnswer-2]]
A: [[code:correctAnswer+1]]
A: [[code:correctAnswer+2]]

[[def:
a0 = random.randrange(-6, 3, 1) # return
a1 = random.randrange(3, 8, 1)  # random
a2 = random.randrange(3, 8, 1)  # numbers

def algorithm(a):
    from sympy import *
    a0=int(a[0])
    a1=int(a[1])
    a2=int(a[2])
    s,t = symbols('s,t')
    s=a0+a1*t
    r = float(solve(s-a2,t)[0])
    return r

global correctAnswer
correctAnswer= algorithm([a0,a1,a2])
]]
```

Fig. 37.3 Example of a ULM question

On the first line we classify both the difficulty and the subject QE::ulm::. Notice that we may use LATEX syntax in the statement and in the alternatives. For instance, comments are toggled with % and math symbols enclosed with $. For a multiple-choice question, right after the statement we include each alternative followed by A:. The first must be the right one, according to the MCTest method. Therefore we have A: [[code:correctAnswer]], where correctAnswer is a variable computed in [[def:, which also defines a function called algorithm.

The algorithm must have all variables in the statement as parameters. The ULM problem asks for the solution of an algebraic equation. For that we use the Python library *sympy* and define two variables: s and t. Afterwards one simply writes the corresponding equation s=a0+a1*t. The result is given by the function *solver* of the library *sympy*. Three random variables a0, a1 and a2 come either in the statement or in the alternatives. That is why we have included the function *random.randrange* from Python.

Finally we must assign algorithm([a0,a1,a2]) to the global variable correctAnswer in order to make it store the correct result, which is then inserted in the first alternative. The wrong alternatives were defined as the solution ±1 and ±2. These are arbitrary values but most

```
QE::matrix:: % question text
Build a matrix $[[code:a0]] \times
[[code:a1]]$ whose elements $(i,j)$ are
$((((i+1) * [[code:a2]]) + ((j+1) *
[[code:a3]])) \mod{100})$. Compute the sum
of the entries of this matrix. Indexes $i$
of rows and $j$ of columns begin with $0$.

% createWrongAnswers([5,10]) - Makes
% 5 different wrong alternatives
% between +/- 10
A: [[code:correctAnswer]]
A: [[code:createWrongAnswers([5,10])]]

[[def:
# code to return a correct answer with the
# following variables
def algorithm(a):
  a0 = int(a[0])
  a1 = int(a[1])
  a2 = int(a[2])
  a3 = int(a[3])
  P = np.zeros((a0,a1))
  for i in range(a0):
    for j in range(a1):
      P[i,j] = (((i+1)*a2)+((j+1)*a3))%100
  return int(P.sum())

# variables used in the question
# produces a random number between 60 and 80
a0=random.randrange(60, 80, 1)
a1=random.randrange(60, 80, 1)
# takes a number at random from a set of three
a2=random.choice([7, 13, 19])
a3=random.choice([11, 17, 23])

global correctAnswer
correctAnswer= algorithm([a0,a1,a2,a3])
]]
```

Fig. 37.4 Example of a parametric question on matrix

importantly is not to repeat alternatives. We can resort to another method that produces wrong answers and avoids repetition, as shown in the next example. Figure 37.5 depicts three random outputs of that ULM problem.

37.4.2 Parametric Question to Handle Matrices

Figure 37.4 illustrates an example for handling matrix to evaluate the students' logical programming skills in a chosen language. Dimensions are large because the students have to solve through a program code. They must compute the sum of the entries of a matrix whose dimensions and elements are both randomly taken while MCTest produces each student's test. Differently from the previous question, this one resorts to a new method called createWrongAnswers([5,10]), which produces five distinct random values between correctAnswer-10 and

1. A car moves on a road with an hourly function $s = 1 + 5t$, where s is given in miles and t in hours. The car passes the mile 3 exactly at:
   A. -0.6   B. 0.4   C. 2.4   D. -1.6   E. 1.4
2. A car moves on a road with an hourly function $s = 1 + 4t$, where s is given in miles and t in hours. The car passes the mile 7 exactly at:
   A. 0.5   B. 3.5   C. 1.5   D. 2.5   E. -0.5
3. A car moves on a road with an hourly function $s = -3 + 6t$, where s is given in miles and t in hours. The car passes the mile 3 exactly at:
   A. -1.0   B. 2.0   C. 3.0   D. 1.0   E. 0.0

Fig. 37.5 Example with three outputs of the parametric ULM

1. Build a matrix 68×75 whose elements (i, j) are $((((i+1) * 19) + ((j+1) * 11)) \mod 100)$. Compute the sum of the entries of this matrix. Indexes i of rows and j of columns begin with 0.
   A. 252348   B. 252350   C. 252349   D. 252354   E. 252346   F. 252355
2. Build a matrix 74×74 whose elements (i, j) are $((((i+1) * 19) + ((j+1) * 11)) \mod 100)$. Compute the sum of the entries of this matrix. Indexes i of rows and j of columns begin with 0.
   A. 270897   B. 270891   C. 270898   D. 270893   E. 270900   F. 270896
3. Build a matrix 67×77 whose elements (i, j) are $((((i+1) * 19) + ((j+1) * 23)) \mod 100)$. Compute the sum of the entries of this matrix. Indexes i of rows and j of columns begin with 0.
   A. 255332   B. 255338   C. 255331   D. 255337   E. 255346   F. 255327

Fig. 37.6 Example of three outputs from the matrix parametric question

`correctAnswer+10`. Figure 37.4 shows the text in TXT that made MCTest produce the three outputs in Fig. 37.6.

37.4.3 Experiments

Our new method to generate parametric questions was applied for the course Introduction to Computer Science (ICS) at our university in the second trimester of 2018. On that occasion we had 167 matriculated students and ICS was a half-distance-learning half-classroom course. In the classroom their presence took place 4 times: opening, first test, project and second test. This course belongs to an interdisciplinary bachelor's programme offered by our university, where the students come from many different backgrounds.

Our university's instruction periods are organized in three trimesters. Here all undergraduate students follow two Interdisciplinary Bachelor's Programmes (IBP): Science&Technology and Science&Humanities abbreviated as BCT and BCH, respectively. On average we have 1,600 freshmen in BCT and 400 in BCH annually. ICS is mandatory for all students, it belongs to BCT and is always scheduled for the third trimester.

Both BCT and BCH take three years, and with an additional year the student can graduate either in Maths, in Physics, or in Computer Science. Right after BCT, with two additional years the student can graduate in one of the nine engineering programmes offered by our institution. Since our students come from many different backgrounds, ICS is devoted to teaching logic programming for daily problems on several subjects in the student's life.

In the classroom we gave tests with parametric questions. The first exam was in hard copy with dissertation questions.

The project and the second exam were given in a computer laboratory, where students had to hand in three program codes by uploading them into a virtual learning platform. Each exam consisted of three questions: of easy, medium and high difficulty. Moreover, each question type came in different versions, as depicted in Figs. 37.4 and 37.6. Notice that the versions do not change the logic programming for solving them.

Hence, MCTest enables us to produce several versions of the same question without repetition. We just need one statement as shown in Figs. 37.4 and 37.6. Thus, for each student a different triple of questions was generated (of easy, medium and high levels of difficulty). Some examples of easy level are illustrated in Fig. 37.6.

37.5 Conclusions

The main contribution of our work is a method to generate parametric questions for exams in hard copy. Both production and generation of multiple-choice questions are free softwares, and the former even open-source. Automatic correction of the multiple-choice answer cards follows the method presented in a previous work. To the best of our knowledge MCTest is the first platform totally devoted to exams in hard copy that includes automatic correction, besides production of tests with random answer keys.

In this paper we explained how to produce parametric questions with MCTest. Though it apparently requires some knowledge of both LaTeX and Python, MCTest offers a folder with several templates that can be used by any teacher and any professor, and they just have to adapt the templates to write their personal questions. Moreover, the MCTest exam generator is open-source, and the several examples of parametric questions available in our repositories can not only help teachers and professors check what we have already done, but they can also contribute with new models of questions. With this interaction we hope to achieve a great facilitation in the arduous task of evaluating large classes.

With MCTest we can also give weekly tests without much effort. This helps track the gradual performance of each student through periodical and personal evaluations, as suggested by Gusev et al. [11]. For example, if a student cannot answer questions about a certain subject, then we include more related to questions in the following exam. Our approach also differs from others in the sense that we focus on exams in hard copy, which include parametric questions instead of quizzes on the web (Gasev's proposal does not include parametric questions either).

In this paper we only gave examples of mathematical and programming questions but an online version of MCTest is under development and it will include a great variety of parametrized questions. But the teacher/professor will always have to prepare them in order to make a reasonable test that covers the programme of the course, with an adequate number of questions, alternatives per question and so on. For this task [12] presents many results of preparing questions based on cognitive model.

In a future work we shall implement a web server endowed with access control, for teachers and professors to store their parametric questions through a friendly Graphical User Interface (GUI). Parametric questions will also take other formats besides multiple-choice. For instance, dissertative and true/false models. Besides Python, solution of parametric questions by other programming languages could be easily encompassed by another feature that will be included in a forthcoming work. For instance, commands

like `[[def:python:...]]`, `[[def:cpp:...]]` and `[[def:java:...]]` will make MCTest generate the corresponding PDF-file according to each specified language.

References

1. Pathak, S., Brusilovsky, P.: Assessing student programming knowledge with web-based dynamic parameterized quizzes. In: Proceedings of ED-MEDIA, pp. 24–29 (2002)
2. Brusilovsky, P., Sosnovsky, S.: Engaging students to work with self-assessment questions: a study of two approaches. ACM SIGCSE Bull. **37**, 251–255 (2005)
3. Gangur, M.: Matlab implementation of the automatic generator of the parameterized tasks, June 2018. [Online]. Available: https://goo.gl/wwATNW
4. Basaran, A., Sezer, G., Özcan, H., Ugurdag, H.F., Argali, E., Eker, O.E.: Smart question (sq): tool for generating multiple-choice test questions. In: Proceedings of the 8th WSEAS International Conference on Education and Educational Technology, EDU'09, pp. 173–177 (2009)
5. Sobre o MEGUA, June 2018. [Online]. Available: http://cms.ua.pt/megua
6. Cruz, P., Oliveira, P., Seabra, D.: Parametrized Problem Databases in Sage (2014)
7. Tutorial do MEGUA, June 2018. [Online]. Available: http://megua.web.ua.pt/tutorial
8. TestMakPro3, June 2018. [Online]. Available: http://www.image-ination.com/testmakePro3
9. AMC – multiple choice questionnaires management with automated marking, June 2018. [Online]. Available: http://auto-multiple-choice.net
10. Kagotani, H., Bréal, F., Bienvenüe, A., Sarkar, A.: Auto multiple choice, June 2018. [Online]. Available: http://download.gna.org/auto-qcm/auto-multiple-choice.en.pdf
11. Gusev, M., Ristov, S., Armenski, G.: Technologies for interactive learning and assessment content development. Int. J. Dist. Educ. Technol. (IJDET) **14**(1), 22–43 (2016)
12. Gierl, M.J., Lai, H., Turner, S.R.: Using automatic item generation to create multiple-choice test items. Med. Educ. **46**(8), 757–765 (2012)

STEM Education Enrichment in NYC

Fangyang Shen, Janine Roccosalvo, Jun Zhang, Yang Yi, Yanqing Ji, Kendra Guo, Ahmet Mete Kok, and Yi Han

38.1 Introduction

The NSF NEST grant is focusing on enhancing the quality of an existing Noyce teacher preparation program in high need schools in Brooklyn and New York City public school districts by offering advanced internship and scholarship opportunities to STEM students. Recently the New York City mayor Bill De Blasio announced that all city public schools will be mandated to offer a Computer Science program to all students by 2025. In line with it, New York state will start requiring a certificate for teaching computer science in the public schools. All this along with the fact that many Mathematics teachers in the NYC public schools are expected to retire in the next years makes this project timely and worthwhile.

Several improvements were applied based upon previous results from the National Science Foundation Noyce scholarship program including: (1) Enhanced internship and scholarship training and early teaching experiences (2) Teacher certification and job placement support (3) Professional development training workshops (4) Improved teacher retention by offering pre-service teacher support for first-year Noyce teachers (5) Noyce scholar and pre-service teacher mentorship using social media tools (6) Expansion of the recruitment of Mathematics teachers from the BMCC's Mathematics Associate's degree program and ensure smooth transfer to City Tech's Mathematics Education Bachelor's degree program (7) Expansion of school network partnerships and an addition of a new science component into the existing internship program.

The outcomes from the first National Science Foundation Noyce program highlighted a need for further mentorship and support for Noyce scholars and teachers which will help to recruit and retain STEM teachers. The contemporary three-tiered model will continue to be enhanced as it designates potential scholars as Noyce explorers, continues to support a selected cohort as Noyce scholars and maintains professional support through their induction into the profession as Noyce teachers. As part of the NEST program, Noyce scholars and teachers participate in various middle school and high school teaching internships, attend summer programs and mentorships involving STEM and Education faculty from City Tech and BMCC. Ultimately, the expanded and enriched NSF project model will help increase the number of NYS certified STEM teachers in high-need school districts upon completion of this project.

The rest of this paper is organized as follows: Sect. 38.2 reviews the literature for this topic; Sect. 38.3 introduces the enriched E-NEST project model to further recruit and train interns to become qualified STEM teachers; Sect. 38.4 presents program data collection and external program evaluations; Sect. 38.5 summarizes the findings of this study and discusses possible directions for future research.

F. Shen (✉) · J. Roccosalvo · K. Guo
Department of CST, NYC College of Technology (CUNY), Brooklyn, NY, USA
e-mail: fshen@citytech.cuny.edu

J. Zhang
Department of Math & CS, University of Maryland, Eastern Shore, Princess Anne, MD, USA

Y. Yi
Department of ECE, Virginia Tech, Blacksburg, VA, USA

Y. Ji
Department of ECE, Gonzaga University, Spokane, WA, USA

A. M. Kok · Y. Han
Borough of Manhattan Community College (CUNY), New York, NY, USA

© Springer Nature Switzerland AG 2019
S. Latifi (ed.), *16th International Conference on Information Technology-New Generations (ITNG 2019)*,
Advances in Intelligent Systems and Computing 800,
https://doi.org/10.1007/978-3-030-14070-0_38

38.2   Literature Review

According to [1], the United States is in dire need of highly effective STEM teachers in classrooms, especially in underserved high-need public school communities in New York City school districts. There were less fully certified Mathematics and Science teachers at the middle school and high school levels in high-need schools in the United States. STEM teacher recruitment and retention is a difficult challenge which directly affects student learning and academic success, especially in Computer Science and Mathematics.

In [2], this research presents how several new teachers are recruited from the path of changing careers soon after being hired. With increasing demand, vacancies are often filled with substitute teachers or those without full certification.

In [3], several studies have demonstrated that the effects of teacher turnover are most negative to students in high-need schools in underserved communities. Teachers tend to leave high-need schools for districts with lower percentages of minority students and higher socioeconomic status.

In [4], this research discusses how teacher turnover rates remain highest for students in grades 6–8. More than 50% of New York City teachers who began teaching in a middle school in the last 10 years left their job within 3 years.

The National Science Foundation has developed the successful NSF Noyce program [5] to help address the need of qualified STEM teachers. The Robert Noyce Teacher Scholarship program seeks to encourage talented Science, Technology, Engineering and Mathematics majors and professionals to become K-12 STEM teachers.

The NSF NEST project model is based upon three broad-based themes on the Engagement, Capacity and Continuity (ECC) trilogy [6]. The Engagement, Capacity and Continuity theory has served as a successful framework for implementing multi-year programs, particularly among low-income underrepresented minorities. ECC has been actively integrated in designing program goals and its respective components.

1. Engagement. Evidence-based information is combined with interactive teaching strategies and activities that motivate and develop a strong induction into teaching inquiry and experiential learning, thereby increasing student engagement, interests and motivation.
2. Capacity. The Teacher Education programs are situated within the successful infrastructure in the Career and Technical Teacher Education-CTTE and Mathematics Education programs at City Tech to increase participants' capacity to gain content knowledge and pedagogical skills in their respective disciplines. Both of these programs lead to initial teacher certification in New York State.
3. Continuity. A dual-discipline specific network system between the Noyce scholars and the CTTE /Mathematics

Education faculty was created to support the continuity of material learned both within the discipline and in educational pedagogy (i.e., Scholars will have both a discipline specific and a teacher education mentor). An online and in-person social support system was created to support student-to-student and faculty-to-student mentoring.

In [7, 8], this research presents the first three-tiered structure of the Noyce project along with mentorship and recruitment strategies implemented in STEM Teacher Education. The results of the project proved that the three-tiered structure was highly successful. An additional seven improvements have been identified and will be executed in the enriched E-NEST project.

38.3   Enriched E-NEST Project Model

The Noyce project overall has greatly impacted students in STEM Education. A diverse population of students is gaining technical and teaching experience in a positive and welcoming learning environment. Students continue to gain confidence in their technical and teaching abilities and faculty are learning how to better introduce, prepare and motivate students. The Noyce internship program has served and provided a strong scholar production which has recruited highly effective STEM teachers and teachers-to-be. The program has recruited STEM teacher candidates who have begun to obtain their degrees and certifications for STEM teaching in either Technology Teacher Education or Mathematics Education fields. The Noyce summer program and internships have improved teaching techniques and pedagogy thus increasing overall impact on student learning in K-12 STEM classrooms. The summer workshops have demonstrated innovative teaching methods while using technological advancement to efficiently deliver the teaching.

The contemporary three-tiered structure has strengthened the current Noyce program. The enriched E-NEST model consists of the following updated three tiers: Noyce Explorers, Scholars and Teachers. (1) Noyce Explorers are associate-level STEM students; (2) Noyce Scholars are baccalaureate-level students in Computer Systems Technology or Mathematics students; and (3) Noyce Teachers consist of post-baccalaureate students who receive mentoring and support as they begin their career as STEM teachers and complete their mandatory teaching years.

Using this structure, new Computer Science and Mathematics teachers were recruited which will help address the shortage of STEM teachers in the United States, especially in New York City high-need public school districts.

Based on program data collection and external program evaluations through surveys and interviews from our NSF

Noyce Scholarship Phase I program, several key findings emerged that will be used in the continued improvement of design of the NEST project. The following seven new improvements of the current model have been identified and plan to include the following:

1. Develop enhanced internship and scholarship training and early teaching experiences.

   Objectives: A two-hour orientation will be required for students as part of their internship and scholarship training. Faculty will discuss program guidelines and procedures in detail about required work. Early teaching experiences will include participating in teaching internships at cooperating schools at various grade levels. This will allow students to gain experience in more than one grade level which will better prepare them to become STEM teachers to different age groups and learners. During these middle school and high school fieldwork placements, students will participate in whole group, small group and one-on-one teaching instruction.

2. Prepare students efficiently with teacher certification and job placement support.

   Objectives: Noyce will collaborate with the NYC Men Teacher Fellows program at City Tech and BMCC. Through both of these programs at City Tech and BMCC, students will be eligible for monthly metro-cards, teacher certification workshops, certification exam assistance and job placement support as well as further faculty mentoring. The project team leaders from Computer Systems Technology, Technology Teacher Education and Mathematics Education will provide group monthly mentorship and individual meetings to support the Noyce scholars on the path to teacher certification and to prepare for job placement.

3. Develop professional development workshops involving STEM and Education content topics.

   Objectives: Professional development workshops specifically training new teachers will be administered by City Tech and BMCC during the Noyce summer workshops. These will include teaching and learning strategies that can be incorporated into the classroom and allow students to make deeper connections among STEM content areas. In addition, students will participate in the Noyce annual conference and other professional conferences.

4. Improve teacher retention by offering teaching support for first-year Noyce teachers.

   Objectives: Post-graduation activities will include participation in both informal and formal sessions through various modes of communication including email, social media and in-class meetings. Reading materials will be assigned to teachers who will engage in open book discussions on online forums.

5. Utilize supplementary social media tools for Noyce scholar and pre-service teacher mentorship

   Objectives: Use varied social media tools to extend networking and provide additional support and resources to Noyce scholars and teachers virtually since social media has become more widespread. Social network sites will include a monthly discussion forum between Noyce scholars, teachers and faculty using tools such as Twitter and LinkedIn. Scholars and teachers will discuss their questions, challenges and offer a peer support system to one another. Students will be responsible for communicating with faculty after graduation as they venture into teaching. Teacher mentorship will include appropriate STEM subject areas.

6. Expand the recruitment of Mathematics teachers.

   Objectives: Recruit Mathematics teachers by stimulating the path of BMCC's Mathematics Associate's degree program to transfer to City Tech's Mathematics Education Bachelor's degree program. This will include a year-round recruitment effort collaboration among BMCC and City Tech faculty. Printed flyers and posters will be distributed throughout both campuses. Electronic recruitment flyers will also be disseminated through college emails to all students as well as at open house orientations.

7. Further develop school network partnerships and add a new science component to the existing internship program.

   Objectives: Expand network with junior high school partnerships to give students an opportunity to intern at the middle school level to supplement training at the high school level. A new science component will be added to allow Noyce explorers to participate in internships and mentorships involving specific Science content faculty in addition to the current choices of Technology or Mathematics fields.

For the enhanced three tiered structure of Noyce Explorer, Scholar and Teacher, see Fig. 38.1 below.

During the Noyce Scholarship application process, students were evaluated and interviewed by the Noyce STEM and STEM Education recruitment committee from both BMCC and City Tech. The interview process helped provide a complete profile of the candidate. Noyce Scholarship candidates are not selected solely based on their academic qualifications, but also by their motivation, determination and zeal for STEM education.

For mentorship, all project members work collaboratively to enhance each scholar and teacher's personal growth and capacity for pursuing STEM teaching. Noyce explorers, scholars and teachers participate in Noyce activities including information sessions, individual mentorship meetings, STEM workshops, peer tutoring, STEM research project presentations, online mentoring, Noyce social events and

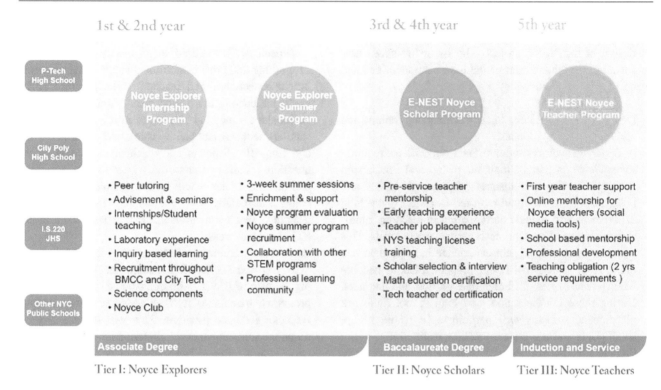

Fig. 38.1 Participant flow-E-NEST program

Noyce summits. The students participate in planned activities as a cohort which facilitate a greater sense of community among their peers. In addition to formal instruction, close mentorship helps scholars to advance their learning and accomplish their teaching goals.

As a result of the current NSF Noyce scholarship program, the Noyce program has built strong school partnerships among high-need middle schools and high schools in NYC and worked collaboratively to accommodate all interns. Careful coordination and placement of the interns has guaranteed the quality of the internship program. As school partners, the cooperating teachers provide mentorship and supervision to students during teaching internships and student teaching and communicate with various departmental CUNY faculty.

38.4 Program Data Collection and External Program Evaluations

During the Fall and Spring semesters, Noyce scholars and explorers accepted Noyce internship positions and were placed in classroom settings, engaged as tutors in peer-led team learning activities, as well as in-house interns. Scholars and explorers were not only able to observe the pedagogies used during the lectures, but also scaffold students by assisting them during in-class laboratory exercises. These students were surveyed at the end of the semester by the external evaluator. Based on survey results, the overall feedback from students was positive. Students very strongly or strongly agreed to the following statements:

- During my internship, I gained a significant amount of instructional experiences and was able to have a positive impact on student learning.
- After completing my internship, I have greater confidence in my ability to help students learn. I am better able to maintain and positive and consistent rapport with students and make course and/or laboratory content understandable for students.

When asked to write comments regarding the NSF Noyce scholarship program on a survey, the response of recent scholars and explorers consisted of the following direct quotes:

- I am very happy for the opportunity and experience I gained.
- It was great having an opportunity to actually help students understand, learn and make a difference.
- Excellent program and hope more students can take advantage.

- During my internship, I gained a lot of real experience of working with students. I feel more comfortable communicating with students and presenting a new topic to a class. I think it is important for future educators to have in-class experience to get a better idea of an ideal classroom setting.
- I really enjoy continuing this program until I graduate. I learned a lot by working with students and gained insights of pedagogical ways that we can use for the students. Students obtain knowledge in different ways that we need to pay a lot of attention in keeping them at the same pace. I would recommend students who are not in the education field to take this internship as a way to learn a new skill of working with students. In this way, they might change their mind to take education as a career. What a wonderful experience.

Noyce scholars were interviewed individually by the external evaluator.

Student 1 comments included: Workshop material is excellent. She would recommend to other students and she did influence one friend to participate in the program. She likes the assignments, group work and discussion. These are helping her become a better listener.

Student 2 comments included: She wanted to teach computer courses in the United States so she applied to City Tech's Technology Teacher Education program. Noyce has been a salvation – scholarship has been great. She will graduate and look for a teaching job then. She has learned to give students time to process questions. She likes the opportunity to work with other students and network. The workshops have increased her interest in becoming a STEM teacher. Special note: this student was extremely grateful for the opportunity this project has provided and cried (tears of emotion, happiness, and joy) at times during the interview.

All students expressed, as a result of the NSF Noyce scholarship program, an increased understanding and interest in becoming a STEM educator.

Program data collection results including student feedback and external program evaluations continue to remain positive. Figures 38.2 and 38.3 below illustrate significant survey responses on student learning and teaching experience given by Noyce students.

The lead instructor for the course allowed me to spend quality time with the students demonstrating techniques, lecturing and responding to their questions.

Answer options	Response percent	Response count
Strongly agree	52.9%	9
Agree	29.4%	5
Neutral	17.6%	3
Disagree	0.0%	0
Strongly disagree	0.0%	0
	answered question	17
	skipped question	1

Fig. 38.3 Teaching experience

Question: After completing my internship, I am better able to make course and/or laboratory content understandable for students.

Question: The lead instructor for the course allowed me to spend quality time with the students demonstrating techniques, lecturing and responding to their questions.

The survey results of the questions in Figs. 38.2 and 38.3 show that most of the interns are gaining instructional and teaching experience.

38.5 Conclusions and Future Work

In this paper, the enriched three-tiered structure of the NSF NEST project model implemented for the recruitment and training of interns to become STEM teachers was described. In addition, program data collection and external program evaluations were presented. The results continue to convey that the program is successful and could be applied to many other similar projects nationwide. For future work, we plan to continue to enhance the three-tiered structure model and continue to modify and collect more experiences and meaningful results for the Noyce project which could greatly contribute to future STEM research.

Acknowledgments This work is supported by the National Science Foundation (Grant Number: NSF 1340007, $1,418,976, Jan. 2014 – Dec. 2019, PI: Fangyang Shen; Co-PI: Mete Kok, Annie Han, Andrew Douglas, Estela Rojas; Project Manager: Janine Roccosalvo; Program Assistant: Nanase Akagami, Kimberly De La Santa).

The Noyce project team would like to thank Prof. Gordon Snyder for his help on the project's evaluation. We also want to thank all faculty and staffs at both City Tech and BMCC who have helped and supported the Noyce project.

After completing my internship, I am better able to make course and/or laboratory content understandable for students.

Answer options	Response percent	Response count
Strongly agree	55.6%	10
Agree	38.9%	7
Neutral	5.6%	1
Disagree	0.0%	0
Strongly disagree	0.0%	0
	answered question	18
	skipped question	0

Fig. 38.2 Student learning

References

1. National Science Board: Science and Engineering Indicators 2018. National Science Foundation (NSB-2018-1), Alexandria, VA (2018)
2. Ingersoll, R., Merrill, L., Stuckey, D.: Seven trends: the transformation of the teaching force, updated April 2014. CPRE Report (#RR-80). Philadelphia: Consortium for Policy Research in Education, University of Pennsylvania (2014)
3. Loeb, S., Ronfeldt, M., Wyckoff, J.: How teacher turnover harms student achievement. Am. Educ. Res. J. **50**(1), 4–36 (2013)

4. Coca, V.M., Marinell, W.H.: Who Stays and Who Leaves? Findings from a Three Part Study of Teacher Turnover in NYC Middle Schools. The Research Alliance for NYC Schools, New York (2013)
5. Robert Noyce Teacher Scholarship Program Solicitation NSF 16-559. National Science Foundation (Sept 2016)
6. Jolly, E., Campbell, P.B., Perlman, L.: Engagement, capacity and continuity: a trilogy for student success. GE Foundation (2004)
7. Shen, F., Roccosalvo, J., et al.: A new approach for STEM teacher scholarship implementation. In: 14th International Conference on Information Technology: New Generations. Springer Publications (2017)
8. Shen, F., Roccosalvo, J., et al.: NSF Noyce recruitment and mentorship. In: 15th International Conference on Information Technology: New Generations. Springer Publications (2018)

Humanoid Soccer Player and Educational Robotic: Development an Architecture to Connect the Dynamixel AX-12A Servo Motor from ROBOTIS to the Raspberry Pi 3B

Gilmar Correia Jeronimo, Paulo Eduardo Bertolucci Consoni, Rodrigo Luis Fialho Gonçalves, Wagner Tanaka Botelho, and Maria das Graças Bruno Marietto

39.1 Introduction

Robotic field has been introduced in many areas, from domestic tasks until industrialized sectors. Nevertheless, the Robotics competitions have been given relevant results, in which students may apply in practice what they learn in theory. It is possible to cite the robots soccer that became a reference point in the last years. The soccer is a dynamic and physic game, that demands real-time control, quick decisions, agile robots, resistant and mainly, intelligent [1].

The studies viability among diverse robots soccer categories are extensive, different areas of knowledge may become a focus on learning. As an example, control theory, mechanic and electronic projects, the design of many robots components, among others. Furthermore, provides an integrated vision, inter and multidisciplinary about necessities that eventually comes to occur in creation and studies of robots.

The idea of robots soccer was first mentioned in 1993 by Professor Alan Mackworth [2] in a paper entitled "*On Seeing Robots*" [3], studying the behavior of robotics agents in the world and their relations with the environment. In this same year a group of Japanese students organized the first robot soccer competition called Robot J-League. In a short time, the researchers started to participate in this competition, changing the name to Robot World Cup Initiative, RoboCup for short [4].

The main target of this paper is to present an architecture that allows the humanoid robot called Robotic Player, plays soccer in the Kid Size category of RoboCup [5]. The Vision System, Electronic Project, and a Java library, known as *AX12-JavA* [6], compose the proposed architecture, besides the Robotic Player that was built using the Bioloid ROBOTIS Premium platform [7].

The Vision System is composed by the rotation Pan/Tilt module with two servo motors SG90. This module is connected directly to the Electronic Project. In order to make the Robotic Player choose the best action during the soccer match, the Logitech c920 [8] webcam, connected to the Raspberry Pi 3B, captures the images in real time. The Open Source Computer Vision (OpenCV) library aims to process these images. Also, the Raspberry Pi 3B [9] and *Raspi2Dynamixel* [6] shield are part of the Electronic Project.

According to [10], robots are becoming an integral component of the society and have great potential in being utilized as an educational technology. Educational Robotics is a transformational tool for learning, computational thinking, coding, and Engineering [11]. Therefore, the proposed architecture also can be used in the teaching-learning process allows students to develop abilities related to the elaboration of hypotheses, finding solutions, establishing relationships, among others.

Some related works are describe in Sect. 39.2. The proposed architecture composed with Bioloid ROBOTIS Premium platform, Logitech c920 webcam, Raspberry Pi 3B, Pan/Tilt module, *Raspi2Dynamixel* [6] shield, *AX12-JavA* library and OpenCV is described in Sect. 39.3. Section 39.4

This work was supported by São Paulo Research Foundation (FAPESP) (2016/26184-9).

G. C. Jeronimo (✉) · P. E. B. Consoni · R. L. F. Gonçalves
Centre of Engineering, Modelling and Applied Social Sciences (CECS), Federal University of ABC (UFABC), Santo André, São Paulo, Brazil
e-mail: gilmar.correia@aluno.ufabc.edu.br;
paulo.consoni@aluno.ufabc.edu.br; rodrigo.fialho@aluno.ufabc.edu.br

W. T. Botelho · M. G. B. Marietto
Centre of Mathematics, Computation and Cognition (CMCC), Federal University of ABC (UFABC), Santo André, São Paulo, Brazil
e-mail: wagner.tanaka@ufabc.edu.br; graca.marietto@ufabc.edu.br

© Springer Nature Switzerland AG 2019
S. Latifi (ed.), *16th International Conference on Information Technology-New Generations (ITNG 2019)*,
Advances in Intelligent Systems and Computing 800,
https://doi.org/10.1007/978-3-030-14070-0_39

presents the results. Finally, the conclusion and future works are considered in Sect. 39.5.

39.2   Related Works

In the beginning, a bibliography survey was made to verify the possibility to make the communication between Raspberry Pi 3B and Dynamixel AX-12A servo motors using Java language. After that, it was verified the need to develop a shield (Raspi2Dynamixel) and a library (AX12-JavA), responsible for this communication.

Thiago Hersan [12] wrote a library in Python language responsible to make the communication between Raspberry Pi 3B and AX-12A servo motors. This library allowed to set torque, position, speed, baud rate and others AX-12A configurations. Therefore, it was used as reference to write the AX12-JavA library in Java language.

Some researchers, such as [13] and [14] use the Dynamixel AX-12A servo motors with the USB2Dynamixel [15] from ROBOTIS, controlling them through a Java library for Raspberry Pi 3B. However, several circuits are used to make their connection, for example, batteries to power the servos and USB extenders, without a direct connection to the controller pins. Besides having an incomplete or not implemented codes.

After analyzing the related works described in this section, some electronic and programming limitations were found, such as the possibility to use different types of sensors, numbers of AX-12A bus channels and power supplies. For these reasons, this paper presents in the next sections the proposed architecture that includes basically the *Raspi2Dynamixel* shield and the *AX12-JavA* library to control the AX-12A servo motors using Java Language.

39.3   Proposed Architecture

Figure 39.1 shows the architecture composed of Electronic Project, Vision System and Robotic Player in ①, ② and ③, respectively. The *AX12-JavA* library in ④ was developed using as reference the libraries implemented in [12, 16]. The main target of this library is to make the communication between Raspberry Pi 3B [17] (a) and Dynamixel AX-12A servo motors (b).

The main target of the proposed architecture shown in Fig. 39.1 is to allow the data collected by the Vision System, i.e. the webcam (e), to be transferred to the Raspberry Pi 3B (a). From these data, the AX-12A servo motors (b) and neck (SG90) servo motors, represented by the Pan/Tilt module (d), must be activated through the functions implemented in *AX12-JavA* library ④. In addition, the servo motors (b) and

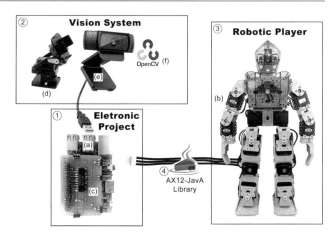

Fig. 39.1 Proposed architecture

(d) are connected to *Raspi2Dynamixel* [6] shield (c) that is also connected to the Raspberry Pi 3B (a).

In order to make the communication between Raspberry Pi 3B and AX-12A servo motors, it was necessary to implement the *AX12-JavA* [6] using as reference the Python code available in [12, 16]. Therefore, *AX12-JavA* can be used by the scientific community that has more familiarity with Java and needs the integration between Raspberry Pi 3B and AX-12A servo motor.

The developed *Raspi2Dynamixel* shield and the *AX12-JavA* library make it possible to expand the programming possibilities and applications of the Bioloid ROBOTIS Premium platform. With this shield, the CM-530 microcontroller from ROBOTIS can be replaced by Raspberry Pi 3B, and it is possible to build other robots that use the AX-12A, different sensors can be used and can plug other shields. In addition, the Raspberry Pi 3B has greater processing power and memory, WiFi module, Bluetooth, USB and others. It's important to highlight that the shield is able to be used in any other application with Raspberry Pi 3B and AX-12A, not limited to Bioloid ROBOTIS Premium. Thus, the *Raspi2Dynamixel* and *AX12-JavA* library are the main contributions of this paper and are available in [6, 18, 19] for free use, adaptation and distribution to the scientific community.

39.3.1   Electronic Project

The Electronic Project in Fig. 39.1 ① is composed by the *Raspi2Dynamixel* [6] shield in (c). It is responsible for the communication between Raspberry Pi 3B (a), and the Dynamixel AX-12A servo motors (b) [20]. These servo motors are used to move the Robotic Player ③ during the soccer match.

The Raspberry Pi 3B in Fig. 39.1 (a), is a single-board computer that allows the integration of all architecture components, as well as to control, through the *Raspi2Dynamixel* [6] (c), the AX-12A servo motors (b) of the Robotic Player ③ and the neck in (d).

The developed *Raspi2Dynamixel* [6] shield in Fig. 39.1 (c) has only one Printed Circuit Board (PCB) that connects all AX-12A (b) and SG90 (d) servo motors cables, sensors, battery and all Raspberry Pi 3B ports in (a), working as an extension of its connectors. It is important to highlight that the Universal Asynchronous Receiver/Transmitter (UART) TX, RX, one General Purpose Input/Output (GPIO), Voltage of Direct Current (VDC) and Ground (GND) ports are used to make the communication with the AX-12A servo motors (b).

39.3.2 Vision System

The Vision System illustrated in Fig. 39.1 ② is composed by Logitech c920 webcam [8] and Pan/Tilt module in (e) and (d), respectively. The module (d) is composed of the Robotic Player ③ neck and its mechanism has two servo motors (SG90) responsible for making the Robotic Player has a field of vision of one hundred and eighty degrees horizontally and ninety vertically. The two SG90 (d) are connected to the shield (c) and are directly controlled by two GPIO ports of the Raspberry Pi 3B (a) and the *AX12-JavA* library ④.

The Vision System ② communicates with Raspberry Pi 3B (a) of the Electronic Project ①. This communication is realized through OpenCV (f) [21].

The OpenCV is one of the largest libraries of computer vision and machine learning, being widely used in both educational and industry sectors. The library provides more than 2500 algorithms, for example, face tracking, object identifications, image filters, among others. It supports Windows, Linux, Mac OS, IOS and Android. Also, it is possible to use C++, C, Python, Java, Matlab, among others [21].

The Logitech c920 webcam in Fig. 39.1 (e) is a Full HD with 1080p that captures images in widescreen. In addition, it can be used in several projects related to computer vision, for example, the humanoid vision system proposed in the AdultSize category of RoboCup [22].

39.3.3 Robotic Player

Figure 39.1 shows the Robotic Player ③ built using the Bioloid ROBOTIS Premium platform [7]. It has eighteen (18) Degrees of Freedom (DOF), each DOF representing one Dynamixel AX-12A [20] servo motor (b). In addition, it has a gyroscope and a Li-Po battery of 11.1 V/1000 mA.

The Dynamixel AX-12A servo motor in Fig. 39.1 (b) is controlled by a gear reduction box and can be programmed in many languages, such as: Java, Python, LabVIEW, C/C++,

and others. Its torque is 15 Kg/cm, 12 V and the speed is 0.196 s per 60 degrees. More information about the AX-12A can be found in [20].

The UART communication protocol is used by AX-12A servo motor to send and receive data serially, in which there are eight (8) data bits, no parity and one (1) stop bit. Bits per second (bps) is a common measure used to show the average rate at which data is transferred between a computer and a data transmission system. The baud rate is the speed at which data is transmitted over a channel. For example, 9600, 38400, 115200 bps, among others.

The Dynamixel AX-12A especially uses a Transistor-Transistor Logic (TTL). The TTL communication remains 0 V and VDC, that represent 5 V. The receive pin RX is set as input, while the TX pin is set as the microcontroller output.

The required data for a message to be sent by the AX-12A or received by Raspberry Pi 3B is described in [20]. These data are important to understand the communication between AX-12A and Raspberry Pi 3B.

The message sent by TX has two values, the LOW logic represented by bit 0, created when the circuit is in $0V$, and the $HIGH$ logic of bit 1, created when the circuit is VDC. A sequence of zeros and ones is sent by TX to create the message. The sequence has a start bit, usually a byte of data converted through the American Standard Code for Information Interchange (ASCII), one parity bit, and one stop bit. The parity bit is used to verify if the transmitted message has not being attenuated [23, 24].

Figure 39.2 shows the required data for a message to be sent by the AX-12A servo motors or received by Raspberry Pi 3B. In order to the Robotic Player performs actions and read information about the AX-12A, the communication has the instruction and status packages, respectively. The AX-12A message transmission characteristics is important to understand the communication between AX-12A and Raspberry Pi 3B. For example, the configuration of the Raspberry Pi 3B serial port to send and receive data through the AX-12A cables connected to the *Raspi2Dynamixel* pins.

The instruction package sends information from the Raspberry Pi 3B to the AX-12A which then carry out the appropriate actions. In this kind of package, the size commands is around one byte, written in hexadecimal base.

The first two initialization commands in Fig. 39.2 are represented by hexadecimal 0XFF, which indicates the beginning of sending an instruction package.

The ID in Fig. 39.2 is the AX-12A identification, and can range from 0X00 to 0XFD shown in decimal by 0 to 254. For example, it is possible to set the position of the servo motor

Fig. 39.2 Message characteristics

number 1 by sending the parameters with de ID equal to 1. Also, the Size is the size of the received message, in byte. If it is five (5), the servo controller needs five (5) bytes to read the sent message.

The rotation speed, toque or baud rate are the Instruction byte defined in the fifth command of Fig. 39.2. They have a pre-defined number in hexadecimal that represents their respective action on the AX-12A. The number is described in [20]. The Parameter byte is the data to be changed. For example, if it is necessary to change the baud rate, the Instruction byte indicates that the baud rate must be changed, and the Parameter byte sets the frequency to be changed. The Check Sum byte is the last command to be sent. It performs a binary sum operation to check if the bytes have not been attenuated or if there is no error in the transmitted information.

39.4   Results

The main target in this section is to describe the development of the *Raspi2Dynamixel* shield shown in Fig. 39.1 (c), based on [25]. The *AX12-JavA* library ④ is also described. It is responsible for the communication between the Raspberry Pi 3B (a) and the Dynamixel AX-12A servo motors (b). Finally, the communication between the *Raspi2Dynamixel*(c), Raspberry Pi 3B (a), Logitech c920 webcam (e), Pan/Tilt module (d) and the Robotic Player ③ is also demonstrated.

39.4.1 *Raspi2Dynamixel* Shield

First, it was necessary to connect the Raspberry Pi 3B in Fig. 39.1 (a) and the AX-12A (b) through the *Raspi2Dynamixel* (c). For this to happen, some properties about the AX-12A have been studied and were described in Sect. 39.3.3. It is important to point out that two GPIO ports of Raspberry Pi 3B (a) can be used to connect two SG90 servo motors of the Pan/Tilt module (d). The gyroscope can also be connected to the GPIO ports, coupled to an analog/digital converter.

In the prototype circuit of Fig. 39.3 developed on a breadboard, the red wires represent the positive pole of the battery (a) and 5V input of the Raspberry Pi 3B (b). The black wires indicate the GND or the negative pole of the battery (a). The yellow, orange and green wires are the servo motors (c) data, TX and RX ports, respectively.

The prototype is composed by the 74LS241E IC of Fig. 39.3(d), which is an octal buffer and line driver. The IC consists of twenty (20) ports, four (4) Transmitter (TX) and four (4) Receiver (RX), eight (8) ports for communication with the servo motors (c), one (1) Voltage of Direct Current (VDC), one (1) Ground (GND) and two (2) ports for data control.

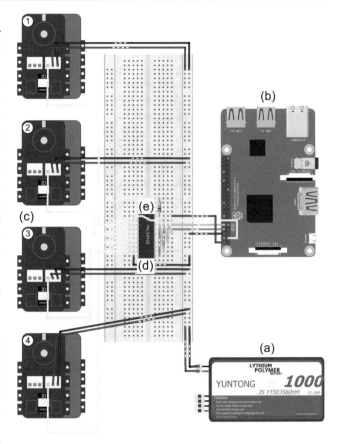

Fig. 39.3 *Raspi2Dynamixel* prototype using four servos lines controlled by a Raspberry Pi 3B

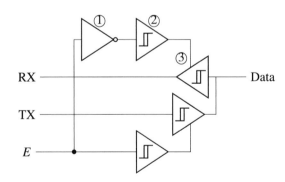

Fig. 39.4 Schematic circuit diagram for one unique dynamixel bus line

The 74LS241E IC is composed by a logic circuit shown in Fig. 39.4. It uses only a NOT port ①, a Schmitt Trigger ② and one Schmitt Trigger with an enable port ③. When the port E, represented as an enable port, has a logic level 1, the TX message available in the Raspberry Pi 3B is transmitted as $DATA$ to the AX-12A servo motors. However, when E has a logic level 0, the RX is enabled to receive servo motors messages.

The AX-12A servo motors in Fig. 39.3 (c) work as a transmission line. Therefore, only four (4) servo motors are connected to the *Raspi2Dynamixel* shield and they are able to control all AX-12A (c) of the Robotic Player ③ shown in Fig. 39.1.

For the best performance of the circuit in Fig. 39.3, one $10k\Omega$ (e) resistor is used at the VDC port of IC. The idea is to control the input circuit voltage. The IC is designed to be a memory address driver, clock and also be able to perform the communication between transmitters and receivers. Therefore, it is a tri-state buffer that accommodates multiple lines in data communication or addresses.

Still according to Fig. 39.3, AX-12A servo motors (①, ②, ③ and ④) in (c) and one Li-Po battery of 11.1 V to power them in (a), represent a group of servo motors of the Robotic Player described in Sect. 39.3.3. The ① and ② are related to the two (2) AX-12A servo motors of the right and left arm of the Robotic Player, respectively. The seven (7) servo motors of each right leg are represented by ③. ④ correspond to the others seven (7) servo motors of the left legs. It is important to point out that the servo motors have a connection to each other, creating a communication channel between them.

39.4.2 Connection Between *Raspi2Dynamixel*, Raspberry Pi, Logitech c920 and Pan/Tilt Module

Figure 39.5 shows the real PCB (a) connected to Raspberry Pi 3B (b). This can also be seen in Figs. 39.1 and 39.3. It was assembled and adapted from [26,27], and was developed using: (i) one (1) double-sided fiberglass board (10×10 cm), responsible to connect the electronic components used to prototype the shield; (ii) fourteen (14) male pins of ninety (90) degrees that connect the AX-12A servo motor cables and the battery (c) to the PCB (a); (iii) four (4) 10-pin female stackable header that connect the PCB to the Raspberry Pi 3B (b); (iv) one (1) $10k\Omega$ resistor (a) regulates the input voltage of 74LS241 IC (a) that was used to perform the communication between the AX-12A servo motors and the Raspberry Pi 3B (b), as recommend by [25–28].

In Fig. 39.5, the Logitech c920 webcam (d) is connected to the Raspberry Pi 3B (b) via USB port in ①. The Pan/Tilt module (e) with its two servo motors SG90 is the neck of the Robotic Player and also can be considered the support of the webcam. The LiPo battery (c) powers the AX-12A servo motors.

The connection between the right and left arms and the right and left legs of the Robotic Player in the *Raspi2Dynamixel* shield (a) is shown in Fig. 39.5 in ②, ③ and ④, ⑤, respectively. The LiPo battery (c) is connected in ⑥ and the webcam (d) with the Pan/Tilt module (e) are connected in ⑦.

Finally, the connection between Raspberry Pi 3B (b) and Logitech C920 webcam (d) was realized through the

Fig. 39.5 Connection between *Raspi2Dynamixel*, Raspberry Pi 3B, Logitech c920, Pan/Tilt module and robotic player

OpenCV library version 3.4.0. The library allowed the use of filter and identification tools of objects.

After some tests with *Raspi2Dynamixel* shield, a study was done on how to translate the code implemented in Python from [12] to Java. After this study, a new library, known as *AX12-JavA* [6], was created and described in the next section.

39.4.3 *AX12-JavA* Library

In order to programming the Raspberry Pi 3B with Java, it was necessary to use the Pi4J [29] library. It is an object-oriented Application Programming Interface (API) that allows using the capabilities of Raspberry Pi 3B functions in Java.

The configuration port of Raspberry Pi 3B is one of the main differences between the library implemented in Python [12] and Pi4J. While Python uses the Raspberry Pi 3B broadcom chip configuration, PI4J has a differentiated configuration, renaming the port as described in [29].

Unlike the *AX12-JavA* library, the Python serial communication works at any baud rate. Therefore, in order to make the communication with Java, it was necessary to change, using the Dynamixel Wizard software [30], the baud rate of the AX-12A servo motors. Initially, it was set to 57,600 bps.

After make all the connections between Electronic Project ①, Vision System ②, Robotic Player ③ and AX12-JavA ④

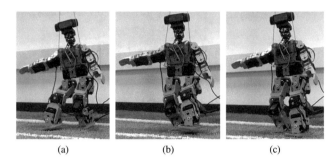

(a)   (b)   (c)

Fig. 39.6 Movements of the robotic player using the architecture proposed in Fig. 39.1 [31]

shown in Fig. 39.1, the legs, arms and neck of the Robotic Player are able to move. Therefore, in order to prove that the proposed architecture works well, Fig. 39.6 shows the movements of the robot from (a) to (c). Also, the movie is available at [31].

39.5 Conclusions and Future Works

The main target of the proposed architecture was to allow the Robotic Player to play soccer in the Kid Size category of RoboCup. It was composed by Electronic Project, Vision System, Robotic Player and *AX12-JavA* library.

The Electronic Project was composed by the *Raspi2Dynamixel* shield and a Raspberry Pi 3B. The Logitech c920 webcam, Pan/Tilt module with two servo motors SG90 and the OpenCV library are related to the Vision System. The Bioloid ROBOTIS Premium platform was used to build the Robotic Player. The webcam was connected to the USB port in Raspberry Pi 3B. The OpenCV allowed that the webcam captured image to be displayed in real-time.

One of the main contributions of this paper was the development of *Raspi2Dynamixel* shield that allowed the efficient connection between Raspberry Pi 3B and AX-12A servo motor, using the *AX12-JavA* library, that was also another contribution.

It is important to highlight that all the information about the *Raspi2Dynamixel* shield and the *AX12-JavA* source code are available for use to other researchers and also contribute to improvements. In addition, it has not found any other shield that can perform the communication between Raspberry Pi 3B and AX-12A servo motors.

As future works, the second version of the *Raspi2Dynamixel* shield should be developed, in order to use an analog sensor, such as accelerometer and gyroscope. In this case, it is necessary to use an Analog-Digital Converter (ADC) with an MCP3008 IC.

The use of Raspberry Pi 3B with *Raspi2Dynamixel* shield allows to extend applications with AX-12A servo motors, not only for robots soccer, but also for other applications, especially in the Educational Robotic field. In this case, the student and also the professor use the creativity to build different types of robots, such as, dog, cat, horse, manipulators, among others. The only limitation is the current that passes through the servo motors, which limits the number of servos used.

The Vision System proposed in this paper can be used to identify different objects in the environment. In the soccer field, the humanoid robot should be able to identify the ball, the field itself, crossbars of the goal, the opponent players and the teammates. Thus, the identification should be performed using image processing through some Artificial Intelligence (AI) technique, such as Neural Network, Fuzzy Logic, Genetic Algorithm, among others. Finally, the humanoid locomotion is an interesting research area to be investigated.

Acknowledgements The authors would like to thank to the São Paulo Research Foundation (FAPESP) (2016/26184-9) for the support of this research.

A special thanks to Thiago Hersan [12] for having developed the AX-12 Python Library for Raspberry Pi. Also, Jesse Merritt [16] who developed a simple module for communicating with Dynamixel AX12 servo motors from Python. These works were important in the development of the architecture proposed in this paper.

References

1. Lund, H.H., Pagliarini, L.: Robot soccer with LEGO mindstorms. In: Asada, M., Kitano, H. (eds.) RoboCup-98: Robot Soccer World Cup II. RoboCup 1998. Lecture Notes in Computer Science, vol. 1604. Springer, Berlin/Heidelberg (1999)
2. Mackworth, A.: Biography. Available in: https://bit.ly/2NmI6UJ (2013). Accessed 17 July 2018
3. Mackworth, A.K.: On seeing robots. In: Computer Vision: Systems, Theory and Applications, pp. 1–13. World Scientific, Singapore (1993). https://doi.org/10.1142/9789814343312_0001
4. RoboCup: The robocup federation. Available in: https://bit.ly/2va76qC (2016). Accessed 12 Sept 2018
5. Humanoid, R.: The official website of the robocup humanoid league. Available in: https://bit.ly/2RwaJ54 (2018). Accessed 28 Sept 2018
6. Jeronimo, G.C., Consoni, P.E.B., Gonçalves, R.L.F., Botelho, W.T.: Ax12-java. Available in: https://bit.ly/2BYeYRC (2018). Accessed 13 June 2018. GitHub Repository
7. Bioloid: Bioloid premium kit – robotis. Available in: https://bit.ly/2NrN5ng (2018). Accessed 01 Aug 2018
8. Logitech: Specifications – logitech c920. Available in: https://bit.ly/2CrWfyZ (2017). Accessed 20 Aug 2018
9. Pi, R.: Raspberry pi. Available in: https://bit.ly/1Jua4qn (2018). Accessed 22 Aug 2018
10. Mubin, O., Stevens, C.J., Shahid, S., Mahmud, A.A., Dong, J.: A review of the applicability of robots in education. J. Technol. Educ. Learn. **1**(209-0015), 13 (2013)
11. Eguchi, A.: Robotics as a learning tool for educational transformation. In: Proceedings of 4th International Workshop Teaching Robotics, Teaching with Robotics & 5th International Conference on Robotics in Education Padova (2014)

12. Hersan, T.: Ax-12 python library (for raspberrypi). GitHub (2014)
13. Skokan, L.: Robotis dynamixel java lib for robotis servos xl-320 and ax12a witn rest api. Available in: https://bit.ly/2y1GDyC (2017). Accessed 01 Oct 2018. GitHub Repository
14. de Vries, R.: Raspberry pi hexabot robot powered by java. Available in: https://bit.ly/2pEGsod (2017). Accessed 12 Aug 2018. Wordpress Sites
15. ROBOTIS: Usb2dynamixel. Available in: https://bit.ly/2Cw20eY (2010). Accessed 01 Jan 2019. ROBOTIS
16. Merritt, J.: A simple module for communicating with dynamixel ax12 servos from python. GitHub (2015)
17. Jain, S., Vaibhav, A., Goyal, L.: Raspberry pi based interactive home automation system through e-mail. In: International Conference on Optimization, Reliabilty, and Information Technology (ICROIT), Feb 2014, pp. 277–280
18. Jeronimo, G.C., Botelho, W.T.: Opencv+raspberry+java. Available in: https://bit.ly/2Qtybij (2018). Accessed 01 Aug 2018. GitHub Repository
19. Jeronimo, G.C., Gonçalves, R.L.F., Consoni, P.E.B., Botelho, W.T.: Ax12-python. Available in: https://bit.ly/2yioHPp (2018). Accessed 10 Sept 2018. GitHub Repository
20. Robotis®: Dynamixel ax-12. Available in: https://bit.ly/2RusuSw (2006). Accessed 03 Sept 2018. Robotis®
21. Bradski, G., Kaehler, A.: Learning OpenCV: Computer Vision with the OpenCV Library. O'Reilly Media, Inc., Sebastopol (2008)
22. Saeedvand, S., Gheibi, M., Saber, A.K., Jafari, M., Abbaszadeh, M.: Irc team description paper 2016 adult-size humanoid robot soccer team (2016)
23. Sparkfun: Rs-232 vs. ttl serial communication. Available in: https://bit.ly/2Nqtmo2 (2010). Accessed 05 Oct 2018. Sparkfun
24. SparkFun: Serial communication. Available in: https://bit.ly/2ax5EHd (2017). Accessed 10 Feb 2018. Sparkfun
25. Hersan, T.: How to drive dynamixel ax-12a servos (with a raspberrypi). Available in: https://bit.ly/1KgPKKC (2015). Accessed 03 Sept 2018. Instructables
26. Savage, J.A.: Arduino and dynamixel ax-12. Available in: https://bit.ly/2QBNUvV (2011). Accessed 03 Oct 2018. Savage Electronics
27. Robottini: Dynamixel ax-12a and arduino: how to use the serial port. Available in: https://bit.ly/2IFKIMM (2011). Accessed 22 Sept 2018. Robottini
28. Oppendijk: Dynamixel ax12 and the raspberry pi. Available in: https://bit.ly/2O8rO77 (2015). Accessed 03 Oct 2018. Oppedijk
29. Pi4J©: The pi4j project: Java i/o library for the raspberry pi. Available in: https://bit.ly/2yavaNh (2017). Accessed 22 Sept 2018. Pi4J©
30. ROBOTIS: Dynamixel wizard. Available in: https://bit.ly/2Qr25U8 (2017). Accessed 01 Jan 2019. ROBOTIS
31. Jeronimo, G.C., Consoni, P.E.B., Gonçalves, R.L.F., Botelho, W.T., Marietto, M.G.B.: Experimental results to prove the communication between the dynamixel ax-12a servo motors from robotis to the raspberry pi 3b. Available in: https://bit.ly/2y10JJ4 (2018). Accessed 01 Jan 2019. YouTube

William Ross and Byeong Kil Lee

40.1 Introduction

To reduce the off-chip memory access, the last level cache (LLC) is usually organized as a shared cache in chip multiprocessors. In a multicore processor, each core has its own private cache, and the next levels of caches can be shared or private to that core. The last level cache is usually a shared cache and an effective utilization of the LLC helps in reducing the effect of costly miss penalty [1]. As the number of cores and threads per core increase, the number of sets utilized at certain point of time is limited, which is leading to cache thrashing. This behavior of cache, when the number of cores and threads increase, can be minimized by using effective data placement strategy which helps in the increase in the usage of cache sets more effectively thereby reducing cache thrashing.

The effect of increased cores and threads can be seen from the graphs in Fig. 40.1. These experiments are performed with multicore configurations, in which a single thread and multiple threads are running on each configuration. The objective of this research is to improve overall performance (e.g., Instructions Per Cycle) by increasing utilization of cache sets.

Data tend to be arranged non-uniformly among the cache sets as the number of cores and threads increase. Usually data that is not shared among the cores is aligned along the page boundaries of main memory [2]. This non-uniform distribution of data causes cache thrashing. Previous studies [2–4] show that a more uniform distribution of data in the cache reduces this effect to an extent. One method to reduce this effect is stack randomization. Stack randomization means it randomly aligns the thread private data in the memory page leading to more cache set usage and more distributed data among the sets. However, this causes data fragmentation because of random organization of data. For example, let us consider a scenario where we have a last level cache with 512 sets and block size 128 Bytes. If the physical address is translated into a cache decode address, same sets are used because the stack data, which is extensively used, is on the page boundaries and during translation the index bits are always the same most of the times. This leads to cache thrashing.

In a cache with 512 sets, index has 9 bits. If the block size is 128 Bytes, the offset has 7 bits as shown in Fig. 40.2. When a 32 bit physical address (which consists of an 8 KB page size) has 13 bits in page offset and the remaining bits as physical page ID is seen from the view point of caches, the index has 6 bits from the first 6 bits of the page offset. Since the data in the boundaries of the pages are accessed more as there are more number of threads, only some number of sets is being used repeatedly. This causes the displacement of a frequently used cache line by another frequently used cache line thereby, causing cache thrashing. A more efficient method should be designed such that the utilization of sets can be more distributed which would minimize cache thrashing and thereby decreasing miss rate and increasing IPC (Instruction Per Cycle).

In this paper, we propose an effective data placement to minimize cache thrashing in last level cache by reducing non-uniform distribution of cache set. In the proposed method, different scheme of bit index is taken into consideration to make more number of sets available which is leading to more uniform distribution of cache data. On 4-core to 32-core configurations, there was an improvement in IPC by up to 10% for all the four replacement policies – LRU (Lease Recently Used), LIP (LRU Insertion Policy), FIFO and random.

W. Ross · B. K. Lee (✉)
University of Colorado, Colorado Springs, CO, USA
e-mail: blee@uccs.edu

© Springer Nature Switzerland AG 2019
S. Latifi (ed.), *16th International Conference on Information Technology-New Generations (ITNG 2019)*,
Advances in Intelligent Systems and Computing 800,
https://doi.org/10.1007/978-3-030-14070-0_40

Fig. 40.1 Experimental results showing the decrease in IPC as the number of cores or threads is increasing (black bar: single-threaded workload, gray bars: multi-threaded workload)

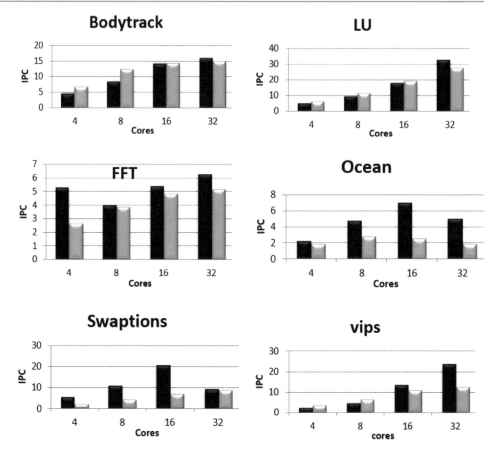

The rest of paper is organized as follows: Sect. 40.2 describes related works. Effective data placement for cache reduction scheme is explained in Sect. 40.3. Section 40.4 shows simulation and observation, and concluding remarks and future works are presented in the last section.

40.2   Related Works

Previous work [2] proposes stack based randomization as a technique to avoid cache thrashing. Stack based randomization involves modifying the thread library in order to align the stack bases randomly in the page but not only on the page boundaries.

According to Meng and Skadron's study [2], private data which is the data accessed by single core only mainly consists of the stack data. As the number of cores and the threads per each core increase, it is seen that more data are accessed from the stacks. It means that the private data is accessed more compared to the shared data that is the data accessed by all the cores. Since the stack bases are all aligned to the page boundaries, it causes a contention among LLC cache sets leading to cache thrashing and non-uniform distribution of data. Stack based randomization has been proposed to this problem.

Data address from the viewpoint of OS

Physical page ID	8KB page offset (13 bits)

Data address from the viewpoint of caches

Tag	Index (9 bits)	Block offset (7)

Fig. 40.2 Address from different viewpoints [2]

Some other replacement policies [5] have also been proposed which tolerate the effect of non-uniform distribution of data like LRSU (Least Recently Used Shared data) and LIP (LRU insertion policy). Since the active set of data consists of private data, it is better to evict shared data first. This policy is called Least Recently Used Shared Data where the least recently used shared data is evicted first instead of the least recently used block from the combination of private and shared data.

Another replacement policy which can tolerate this non-uniform distribution of data to an extent is Least Insertion Policy (LIP). In the LIP policy, the incoming block is placed in the Least Recently Used position instead of Most Recently Used position. It is moved to the Most Recently Used (MRU) position only on the second access. This way the cache block which is used only once can be evicted without keeping it for long in the cache. For the shared data that are only accessed once, remain in the LRU position and get evicted during the next replacement so that the private data can reside in the cache. But in some cases, LIP [6] can cause an increase in cache thrashing because since the cache is an inclusive cache, there might arise a case where the data is in L1 and the LLC copies are not used. This data gets evicted if it is in LRU position which is not wanted.

40.3 Effective Data Placement for Cache Thrashing Reduction

40.3.1 Page Sizing

As we discussed earlier, the cache thrashing problem is due to the non-uniform distribution of data among the cache sets. To efficiently use the sets in last level cache, we change the way the address in the viewpoint from the cache to solve the cache thrashing problem to an extent.

While translating the physical address to the cache address, firstly the cache with the original index is looked upon to find the original set. If there is an available place in that set, then it will be placed there. If the set is full, the data can be placed at new location by adding a new logic which is required to accommodate the translation. In the new logic, the index bits are shifted left by 6 bits while translating the address (e.g., 32 bit address). In this case, the location overlap can be avoided. The least significant 7 bits are used as offset and most significant 10 bits and 6 bits (12:7 bits) are concatenated to produce the tag. This will lead to more distribution among the sets in an LLC. Also, it does not have any fragmentation issues mentioned in the previous approaches. This address swizzling scheme is shown in Table 40.1 for a cache with 512 sets and 128 Byte cache line.

40.3.2 Address Swizzling and Placement

The procedure is that firstly the data will be arranged in the cache in the regular way. When the set is full for a given address, then the address will be swizzled and a new set

location will be calculated. If this new set is available or vacant, then the data will be kept there. If that set is also full, then eviction happens by removing the Least Recently Used element in the original set. This method certainly decreases the miss rate and increases the IPC by increasing utilization of the cache storage. This method works for different replacement policies like LRU, LIP, FIFO and random and it showed significant improvement in IPC as well.

The Table 40.2 shows the tag, index and offset bits for different configurations as we increase the number of cores from 4 to 32, the cache size is also increased to ensure that high miss rate which can be seen in the regular scenario is not due to the low cache size [3]. Therefore, the number of sets will differ accordingly and will have different number of bits for each configuration for index, tag and offset.

In order to get the required bits for tag and index, some bit manipulation must be done. Figure 40.3 is the part of source code which has been used to implement the bit manipulations for a 16-core simulation.

If all the blocks in the set which is pointed by the original index is full, and also the set which is pointed by the new index is full then the question of eviction arises. Experimental results show that replacing the LRU block or the block which needs to be replaced by whichever replacement policy we use, that block must be from the original set but not the new set pointed after the swizzle logic, will only show more improvement in the reduction of miss rate. The flow chart in Fig. 40.4 shows the flow of the procedure.

Table 40.1 Bit information for original and proposed method

	Original value	New value
Tag	[31–16]	[31–22]|[12–7]
Index	[15–7]	[21–13]
Offset	[6–0]	[6–0]

Tag2	Index	Tag1	Offset
(10bits)	*(9 bits)*	*(6 bits)*	*(7bits)*

Table 40.2 Tag, index and offset bits of different configurations for both original and proposed methods

Core# (Cache size)/Bit information		Tag bits	Index bits	Offset bits
4-core (128K)	Original	31–13	13–7	6–0
	New	**(31–20)|(12–7)**	**19–13**	**6–0**
8- core (256K)	Original	31–14	14–7	6–0
	New	**(31–21)|(12–7)**	**20–13**	**6–0**
16- core (512K)	Original	31–15	15–7	6–0
	New	**(31–22)|(12–7)**	**21–13**	**6–0**
32- core (1024K)	Original	31–16	16–7	6–0
	New	**(31–23)|(12–7)**	**22–13**	**6–0**

```
index_or_set = ((addr & 0x3fe000) >> 0xd);   //new index
*set = index_or_set;
temp_tag1 = ((addr & 0xffc00000) >> 0x16);
temp_tag2 = ((addr & 0x1f80) >> 7);
temp_tag2 = (temp_tag2 << 0xa);
temp_tag = temp_tag1 | temp_tag2;             //new tag
*tag = temp_tag;
```

Fig. 40.3 Part of the source code for bit manipulation

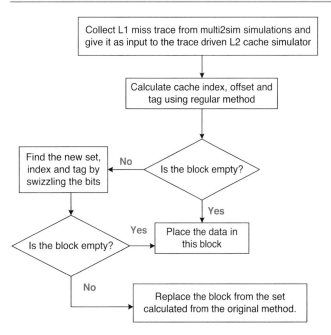

Fig. 40.4 Flowchart showing the memory access flow in proposed method

Table 40.3 Cache configuration

Cache size	16 KB L1 instruction cache (16-way)
	16 KB L1 data cache (16-way)
	L2 cache size: depends on the number of cores
	(8-way set associative)
	(For example, 32-core system has 1024KB L2 cache and 4-core system has a 128KB L2 cache.)
Latency	L1 cache: 3 cycles
	L2 cache: 30 cycles
	Main memory: 50 cycles
Main memory page size	8 KB

40.4   Simulation and Observations

Multi2sim 3.2 [7] simulator is used to generate L1 miss traces and a trace driven L2 simulator was used to calculate the miss rate for the last level cache (LLC). Since the current microprocessors are based on complex designs, Multi2sim simulation framework which models major components of incoming systems is used for the simulations. The tool is configured to get the miss traces for various configurations. Table 40.3 shows experimental setup for this experiment.

Three benchmarks from SPLASH2 benchmark suite [8] and three benchmarks from PARSEC benchmark suite [9–11], totally six were taken to examine the results. SPLASH2 is the most commonly used suite for scientific studies of parallel machines with shared memory. PARSEC is a collection of multithreaded benchmarks which is widely used by parallel workloads researchers. The description for each benchmark which was used for simulations are shown in Table 40.4.

Table 40.4 Workloads used for simulation

Benchmark	Description
Bodytrack	Computer vision (PARSEC 2.1) Problem size:4 frames, 4000 particles
LU	LU decomposition (SPLASH 2). Dense linear algebra, alternating row-major and column-major computation Input: a 300 × 300 matrix
FFT	Signal processing (SPLASH 2) Problem size: 4,194,304 data points
Swaptions	Financial analysis (SPLASH 2) Problem size: 64 swaptions, 20,000 simulations
Ocean	High performance computing (PARSEC 2.1) Problem size:514 × 514 grid
Vips	Media processing (PARSEC 2.1) Problem size: 1 image, 2662 × 5500 pixels

A mix of benchmarks from both PARSEC 2.1 and SPLASH2 benchmark

For each benchmark, miss information of L1 cache have been collected by changing the number of cores with eight threads running on each core in Multi2sim. Also, the L2 cache size has been changed accordingly. IPC, number of instructions and number of cycles have been noted down. Then miss traces were given as input to the trace driven L2 cache simulator which gave the miss rate. Simulations were performed for different configurations and different replacement policies.

Number of cycles saved for each benchmark are observed and shown in Fig. 40.5. The number of cycles which are saved are investigated for several configurations: 4-core, 8-core, 16-core and 32-core configurations with eight threads per core. Several replacement policies including LRU, LIP, FIFO and Random are also examined.

Observations The swizzling logic distributes cache blocks more uniformly among the cache sets. This can be observed from Fig. 40.6 for swaptions benchmark. There will be definitely an improvement in the IPC because a lot of cycles have been saved using the proposed method. Since the decrease in the number of misses, decreases the penalty of L2 cache miss which is due to the time taken to fetch the block from main memory which has around 50 cycles miss penalty (assumed for simulation). Also, it is seen that there is an efficient usage of all the sets and there is a more uniform distribution of data. For example, the sets which are being frequently accessed will be accessed less. Instead, the other sets which were not used frequently are having more number of accesses. In Fig. 40.6, it is seen that the set number 192 which was accessed for 1,603,754 times is now being reduced to 1,199,746 times. Those reduced accesses are distributed among other cache sets. This improvement in cache set utilization has the cache thrashing reduced to an extent, and overall performance can be increased with lower cache miss rate. More hardware

Fig. 40.5 # of cycles saved for various replacement policies and configurations

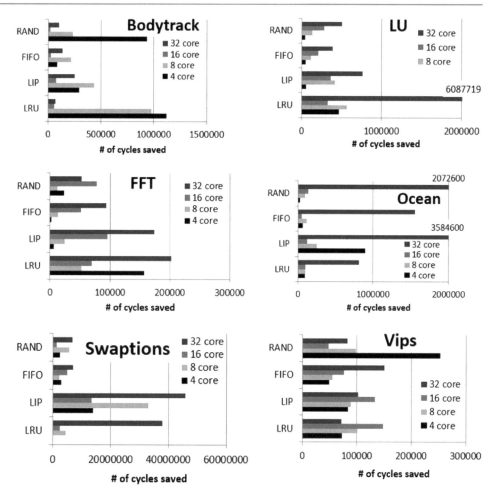

(e.g., MUX, comparator for comparison) is required for writing and reading data, but there is no increase in data array or tag array memory which is demanding more hardware cost and operational power. The data access time will be same if parallel searching and parallel comparison operations are applied.

40.5 Conclusion

The data will be non-uniformly distributed among the sets when there is an increase in number of cores and number of threads. Only some cache sets are being frequently used while the others are not being accessed that much. These cache sets which are being accessed repeatedly are prone to cache thrashing. This problem is due to the alignment of the mostly accessed data which is private data on the page boundaries. When the physical address is decoded and seen from the viewpoint of caches, there is a probability of using a limited number of sets. A new method is proposed in which different index bits are taken into consideration so that more number of sets will be available leading to a more uniform distribution of data.

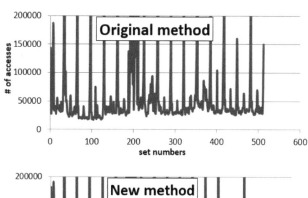

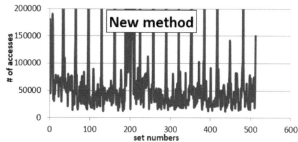

Fig. 40.6 Distribution of set utilization (zoom-in version)

Future work can be done in terms of further decreasing the miss rate and come up with compromised replacement policies towards the proposed data swizzling scheme.

References

1. Patterson, D.A., Hennessy, J.L.: Computer Architecture – A Quantitative Approach. Fifth edition, Morgan Kaufmann (2006)
2. Meng, J., Skadron, K.: Avoiding cache thrashing due to private data placement in last level cache for manycore scaling, in ICCD (2009)
3. Qureshi, M.K., Thompson, D., Patt, Y.N.: The V-Way cache: demand-based associativity via global replacement. International Symposium on Computer Architecture (2005)
4. Jouppi, N.P.: Improving direct-mapped cache performance by the addition of a small fully-associative cache and prefetch buffers. SIGARCH Comput. Archit. News. 18(3a), 364–373 (1990)
5. CPU Cache. Available from: http://en.wikipedia.org/wiki/CPU_cache
6. Qureshi, M.K., Jaleel, A., Patt, Y.N., Steely, S.C. Jr., Emer, J.: Adaptive insertion policies for high performance caching. In: International Conference on Parallel Architectures and Compiler Techniques (PACT) (2008)
7. Ubal, R., et al.: Multi2Sim: A Simulation Framework to Evaluate Multicore-Multithreaded Processors, In: Proceedings of the 19th International Symposium on Computer Architecture and High Performance Computing (2007)
8. Woo, S.C., Ohara, M., Torrie, E., Singh, J.P., Gupta, A.: The SPLASH-2 programs: characterization and methodological considerations. In: Proceedings of the 22nd International Symposium on Computer Architecture, pp. 24–36 (June 1995)
9. Bienia, C., Kumar, S., Li, K.: PARSEC vs. SPLASH2: a quantitative comparison of two multithreaded benchmark suites on chip multiprocessors. In: IEEE International Symposium on Workload Characterization (Sept 2008)
10. Bienia, C., Li, K.: PARSEC 2.0: a new benchmark suite for chip-multiprocessors. Workshop on Modeling, Benchmarking and Simulation (June 2009)
11. Bienia, C.: Benchmarking Modern Multiprocessors. Ph.D. Thesis, Princeton University (Jan 2011)

Data Imputation with an Improved Robust and Sparse Fuzzy K-Means Algorithm

41

Connor Scully-Allison, Rui Wu, Sergiu M. Dascalu, Lee Barford,
and Frederick C. Harris Jr.

41.1 Introduction

With any dataset collected in large volumes and at high velocities, missing data will occur. For the Nevada Research Data Center(NRDC), a Nevada-based data management center, the occurrence of missing data can be frustrating [1]. Data holes hinder data analytics by reducing usable data. Oftentimes, entire rows of downloaded measurements are thrown out by data users to simplify pre-processing. Any loss or lack of usable data slows down research being done by various institutions on issues like water conservation and global warming. This, in turn, reduces the production of actionable intelligence which can be used to drive institutional policies on these matters.

Myriad phenomena can poke holes in otherwise valid datasets, and can result in long sequential strings of missing data. When a large series of data points goes missing, it makes data repair that much harder than if scattered holes were found. Specifically, sequences of missing data preclude the use of intuitive varieties of data imputation, like interpolation. With interpolation, a singular hole in time series data can be filled by finding the mean between the preceding and following data points and inserting that value. While not foolproof, methods like this can be simple and effective, however, they start to fail as the gaps between known data points get wider.

To surmount this problem more advanced imputation techniques are required. Many approaches exist to accurately fill data holes, however one of the most effective techniques comes from the domain unsupervised machine learning: clustering, specifically Fuzzy K-Means (FKM) clustering [2]. By using multivariate data from other sites and sensors, a missing value can be assigned into multiple clusters with similar values according to the similarity of other values, or "features", collected at the missing timestamp. Even in the occurrence of large strings of missing data points, accurate clustering is still possible and provides sufficient contextual information for imputing data.

This method comes with a downside however. Fuzzy K Means clustering is a notoriously slow algorithm, especially when compared with other common imputation methods [2]. FKM often requires hundreds of iterations of tens of thousands of euclidean distance comparisons. If naively implemented, this algorithm can take hours to separate 40,000 vectors of floating point measurements into meaningful groups.

To address this, we propose to leverage the power of GPUs to enable the massive concurrent processing of embarrassingly parallel distance and optimization calculations used by FKM Clustering. Specifically, this paper proposes modifications to a specific Fuzzy K Means algorithm, called "Robust and Sparse Fuzzy K Means". This algorithm ensures that robust and accurate clustering occurs with any number of GPUs. This paper shows that our proposed *improved Robust and Sparse Fuzzy K-Means* (iRSFKM) algorithm

C. Scully-Allison (✉) · S. M. Dascalu · F. C. Harris Jr.
Department of Computer Science and Engineering,
University of Nevada, Reno, NV, USA
e-mail: cscully-allison@nevada.unr.edu; dascalus@cse.unr.edu;
fred.harris@cse.unr.edu

R. Wu
Department of Computer Science, East Carolina University,
Greenville, NC, USA
e-mail: wur18@ecu.edu

L. Barford
Department of Computer Science and Engineering,
University of Nevada, Reno, NV, USA

Keysight Laboratories, Keysight Technologies, Reno, NV, USA
e-mail: lee.barford@ieee.org

© Springer Nature Switzerland AG 2019
S. Latifi (ed.), *16th International Conference on Information Technology-New Generations (ITNG 2019)*,
Advances in Intelligent Systems and Computing 800,
https://doi.org/10.1007/978-3-030-14070-0_41

provides accuracy results sufficient for the imputation needs of the NRDC with up to 180× speedup over an optimized implementation of the original.

The remainder of this paper is organized as follows: Sect. 41.2 describes the background of data imputation, FKM clustering, and the use of GPUs in these domains. Section 41.2 also describes in more significant detail the background required to understand the RSFKM algorithm which this paper modifies. Section 41.3 describes the details of implementing and adapting the RSFKM algorithm to one GPU and multiple GPUs. Section 41.4 outlines the experiments run and results which validate the iRSFKM implementation on one GPU and multiple GPUs. Section 41.5 concludes the paper and outlines avenues and opportunities for future research based on this work.

41.2 Background Related Works

"Missing data imputation" describes the process of filling in holes that occur in sufficiently large data sets [3]. To ensure valid statistical analyses with a dataset these holes must be filled with accurate estimations of "ground truth" values. Although many simple and effective statistical imputation methods exist, like K Nearest Neighbors which takes a simple mean of "closely related" data points [4], not every dataset is as sympathetic to these approaches as others. Accordingly, when choosing approaches to data imputation, multiple factors must be considered. The randomness of missing data and the structure of the dataset being imputed are a few examples of this.

The work of Schmitt et al. explores these factors in detail with a comparative analysis of six common methods of data imputation [2]. In this paper, the authors compare the imputation methods of a simple Mean, KNN, fuzzy K-means, singular value decomposition (SVD), Bayesian principal component analysis (bPCA) and multiple imputations by chained equations (MICE). Using a quantitative analysis gauging the accuracy of imputed data on multiple benchmark data sets, the authors largely concluded that – for both large and small datasets – FKM provided accurate results at the cost of a very poor execution time. This trade-off indicates a clear opportunity for optimization with the aid of modern GPU programming techniques.

Fuzzy K-Means clustering algorithms are modifications of traditional K-Means clustering. These algorithms exploit fuzzy set theory to define membership as a percentage allowing for a data point to have membership in multiple clusters [5]. This approach to clustering has proven to be very effective for applications in the domains of computer vision and pattern recognition [6, 7]. Data imputation as an application of FKM has been explored by many authors [8, 9], but in the recent past, "pure" FKM approached to

Data Imputation fallen out of vogue, yielding to hybrid approaches with other machine learning techniques [10, 11]. Accordingly, there currently exists a research gap in the area of utilizing a standalone FKM algorithm for Data Imputation.

Finally, the concept of leveraging GPUs to accelerate the time consuming process of clustering is not a new one. A handful of publications have been produced exploring how best GPUs can be leveraged for clustering. However, many of these publications are now outdated [12, 13], having been released very shortly after the introduction of NVIDIA's GPU processing framework, CUDA. Accordingly, they do not reflect significant changes to GPU hardware and software. Changes which certainly affect the structure of proposed algorithms and implementations. Additionally, more prominent contemporary articles do not consider applications of data imputation and, more importantly, do not create Multi-GPU implementations of their algorithms [14]. We assert that the inclusion of Multiple GPUs is a substantial contribution of this paper over prior work.

41.2.1 Fuzzy K Means

The traditional approach to Fuzzy K-Means, as introduced by Dunn [5], is relatively simple. A set $\mathbf{X}$ of n objects can be grouped into c clusters with membership coefficients $\mathbf{U}$ defined by centroids in matrix $\mathbf{V}$ with the following algorithm:

while Not *Converge* **do**
    Compute centroids $\mathbf{V}$ via (1)
    Compute coefficients of memberships for U via (2)
end while

The equations referenced in the above algorithm follow here with a brief explination:

$$v_k = \frac{\sum_x w_k(x)^m x}{\sum_x w_k(x)^m} \tag{41.1}$$

$$\sum_{i=1}^{n} \sum_{j=1}^{c} u_{ij} = \frac{1}{\sum_{k=1}^{c} (\frac{\|x_i - v_j\|}{\|x_i - v_k\|})^{\frac{2}{m-1}}} \tag{41.2}$$

u_{ij} in the above equation describes a specific cell of membership matrix $\mathbf{U}$ which defines the membership of object x_i in centroid v_j.

41.2.2 Robust and Sparse Fuzzy K-Means

The fundamental algorithm modified and used in this work is derived from Xu et al. [15]. This algorithm, called "Robust and Sparse Fuzzy K-Means Clustering," makes several adjustments to traditional Fuzzy K Means Algorithms to enforce "robustness" and "sparsity" of the clustering. In this context, "robustness" refers to a characteristic of centroids which mitigates outlier influence in updating their mean value. "Sparsity" refers to a cluster's membership characteristics. A membership vector is sparse if a data point is wholly a member of only one cluster and no others. It is the authors' assertion that there exists an optimal sparsity for each data point where it belongs to a few clusters but not others.

The algorithm proposed by Xu et al. modifies the core FKM algorithm expressed above with the modification of Eqs. (41.1) and (41.2). Several supplementary equations were also introduced by the authors of this paper to enforce robustness and sparsity. The following equations replaces (41.2) and (41.1), respectively:

$$\min_{\substack{u^i 1 = 1, \\ u^i \geq 0}} \|u^i - \widetilde{h}^i\|_2^2 \tag{41.3}$$

This equation optimizes membership for a value denoted by row u_i, where $1 \leq i \leq n$, and n is the number of rows in our data matrix. Equation (41.3) is a wholly independent sub problem for each line. So with n values this minimization can be performed entirely in parallel with n cores. $\widetilde{h}^i$ is an auxiliary variable used to enforce sparsity that is stored in an auxiliary $\mathbf{H}$ matrix. The next equation updates the centroids and replaces (41.1):

$$v_k = \frac{\sum_{i=1}^{n} s_{ik} u_{ik} x_i}{\sum_{i=1}^{n} s_{ik} u_{ik}} \tag{41.4}$$

s_{ik} in this equation is an auxiliary variable which enforces robustness. If a euclidean distance between a centroid and data vector is within a user defined threshold then s_{ik} will be defined as the reciprocal of that distance, which reduces the weight of farther points. If its outside the threshold then s_{ik} will be defined as 0. This prevents outliers from influencing centroid updates. u_{ik} describes the membership percentage of data vector i in cluster k, and x_i describes data vector i out of n values being clustered. As many calculations in this algorithm require numerous auxiliary matrices and sum reductions plentiful opportunities for parallelization could be found.

41.2.3 Imputation Method

Data imputation with Fuzzy K-Means algorithms is relatively simplistic. After all data objects have been clustered, a missing data value j for a specific data vector x_i can be filled in with the following equation [8]:

$$x_{i,j} = \sum_{k=1}^{V} U(x_i, v_k) * v_{k,j} \tag{41.5}$$

To explain this formula in greater detail, $U(x_i, v_k)$ describes a membership value for a particular vector x_i in a cluster v_k. This value between, 0 and 1, can be used as a weight to scale how much a given centroid k should influence the sum which yields our missing value. The sum itself is the summation of the value at feature j in each centroid. With the weights, this should accurately place our missing value between all centroids of which it has some membership, this should also accurately reflect it's true value.

41.3   Methods

Four iterations of the improved RSFKM algorithm were implemented for experimentation and analysis. First, a standard CPU implementation of the RSFKM was implemented using a generic solver for Eq. (41.3). This implementation was developed to provide a clear baseline of the existing algorithm for timing purposes. Next, the sequential CPU-only implementation was optimized with the introduction of library free convex optimizer in lieu of the generic solver. From there, a single GPU implementation was introduced and optimized for speed and overhead efficiency. Finally, to account for restrictions on GPU memory and produce more consistent results, a multi-GPU iteration of this software was implemented.

41.3.1 Convex Optimization

In the work by Xu et al., the authors reference the use of "the technique" utilized by Huang et al. [16] to solve (41.3). To speed up development time for this baseline code and to promote reproducibility among computer scientists who are not intimately familiar with convex optimization, we diverged from the method in Xu et al. and utilized ECOS convex optimization solver [17] through the CVXpy interface [18]. Unfortunately, the introduction of this generic solver cased problems for plans of a GPU adaptation of this algorithm.

Presently, no generic solver currently exists which leverages the power of GPUs to speed up its numerous and dense calculations. Since preliminary timings of this solver indicated that this was a substantial bottleneck in the RSFKM algorithm, a workaround was necessary to enable the use of GPU architectures with RSFKM. To solve this problem, a library free solver, comprised of optimized C code generated by CVXGEN [19], was integrated into a modified sequential implementation of RSFKM. This improved implementation was significantly faster and provides a best case baseline to measure GPU versions against.

The code generated by CVXGEN was much more sympathetic to a GPU implementation of RSFKM compared to CVXpy as it was library free and written in C, which directly maps to CUDA. However, this code also introduced new problems hindering a successful GPU implementation. First, it generates code that only works on a pre-defined, constant vector size. In our case, that means it only works for a fixed number of centroids. Next, as this generated code was optimized for embedded systems, many frequently accessed variables were set at global scope. As this code would have to be adapted to run sequentially on a single thread (but with multiple instances running in parallel across multiple cores) this scoping creates problems. CUDA doesn't have an equivalent global scope that exists independently for each thread, so a means to trick the architecture was required.

41.3.2 GPU Adaptation

As alluded to in Sect. 41.3.1, the approach of prioritizing the adaptation of optimization code to GPUs came from two directions. First, the mathematical minimization formula for membership is fundamentally independent, indicating that calculations for membership vector u_i can occur at the same time as u_{i+1}. This makes the corresponding algorithm embarrassingly parallel. Second, the significant bottleneck of this minimization problem demanded resolution before other approaches to optimization and parallelization could be considered.

The problem of the globally scoped variables was solved first. To properly adapt the generated C code, a method was devised to maintain the structure of the program as developed to run on a CPU. To maintain the illusion of global scope, per-thread, the entirety of generated code, approximately 1500 lines, was encapsulated into a class-like struct. Encasing our C code into a struct allowed all required variables to be treated like private members which could be freely and safely accessed by the component functions which performed the minimization solving. The struct functions themselves were labeled as "__device__" functions and could then be called freely by individual threads from the "Update Membership" kernel.

As can be seen in Fig. 41.1, the modified kernel was called on n blocks, each using one thread. Although this configuration seems naive, attempts to call this sequential code with multiple threads per block, in multiples of 32 to take advantage of warp efficiency, saw much slower results than the one-thread-per-block approach. Furthermore, very few threads-per-block could be instantiated due to a lack of register space. This is likely due to the size of the solver struct being loaded to run on the GPU. By using this structure, our solver could make the best use of GPU resources to perform the embarrassingly parallel computations promised by Eq. (41.3) in the most straightforward manner possible.

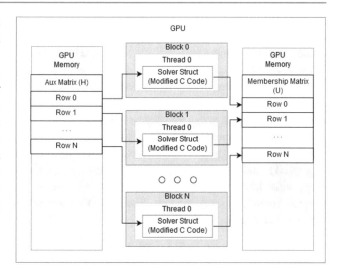

Fig. 41.1 A diagram showing a basic breakdown of the implementation and operation of the core solver implemented as a CUDA struct from modified C code. This diagram shows the mapping of the solver to each block and indicates how the problem is decomposed, where each line in our membership matrix is solved in parallel on the GPU. The decision to call this solver kernel on one thread per block came from slowdown observed with more than one thread per block

41.3.3 Further Optimization

After converting the primary bottleneck of the minimization solver into GPU code, many other opportunities for optimization became evident. In addition to the aforementioned "Update Membership" function, there existed three other computationally intensive functions which comprised the core functionality of this algorithm. Those functions are **build_h_matrix**, **find_centroids**, **update_S**. Generally, these functions involved some kind of matrix manipulation with a sum reduction and could be easily mapped one matrix cell to a block with multiple threads doing reduction and simple addition work in parallel. Shared memory was used for each of these functions as a buffer to hold summed data, pre-reduction, for fast access and simplicity of implementation.

Through rewriting these functions in CUDA to run on the GPU, and adding supplemental functions to store and calculate auxiliary scalars between iterations, the entirety of this program (within a single iteration) was successfully configured to run on the GPU. We successfully minimized overhead with this full conversion by initializing all derived matrices, **U**, **V**, **H**, and **S** in global memory on the GPU. Each of these matrices can be maintained on the GPUs main memory between the core kernel calls which perform the computations of our clustering algorithm. Although highly efficient, this memory configuration caused problems in terms of space efficiency which are addressed by the multi-GPU implementation of this algorithm.

The only matrix which could not be initialized on the GPU was the original data matrix $\mathbf{X}$, because its contents are read from the disk. This however does not significantly impact overall run time because the overhead of transferring this data to the GPU occurs only once before clustering begins. After this, the only continuous data transfer occurs at the end of each iteration when our centroids, $\mathbf{V}$, are transferred back down to the CPU to check for convergence. The overhead of this transfer is very minimal however, as the number of centroids is generally going to be relatively small to provide meaningful clustering. The final overhead from data movement between host and device comes from the output of the membership matrix $\mathbf{U}$. Like our data matrix, this matrix transmits between GPU to CPU only one time after all the iterations of this clustering algorithm have concluded. Because of this, the overhead of this retrieval can be considered negligible.

41.3.4 Multi-GPU

The need for a multi-GPU implementation of this algorithm was driven by the memory limitations of our single-GPU implementation. **calculate_centroids** requires a substantial sum reduction resulting from a matrix multiplication between the data matrix and an auxiliary matrix which holds a regulating scalar. After multiplication, but before reduction, result vectors are stored in shared memory assigned to a block. This requires our shared buffer to hold n spaces of size 8 bytes for the double precision values. CUDA provides a max of 48 KB per block. Accordingly, this only allows for approximately 6,000 double precision values to be processed by this function. At one point in this algorithm, this function handles every data value we intend to cluster. With the space limitations, this indicates that we cannot cluster more than 6,000 values with this program.

Although there are certainly many approaches and solutions to this problem, we chose to leverage data decomposition and split our calculations across multiple GPUs. By randomly sampling data points and distributing them across GPUs we can accurately cluster up to 48,000 data points in the same amount of time it would take for 6000. Although the memory limitation remains, per-gpu, it becomes significantly less impactful with each GPU we add. This horizontal scaling is not significantly impacted by overhead due to the design of our per-gpu algorithm, where all substantial matrices are instantiated and maintained on each GPU's main memory. Data transfer occurs minimally in between interactions to check for convergence and at the end to transfer back results for imputation.

41.3.5 Data Imputation

The data imputation algorithm was implemented in Python as a collection of methods which performed basic pre-processing, called the above detailed GPU clustering algorithm, removed data for imputation and performed the imputation algorithm specified in Eq. (41.5) using the centroids and returned membership matrix retrieved from the GPU. The implemented imputation method checked its accuracy using a simple RMSE algorithm finding the difference between the actual values removed from the dataset earlier against those which were imputed.

41.4 Experiments and Analysis

To evaluate the improved RSFKM algorithm on a single GPU and multiple GPUs, timings were collected on repeated clustering of a set of environmental sensor data downloaded from the Nevada Research Data Center's website [1]. Preliminary experiments showed that changing the number of features on our data and adjusting the number of centroids did very little to influence overall runtimes for either CPU or GPU implementations of this algorithm. So, for all the following experiments, the number of centroids used is fixed at 15 and the number of features is fixed at 11.

41.4.1 Experimental Setup

The following experiments were performed on a remote server containing 24 2.00 Ghz Intel Xeon CPUs connected over two PCIe buses to 8 GeForce GTX 1080 GPUs. The GTX 1080s are grouped 4 cards to a single bus. Each GPU has an available memory of 8 GB with 6 KB of shared memory available per block. Each GPU implements the Pascal architecture which provides support for advanced processing features like unified memory.

As an initial proof of concept, the clustered data set was a selection of 550,000 data points of time series data organized into 50,000 vectors with 11 features. Each vector represented a per-minute log of autonomously collected meteorological data with each feature representing a measurement collected at that minute from a distinct sensor. Table 41.1 shows an excerpt of the data set used.

41.4.2 Data Imputation

Although not explored in great detail for this paper due to constraints of space, this algorithm performed well in

Table 41.1 An excerpt of environmental data used for bench marking the effectiveness of the improved RSFKM clustering algorithm. The values depicted here are humidity and temperature measurements collected from many sites

1/25/2018 1:19	−5.678	−6.482	−6.499	−6.455	0.491	
1/25/2018 1:20	−5.654	−6.474	−6.499	−6.452	0.492	...
1/25/2018 1:21	−5.697	−6.462	−6.479	−6.452	0.494	
1/25/2018 1:22	−5.774	−6.481	−6.499	−6.449	0.494	
1/25/2018 1:23	−5.788	−6.491	−6.515	−6.472	0.494	
1/25/2018 1:24	−5.793	−6.503	−6.519	−6.478	0.492	
1/25/2018 1:25	−5.732	−6.515	−6.538	−6.492	0.494	

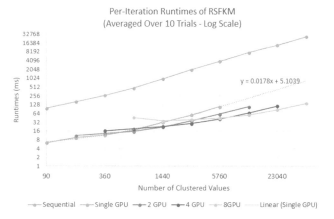

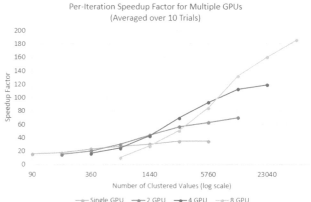

Fig. 41.2 A line plot showing the runtimes of the RSFKM algorithm running sequentially on a CPU and in parallel on one, two, four and eight GPUs. The runtimes are per iteration of the clustering algorithm and averaged over ten trials. The dotted line indicates a trendline which shows the projected runtimes of a single GPU experiment. The equation next to the trendline was used to derive runtime values for efficiency calculations

Fig. 41.3 A line plot showing the per-iteration speedup factor of singular and multi-GPU experiments run with the RSFKM algorithm. For larger numbers of clustered values, the addition of more GPUs produces near equivalently scaled speedup factors

providing reasonably accurate data imputation. With RMSE values as low as 0.18 the GPU implementation of this algorithm provided accurate clustering on par with the CPU implementation. Since the applications and datasets were very different between this paper and those tested by Xu et al. [15], a more detailed imputation experiment will be required and explored in future work.

41.4.3 Timings

Fuzzy K Means clustering, is a variable length iterative algorithm that does not converge uniformly at a set number of iterations. Between runs where the same data set is organized into the same number of clusters, convergence can occur in half as many iterations as expected. Accordingly, this makes collecting timings of full program runs problematic, because times can vary wildly. To solve this problem the runtimes collected and manipulated in the following sections were calculated per-iteration by dividing the overall runtime of this algorithm by the number of iterations taken. Using this timing scheme significantly reduced outliers in our timing data and produced meaningful results.

The raw per-iteration timings, shown in Fig. 41.2, show a comparison between the sequential implementation of

RSFKM and our modified GPU implementation, on one, two, four and eight GPUs. The sequential runtimes, indicated by the grey line, indicate a clear linear curve which is to be expected by the *O(NVF)* complexity of our algorithm. As *V*, the number of centroids, and *F*, the number of features in the data vectors, are fixed at relatively small sizes compared to *N*, the algorithm becomes linear as *N* becomes sufficiently large.

At a glance its evident that GPU implementations were very successful in reducing the runtimes compared to the sequential algorithm. On this logarithmically scaled graph, per-iteration runtimes of GPU implementations remain relatively level as data inputs scale from 90 vectors to approx. 40,000, ending with runtimes of 128 ms up from 6 ms. By comparison, across the same spread of processed data points the CPU implementation grows significantly in runtimes, from 128 ms to almost 32,000 ms. The difference between these runtimes is so severe that when rendered with the CPU runtimes, without a logarithmic scaling, the GPU runtimes cannot be seen as anything other than a multicolored horizontal line on the bottom of the graph.

The significance of this difference in runtimes is further reinforced by Fig. 41.3 which shows the speedup factor of various GPU timings when compared with the sequential timings. The speedup factor, *S*, was calculated with the following equation: $S = t_s/t_p$, with t_s and t_p representing sequential and GPU runtimes respectively. Overall, the

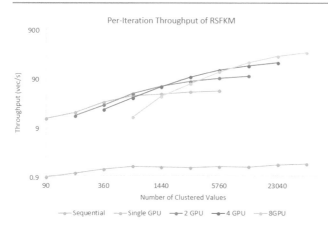

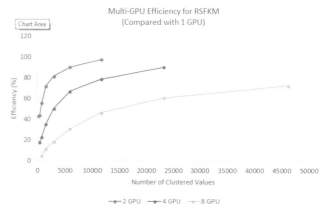

Fig. 41.4 A line plot showing per-iteration throughput of the RSFKM algorithm as run on CPU, one, two, four and eight GPUs. Throughput for this paper was expressed in terms of vectors/ms and was calculated as $thp = n/ms$ where n is the number of vectors input for clustering. GPU throughput handily outstrips sequential throughput by a wide amount, even when processing a small number of values

Fig. 41.5 A line plot showing the relative efficiency of multi gpu implementations vs. single GPU implementation. Calculated as $E = t_{gpu}/(t_{gpuN} * N) * 100$, where N is the number of GPUs being used, this graph indicates how much meaningful work each GPU is performing

speedup was very promising with a low of around $10\times$ speedup for 8 GPUs (with a small number of values) and a high of over $180\times$ speedup for 8 GPUs. Speedup was shown to be more significant as more GPUs were added and additional data points could be processed. This increase in speedup is certainly consistent with expectations because there is very little cost to calculations performed with two GPUs, compared to one due to a lack of communication overhead. Broadly this indicates that when the fundamental overhead of CPU/GPU communication is overcome by data processing then speedup scales with the number of GPUs utilized. Analysis of throughput and efficiency reinforce the conclusions made in reference to timings and speedup. In Fig. 41.4, its shown that overall throughput increases with more GPUs but only for sufficiently large amounts of data. For smaller amounts of clustered data we see a general dropoff of throughput when compared with fewer GPU configurations, again likely due to increased overhead that comes with the addition of multiple GPUs.

Similarly, Fig. 41.5 most clearly shows that at lower values very little work is being done on each GPU. As more GPUs are added the lack of work being done is exacerbated dramatically. For 360 values, 2 GPUs get near 40% efficiency with below 10% 8 GPUs. This difference in efficiency drops off dramatically however as more data is processed by each GPU. We do see that overall efficiency still drops off with the addition of more GPUs despite increased throughput and speedup. Again, this is likely due to increased competition for bus access which is harder to offset as more GPUs are added.

Within the GPU runtimes themselves some characteristics of these lines should be noted. First, there is a trendline on the single GPU runtimes. As explained in Sect. 41.3, space

limitations prevented successful operation of this program on one GPU with more than 6000 initial values. A trendline enables us to visualize and estimate runtimes going out towards the max processed 40,000 values. This trendline and its derived data will help provide efficiency data for our multi-GPU implementations.

It should be noted that with the two, four and eight GPU experiments in Fig. 41.2 a bow-like curvature can be seen. This is almost certainly due to the overhead of transferred data between our multiple GPUs and the CPU. Since only two PCIe buses connect GPUs and CPUs the overhead of data transfer becomes more pronounced as GPUs are added and each GPU must compete to transfer data. And, even though there is relatively little data transferred between the GPU and CPU with this algorithm, there is certainly enough, especially with the per iteration transfer of our centroids, that slowdowns occur when there is not a sufficient amount of data for each GPU to process. The downward curve occurs as a result of more data being computed on each GPU and causing less frequent calls back to the CPU to check for convergence. This in turn, reduces simultaneous demand for the limited bus access.

41.5 Conclusion and Future Works

This paper presented improvements to an existing Robust and Sparse Fuzzy K Means algorithm, that allowed for the entirety of its processing to be performed on a single GPU or multiple GPUs. This improved algorithm was demonstrated to provide accurate imputation and facilitate much faster clustering through numerous experiments performed on environmental time-series data acquired from the NRDC.

Specifically, this paper demonstrated that robust clustering can be performed on a single GPU with as much as

34 times speedup. It was also shown that this algorithm scales very well horizontally with allowing for the use of up to 8 GPUs, resulting in as much as 185 times speedup over sequential methods. It was also demonstrated that this algorithm was very efficiently designed to minimize communication overhead between CPU and GPU, with efficiencies as high as 97% shown with two GPUs.

In developing this GPU-based iRSFKM algorithm many avenues for continued development and research appeared. First, adjustments to current memory organization in this algorithm which are needed to facilitate each GPU handling more than 6000 data points. When a key goal of graphics processing is the ability to manipulate large amounts of data quickly, this restriction significantly limits the applicability of iRSFKM (even with the help of additional GPUs).

A further area of research which expands on this work would certainly be a much more comprehensive exploration of the effectiveness of data imputation with this algorithm on environmental sensor data. A better curated and preprocessed data set could provide significantly better results for data imputation than what was shown in this paper. Additionally, the data imputation algorithm in this paper was implemented for CPU only and can be adapted for GPUs. By implementing code which performs the imputation itself in CUDA the overall algorithm could run much faster and be better encapsulated into a distributable package for active use by data scientists in need of fast, effective imputation.

Acknowledgements This material is based upon work supported by the National Science Foundation under grant number IIA1301726. Any opinions, findings, and conclusions or recommendations expressed in this material are those of the authors and do not necessarily reflect the views of the National Science Foundation.

References

1. NRDC: Nevada research data center [Online]. Available: http://sensor.nevada.edu/NRDC/ pp. 1–6 (2018). Last accessed 12 Dec 2018
2. Schmitt, P., Mandel, J., Guedj, M.: A comparison of six methods for missing data imputation. J. Biometrics Biostat. **6**, 1–6 (2015). https://www.omicsonline.org/open-access/a-comparison-of-six-methods-for-missing-data-imputation-2155-6180-1000224.php?aid=54590
3. Soley-Bori, M.: Dealing with missing data: key assumptions and methods for applied analysis. Department of Health Policy and Management, School of Public Health, Boston University, Technical Report 4, May 2013
4. Beretta, L., Santaniello, A.: Nearest neighbor imputation algorithms: a critical evaluation. BMC Med. Inform. Decis. Mak. **16**(3), 74 (2016)
5. Dunn, J.C.: A fuzzy relative of the isodata process and its use in detecting compact well-separated clusters. J. Cybern. **3**(3), 32–57 (1973)
6. Banerjee, T., Keller, J.M., Skubic, M., Stone, E.: Day or night activity recognition from video using fuzzy clustering techniques. IEEE Trans. Fuzzy Syst. **22**(3), 483–493 (2014)
7. Valafar, F.: Pattern recognition techniques in microarray data analysis: a survey. Ann. N. Y. Acad. Sci. **980**(1), 41–64 (2002)
8. Li, D., Deogun, J., Spaulding, W., Shuart, B.: Towards missing data imputation: a study of fuzzy k-means clustering method. In: Tsumoto, S., Słowiński, R., Komorowski, J., Grzymała-Busse, J.W. (eds.) Rough Sets and Current Trends in Computing, pp. 573–579. Springer, Berlin/Heidelberg (2004)
9. Liao, Z., Lu, X., Yang, T., Wang, H.: Missing data imputation: a fuzzy k-means clustering algorithm over sliding window. In: Sixth International Conference on Fuzzy Systems and Knowledge Discovery, 2009. FSKD'09, vol. 3, pp. 133–137. IEEE (2008)
10. Tang, J., Zhang, G., Wang, Y., Wang, H., Liu, F.: A hybrid approach to integrate fuzzy c-means based imputation method with genetic algorithm for missing traffic volume data estimation. Transp. Res. Part C: Emerg. Technol. **51**, 29–40 (2015)
11. Azim, S., Aggarwal, S.: Hybrid model for data imputation: using fuzzy c means and multi layer perceptron. In: 2014 IEEE International Advance Computing Conference (IACC), pp. 1281–1285, Feb 2014
12. Shalom, S.A.A., Dash, M., Tue, M.: Efficient k-means clustering using accelerated graphics processors. In: Song, I.-Y., Eder, J., Nguyen, T.M. (eds.) Data Warehousing and Knowledge Discovery, pp. 166–175. Springer, Berlin/Heidelberg (2008)
13. Shalom, S.A.A., Dash, M., Tue, M.: Graphics hardware based efficient and scalable fuzzy c-means clustering. In: Proceedings of the 7th Australasian Data Mining Conference – Volume 87, AusDM'08, Darlinghurst, pp. 179–186. Australian Computer Society, Inc. (2008)
14. Al-Ayyoub, M., Abu-Dalo, A.M., Jararweh, Y., Jarrah, M., Sa'd, M.A.: A gpu-based implementations of the fuzzy c-means algorithms for medical image segmentation. J. Supercomput. **71**(8), 3149–3162 (2015)
15. Xu, J., Han, J., Xiong, K., Nie, F.: Robust and sparse fuzzy k-means clustering. In: Proceedings of the Twenty-Fifth International Joint Conference on Artificial Intelligence, IJCAI'16, pp. 2224–2230. AAAI Press (2016)
16. Huang, J., Nie, F., Huang, H.: A new simplex sparse learning model to measure data similarity for clustering. In: Proceedings of the 24th International Conference on Artificial Intelligence, IJCAI'15, pp. 3569–3575. AAAI Press (2015)
17. Domahidi, A., Chu, E., Boyd, S.: ECOS: an SOCP solver for embedded systems. In: European Control Conference (ECC), pp. 3071–3076 (2013)
18. Diamond, S., Boyd, S.: CVXPY: a Python-embedded modeling language for convex optimization. J. Mach. Learn. Res. **17**(83), 1–5 (2016)
19. Mattingley, J., Boyd, S.: Cvxgen: a code generator for embedded convex optimization. Optim. Eng. **13**(1), 1–27 (2012)

Syed Zawad, Feng Yan, Rui Wu, Lee Barford, and Frederick C. Harris Jr.

42.1 Introduction

The field of quantum information explores the possibilities of exploiting the laws of quantum mechanics to gain benefits in computational complexities that are otherwise largely problematic for classical computers to solve. Quantum computing involves using the superposition principle to carry out computational tasks in a more efficient way than is possible with devices governed by classical physics. There is a broad range of longstanding problems in strongly correlated systems, and quantum computing has shown great prospect in solving them. This has encouraged a significant increase in research, especially over the last decade. The tools required to design, build, and implement these quantum systems [1, 2] have seen rapid development and are still being developed, currently reaching some very sophisticated levels [3]. Research breakthroughs in ultra-cold atoms and photons have become more commonplace.

However, building quantum computers represents an immense technological challenge and, at present, the quantum hardware is only available in research labs. Under these circumstances quantum simulators have become valuable instruments in developing and testing quantum algorithms and in the simulation of physical models used in the implementation of a quantum processor. Simulating a quantum computer on a classical computer is a computationally hard problem. According to Feynman's paper [1] classic computers will never be able to perform simulations of full behavior of a quantum system in a polynomial time. Because of the exponential behavior of quantum systems, simulating them on conventional computers requires an exponential amount of operations and storage. Parallelization alleviates this problem, allowing the simulation of more qubits at the same time or the same number of qubits to be simulated in less time.

This parallelism can be achieved through the use of Graphics Processing Units. Modern graphics processing units (GPUs) have been at the leading edge of increasing chip-level parallelism for some time. While originally designed to perform calculations for graphics in video editing and video games, they are now being widely used for general purpose programming in many fields which require intense parallel computations. For this paper, we demonstrate that GPUs are a viable candidate for simulating quantum computers in terms of both accuracy and speed of execution. We also show how the GPU accuracy changes with changes in its floating point precision, and how that can be used to simulate accurate models of decaying quantum gates. This paper makes three contributions -

- *Develop a GPU-based Quantum Simulation framework to accurately simulate the application of Quantum Gates –* The simulation uses the Clifford gates [4] model, and is implement in C and CUDA. It provides the linear algebra calculations required to model the fundamentals of the Clifford gates system. The framework has support for multiple GPUs.
- *Optimize quantum simulator calculations –* Certain linear algebra functions such as transpositions and matrix inner

S. Zawad · F. Yan · F. C. Harris Jr. (✉)
Department of Computer Science and Engineering,
University of Nevada, Reno, NV, USA
e-mail: szawad@nevada.unr.edu; fyan@unr.edu;
fred.harris@cse.unr.edu

R. Wu
Department of Computer Science, East Carolina University,
Greenville, NC, USA
e-mail: wur18@ecu.edu

L. Barford
Department of Computer Science and Engineering,
University of Nevada, Reno, NV, USA

Keysight Laboratories, Keysight Technologies, Reno, NV, USA
e-mail: lee.barford@ieee.org

© Springer Nature Switzerland AG 2019
S. Latifi (ed.), *16th International Conference on Information Technology-New Generations (ITNG 2019)*,
Advances in Intelligent Systems and Computing 800,
https://doi.org/10.1007/978-3-030-14070-0_42

products can be considered redundant for some parts of the quantum computing model. This paper discusses how they can be removed or optimized to speed up the execution for quantum bit calculations as well as how and where to reduce the problem size.

- *Apply Randomized Benchmarking* – Randomized Benchmarking is currently one of the more popular tests which are used to analyze the fidelity of hardware quantum gates. It measures how accurate a certain hardware is in terms of performing quantum calculations. We apply this to our GPU implementation and show that GPUs can be an accurate simulator. We also vary the precision of the calculations manually and show how a varying precision can accurately model the gate incoherence present in real systems.

The rest of this paper is organized as follows: In Sect. 42.2 we talk about the background of our quantum model. We provide a short explanation of the randomized benchmarking algorithm and give a brief description of the NVidia CUDA architecture. In Sect. 42.3 we review the current state of research in the application of quantum computing simulations for GPUs and explain how our research fits there. In Sect. 42.4 we discuss our implementations details, our optimizations and parallel strategies. For Sect. 42.5, we provide the results of our optimizations and parallel implementations. We then compare and discuss our observations, and draw our conclusions and present some future work in Sect. 42.6.

42.2 Background

42.2.1 Quantum Computing Model

The concept of using Quantum Computer simulations in a classical computer were an extension of Feynman [1]. The application of the principles of quantum physics in the computer area led to the concept of quantum computer, in which the data isn't stored in bits like in the conventional memory, but as a combined state of several systems with 2 qubit states. Nowadays, the most common model involves using Pauli matrices and Clifford groups to perform computations for quantum processes, consistent with the Gottesman – Knill theorem [5, 6].

The fundamental concept, or representation of a quantum system revolves around the *qubit*. The qubit can be considered as the equivalent bit representation of a quantum system. While a classical computer uses 0's or 1's to represent data, a quantum system uses the qubit. The qubit is represented as a square matrix of the dimensions 2^n-by-2^n, where n is the

$$
\begin{array}{c c c c c}
 & 00 & 01 & 10 & 11 \\
00 & \begin{bmatrix} 0.1 & 0 & 0 & 0 \\ 01 & 0 & 0.2 & 0 & 0 \\ 10 & 0 & 0 & 0.3 & 0 \\ 11 & 0 & 0 & 0 & 0.4 \end{bmatrix}
\end{array}
$$

Fig. 42.1 A 2-bit qubit density matrix representation

$$
\begin{bmatrix} 0 & 1 \\ 1 & 0 \end{bmatrix}
$$

Fig. 42.2 X (qubit NOT) gate. It switches the probabilities between the bit combinations

number of bits in the qubit. Figure 42.1. Shows the density matrix of a 2-bit qubit.

Here, the rows and columns represent the possible bit combinations, and the corresponding diagonal values represent the probability with which the bit combination will be returned when the qubit state is read. The density matrix contains real and imaginary values. The density matrices must be have sum of diagonal elements to 1, must be a Hermetian and a positive semi-definite matrix.

Similarly, as with classical computers, quantum computers also have gates which transform the probabilities of the bit combinations in a qubit. They are also represented by 2-by-2 density matrices, and are complex and unitary. Figure 42.2 shows the qubit representation of an X gate (also called the qubit *NOT* gate).

A 2-by-2 gate, however, can only be applied to a single qubit density matrix. In order to be applied to higher order matrices, the gate must be expanded to equal the dimensions of the qubit itself. This is done via creating the appropriate Kronecker product of the gate. The formula for creating an n-qubit gate is as follows –

$$
\underbrace{\underbrace{I \otimes \cdots \otimes I}_{i-1 \text{ times}} \otimes G \otimes \underbrace{I \otimes \cdots \otimes I}_{n-i \text{ times}}}_{n \text{ times total}}
$$

where i is the ith qubit that the gate will apply to, n is the number of qubits, I is a 2-by-2 identity matrix, G is the 2-by-2 gate matrix, and $\otimes$ represents the kronecker product function. The above equation results in a 2^n-by-2^n matrix which can be applied to the 2^n-by-2^n qubit density matrix. Application of qubit gates to qubits is done by –

$$
Q_{\text{new}} = G^\dagger Q G
$$

Here, Q_{new} is the newly transformed qubit resulting from the application of gate G to the original qubit Q. $G^\dagger$ is the Hermetian of the gate G.

42.2.2 Randomized Benchmarking

The purpose of the Randomized Benchmarking method is to find how accurate the computations are after applying the gates to a certain initial starting qubit. What it measures is the "proximity" of the result after the application of gates. This "proximity" can be considered as how erroneous the gates are and is known as the *infidelity*. Algorithm 42.1 shows the code for Randomized Benchmarking.

Algorithm 42.1

Randomized Benchmarking

```
Input:
  Number of Runs N
  Number of Gates M.
  Initial Qubit Qubit
Output:
Fidelity
1 forn: 1 to Ndo
2    form: 1 to Mdo
3Gₘ = getRandomGate();
4       Qubit = applyGate(Gₘ, Qubit);
5GateStack.push(Gₘ);
6    end
7    form: M to 1do
8Gₘ = hermetian(GateStack.pop());
9Qubit = applyGate(Gₘ,Qubit);
10   end
11fidelity = calculateFidelity(Qubit);
12 end
```

The algorithm takes in the number of test runs N, the number of gates to apply per run M and the initial qubit *Qubit* as inputs. For every run, M gates are chosen at random and applied one after the other. These gates are then stored in a LIFO queue. The initial Qubit undergoes a series of probability transformations due to the application of these gates that results in a completely different set of probabilities. After the first inner loop, the initial Qubit has been destroyed. The next inner loop pops the queues from the LIFO Queue and applies the Hermetian transformation to the gate. One of the gate properties is that the conjugate transpose of the gates is equal to itself, and that they are unitary gates. This means that by applying a gate to a Qubit and applying the Hermetian of the gate to the resulting qubit will result in the Qubit reverting back to its original state. As such, the application of a sequence of gates and then the Hermetians of those gates should ideally result in the initial Qubit. Practically, however, quantum hardware does not give the exact initial state. Due to noise in gate hardware, the exact calculations are rarely accurate, and this becomes apparent after the application of the gates and their Hermetians. It is expected that as more gates are applied, the "further" the final state is from the initial state.

42.2.3 NVidia CUDA and GPUs

The Compute Unified Device Architecture (CUDA) library, developed by NVIDIA, is a software and hardware architecture that enables the users to harness the high counts of parallel processing power of the recent NVIDIA graphics cards. From the hardware perspective, the GPU consists of several multiprocessors working in a SIMT (Single Instruction Multiple Thread) fashion, each of them containing a certain number of streaming processors. In order to develop GPU-enabled applications, programmers can make use of various programming languages: C/C++ for CUDA, OpenCL, Fortran or DirectCompute. However, CUDA is the proprietary library provided by the hardware developers themselves, and thus provides many functionalities that allow users to fully utilize the hardware. The primary difference between a GPU and a CPU is that GPUs contain a high number of less powerful cores while CPUs contain a few number of highly powerful cores. Other than that, the other properties are similar. GPUs contain its own memory spaces; (1) Global memory – Data stored in global memory is visible to all threads within the application (including the host), and lasts for the duration of the host allocation. (2) Shared Memory – Data stored here is visible to all threads within the block that allocates it. This type of memory allows for inter-thread communication to occur and permits the sharing of data between threads. This is also faster than the global memory. (3) Local memory – Local memory has the same properties as normal registers, but has a larger available memory size but it is much slower. Apart from these, there are also Texture and Constant memory available, but it is not relevant in this context. CUDA provides a hierarchical execution model for execution of its threads. The abstraction is done at four levels. These abstractions are (in order of highest abstraction to the lowest) – grids, blocks, warps and threads.

42.3 Related Work

The concept of Quantum Systems was first put forth by Feynmen [1]. The paper emphasized the complexity of simulating quantum systems using classical computers. A well-controlled system can be built from the bottom up, and by doing so, one could create a computer whose constituent parts are governed by quantum dynamics generated by a desired Hamiltonian. However, at that time period, the computational power required to even describe the quantum system which scales exponentially with the number of its qubits was practically infeasible. Additionally, delving this deeply into the properties of this system led to the discovery of difficult to compute properties of a quantum many-body model, such as the nature of quantum-phase diagrams. This initial proposed model is known as the "quantum simulator".

Since then, there have been great strides towards feasible quantum computers, even though much work is yet to be done to make quantum computers mainstream. As such, quantum computing simulation libraries are still what drives the majority of research. These simulators come in different variations and levels of complexity [7], each catering to a specific need. Sequential quantum simulators are many, and have a good variety of representations and contains different types of simulators [8]: quantum programming languages (QCL, Q language, Quantum Superpositions, QuBit), quantum compilers (Qubiter, GQC), quantum circuits simulators (QCAD, QuaSi, Libquantum), quantum hardware emulators (QCE, QSS) and purely pedagogical software (quantum Turing machine simulator, Quantum Search Simulator, Shor's algorithm simulator). The need for parallel simulators emerges due to the super-exponential nature of quantum computation. It is extremely time consuming for classical computing devices simulate it. The first parallel simulator was developed by Obenland and Despain [9], but was based on the physical model of a very specific model and so did not see mainstream use. Since then, there have been many parallel CPU implementations, [9, 10]. However, they all fall short due to lack of scalability. The latest approach that researchers are taking to further optimize the simulators are through the GPUs. Much work has already been done on simulating specifics applications [11, 12]. Similarly, several generic single GPU and distributed GPU systems for quantum simulators have been developed and see widespread use [13, 14].

One of the key problems with quantum systems are that they have degrading accuracy [15, 16]. While many companies are racing to become the first to develop the first real quantum computer, one of the major hurdles of noisy systems still exist. To facilitate the work in this field, researchers have come up with standards to measure and standardize the testing process [17–21]. The industry standard so far has been QST and GST, but [22] argues that they are both too slow and bad at scaling. To counter this, they propose a new type of testing called Randomized Benchmarking. Since then, there have been quite a few experiments where RB has been used to good effect [22–24]. The scalability, simplicity and ease of implementation have made RB one of the currently favored testing methods for quantum hardware. While much work regarding quantum simulators have been done, there has been no research into how RB will perform on GPUs. Since GPUs are now the go-to implementations for quantum simulators, we need to find a means of simulating the deteriorating effect of applying gates to qubits.

42.4 Implementation

One of our key motivations is to be able to generate the decaying effect shown by real quantum gates on the GPU. We do this by manipulating the floating point precision of the calculations of the operations. We do this via the formula –

$$R_{factor} = 1 \ll M$$

$$Value = \frac{round\left(Value_{original} * R_{factor}\right)}{R_{factor}}$$

Here, M is the precision in bits, and $Value_{original}$ is the actual value with full 64-bit precision. This function is applied to all mathematical calculations done in every thread. For our implementation, we start by setting the probability of all qubits being 0 to 1.0. In other words, our Q_{old} [0][0] is set to 1.0. With full precision, we expect that the value for Q_{old} [0][0] will be 1.0 after going through Randomized Benchmarking. With less precision, we will find that the value will deviate from 1.0. *Fidelity* is the measure of how much the final value deviates. This is given by –

$$Fidelity = 1 - conj\left(Q\left[0\right]\left[0\right]\right) * Q\left[0\right]\left[0\right]$$

Q is the final qubit after applying Randomized Benchmarking. The closer *Fidelity* is to 1.0, the less the noise is in the system. For our experiments, we vary the precision and measure the corresponding *Fidelity* with a range of gates. Next, we talk about the two facets to the implementation of our framework. First we focus on reducing the number of computations for the operations while maintaining the accuracy of the results. Then we focus on how to implement a parallelized version of the algorithm for deployment on the GPU.

42.4.1 Optimizations

For the linear algebra operations involved in applying the gates to the quantum bit and for calculating the Hermetians, the number of calculations were reduced by taking advantage of the matrix properties of the gates and the qubits. These properties enable the reduction of the problem size due to rendering most of the computations necessary redundant. Specifically, three major algorithmic changes were done.

The first change was to only calculate the Hermetian of gates (line 8 of Algorithm 42.1) for specific gates. The traditional way of applying gates in quantum simulators use the transposed conjugate of the gate matrix and the gate matrix itself. The original qubit is then right multiplied by the transpose-conjugate and left multiplied by the original gate. These inner products are done in sequence to each other. It should also be noted that for multi-qubit systems the multi-qubit gate is first generated by the Kronecker product and then the conjugate transpose is applied. However, this conjugate transpose step can be completely ignored due to the fact that gate matrices must be Hermetian matrices. In other words, one of the properties that a matrix must fulfill to be a gate is that it must be a conjugate transpose of itself. Therefore, the Clifford group of gates (X, Y, H, Z) used are all Hermetians, meaning that they do are not required to undergo the Hermetian transformation. This change effectively removes one full computational step of the $O(2^{2n+1})$, where n is the number of qubits. The second optimization done involves using the Eigen property of the gates that reduces the computations to only using the upper half of the square matrices. Algorithm 42.1 can be unwound in the form of the equation –

$$G_{right} = G_1 G_2 \dots G_{N-1} G_N G_N^H G_{N-1}^H \dots G_2^H G_1^H$$

$$G_{left} = G_1 G_2 \dots G_{N-1} G_N G_N^H G_{N-1}^H \dots G_2^H G_1^H$$

$$Q_{new} = G_{right} Q_{old} G_{left}$$

where N is the total number of gates. However, we know that the Qubit Q_{old} can be decomposed as –

$$Q_{old} = L L^T$$

where L is the Eigen decomposition of Q_{old}. So now we can reduce the original equations to

$$G_{right} = G_1 G_2 \dots G_{N-1} G_N G_N^H G_{N-1}^H \dots G_2^H G_1^H L$$

$$G_{left} = G_1 G_2 \dots G_{N-1} G_N G_N^H G_{N-1}^H \dots G_2^H G_1^H L^T$$

$$Q_{new} = G_{right} G_{left}$$

When calculating the infidelity, we are only interested in the first diagonal value of the resulting matrix, i.e. $Q_{new}[0][0]$. Therefore, instead of calculating the full matrix, we may simply get the inner product of the first column of G_{right} and the first row of G_{left}. Thus, the total number of calculations of the full algorithm sequence can effectively be reduce to half, which contributes greatly to the reduction of the execution time of the complete run.

The last and most impactful of the optimizations is the complete removal of the calculation of the Kronecker product. The calculation of the Kronecker product takes place in line 3 of Algorithm 42.1. While getting the random gate, a random qubit is chosen from the N qubits. Then a random gate from among the Clifford groups is chosen and then the appropriate number of Kronecker multiplications are done to get the full N-qubit gate. Generating this gate for every step of the algorithm is time consuming, and any reduction here will reduce the total execution time greatly. The eventual product of the Kronecker multiplications is a full n-qubit gate, which is then multiplied against the qubit matrix. However, the n-qubit gate is very sparse, meaning that most of the values need not be generated via Kronecker at all. Additionally, given the index, the exact value of the Kronecker product can be deduced due to the structured expansion that Kronecker multiplications result in.

42.4.2 CUDA Implementation

As shown in Algorithm 42.1, there is one outer loop and two inner loops. The inner loops are involved in the actual calculations while the outer loop is based on the number of runs. The two inner loops contain the inner product of the matrices. However, in order for the calculations to be accurate, these operations must be applied in sequence. Therefore, the scope for parallelization is within the operation calculation itself. The operations to be parallelized here are the Hermetian, the optimized Kronecker products (n-qubit gate generation) and the inner product of the qubits and the gate matrices. The memory was allocated using CUDA's unified memory.

For parallelizing the inner product for this simulation scheme, the qubit state matrix is partitioned in sets of fixed dimensions and assigned to CUDA blocks, where the sub-vectors are processed in parallel on a SIMT (single Instruction Multiple Thread) fashion. The index multiplications are done in parallel threads. The final sum is done by reduction, where every summation between two values are done in parallel. The major bottleneck here is during the synchronization of the threads; every sum step needs a blocking call for the previous summing step so that the values required for the sum have been correctly calculated. For a matrix inner product of two square matrices of the dimensions 2^n-by-2^n, we will end up with a new matrix of dimensions 2^n-by-2^n. For maximum parallelism, every index is calculated in a separate independent thread. The multiplication and reduction are also done in independent threads. During deployment of the threads, the number of threads per CUDA block was kept at the maximum possible number of threads (i.e. 1024) for the GPU experimented on. This is not enough for large qubit sizes, so the number of blocks used were set such that –

$$number\ of\ blocks = \frac{2^{2n}}{1024}$$

This ensures that there are always enough parallel threads to get the maximum parallelism possible. There were also checks in place to ensure that the number of threads were multiples of 32 in order to keep the number of threads consistent with the warp sizes.

For the Hermetian calculations, the conjugate and the transpose functions on a single matrix index were merged to form a single operation where given an index, the transpose of the index is calculated by getting the value of the "mirror" index and then the conjugate of that value calculated and stored. Each index is run in a separate thread, with the threads distributed in the same manner as for the inner product. The dynamic qubit gate generation without the use of Kronecker products is also done on the GPU. The gate is generated and stored in the GPU's global memory, so there is no need to transfer the large gates from the host to the device and vice versa. The function `applyGate()` from Algorithm 42.1 deploys a single thread for the each of the 2^{2n} values in the matrix. This is run once when a random gate is generated (line 3 Algorithm 42.1). The execution time for the dynamically generated gate instead of doing the Kronecker products is further beneficial since the Kronecker product calculations, had they been done on the GPU, would have required thread synchronization for every loop iteration. However, the dynamic generation requires no such bottleneck and the gates require much less time to be generated. The multi-GPU implementation is done for the outer loops. Note that the final value for fidelity is calculated for every run of the outer loop, and is independent of the rest of the other calculations. Therefore, the outer loop can also be deployed in parallel threads. We take advantage of this fact by deploying the different outer loops in different GPUs. The devices never need to be synchronized.

42.5   Experiment and Results

The test bed used for the experiments contain 4 NVidia GTX 1080s. The CPU is Intel(R) Xeon(R) CPU E3-1225 v3, 3.20GHz with 4 Cores and has 64 GB of RAM. The operating system used was Linux's Ubuntu 16.06. The timings were taken using CUDA's event synchronization library functions. The values used are means from 20 runs. Profiling was done using NVidia's profiler. Figure 42.3 shows the pure execution time for the single GPU implementations, including the computation reduction optimizations. The log-scaled graph shows the clear benefits in performance of the GPU's parallelized version. We can see that for smaller number of qubits, the sequential version is better since the overhead of moving the qubit from the CPU to the GPU is too large to make up for the execution time reduction. However, at around the 5th qubit, the execution time of the sequential version starts taking more time than the parallel version. By the 13th qubit, the total speedup is around 400 times.

The throughput graph (Fig. 42.4) shows that up to the 7th qubit there is a clear benefit from parallelism but plateaus out after that. The bulk of the computations among all the steps are taken up by the matrix multiplications. The synchronization required for the sum reduction becomes a significant bottleneck at that stage, resulting in a flat throughput. Figure 42.5 presents the execution time of the multi-GPU implementation of the Randomized Benchmarking algorithm. As expected, the 2-GPU implementation is clearly almost twice as better as the single GPU case.

However, the addition of more GPUs do not clearly benefit at the same scale, as can be see for the 3 and 4 GPU cases. As the number of GPUs increase, the benefits of adding more GPUs tend to decrease. The overall execution time does decrease with increasing number of runs, and it scales linearly. Ideally, with 4 GPUs we should expect a

Fig. 42.3 Sequential versus parallel execution times

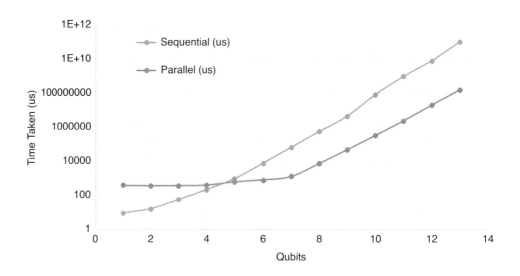

Fig. 42.4 Sequential versus
parallel throughput

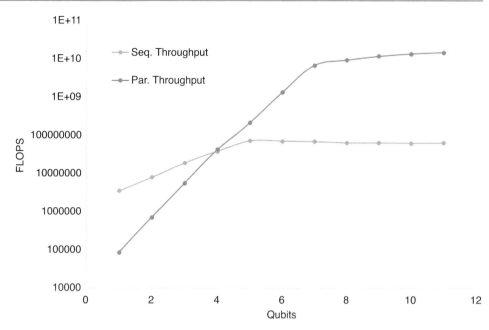

Fig. 42.5 Multi-GPU execution
time

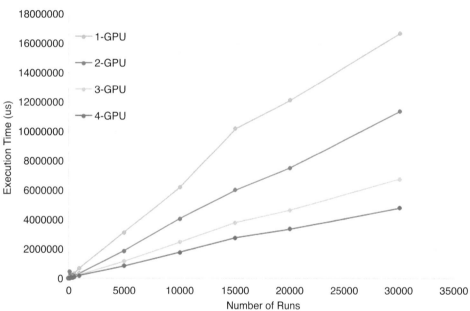

speedup of 4 times if the scale was perfectly linear. The
graphs show that this is not the case, and we have a speed
up of around 3.6 times with 4 GPUs. This phenomenon can
be explained by the fact that while pure execution time may
speed up fourfold, the overall execution time given here also
includes the data transfer time from host to device and vice
versa, which does not scale linearly. Thus a sub-linear effect
is introduced and the benefits of using multiple GPUs are
somewhat diminished. Nonetheless, the speed up achieved is
90% of the maximum expected.

The final set of experiments performed focuses on the
evaluation of the GPU's calculated fidelity based on the
varying bit precision. The full implementation of the system
was done using double precision. We have observed that
the fidelity largely stays the same throughout the bit values
between 64 and 10 bits, no matter how many gates are
applied after each other. Once we started seeing decreasing
fidelity, we varied the number of gates and the qubit size
to understand how the fidelity changes against them. Figure
42.6 shows how the fidelity is affected by an increasing
number of gates. The data shown is the calculated mean and
is using a precision of 9 bits. The trend here shows a strong
inverse correlation between the increasing number of gates
applied and the fidelity. This is consistent with the findings
in [16, 18], where real quantum systems show a tendency to
deteriorate with higher number of gates.

Fig. 42.6 Fidelity vs. number of gates applied with varying qubits

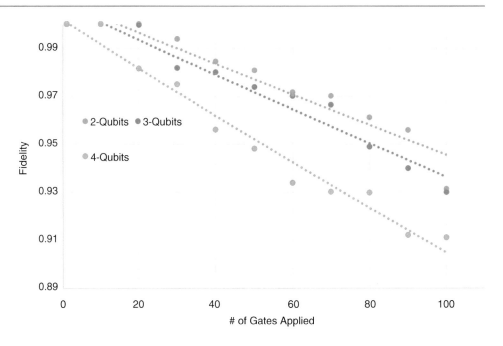

Another observation here is that for higher qubits, the fidelity deteriorates faster than for smaller qubit sizes. This is due to the fact that larger qubits undergo more computations and are thus more affected by round off errors occurring due to coarser precisions.

42.6 Conclusion

We introduce a set of new linear algebra optimizations that reduce the problem sizes of quantum computing calculations that drastically reduce execution time as well as make it easy to parallelize. These should be generic enough to be applicable for all quantum computing systems since these reductions were done at the most basic levels which are required by virtually all quantum simulators. We have conducted experiments based on which we can conclude that Quantum simulators benefit greatly from their usage of the parallelism afforded by multi-GPU systems in addition to the reduction in problems size, achieving a speedup of more than 400 times. Additionally, simply by changing the precision we were able to simulate the decaying effect of real quantum systems, making it a lightweight solution to a complicated simulation problem. While this is a sufficient solution currently, future work can involve usage of Gaussian noise to make gate decay simulation more faithful to how actual Quantum systems perform.

Acknowledgments This material is based in part upon work supported by the National Science Foundation under grant number IIA- 1301726. Any opinions, findings, and conclusions or recommendations expressed in this material are those of the authors and do not necessarily reflect the views of the National Science Foundation.

References

1. Feynman, R.: Simulating physics with computers. Int. J. Theor. Phys. **21**(6–7), 467–488 (1982)
2. Buluta, I., Nori, F.: Quantum simulators. Science. **326**(5949), 108–111 (2009)
3. Lin, Y.-J., Compton, R.L., Jimenez-Garcia, K., Porto, J.V., Spielman, I.B.: Synthetic magnetic fields for ultracold neutral atoms. Nature. **462**(7273), 628–632 (2009)
4. Bravyi, S., Kitaev, A.: Universal quantum computation with ideal Clifford gates and noisy ancillas. Phys. Rev. A. **71**(2), 022316 (2005)
5. Gottesman, D.: The heisenberg representation of quantum computers. In: Proceedings of the 22nd International Colloquium on Group Theoretical Methods in Physics, pp. 32–43. International Press, Cambridge (1999)., p. 23
6. Aaronson, S., Gottesman, D.: Improved simulation of stabilizer circuits. Phys. Rev. A. **70**(5), 052328 (2004)
7. Eason, G., Noble, B., Sneddon, I.N.: On certain integrals of Lipschitz-Hankel type involving products of Bessel functions. Philos. Trans. R. Soc. Lond. A. **247**, 529–551 (1955)
8. Quantiki: List of QC simulators. www.quantiki.org/wiki/list-qc-simulators (25 Apr 2018)
9. Obenland, K.M., Despain, A.M.: A parallel quantum computer simulator. *arXiv* preprint quant-ph/9804039 (1998)
10. Brandl, M.F.: A quantum von Neumann architecture for large-scale quantum computing in systems with long coherence times, such as trapped ions. *arXiv* preprint arXiv:1702.02583 (2017)
11. Ufimtsev, I.S., Martinez, T.J.: Quantum chemistry on graphical processing units – strategies for two-electron integral evaluation. J. Chem. Theory Comput. **4**(2), 222–231 (2008)
12. Maia, J.D., Urquiza Carvalho, G.A., Mangueira Jr., C.P., Santana, S.R., Cabral, L.A., Rocha, G.B.: GPU linear algebra libraries and GPGPU programming for accelerating MOPAC semiempirical quantum chemistry calculations. J. Chem. Theory Comput. **8**(9), 3072–3081 (2012)

13. Amariutei, A., Caraiman, S.: Parallel quantum computer simulation on the GPU. In: Proceedings of the 15th International Conference on System Theory, Control, and Computing (ICSTCC), pp. 1–6. IEEE, New York (2011). https://ieeexplore.ieee.org/stamp/stamp.jsp?arnumber=6085728

14. Gutierrez, E., Romero, S., Trenas, M.A., Zapata, E.L.: Parallel quantum computer simulation on the CUDA architecture. In: Bubak, M., van Albada, G.D., Dongarra, J., Sloot, P.M.A. (eds.) Computational Science – ICCS 2008. ICCS 2008. Lecture Notes in Computer Science, vol. 5101, pp. 700–709. Springer, Berlin/Heidelberg

15. Deutsch, D., Jozsa, R.: Rapid solution of problems by quantum computation. Proc. R. Soc. Lond. A. **439**(1907), 553–558 (1992)

16. Shor, P.W.: Scheme for reducing decoherence in quantum computer memory. Phys. Rev. A. **52**(4), R2493 (1995)

17. O'Brien, J.L., Pryde, G.J., Gilchrist, A., James, D.F.V., Langford, N.K., Ralph, T.C., White, A.G.: Quantum process tomography of a controlled-NOT gate. Phys. Rev. Lett. **93**(8), 080502 (2004)

18. Knill, E.: Quantum computing with realistically noisy devices. Nature. **434**(7029), 39 (2005)

19. Reichardt, B.W.: Quantum universality by state distillation, *arXiv* preprint quant-ph/0608085 (2006)

20. Raussendorf, R., Harrington, J.: Fault-tolerant quantum computation with high threshold in two dimensions. Phys. Rev. Lett. **98**(19), 190504 (2007)

21. Mohseni, M., Rezakhani, A.T., Lidar, D.A.: Quantum-process tomography: resource analysis of different strategies. Phys. Rev. A. **77**(3), 032322 (2008)

22. Knill, E., Leibfried, D., Reichle, R., Britton, J., Blakestad, R.B., Jost, J.D., Langer, C., Ozeri, R., Seidelin, S., Wineland, D.J.: Randomized benchmarking of quantum gates. Phys. Rev. A. **77**(1), 012307 (2008)

23. Magesan, E., Gambetta, J.M., Emerson, J.: Scalable and robust randomized benchmarking of quantum processes. Phys. Rev. Lett. **106**(18), 180504 (2011)

24. Ryan, C.A., Laforest, M., Laflamme, R.: Randomized benchmarking of single-and multiqubit control in liquid-state NMR quantum information processing. New J. Phys. **11**(1), 013034 (2009)

A Memory Layout for Dynamically Routed Capsule Layers

A Memory Layout for Dynamically Routed Capsule Layers

43

Daniel A. Lopez, Rui Wu, Lee Barford, and Frederick C. Harris Jr.

43.1 Introduction

Capsule Networks are a novel interpretation of neural networks where the perceptron model has been expanded to process vectors rather than scalars [12]. In this model, the orientation of the output vector corresponds to pose parameters of the feature being classified and the length of the vector corresponds to the probability of the feature being detected in the input. Analogously to a conventional neural network, the capsule outputs from one layer (after undergoing some dimensionality transformation via transformation matrices) to the next are compiled via some weighted sum-reduction algorithm. The weights are computing by a routing-by-agreement algorithm which, similarly to K-means [7], strengthens connections between vectors whose outputs tend to cluster together, thus exploiting the rareness of "agreeing" vectors in higher dimensions found in discriminatory learning [4]. The original theorists of this algorithm for dynamic routing stress that this straightforward method is Although the authors of [12] present this straightforward method, they also stress that there are many different ways to produce a similar output such as in [6].

The presented CapsNet architecture in [12] features a convolutional layer, followed by a capsule layer, the latter of which contains a capsule for each feature class defined. The dynamic routing algorithm occurs between the two layers in a network, where the initial "output" vectors of the first capsule layer are composed of the same pixel value from many filter outputs from the previous convolutional layer. In machine learning libraries, individual tasks are distributively organized to accommodate many different devices and easy scaling from a single machine to a distributed system [1]. However, when implementing Capsule Networks, memory resources are wasted when the tensor representing the cube output from the convolutional layer is transformed via these tasks to a set of 8D vectors. Moreover, during reconstruction error generation, although tasks are modeled alongside their dependencies in a graph-based representation in an effort hide latency, redundant or unnecessary tasks may accidentally execute. This GPU acceleration method does not compute the reconstruction error.

The number of vector channels in the reshaping of the convolutional is varied, and thus, the number of filters in the PrimaryCaps layer is d_l times multiplied by the number of vector channels. This paper proposes a set of CUDA kernels which aim to optimize the operations of a capsule network.

The rest of this paper is structured as follows: Background and related work is covered in Sect. 43.2. Section 43.3 presents the methodology from data allocation through algorithmic definitions all the way to loss and activation functions. Results are presented in Sect. 43.4 including timings, speedup and throughput calculations. Conclusions and future work follow in Sect. 43.5.

D. A. Lopez · F. C. Harris Jr. (✉)
Department of Computer Science and Engineering,
University of Nevada, Reno, NV, USA
e-mail: daniellopez@nevada.unr.edu; fred.harris@cse.unr.edu

R. Wu
Department of Computer Science, East Carolina University,
Greenville, NC, USA
e-mail: wur18@ecu.edu

L. Barford
Department of Computer Science and Engineering,
University of Nevada, Reno, NV, USA

Keysight Laboratories, Keysight Technologies, Reno, NV, USA
e-mail: lee.barford@ieee.org

© Springer Nature Switzerland AG 2019
S. Latifi (ed.), *16th International Conference on Information Technology-New Generations (ITNG 2019)*,
Advances in Intelligent Systems and Computing 800,
https://doi.org/10.1007/978-3-030-14070-0_43

43.2 Background and Related Work

Convolutional layer operations are a widely attacked problem for two reasons: the resulting abstraction and how easily it lends itself to parallelization. The abstraction provided helps facilitate higher order algorithms map single dimensional or stream based data into multidimensional data. Its distributed nature yielded many multi-GPU optimizations in recent years for machine learning libraries [2, 3, 9]. However, this generalization has cornered advancements to be accompanied with structural overhead limiting full optimization. This work addresses the short comings of a specific use case: the interface between the convolutional layer and an immediate capsule layer.

43.2.1 Existing Tensor Flow Implementations

TensorFlow provides an API to a model centric framework; a model is defined as a directed acyclic graph of tasks that eventually lead to a loss function which is minimized through a solver [1]. This generalization allows it to attempt to complete tasks that are independent of one another in parallel, as well as distributing a task across many nodes and potential multi-core devices. Generalized as it may be, this has serious pitfalls, as resources may be wasted on needless tasks, simply because a later dependent task decides not to used based on other input. The framework seizes the control flow and may waste computational power on values that are thrown out [1]. However, the solvers in this framework still enable a Tensorflow implementation to reach around 90% accuracy within hours sequentially, and within minutes with GPU computation [5, 8, 11].

Moreover, TensorFlow back-end computation is accelerated by cuDNN, a deep neural network library.

GPU-accelerated primitives specific for convolutional neural networks (as well as other deep neural network formats) are provided in cuDNN, a library released by NVIDIA [2, 4]. These primitive include layer definitions forward and backward propagation for convolutional layers which are accelerated through very specific indexing filter values allocated to these layer. As with normal batching techniques, the first most dimension of their tensor primitive is allocated for the batch size in all tensors used in computation [4].

The host-only API provided enables the initialization and utilization of the filter kernel with no human interaction,

enabling speedup for their back-end processing. Although it is generally not required to manually update or check the individual values in these tensor, the authors found it easier to implement a simple, straightforward convolutional pass. In this implementation, the input (image), filter, and output tensors are allocated sequentially.

An admittedly performance-hindering method; back propagation is done with the use of built in "atomicAdd" functions, since the same filter value has an effect on multiple output values during forward propagation and thus, gets the same influence in back propagation [13]. In Sect. 43.5, using these primitives is considered.

43.2.2 Computational Walk through

The CapsNet Architecture is defined as a convolutional layer (PrimaryCaps), followed by a capsule layer, which reinterprets the output of the convolutional layer as its own output, followed by a smaller capsule layer (DigitCaps). The principle novel computation in capsule networks lies in the operations between capsule layers during the forward propagation stage. Here, lower-level, lower-dimensional capsules undergo dimensional transformation and a dynamically weighted reduction to become the output of a higher-level, higher-dimensional capsule network. This transformation is analogous to multiplying individual scalars by weights in feed-forward, fully-connected layers.

The lower-dimensional capsules has "outputs" by the feature map output of a convolutional layer with ReLU activation. In PrimaryCaps, the number of filters must be divisible by the lower dimension to reshape the feature map outputs as vector maps. For referential integrity, the lower dimension, d_l, will be 8, and the higher dimension, d_{l+1} will be 16. Therefore, depth wise, d_l sized increments may be considered a vector map, and the number of lower level capsules is d_l times the number of vector channels. The number of vector maps (and consequently, the number of lower-level capsules) are varied in this paper to study the speed up effectiveness. Furthermore, these methods will use the MNIST data set of hand written digits of 28×28 pixels.

Forward Propagation

First, the lower level capsules, j, produce an output vector, $\hat{\mathbf{u}}_{j|i}$, for each higher level capsule, i, once for each possible output class, estimating the parameters of *their* output vector, $\mathbf{v}_i$; which is transformed to the dimensional space of the higher layer by an evolved transformation matrix, $\mathbf{W}$, such that $\hat{\mathbf{u}}_{j|i} = \mathbf{W}_{ij}\mathbf{u}_i$. Each higher level capsule then computes a dynamic weighted sum of these vectors as their output, $\mathbf{v}_j$.

For a vector to calculate its output, the weighted sum result, $\mathbf{s}_j$, undergoes a vector squishing activation function, $\mathbf{v}_j = \frac{\|\mathbf{s}_j\|^2}{1+\|\mathbf{s}_j\|^2}\frac{\mathbf{s}_j}{\|\mathbf{s}_j\|}$. This is analogous to the sigmoid activation functions usually applied onto the weighted sums in traditional capsule networks.

The dynamic weights, $\mathbf{c_i}$, are updated by computing the log prior probabilities, $\mathbf{b_i}$, which are iteratively updated. During each iteration, the probabilities, $\mathbf{b_i}$ are incremented by the scalar product of the activated weighted sum, $\mathbf{v}_j = squash(\mathbf{s}_j)$, $\mathbf{s}_j = \sum_i c_{ij}\hat{\mathbf{u}}_{j|i}$, and the vector in question, $\hat{\mathbf{u}}_{j|i}$.

The final capsule vectors outputted encode the probability of the existence of that feature in the length of the vector, while encoding the instantiation parameters of the pose of that feature in the orientation of the vector. The orientation parameters are heavily determined by the transformation matrix, which are optimized using μ momentum, instead of the Adam optimizer. The network will classify an image as being part of class i from k classes with the maximum length, $\|\mathbf{v_i}\|$.

Back Propagation

In back propagation, a corresponding error value is computed as a factor of the length of the vectors. This error value is the combination of the gradient of the loss function, multiplied by the derivative of the vector squashing activation function, effectively representing the error gradient towards which the free $\mathbf{W}$ values scattered throughout the network collectively inch towards. These are weighted by their respective c values to produce the error gradients, $\delta\mathbf{u}_{j|i}$, for all i output capsule vectors.

The error gradients are then transformed into the dimension of the lower level capsules, by being multiplied by the transpose of the original transformation matrix, $\mathbf{W}_{ij}^T$. Before then, however, a matrix product of these error gradients, and the original inputs to the network become part of the $\Delta\mathbf{W}_{ij}$. The output of the lower level capsules, however, are in truth, the rearranging of the output of the earlier convolutional layer. Back propagation occurs in this layer as normal.

Sequential Batch Updating

Given the time it takes to train a network, and the potential bias the ordering of the training examples gives to the network, different techniques were created in order to speed up processing and reduce potential bias and equalize the change all training examples provide. The latter is important to greatly increase the chances the network will converge to a more global optima. Mini-batching is one such technique.

In mini-batching, a batch of input images are provided to the network, where forward, and subsequently, back propagation are computed in parallel to one another. Afterwards, the resulting $\Delta\mathbf{W}$'s are reduced from all these "layers" to provide one main error, by which the network is updated. This paradigm is used in machine learning frameworks such as TensorFlow.

All matrices and vectors used in these computations are allocated as 1D arrays and indexed very precariously in the proposed method. Traditional mini-batching compilations of images would include higher complexities in these indexing. Moreover, the reduction of these $\Delta\mathbf{W}$'s from multiple devices would increase the communication needed between the host and all potential devices, thereby reducing potential scaling benefits. Therefore, this method does not use this technique, but rather computes forward and back propagation for each image sequentially, accumulating the error in $\mathbf{W}$ and then applying it to $\mathbf{W}$ at the end.

43.3   Methodology

The network is trained with several forward and back propagation passes for images from a training data set with periodic weight updates. To go through all data points is an epoch, and several epochs are performed in an effort to minimize the overall network loss, and conversely, maximize accuracy [4,6,12]. Network accuracy evaluation is done after training, where a different testing data set is used to eliminate biased estimations.

43.3.1 Data Allocation

All data structures are allocated using Unified memory, where data movement is optimized by the device scheduler. Since everything is allocated with 1D arrays, memory management and organizing is highly significant, and this method presents one way to arrange the data.

Data may be thought of as a $k \times t$ grid of potentially multi-dimensional elements in row-major ordering, where an element i, j corresponds to class i and lower level capsule j. An

$$
\begin{vmatrix}
\mathbf{W}[0,0] & \mathbf{W}[0,1] & \dots & \mathbf{W}[0,k] \\
\mathbf{W}[1,0] & \mathbf{W}[1,1] & \dots & \mathbf{W}[0,k] \\
\dots & \dots & \dots & \dots \\
\mathbf{W}[t,0] & \mathbf{W}[t,1] & \dots & \mathbf{W}[t,k]
\end{vmatrix}
\otimes
\begin{vmatrix}
\mathbf{u}[0] & \mathbf{u}[0] & \dots & \mathbf{u}[0] \\
\mathbf{u}[1] & \mathbf{u}[1] & \dots & \mathbf{u}[0] \\
\dots & \dots & \dots & \dots \\
\mathbf{u}[t] & \mathbf{u}[t] & \dots & \mathbf{u}[t]
\end{vmatrix}
=
\begin{vmatrix}
\hat{\mathbf{u}}[0,0] & \hat{\mathbf{u}}[0,1] & \dots & \hat{\mathbf{u}}[0,k] \\
\hat{\mathbf{u}}[1,0] & \hat{\mathbf{u}}[1,1] & \dots & \hat{\mathbf{u}}[0,k] \\
\dots & \dots & \dots & \dots \\
\hat{\mathbf{u}}[t,0] & \hat{\mathbf{u}}[t,1] & \dots & \hat{\mathbf{u}}[t,k]
\end{vmatrix}
$$

Fig. 43.1 The $d_{l+1} \times d_l$ transformation matrices, shown in the left most tensor, are multiplied element wise with the d_l dimensional outputs from the lower level capsules. These outputs are stored column-wise in the middle tensor, but are duplicated by column-wise for each higher-level capsule. The Hadamard product of these tensor produces d_{l+1} sized vector inputs to the higher level capsules in the right most tensor

example of the data layout may be found in Fig. 43.1, where $\hat{\mathbf{u}}$ is being created for each higher-level capsule column-wise, from each lower-level capsule row-wise. The lower level capsule outputs, $\mathbf{u}_{i,j}$, are represented in the middle tensor, and are duplicated along each column. Although this is potential memory storage waste during forward propagation, the extra storage will be used to save appropriate $\delta\mathbf{u}_{i,j}$ during back propagation.

For the capsule layer interface operations, there are a total of two 1D element grids, **b** and **c**, three vector-element grids, **u**, $\hat{\mathbf{u}}$, and **v**, (**v** has a height of 1, and shares the dimensionality of $\hat{\mathbf{u}}$) and three matrix-element grids, **W**, $\Delta\mathbf{W}$, and $\mathbf{W}_{\text{velocity}}$, the latter two of which are used for updating.

The preceding convolutional layer, however, requires simpler, sequentially indexed (channel, then depth if applicable, then height, then width) of 3 and 4 dimensional arrays. These are required for the input, **x**, the output, $\hat{\mathbf{x}}$, and the filters of the arrays. Much like the **W** in the capsule layer, the filters have two other equally sized companion arrays, to hold the errors, and the velocities required in momentum updating.

43.3.2 Algorithmic Definitions

In Procedure 43.1, forward propagation is given the image as a vector, **x**, and requires the use of the dynamic routing procedure defined in [12].

The *Rearrange* method interprets the output tensor from the convolutional layer as a list of vectors as captured down the output depths. These vectors are then undergo the vector squash activation function, the non-linear function which facilitates discriminatory learning by scaling vector lengths close to zero and long vectors closer to one. Afterwards, as is illustrated the middle tensor shown in Fig. 43.1, these vectors are duplicated along the "columns", representing each of the DigitCaps classes. The Hadamard product, represented

by $\otimes$, then produces distinct $\hat{\mathbf{u}}_i, j$ used in dynamic routing.

Algorithm 43.1 Forward Propagation

1: **procedure** FP(**x**)
2:    $\hat{\mathbf{x}} \leftarrow PrimaryCaps.FP(\mathbf{x})$
3:    $\mathbf{u} \leftarrow Duplicate(Activate(Rearrange(\hat{\mathbf{x}})))$
4:    $\hat{\mathbf{u}} \leftarrow \mathbf{W} \otimes \mathbf{u}$
5:    **return** $Routing(\hat{\mathbf{u}}, 3, 2)$ ▷ This is defined in [12]

On the other hand, during back propagation, the corresponding label to the vector, $y_\mathbf{x}$, is provided to calculate the error functions. This is performed by the *DerivativeActivationAndLoss* kernel. With the remaining **c** values which were set during dynamic routing, a $\delta\mathbf{u}$, corresponding to lower level capsules are generated by multiplying **v** with the corresponding transpose transformation matrix, $\mathbf{W}^T$, and scaled with **c**. Before continuing, however, $\Delta\mathbf{W}$ needs to be incremented by the matrix product of the previous input $\mathbf{u}^T$ and the error gradient for the output, $\delta\mathbf{v}$. After $\delta\mathbf{u}$ has been calculated for all j along the columns, they are reduced to the left, to compile all the error gradients proposed by each higher level capsule, before having each undergo another "unsquashing". This inverse activation is performed to match the initial activated squashing done during forward propagation. Finally, these vectors are rearranged and handed back to the convolutional layer for tradition convolutional back propagation.

Algorithm 43.2 Back Propagation

1: **procedure** BP($y_\mathbf{x}$)
2:    $\delta\mathbf{v} \leftarrow DerivativeActivationAndLoss(\mathbf{v}, y)$
3:    $\delta\mathbf{u}_{ij} \leftarrow \delta\mathbf{c}_{ij}\mathbf{W}_{ij}^T\delta\mathbf{v}_j$
4:    $\Delta\mathbf{W}_{ij} \leftarrow \Delta\mathbf{W}_{ij} + \delta\mathbf{v}_j\mathbf{u}_{ij}^T$
5:    $\delta\hat{\mathbf{x}} \leftarrow DerivativeActivation(ColReduction(\delta\mathbf{u}))$
6:    **return** $PrimaryCaps.BP(\delta\hat{\mathbf{x}})$

$$squash(\mathbf{s}_j) = \frac{\|\mathbf{s}_j\|^2}{1 + \|\mathbf{s}_j\|^2} \frac{\mathbf{s}_j}{\|\mathbf{s}_j\|} \tag{43.1}$$

$$\frac{\partial}{\partial \|\mathbf{s}_j\|}[squash] = \frac{2\|\mathbf{s}_j\|}{(\|\mathbf{s}_j\|^2 + 1)^2} \frac{\mathbf{s}_j}{\|\mathbf{s}_j\|} \tag{43.2}$$

$$L_k = T_k \max(0, m^+ - \|\mathbf{v}_k\|)^2 + \lambda(1 - T_k) \max(0, \|\mathbf{v}_k\| - m^-)^2 \tag{43.3}$$

$$\frac{d}{d\|\mathbf{v}_k\|}[L_k] = \begin{cases} -2T_k(m^+ - \|\mathbf{v}_k\|) & \|\mathbf{v}_k\| < m^+, \|\mathbf{v}_k\| \leq m^- \\ 2\lambda(T_k - 1)(m^- - \|\mathbf{v}_k\|) & \|\mathbf{v}_k\| \geq m^+, \|\mathbf{v}_k\| > m^- \\ 2(\lambda(T_k - 1)(m^- - \|\mathbf{v}_k\|) + T_k(\|\mathbf{v}_k\| - m^+)) & \|\mathbf{v}_k\| < m^+, \|\mathbf{v}_k\| > m^- \end{cases} \tag{43.4}$$

Eq. 43.1 shows the original activation algorithm proposed, whereas its derivative is shown in Eq. 43.2. Equation 43.3 is the loss function proposed by [12] as a function of the length of output vector. Equation 43.4 is the derivative of the loss function with respect to the length of the instantiation vector. Note the normalization factor, $\frac{\mathbf{s}_j}{\|\mathbf{s}_j\|}$ remains present in both equations. Constant hyper-parameters: $m^+ = 0.9$, $m^- = 0.1, \lambda = 0.5$

43.3.3 Loss and Activation Functions

The error function applied to the back vectors include the derivative of two functions, the loss function aforementioned and the partial derivative of the vector activating-squash function. The loss function and its derivative include T_k which is set if an instance of class k is present in the image. The original vector activation squash-function and its corresponding derivative may be seen in Eqs. 43.1 and 43.2.

The activation function multiplies a non-linear squashing scalar with a normalized vector. The partial derivative function is determined by the scalar portion of the activation function, and effectively scales the error from the loss function to the length of the output instantiation vector. The original vector loss function may be seen in Eq. 43.3 and its appropriate loss function is Eq. 43.4.

43.4   Results

To measure the given speed up of these methods, capsule networks were generated with varying numbers of vector channels. This effectively changes the number of lower level capsules by multiples of the convolutional layer's height and width, which were set to 6 for referential integrity's sake.

The time taken to perform forward propagation, backward propagation and epoch timings are reported, along with the equivalent speed up and throughput calculations.

Varying the number of channels in the tensor also helps better profile the program better through its throughput measurements. Reported below are the time taken to perform forward propagation, backward propagation and epoch timings, along with the equivalent speed up and throughput calculations. All timings seen in Figs. 43.2 and 43.3 are an average 30 statistical runs, after removing the highest and lowest outliers. To ensure sequential optimization, the sequential version uses the Armadillo library for linear algebra operations (matrix-vector multiplication) [14].

In Fig. 43.2, back propagation is shown to have a higher percentage of data computation rather than data movement, since data movement is not a hindering factor on CPU-based operations. Parallel timings in Fig. 43.3 further illustrate the GPU bottleneck since back propagation is faster than forward propagation. The single image trend lines show the computation time of processing an entire image, combining forward and back propagation.

The sequential version of these methods are completed with the help of the Armadillo library for linear algebra operations (matrix-vector multiplication) [14]. Figure 43.2 shows, however, back propagation is shown to be more computationally intensive, since data movement is not a hindering factor on CPU-based operations. Parallel timings in Fig. 43.3 further illustrate this point, as near linear trends are also shown, similar to the sequential version, but back propagation is clearly faster than forward propagation, although not by much. However, the gap between forward propagation and back propagation steadily increases over larger capsule layer sizes, despite the amount of data being transferred being the same. This indicates the back propagation problem scales better to GPUs as opposed to forward

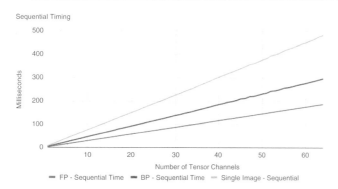

Fig. 43.2 Forward propagation takes less time than back propagation in the sequential version of these methods since there is no data movement (despite the inner loop found in Dynamic Routing [12]. This indicates back propagation is more computationally intensive. These methods are not multi-threaded

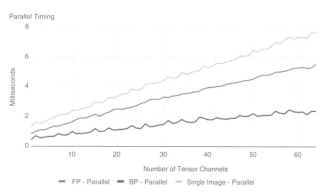

Fig. 43.3 Back propagation is clearly faster than forward propagation due to lack of communication overhead in data movement. However, the steadily increasing gap between the two also indicates computation overhead in back propagation scales better with these methods than forward propagation

propagation. This comparison is made more apparent since forward propagation adds initial data transfer overhead.

43.4.1 Speedup

To measure the speedup, capsule networks were generated with varying numbers of vector channels. This changes the lower level capsule layer size, dependent on the convolutional layer height and width, both set to 6 for the sake of referential integrity. Thus, the number of rows in the **u** grid (and other grids) is based off the height (6), the width (6), and the tensor channel size. For the equal tensor channel values found in [12], forward propagation obtained 32x speedup and back propagation obtained 116x speedup. These methods were able to obtain up to 33x speed up for forward propagation and 130x speed up for back propagation procedures alone (Fig. 43.4).

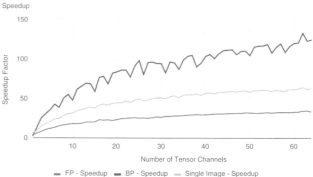

Fig. 43.4 Speedup of these methods start to slow down between 10–15 tensor channels (360–540 lower level capsules) as these methods increase due to Amdahl's law. Note that back propagation, which only communicates resulting $\mathbf{v}_j$ errors back to the host, a constant $k \times d_{l+1} = 160$ values, has higher speedup than forward propagation, which requires the movement of a 28×28 sized image from the MNIST dataset [10]

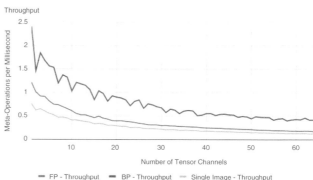

Fig. 43.5 The throughput is measured by a factor the number of floating points required to compute (not the operations) at the variable layer divided by the amount of time taken to complete the meta-operation. These floating points are the ones for the interim layer only

43.4.2 Throughput

Traditionally, efficiency is calculated in multi-CPU application to measure resource and memory exploitation in distributed algorithms. However, GPU speedup is accompanied with throughput; how many floating point operations (FLOPS) are computed in a given amount of time. For GPUs, throughput focuses on the bandwidth of the data flow rather than hardware architecture; an more appropriate, important distributed algorithm metric. For the graph seen in Fig. 43.5, throughput was computed solely on the amount of computation used during the variable layer, the convolutional output to tensor, for easy comparison. In forward and back propagation, this equates to a $6 * 6$ grid, multiplied by the appropriate number of depth channels, divided by the processing time in seconds. This is similar for single image processing, where throughput is obviously slower due to the concatenation of these two operations. The warp-based valleys found during the speed up of the program also make an impact here; computation is less efficient when hardware resource allocation is not optimized. There is no surprise that back propagation can produce a higher throughput than forward propagation, even when the hidden layers are increased. Contributing to this advantage are lack of required CPU-to-GPU data communications and no iterative dynamic routing necessary.

43.5 Conclusion and Future Work

This paper illustrates data layout methods, and shows their effectiveness in CUDA based architectures for Capsule Networks. Although cumbersome, these methods provide increased manipulation of lower level arrays while maintain high level grid abstraction intact; increasing flexibility than machine learning libraries, such as TensorFlow.

Though other major batch-based techniques were not used, single data point processing was increased almost $20\times$ for 40 vector-tensor channels, equating to 1440 lower level capsule, each having unique 8×16 ($d_l \times d_{l+1}$) weight matrices. Increasing another dimension of complexity may become a single GPU issue, given the data limits and how the size of these transformation $\mathbf{W}$'s grows. However, in a future, heterogeneous distributed version of these methods, data points may be partitioned per device, and $\Delta\mathbf{W}$'s may be reduced between these devices, as a single point of communication during batch updating.

43.5.1 Using cuDNN Primitives

The methods in this paper involves low-level organization of the equivalent weight elements between capsule networks, and the rearrangement of incoming convoluted vector data from the convolution layer. Although cuDNN primitive use very specific ways of arranging their tensors in memory without a clear way of accessing the values in the convolutional layer filters themselves, their use as a convolutional layer to this method could aid in potential future speedup. However, the tensor indexing for individual values must be either transformed to reshape as the tensor shown here. To lower kernel call overhead even further, some of the original computation kernels may be rewritten to output into the required tensor shape.

43.5.2 Parallel Mini-Batching

To achieve multiple image processing for each forward and back propagation step, an extra dimension in the proposed arrays could be used to hold the same information for varying inputs. After a backward pass, all values for $\Delta\mathbf{W}$ would have to be reduced along this new axis, however, before being applied to $\mathbf{W}$. The complexity of this reduction problem increases depending on both the solver and potential distributed systems, for memory restrictions, and bottleneck data movement overhead, respectively.

43.5.3 Multiple Parameter Varying

In efforts to remain truthful to source material, d_l was set to 8 and d_{l+1} was set to 16. These values seem to be arbitrary, as no real explanation as to the dimensionality is mentioned. However, as the orientations of the vectors in these space represent the instantiation of the perceived feature in each capsule, it could be argued that the limited space helps the vector "point" in the most likely direction. Therefore, increasing these values will most likely give more space and wiggle room for these outputs. Nevertheless, the $\mathbf{W}$ matrices contain $d_l \times d_{l+1}$ degrees of freedom; increasing this may add too much volatility to the space of transformations.

Acknowledgements This material is based in part upon work supported by the National Science Foundation under grant number IIA-1301726. Any opinions, findings, and conclusions or recommendations expressed in this material are those of the authors and do not necessarily reflect the views of the National Science Foundation.

References

1. Abadi, M., Agarwal, A., Barham, P., et al.: Tensorflow: large-scale machine learning on heterogeneous distributed systems. CoRR, abs/1603.04467 (2016)
2. Chetlur, S., Woolley, C., Vandermersch, P., Cohen, J., Tran, J., Catanzaro, B., Shelhamer, E.: cudnn: efficient primitives for deep learning. CoRR, abs/1410.0759 (2014)
3. Coates, A., Huval, B., Wang, T., Wu, D.J., Ng, A.Y., Catanzaro, B.: Deep learning with cots hpc systems. In: Proceedings of the 30th International Conference on International Conference on Machine Learning – Volume 28, ICML'13, pp. III–1337–III–1345. JMLR.org (2013)
4. Di, W., Bhardwaj, A., Wei, J.: Deep Learning Essentials: Your Hands-on Guide to the Fundamentals of Deep Learning and Neural Network Modeling, 1 edn. Packt Publishing, Birmingham (2018)
5. Hecht-Nielsen, R.: Iii.3 – theory of the backpropagation neural network*. In: Wechsler, H. (ed.) Neural Networks for Perception, pp. 65–93. Academic Press, Inc., San Diego, CA (1992). https://doi.org/10.1016/B978-0-12-741252-8.50010-8

6. Hinton, G.E., Sabour, S., Frosst, N.: Matrix capsules with EM routing. In: International Conference on Learning Representations (2018)

7. Kanungo, T., Mount, D.M., Netanyahu, N.S., Piatko, C.D., Silverman, R., Wu, A.Y.: An efficient k-means clustering algorithm: analysis and implementation. IEEE Trans. Pattern Anal. Mach. Intell. **24**(7), 881–892 (2002)

8. Kingma, D.P., Ba, J.: Adam: a method for stochastic optimization. CoRR, abs/1412.6980 (2014)

9. Krizhevsky, A., Sutskever, I., Hinton, G.E.: Imagenet classification with deep convolutional neural networks. In: Pereira, F., Burges, C.J.C., Bottou, L., Weinberger, K.Q. (eds.) Advances in Neural Information Processing Systems, vol. 25, pp. 1097–1105. Curran Associates, Inc., Red Hook, NY (2012)

10. LeCun, Y., Cortes, C.: MNIST handwritten digit database. http://yann.lecun.com/exdb/mnist/ (2010). Last Accessed 01 Sept 2019

11. Minai, A.A.: Acceleration of backpropagation through learning rate and momentum adaptation. In: Proceedings of the International Joint Conference on Neural Networks, pp. 676–679 (1990)

12. Sabour, S., Frosst, N., Hinton, G.E.: Dynamic routing between capsules. CoRR, abs/1710.09829 (2017)

13. Sanders, J., Kandrot, E.: CUDA by Example: An Introduction to General-Purpose GPU Programming, 1st edn. Addison-Wesley Professional, Boston, Mass (2010)

14. Sanderson, C.: Armadillo C++ Linear Algebra Library. https://doi.org/10.5281/zenodo.55251, June 2016. Last Accessed Date 01 Sept 2019

Image Processing Using Multiple GPUs on Webcam Image Streams

44

Hannah Munoz, Sergiu M. Dascalu, Rui Wu, Lee Barford, and Frederick C. Harris Jr.

44.1 Introduction

Analysis of image series is used in various research areas, from vehicle speed monitoring to land coverage recognition via satellites [1, 2]. Remote sensing is a vastly expanding field that has begun to welcome webcams as a cheap solution to collecting a different type of data from their research sites [3]. Webcams utilized in remote sensor networks can acquire large amounts of images by taking pictures minute or hourly, making it hard for a person to analyze image streams in a timely manner. This project aims to find the best method of performing image processing using OpenCV on high volumes of low resolution images. By varying the number of images in the stream, and the size of the images, we can determine the best method of processing image streams in minimal time.

The Nevada Research Data Center (NRDC) manages the data from several remote sensor network projects in Nevada [4]. Most of the supported projects deploy web cameras in their research sites. The Walker Basin Hydroclimate was chosen as our data set due to the southern camera's clear view of the sky and vast portion of land without sagebrush or forest [4]. The Walker Basin Hydroclimate project began

H. Munoz · S. M. Dascalu · F. C. Harris Jr. (✉)
Department of Computer Science and Engineering,
University of Nevada, Reno, NV, USA
e-mail: hannahmunoz@nevada.unr.edu; dascalus@cse.unr.edu;
fred.harris@cse.unr.edu

R. Wu
Department of Computer Science, East Carolina University,
Greenville, NC, USA
e-mail: wur18@ecu.edu

L. Barford
Department of Computer Science and Engineering,
University of Nevada, Reno, NV, USA

Keysight Laboratories, Keysight Technologies, Reno, NV, USA
e-mail: lee.barford@ieee.org

in 2012 and collects images hourly. The project has collected over 17,000 landscape images in the Western Great Basin. We developed snow and cloud estimation algorithms. Each algorithm was applied to a small section within an image, called a regions of interest (ROI).

The Nevada Research Data Center (NRDC) manages the data from several remote sensor network projects in Nevada [4]. Six of the research sites deploy web cams as a type of sensor. The Walker Basin Hydroclimate Project's Rockland Summit research site was chosen as our data set due to the southern camera's clear view of the sky and vast portion of land without large sagebrush or forest obscuring the view [4]. Rockland Summit has 20 different camera angles, using pan-tilt-zoom presets in a Canon VB-H41 on site. The camera is set up to take HD pictures in 60-min intervals from 10 AM PST to 5 PM PST. Images are taken in JPEG format with a resolution of 960×540. The southwest camera angle that was chosen has the best field of view for our use and does not have any sun interference. By using image analysis to track snow and cloud coverage in an area, time spent analyzing collected data can be reduced. The project has collected over 17,000 landscape images in the Western Great Basin.

Image processing is a method of analyzing and changing images [5]. Commonly, it is used for improving an image's quality. OpenCV is a computer vision/image processing library for various language and operating systems [6]. With over 2,500 available algorithms and 2.5 million downloads, OpenCV is widely used in many types of computer vision projects [7]. OpenCV uses matrices as a way of storing images, where each element of the matrix is a pixel in an image. Most computer vision algorithms work by analyzing pixel intensity values in a image. A pixel's intensity it's color value. Usually, a digital image has three color channels, red, blue, and green, each of which has its own intensity value. Since images can have thousands of pixels, computation on large images can become costly. However, OpenCV's algorithms have been highly optimized and as such have significantly outperformed other popular computer vision libraries [8].

In 2011, OpenCV introduced Graphical Processing Unit (GPU) accelerated algorithms [9]. Current goals of OpenCV GPU project is to provide a GPU computer vision framework consistent with the CPU functionality, achieve high performance with the GPU algorithms, and to complete as many algorithms as possible so image analysis can be done completely on the GPU [9]. OpenCV's GPU implementation was written using CUDA, so developers could take advantage of previously developed CUDA libraries.

Algorithms developed for GPU often differ than algorithms implemented on CPU. GPUs can do hundreds of independent calculations simultaneously, which lends to immense speedup over CPU implementations. Because of the architecture differences between CPU and GPU, there is no guarantee that algorithms implemented on the GPU will necessarily be faster than those done on the CPU [10]. There can be many factors, such as image size or algorithm technique, that could increase execution time on GPU compared to CPU. If there are any dependencies between pixels in an algorithm, such as a blurring mask, the time spent trying to sync pixel values could attribute to low performance [11].

There is currently no multiple GPU (multi-GPU) support for the GPU algorithms beyond a method to manually switch GPU devices [12]. OpenCV does not recommend using multiple GPUs on smaller images, as added overhead of data transfers between GPUs could negate any speedup achieved from using GPU. However, there is no mention on whether multiple GPUs could attain speedup on streams of images.

In this paper, we compare execution times, speedup factors, and throughput between various implementations of low-level image processing algorithms on image series. This paper is structured as follows: Sect. 44.2 review work previously done on the subject; Sect. 44.3 discusses the detection algorithms applied to the image series; Sect. 44.4 discusses differences between CPU, GPU, and multiple GPU implementations; Sect. 44.5 discusses results; and Sect. 44.6 contains our conclusions and goals for future development.

44.2   Related Works

Hwang et. al's work in real-time image processing on high-resolution images shows that by offloading image computation to the GPU, images can be analyzed for simple operations in real time [13]. By simply thresholding pixel differences against a given value, items in the foreground of an image can be detected. Image pre-processing is done on the CPU before the image is sent to the GPU for further analysis using OpenCV. No run time comparison on

the proposed algorithm were done. In contrast, our project compares and discusses execution time between CPU and GPU implementations of image processing techniques on image streams.

Agrawal et. al's paper implements a GPU version of a saliency model performing in real time [14]. Saliency is a pre-processing technique which must be done quickly. On CPU, the processes is slow, however, by exploiting explicitly paralleled parts of the method onto the GPU, a speedup by a factor of almost 600 was achieved. Agrawal's et. al's implementation outperformed OpenCV by a speedup factor of almost 300. This was only done for a single image, however, not a stream of images as our results show.

Wang et. al's work proposes a multi-GPU accelerated version of two popular algorithms for satellite image processing [15]. They discuss the difference between two popular methods of doing multi-GPU image processing. Namely, multiple image at once or division of a single image among GPU. They compared CPU implementation to GPU and multi-GPU. Speedups of over 100 were achievable on both proposed algorithms. Our project tries to develop similar, but with much lower resolution images.

44.3   Algorithms

Two low-level detection algorithms for snow and cloud coverage were developed for testing. Figure 44.1 shows an example image from our data set that the cloud and snow algorithms were developed on, as well as the regions of interest chosen for testing.

44.3.1   Cloud Coverage

Algorithm 44.1 Cloud coverage algorithm

1: Split Image into BGR Color Streams
2: **for** Pixel i in Image **do**
3:    **if** $Blue_i - Red_i > Threshold_c$ **then**
4:        i is Cloud
5:    **else**
6:        i is Sky
7:    **end if**
8: **end for**

The cloud coverage algorithm was used as our low computation test using only one color channel: blue, green, red (BGR). As such, it does very little calculations to determine cloud coverage. A comparison between a ROI from the data set and the results of the algorithm are in Fig. 44.2. Algorithm 44.1 was based off a hybrid threshold algorithm previously developed by Li [16]. Little computational work is done in determining cloud coverage, however it was de-

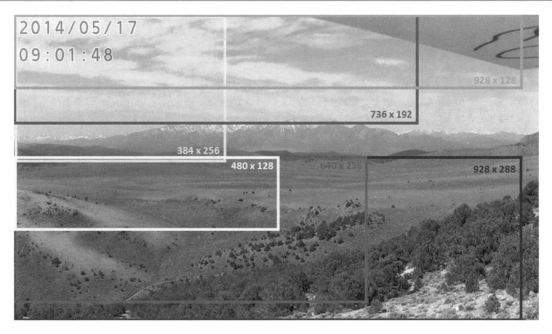

Fig. 44.1 The various region of interest on which the cloud and snow algorithms were applied

Fig. 44.2 A comparison of the cloud coverage algorithm's output

Fig. 44.3 A comparison of the snow coverage algorithm's output

termined to be is very accurate with a threshold value of 32. Lines 3 and 4 of Algorithm 44.1 are completely independent and can be run simultaneously on GPU causing speedup.

44.3.2 Snow Coverage

The snow coverage algorithm requires converting color streams, blurring, and several elemental operations making its execution a bit more intensive than the cloud algorithm. Because some of these operations rely on pixels related to them, the actions cannot always be done in parallel. The snow detection algorithm was based off work previously done by Salvatori [17]. In the version implemented $Thresold_l$ is set to 20 and $Thresold_b$ is set to 127. Figure 44.3 shows how the output of Algorithm 44.2 compares to the original image.

Salvatori's algorithm did not have high accuracy results when tested on our dataset. This may be because the im-

---

Algorithm 44.2 Snow coverage algorithm

1: Split Image into RGB Color Channels
2: Convert Image to HLS Color Space
3: Split Image into HLS Color Channels
4: Gaussian Blur (5x5) Hue Channel
5: **for** Pixel i in Image **do**
6:    **if** $Hue_i > Thresold_h$ **then**
7:       **if** $Blue_i > Thresold_b$ **then**
8:          i is Snow
9:       **end if**
10:   **else**
11:      i is Ground
12:   **end if**
13: **end for**

---

age set tested in Salvatori's project was from mountainous regions, where as our image set is from a desert. Salvatori's algorithm had problems distinguishing snow from sand.

We were able to fix this problem by adding a test of another color channel. The difference in pixel intensity in the hue, saturation, and luminance (HSL) color space for snow and sand pixel was significant enough to distinguish between the two. This helped improved the accuracy of our snow coverage algorithm.

Although some functions must have algorithm changes to increase performance of the GPU, our implemented CPU algorithms did not need to be changed. The only difference between implementations is that the GPU algorithms have an added overhead of transferring the image to and from the graphics card. All developed algorithms were made to be as similar to the CPU implementation as possible. This allowed for more accuracy between comparisons of the algorithms. OpenCV's design allows for an almost direct correlation between its CPU and GPU code. Figure 44.1 shows regions of interest chosen for testing.

44.4 Implementations

44.4.1 Hardware

The algorithms were applied on a Ubuntu 16.04 machine with eight GeForce GTX 1080 (Pascal Architecture v6.1) graphics cards, two Intel Xeon Processors E5-2650 CPU and 64GB of RAM. Each card has 8114 MiB of memory, 65536 bytes of constant memory, and 49152 bytes of shared. Concurrent copying and execution is enabled with 2 copy engines running.

44.4.2 CPU

The CPU implementation for Algorithms 44.1 and 44.2 are applied during separate runs to each image in the series. Image sizes and number of images in a series were varied over five runs and an average execution time was calculated in Fig. 44.4.

Both algorithms run in linear time; as the number of images in a series increases so does the execution time. The image size does affect execution time, as seen from the increase in execution times in Cloud 736×128 and Snow 928×288, but not until a significant amount of images are added to the series. Algorithm 44.2 has higher execution times than Algorithm 44.1 because there are more steps that need to be done during it.

44.4.3 GPU

Similarly, the GPU algorithm was implemented using OpenCV's GPU algorithm. In this implementation, after the image is read in from the file, it must be "uploaded" to the GPU. This take a significant amount of time initially to set up. To try to alleviate the initial overhead, a 1×1, single channel GPU matrix is created before processing the image series. This reduced the execution time of a single image using Algorithm 44.2 from an average of 755.822 to 23.356 ms; 32 times speedup. Once either algorithm has been applied to the image series, the resulting image is "downloaded" from the GPU to the CPU and the coverage percentage is pushed to a vector. The average execution time over 5 runs can be seen in Fig. 44.5.

When moved to the GPU, OpenCV's matrices are divided up into $(c+15)/15$ by $(r+15)/15$ grids where c and r are the columns and rows in the image. Each block in the grid has 16 threads in both the x and y direction, resulting in 256 threads per block. Since 256 is divisible by 32, it takes advantage of the GPU's warp scheduling to increase speedup.

One the GPU, all the execution times begin to converge together. The snow coverage algorithm has a significant amount of speedup. However, the cloud coverage algorithm actually takes longer on the GPU than it did on the CPU.

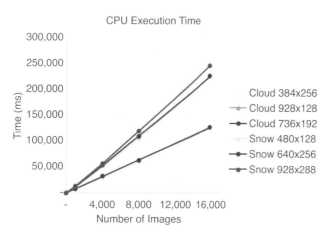

Fig. 44.4 The execution time in milliseconds of both the cloud and snow algorithm run on the CPU

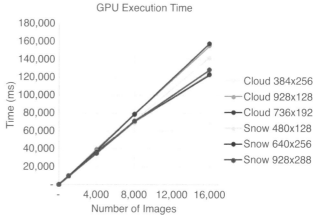

Fig. 44.5 The execution time in milliseconds of both the cloud and snow algorithm run on the GPU

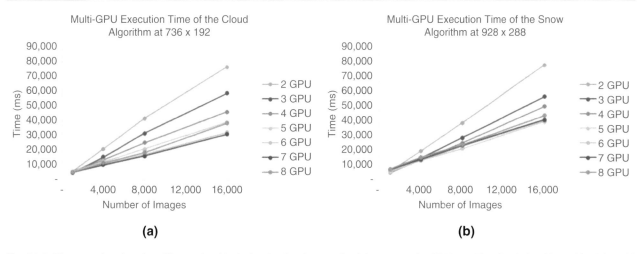

Fig. 44.6 The execution time in milliseconds of both the cloud and snow algorithm run on the GPU. (**a**) The cloud algorithm with 736×192 ROI. (**b**) The snow algorithm with 928×288 ROI

This is because the added overhead of sending the images to the GPU and back takes more time than execution on the GPU could decrease. In other words, there is not enough computations done to justify sending it to the GPU.

44.4.4 Multiple GPU

The multiple GPU implementation utilizes threads to control multiple GPUs at once. Each thread is given all the file names of the images in the series and the ROI's height and width. The number of images in the series is divided among available GPUs. Each thread then runs the previously implemented single GPU algorithms on their own GPU stream. This allows for all image analysis to happen concurrently and images are uploaded and downloaded from the GPU asynchronously. The average execution time over 5 runs can be seen in Fig. 44.6. For simplicity's sake, Fig. 44.6 shows the largest of the ROI run for both algorithms. The full data set, however, is available [18].

As the number of GPUs increases, the execution time decreases. Even in the cloud detection algorithm, which did not achieve speedup when run on a single GPU, decreases its run time when more GPUs are added. Once 8 GPUs are added, however, execution time begins to increase again. Once again, the issue is the overhead resulting from sending the images to the GPUs. The amount of work being done on each GPU is not efficient and the images are going to so many different GPUs that they are bogging down the PCIe bus that connects the GPUs together.

44.5 Results

While Algorithm 44.1 does not increase speedup on a single GPU, speedup is gained during multi-GPU computing.

Figure 44.7a shows the speedup factor of the 736×192 ROI over 1 to 8 GPUs. For GPUs 2 through 7 a steady increase of speedup occurs. At 8 GPUs, speedup begins to decrease. A discussed previously, this is because there is too much data transfer happening between GPUs.

Algorithm 44.1's throughput in Fig. 44.8a is quite clear. As the number of GPUs increases, up to 7 GPUs, the throughput increases as well.

Algorithm 44.2's speedup graph in Fig. 44.7b is a little different. A steady increase in speedup factor occurs from the use of 1 to 5 GPUs, not 7 GPUs. This is because of the extra work load that needs to be done to calculate snow coverage in the region. At a smaller number of images, anything less than 4,000, it becomes confusing to determine the most efficient number of GPU to use. The speedup for such a small amount of images does not increase with more GPUs. This could be because the machines the algorithms were run on were either closer or farther from each other, thus affecting execution time.

Algorithm 44.2's throughput in Fig. 44.8b is similar, in that it reflects trends previously discussed in its speedup graph. The throughput also increases as more images are added to the series, because large amounts of images can be more efficiently handled on more GPUs.

Figure 44.7b displays a superlinear speedup characteristic of Algorithm 44.2. This is an expected characteristic, as OpenCV algorithms can have up to $100\times$ speedup over their original implementation [9].

44.6 Conclusion and Future Work

OpenCV is a popular image processing and computer vision library. Although they have a expansive GPU-based algorithms library, they have yet to implement algorithms that

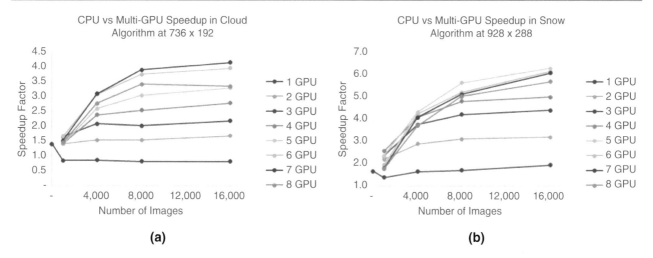

Fig. 44.7 The speedup factor of the largest ROI run on CPU, GPU, and multiple GPUs. (**a**) The cloud algorithm with 736×192 ROI. (**b**) The snow algorithm with 928×288 ROI

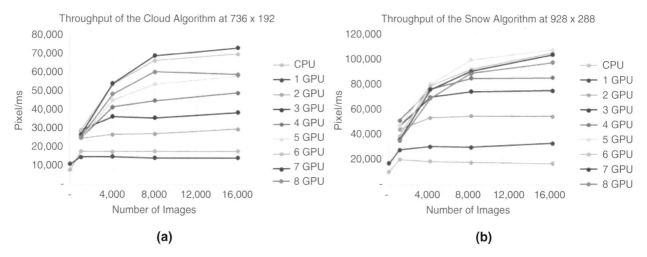

Fig. 44.8 The throughput of the largest ROI run on CPU, GPU, and multiple GPUs. (**a**) The cloud algorithm with 736×192 ROI. (**b**) The snow algorithm with 928×288 ROI

utilize multiple GPUs. In this paper, we have implemented low and high work intensity image processing algorithms on a large image series to determine when it's appropriate to do image processing on CPU, single GPU, or multiple GPUs. It was determined that high work intensity algorithms on image series could achieve speedup on single GPU, however low intensity algorithms could not. Both algorithm types could achieve speedup using multiple GPUs with an increase in throughput with larger image series. There is, however, a maximum amount of GPUs that can be used before speedup begins to decrease.

Further work could be done to determine if multiple GPUs should be used to analysis images in real time from a video stream. The algorithms developed in this paper could be tested on live cameras to see if they can calculate coverage in real time. This could be helpful for use in citizen science projects, where citizens can use their own webcams to help scientists in their research.

Acknowledgements This material is based in part upon work supported by the National Science Foundation under grant number IIA-1301726. Any opinions, findings, and conclusions or recommendations expressed in this material are those of the authors and do not necessarily reflect the views of the National Science Foundation.

References

1. Gerát, J., Sopiak, D., Oravec, M., Pavlovicová, J.: Vehicle speed detection from camera stream using image processing methods. In: Proceedings of ELMAR-2017 – 59th International Symposium ELMAR-2017, pp. 201–204 (2017)
2. Absardi, Z.N., Javidan, R.: Classification of big satellite images using hadoop clusters for land cover recognition. In 2017 IEEE 4th International Conference on Knowledge-Based Engineering and Innovation (KBEI), pp. 600–603 (2017)
3. Jacobs, N., Burgin, W., Fridrich, N., Abrams, A., Miskell, K., Braswell, B.H., Richardson, A.D., Pless, R.: The global network of outdoor webcams: properties and applications. In: Proceedings of the 17th ACM SIGSPATIAL International Conference

on Advances in Geographic Information Systems, ser. GIS'09, pp. 111–120. ACM, New York (2009). [Online]. Available: http://doi.acm.org/10.1145/1653771.1653789

4. Nevada research data center. [Online]. Available: https://sensor.nevada.edu/NRDC/

5. Sonka, M., Hlavac, V., Boyle, R.: Image Processing, Analysis, and Machine Vision. Cengage Learning (2014). [Online]. Available: https://books.google.com/books?id=QePKAgAAQBAJ

6. Opencv library. [Online]. Available: https://opencv.org/

7. Culjak, I., Abram, D., Pribanic, T., Dzapo, H., Cifrek, M.: A brief introduction to OpenCV. In: Proceedings of the 35th International Convention MIPRO, pp. 1725–1730 (2012)

8. Matuska, S., Hudec, R., Benco, M.: The comparison of CPU time consumption for image processing algorithm in matlab and opencv. In: Proceedings on 9th International Conference – 2012 ELEKTRO, pp. 75–78 (2012)

9. Cuda. [Online]. Available: https://opencv.org/platforms/cuda.html

10. Saahithyan, V., Suthakar, S.: Performance analysis of basic image processing algorithms on GPU. In: 2017 International Conference on Inventive Systems and Control (ICISC), pp. 1–6 (2017)

11. Ke, J., Bednarz, T., Sowmya, A.: Optimized GPU implementation for dynamic programming in image data processing. In: 2016 IEEE 35th International Performance Computing and Communications Conference (IPCCC), pp. 1–7 (2016)

12. Gpu module introduction. [Online]. Available: https://docs.opencv.org/2.4/modules/gpu/doc/introduction.html

13. Hwang, S., Uh, Y., Ki, M., Lim, K., Park, D., Byun, H.: Real-time background subtraction based on gpgpu for high-resolution video surveillance. In: Proceedings of the 11th International Conference on Ubiquitous Information Management and Communication, ser. IMCOM'17, pp. 109:1–109:6. ACM, New York (2017). [Online]. Available: http://doi.acm.org/10.1145/3022227.3022335

14. Agrawal, R., Gupta, S., Mukherjee, J., Layek, R.K.: A GPU based real-time cuda implementation for obtaining visual saliency. In: Proceedings of the 2014 Indian Conference on Computer Vision Graphics and Image Processing, ser. ICVGIP'14, pp. 1:1–1:8. ACM, New York (2014). [Online]. Available: http://doi.acm.org/10.1145/2683483.2683484

15. Wang, M., Fang, L., Li, D., Pan, J.: Using multiple gpus to accelerate MTF compensation and georectification of high-resolution optical satellite images. IEEE J. Sel. Top. Appl. Earth Obs. Remote Sens. **8**(10), 4952–4972 (2015)

16. Li, Q., Lu, W., Yang, J.: A hybrid thresholding algorithm for cloud detection on ground-based color images. J. Atmos. Ocean. Technol. **28**(10), 1286–1296 (2011)

17. Salvatori, R., Plini, P., Giusto, M., Valt, M., Salzano, R., Montagnoli, M., Cagnati, A., Crepaz, G., Sigismondi, D.: Snow cover monitoring with images from digital camera systems. Ital. J. Remote Sens. **43**, 6 (2011)

18. Project timings. [Online]. Available: https://github.com/hannahmunoz/MultiGPUOpenCV/blob/master/FinalRuntimes.xlsx

A CUDA Algorithm for Two-Dimensional Detrended Fluctuation Analysis

Vanessa Cristina Oliveira de Souza, Reinaldo Roberto Rosa, and Arcilan Trevenzoli Assireu

45.1 Introduction

Detrended Fluctuation Analysis (DFA) [1] has been used repeatedly in studies that seek to understand the scaling regime governing systems and processes in different areas. DFA has the advantage of tackling two critical problems in time series studies: trend and short series. The method is, in fact, a Hurst exponent gauge [2] optimised for non-stationary signal analysis. The objective of DFA is, therefore, to quantify long-range correlation (memory), if any, in the time series, characterising it as persistent or anti-persistent and revealing its intensity.

In 2002, DFA was generalised to characterise multifractal signals [3], and in 2006, both DFA and multifractal DFA (MF-DFA) were expanded to operate not only on one-dimensional time series, but also on any other dimensions, especially two-dimensional signals (2D-DFA and 2D-MF-DFA), which include the observation of patterns from digital images [4]. In addition, several other generalisations have been proposed over the last few years [5].

The notion of persistence, when applied to two-dimensional structural patterns, may reveal characteristics related to the fluctuation of the roughness of these surfaces. This is because the more persistent the signal, the more

V. C. O. de Souza (✉)
Institute for Mathematics and Computing, Federal University of Itajubá, Itajubá, Minas Gerais, Brazil
e-mail: vanessasouza@unifei.edu.br

R. R. Rosa
Laboratory for Computing and Applied Mathematics, National Institute for Space Research, São José dos Campos, SP, Brazil
e-mail: reinaldo@lac.inpe.br

A. T. Assireu
Institute for Natural Resources, Federal University of Itajubá, Itajubá, Minas Gerais, Brazil
e-mail: arcilan@unifei.edu.br

constant the series is, and vice versa [1]. Therefore, 2D-DFA and 2D-MF-DFA have been particularly applied in studies that evaluate surface roughness, considering digital images that reveal spatial patterns produced through an underlying dynamic. And, although they are still little explored, they have been used successfully in different applications [6–9].

A limitation of the use of 2D-DFA and 2D-MF-DFA is computational complexity (close to $O(n^3)$). Depending on the type of application, the processing time of these algorithms may make it unfeasible to use [7, 8]. However, as will be observed in Sect. 45.2, the methods are highly parallelisable, since the same computation is independently performed in different sub-matrices of the digital image. This finding guided the idea of parallelising the 2D-DFA and 2D-MF-DFA algorithms.

In this context, the objective of this work is to present a parallel implementation of the 2D-DFA and 2D-MF-DFA methods using General Purpose Graphics Processing Units (GPGPUs) and the parallel computing software platform using NVIDIA GPUs, termed CUDA (*Compute Unified Device Architecture*), through the PyCUDA Wrapper.

Two concepts are important in GPU programming: data parallelism and independence. These attributes allow the same computing task to be applied to different data streams, which depend very little on each other, i.e., they allow the simultaneous execution of the same instruction in different portions of a data set. These characteristics are found in the 2D-DFA and M2D-DFA algorithms, leaving them ready to be programmed using GPU/CUDA.

It is not the objective of this paper to describe the architectures of GPU cards nor of the CUDA platform. Other literature can be consulted for this purpose [10, 11].

The contribution of this work is based on presenting and discussing the GPU/CUDA parallelisation of the 2D-DFA and 2D-MF-DFA algorithms, showing its feasibility, especially for real-time applications.

© Springer Nature Switzerland AG 2019
S. Latifi (ed.), *16th International Conference on Information Technology-New Generations (ITNG 2019)*,
Advances in Intelligent Systems and Computing 800,
https://doi.org/10.1007/978-3-030-14070-0_45

This work is organised as follows: in Sect. 45.2, the 2D-DFA algorithm is presented. Section 45.3 presents the modelling used for the parallelisation of the algorithms using GPU/CUDA. Section 45.4 presents the work methodology. In Sect. 45.5, the results and discussions are presented. In Sect. 45.6, the final considerations are presented.

45.2 Two-Dimensional Detrended Fluctuation Analysis (2D-DFA)

Detrended fluctuation analysis is one of the techniques used to characterise stochastic fractal signals, which aims to estimate the scaling exponent α. DFA was originally defined by [1] for gene sequence analysis and has since become a robust technique for quantifying correlations in non-stationary time series. Since then, the one-dimensional version has been used in different applications. The main advantage of the DFA method is that it can eliminate trends from different orders, distinguishing trends from intrinsic long-range fluctuations in the data [12]. Reducing the effects of trends on the data, DFA enables a more accurate estimation of the scaling exponent and *crossover point* positions, which show a change in the regime of the analysed series.

Further details of the algorithm, a deep discussion regarding the significance of its results and some examples of application can be seen in [13, 14].

There are two approaches to implementing detrending fluctuation analysis in two-dimensional (2D-DFA) images: those in which one-dimensional DFA is calculated in different rows and columns of the image and subsequently averaged; and the approach where 1D-DFA was in fact generalised to work with more than two dimensions. In this work, we are interested in the second approach, but it must be pointed out that some initiatives of the first approach arose precisely to the detriment of the computational cost of 2D-DFA generalisation.

45.2.1 Implementation of the 2D-DFA Algorithm

The implementation of the 2D-DFA method is an adaptation of the DFA method for a one dimensional temporal series and is described below, as defined in [4]. In the 2D adaptation, the matrix is subdivided into submatrices that range from 6×6 pixels to $n \times n$ pixels, where n is the lowest value between (M/4) and (N/4), where M and N correspond to the number of lines and columns of the matrix respectively.

This partition constitutes the analysed scales and is controlled by the variable s. After the subdivision of the matrix, it is integrated in accordance with Eq. 45.1.

$$y_{v,w}(i, j) = \sum_{k_1}^{i} \sum_{k_2}^{j} X_{v,w}(k_1, k_2) \quad (45.1)$$

The local trend is removed for each sub-matrix. The least squares method is employed by adjusting the surface by one of the surface equations defined in [4]. Following this, the residual matrix between the sub-matrix (y) and the adjusted surface (y) is calculated (Eq. 45.2).

$$z_{v,w}(i, j) = y_{v,w}(i, j) - \widetilde{y}_{v,w}(i, j) \quad (45.2)$$

The local fluctuation function for each scale s ($6 \leq s \leq \min((M,N)/4)$) is calculated by means of Eq. 45.3.

$$F^2(u, v, s) = \frac{1}{s^2} \sum_{i=1}^{s} \sum_{j=1}^{s} \left(z_{v,w}(i, j)\right)^2 \quad (45.3)$$

where: u, v identify the submatrices for s scale.

When M and/or N are not multiples of s, it is necessary to restart the calculation of the other three corners of the matrix and $F^2(u, v, s)$ is an average of the four values obtained for the size of s. This calculation is important to ensure that the entire length of the image is analysed. On the other hand, the computational cost of the method increases considerably.

The steps depicted by Eqs. 45.1, 45.2 and 45.3 are repeated until s reaches its upper limit. After, the global fluctuation function is calculated as an average (*root-mean-square*) of the local fluctuation functions, by means of Eq. 45.4.

$$F_2(s) = \left(\frac{1}{p * q} \sum_{i=1}^{p} \sum_{j=1}^{q} F^2(u, v, s)\right)^{1/2} \quad (45.4)$$

where: p = M/s and q = N/s.

Thus by varying the value of s, it is possible to determine the relation of the scale between the fluctuation of the $F_2(s)$ function and the size of scale s, by means of Eq. 45.5:

$$F(s) \sim s^{\alpha} \quad (45.5)$$

The exponent for fluctuation scaling (α) is thus calculated as the slope of the straight line of the log-log graph between $F_2(s)$ and s.

In 2D-DFA, stationary signals return values of α between 0 and 1, while non-stationary signals return values between 2 and 3. Values of α between 1 and 2 indicate that the signal is not self-similar. Understanding the significance of the α parameter value helps to understand not only the correlation strength of the analysed signal, but also the characteristics of the process that originated it.

In multifractal DFA, instead of analysing only one scaling exponent in the calculation of the global fluctuation function, it is evaluated on several orders, given by the parameter q (Eq. 45.6, for values of q other than zero) [2].

$$F_q(s) = \left(\frac{1}{M_s N_s} \sum_{u=1}^{M_s} \sum_{v=1}^{N_s} [F(u, v, s)]^q \right)^{1/q} \quad (45.6)$$

45.3   Method

In this work, the parallelisation in a GPU was used. Each card has several multi-processors, each consisting of some *streaming* processors. These processors are responsible for running all *threads*[1] in parallel.

When a CPU code (also called *host* code) triggers the so-called *kernel* function, the CUDA system generates a grid of *threads*, which are organised in a two-tiered hierarchy: block and thread. All of the blocks have the same size, i.e., the same amount of threads, specified in the host code. Each block receives a data set, which is processed by the threads [10].

As the sequential algorithms had already been implemented in Python, it was decided to use the PyCUDA *wrapper* [11], which connects a Python programme to CUDA API functions, which is written in C. The Python compiler, version 3.4, numpy 1.11 and scipy 0.17 was used.

The empirical tests were performed on a desktop machine, with Core i7–950 processor, 3.07 GHz, 8 GB of RAM, running Linux CentOS 7.2.15.11, fully dedicated. The GPU used was NVIDIA GeForce GTX 460, compute capability 2.1 and CUDA toolkit 7. The main features of the GPU are summarised in Table 45.1.

For the validation of the method implemented, images of fBm fractals,[2] generated in Matlab with the software Fraclab [15] were used. The matrices generated were square,

Table 45.1   Features of the GeForce GTX 460 card

Maximum number of threads per SM	1,536
Maximum number of blocks per SM	8
Maximum number of threads per block	1,024
Amount of CUDA cores	336
Number of multiprocessors	7
Maximum number of concurrent threads	10,752
Overall memory size	963 kB
Shared memory size per block	48 kB

SM streaming multiprocessor

---

[1]Parallel processing units.

[2]Fractional Brownian motions: non-stationary signals.

with different dimensions and Hurst exponents. The processing time of the parallel and sequential versions of the 2D-DFA and 2D-MF-DFA algorithms were compared for different matrix sizes. The code for the sequential versions of the algorithms can be accessed at https://github.com/vanessavcos/2D-DFA. The dataset can be accessed at https://doi.org/10.13140/RG.2.2.22950.78409.

The parameter s, which defines the size of the sub-matrices, varied from 6 to $n/4$ in the analyses, n being the dimension of the matrix. The detrending of the matrix used the surface given by the following plane equation $\widetilde{y}_{v,w}(i, j) = ai + bj + c$.

45.4   Parallelisation of 2D-DFA and 2D-MF-DFA Using GPU/CUDA

In the first step of the 2D-DFA algorithm, the input matrix is divided into sub-matrices. The main idea of the parallel approach is that the algorithm steps represented by Eqs. 45.1, 45.2 and 45.3 (Sect. 45.2) were performed in parallel between the sub-matrices.

The first important observation is that steps 1, 2 and 3 are not, in themselves, parallelisable. The reasons will be discussed below. So the main issue is that the parallelisation of the method occurred, in practice, through the parallelisation of the loop between the sub-matrices. The main points of this approach will be discussed below.

The step depicted by Eq. 45.1 of the algorithm is to obtain the surface profile, via cumulative sum. The cumulative sum in the GPU is a well-studied problem, being a strictly sequential problem. Some approaches are cited in the literature and even implemented in some CUDA libraries [16]. However, the functions implemented by libraries are *kernel* functions and cannot be called within the GPU code. In this case, the gain from using a library function would be small, since the data transfer would be large. So it was decided to calculate the cumulative sum inside the GPU. The strategy used was sequential, i.e., a single *thread* allocated to perform this task.

In order to obtain the detrended matrix (Eq. 45.2), LU factorisation was used. In the resolution of a system $Ax = b$, using LU factorisation, we have $A *A^T \equiv A$; $A \equiv L * U$ e $A^T * B \equiv b$. To solve the system, it is necessary that: $L * y = b$; $U * x = y$. As for the entire scale s, all sub-matrices share the same L and U matrices, they are computed in the *host* code in python and sent to the GPU. In the CUDA code, three threads were assigned to calculate $A^T * B \equiv b$, given that multiplication is also a sequential task.

A single thread executes the equations $L * y = b$; $U * x = y$ and finds parameters a, b, and c that solve the system $\widetilde{Y}_{v,w}(i, j) = ai + bj + c$. Finally, allocating one thread per line, the difference is calculated between $Y_{v, w}$ and $\widetilde{Y}_{v,w}$ in order to obtain the detrended matrix and the local fluctuation function (Eq. 45.3) is computed and returned to the host.

Therefore, threads are triggered for each block, one for each row of the submatrix. This option was taken because, as stated above, there is not much parallelisation inside the process and many threads would be idle throughout almost the entire data processing. In addition, in the architecture used, each block has a limit of 1024 threads, which means that it is not possible to trigger one thread per cell in matrices larger than 32×32. In addition to calculating the local function, the other threads are used to cooperatively copy the data from the global memory to the shared memory.

By limiting the size of the shared memory of the card used, when the scale s is greater than 110, a second kernel function is triggered, where the data remains in the GPU's global memory throughout processing.

Figure 45.1 shows the kernel function call termed '*fit*', where:

- *d-L*: non-zero parameters of matrix L
- *d-U*: non-zero parameters of matrix U
- *d-mat* is the data matrix
- *s* is the current scale
- *tam* is the size of the data matrix

The kernel function is called for each scale s. The kernel parameters are defined with the variable $blockSize = (s, 1, 1)$ and $grid = \left(\left\lfloor \frac{m}{s} \right\rfloor, \left\lfloor \frac{m}{s} \right\rfloor, 1 \right)$, i.e., the grid represents the sub-division of the matrix into the sub-matrices of size s. And, as already mentioned, each block has s threads. The function returns the d_vetF matrix, which stores the local fluctuation function for each submatrix of the scale s and, on top of which, the global fluctuation function is calculated.

```
#launch kernel
#get the kernel function from the compiled module
kernel = mod.get_function("fit")
#call the kernel on the card
kernel(
    #inputs
    d_L, d_U, d_mat,np.int32(s), np.int32(tam),
    #output
    d_vetF,
    # kernel parameters
    grid= grid, block=blockSize,
    #size of shared memory
    shared = sizeofSharedMemoryinBytes)
```

Fig. 45.1 Excerpt from the PyCUDA code that triggers the kernel function termed 'fit', for the parallel calculation of the 2D-DFA

The complete parallel code can be accessed at https://github.com/vanessavcos/DFA2D-PARALLEL.

The implementation of the Parallel Multifractal 2D-DFA followed the same approach as the monofractal version. What differentiates the versions is the global function calculation, which was not parallelised.

45.5 Results and Discussion

The tests indicated that the parallel version improved the processing time considerably, both in the monofractal version (Table 45.2) and in the multifractal version (Table 45.3), without affecting the correctness of the results. The reported times are in seconds and were counted from the start of both procedures, i.e., also considering the processing time on the host and the data transfers in the parallel versions. The speed up was calculated as the division of sequential time by parallel time.

The tests confirm that the increase in time is related to the increase of the matrix size. The sequential implementation for 2048×2048 matrices already consumes a high processing time (approximately 5 minutes), making it impossible to use applications with large databases. This is because, when processing an image every 5 minutes, only 288 images would be processed per day (24 hours). The parallel approach

Table 45.2 Comparison of processing time between sequential python and parallel implementations in PyCuda for Monofractal 2D-DFA

Matrix size	Sequential time (s)	Parallel time (s)	Speed up
64	0.09107 ± 0.0015	0.00909 ± 0.0005	10.1037
128	0.48455 ± 0.0025	0.02337 ± 0.0009	20.7370
256	2.35751 ± 0.0103	0.07386 ± 0.0015	31.9179
512	11.58794 ± 0.0715	0.52457 ± 0.0029	22.0905
1024	59.90978 ± 0.1484	3.03094 ± 0.0389	19.7660
2048	322.39272 ± 1.892	27.5368 ± 0.03888	12.9642
4000	2443.151 ± 3.7664	320.7264 ± 0.0550	7.6176

Times are given in seconds. Average of 50 repetitions

Table 45.3 Comparison of processing time between sequential python and parallel implementations in PyCuda for Multifractal 2D-DFA

Matrix size	Sequential time (s)	Parallel time (s)	Speed up
64	0.3444 ± 0.001	0.1691 ± 0.003	2.0356
128	1.1235 ± 0.007	0.2843 ± 0.003	3.9511
256	4.4307 ± 0.016	0.6015 ± 0.004	7.3662
512	19.2487 ± 0.043	1.8361 ± 0.006	10.4835
1024	89.9749 ± 0.1977	6.9600 ± 0.012	12.9273
2048	479.3076 ± 0.8609	40.5287 ± 0.03	11.8264
4000	2875.504 ± 29.97	344.4933 ± 0.188	8.3471

Times are given in seconds. Average of 50 repetitions

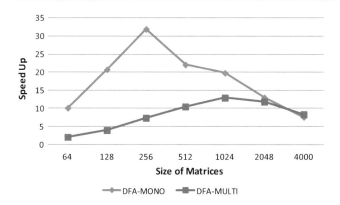

Fig. 45.2 Comparison of the speed up between the parallel monofractal and multifractal approaches

45.6 Final Considerations

This work addressed a critical point of the 2D-DFA and 2D-MF-DFA algorithms, related to their performance. Parallelisation of the method using GPGPU-CUDA significantly reduced processing time. It is important to note that parallelisation occurred more in the simultaneous processing of several submatrices, than in the parallelisation of the internal procedure in each submatrix. Other modelling, especially in the case of the cumulative sum, can further enhance gain, which is essential in large data flow environments or in scenarios where image processing needs to be rapid. Another point is that the GPU card used in this work has limitations that culminated in some adaptations of the code. Even so, the gain was substantial.

The availability of this program, developed in PyCUDA, will facilitate studies on scaling regime governing systems and lends itself to comparative studies of long-range correlation in time series across the globe.

reduced this time to approximately 27 seconds, enabling the processing of 3200 images in 1 day.

It is also worth noting that in the work [7] the time to process a 4080 × 3072-pixel image using the sequential implementation was greater than 3 days. The sequential implementation of the 2D-DFA conducted in this work took, on average, 6.78 hours for a 4000 × 4000-pixel image. This time was reduced to about 5 min in the parallel version. In fact, for real-time applications and those classified as Big Data, the sequential approach may become impractical [7, 8].

In Table 45.2, it is assessed that, although the speed up is good, there is a drop in parallel performance for bigger matrices (larger than 512 × 512-pixel). The explanation for this is the shared memory limitation of the GPU card used. As previously stated, when the submatrix size is greater than 110 × 110 pixels, a second kernel function is called and the data resides in the GPU's global memory throughout processing. It is believed that by using more robust cards, the performance for larger images will be more significant. This same behaviour is observed for the multifractal algorithm (Table 45.3). However, for the multifractal version, the speed up was lower than the monofractal version. The explanation is that calculating the global fluctuation function at different scales is performed in the host code and not in the GPU. Thus, the impact of parallelisation was lower for this algorithm. The graph in Fig. 45.2 illustrates the speed up behaviour for both algorithms. From the graph in Fig. 45.2, it can be observed that the increase in the speed up of the multifractal algorithm is more constant and smoother than for the monofractal algorithm, which also exhibits more abrupt drops in speed up when compared to the multifractal version.

References

1. Peng, C., et al.: Long-range correlations in nucleotide sequences. Phys. Rev. E. **49**, 1685–1689 (1994)
2. Hurst, H.: Long term storage capacity of reservoirs. Trans. Am. Soc. Civ. Eng. **116**, 770–799 (1951)
3. Kantelhardt, J.W., et al.: Multifractal detrended fluctuation analysis of nonstationary time series. Physica A. **316**(1–4), 87–114 (2002)
4. Gu, G.F., Zhou, W.X.: Detrended fluctuation analysis for fractals and multifractals in higher dimensions. Phys. Rev. **74**(6), 1–7 (2006)
5. Xiong, H., Shang, P.: Detrended fluctuation analysis of multivariate time series. Commun. Nonlinear Sci. Numer. Simul. **42**, 12–12 (2017)
6. Barrera, E., et al.: Correlation of optical properties with the fractal microstructure of black molybdenum coatings. Appl. Surf. Sci. **256**(6), 1756–1763 (2010)
7. Yeh, R.G., et al.: Two-dimensional matrix algorithm using detrended fluctuation analysis to distinguish Burkitt and diffuse large B-cell lymphoma. Comput. Math. Methods Med. **2012**, 947191 (2012)
8. Freitas, R.M.: Laboratório virtual para visualização e caracterização do uso e cobertura da terra utilizando imagens de sensoriamento remoto. 235 p. Ph.D. dissertation – National Institute for Space Research (INPE), Brazil (2012)
9. Alpatov, A.V.: Revealing the surface interface correlations in a-Si:H films by 2D detrended fluctuation analysis. Semiconductors. **47**(3), 365–371 (2013)
10. NVIDIA: NVIDIA CUDA programming documentation. Available from Internet: http://docs.nvidia.com/cuda/ (2016)
11. Klöckner, A.: PyCuda. Available from Internet: https://mathema.tician.de/software/pycuda/ (2017)

12. Kantelhardt, J.W., et al.: Detecting long-range correlations with detrended fluctuation analysis. Physica A. **295**, 441–454 (2001)

13. Souza, V.C.O., Assireu, A.T.: Detrended fluctuation analysis of spatially extended digital surfaces: the classification process of 1/f noise and computacional performance. J. Comput. Interdiscip. Sci. **24**, 25 (2016)

14. Souza, V.C.O.: Análise computacional de padrões estruturais não-lineares a partir de imagens digitais com estudos de caso em ciências ambientais e espaciais. Ph.D. dissertation, National Institute for Space Research (INPE), Brazil (2017)

15. Inria. FRACLAB: A fractal analysis toolbox for signal and image processing [online]. Available at: http://fraclab.saclay.inria.fr/ (2016)

16. Harris, M., Sengupta, S., Owens, J.D.: Parallel prefix sum (scan) with CUDA. GPU Gems. **3**(39), 851–876 (2007)

Xiang Li, A. Grant Schissler, Rui Wu, Lee Barford, and Frederick C. Harris Jr.

46.1 Introduction

Many stochastic simulations cannot be accurately represented by independent and identically distributed samples. Realistic simulations must include correlations and allow for heterogeneous marginal distributions, modeled using a correlation matrix. For example, Biological processes are often highly correlated, involving many complex interactions. Applications in integrative analysis of multiple data sets has been recently researched [1]. One of the pitfalls of not accounting for correlations was researched by Gatti et al. [2]. They found significantly higher rates of false positives in gene set testing when independent genes were inappropriately assumed. In order to accurately represent this data, a multivariate (correlated) structure must be used. The expression of genes is also heterogeneous, that is not all genes follow the same distribution. Using the NORTA method described in this paper, we are able to simulate

random vectors whose components follow arbitrary distributions marginally while jointly reflecting a specified correlation matrix.

The ability to simulate high dimensional correlated data has applications in machine learning as well. For example, Wilkins, Morris, and Boddy researched methods of using backpropagation neural networks to identify marine phytoplankton using multivariate flow cytometry data [3]. In identification problems, lower dimensionality leads to increased overlap between classes. For higher accuracy, it becomes necessary to use the full multivariate nature of the data to distinguish between classes. Another common simulation method is the Monte Carlo Method which was used recently by Russkova to model weather phenomena [4].

There are numerous uses of multivariate statistical analysis, ranging from RNA sequence analysis to health records [5]. More recently Often complex real world system are composed of dependent variables that cannot be treated as independent. The use cases are extremely broad and as computing power continues to increase, use of high performance algorithms for preforming these statistical analyses will become more and more commonplace.

In general, methods for generating correlated random vectors can be classified into three categories [6]:

(1) Analytic methods using conditional distributions.
(2) Numerical approaches including accept/reject methods.
(3) Transformation of univariate vectors.

Numerous approaches researched in the first two categories are limited to bivariate distributions and can only be used for generation of variables that share a common distribution [6].

NORTA falls into category 3. The advantage of employing this method is in its simplicity, propensity for parallelization, and broad applications. Algorithms in this category are able to use the appropriate set of marginal probability

X. Li · F. C. Harris Jr. (✉)
Department of Computer Science and Engineering,
University of Nevada, Reno, NV, USA
e-mail: xli@nevada.unr.edu; fred.harris@cse.unr.edu

A. G. Schissler
Department of Mathematics and Statistics, University of Nevada,
Reno, NV, USA
e-mail: aschissler@unr.edu

R. Wu
Department of Computer Science, East Carolina University,
Greenville, NC, USA
e-mail: wur18@ecu.edu

L. Barford
Department of Computer Science and Engineering,
University of Nevada, Reno, NV, USA

Keysight Laboratories, Keysight Technologies, Reno, NV, USA
e-mail: lee.barford@ieee.org

© Springer Nature Switzerland AG 2019
S. Latifi (ed.), *16th International Conference on Information Technology-New Generations (ITNG 2019)*,
Advances in Intelligent Systems and Computing 800,
https://doi.org/10.1007/978-3-030-14070-0_46

distributions and a correlation matrix at the cost of partially specifying the joint distributions [6]. This technique offers the benefit of being general, while avoiding solving complex equations [7]. The NORTA method transforms elements from a multivariate standard normal distribution to the desired marginal distributions [6]. A more detailed description of the algorithm is given in Sect. 46.3.

For large and dense matrices, the number of operations scale with the number of elements, potentially leading to compute time bottlenecks. Analysis of the bottlenecks of a sequential implementation is discussed in Sect. 46.4. Therefore, implementation through parallel computing algorithms on the GPU becomes attractive. Due to the Single Instruction Multiple Data (SIMD) architecture of GPUs, operations can be performed in parallel thus providing a speedup over a sequential implementation.

The rest of the paper is organized as follows. Section 46.2 provides a brief summary of statistical background relevant to the main algorithm used, as well as an introduction to the GPU architecture, followed finally by a discussion of unified memory. Section 46.3 discusses the NORTA algorithm implemented in this paper. Section 46.4 reviews the sequential implementation written in R. Section 46.5 reviews the GPU implementation. Section 46.6 includes the testing methodology and experimental results followed by a discussion in Sect. 46.7. Section 46.8 presents the conclusion and future work.

46.2 Background

46.2.1 Statistical Background

The first topic to discuss is the correlation matrix. This matrix is positive semi-definite symmetric matrix with dimensions $d \times d$, where d corresponds to the number of variables in a multivariate vector whose correlations are represented by the matrix. Each entry in the matrix, with indices (i, j), represents the Pearson correlation coefficient $\rho(X_i, X_j)$ between random variables X_i, X_j given by the equation [8]:

$$\rho_{X_i, X_j} = \frac{E[(X_i - \mu_{X_i})(X_j - \mu_{X_j})]}{\sigma_{X_i} \sigma_{X_j}} \quad (46.1)$$

E is the expected value. μ_{X_i}, μ_{X_j} are the means and $\sigma_{X_i}, \sigma_{X_j}$ are the standard deviations of X_i, X_j respectively.

The matrix is symmetric because correlation between X_i and X_j is the same as the correlation between X_j and X_i. Also the main diagonal is composed of d 1s, reflecting the fact that a random variable is perfectly correlated with itself. Essentially, a correlation matrix represents the relationship between variables in a multivariate vector.

The Cholesky decomposition is used because it is an efficient method for decomposition of an input correlation matrix (and, in general any semi-positive definite matrix) [9]. The Cholesky decomposition provides two factors: one lower and one upper triangular $n \times n$ matrix. Multiplication using one of these factors onto a matrix of independent random variables will induce a specified correlation [10].

For convenience and ease of simulation, the NORTA algorithm begins with independent and identically distributed normal vectors. Most pseudo-random number generation software include the ability to simulate random normal variables. Further, we choose to start with normal vectors to begin our stochastic simulation to align with the original NORTA algorithm. One could start from other distributions, such as marginally uniform random variables. The important part is that we begin with any suitable random variable that can be easily transformed into a copula by applying the distribution's inverse cumulative distribution function (CDF). A copula is a multivariate distribution with uniform marginals and an arbitrary correlation matrix [11]. Importantly, Sklar's theorem [12] shows that any joint distribution has copula representation—providing a guarantee for the success of the NORTA algorithm.

The inverse (probability) transform is a general method of transforming one random variable into another. The transformation begins with values (usually a vector) obtained from one distribution with known CDF. Then the known CDFs are applied to these values to obtain values that have a uniform (on the interval 0 to 1) distribution. These values can be thought of as probabilities that a value from the original distribution was less than or equal to the original, observed value. Then these probabilities can be transformed into another target distribution by applying the inverse CDF of the target distribution. See Rizzo 2007 for details [10].

46.2.2 GPU Architecture

GPUs feature single instruction stream multiple data stream architecture that is optimized for data-parallel computations. This architecture works well for applications running a single instruction set over many different data elements. Compared to traditional central processing unit (CPU) architectures, the GPU is throughput focused. A GPU accomplishes this by having many compute cores that are able to process thousands of calculations simultaneously. While a GPU is conventionally used for graphics displays, in the past decade or so, there has been a large growth in general purpose GPU computing (GPGPU) [13]. Nvidia has created their own API for GPGPU programming using their chips. This language/API is an extension of the ANSI C language and is called CUDA which stands for the Compute Unified Device Architecture. Functions that are to be ran on the device

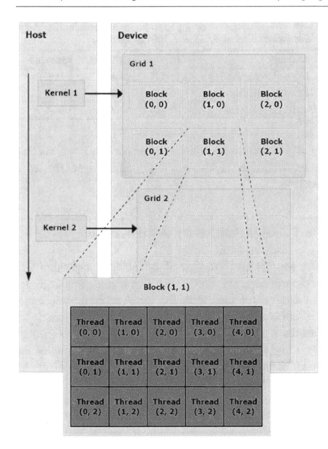

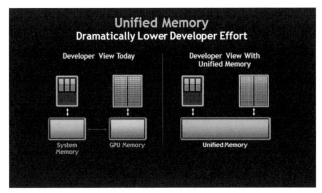

Fig. 46.2 Nvidia's Unified Memory combines physically distinct CPU and GPU memory into a single memory address space and handles data migration between the two automatically [15]

Fig. 46.1 Nvidia thread block structure: Each kernel corresponds to a grid. Each parallel invocation of the kernel corresponds to a block, and each block can be further divided into threads [13, 14]

(GPU) are called kernels. Each thread launches an instance of the kernel to run.

Kernels on the graphics processor is organized into grids of blocks of threads. Figure 46.1 visualizes this structure. Threads in the same block are run on the same streaming multiprocessor, which is the Nvidia compute unit. The advantage of using a GPU in parallel computing and high performance computing lies in the fact that Nvidia GPU's are commodity hardware and readily available for purchase. They are also fairly inexpensive and easy to setup compared to the more traditional computer cluster that are often used in high performance computing (HPC). With a GPU, high performance computing becomes more accessible, being able to speedup scientific computation by a significant amount without having the cost of setting up a CPU cluster. The GPUs used in this work were a Tesla P100 and an Nvidia GTX 1080 which are part of the Pascal architecture family.

46.2.3 Unified Memory

CUDA tool kit 6 introduced unified memory to the CUDA ecosystem [15]. CUDA programming up until this point re-

quired manual memory manipulation between the device and the host (CPU). With unified memory, the programmer no longer has to explicitly move data between memories. Memory management is handled by the CUDA backend and is abstracted from the programmer. This is illustrated in Fig. 46.2.

The Pascal architecture, which is the architecture of GPUs used for this paper, leverages a built-in hardware Page Migration Engine to handle page faults and data migration [16]. Unified memory is especially useful when the data set is too large to fit on the device. Without unified memory, the developer would have to manually chunk their data and transfer it between CPU and GPU. In the case of this project, the largest input matrices are about 3.4 GB for double precision numbers. This is quite large and can pose memory issues when used with common GPUs with 4 GB and less memory, especially with larger values of simulation replicates, n. Although unified memory reduces the complexity of CUDA code, it is relatively new and not optimized yet. Therefore, it is reasonable to speculate that the algorithm could run faster if using manual data migration.

46.3 NORTA Algorithm

Algorithm 46.1 describes the NORTA algorithm for the generation of a $d \times 1$ random vector X. This algorithm is extended to an $d \times n$ matrix, where n is the desired number of simulation replicates obtained by repeating the procedure. For a more in-depth explanation of the NORTA algorithm, refer to Cario [7].

The NORTA algorithm is divided into four steps (lines 2–5). The first step (line 2) is the Cholesky decomposition of the input correlation matrix, Σ_z, into a product of an upper triangular and it's conjugate, M and M' respectively. The input symmetric matrix has dimension $d \times d$. This step is the precursor to inducing the specified correlation matrix. The next step (line 3) is to generate a $d \times 1$ independent and

Algorithm 46.1 NORTA algorithm

1: **procedure** NORTA
2:　　Produce M of Σ_z sothat $MM' = \Sigma_z$.
3:　　Generate $W = (W_1, W_2, \ldots, W_d)' \leftarrow d \times 1 \; vector$.
4:　　Set Z by $Z \leftarrow MW$.
5:　　Return X where $X_i \leftarrow F_{X_i}^{-1}[\Phi(Z_i)]$, $i = 1, 2, \ldots, d$.

Table 46.1 Timings comparing sequential R and parallel CUDA implementations for $d = 20501$, $n = 1094$ (in seconds)

Functions	R (s)	CUDA (s)	Speedup
Cholesky decomposition	1,058.27	11.440	92.5
Random normal gen.	1.64	0.071	23.1
Matrix multiplication	316.16	3.730	84.8

identically distributed normal vector W. Then (line 4), W is applied the proper correlations using a matrix multiplication between M and W to return Z, a multivariate normal vector. In the last step (line 5), each element of Z is applied $\Phi(x)$, which is the standard normal CDF. Then, the inverse CDF of desired marginal distribution, also known as the quantile function, $F_{X_i}^{-1}(x)$ is applied. This will return, X, the final transformed vector, which is repeated n times to get the simulation matrix.

Importantly, we deviate from the original NORTA algorithm in that we do not adjust the input correlation matrix to get a (near) exact target correlation matrix from our final simulation data set. Instead, we use the target correlation as input and forgo the massive, brute force search for an adjusted input correlation matrix suggested by Cairo and Nelson [7]. Later, we assess the loss of accuracy associated with using the target correlation matrix by comparing to the publicly available COPULA R package [17].

46.4　Sequential Implementation

The sequential version of this simulation program was implemented using the R programming language. R is commonly used by statisticians and has many built-in statistical functions. This made the implementation of the inverse transformation much easier as R has support and optimization for all of the distributions used for this project are available. The random number generator, quantile, and distribution functions of each type of distribution were used for the sequential implementation. In addition to statistical functions, R also features support for working with dense and sparse matrix math. The Cholesky decomposition, matrix multiplication and random matrix generation were also handled through built-in functions. A text file containing the correlation matrix and a text file specifying marginal distributions with parameters were the inputs to the program. The output is a text file containing the final transformed simulation matrix.

The Cholesky decomposition, random number generation, matrix multiplication, and inverse transformation steps of the algorithm were expected to benefit the most from GPU parallelization because, due to the number of independent operations, the computation time scales with size of matrix for sequential implementations. Results from bottleneck analysis in Table 46.1 confirms this. Analysis of the inverse

transformation step was not included because it was not able to be parallelized in this study due to difficulties explained in Sect. 46.5.

46.5　GPU Parallel Implementation

The parallel GPU version of the NORTA algorithm was implemented using the CUDA programming language. Input and output data were the same as in the sequential implementation described above. The matrix_read_in function and distribution_list_read_in functions were implemented sequentially in the GPU version of the code. Memory management was handled automatically by CUDA unified memory.

Several official CUDA APIs from Nvidia were used. They were: CuBLAS, CuRAND, and CuSOLVER [18–20]. Library functions abstracted out the notion of kernels. The Cholesky decomposition was implemented using the potrf function from CuSOLVER. This function performs the Cholesky factorization and returns an upper or lower triangle in column major order depending on arguments specified and uses double precision. Matrix multiplication was implemented using the gemm function from CuBLAS. gemm performs matrix multiplication given two input matrices and stores them in an output matrix specified in the arguments. The output matrix used was one of the original input matrices for better memory efficiency. Double precision was used for this function as well. CuRAND was used for the generation of the random normal matrix. The pseudo random number generator was used out of this library using the default generator to generate random numbers in double precision sampled from a standard normal distribution.

For the inverse transformation function, the C++ stats library StatsLib made by Keith O'Hara was used [21]. Distributions supported are: Beta, Cauchy, Exponential, F, Normal, Log Normal, Logistic, Poisson, t, Uniform, and Weibull. This library was chosen because it was one of few C++ statistical libraries found that imitated the R distributions library. Implementing distribution functions is a complex process so due to time and technical ability constraints, distributions were not written directly in CUDA. For testing purposes, only the Poisson distribution was used. The reasoning was that the input data came from RNA sequence data may be modeled by Poisson. Due to difficulties stemming from incompatibilities between C++ and CUDA, the inverse transformation function was also implemented on the CPU.

46.6   Experiments and Results

The CUDA libraries and functions used in this study treat matrices as column major instead of the C style row major. This causes most functions such as the Cholesky decomposition and the matrix multiplication routine to return the transpose of the expected result. This implicit transpose is caused by the conversion from row major to column major. Care must be taken to get the correct results out of these routines. Unified memory was used to store all arrays. This allows arrays to be larger than the memory of the GPU. Timings were captured for various n and d values. Speedup was calculated between GPU version and sequential version. Speedup vs another popular R library for multivariate statistical simulation called COPULA was also captured. Refer to Yan for details of the COPULA R package [17].

Once the implementations discussed in Sects. 46.4 and 46.5 were finished, timings for the various modules of each implementation were done in order to see how the R and CUDA computations compared. Timings for the sequential R column shown in Table 46.1 were done on a Macbook Air with an Intel Core i5 processor at 1.3 Ghz and 4 gb of memory. The CUDA column in Table 46.1 was performed on a CUBIX box with dual 12 core Intel Xeon CPUs running at 2.00 GHz. This machine has 8 Nvidia GeForce GTX 1080 GPUs but only one was used in the computation of this project. A different machine was used than the timing data collected for sequential R column due to the fact that the Macbook Air used did not have a CUDA compatible graphics card.

Speedup and timing data comparing sequential, GPU, and COPULA methods in Figs. 46.3, 46.4, and 46.5 were collected

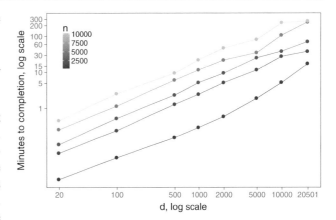

Fig. 46.4   Sequential NORTA timings (R version) with increasing n and d

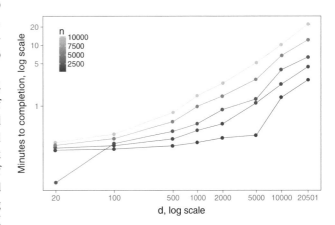

Fig. 46.5   GPU-NORTA timings with increasing n and d

on a machine with an Intel Xeon Gold 6126 CPU running at 2.60 GHz. The GPU was a Nvidia Tesla P100. Using this machine, and CUDA Unified Memory, we were able to overcome memory limitations on the Macbook Air for larger values of n.

46.7   Discussion

Table 46.1 summarizes the timings of the functions that were parallelized from sequential R version to GPU CUDA version and provides the speedups. Correlation matrix and distribution file read in timings are not included since they were done sequentially in both cases. There is potential to improve these steps, for instance, by using binary files instead of text files for input data. The inverse transformation step was not included either for reasons discussed in Sect. 46.5. The other steps: Cholesky decomposition, matrix multiplication, and random number generation were all responsive to parallelization, and are optimized by Nvidia in

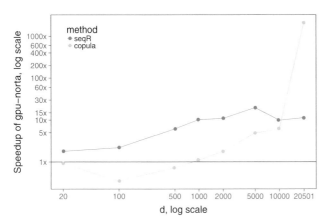

Fig. 46.3   Speedup of GPU-NORTA vs Sequential NORTA (R version) and GPU-NORTA vs Sequential COPULA (for $n = 2,000$)

Table 46.2 Quadratic losses of sequential and GPU implementation at various d and using $n = 2000$

d	1,000	2,000	5,000	10k	20,501
GPU-NORTA	1,275	5,060	7,554	10,225	9,704
R-NORTA	1,232	4,825	7,720	10,571	10,572
COPULA	1,131	4,853	7,524	10,272	10,426

their cuSOLVER, cuBLAS, and cuRAND libraries. Speedup of the decomposition step resulted in a 92.5× increase. The random normal generation increased by a factor of 23.1 and matrix multiplication increased by a factor of 84.8 over the sequential version.

Figure 46.3 shows the speedup of GPU-NORTA compared to both the sequential NORTA and sequential COPULA methods. Speedup data was collected using $n = 2,000$ and various d. Speedups were calculated for the whole program running, including data read in and inverse transformation. GPU-NORTA performed faster than sequential NORTA at all d sizes. The speedup peaks (19.6x) at $d = 5,000$ for sequential R comparison and a dramatic increase (2093×) at $d = 20,501$ for comparison against COPULA package. The spike in COPULA is because computation time does not scale linearly.

Figure 46.4 shows the total run time of the sequential version for various n and d dimensions and Fig. 46.5 shows the total run time for the GPU-NORTA implementation. The GPU-NORTA version took less than 30 min to run for the largest set tested, compared to over 200 min for the sequential version. Even with non-optimized read in functions and sequential inverse transformation, GPU-NORTA had notable speedup that increases with increasing data size compared to optimized R code and COPULA packages.

Table 46.2 displays the quadratic loss between the simulated correlation matrix $\hat{R}$ from the input (target) correlation, R for $n = 2,000$. Quadratic loss is given by the expression $||\hat{R}R^{-1} - I||_2$ and we compare our proposed GPU-NORTA to the R-NORTA and COPULA implementation. Both our GPU-NORTA and R-NORTA forgo the grid search for an adjusted input matrix, whereas the COPULA approach does not require such an adjustment a priori. In our limited studies, at high dimension, we found no substantive loss in accuracy and can safely avoid the search suggested by Cairo and Nelson [7]. Future studies, however, are needed to fully assess the effects.

The speedup of the decomposition and matrix multiplications led to a huge gain which helped to off balance the slow run times of sequential read in function and inverse transformation function. There is less significant speedup of the random normal matrix generation, but it still provided a 23.1 times speed.

46.8 Conclusion and Future Work

In this paper, we have introduced a GPU accelerated version of simulating multivarate data with input correlation matrix and arbitrary marginal distributions. This parallel GPU-NORTA algorithm exhibited speedups over a sequential NORTA version and a sequential COPULA version both written in R. The analysis summarized in Table 46.1 revealed that the Cholesky and matrix multiplication steps took significant portions of computation time in the sequential version. Fortunately, these functions have attractive parallelization potential and thus had significant speedups over the sequential counterpart. The inverse transformation also has parallelization potiential, but was not implemented in this paper. Future work needs to be done in implementing the statistical functions required for the inverse transformation into CUDA device code. This allows the last step to be parallelized, which should result in more dramatic speedups than what was obtained in this study.

The issue concerning the necessity of adjusting input correlation matrix warrants further study at various dimensions and for other marginal distributions. Our GPU-NORTA implementation could be improved with the ability to allow a user to conduct the search resulting in greater simulation accuracy.

For larger data sets than what was used in this study, this algorithm has the potential to be extended to utilize multiple GPUs. The CUDA API has support for multi-GPU applications [13].

Further speedup could be possible from manual memory management with the trade off of added complexity. The question of whether or not the speedups are worth the complexity trade off needs further exploration. However, results show that even with inefficiencies associated with Unified Memory, the GPU implementation is still faster than both R-NORTA and COPULA versions. With future releases of the CUDA toolkit, Unified Memory management is expected to become more optimized.

Acknowledgements This material is based upon work supported by the National Science Foundation under grant number IIA1301726. Any opinions, findings, and conclusions or recommendations expressed in this material are those of the authors and do not necessarily reflect the views of the National Science Foundation.

References

1. Pucher, B.M., Zeleznik, O.A., Thallinger, G.G.: Comparison and evaluation of integrative methods for the analysis of multilevel omics data: a study based on simulated and experimental cancer data. Brief. Bioinform. 1–11

(2018). [Online]. Available: https://academic.oup.com/bib/advance-article/doi/10.1093/bib/bby027/4982568

2. Gatti, D.M., Barry, W.T., Nobel, A.B., Rusyn, I., Wright, F.A.: Heading down the wrong pathway: on the influence of correlation within gene sets. BMC Genomics **11**(1), 574 (2010). [Online]. Available: https://doi.org/10.1186/1471-2164-11-574

3. Wilkins, M.F., Morris, C., Boddy, L.: A comparison of radial basis function and backpropagation neural networks for identification of marine phytoplankton from multivariate flow cytometry data. Bioinformatics **10**(3), 285–294 (1994). [Online]. Available: http://dx.doi.org/10.1093/bioinformatics/10.3.285

4. Russkova, T.V.: Monte Carlo simulation of the solar radiation transfer in a cloudy atmosphere with the use of graphic processor and NVIDIA CUDA technology. Atmos. Oceanic Opt. **31**(2), 119–130 (2018). [Online]. Available: https://link-springer-com.unr.idm.oclc.org/content/pdf/10.1134/S1024856018020100.pdf

5. Häyrinen, K., Saranto, K., Nykänen, P.: Definition, structure, content, use and impacts of electronic health records: a review of the research literature. Int. J. Med. Inform. **77**(5), 291–304 (2008). [Online]. Available: https://www.sciencedirect.com/science/article/pii/S1386505607001682

6. Niaki, S.T.A., Abbasi, B.: Generating correlation matrices for normal random vectors in NORTA algorithm using artificial neural networks. J. Uncertain Syst. **2**(3), 192–201 (2008). [Online]. Available: http://www.worldacademicunion.com/journal/jus/jusVol02No3paper04.pdf

7. Cario, M.C., Nelson, B.L.: Modeling and generating random vectors with arbitrary marginal distributions and correlation matrix. Northwestern University, Technical Report (1997). [Online]. Available: http://citeseerx.ist.psu.edu/viewdoc/summary?doi=10.1.1.48.281

8. Casella, G., Berger, R.L.: Statistical Inference. Duxbury, Pacific Grove (2002)

9. Strang, G.: Introduction to Linear Algebra. Cambridge Press, Wellesley (1993)

10. Rizzo, M.L.: Statistical Computing with R. Chapman and Hall/CRC (2007). [Online]. Available: https://www.taylorfrancis.com/books/9781420010718

11. Genest, C., Mackay, J.: The joy of copulas: bivariate distributions with uniform marginals. Am. Stat. **40**(4), 280–283 (1986). [Online]. Available: https://www.tandfonline.com/doi/abs/10.1080/00031305.1986.10475414

12. Sklar, M.: Fonctions de répartition à n dimensions et leurs marges. Publ. Inst. Statist. Univ. Paris **8**, 229–231 (1959)

13. Sanders, J., Kandrot, E.: CUDA by example: an introduction to general-purpose GPU programming, 1st edn. Addison-Wesley Professional, Upper Saddle River (2010)

14. Nobile, M.S., Cazzaniga, P., Besozzi, D., Pescini, D., Mauri, G.: cuTauLeaping: a GPU-powered Tau-leaping stochastic simulator for massive parallel analyses of biological systems. PLoS One **9**(3), e91963 (2014). [Online]. Available: http://dx.plos.org/10.1371/journal.pone.0091963

15. Harris, M.: Unified memory in CUDA 6 (2013). [Online]. Available: https://devblogs.nvidia.com/unified-memory-in-cuda-6/

16. Harris, M.: CUDA 8 features revealed: pascal, unified memory and more (2016). [Online]. Available: https://devblogs.nvidia.com/cuda-8-features-revealed/

17. Yan, J.: Enjoy the joy of copulas: with a package copula. J. Stat. Softw. **21**(4), 1–21 (2007). [Online]. Available: http://www.jstatsoft.org/v21/i04/

18. cuSOLVER::CUDA Toolkit Documentation (2018). [Online]. Available: https://docs.nvidia.com/cuda/cusolver/index.html

19. cuRAND::CUDA Toolkit Documentation (2018). [Online]. Available: https://docs.nvidia.com/cuda/curand/notices-header.html#notices-header

20. cuBLAS::CUDA Toolkit Documentation (2018). [Online]. Available: https://docs.nvidia.com/cuda/cublas/index.html

21. O'Hara, K.: StatsLib (2018). [Online]. Available: https://github.com/kthohr/stats

A GPU-Based Smith-Waterman Approach for Genome Editing

47

Luay Alawneh, Mohammad Shehab, Mahmoud Al-Ayyoub, and Yaser Jararweh

47.1 Introduction

Genome editing technologies enable scientists to alter any organism's DNA for the prevention and treatment of human diseases [1]. Genome editing is relatively new and is currently being experimented with on cells and animal models to determine the applicability of this approach on humans. Research in genome editing is being conducted on various sets of diseases such as single-gene disorders (cystic fibrosis and hemophilia) and is extended to more complex diseases such as cancer, human immunodeficiency virus (HIV), heart diseases and mental illness.

Changing the genome structure includes adding, removing and replacing particular locations in the genome. There exist several approaches for genome editing such a ZFN, TALEN and CRISPR-based technologies [2]. In this paper, we focus our study on CRISPR-Cas9 (clustered regularly interspaced short palindromic repeats and CRISPR-associated protein 9) [2]. CRISPR-Cas9, discovered in the 1980s [3], is an adaptive immunity system for fighting invading genetic material. The CRISPR-Cas9 technique was inspired by a natural genome editing system found in bacteria where the latter capture snippets of DNA from invading viruses to create DNA segments known as CRISPR arrays. These arrays are then used by the bacteria as a signature to prevent the same viruses (or similar ones) from attacking again. The bacteria will use these arrays to produce RNA segments to target the viruses' DNA. Finally, the bacteria use Cas9 or similar enzymes to split the DNA apart and disable the virus. Scientists can use the same approach for genome editing for making changes to the DNA sequence. Although this approach may incur ethical concerns when applied on humans, however, it is very promising when used for treatment and prevention of diseases supported by its efficiency, accuracy and feasibility.

A major line of research within Bioinformatics aims at reducing the cost of medical research by using advanced techniques that help in detecting specific patterns in DNA sequences which correspond to unwanted cells such as Cancer [4]. Genome editing tools, such as CRISPR, are used to detect these optimal patterns. These tools need alignment methods to calculate the maximum number of gaps and the cost that biologists set for the gap [3]. In this paper, we propose to use the Smith-Waterman (SM) algorithm [5] as local sequence alignment algorithm for this purpose. SM algorithm is a dynamic programming technique proposed to calculate the similarity between two strings of nucleic acid sequences or protein sequences. It is suitable to be implemented and executed in parallel which makes it fit to run on GPUs (Graphic Processing Units). Our approach targets detecting CRISPR patterns using the SM algorithm.

In this paper, we present four different implementations[1] of the string similarity algorithm and provide performance evaluation when tested on five different sequences of various lengths. The results show that using GPU will significantly improve the performance of the Smith-Waterman algorithm.

The rest of the paper is organized as follows. Section 47.2 discusses the related work. In Sect. 47.3, we detail the proposed approach. Section 47.4 presents the analysis results. Section 47.5 concludes the paper with a glimpse on the future directions.

L. Alawneh (✉) · M. Al-Ayyoub · Y. Jararweh
Jordan University of Science & Technology, Irbid, Jordan
e-mail: lmalawneh@just.edu.jo; maalshbool@just.edu.jo; yijararweh@just.edu.jo

M. Shehab
Concordia University, Montreal, QC, Canada
e-mail: m_shehab@live.concordia.ca

[1]Implementations and data can be found at: https://github.com/M12Shehab/Dynamic-parallelism-technology-cuda-needleman-wunsch-and-smith-waterman.

© Springer Nature Switzerland AG 2019
S. Latifi (ed.), *16th International Conference on Information Technology-New Generations (ITNG 2019)*,
Advances in Intelligent Systems and Computing 800,
https://doi.org/10.1007/978-3-030-14070-0_47

47.2   Related Work

This section presents studies that target efficient detection of CRISPR patterns and also sheds light on the use of GPU in bioinformatics.

Biswas et al. [6] implemented the CRISPRDetect algorithm to detect CRISPR patterns from RNA repeats in bacteria. The tool needs Genomic sequence, pattern size, number of minimum repeated patterns and the maximum number of gaps as input from the user. Then, the tool calculates the quality of score based on these inputs and saves the predicted location of the pattern that will then be used in the genome editing.

Bland et al. [7] implemented CRISPR Recognition Tool (CRT) to directly detect repeats from a DNA sequence. According to the authors, CRT showed significant improvement in terms of CRISPR identification with respect to accuracy and performance when compared to the other approaches such as Patscan [7]. Additionally, the authors recommended using CRT in conjunction with Pilercr [7] as both approaches complement each other in terms of performance and accuracy.

Bo et al. [8] proposed a new algorithm based on automata. Their implementation is designed and tested for different hardware platforms. The implementation is suitable for CPU, GPU, FPGA and AP platforms. The results showed that FPGA achieves significant speedup when compared to the other platforms. Moreover, this algorithm proved not appropriate for GPUs.

Several approaches use GPUs (Graphic Processing Units) to improve the performance of various bioinformatics techniques. Vouzis et al. [9] used the GPU to accelerate the Basic Local Alignment Search Tool (BLAST) tool. Wilton et al. [10] improved the performance of sequence alignment (A21 to compare similar protein sequences) by using a sorting method with reduction in order to make the implementation suitable for parallel programming. Their GPU-based approach maintained the same accuracy of the original approach.

Alawneh et al. [11] parallelized a Shannon-entropy sequence segmentation approach. Their GPU parallel approach provided 6 times faster execution than the CPU-based approach although the original algorithm incurs high dependencies. Liu and Shmidt [12] presented a dynamic programming approach that performs optimal alignments for short DNA sequences. Their approach, a GPU-accelerated pairwise sequence alignment algorithm (GSWABE), used the Smith-Waterman to perform all-to-all pairwise global, semi-global, and local alignments. Zhu et al. [13] proposed to speed up the MAFFT multiple sequence alignment algorithm [14] using the power of GPU. In [15], a parallel implementation of the Smith-Waterman algorithm on SLI NIVIDA technology was

introduced. To achieve parallelization, the authors adopted the diagonal implementation of the Smith-Waterman. They conducted their experiments on an NVIDIA card with 4 GPUs where it provided 45 times faster execution when compared to the sequential implementation. The main drawback of this work is that it only compares a pair of sequences and does not consider searching in large databases of protein sequences.

47.3   Approach

CRISPR-Cas9 represents DNA sequences in bacteria used to detect and destroy DNA from similar viruses during subsequent attacks. The sequences contain snippets of DNA from attacking viruses. CRISPR-Cas9 plays an important role in the immune system of the prokaryotic cells [16]. Each sequence has a specific pattern location which cannot be recovered if the sequence is broken. Searching for this pattern has two main problems. Keeping the number of gaps constant and searching for patterns in long sequences.

Smith-Waterman is a local alignment algorithm used to calculate similarity between a substring and a long sequence with gaps and cost for each gap are detected [5]. In this research, we used the Smith-Waterman algorithm to achieve the goal of searching for CRISPR patterns in DNA sequences. We set the maximum number of gaps and the cost for each gap as dynamic arguments for the system.

The user enters the pattern query and set the two main arguments (maximum number of gaps and cost for each gap). Then, our system starts searching for snippets and save them for experts. To accelerate the execution time for the searching process, we utilize the capabilities of GPU. We then compare the performance gain in GPU with respect to the performance when run on the CPU.

In the following, we present the Smith-Waterman algorithm followed by a presentation of the four different implementations on CPU and GPU.

47.3.1 Smith Watermann Algorithm

The Smith–Waterman is a dynamic programming algorithm which performs local sequence alignment to determine similar regions between two strings (nucleotide or protein sequences). Similarity in nucleotide or protein sequences indicates that the compared sequences are related. Moreover, local alignment can be used to find similarity between regions (subsequences) of the same sequence.

The Smith-Waterman algorithm identifies optimal local alignments between two sequences which mean that it finds a segment in each sequence where their alignment indicates the maximal similarity scoring. Thus, the algorithm uses a scoring matrix to determine the level of similarity, where a zero score means no similarity. The matrix dimensions are $(M + 1) \cdot (N + 1)$ where M is the length of the first string S_1 and N is the length of the second string S_2. The extra row r_0 and the extra column c_0 are added at index 0 in the generated matrix and their values are set to 0. Equation 47.1 is used to calculate the values in each cell based on the score that is calculated using Algorithm 47.1. The scoring values that are used in our implementation are match = 2, mismatch = 0 and gap = −1. Simply, the cell score is initially determined if the two compared characters match or mismatch. Algorithm 47.1 calculates the score value for each cell in matrix H based on the values in the neighboring cells (upper, left and upper-left corner). It should be noted that a sequence alignment algorithm may be implemented by either sequentially or diagonally traversing the matrix. The problem with using sequential traversal is that the dependencies among the matrix cells are very strong [10].

Algorithm 47.1

Smith-Waterman (Sequential Traversal)

```
1.  SmithWaterman(S₁, S₂, N, M)
2.  Def score, gaps
3.  Def match = 1, mismatch = 0, gap = -1
4.  For I = 0 To N - 1 Do   H[I][0] = 0   End For
5.  For J = 0 To M - 1 Do   H[0][J] = 0   End For
6.  For I = 0 To N - 1 Do
7.         For J = 0 To M - 1 Do
8.                 If S1[I] EQ S2[J] Then
9.                         score = match
10.                Else
11.                        score = mismatch
12.                End If
13.                H[I][J] = Equation_1 (I, J,
                       score, gap)
14.                gaps [row][col] = get number of
                                    gaps
15.        End For
16. End For
```

$$H_{i,j} = \max \begin{cases} 0 \\ H_{i-1,j-1} + match \\ H_{i-1,j-1} + mismatch \\ H_{i,j-1} + gap \\ H_{i-1,j} + gap \end{cases} \quad (47.1)$$

When the calculation of the final scores for each cell ($H_{i,j}$) is determined, the highest scoring alignment can be determined by backtracking the matrix from the highest scoring cell until a cell with a zero score is reached.

Figure 47.1 shows two possible alignments using the matrix that is populated using the Smith-Waterman algorithm.

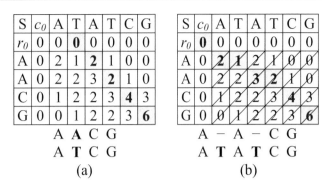

Fig. 47.1 Smith-Waterman Example (score = 6). (**a**) Alignment 1. (**b**) Alignment 2

Figure 47.1a shows the first alignment with no gaps (one mismatch; insertion or deletion of character in the corresponding sequence) while Fig. 47.1b shows the alignment with two gaps. Thus, the gap cost in Fig. 47.1b is 2 while in Fig. 47.1a is *zero*. Figure 47.1b shows that the matrix was constructed using diagonal traversal according to Algorithm 47.2. It is clear that as the diagonal line gets longer, the score for the cells in the diagonal line can be calculated independently. Thus, diagonal traversal is convenient in case of parallelization.

Algorithm 47.2 presents the diagonal traversal implementation of the Smith-Waterman algorithm.

Algorithm 47.2

Smith-Waterman (Diagonal Traversal)

```
1.  SmithWaterman(S₁, S₂, N, M)
2.  Def score, gaps
3.  Def match = 1, mismatch = 0, gap = -1
4.  For I = 0 To N - 1 Do   H[I][0] = 0   End For
5.  For J = 0 To M - 1 Do   H[0][J] = 0   End For
6.  For slice = 0 To M + N - 1 Do
7.         If slice < M Then
8.             Z = 0
9.         Else
10.            Z = slice - M + 1
11.        End If
12.        For j = Z To slice - 1 Do
13.            row = slice - j
14.            col = j
15.            If S1[row] EQ S2[col] Then
16.                    score = match
17.            Else
18.                    score = mismatch
19.            End If
20.            H[row][col] = Equation_1 (row,
                   col, score, gap)
21.            gaps [row][col] = get number
of gaps
22.        End For
23. End For
```

The Smith-Waterman has a quadratic computational complexity which hinders its applicability for large sequences. In the following, we present the string matching approach and the four versions that will be used for searching for CRIPR-Cas9 patterns. A comparison between the four versions is provided in Sect. 47.4.

47.3.2 String Matching Algorithm

Searching for similar CRISPR patterns is assisted using the Smith-Waterman algorithm (diagonal version). The CRISPR pattern is compared to the different segments in the DNA sequence using a sliding window. The number of *gaps* determines whether to consider the alignment a potential match or not. The user predetermines the maximum number of gaps allowed using the *MAXGAPS*. Algorithm 47.3 will try to match the pattern starting at each position in the target sequence. If the number of *gaps* is larger than the *MAXGAPS* value, then this alignment is skipped and a new alignment will be examined at the next position in the target sequence.

Algorithm 47.3
String Matching – Sliding Window

```
1.  Def  MAXGAPS
2.  P: Pattern
3.  S: Sequence
4.  M = len(P)
5.  N = len(S)
6.  For I = 0 To N Do
7.       gaps = SmithWaterman(S_{I,M-1}, P, M, M)
8.       If gaps > MAXGAPS Then
9.            No alignment starting at I
10.      Else
11.           Add S_{I,M-1} as possible CRISPR region
12.      End If
13. End for
```

We have implemented four different versions of Algorithm 47.3 as follows.

1. *Sequential CPU.* This version is implemented using the diagonal traversal technique in order to be able to provide fair comparison with the parallel implementation. The implementation is executed in the CPU side only with a single thread.
2. *Parallel CPU.* The CPU implementation is based on OpenMP [17]. Algorithm 47.3 is modified by having each thread calculate the score for a cell in the matrix at each diagonal concurrently. In the CPU, the number of possible threads is limited to the number of cores.
3. *Parallel GPU.* This version is proposed with multithreading in the GPU side using CUDA. The advantage of using the GPU is that we can launch a larger number of threads at the same time in order to calculate the score and gaps in the Smith-Waterman matrix. GPUs are capable of

running thousands of threads in parallel which make them very appropriate for solving the local sequence alignment problem.

4. *Dynamic Hybrid CPU-GPU.* This last version is implemented to capitalize the power of the new NIVIDA technology using dynamic parallelism [18]. This allows us to utilize the parallel power of CPUs and GPUs. Thus, each CPU thread will have its own alignment operation on the GPU. Using dynamic parallelism, the CPU works in parallel to send multiple queries to the GPU that in turn works on the Smith-Waterman algorithm which saves transfer time between CPU and GPU. The Hybrid implementation needs to reduce the number of data transfers between the CPU and the GPU's main memory which is known as the memory management challenge in hybrid CPU-GPU implementations. The dynamic parallelism is supported only with GPUs of Kepler architecture and compute capability 3.5 or more. Thus, this hybrid version is limited for certain GPUs only.

47.4 Experimental Results

In this paper, we tested our approach on five different randomly generated sequences that resemble DNA structure. Table 47.1 shows the execution times for the four different implementations on each sequence. The experiments were conducted on a Windows 10 machine (32 GB RAM) with Core i7 (7th generation) processor and GTX 970 GPU (8 GB Ram) running CUDA 8.

It is clear that the sequential implementation is the slowest while a slight improvement is achieved using the parallel CPU implementation. The GPU implementation outperforms the two CPU implementations. The Parallel GPU is 5 times faster than the parallel CPU and 18 times faster than the sequential version. The parallel CPU is 4 times faster than the sequential version. However, the synergy between the CPU and the GPU provided the best performance (hybrid approach) since we are able to utilize a larger number of threads in the GPU.

The number of GPU threads in the hybrid approach, in its best case, will be multiplied by the number of CPU

Table 47.1 Smith-Waterman execution time (seconds)

Sequence length	Implementation			
	Sequential	Parallel CPU	Parallel GPU	Hybrid CPU-GPU
100	2.50	0.65	0.13	0.11
1000	72.00	17.89	3.72	2.89
10,000	2102.23	524.66	108.85	84.00
100,000	60,972.45	15243.20	3034.86	2438.30
1000,000	1,768,202.50	442,050.90	85,863.91	70,729.50

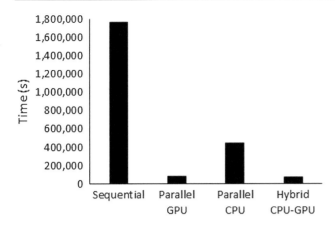

Fig. 47.2 Execution evaluation (1,000,000 length)

threads (four threads in our experiments). The hybrid CPU-GPU implementation achieves the best performance gain with an average of 24 times faster than the sequential version. Furthermore, the hybrid implementation is 6 times faster than the parallel CPU version and only 1.3 times faster than the parallel GPU.

Figure 47.2 shows the execution times of the four different implementations on the 1,000,000 sequence. It is clear that the sequential approach is inferior for this kinds of problems. However, the parallel approaches should be tested on very large DNA sequences (millions) in order to measure their scalability.

47.5   Conclusion and Future Work

CRISPR is a mechanism used by bacteria as an immunity system. This system eliminates genetic material from virus attack. It needs to detect pattern locations in DNA sequences and then use them to deactivate RNA of viruses.

Biologists utilize this technique as treatment for Cancer and to deactivate HIV virus. To save time and cost, biologists need computer algorithms to predict patterns from DNA sequences. Many tools and algorithms are implemented for this issue. However, some of those algorithms need to improve their execution time and it are not suitable to be executed with GPUs.

In this paper, we presented four different implementations of the Smith-Waterman algorithm. We compared the performance of the Sequential (single-threaded on CPU), parallel CPU, parallel GPU and hybrid CPU-GPU implementations (implementations and data used in this paper are available on GitHub[2]). The hybrid CPU-GPU implementation gets the

best results with performance gain with around 25 times faster than sequential version.

We are currently working on testing the approach on real virus DNAs. In the future, we are planning to test this approach on large DNA sequences in order to prove the applicability and scalability of the hybrid parallel approach.

Acknowledgment   The authors would like to thank the Deanship of Research at Jordan University of Science and Technology for funding this work (grant number 20170396).

References

1. Maeder, M.L., Gersbach, C.A.: Genome-editing technologies for gene and cell therapy. J. Mol. Ther., Elsevier. **24**(3), 430–446 (2016)
2. Gaj, T., Gersbach, C.A., Barbas, C.F.: ZFN, TALEN, and CRISPR/Cas-based methods for genome engineering. Trends Biol. **31**(7), 397–405 (2013)
3. Horvath, P., Barrangou, R.: CRISPR/Cas, the immune system of bacteria and archaea. Science. **327**, 167–170 (2010)
4. Li, J., Zhang, Z., Rosenzweig, J., Wang, Y.Y., Chan, D.W.: Proteomics and bioinformatics approaches for identification of serum biomarkers to detect breast cancer. Clin. Chem. **48**(8), 296–304 (2002)
5. Rognes, T.: Faster Smith-Waterman database searches with intersequence SIMD parallelisation. BMC Bioinform. **12**(1), 221 (2011)
6. Biswas, A., Staals, R.H., Morales, S.E., Fineran, P.C., Brown, C.M.: CRISPRDetect: a flexible algorithm to define CRISPR arrays. BMC Genomics. **17**, 356 (2016)
7. Bland, C., Ramsey, T.L., Sabree, F., Lowe, M., Brown, K., Kyrpides, N.C., Hugenholtz, P.: CRISPR recognition tool (CRT): a tool for automatic detection of clustered regularly interspaced palindromic repeats. BMC Bioinform. **8**, 209–216 (2007)
8. Bo, C., Dang, V., Sadredini, E., Skadron, K.: Searching for potential gRNA off-target sites for CRISPR/Cas9 using automata processing across different platforms. In: IEEE International Symposium on High Performance Computer Architecture (HPCA), pp. 737–748 (2018)
9. Vouzis, P.D., Sahinidis, N.V.: GPU-BLAST: using graphics processors to accelerate protein sequence alignment. Bioinformatics. **27**(2), 182–188 (2011)
10. Wilton, R., Budavari, T., Langmead, B., Wheelan, S.J., Salzberg, S., Szalay, A.: Faster sequence alignment through GPU-accelerated restriction of the seed-and-extend search space. bioRxiv (2014)
11. Alawneh, L., Rawashdeh, E., Al-Ayyoub, M., Jararweh, Y.: GPU parallelization of sequence segmentation using information theoretic models. Simul. Model. Pract. Theory. **86**, 11–24 (2018)
12. Liu, Y., Schmidt, B.: GSWABE: faster GPU-accelerated sequence alignment with optimal alignment retrieval for short DNA sequences. Concurr. Comput. **27**(4), 958–972 (2015)
13. Zhu, X., Li, K., Salah, A., Shi, L., Li, K.: Parallel implementation of MAFFT on CUDA-enabled graphics hardware. IEEE/ACM Trans. Comput. Biol. Bioinform. **12**(1), 205–218 (2015)
14. Katoh, K., Toh, H.: Recent developments in the MAFFT multiple sequence alignment program. Brief. Bioinform. **92**, 86–98 (2008)

---

15. Khajeh-Saeed, A., Poole, S., Perot, J.B.: Acceleration of the smith-waterman algorithm using single and multiple graphics processors. J. Comput. Phys. **229**(11), 4247–4258 (2010)

16. Ratha, D., Amlinger, L., Rathc, A., Lundgren, M.: The CRISPR-Cas immune system: biology, mechanisms and applications. Biochimie. **117**, 119–128 (2015)

17. Tian, X., Bik, A., Girkar, M., Grey, P., Saito, H., Su, E.: Intel OpenMP C++/Fortran compiler for hyper-threading technology: implementation and performance. Intel Technol. J. **6**(1), 1–11 (2002)

18. Jones, S.: Introduction to dynamic parallelism. In: GPU Technology Conference Presentation S, vol. 338 (2012)

## Part VI

Biostatistical Analysis, Wavelet Analysis, and Verified Numerical Computations

The SMp(x or y;PXmin,Xmax,ML,p1,p2,Max)
a Probabilistic Distribution, or a Probability
Density Function of a Random Variable X

48

Terman Frometa-Castillo

48.1 Introduction

The SP/Es, contrary to the deterministic ones, do not have a unique value of their outcomes but a range of values that are evaluated with the likelihood of their occurrences. The mean and/or the most likely value are used to make decisions involving SP/Es.

There is an inappropriate use of the normal cumulative distribution function (NCDF) in the normal tissue complication probability (NTCP) models in [1]. The NCDF is used to evaluate the probabilities of random variable X in a function of x ($P_X = f(x)$), whereas the NTCP is a function of a different variable y. Lyman-Kutcher-Burman (LKB) NTCP(Deff) is a NCDF, so it calculates the probabilities of Deff $\leq$x if Deff follows a normal distribution. It is completely different from its formulation, where one must calculate the NTCP function of Deff. The deficient use of the NCDF has led to deficient-sigmoid functions, e.g., in [2], which are less complex logistic functions but have the same sigmoid forms.

The SMp is a modeling and simulation project based on the most appropriate treatment, modeling and evaluations of many SP/Es in the fields, e.g., radiobiology, where the SMp has been initially applied. The SMp determines the probabilities using its probabilistic-mechanistic models or computer simulators, as described in [3].

Based on strong probabilistic foundations, the SMp will provide suitable PFs to describe and predict many SP/Es to simplify, increase the precision, and unify the treatments and models for these SP/Es through calculation methodologies using new PFs and simulation tools. The SMp functions will represent new PFs, where the probability of a random discrete/continuous variable X can be expressed as $P_X = p(y)$, where the independent variable y, instead of x.

T. Frometa-Castillo (✉)
Oncology Hospital of Santiago de Cuba, Chicago, IL, USA

The current PDFs, e.g., Weibull density function, produce bell-share symmetric or asymmetric functions or decreasing curves. These PDFs can be reproduced using the proposed function, which is possible with an appropriate selection of its parameters. Many current PDFs have similar forms with SMp(y) of [4].

SMp(x) has six parameters, but based on probabilistic conditions and according its use, one or more parameters can be obtained from others, e.g. The SMp(x) has only three independent parameters in its role of normal distribution. Also the parameters Xmin, ML and Xmax are of easy selection.

Nowadays it is habitual that PFs of a random variable X are expressed in function of variable independent x. The proposed PF can be expressed in function of x or a different independent variable y.

The SMp(x) would be a unique simple-probabilistic function that would play the role of some current PDs and PDFs, such as Poisson, Gaussian, log-normal and Weibull. Despite SMp function has six parameters; five or less of them are independent, and this simple function can play role of many complex PFs with one or more independent parameters.

The advantages and disadvantages of the SMp function are described in Table 48.1.

48.2 Methodology

The SMp has classified in three types (P1, P2 and P3) the behaviors of the mean values of the following SP/Es. P1: For values of the independent variable (in radiotherapy, the dose that is translated to absorbed energy) < a threshold value (TV) the process is 0%-deterministic, i.e., it will never occur. For values $\geq$ this threshold, the process is increased when the independent variable (IV) increases from 0% to 100% and is stochastic until a determined value of IV, where the

S. Latifi (ed.), *16th International Conference on Information Technology-New Generations (ITNG 2019)*,
Advances in Intelligent Systems and Computing 800,
https://doi.org/10.1007/978-3-030-14070-0_48

Table 48.1 Disadvantages and advantages of the SMp(x)

Disadvantages	Advantages
1. Not recognized yet	1. One of the few functions that are pure-probabilistic functions (PFs); i.e. that defines the stochastic and non-stochastic regions.
2. It has two steps	2. It can play the role of current PFs, such as the normal and Poisson distributions, and even overcome their deficiencies and limitations. Also, in these roles the number of independent parameters is ≤ 5.
3. It has six parameters	3. It can play the role of the three SMp types (P1, P2 and P3) if the independent variable x is replaced for y
	4. Except its powers, other parameters are of easy selection given the strong-probabilistic foundations of this function.
	5. For its probabilistic conditions, at least one of the six parameters is dependent of the others.
	6. It is less complex than some current PFs.
	7. It can be used for modeling SP/Es that have not been modeled yet, like the PDD curves

process becomes 100%-deterministic, i.e., it certainly occurs. P2: For values of IV < a TV, the process is 0%-deterministic; for values ≥ this threshold, the process is increased when IV increases from 0% to a maximum value; then, it begins to decrease to 0%, where the process is 0%-deterministic when IV increases. P3: For values of IV < a TV, the process is 100%-deterministic; with an increase of IV, it is stochastic and decreases from 100% to 0%, where the process is 0%-deterministic when IV increases. An example of the application of the SMps P1, P2 and P3 is the mean behaviors for several homogeneous populations Vs. Different levels of alcohol in blood follow these three SP/Es: death by alcohol poisoning, sublethal damage, and without effects.

48.2.1 SMp(x or y;PXmin,Xmax,ML,p1,p2,Max) Function

The SMp proposes Eq. (48.1), which can be used as a PD or PDF. Be X a random variable depending an independent variable (IV) x or y, then

$$SMp(IV) = \begin{cases} \left(\frac{IV - PXmin}{ML - PXmin}\right)^{p1} Max & Xmin \leq IV \leq ML \\ \left(\frac{Xmax - IV}{Xmax - ML}\right)^{p2} Max & ML < IV \leq Xmax \end{cases}$$

(48.1)

The SMp(IV) = 0 in IV < Xmin, and IV > Xmax. Where *PXmin*: If *PXmin* ≥ 0 *PXmin* = Xmin. If *PXmin* < 0 SMp (0) > 0 and *Xmin* = 0; ML: *IV* where SMp(x) = Max.; *Xmax*: Upper limit value of IV in the stochastic region. *Xmax* > ML; *Xmin*: Minimum value can take X. *Xmin* < ML; *Max*: Maximum of this model; *p1, p2*: Powers of this function *(p1 > 0 and p2 > 0)*; and *IV* ϵ [0;∞).

SMp(x) is a six-parameter function; and for its PDF condition expressed in Eq. (48.2), at least one of its six parameters depends on the others. Ex: *Max* can be determined with Eq. (48.3) as follows:

$$\int_{Xmin}^{Xmax} SMp(x; PXmin, Xmax, ML, p1, p2, Max) = 1$$

(48.2)

$$Max = \frac{1}{\int_{Xmin}^{Xmax} SMp(x; PXmin, Xmax, ML, p1, p2)}$$

(48.3)

When SMp(x) is used as a PD of a random discrete variable, for its PD condition in Eq. (48.4), at least one of its six parameters depends on the others. Ex. *Max* can be determined with Eq. (48.5) as follows:

$$\sum_{Xmin}^{Xmax} SMp(x; PXmin, Xmax, ML, p1, p2, Max) = 1$$

(48.4)

$$Max = \frac{1}{\sum_{Xmin}^{Xmax} SMp(x; PXmin, Xmax, ML, p1, p2)}$$

(48.5)

48.2.2 SMpxtools

The MATLAB code and its description are available in [5].

In the application developed by this study, the probabilities are calculated using the SMp CDF and other probabilistic relationships, such as follows:

$$SMp \, CDF(x) = \int_{Xmin}^{x} SMp\left(x\right)) \, dx$$

(48.6)

$$P(X <= a) = SMp \, CDF(a)$$

(48.7)

$$P(X > a) = 1 - P(X <= a) \qquad (48.8)$$

$$P(a <= X <= b) = SMp \, CDF(b) - SMp \, CDF(a) \qquad (48.9)$$

$$P(X > a \ or \ X > b) = 1 - P(a <= X <= b) \qquad (48.10)$$

where a and $b \, \epsilon \, [0; \infty)$ and $a < b$.

48.3   Results and Discussion

SMp(x) can play the role of some current PDFs or PD, such as Normal with μ-5σ > 0 if p1 = p2, p1 > 1 and PXmin $\geq$ 0; Poisson if p1 > 1 or p1 = 0, p2 > 1 and PXmin < 0, or p1 > 1, p2 > 1 and PXmin $\geq$ 0; Binomial if p1 > 1, p2 > 1 and PXmin $\geq$ 0, as well as variable x is replaced by k.; exponential if p2 > 1 and ML = 0; Log normal if p1 > 1, p2 > 1 and PXmin $\geq$ 0; Rayleigh if p1 < 1, p2 < 1 and PXmin = 0, or p1 < 1, p2 > 1 and PXmin = 0; and Weibull p1 > 1, p2 > 1 and PXmin$\geq$0, or p2 > 1 and ML = 0.

Eq. (48.1) can generate the SMp P1 and P3 models if independent variable x is replaced by y, and the conditions for these reproductions are for SMp P1: p2 = 0, ML = XmaxP1, Max = 1 and Xmax = max(y); and SMp P3: p1 = 0, ML = XminP3, PXmin = 0, Max = 1 and Xmax = XmaxP3. Where *XmaxP1: Minimum value of independent variable y for SMp P1 = 100%. max(y): Maximum of y. XminP3: Maximum value of y for SMp P3 = 100%. XmaxP3: Minimum value of y for SMp P3 = 0%.* Ex. LKB NTCP (Deff;TD50 = 65 Gy, m = 0.14, n = 0.25) taken from [6] can be represented as SMp NTCP (Dmax;45 Gy, 85 Gy, 0.9).

In addition to its previous use, the SMp(x) function can be used in some SP/Es that have not been probabilistically modeled. Ex. 1 the percentage depth dose (PDD) curves in water for a $10 \times 10 \, cm^2$ field at a source-surface distance of 100 cm for cobalt-60 gamma rays, i.e., the PDD(x;10,100,Co) of [7]. This PDD can be modeled as SMp(x;−4,30,0.7,5,1.5,1). Ex. 2 Pharmacokinetic data taken from figure of [8] as SMp(t;0,18,2.7,0.7,2.3,1).

Table 48.1 shows the disadvantages and advantages of the SMp(x) function.

48.3.1 Normal Probability Density Functions

The normal distribution (ND) is an exponential function with quadratic dependence of its independent variable x, which produces bell-shape symmetric functions, whose expression is as follows:

$$ND(x; \mu, \sigma) = \frac{1}{\sqrt{2\pi\sigma^2}} e^{-\frac{(x-\mu)^2}{2\sigma^2}} \qquad (48.11)$$

where μ: Mean; and σ: Standard deviation.

The ND is a non-PF because this does not define the stochastic region and is not wholly appropriate for predicting P(X $\leq$ x), since although X must be $\geq$0, P(X < 0) > 0, and P = 0 in x = $\pm\infty$, which generally can be unrealistic. Although a symmetric interval from μ can be estimated, the ND is symmetric in the interval $(-\infty; \infty)$. Nonetheless, the usefulness of this distribution is conditioned by its similarities with Eq. (48.1); its acceptable approximation is determined by the following equation:

$$\int_{Xmin}^{Xmax} ND(x; \mu, \sigma) \approx 1 \qquad (48.12)$$

Although SMp(x) has six parameters, when it is used as a "ND" with $\mu - 5\sigma > 0$, there will be only three independent parameters: *PXmin, ML,* and *p1,* since we treat a symmetric function; then the following:

$$p2 = p1 > 1 \qquad (48.13)$$

$$ML = \mu \qquad (48.14)$$

$$Xmax = 2ML - PXmin \qquad (48.15)$$

$$\int_{Xmin}^{ML} SMp(x)dx = \frac{1}{2} \qquad (48.16)$$

$$Max = \frac{(p1 + 1)}{2(ML - PXmin)} \qquad (48.17)$$

Table 48.2 Similarities and differences between the NCDF and SMp-normal CDF

Aspect	NCDF	SMp-normal CDF
Behavior function of its independent variable	Sigmoidal	Sigmoidal
Definition of the stochastic region (SR)	Not clearly defined. The SR can be obtained only by estimation	Well-defined
Shapes of its PDF curves	Only bell-shape	Bell, parabolic, and triangular-shape
[*]*Number of parameters*	2	6
[*]*Number of independent parameters*	2	3
Analytical solution (AS)	None	Simple AS from an integral expression
[*]*Year of its introduction or discovery*	1809	Not recognized yet
Math complexity	High	Low

The SMp-normal: When the SMp(x plays the role of the normal distribution. The advantages of the NCDF over SMp-normal CDF appear in the Aspects pointed out with (*)

The ND(x;6,1) can be represented as SMp(x;2.47,9.53,6, 2.25,2.25,0.43).

The NCDF is used to calculate the probabilities of a continuous random variable that follows the well-known Gauss distribution. The cumulative distribution function (CDF) of SMp(x), particularly in its role of the normal distribution SMp-normal, can also be used for these calculations. Table 48.2 shows a comparison between two CDFs.

48.3.2 Probability Distribution of a Discrete Random Variable

Binomial Distribution (BD)

The binomial theorem in its most general case, i.e., the binomial series of [9], states that for any positive integer n, the nth power of the sum of two numbers a and b may be expressed as follows:

$$(a+b)^n = \binom{n}{0} a^0 b^{n-0} + \binom{n}{1} a^1 b^{n-1} + \cdots + \binom{n}{n} a^n b^{n-n} \tag{48.18}$$

The BD is used to calculate the probability P(k;n,p) of k successes in n trials with a parameter p (success probability in each trial) in a determined value of a random variable X = x. For its mathematic origin, the exact BD represents only one of too many possible distributions of P(k;n,p). *BD(k;n,p)* is derived by evaluating $a = p$ and $b = 1 - p$ in Eq. (48.18), where $(a + b)^n = 1$.

When SMp(x) is used as a "BD", there are only four independent ones: *PXmin, Xmax (Xmax ≥ n), p1 and p2*, since *ML ≈ np (ML ∈ N)*, and *Max* is calculated with Eq. (48.5) but with a sum from 0 to n. The BD(k;10,0.3) can be represented as SMp(k;−0.5,11,3,1.1,4,0.299).

BD(k;n,p) is modeled with the SMp(k) function only to show the usefulness of this. The BD is only one distribution of the P(k:n,p) ones, and is obtained from a mathematic exercise based on a mathematic theorem, which does not have any probabilistic relationship. For 400 years, this exercise has changed what one should consider for the SP/Es characterized with parameter p in n trials; its trend (τ) is equal to *np*. The determination of any distribution P(k;n,p), such as SMp(k) and BD(k;n,p), is unnecessary. The introduction of the BD has represented a new and complex mathematic expression and generated some confusion. One should consider important aspects of the SP/Es, such as its τ, mean and most likelihood values, and SMp distributions.

Because the probability is the extent to which an event is likely to occur, which is measured by the ratio of the favorable cases to the total number of possible cases or the ratio of the number of results of an effect or process and its initial number. If a SP/E is characterized with probability p, the most important issue is evaluating its τ in n trials, which is determined as $τ = np$. For example, if a SP/E has p = 40% in 12 trials, τ = 4.8, i.e., approximately five successes occur. By creating new probabilities that depend on p, like BD(k;n,p) it has little probabilistic importance or none, and represents the creation of a complex mathematic expression and even generates confusions.

Poisson Distribution

"The Poisson distribution (PD) is a probability distribution of a discrete random variable that stands for the number (count) of statistically independent events, occurring within a unit of time or space. Given the expected value μ of the variable X, the probability function is defined as" [10] as follows:

$$PD(x; \mu) = P\left(X = x\right) = \frac{e^{-\mu} \mu^x}{x!} \tag{48.19}$$

The PD is a limiting case of the BD for n → ∞. In fact, the well-known PD is a non-PF because this does not define the stochastic region and is not completely appropriate for predicting P(X = x), since P(X ≥ 0) > 0% and P = 0% in x = +∞, which generally can be unrealistic. Nonetheless, the usefulness of this PD is conditioned to its similarities with the PFs, e.g., Eq. (48.1); its acceptable approaches are determined by the following equations:

$$\mu \cong Xminp_{min} + x_2p_2 + \cdots + Xmaxp_{max} \quad (48.20)$$

$$\sum_{Xmin}^{Xmax} PD(x;\mu) \cong 1 \quad (48.21)$$

where *Xmin* is the minimum value can take the random variable X; *Xmax* is the maximum value can take X; and p_i is the probability of occurrence for each value can take X.

The PD was determined as follows:

$$\lim_{n\to\infty} BD(k;n,p) = PD(k;\lambda) \quad (48.22)$$

and used in $P(X = x) = p(x)$ as PD(x;μ) with some limitations because of its characteristics, e.g., some values of x have the same probability for determined parameters: *PD(0;1) = PD(1;1) = 0.368*, and *PD(1;2) = PD(2;2) = 0.27*; μ is always the value of X with the major likelihood if μ is an integer. Here, μ is the expect value of X, and λ = np.

When SMp(x) is used as a "PD", there will be only four independent ones: *PXmin, Xmax, p1* and *p2* because of *ML = μ*, and *Max* is calculated with Eq. (48.5); then, a discrete random variable X that follows a PD(x;1) can be described using SMp(x;−1,6.8,1,0,4,0.368).

48.3.3 SMptools Results

The first version of a computer application that will significantly benefit researchers and students, who widely use the normal, Poisson, binomial and other standard distributions, and models such as the NTCP, linear-quadratic cell survival, percentage depth dose, and pharmacokinetic.

Some results of the probabilities calculated by the SMp application are compared with the results of [11] with notably acceptable approximation.

The works on the SMp application enable one to develop codes for the SMp CDF, which provides an analytical solution (AS) of the SMp(x) integral expression. This is one of the advantages of SMp(x) compared to PFs, e.g., a well-known normal that does not have an AS as discussed in [12].

References

1. Report of AAPM TG 166. The Use and QA of Biologically Related Models for Treatment Planning (2012)
2. Gay, H.A., Niemierko, A.: A free program for calculation EUD-based NTCP and TCP in external beam radiotherapy. Phys. Med. **23**, 115–125 (2007)
3. Frometa-Castillo, T.: The statistical models project (SMp) normal tissue complication probability (NTCP) model and parameters. Am. J. Appl. Math. Stat. **5**(4), 115–118 (2017)
4. Frometa-Castillo, T.: The statistical models project (SMp) for evaluation of biological radiation effects. Am. J. Appl. Math. Stat. **5**(4), 119–124 (2017)
5. https://gitlab.com/tfrometa/SMpx/blob/master/
6. Mayo, C., Martel, M.K., Marks, L.B., et al.: Radiation dose-volume effects of optic nerves and chiasm. Int. J. Radiat. Oncol. Biol. Phys. **76**(3, Supplement), S28–S35 (2010)
7. Podgorsak, E.B. (ed.): Review of Radiation Oncology Physics: A Handbook for Teachers and Students. International Atomic Energy Agency, Vienna (2003)
8. Professor Joel Tarning. https://www.ndm.ox.ac.uk/principal-investigators/researcher/joel-tarning
9. Weisstein, E.W.: Binomial theorem. Wolfram MathWorld. http://mathworld.wolfram.com/BinomialTheorem.html
10. Letkowski, J.: Applications of the Poisson probability distribution. Western New England University. http://www.aabri.com/SA12Manuscripts/SA12083.pdf8
11. Normal distribution. http://onlinestatbook.com/2/calculators/normal_dist.html
12. Vazquez-Leal, H., Castaneda-Sheissa, R., Filobello-Nino, U., et al.: High accurate simple approximation of normal distribution integral. Math. Probl. Eng. **2012**, Article ID 124029, 22 pages. https://doi.org/10.1155/2012/124029

Application of the Multi-Dimensional Hierarchical Mixture Model to Cross-Disorder Genome-Wide Association Studies

Takahiro Otani, Jo Nishino, Ryo Emoto, and Shigeyuki Matsui

49.1 Introduction

Genome-wide association studies (GWASs) across disorders are among the most promising types of studies for identifying genetic components that are either shared by or differentiate multiple disorders. Existing GWASs identify several hundred loci associated with multiple disorders [1–4]; however, only a modest fraction of the total heritability estimated from twin and family research can be explained by these results, which is called the missing heritability problem [5]. From a statistical perspective, the use of single-variant association tests with a very conservative genome-wide significance threshold [6] of $p < 5 \times 10^{-8}$ is one of the most crucial issues. Although several consortiums have conducted extremely large-scale GWASs (sample size $>10^5$) to increase the statistical power, they were able to detect only small numbers of additional significant variants. While efforts to obtain adequate numbers of samples are naturally crucial, the development of more efficient analysis strategies is equally important.

We developed an alternative multi-subgroup gene screening method [7, 8] based on a multidimensional semi-parametric hierarchical mixture model [9, 10]. This method efficiently reveals the underlying effect size distributions of single nucleotide polymorphisms (SNPs) associated with disorders based on summary association statistics (per-allele effect sizes together with their standard errors). One can evaluate the existence of shared or disorder-specific SNPs from such a distribution rather than by identifying individual SNPs by conducting association tests with a conservative genome-wide significance criterion. Furthermore, effect size estimates adjusted for overestimation errors arising from association analyses, called the winner's curse phenomenon [11], can be obtained based on the resulting distributions. The adjusted estimates can serve as fundamental information for developing appropriate therapeutic strategies. In addition, an efficient test was developed using the optimal discovery procedure (ODP) [12, 13], which can provide optimal variant ranking as well as being the most powerful test for detecting disorder-related SNPs with control of multiplicity, e.g., the false discovery rate (FDR). Previous studies [7, 8] have shown that the ODP outperforms traditional multiple testing methods such as the q value procedure [14]. In this study, we employed the newly available method to analyze large-scale cross-disorder GWAS data for schizophrenia and bipolar disorder [4].

49.2 Materials and Methods

49.2.1 GWAS for Bipolar Disorder and Schizophrenia

We examined the data from a recently published case-control GWAS for schizophrenia and bipolar disorder [4] involving 53,555 cases (33,426 schizophrenia and 20,129 bipolar disorder) and 54,065 controls to investigate the genetic factors contributing to the shared and disorder-specific symptoms. A summary of the association data is available at http://www.med.unc.edu/pgc/. Figure 49.1 presents the results of the association tests for schizophrenia and bipolar disorder in the form of Manhattan plots.

T. Otani (✉) · R. Emoto · S. Matsui
Department of Biostatistics, Nagoya University Graduate School of Medicine, Nagoya, Japan
e-mail: otani@med.nagoya-u.ac.jp;
emoto.ryo@b.mbox.nagoya-u.ac.jp; smatsui@med.nagoya-u.ac.jp

J. Nishino
Medical Science Mathematics, Medical Research Institute, Tokyo Medical and Dental University, Tokyo, Japan
e-mail: jonimesm@tmd.ac.jp

© Springer Nature Switzerland AG 2019
S. Latifi (ed.), *16th International Conference on Information Technology-New Generations (ITNG 2019)*,
Advances in Intelligent Systems and Computing 800,
https://doi.org/10.1007/978-3-030-14070-0_49

Fig. 49.1 Manhattan plots from association analyses for (**a**) schizophrenia and (**b**) bipolar disorder [1]. $-\log_{10}p$ for each SNP (y axis) is plotted against the chromosomal position (x axis). The red lines denote the genome-wide significance level ($p = 5 \times 10^{-8}$)

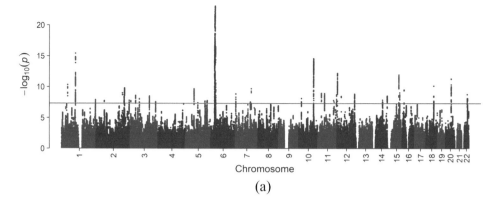

(a)

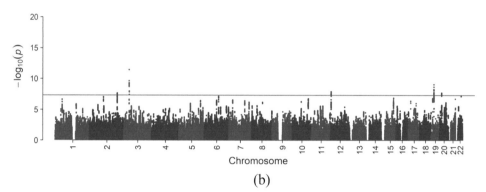

(b)

In this study, we used 2,201,105 SNPs with very high imputation quality (INFO > 0.99) for the analysis using the screening method. Figure 49.2 shows a scatter plot of the effect size estimates (logarithmic odds ratios) for all SNPs. The distribution may involve two types of nuisance factors: (i) SNPs that are not associated with disorders and (ii) random variations irrespective of association with disorders. We attempted to estimate the underlying effect size distribution of the disorder-related SNPs by eliminating these factors using a hierarchical mixture model for the entire SNP set.

49.2.2 Workflow of the Gene Screening Method

We used the multi-subgroup gene screening method [7, 8] (Fig. 49.3). In this method, association analyses are first independently conducted for each disorder. As a result, summary statistics, i.e., effect size estimates and their standard errors for each SNP, are obtained for each disorder group. Using the summary statistics, this method reveals the proportion of SNPs associated with any disorder and the underlying effect size distribution of SNPs across disorders via empirical Bayes estimation under the multidimensional hierarchical mixture model. Furthermore, posterior association probabilities and effect size estimates adjusted for gene selection

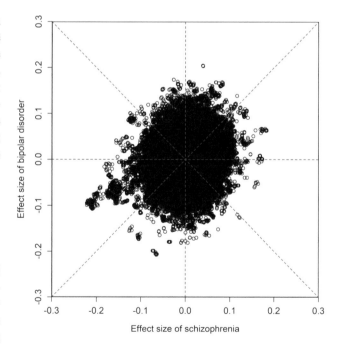

Fig. 49.2 Scatter plot of effect size estimates derived from association analyses. The x and y axes represent the effect sizes for schizophrenia and bipolar disorder, respectively

errors and overestimation can be obtained for each SNP based on the estimated distribution.

Fig. 49.3 Workflow of the gene screening method using multidimensional hierarchical mixture models. The top panels show the genotypic data and case statuses of the schizophrenia and bipolar disorder groups, whose sample sizes are denoted as N_0 and N_1, respectively. M is the total number of SNPs. Genotypes are represented by the counts of minor alleles. The middle panels show the summary statistics, specifically, the estimated effect sizes $b^{(0)}$ and $b^{(1)}$ and their standard errors for the main effects of SNPs derived from association analyses using regression models without interaction terms. The bottom panels show the denoised effect size distribution of SNPs associated with the disorders and the adjusted effect size estimates for each SNP. Significant SNPs were detected by the optimal discovery procedure based on the distribution and posterior probabilities

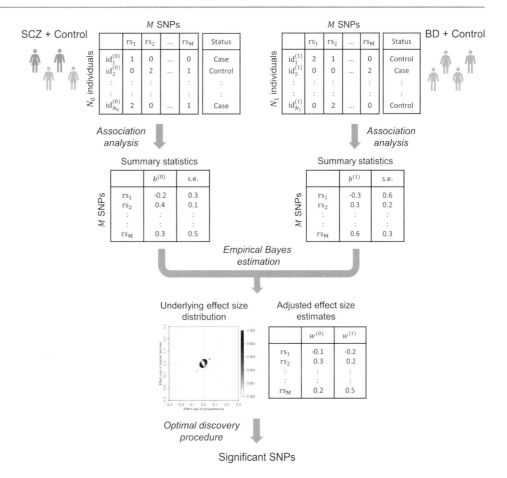

49.2.3 Multidimensional Hierarchical Mixture Model

We used an empirical Bayes framework [7, 8] to estimate the underlying SNP distribution. Suppose that M SNPs are simultaneously tested to determine whether each is associated with disease risk. Of these SNPs, M_0 are truly "null" and are not associated with the outcomes, and $M_1 = M - M_0$ SNPs are truly "non-null" and are associated with the diseases.

We defined an effect size vector $\boldsymbol{\beta}_j = \left(\beta_j^{(0)}, \beta_j^{(1)}\right)$ for SNP j as the logarithmic odds ratio for the minor allele under an additive genetic model, where $\beta_j^{(0)}$ and $\beta_j^{(1)}$ are the logarithmic odds ratios for schizophrenia and bipolar disorder, respectively. As an estimate of $\boldsymbol{\beta}_j$, we considered the maximum likelihood estimate $\boldsymbol{b}_j = \left(b_j^{(0)}, b_j^{(1)}\right)$, i.e., the multivariate logistic regression coefficient.

We assumed a mixture model for the $\boldsymbol{b}_j$:

$$f\left(\boldsymbol{b}_j, \boldsymbol{\Sigma}_j\right) = \pi f_0 \left(\boldsymbol{b}_j, \boldsymbol{\Sigma}_j\right) + (1 - \pi)\, f_1 \left(\boldsymbol{b}_j, \boldsymbol{\Sigma}_j\right),$$

where f_0 and f_1 are the density functions of $\boldsymbol{b}$ for the null and non-null SNPs, respectively, and $\boldsymbol{\Sigma}_j = \mathrm{diag}\left(V_j^{(0)}, V_j^{(1)}\right)$ is a covariance matrix consisting of empirical variances (squared standard errors) from an association analysis of a particular

variant for each disorder. We considered different modeling assumptions for the two components. For the null component, we assumed f_0 to have the normal distribution $N(\boldsymbol{0}, \boldsymbol{\Sigma}_j)$. For the non-null component of interest, a hierarchical model was assumed. Specifically, for a non-null SNP j, we assumed.

$$\boldsymbol{b}_j \mid \boldsymbol{\beta}_j \sim N\left(\boldsymbol{\beta}_j, \boldsymbol{\Sigma}_j\right) \text{ and } \boldsymbol{\beta}_j \sim g_1\left(\cdot\right).$$

In the first level of this model, given an SNP-specific effect size $\boldsymbol{\beta}_j$, $\boldsymbol{b}_j$ follows a normal distribution. In the second level, the SNP-specific $\boldsymbol{\beta}_j$ follows the distribution g_1.

The marginal distribution f_1 is given by

$$f_1\left(\boldsymbol{b}, \boldsymbol{\Sigma}\right) = \int_{-\infty}^{\infty} g_1\left(\boldsymbol{\beta}\right) \varphi_{\boldsymbol{\beta}, \boldsymbol{\Sigma}}\left(\boldsymbol{b}\right) \mathrm{d}\boldsymbol{\beta},$$

where $\phi_{\boldsymbol{\mu}, \boldsymbol{\Sigma}}(\cdot)$ is the density function of the multivariate normal distribution with mean $\boldsymbol{\mu}$ and covariance matrix $\boldsymbol{\Sigma}$.

49.2.4 Expectation-Maximization (EM) Algorithm

We estimated the parameters π and g_1 using an EM algorithm, as was done in previous studies [7–10]. We considered a nonparametric estimate of the prior distribution g_1 in which the estimate was supported by fixed discrete mass points $\mathbf{P} = (p_{kl})_{1 \leq k \leq B, 1 \leq l \leq B}$ at a series of nonzero points $\mathbf{T} = (t_{kl})_{1 \leq k \leq B, 1 \leq l \leq B}$, where the zero point was skipped because we considered non-zero effects for non-null SNPs, and an estimate of the marginal distribution f_1 was calculated using summations rather than integrals.

Let γ_j be the unknown indicator variable for null/non-null status for SNP j, such that $\gamma_j = 1$ if SNP j is non-null and $\gamma_j = 0$ otherwise. The prior probability of being non-null is $P(\gamma_j = 1) = 1 - \pi$ and the posterior probability is $P(\gamma_j = 1 | \boldsymbol{b}_j = \boldsymbol{b}, \boldsymbol{\Sigma}_j = \boldsymbol{\Sigma}) = (1 - \pi)f_1(\boldsymbol{b}, \boldsymbol{\Sigma})/f(\boldsymbol{b}, \boldsymbol{\Sigma})$ respectively. The posterior probability was estimated in the expectation step as

$$\tau_j = \frac{(1 - \widehat{\pi}) \, \widehat{f}_1 \left(\boldsymbol{b}_j, \boldsymbol{\Sigma}_j \right)}{\widehat{\pi} \, f_0 \left(\boldsymbol{b}_j, \boldsymbol{\Sigma}_j \right) + (1 - \widehat{\pi}) \, \widehat{f}_1 \left(\boldsymbol{b}_j, \boldsymbol{\Sigma}_j \right)},$$

where $\widehat{\pi}$ and $\widehat{f}_1$ are the empirical estimates of π and f_1. f_1 was also estimated as

$$\widehat{f}_1 (\boldsymbol{b}, \boldsymbol{\Sigma}) = \sum_{k=1}^{B} \sum_{l=1}^{B} p_{kl} \varphi_{t_{kl}, \boldsymbol{\Sigma}} (\boldsymbol{b}).$$

The parameter $\mathbf{P}$ (for g_1) and the mixing parameter π were estimated in the maximization step as.

$$p_{kl} = \frac{1}{\sum_{j=1}^{M} \tau_j} \sum_{j=1}^{M} \tau_j$$

$$\times \frac{p_{kl} \varphi_{t_{kl}, \boldsymbol{\Sigma}_j} \left(\boldsymbol{b}_j \right)}{\sum_{k'=1}^{B} \sum_{l'=1}^{B} p_{k'l'} \varphi_{t_{k'l'}, \boldsymbol{\Sigma}_j} \left(\boldsymbol{b}_j \right)} \quad (k, l = 1, \ldots, B).$$

and

$$\widehat{\pi} = \frac{1}{m} \sum_{j=1}^{m} \left(1 - \tau_j \right),$$

respectively.

49.2.5 Adjusted Effect Size Estimates

The estimated effect sizes derived from association analysis contain two types of errors: (i) selection errors due to incorrect selection of variants with no association and (ii) overestimation errors. Using the estimated underlying effect size distribution, the effect size estimates adjusted for these errors can be obtained [15]. Thus, the selection error was adjusted by the estimated posterior probability of association τ_j, and the overestimation error was adjusted by using an unconditional mean for the effect size of SNP j. The posterior

mean for the effect size of SNP j is.

$$\mathrm{E} \left(\boldsymbol{\beta} | \gamma_j = 1, \boldsymbol{b}_j = \boldsymbol{b}, \boldsymbol{\Sigma}_j = \boldsymbol{\Sigma} \right)$$

$$= \int \boldsymbol{\beta} f \left(\boldsymbol{\beta} | \gamma_j = 1, \boldsymbol{b}_j = \boldsymbol{b}, \boldsymbol{\Sigma}_j = \boldsymbol{\Sigma} \right) \mathrm{d}\boldsymbol{\beta},$$

which can be estimated as $\widehat{\mathrm{E}} \left(\boldsymbol{\beta} | \gamma_j = 1, \boldsymbol{b}_j = \boldsymbol{b}, \boldsymbol{\Sigma}_j = \boldsymbol{\Sigma} \right)$ by plugging in the estimate of f. By combining this conditional mean, given $\gamma_j = 1$, with the posterior probability of being non-null, the unconditional mean as an index for effect size for SNP j can be derived as.

$$\mathrm{E} \left(\boldsymbol{\beta} | \boldsymbol{b}_j = \boldsymbol{b}, \boldsymbol{\Sigma}_j = \boldsymbol{\Sigma} \right) = \mathrm{Pr} \left(\gamma_j = 1 | \boldsymbol{b}_j = \boldsymbol{b} \right)$$

$$\mathrm{E} \left(\boldsymbol{\beta} | \gamma_j = 1, \boldsymbol{b}_j = \boldsymbol{b}, \boldsymbol{\Sigma}_j = \boldsymbol{\Sigma} \right),$$

and the adjusted effect size estimate $\boldsymbol{w}_j$ can be obtained by plugging in the hyperparameter estimates as.

$$\boldsymbol{w}_j = \widehat{\mathrm{E}} \left(\boldsymbol{\beta} | \boldsymbol{b}_j = \boldsymbol{b}, \boldsymbol{\Sigma}_j = \boldsymbol{\Sigma} \right) = \tau_j \widehat{\mathrm{E}} \left(\boldsymbol{\beta} | \gamma_j = 1, \boldsymbol{b}_j = \boldsymbol{b}, \boldsymbol{\Sigma}_j = \boldsymbol{\Sigma} \right).$$

49.2.6 Detecting Disease-Related SNPs

To detect the SNPs associated with the outcomes, some SNP-specific indices were defined. We used an ODP statistic under the empirical Bayes framework based on the hierarchical Bayesian models derived by Noma and Matsui [13] to screen disease-related SNPs. Adapting their results to the hierarchical mixture model, the ODP statistic becomes

$$R_{\mathrm{ODP}} (\boldsymbol{b}, \boldsymbol{\Sigma}) = \frac{\widehat{f}_1 (\boldsymbol{b}, \boldsymbol{\Sigma})}{f_0 (\boldsymbol{b}, \boldsymbol{\Sigma})}.$$

Multiple hypothesis testing involving estimation of the FDR in the Bayesian sense was conducted as follows. Let k be the number of tests called significant. We first calculated R_{ODP} for each SNP. Then, we ranked the SNPs in order of decreasing R_{ODP}, so that $j = 1, \ldots, k$ represented the tests called significant. The FDR of the significant results was estimated as

$$\widehat{\mathrm{FDR}} = \frac{1}{k} \sum_{j=1}^{k} \left(1 - \tau_j \right).$$

49.3 Results

We fitted the multidimensional hierarchical mixture model to the effect size estimates (Fig. 49.2). We set a series of nonzero mass points $(-0.3, -0.295, \ldots, -0.005, 0.005, \ldots, 0.295, 0.3)$ for $\beta_j^{(0)}$ and $\beta_j^{(1)}$ to support the effect size

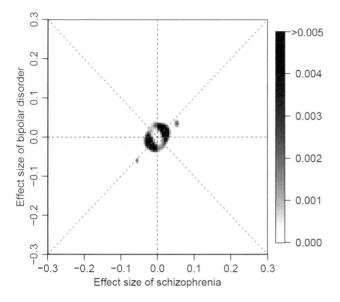

Fig. 49.4 Effect size distribution of SNPs associated with schizophrenia and/or bipolar disoder. The x and y axes represent the effect sizes for schizophrenia and bipolar disorder, respectively

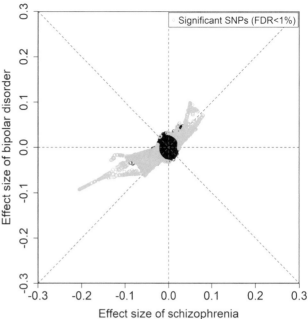

Fig. 49.5 Scatter plot of the adjusted effect size estimates derived from the hierarchical mixture model. The x and y axes represent the effect sizes for schizophrenia and bipolar disorder, respectively. The green points indicate the SNPs detected by the ODP (FDR < 1%)

distribution g_1 for non-null variants. The proportion of null SNPs was estimated as $\widehat{\pi} = 0.662$, indicating that a large fraction of the SNPs were associated with the disorders, while almost all of the SNPs had small effect sizes (odds ratios <1.05). Fig. 49.4 shows the estimated effect size distribution of non-null SNPs g_1. There are two large peaks around $(\beta^{(0)}, \beta^{(1)}) = (0.02, 0.02)$ and $(\beta^{(0)}, \beta^{(1)}) = (-0.02, -0.02)$. These might be related to genetic components shared by the two disorders. In addition to the large peaks, there are several small peaks that deviate from the low-effect area. All of these are in the first and third quadrants, suggesting a shared genetic basis between the two disorders.

Fig. 49.5 shows a scatter plot of the adjusted effect size estimates w_j for all SNPs. Here, 38,880 SNPs detected through the ODP (FDR < 1%) are highlighted; note that the number of independently associated SNPs should be much smaller since some SNPs are in linkage disequilibrium. In addition to the SNPs that might be related to shared genetic components, others with non-zero effects only on one disorder and others with opposite effects on the two disorders were detected. These might be related to disorder-specific genetic components. On the contrary, the qvalue procedure [14] yielded 30,761 SNPs with the same FDR level.

49.4   Discussion

Recent genetic and epidemiological studies have demonstrated substantial overlap between schizophrenia and bipolar disorder [16, 17]. The estimated effect size distribution

(Fig. 49.4) suggests that most SNPs have common effects on both disorders, which is consistent with the results of past studies. This similarity could indicate the difficulty of discriminating the two diseases. On the other hand, a few SNPs with comparably small effect sizes also exist, which could reflect differences between these diseases. Although biological investigations and validation studies for these results are required, these findings could provide important information for understanding the biology of psychiatric disorders and improving diagnosis and treatment based on genetic features.

Hierarchical mixture models are not limited to two-dimensional settings. One can consider an extended model with three or more dimensions to perform cross-disorder analysis with three or more disorders, such as in [1–3].

While we demonstrated cross-disorder analysis in this study, our approach can also be applied to single-disorder analyses involving sample subgroups. In the context of randomized clinical trials with genomic data, for example, this method can be used to explore the existence of predictive genes that are differently associated with outcomes between subgroups (treatment and control groups in a trial) and prognostic genes that are similarly associated with outcomes irrespective of subgroups [7, 8]. Application to observational studies is also expected, although one should consider confounding factors and sample overlap.

Acknowledgment This work was supported by CREST, the Japan Science and Technology Agency (JPMJCR1412), and JSPS KAKENHI (JP16H06299).

References

1. Pickrell, J.K., Berisa, T., Liu, J.Z., et al.: Detection and interpretation of shared genetic influences on 40 human traits. Nat. Genet. **48**(7), 709–727 (2016)
2. Sivakumaran, S., Agakov, F., Theodoratou, E., et al.: Abundant pleiotropy in human complex diseases and traits. Am. J. Hum. Genet. **89**, 607–618 (2011)
3. Cross-Disorder Group of the Psychiatric Genomics Consortium: Identification of risk loci with shared effects on five major psychiatric disorders: a genome-wide analysis. Lancet. **381**(9875), 1371–1379 (2013)
4. Bipolar Disorder and Schizophrenia Working Group of the Psychiatric Genomics Consortium: Genomic dissection of bipolar disorder and schizophrenia, including 28 subphenotypes. Cell. **173**(7), 1705–1715.e16 (2018)
5. Manolio, T.A., Collins, F.S., Cox, N.J., et al.: Finding the missing heritability of complex diseases. Nature. **461**, 747–753 (2009)
6. Dudbridge, F., Gusnanto, A.: Estimation of significance thresholds for genomewide association scans. Genet. Epidemiol. **32**, 227–234 (2008)
7. Matsui, S., Noma, H., Qu, P., et al.: Multi-subgroup gene screening using semi-parametric hierarchical mixture models and the optimal discovery procedure: application to a randomized clinical trial in multiple myeloma. Biometrics. **74**(1), 313–320 (2017)
8. Otani, T., Noma, H., Sugasawa, S., et al.: Exploring predictive biomarkers from clinical genome-wide association studies via multidimensional hierarchical mixture models. Eur. J. Hum. Genet. **27**, 140–149 (2018). https://doi.org/10.1038/s41431-018-0251-y
9. Matsui, S., Noma, H.: Estimating effect sizes of differentially expressed genes for power and sample-size assessments in microarray experiments. Biometrics. **67**(4), 1225–1235 (2011)
10. Nishino, J., Kochi, Y., Shigemizu, D., et al.: Empirical Bayes estimation of semi-parametric hierarchical mixture models for unbiased characterization of polygenic disease architectures. Front. Genet. **9**, 115 (2018)
11. Ferguson, J.P., Cho, J.H., Yang, C., Zhao, H.: Empirical Bayes correction for the Winner's Curse in genetic association studies. Genet. Epidemiol. **37**, 60–68 (2013)
12. Storey, J.D., Dai, J.Y., Leek, J.T.: The optimal discovery procedure for large-scale significance testing, with applications to comparative microarray experiments. Biostatistics. **8**, 414–432 (2007)
13. Noma, H., Matsui, S.: Empirical Bayes ranking and selection methods via semiparametric hierarchical mixture models in microarray studies. Stat. Med. **32**, 1904–1916 (2013)
14. Storey, J.D., Tibshirani, R.: Statistical significance for genomewide studies. Proc. Natl. Acad. Sci. **100**(16), 9440–9445 (2003)
15. Matsui, S., Noma, H.: Estimation and selection in high-dimensional genomic studies for developing molecular diagnostics. Biostatistics. **12**, 223–233 (2011)
16. Cross-Disorder Group of the Psychiatric Genomics Consortium: Genetic relationship between five psychiatric disorders estimated from genome-wide SNPs. Nat. Genet. **45**, 984–994 (2013)
17. Lichtenstein, P., Yip, B.H., Björk, C., et al.: Common genetic determinants of schizophrenia and bipolar disorder in Swedish families: a population-based study. Lancet. **373**, 234–239 (2009)

A Model-Based Framework for Voxel and Region Level Inferences in Neuroimaging Disease-Association Studies

Ryo Emoto, Atsushi Kawaguchi, Takahiro Otani, and Shigeyuki Matsui

50.1 Introduction

Disease-association analyses using neuroimaging data, such as brain magnetic resonance imaging (MRI) data, have contributed to the understanding of the underlying mechanisms of disease and also to developing disease diagnostics. A widely used approach to the association analysis is the cluster-level inference [1] that defines clusters of contiguous voxels based on a certain thresholding rule in order to effectively reduce the dimension of a large number of voxels, incorporating the fact that clusters of contiguous voxels are usually co-activated. Such as cluster-level inference, however, has been criticized in terms of several aspects, including arbitrariness in forming clusters [2] and lack of information about the nature or profile of association, which may warrant further evaluation of individual voxels within clusters.

An alternative approach to association analysis is to test every voxel individually, without any initial clustering using a particular thresholding rule. In such a voxel-level inference, control of false positives, e.g., in terms of false discovery rate (FDR) [3], becomes more important because enormous numbers of voxels will be tested simultaneously. Conventional FDR-controlling procedures (e.g., [4]), however, may suffer from loss of efficiency [5] because they assume independency among hypotheses or voxels. In order to capture

dependency structure among contiguous voxels, Shu et al. [6] employed a hidden Markov random field model and developed a multiple testing procedure based on a local index of significance [5]. With this method, contiguous voxels may be more prone to rejection than in conventional, voxel-level multiple testing procedures.

However, the existing cluster-level and voxel-based inferences described above may not provide a direct inference for "neurologically-defined" brain areas or known regions that have structural or functional features, such as those defined in the region of interest (ROI) analysis (e.g., [7]). In addition, there is no established method for estimating the magnitude of associations or effect sizes for detected voxels or brain regions that effectively solves the overestimation problem owing to a selection bias by picking up voxels or regions with the greatest effect sizes, possibly due to random errors (e.g., [8, 9]). In order to address these issues, we propose a framework for modelling effect sizes of whole-brain voxels. More specifically, we integrate a methodology of hierarchical modeling of summary statistics from the whole set of features [10, 11] with a hidden Markov random field models for voxel-level multiple testing under dependency [6]. Based on the integrated whole-brain model, we obtain an estimate of the local index of significant (LIS) for FDR control and shrinkage effect size estimates for individual voxels. Furthermore, we propose a method for ROI analysis based on the same whole-brain, voxel-based model. This FDR controlling method allows for ROI-level multiplicity adjustment.

R. Emoto (✉) · T. Otani · S. Matsui
Department of Biostatistics, Nagoya University Graduate School of Medicine, Showa-ku, Nagoya, Japan
e-mail: emoto.ryo@b.mbox.nagoya-u.ac.jp;
otani@med.nagoya-u.ac.jp; smatsui@med.nagoya-u.ac.jp

A. Kawaguchi
Faculty of Medicine, Center for Comprehensive Community Medicine, Saga University, Saga, Japan
e-mail: akawa@cc.saga-u.ac.jp

50.2 Proposed Methods

We suppose a situation where disease and normal control subjects are compared without any covariates for simplicity. For a binary disease status variable, we suppose a class

© Springer Nature Switzerland AG 2019
S. Latifi (ed.), *16th International Conference on Information Technology-New Generations (ITNG 2019)*,
Advances in Intelligent Systems and Computing 800,
https://doi.org/10.1007/978-3-030-14070-0_50

label has a value either 1 or 2, for example, disease or normal. Let n_1 and n_2 be the numbers of disease and normal control subjects, respectively, and $n = n_1 + n_2$ be the total number of subjects. We suppose that the voxel values are observed after spatial normalization (e.g.,[2]) that adjust for differences in size or shape of the observed image for each subject and a further normalization to ensure normality for the voxel values across subjects within class. Let S be the set of all voxels in the neuroimaging data and m denote the number of voxels in S. We consider the whole-brain model for the summary statistic of associations in order for direct interpretation of effect size for individual voxels without dependency on the sample size. As the effect size, the standardized mean difference for a voxel is defined by $\delta_s = (\mu_{1s} - \mu_{2s})/\sigma_s, s \in S$, where μ_{1s} and μ_{2s} are the mean of voxel s for classes 1 and 2, respectively, and σ_s is the common standard deviation for voxel s across classes. As the summary statistic of associations, an estimate of δ_s is expressed as

$$Y_s = \frac{\bar{\mu}_{1s} - \bar{\mu}_{2s}}{\hat{\sigma}_s}, \qquad (50.1)$$

where $\bar{\mu}_{1s}$ and $\bar{\mu}_{2s}$ are sample means of voxel values in two groups and $\hat{\sigma}_s^2$ is an estimator of the common within-class variance. This statistic is equivalent to a two-sample t-statistic, except for the sample size term. Thus, Y_s can be easily calculated from the t-value provided by software such as SPM. Let $Y = \{Y_s : s \in S\}$ be the vector of Y_s for all voxels. In incorporating additional covariates, corresponding effect size becomes a coefficient of disease status variable in a general linear model.

We assume a hidden Markov random field model [6] for Y. Let $\Theta = \{\Theta_s : s \in S\} \in \{0, 1\}^m$ be a set of latent variables, where $\Theta_s = 0$ if the voxel s is null (i.e., no association with disease) and $\Theta_s = 1$ otherwise (i.e., association with disease). The dependence structure across contiguous voxels is modeled assuming that a realization θ of the latent variable Θ is generated from the following Ising model with two parameters $\gamma = (\gamma_1, \gamma_2)^T$,

$$P(\theta; \gamma) = \frac{\exp\{\gamma^T H(\theta)\}}{C(\gamma)}, \qquad (50.2)$$

where $H(\theta) = \left(\sum_{(s,t) \in S_1} \theta_s \theta_t, \sum_{s \in S} \theta_s\right)^T$ and $C(\gamma)$ is the normalizing constant. In the vector $H(\theta)$, the first component pertains to a summation over all pairs of contiguous voxels, S_1, and the second component to a summation over all voxels, S.

We assume that the statistics Y_s's are mutually independent, given the latent status $\Theta = \theta$,

$$f(y|\theta) = \prod_{s \in S} f(y_s|\theta_s). \qquad (50.3)$$

For the component, $f(y_s|\theta_s)$, we define $f_0(y_s) = f(y_s|0)$ as the null density function and $f_1(y_s) = f(y_s|1)$ as the non-null density function. We assume the distribution of Y_s as the mixture of null and non-null distributions,

$$f(y_s) = \pi_s f_0(y_s) + (1 - \pi_s) f_1(y_s), \qquad (50.4)$$

where $\pi_s = \Pr(\Theta_s = 0)$.

For the null voxels, we assume f_0 to be a normal distribution, $N(0, c_n^2)$, where $c_n = \sqrt{n/n_1 n_2}$, based on asymptotic normality for Y_s. For the non-null voxels, we assume the hierarchical structure with two levels:

$$\begin{aligned} Y_s|\delta_s, \Theta_s = 1 &\sim N(\delta_s, c_n^2), \\ \delta_s &\sim g(\cdot). \end{aligned} \qquad (50.5)$$

At the first level, the conditional distribution of Y_s given effect size δ_s is normal with mean δ_s and variance c_n^2, again, based on asymptotic normality for Y_s. At the second level, the voxel-specific effect size δ_s has an effect size distribution, g. We consider a non-parametric specification for the effect size distribution, g, because the information about the effect size distribution is generally limited reflecting the exploratory nature of disease-association studies using neuroimaging data with a large number of voxels.

For estimation, we employ an approximation method [12, 13] that g has discrete probabilities $p = (p_1, \ldots, p_B)$ at each mass point $t = (t_1, \ldots, t_B)$,

$$g(t_b; p) = p_b, \quad b = 1, \ldots, B, \qquad (50.6)$$

where B is a sufficiently large number of mass points and discrete probability p_b satisfies $p_1 + \cdots + p_B = 1$. In practice, we set $B = 200$, following the guideline by Shen and Louis (1999) [12]. The mass points t may support a possible range of Y and $t_b \neq 0$ for any b.

From Eqs. (50.5) and (50.6), the marginal non-null distribution of Y_s, f_1, can be expressed as a mixture of normal distributions,

$$f_1(y; p) = \sum_{b=1}^{B} p_b \phi(y; t_b, c_n^2), \qquad (50.7)$$

where $\phi(\cdot; \mu, \sigma^2)$ represents the density function of normal distribution, $N(\mu, \sigma^2)$.

The the set of all parameters $\boldsymbol{\varphi} = \left(\boldsymbol{\gamma}^{\mathrm{T}}, \boldsymbol{p}^{\mathrm{T}}\right)^{\mathrm{T}}$ including the parameter in the Ising model and the hierarchical model can be estimated by an expectation-maximization algorithm.

50.2.1 Voxel-Level Inference

In voxel-level inference, for all voxel $s \in S$, we test the null hypothesis $H_{0s} : \theta_s = 0$. We consider the multiple testing method to control FDR and the method to estimate the voxel specific effect size.

We employ the local index of significance (LIS) [5] to estimate FDR to incorporate dependence structure across voxels. As a function of the parameter, $\boldsymbol{\varphi}$, the LIS is defined as the posterior probability that the voxel is null given all Y_s's,

$$\mathrm{LIS}_s(\boldsymbol{y}) = \Pr(\Theta_s = 0 | \boldsymbol{Y} = \boldsymbol{y}; \boldsymbol{\varphi}). \quad (50.8)$$

Note that the LIS corresponds to the local false discovery rate [14] when independence across voxels is assumed. Let $\mathrm{LIS}_{(1)}(\boldsymbol{y}) \leq \cdots \leq \mathrm{LIS}_{(m)}(\boldsymbol{y})$ represent a series of ordered LIS across voxels and let $H_{(i)}$ be the null hypothesis on the voxel corresponding to $\mathrm{LIS}_{(i)}(\boldsymbol{y})$. The oracle LIS procedure [5,6] determines rejected voxels using the following rule:

$$\mathrm{let}\, k = \max \left\{ i : \frac{1}{i} \sum_{j=1}^{i} \mathrm{LIS}_{(j)}(\boldsymbol{y}) \leq \alpha \right\},$$
$$\quad (50.9)$$

then reject all $H_{(i)}, i = 1, \ldots, k$.

This procedure controls FDR level at α. Since the parameters, $\boldsymbol{\varphi}$, is unknown, a plug-in estimator, $\widehat{\mathrm{LIS}}_s(\boldsymbol{y}) = \Pr(\Theta_s = 0 | \boldsymbol{y}; \hat{\boldsymbol{\varphi}})$, is used. This probability, $\widehat{\mathrm{LIS}}_s(\boldsymbol{y}) = \Pr(\Theta_s = 0 | \boldsymbol{y}; \hat{\boldsymbol{\varphi}})$, can be calculated using Gibbs sampler from the distribution of $\boldsymbol{\Theta} | \boldsymbol{Y}$ [6],

$$\Pr(\boldsymbol{\theta} | \boldsymbol{y}; \hat{\boldsymbol{\varphi}}) \propto \exp \left[\hat{\gamma}_1 \sum_{(s,t) \in S_1} \theta_s \theta_t \right.$$
$$\left. + \sum_{s \in S} \left\{ \hat{\gamma}_2 - \log f_0(y_s) + \log f_1(y_s; \hat{\boldsymbol{p}}) \right\} \theta_s \right]. \quad (50.10)$$

The estimation of effect sizes for selected voxels is important for evaluating their biological or clinical significance. Of note, the naive estimator given by $\tilde{\delta}_s = Y_s$ generally overestimates the true effect size (absolute δ_s) for the selected "top" voxels with the highest statistical significance. This estimation bias reflects the selection bias inherent in selecting voxels with the largest absolute Y_s that is due to random variation. This bias is reduced by shrinkage estimation for

selected voxels. Specifically, we extended posterior indices originally developed in the case of independent Y_s's by [11] to the case of dependent Y_s's.

Since the effect size under the null hypothesis is zero, the posterior mean of the effect size of the voxel s is given by

$$E\left[\delta_s | \boldsymbol{y}; \boldsymbol{\varphi}\right] = E\left[\delta_s | y_s, \Theta_s = 1; \boldsymbol{\varphi}\right] \Pr(\Theta_s = 1 | \boldsymbol{y}; \boldsymbol{\varphi}).$$

We can then estimate the effect size using the following posterior indices,

$$\hat{\delta}_s = d_s \ell_s, \quad (50.11)$$

where d_s and ℓ_s are plug-in estimators, $d_s = E\left[\delta_s | y_s, \Theta_s = 1; \hat{\boldsymbol{\varphi}}\right]$ and $\ell_s = \Pr(\Theta_s = 1 | \boldsymbol{y}; \hat{\boldsymbol{\varphi}}) = 1 - \widehat{\mathrm{LIS}}_s(\boldsymbol{y})$.

Since the posterior probability $f(\delta | y_s, \theta_s = 1; \boldsymbol{p})$ is given by

$$f(\delta | y_s, \theta_s = 1; \boldsymbol{p}) = \phi(y_s; \delta, c_n^2) g(\delta; \boldsymbol{p}) / f_1(y_s; \boldsymbol{p}),$$

the plug-in estimator of posterior mean of δ_s for a non-null voxel s is expressed as

$$d_s = \sum_{b=1}^{B} t_b f(t_b | y_s, \theta_s = 1; \hat{\boldsymbol{p}})$$
$$= \sum_{b=1}^{B} t_b \hat{p}_b \phi(y_s; t_b, c_n^2) / \sum_{b=1}^{B} \hat{p}_b \phi(y_s; t_b, c_n^2).$$

The probability, ℓ_s, depends on the multiple testing index, $\widehat{\mathrm{LIS}}_s(\boldsymbol{y})$, which is calculated using Gibbs sampler from the distribution of $\boldsymbol{\Theta} | \boldsymbol{Y}$, presented by Eq. (50.10).

50.2.2 ROI Analysis with Voxel-Level Multiplicity Adjustment

In ROI analysis, we are interested in the regions rerated with disease. Let $R_1, \ldots, R_L$ be all the ROIs. For detecting disease-associated ROIs, we consider an FDR procedure based on the average of LIS in each ROI. Let $Q_l = \sum_{s \in R_l} \mathrm{LIS}_s(\boldsymbol{y}) / r_l$, where r_l is the number of voxels in R_l ($l = 1, \ldots, L$) and $Q_{(1)} \leq \cdots \leq Q_{(L)}$ represent a series of ordered $Q_1, \ldots, Q_L$. Let $R_{(l)}$ be the ROI corresponding to $Q_{(l)}$ and $r_{(l)}$ be the number of voxels in $R_{(l)}$. We can derive the following LIS procedure for ROI analysis adjusting voxel-level multiplicity,

$$\mathrm{let}\, k = \max \left\{ i : \left(\sum_{l=1}^{i} r_{(l)} Q_{(l)} / \sum_{l=1}^{i} r_{(l)} \right) \leq \alpha \right\},$$

then reject all H_{0s} s.t. $s \in R_{(i)}, i = 1, \ldots, k$. (50.12)

With rejection, we declare that all $R_{(i)}, i = 1, \ldots, k$ are related to disease. Again, since the parameters, $\boldsymbol{\varphi}$, is unknown, a plug-in estimator, $\hat{Q}_l = \sum_{s \in R_l} \widehat{\mathrm{LIS}}_s(\boldsymbol{y})/r_l$, is used.

50.2.3 ROI Analysis with ROI-Level Multiplicity Adjustment

We note that the procedure in Sect. 50.2.1 pertains to the association of L ROIs with disease, but controls FDR against m null hypotheses. We can also propose another method for ROI analysis that controls FDR against ROI-based L null hypotheses. In this method, the null hypothesis, H_{0l}^* : there are no disease-related voxels in R_l, is tested for all ROIs. Let Φ_l be a variable representing disease-related status for all ROIs, where $\Phi_l = 0$ if the ROI l is null and $\Phi_l = 1$ otherwise. We can define LIS for all ROIs,

$$\mathrm{LIS}_l^*(\boldsymbol{y}) = \Pr(\Phi_l = 0 | Y = y; \boldsymbol{\varphi}).$$

Since the event $\Phi_l = 0$ means the event for all $s \in R_l$, $\Theta_s = 0$,

$$\mathrm{LIS}_l^*(\boldsymbol{y}) = \Pr\left(\bigcap_{s \in R_l} \{\Theta_s = 0\} | Y = y; \boldsymbol{\varphi}\right).$$

Let $\mathrm{LIS}_{(1)}^*(\boldsymbol{y}) \leq \cdots \leq \mathrm{LIS}_{(L)}^*(\boldsymbol{y})$ represent a series of ordered LIS_l^* across ROIs and let $H_{(i)}^*$ be the null hypothesis (representing no association with disease) on the voxel corresponding to $\mathrm{LIS}_{(i)}^*(\boldsymbol{y})$. We can derive a LIS procedure for ROI analysis adjusting ROI-level multiplicity,

$$\mathrm{let} \, k = \max \left\{ i : \frac{1}{i} \sum_{l=1}^{i} \mathrm{LIS}_{(l)}^*(\boldsymbol{y}) \leq \alpha \right\}, \quad (50.13)$$

then reject all $H_{(i)}^*, i = 1, \ldots, k$.

Again, since the parameters, $\boldsymbol{\varphi}$, is unknown, a plug-in estimator, $\widehat{\mathrm{LIS}}_{(j)}^*(\boldsymbol{y}) = \Pr(\Phi_l = 0 | Y = y; \hat{\boldsymbol{\varphi}})$, is used.

In the ROI analyses given in Sects. 50.2.2 and 50.2.3, we can evaluate effect sizes for the significant ROIs based on the distribution of voxel-level effect size estimates within ROI, $\hat{\Delta}_l = \sum_{s \in R_l} \hat{\delta}_s/r_l$.

50.3 Application

Alzheimer's disease is one of the most common neurodegenerative disorders that cause dementia with brain atrophy. We adapted our methods to quantify brain atrophy using a dataset of MRI images from the Open Access Series of Imaging Studies (OASIS) [15] from January 2011. Intensity values on $m = 185, 405$ voxels were observed for $n_1 = 27$

Alzheimer's disease patients and $n_2 = 28$ healthy controls ($n = n_1 + n_2 = 55$). The summary statistic in incorporating single covariate of the total brain volume was calculated from a t-statistic in testing a coefficient of disease status variable equal to zero in the general linear model. Here, the calculation of the t-statistic was performed using the SPM package.

When we fit our statistical models (see Sect. 50.2), the whole brain image was divided into 116 ROIs based on the Automated Anatomical Labeling (AAL). Since substantial heterogeneity in effect size was presumed across ROIs, we separately fit our model for each ROI to obtain more accurate voxel-level estimates within each ROI. We then obtained the effect size estimate, $\hat{\delta}_s$, in Eq. (50.11) and the LIS statistic, $\widehat{\mathrm{LIS}}_s(\boldsymbol{y})$, in Sect. 50.2 for individual voxels within each ROI based on the estimated model within the ROI. To detect disease-associated voxels for the criterion of FDR $\leq 5\%$, all the LIS values were pooled across ROIs and ordered to determine rejection of voxels based on the criterion in Eq. (50.8), which corresponds to a pooled LIS procedure by [16].

Figure 50.1a, b display significant voxels at FDR $\leq 5\%$ by the pooled LIS procedure and their effect size estimates, $\hat{\delta}_s$, in Eq. (50.11), based on the ROI-specific estimated models. We obtained similar conclusion when performing the estimation based on the t-distribution, rather than the normal distribution, for the sampling distribution of Y_s to cope with the small sample size. Figure 50.1a indicates that so much of the brain was significantly different between disease and control subjects that it is not very informative. By contrast, Fig. 50.1b indicates some variation in effect size among the significant voxels. The regions that contain the detected voxels with the great effect sizes were consistent with those reported by a large-scale meta-analysis [17].

| (a) Rejected voxels | (b) Estimated effect size |

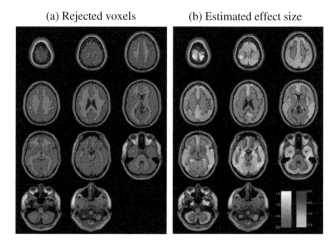

Fig. 50.1 Application of the method for voxel-level inference to Alzheimer's disease. Panel (**a**) displays rejected voxels for the nominal FDR level of 0.05. Panel (**b**) displays effect size estimates for all voxels

Table 50.1 Rejected ROIs by the method for ROI analysis with voxel-level multiplicity adjustment in application to Alzheimer's disease

Index	Name	Lobe	r_l	$\hat{\Delta}_l$	$\hat{Q}_l$
83	TPOsup.L	Temporal	1285	0.952	0.001
38	HIP.R	Limbic	946	1.075	0.002
59	SPG.L	Parietal	2065	0.844	0.003
84	TPOsup.R	Temporal	1338	1.041	0.003
37	HIP.L	Limbic	932	1.125	0.003
39	PHG.L	Limbic	978	0.929	0.004
87	TPOmid.L	Temporal	755	0.888	0.005
41	AMYG.L	Limbic	220	1.184	0.006
40	PHG.R	Limbic	1132	0.928	0.010
82	STG.R	Temporal	3141	0.895	0.015
81	STG.L	Temporal	2296	0.828	0.017
55	FFG.L	Occipital	2310	0.996	0.019
62	IPL.R	Parietal	1345	0.849	0.019
17	ROL.L	Frontal	990	0.636	0.023

We defined ROIs as the regions divided by AAL and performed ROI analysis. Using the effect size estimate, $\hat{\delta}_s$, and the LIS statistic, $\widehat{\mathrm{LIS}}_s(\boldsymbol{y})$, obtained by voxel-level inference, we calculated $\hat{Q}_l$ and $\hat{\Delta}_l$. Table 50.1 shows significant ROIs at FDR $\le 1\%$ by the method for ROI analysis with voxel-level multiplicity adjustment.

50.4 Discussion

We have proposed a framework for modelling effect sizes of whole-brain voxels, through integrating a hierarchical modeling of summary statistics from the whole set of features [10, 11] and a modeling in the hidden Markov random field to accommodate dependency in neuroimaging data [6]. With the whole-brain model, we can derive efficient multiple testing procedures with FDR control and also, less biased, shrinkage estimates for individual voxels. Furthermore, with the same whole-brain model, we can derive a procedure for prespecified brain regions, such as those in the ROI analysis with ROI-level multiplicity adjustment. For multiplicity adjustment, we proposed both voxel-based and ROI-based multiple testing procedures in Eqs. (50.9), (50.12) and (50.13). The voxel-based procedure in Eq. (50.9) can detect new regions that are independent of the known ROIs. This may be particularly usefull for exploratory research. Meanwhile, for the purpose of ROI analysis, when a related region is a small part of a ROI, the procedure in Eq. (50.12) could fail to detect the signal, because this procedure is based on the average of LIS in each ROI. In such a situation, the procedure in Eq. (50.13) may be a better choice since this employs the strict hypotheses, H_{0l}^*. It should be noticed that this is sensitive to the parcellation to define ROIs, that is,

when one disease-related voxel in a ROI is labeled as another ROI, the labeled ROI could be incorrectly detected. We note that the effect size index in Eq. (50.1) allows for evaluation without dependency of sample size. Such a feature may be particularly useful for comparing effect size estimates across different studies with distinct sample sizes. Also, we note that our procedures can be easily adapted to incorporate baseline covariates via regression models.

Lastly, another important advantage of the proposed framework with a whole-brain, voxel-based model is that it allows for power and sample size calculations of neuroimaging, disease-association studies. This will be a subject for future report.

Acknowledgements This research was supported by a Grant-in-Aid for Scientific Research (16H06299) and JST-CREST (JPMJCR1412) from the Ministry of Education, Culture, Sports, Science and Technology of Japan. We appreciate Open Access Series of Imaging Studies (OASIS) supported under grants: P50 AG05681, P01 AG03991, R01 AG021910, P20 MHO71616, and U24 RR021382 in using Alzheimer's disease data in Sect. 50.3.

References

1. Chumbley, J.R., Friston, K.J.: False discovery rate revisited: FDR and topological inference using Gaussian random fields. Neuroimage **44**(1), 62–70 (2009)
2. Poldrack, R.A., Mumford, J.A., Nichols, T.E.: Handbook of functional MRI data analysis. Cambridge University Press, Cambridge (2011)
3. Benjamini, Y., Hochberg, Y.: Controlling the false discovery rate: a practical and powerful approach to multiple testing. J. R. Stat. Soc. Ser. B Methodol. **57**(1), 289–300 (1995)
4. Genovese, C., Wasserman, L.: Operating characteristics and extensions of the false discovery rate procedure. J. R. Stat. Soc. Ser. B Stat. Methodol. **64**(3), 499–517 (2002)
5. Sun, W., Cai, T.T.: Large-scale multiple testing under dependence. J. R. Stat. Soc. Ser. B Stat. Methodol. **71**(2), 393–424 (2009)
6. Shu, H., Nan, B., Koeppe, R.: Multiple testing for neuroimaging via hidden Markov random field. Biometrics **71**(3), 741–750 (2015)
7. Poldrack, R.A.: Region of interest analysis for fMRI. Soc. Cogn. Affect. Neurosci. **2**(1), 67–70 (2007)
8. Reddan, M.C., Lindquist, M.A., Wager, T.D.: Effect size estimation in neuroimaging. JAMA Psychiat. **74**(3), 207–208 (2017)
9. Lindquist, M.A., Mejia, A.: Zen and the art of multiple comparisons. Psychosom. Med. **77**(2), 114–125 (2015)
10. Efron, B.: Empirical Bayes estimates for large-scale prediction problems. J. Am. Stat. Assoc. **104**(487), 1015–1028 (2009)
11. Matsui, S., Noma, H.: Estimating effect sizes of differentially expressed genes for power and sample-size assessments in microarray experiments. Biometrics **67**(4), 1225–1235 (2011)
12. Shen, W., Louis, T.A.: Empirical Bayes estimation via the smoothing by roughening approach. J. Comput. Graph. Stat. **8**(4), 800–823 (1999)
13. Matsui, S., Noma, H.: Estimation and selection in high-dimensional genomic studies for developing molecular diagnostics. Biostatistics **12**(2), 223–233 (2011)
14. Efron, B.: Large-Scale Inference: Empirical Bayes Methods for Estimation, Testing, and Prediction. Institute of Mathemati-

cal Statistics Monographs. Cambridge University Press, Leiden (2010)

15. Marcus, D.S., Wang, T.H., Parker, J., Csernansky, J.G., Morris, J.C., Buckner, R.L.: Open Access Series of Imaging Studies (OASIS): cross-sectional MRI data in young, middle aged, non-demented, and demented older adults. J. Cogn. Neurosci. **19**(9), 1498–1507 (2007)

16. Wei, Z., Sun, W., Wang, K., Hakonarson, H.: Multiple testing in genome-wide association studies via hidden Markov models. Bioinformatics **25**(21), 2802–2808 (2009)

17. Chapleau, M., Aldebert, J., Montembeault, M., Brambati, S.M.: Atrophy in Alzheimer's disease and semantic dementia: an ALE meta-analysis of voxel-based morphometry studies. J. Alzheimers Dis. **54**(3), 941–955 (2016)

Hajime Omura and Teruya Minamoto

51.1 Introduction

Gait recognition is a biometric authentication method for identifying an individual that is focused on his/her manner of walking. In contrast to biometric authentication methods that use a fingerprint or an iris, where the subject must be close to the sensor for recognition processing, gait recognition methods can recognize a human being at a long distance from the monitoring camera video. Gait recognition methods authenticate the individual person based on the gait features of the object person extracted from the image sequence obtained from the monitoring camera video.

Most existing gait recognition methods use gait features based on the silhouette image sequence. Han and Bhanu [2] proposed a gait representation called the gait energy image (GEI), which is obtained by averaging the silhouette image sequence over a gait period. Makihara et al. [4] extracted the time-independent frequency domain feature from the gait silhouette volume. Bashir et al. [1] proposed a gait representation called the gait entropy image (GEnI), which is obtained by computing the temporal entropy from the silhouette image sequence of a gait period. In [3], Lam et al. proposed a gait representation called the gait flow image (GFI), which is computed from the optical flow in the silhouette image sequence. However, the gait recognition accuracy decreases if the resolution of the silhouette image sequence is low.

The purpose of this paper is to present a new gait feature extraction method that uses the dyadic wavelet transform (DYWT) and structural similarity (SSIM) for gait recognition. To authenticate an individual person, our method

focuses on the similarity of the structural features in each frame of the silhouette image sequence of a gait period. We use SSIM [8] to compute this similarity. SSIM is an objective image quality assessment method for quantitatively measuring image quality by computing the structural features extracted from the objects in the scene. A gait feature extraction method that uses the SSIM technique does not exist.

The remainder of this paper is organized as follows. In Sect. 51.2, we briefly describe the DYWT and the wavelet transform modulus (WTM). In Sect. 51.3, we describe the SSIM technique used for extracting the gait features. In Sect. 51.4, we present a new method for extracting the gait features. We present our experimental results in Sect. 51.5 and conclude the paper in Sect. 51.6.

51.2 Dyadic Wavelet Transform

Let us represent the original image by $C^0[m, n]$, which represents samples of a normalized discrete image defined on a two-dimensional lattice of finite extent. In [5,6], as is well-known, the DYWT for this image based on the "algorithme à trous" is given by

$$C^{j+1}[m, n] = \sum_k \sum_l h[k]h[l]C^j_{k,l}[m, n],$$

$$D^{j+1}[m, n] = \sum_k \sum_l g[k]h[l]C^j_{k,l}[m, n],$$

$$E^{j+1}[m, n] = \sum_k \sum_l h[k]g[l]C^j_{k,l}[m, n], \quad (51.1)$$

$$F^{j+1}[m, n] = \sum_k \sum_l g[k]g[l]C^j_{k,l}[m, n].$$

Here, $C^j_{k,l}[m, n] = C^j[m + 2^j k, n + 2^j l]$, h is a low-pass filter, and g is a high-pass filter. More precisely, $C^j[m, n]$,

H. Omura (✉) · T. Minamoto
Department of Information Science, Saga University, Saga, Japan
e-mail: ohmurah@ma.is.saga-u.ac.jp; minamoto@ma.is.saga-u.ac.jp

© Springer Nature Switzerland AG 2019
S. Latifi (ed.), *16th International Conference on Information Technology-New Generations (ITNG 2019)*,
Advances in Intelligent Systems and Computing 800,
https://doi.org/10.1007/978-3-030-14070-0_51

$D^j[m, n]$, $E^j[m, n]$, and $F^j[m, n]$ are the low-frequency components, the horizontal high-frequency components, the vertical high-frequency components, and the high-frequency components in both directions, respectively. The indices m and n indicate the components' locations in the horizontal and vertical directions, respectively.

We now explain the WTM using the DYWT. We define the WTM as

$$WM^j[m, n] = \sqrt{(D^j[m, n])^2 + (E^j[m, n])^2 + (F^j[m, n])^2}$$

Figure 51.1 shows an example obtained by applying the WTM to a part of the silhouette image sequence.

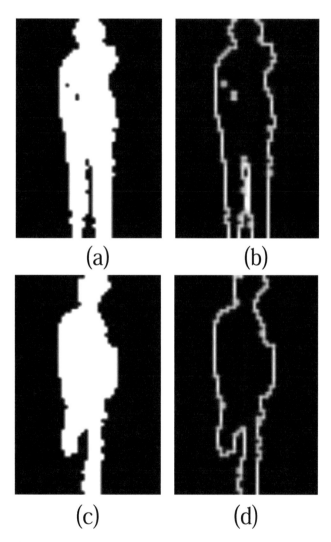

(a)   (b)

(c)   (d)

Fig. 51.1 (**a**) and (**c**) Part of the silhouette image sequence. (**b**) and (**d**) Images obtained by applying the wavelet transform modulus to (**a**) and (**c**), respectively

51.3 Structural Similarity

In this section, we explain the SSIM method [8]. The SSIM is evaluated by computing and comparing three elements of two nonnegative images: luminance, contrast, and structure. The range of the SSIM value is [0, 1], where a value of 1 means that the similarity is of the best quality and a value of 0 means it is of the poorest quality.

We assume that $x = (x_1, x_2, \cdots, x_N)$ and $y = (y_1, y_2, \cdots, y_N)$ are the original image and reference image, respectively; the size of each image is $N = M_1 \times M_2$. First, the luminance element $l(x, y)$ is defined by

$$l(x, y) = \frac{2\mu_x\mu_y + C_1}{\mu_x{}^2 + \mu_y{}^2 + C_1},$$

where $\mu_x = (\sum_{i=1}^N x_i)/N$ and $\mu_y = (\sum_{i=1}^N y_i)/N$, and the role of C_1 in avoiding instability in the case where $\mu_x^2 + \mu_y^2$ is very close to zero is important. If C_1 is not included, $l(x, y)$ becomes the infinite value. Assuming that L is the dynamic range of the pixel values (255 for 8-bit grayscale images) and $K_1 \ll 1$ is a small constant, we choose $C_1 = (K_1 L)^2$ and $K_1 = 0.01$ by default.

Next, a contrast element $c(x, y)$ is defined as

$$c(x, y) = \frac{2\sigma_x\sigma_y + C_2}{\sigma_x{}^2 + \sigma_y{}^2 + C_2},$$

where $\sigma_x^2 = \left(\sum_{i=1}^N (x_i - \mu_x)^2\right)/(N - 1)$, $\sigma_y^2 = \left(\sum_{i=1}^N (y_i - \mu_y)^2\right)/(N - 1)$, $C_2 = (K_2 L)^2$, $K_2 \ll 1$, and $K_2 = 0.03$ by default.

A structure element $s(x, y)$ is defined as

$$s(x, y) = \frac{\sigma_{xy} + C_3}{\sigma_x\sigma_y + C_3},$$

where $\sigma_{xy} = \left(\sum_{i=1}^N (x_i - \mu_x)(y_i - \mu_y)\right)/(N - 1)$, and C_3 is a constant.

Then, the SSIM between images x and y is defined as

$$SSIM(x, y) = [l(x, y)]^\alpha \cdot [c(x, y)]^\beta \cdot [s(x, y)]^\gamma,$$

where $\alpha > 0$, $\beta > 0$, $\gamma > 0$. A specific form of the SSIM is obtained by setting $\alpha = \beta = \gamma = 1$ and $C_3 = C_2/2$:

$$SSIM(x, y) = \frac{(2\mu_x\mu_y + C_1)(2\sigma_{xy} + C_2)}{(\mu_x{}^2 + \mu_y{}^2 + C_1)(\sigma_x{}^2 + \sigma_y{}^2 + C_2)}.$$

51.4 Gait Feature Extraction Method

In this section, we explain the proposed method, which detects a gait period and extracts the gait features.

51.4.1 Detection of Gait Period

In this subsection, we describe the detection of the gait period. According to the authors of [4], to detect the gait period first the following normalized autocorrelation is applied to the input silhouette image sequence.

$$C(N) = \frac{\sum_{x,y} \sum_{n=0}^{N_{total}-N-1} Seq(x,y,n)Seq(x,y,n+N)}{\sqrt{\sum_{x,y} \sum_{n=0}^{N_{total}-N-1} Seq(x,y,n)^2} \sqrt{\sum_{x,y} \sum_{n=0}^{N_{total}-N-1} Seq(x,y,n+N)^2}}.$$

Here, $C(N)$ is the autocorrelation for N frame shifts, $Seq(x,y,n)$ is the silhouette image at the n-th frame, and N_{total} is the number of total frames in the image sequence. Then, the gait period N_{gait} is computed as

$$N_{gait} = \arg \max_{N \in [20,40]} C(N). \quad (51.2)$$

Here, from our experience, we set the range of N to be [20, 40].

We extract the silhouette image sequence of a gait period using N_{gait} from the input silhouette image sequence.

51.4.2 Gait Feature Extraction

(1) Apply the WTM to each frame in the input silhouette image sequence $Seq(x, y, N_{gait})$ of a gait period and obtain $F(x, y, N_{gait})$.
(2) Apply the following equation to $F(x, y, N_{gait})$.

$$SumSeq(x, y, k) = \sum_{n=1}^{k} F(x, y, n),$$

where $k = 1, \cdots, N_{gait}$.
(3) Compute the SSIM between $SumSeq(x, y, k)$ as

$$Feat_i$$

$$= SSIM(SumSeq(x, y, i), SumSeq(x, y, i+1)).$$

Here, $i = 1, \cdots, N_{gait} - 1$.
(4) Apply moving average to $Feat_i$ and obtain the $mFeat_i$.
(5) Normalize $mFeat_i$ in the range between 0 and 1, and obtain $NmFeat_i$.

(6) The mean of $NmFeat_i$ is computed, and $SSIM_{feat}$ is defined as

$$SSIM_{feat} = \frac{1}{N_{gait}-1} \sum_{i=1}^{N_{gait}-1} NmFeat_i.$$

51.5 Experiment

To evaluate the proposed method, we conducted an experiment in which its performance was compared with that of the GEI method using the datasets of the OU-ISIR database [7]. Figure 51.2 shows the procedure of the performance evaluation experiment. In this experiment, we used the gait data of 100 subjects with respect to the view direction 0° in this dataset. The data of each subject were composed of two types of silhouette image sequence: probe and gallery. In addition, we resized these silhouette image sequences to 64×44, 32×22, and 16×11 so that the aspect ratio could be preserved. As the index of the evaluation, we used the cumulative matching characteristics (CMC) curve.

51.5.1 Computation of the Similarity

In this subsection, we describe the computation of the similarity for the performance evaluation experiment. We employed the 1-norm as the computation of the similarity. Here, if the size of the feature vector of the probe and gallery image sequences was different, the processing was applied so that the sizes of two feature vectors were the same. More precisely, if the vector size was large, the size of the images was reduced by applying downsampling.

We assumed that F_P and F_G are the $SSIM_{feat}$ of the probe and gallery image sequences, respectively, obtained by the

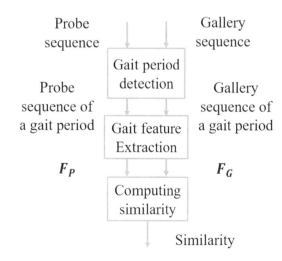

Fig. 51.2 Procedure of the performance evaluation experiment

gait feature extraction method described in Sect. 51.4. We
define the similarity $D(F_P, F_G)$ as

$$D(F_P, F_G) = |F_P - F_G|.$$

However, the similarity $D(GEI_P, GEI_G)$ with respect to
the GEI method for an image size $M_1 \times M_2$ was computed
as

$$D(GEI_P, GEI_G) = \sum_{m=1}^{M_1} \sum_{n=1}^{M_2} |GEI_P(m, n) - GEI_G(m, n)|.$$

51.5.2 Experimental Results

Figures 51.3, 51.4, and 51.5 show the CMC curve of the
proposed and the GEI method for the gait data of 100
subjects in the case where the silhouette image size was
64×44, 32×22, and 16×11. The CMC represents the
authentication rates within the top N ranks.

In Figs. 51.3 and 51.4, it can be seen that the curve of
the GEI method is located higher than that of the proposed
method; that is, the performance of the GEI method is
superior to that of the proposed method. In Fig. 51.5, it can be
seen that the curve of the GEI method is also located higher
than that of the proposed method; that is, the performance of
the GEI method is superior to that of the proposed method.
However, the authentication rate of the proposed method
in the case where the silhouette image sequence size was
16×11 increased in comparison with the case where the
size was 64×44 and 32×22.

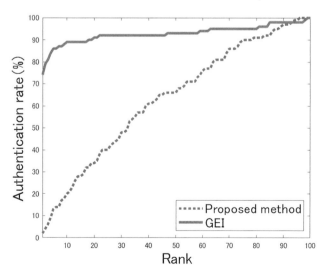

Fig. 51.3 Proposed vs. the gait energy image method in the case where
the silhouette image size is 64×44

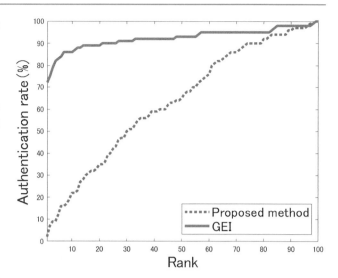

Fig. 51.4 Proposed vs. the gait energy image method in the case where
the silhouette image size is 32×22

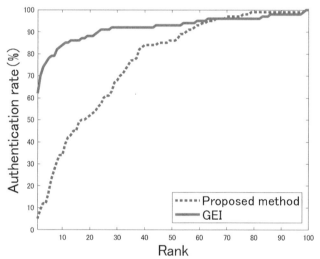

Fig. 51.5 Proposed vs. the gait energy image method in the case where
the silhouette image size is 16×11

51.6 Conclusion

We proposed a gait feature extraction method that uses the
DYWT and SSIM for gait recognition. The use of the SSIM
in image quality assessment to extract gait feature is novel.
According to the experimental results, the performance of
the GEI method is superior to that of the proposed method.
However, if the resolution of the silhouette image is low, the
authentication rate increases. Moreover, in terms of the CMC
curve for a silhouette image size of 16×11, the authentication
rate, at Rank $= 70$, of the proposed method is superior to
that of the GEI method. Therefore, for authenticating a low-
resolution sequence, it may be effective to use a gait feature

extracted using SSIM. The development of a gait recognition method using SSIM remains the topic of a future study.

References

1. Bashir, K., Xiang, T., Gong, S.: Gait recognition using gait entropy image. In: Proceedings of the 3rd International Conference on Imaging for Crime Detection and Prevention, pp. 1–6 (2009)
2. Han, J., Bhanu, B.: Individual recognition using gait energy image. IEEE Trans. Pattern Anal. Mach. Intell. **28**(2), 316–322 (2006)
3. Lam, T.H.W., Cheung, K.H., Liu, J.N.K.: Gait flow image: a silhouette-based gait representation for human identification. Pattern Recogn. **44**(4), 973–987 (2011)
4. Makihara, Y., Sagawa, R., Mukaigawa, Y., Echigo, T., Yagi, Y.: Gait recognition using a view transformation model in the frequency domain. Lect. Notes Comput. Sci. **3953**, 151–163 (2006)
5. Mallat, S.: A Wavelet Tour of Signal Processing. Academic, Amsterdam/London (2009)
6. Minamoto, T., Ohura, R.: A blind digital image watermarking method based on the dyadic wavelet transform and interval arithmetic. Appl. Math. Comput. **226**, 306–319 (2014)
7. The OU-ISIR Gait Database, Multi-View Large Population Dataset (Osaka University). http://www.am.sanken.osaka-u.ac.jp/BiometricDB/GaitMVLP.html
8. Wang, Z., Bovik, A.C., Sheikh, H.R., Simoncelli, E.P.: Image quality assessment: from error visibility to structural similarity. IEEE Trans. Image Process. **13**(4), 600–612 (2004)

Automatic Detection of Early Gastric Cancer from Endoscopic Images Using Fast Discrete Curvelet Transform and Wavelet Transform Modulus Maxima

Yuya Tanaka, Kohei Watarai, and Teruya Minamoto

52.1 Introduction

As reported in [1], it is estimated that more than 1 million people worldwide were newly diagnosed with gastric cancer in 2018. Gastric cancer is divided into Stages 1 to 4. The different stages are distinguished by the extent to which the cancer has grown and spread according to diagnosis. According to [2], the five-year survival rate of Stage 3A gastric cancer patients is 54%, whereas that of Stage 1 gastric cancer patients is 94%. Therefore, it is very important to detect gastric cancer at an early stage.

However, doctors cannot easily detect abnormal regions, that is, early stage gastric cancer regions, in an endoscopic image obtained by a endoscopic examination, in which a white light is used in general. This is because the appearance of abnormal regions and other regions, called normal regions in this paper, are very similar. In addition, diagnostic criteria for early stage gastric cancer based on endoscopic images have not been established, and thus, endoscopic diagnosis depends on the subjective diagnosis of each doctor and his/her skill and experience. Therefore, to support the doctor's diagnosis, the development of a detection method for early stage gastric cancer from endoscopic images is required.

In [3] and [4], an early stage gastric cancer detection method from an endoscopic image was proposed. However, in these studies, endoscopic images captured in narrow band imaging (NBI) mode [5] were targeted. The NBI mode is, however, installed only in the Olympus endoscope system. The most frequently used endoscopy produces endoscopic images captured with white light. However, as compared to the NBI mode, there are few studies on the detection of early stage gastric cancer from endoscopic images captured with white light.

In [6–8], and [9], an early stage cancer detection method using wavelet transforms was proposed. A detection method for early stage gastric cancer using the wavelet transform modulus maxima (WTMM) was proposed in [7]. Further, many types of studies have been conducted on medical image analysis using the curvelet transform, such as those reported in [10] and [11]. The results show that the curvelet transform may also be a useful tool for the detection of early stage gastric cancer. However, few studies exist on the detection of early stage gastric cancer using the curvelet transform.

The aim of this study was to develop a new method for detecting early stage gastric cancer from endoscopic images. We combine the wavelet transform, which is sensitive in the horizontal and vertical directions, and curvelet transform, which is sensitive to curves, to enhance the features of abnormal regions. We also show that our new method is superior to our previous method.

The remainder of this paper is organized as follows. We briefly introduce the dyadic wavelet transform (DYWT) and WTMM in Sect. 52.2. In Sect. 52.3, we discuss the fast discrete curvelet transform (FDCT). In Sect. 52.4, we explain the features of abnormal regions when the FDCT and WTMM are used. In Sect. 52.5, we propose a new method of early gastric cancer detection from endoscopic images captured with white light. We present the experimental results in Sect. 52.6 and conclude this paper in Sect. 52.7.

52.2 Preliminaries

52.2.1 Dyadic Wavelet Transform

The DYWT has the shift-invariant property, and the size of each frequency component is the same as in the original

Y. Tanaka (✉) · K. Watarai · T. Minamoto
Department of Information Science, Saga University, Saga, Japan
e-mail: tanakay@ma.is.saga-u.ac.jp; wataraik@ma.is.saga-u.ac.jp;
minamoto@ma.is.saga-u.ac.jp

© Springer Nature Switzerland AG 2019
S. Latifi (ed.), *16th International Conference on Information Technology-New Generations (ITNG 2019)*,
Advances in Intelligent Systems and Computing 800,
https://doi.org/10.1007/978-3-030-14070-0_52

image. According to [8, 12], the DYWT for images is given by

$$C^{m+1}[i, j] = \sum_{l=-\infty}^{+\infty} \sum_{k=-\infty}^{+\infty} h[l]h[k]C^m[I, J],$$

$$D^{m+1}[i, j] = \sum_{l=-\infty}^{+\infty} \sum_{k=-\infty}^{+\infty} h[l]g[k]C^m[I, J],$$

$$E^{m+1}[i, j] = \sum_{l=-\infty}^{+\infty} \sum_{k=-\infty}^{+\infty} g[l]h[k]C^m[I, J],$$

$$F^{m+1}[i, j] = \sum_{l=-\infty}^{+\infty} \sum_{k=-\infty}^{+\infty} g[l]g[k]C^m[I, J].$$

for $m \geq 0$. Here, $C^0[i, j]$ is the original image, $I = i + 2^m l$, $J = j + 2^m k$, h is a low-pass filter, and g is a high-pass filter. Thus, $C[i, j]$, $D[i, j]$, $E[i, j]$, and $F[i, j]$ are the low frequency component, the vertical high frequency component, the vertical high frequency component, and the horizontal high frequency component in both directions, respectively. The indices i and j indicate the component's location in the vertical and horizontal directions, respectively. The low frequency component is the smoothed approximate image and, the three high frequency components are the edge information.

52.2.2 Wavelet Transform Modulus Maxima

The WTMM method was developed by Mallat et al. [12, 13], and the wavelet transform modulus (WTM) was originally defined by

$$M_o^m[i, j] = \sqrt{|D^m[i, j]|^2 + |E^m[i, j]|^2},$$

which represents the multiscale edge maps of the original image. The WTMM are defined as locations $[i, j]$ where the modulus image $M_o^m[i, j]$ has a local maximum, along the gradient direction given by

$$A_o^m[i, j] = \begin{cases} \alpha_o^m[i, j] & \text{if} D^m[i, j] \geq 0 \\ \pi + \alpha_o^m[i, j] & \text{if} D^m[i, j] < 0 \end{cases}$$

with

$$\alpha_o^m[i, j] = \tan^{-1}\left(\frac{E^m[i, j]}{D^m[i, j]}\right).$$

In contrast, we define the WTM by the following:

$$W_o^m[i, j] = \sqrt{|D^m[i, j]|^2 + |E^m[i, j]|^2 + |F^m[i, j]|^2}.$$

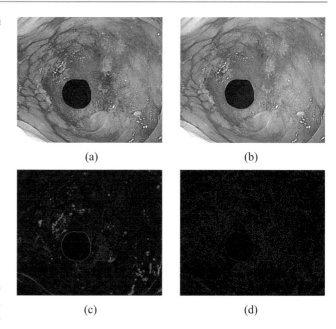

(a)   (b)

(c)   (d)

Fig. 52.1 Original image and image obtained by WTMM. (**a**) Original image. (**b**) Grey scale image. (**c**) WTM image. (**d**) WTMM image

The WTM detects edges in three directions in an image.

We also define the WTMM as locations $[i, j]$ where the modulus image $WM^m[i, j]$ is locally maximum, along the gradient direction given by $A_o^m[i, j]$.

We set a threshold:

$$W_M^m[i, j] = \begin{cases} 1 & \text{if } W_o^m[i, j] \text{ is locally maximum} \\ & \text{and } W_o^m[i, j] \geq \mu + \sigma \\ 0 & \text{otherwise.} \end{cases}$$

Here, μ is the mean value of the image and σ is its standard deviation.

Figure 52.1 shows an example of a WTMM image. As seen in this figure, in the WTMM image generated by the threshold the edges can be located.

52.3 Curvelet Transform

52.3.1 Fast Discrete Curvelet Transform

In [12], curvelet frames, introduced by Candès et al. [14], were used to construct sparse representation for images including geometrically regular edges. As in directional wavelets, curvelet frames are obtained by rotating, dilating, and translating basic waveforms. However, curvelets have a very elongated support obtained with a parabolic scaling using different scaling factors along the width and length of the curvelet. The direction sensitivity of these anisotropic waveforms is considerably greater than that of directional wavelets.

It was shown in [14] that the digital curvelet transform for images is given by

$$c(j, l, k) = \sum_{0 \le t_1, t_2 < n} f[t_1, t_2]\overline{\phi_{j,l,k}[t_1, t_2]}.$$

Here, $c(j, l, k)$ is the set of obtained curvelet coefficients, $f[t_1, t_2]$ is a Cartesian array, t_1 and t_2 represent the position of the element of the array, $\phi_{j,k,l}$ is the digital curvelet waveform, and j, k, and l represent the scale, orientation, and spatial position, respectively.

In [14], implementation of the FDCT by wrapping was proposed. The algorithm is as follows.

(1) Apply the 2D fast Fourier transform (FFT) and obtain Fourier samples $\hat{f}[n_1, n_2]$, $-n/2 \le n_1, n_2 < n/2$
(2) For each scale j and angle l, form the product $\tilde{U}_{j,l}[n_1, n_2]\hat{f}[n_1, n_2]$. Here, $U_{j,l}$ is the frequency window.
(3) Wrap this product around the origin and obtain

$$\tilde{f}_{j,l}[n_1, n_2] = W(\tilde{U}_{j,l}\hat{f})[n_1, n_2],$$

where the ranges for n_1 and n_2 are respectively $0 \le n_1 < L_{1,j}$ and $0 \le n_2 < L_{2,j}$ (for θ in the range $(-\pi/4, \pi/4)$), $L_{1,k} \sim 2^j$, $L_{2,j} \sim 2^{j/2}$.
(4) Apply the inverse 2D FFT to each $\tilde{f}_{j,l}$, thereby collecting the discrete coefficients $c(j, l, k)$.

The computational complexity of this algorithm is $O(n^2 \log n)$.

52.3.2 Proposed Method

We propose a method that incorporates the WTMM idea in the FDCT. When the FDCT is applied to an image, a curvelet coefficient of complex type is obtained. Then, only imaginary components are extracted and the inverse fast discrete curvelet transform (IFDCT) is applied to obtain a real type reconstructed image. The visualization of the absolute values of the component obtained by this inverse transform is as shown in Fig. 52.2. We define this image as a CTM (Curvelet Transform Maxima) image. Next, we find the local maximum of the CTM image. We set a threshold:

$$C_M[i, j] = \begin{cases} 1 & \text{if } C_o[i, j] \text{ is locally maximum} \\ & \text{and } C_o[i, j] \ge \mu + \sigma \\ 0 & \text{otherwise.} \end{cases}$$

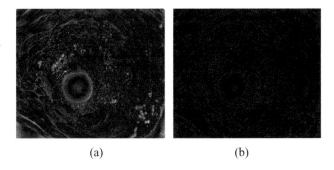

(a)   (b)

Fig. 52.2 Image obtained by CTMM. (**a**) CTM image. (**b**) CTMM image

Here, μ is the mean value of the image and σ is its standard deviation, and C_o is a CTM image.

We define the image obtained by this process as a CTMM (Curvelet Transform Maxima) image. Figure 52.2 shows the CTMM image generated from the gray scale image in Fig. 52.1. As can be seen, the CTMM image also shows that the edges can be detected.

52.4   Preliminary Experiment

As a preliminary experiment, we generated enhanced images from the WTMM and CTMM images. Then, in order to analyze the features of abnormal regions, we compared the average value of the region corresponding to the abnormal region with that of the entire image.

The procedure of the preliminary experiment was as follows.

(1) Convert the endoscopic image to the CIE L*a*b* color space. In this experiment, we used the L* (lightness) and the a* (complementary color) component.
(2) Clip the image size of the two images to 640 × 640 pixels. Then, emphasize the contrast by applying histogram equalization, and create a composite image from the two images using

$$IC[i, j] = \sqrt{IL^2[i, j] + IA^2[i, j]}.$$

Here, IL and IA are images of the L* and the a* component, respectively, which are contrast-enhanced, IC is a composite image, and i and j are the positions of the elements in the horizontal direction and the vertical direction of the image, respectively.
(3) Compute the WTM image by applying the DYWT twice to the image of the a* component. In this experiment, we used the DYWT with the coefficients of filters computed from the quadratic spline wavelets and the scaling functions as descibed in [12].

Table 52.1 Average values of the entire image and the average value only the block of the abnormal region

Average value/Endoscopic image	Image 1	Image 2
Average value of the WTMM image	0.3640	0.5651
Average value of blocks with only abnormal regions in the WTMM image	0.0748	0.2371
Average value of the CTMM image	0.4763	0.3109
Average value of blocks with only abnormal regions in the CTMM image	0.1185	0.1088

(4) Find the local maximum for the WTM image. Then, generate the WTMM image by threshold processing.
(5) Apply the FDCT to the composite image obtained by Step 2.
(6) Extract the imaginary components of the curvelet coefficient and apply contrast limited adaptive histogram equalization (CLAHE) to enhance the contrast. In addition, set all the real components of the curvelet coefficient to 0.
(7) Apply the IFDCT to calculate the absolute value and generate the CTM image.
(8) Find the local maximum for the CTM image. Then, generate the CTMM image by threshold processing.
(9) Divide the WTMM and CTMM images into non-overlapping blocks of size 64 × 64 pixels. In this experiment, the image was divided into 100 blocks.
(10) Compare the average value of the entire image and the average value of only the block of the abnormal region.

Table 52.1 presents the results obtained in the preliminary experiments. They show that the average value of the region corresponding to abnormal regions is smaller than that of the entire image. This means that there is a characteristic difference in the edge information in abnormal regions and normal regions, and there is a threshold for detecting abnormal regions.

52.5 Algorithm for Early Gastric Cancer Detection

On the basis of the preliminary experiment, we present an algorithm for detecting abnormal regions as follows

(1) Perform Step 1 to Step 9 described in Sect. 52.4.
(2) For each of the two images divided into blocks, compute the representative value $R[i, j]$ of each block which is the average pixel value of the block.
(3) For each of the two images divided into blocks, compute the average pixel value μ and the standard deviation σ of the representative values of all blocks.

(4) For each of the two images divided into blocks, generate the enhanced image O_{bin} using the relations

$$O_{bin}[i, j] = \begin{cases} 1 & \text{if } R[i, j] \leq \mu - \sigma \\ 0 & \text{otherwise.} \end{cases}$$

(5) Compute the Hadamard product of the two binarized images, and generate an enhanced image.

52.6 Experimental Results

Our results were evaluated from the perspective of false positives. That is, even if normal regions were determined to be abnormal regions, it was deemed that the detection method succeeded provided that abnormal regions were not overlooked. The areas detected as abnormal regions by our method are shown in white in the resulting images. As shown in Figs. 52.3, 52.4, 52.5, 52.6, and 52.7, we found that the white areas of the image obtained using our method contained the abnormal regions marked by the doctor. According to Table 52.2, as compared to our previous method [7], in our new method in fewer areas are normal regions judged to be abnormal regions. This shows that the proposed method is more accurate than the previous one. This is considered to occur because it is possible to capture the features of the abnormal part more effectively by combining a strong curvelet transform on the curve and a strong wavelet transform in the horizontal and the vertical directions.

However, instances remain where the normal regions are detected as abnormal regions. This is considered to occur because the threshold value of the image being examined

Fig. 52.3 Endoscopic image in which the region of early stage gastric cancer was marked by a doctor, the detection result of our new method, the detection result of our previous method (WTMM only), and the detection result using only CTMM (from left to right)

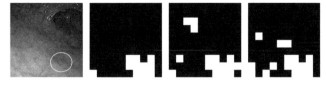

Fig. 52.4 Endoscopic image in which the region of early stage gastric cancer was marked by a doctor, the detection result of our new method, the detection result of our previous method (WTMM only), and the detection result using only CTMM (from left to right)

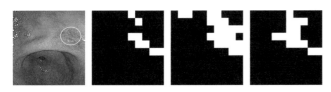

Fig. 52.5 Endoscopic image in which the region of early stage gastric cancer was marked by a doctor, the detection result of our new method, the detection result of our previous method (WTMM only), and the detection result using only CTMM (from left to right)

Fig. 52.6 Endoscopic image in which the region of early stage gastric cancer was marked by a doctor, the detection result of our new method, the detection result of our previous method (WTMM only), and the detection result using only CTMM (from left to right)

Fig. 52.7 Endoscopic image in which the region of early stage gastric cancer was marked by a doctor, the detection result of our new method, the detection result of our previous method (WTMM only), and the detection result using only CTMM (from left to right)

Table 52.2 The average of the number of blocks in which normal regions were detected as abnormal regions for each method

	The average number of blocks
Our new method	8.2
Our previous method (WTMM only)	14.2
CTMM only	11.9

is subjected to threshold processing using the average and standard deviation. In order to detect only abnormal regions without misdetecting a normal region as an abnormal region, the threshold must be set appropriately.

52.7 Conclusion

We proposed an automatic detection method of early stage gastric cancer from endoscopic images using the FDCT and WTMM. The experimental results show that it is possible to detect the area containing abnormal regions. Moreover, our

method finds fewer areas where normal regions are judged to be abnormal than our previously method. This would lead to diagnostic support of doctors performing an endoscopic examination.

In future work, a threshold that allows only abnormal regions to be detected will be investigated. We will consider using not only the dynamic threshold but also a hard threshold. Further, we would like to emphasize the features of abnormal regions using additional analysis methods.

References

1. WCRF: Stomach cancer statistics. https://www.wcrf.org/dietandcancer/cancer-trends/stomach-cancer-statistics
2. American Cancer Society: Stomach Cancer Survival Rates. https://www.cancer.org/cancer/stomach-cancer/detection-diagnosis-staging/survival-rates.html
3. Kaise, M.: Advanced endoscopic imaging for early gastric cancer. Best Pract. Res. Clin. Gastroenterol. **29**, 575–587 (2015)
4. Song, M., Ang, T.L.: Early detection of early gastric cancer using image-enhanced endoscopy: current trends. Gastrointest. Interv. **3**, 1–7 (2014)
5. Gono, K., Obi, T., Yamaguchi, M., Ohyama, N., Machida, H., Sano, Y., Yoshida, S., Hamamoto, Y., Endo, T.: Appearance of enhanced tissue features in narrow-band endoscopic imaging. J. Biomed. Opt. **9**(3), 568–577 (2004)
6. Hu, T., Lu, Y.H., Cheng, C.G., Sun, X.C.: Study on the early detection of gastric cancer based on discrete wavelet transformation feature extraction of FT-IR spectra combined with probability neural network. Spectroscopy **26**(3), 155–165 (2011)
7. Tanaka, Y., Minamoto, T.: Detection of early gastric cancer from endoscopic images using wavelet transform modulus maxima. Adv. Intell. Syst. Comput. **738**, 589–594 (2018)
8. Ohura, R., Omura, H., Sakata, Y., Minamoto, T.: Computer-aided diagnosis method for detecting early esophageal cancer from endoscopic image by using dyadic wavelet transform and fractal dimension. Adv. Intell. Syst. Comput. **448**, 929–938 (2016)
9. Matsunaga, H., Omura, H., Ohura, R., Minamoto, T.: Daubechies wavelet-based method for early esophageal cancer detection from flexible spectral imaging color enhancement image. Adv. Intell. Syst. Comput. **448**, 939–948 (2016)
10. Liu, G., Yan, G., Kuang, S., Wang, Y.: Detection of small bowel tumor based on multi-scale curvelet analysis and fractal technology in capsule endoscopy. Comput. Biol. Med. **70**(1), 131–138 (2016)
11. Dhahbi, S., Barhoumi, W., Zagrouba, E.: Breast cancer diagnosis in digitized mammograms using curvelet moments. Comput. Biol. Med. **64**(1), 79–90 (2015)
12. Mallat, S.: A Wavelet Tour of Signal Processing: The Sparse Way, 3rd edn. Academic Press, Amsterdam/London (2008)
13. Mallat, S., Zhong, S.: Characterization of signals from multiscale edges. IEEE Trans. Pattern Anal. Mach. Intell. **14**(7), 710–732 (1992)
14. Candès, E., Demanet, L., Donoho, D., Ying, L.: Fast discrete curvelet transforms. Multiscale Model. Simul. **5**(3), 861–899 (2006)

## Part VII

Workshop on Next Generation Infrastructures:
Health, Automotive, Mobility

Adriano M. S. Lima and Methanias Colaço Júnior

53.1 Introduction

Over the years, various administrative theories have been created to improve corporate performance. Some examples are: Organization, Systems and Methods (OSM) [1], Total Quality Management (TQM) [2], Organizational Reengineering [3], and others. One point of intersection between such approaches was the concept of Business Process, which [4] defines as a flow of coordinated tasks, activated by specific events, conducted by participants who act with data, information and knowledge to reach a specific goal. According to [5], the synchronism between processes makes the organization aware of the needs of the customers, guarantees the quality and productivity, and provides greater agility and objectivity in decisions, making the company more competitive. For the perfect functioning of this strategy, processes need to be managed transversally across the sectors, generating the concept of Business Process Management, which [5] defines as a key part of developing enterprise information systems.

Industry 4.0 is a term supposedly used first to describe a high-tech strategy proposed by the German government, and now commonly used to refer to the development of cyber-physical systems (CPS) and dynamic data processes that use large amounts of data to drive intelligent machines [6]. This new revolution will not have the desired success either, if all the requirements for a complete business transformation are not met. For [7], the challenges of industry 4.0 are structured around three important aspects: people, organizational struc-

tures and technology, in which the most important aspect is the interdependence among them for the success of the strategy.

According to [8], the Strategic Planning is an administrative process, usually carried out by the highest levels of the organization, which includes establishing the objectives and means to achieve them, observing the internal and external, present and future environments. For [9], people are essential for the organizational change process, the development of changeable environments and the sharing information, gaining competitive advantages and improvements in the processes. In this sense, it is important to emphasize that, in order for such synchronicity in the organizational processes to happen, it is necessary that people also be strategically aligned, since they are the fundamental pillar for the success of an organization. In order to guarantee operational and marketing profits in companies, a technological analysis must be carried out on the opportunities for the digital transformation of each process. The objective of this article is to present the results of the construction and experimentation of a management methodology that integrates the areas of Strategic Planning, BPM, Digital Transformation and People Management, in an agile way, through an *in vivo* experiment in the goods receipt process of a food distributor company. In previous works, such as, [10–12] methodologies for the implementation of process management were used, but none of them integrated the four areas described above and the results of the implementation of the methodology were not considered via an in vivo experiment with the statistical analysis of the results obtained.

The remainder of the paper is structured as follows. Section 53.2 presents the research methodology. Section 53.3 presents the proposed management methodology and its brief description. In Sect. 53.4, the related works are presented. In Sect. 53.5 we find the definition and the planning of the experiment. In Sect. 53.6, the operation of the experiment

A. M. S. Lima (✉)
Department of Computer Science (DCOMP), Federal University of Sergipe (UFS), Aracaju, Sergipe, Brazil

M. C. Júnior (✉)
Federal University of Sergipe (UFS), Aracaju, Sergipe, Brazil
e-mail: methanias@ufs.br

© Springer Nature Switzerland AG 2019
S. Latifi (ed.), *16th International Conference on Information Technology-New Generations (ITNG 2019)*,
Advances in Intelligent Systems and Computing 800,
https://doi.org/10.1007/978-3-030-14070-0_53

is presented. Section 53.7 contains the results of the experiment. Finally, Sect. 53.8 presents the conclusion and future works.

53.2 Research Methodology

The methodology used for the work in question consists, in terms of classification, of an exploratory and explanatory research [13], with quantitative connotation, in which the method used was experimental, based on [14]. Summarizing, the experiment was performed in 15 steps. At the end of the experiment, the metrics used as a basis for evaluating the hypotheses were analyzed. At that specific moment, two metrics were analyzed: (a) process execution time – to verify the time spent for the execution of an instance of the receiving process; (b) goods receipt errors – number of errors found in the goods receipt process. The experiment in question is detailed in Sects. 53.5, 53.6 and 53.7.

53.3 BTrans Management Methodology

The Business Transformation Management Methodology (BTrans) was developed in ten phases. Each phase will be explained in detail for a better understanding of the results of this work in the item 6.

53.3.1 Implementation of the BTrans Methodology

The implementation of the methodology is completely agile, that is, instead of carrying out each proposed phase through all the processes in the company, a prioritization of processes is conducted in Phase III and one process is chosen at a time, to individually pass through Stages IV–X. Each new chosen process must always follow such cycle.

The estimated average time for the transformation of the first selected process is three sprints of 15 days, considering a total of 12 hours of work per week. The other processes can be transformed into an average of two sprints with the same time and effort.

53.4 Related Works

There are currently several studies related to the implementation of BPM project methodologies. In [10], the methodology was based on the understanding of processes and error reduction to conduct a BPM case study in a public company. The work showed efficiency in the design of the processes and the reduction of errors, but it was not

implemented in an agile way, it did not show an alignment with the strategy and it did not give the improvement to the people management. In [11], for example, a CBOK-based methodology for sales process modeling was used. The process was modeled and accomplished the standardization of activities, but the improvements were not scientifically presented. In [12], a study to identify the management model used in the units of a public university is conducted, aiming to understand why public management does not function efficiently. The conclusion was that the lack of a process-oriented strategic management was responsible for most of the problems in the organization. In summary, to date, no studies have been found describing a complete methodology for the implementation of BPM projects focused on strategy, people management and digital transformation, not even with an agile implementation. The description of this methodology, along with the results collected via the *in vivo* experiment, can help other companies replicate the model to obtain great benefits for the organizational strategy.

53.5 Definition and Planning of the Experiment

In this section, and in the next two ones, our work is presented as an experimental process. It follows the guidelines of Wohlin et al. in [15]. This section will focus on development of the objective and the planning of the experiment.

53.5.1 Goal Definition

The objective was formalized using the GQM model proposed by Basili [16]: **Evaluating** a management methodology that combines BPM, strategic planning, digital transformation and people management, in an agile way, **in order to** measure the results obtained with the proposed model against the traditional, non-process-oriented approach, **with respect to** the efficiency and effectiveness of critical business processes and **from the perspective** of professionals involved in business management, in the context of a food distributor company.

53.5.2 Planning

Context Selection
The *in vivo* experiment was conducted at a food distributor company, in which the authors of this article are consultants. At the time the research was conducted, the company did not have a strategic planning formalized and disseminated among all employees yet, but the strategic objectives were identified through interviews with the partners.

Dependent and Independent Variables

The independent variables of the experiment are: The traditional, non-process-oriented model and the proposed model;

The dependent variables of the experiment are: The waiting time of the transportation vehicle for the delivery of the cargos and the percentage of deliveries with errors

Formulation of Hypotheses

The research questions for the experiment, which need to be answered, are the following: (1) Can the implementation of the proposed model reduce the execution time of the prioritized process? (2) Can the implementation of the proposed model reduce the number of errors during the execution of the prioritized process?

Two metrics will be used as dependent variables for evaluating the questions above: (1) Average time to conduct the process; (2) Calculated average errors, considering: (a) Number of cargos without previous registration of the orders in the ERP; (b) Number of cargos with differences not corrected by the purchasing department until the arrival of the goods. Having the objectives and metrics defined, the following hypotheses will be considered:

Hypothesys 1

- H_{0time}: The traditional management practices and the proposed model have the same execution time for the prioritized process ($\mu_{TraditionalExecTime} = \mu_{BTransExecTime}$).
- H_{1time}: The proposed model has a shorter execution time for the prioritized process than traditional practices ($\mu_{TradicionalExecTime} > \mu_{BTransExecTime}$).

Hypothesys 2

- $H_{0errors}$: Traditional management practices guarantee the same number of errors for the prioritized process as the proposed model ($\mu_{TraditionalExecErrors} = \mu_{BTransExecErrors}$).
- $H_{1errors}$: The proposed model guarantees a smaller number of errors for the prioritized process than traditional practices ($\mu_{TraditionalExecErrors} > \mu_{BTransExecErrors}$).

Selection of Participants

For the composition of the participants in the experiment, 12 collaborators of the logistics department were selected: a logistics coordinator, a receiver and 10 stock assistants. As the goods receipt process relates to other sectors, it is always normal to have interactions with the purchasing, and the financial and fiscal teams.

Design of the Experiment

The macro business process of the company will be designed to discovery the main business processes. According to the strategic planning, the main strategic objectives will be identified. The processes of the macro business process will be prioritized in relation to the strategic objectives, determining the process to be selected. The strategic pain points related to the selected process will be written as user stories. The selected process will be drawn (AS-IS diagnostic) and the times, profile of the people and errors in the traditional model will be recorded. The gaps in the process will be recorded and prioritized with the digital transformation to define which one can be immediately corrected. The new design of the process will be elaborated (TO-BE model) based on the prioritized gaps, defining the owner performance indicators of the process. The HR adjustments will be made, the process will be implemented and the new times and errors associated with the process activities will be recorded in the new model. The results will be statistically compared.

Instrumentation

The *in vivo* experiment was carried out in the real working environment of the food distributor company, taking advantage of a project contracted by the client. Improvements in the goods receipt process were defined by the consultants and all activities were performed directly by the selected participants, without any interference from the authors of this article.

The tools used are the BTrans methodology described in Sect. 53.3.1, the Bizagi Modeler for the design of the process, Excel for the prioritization of processes and gaps, Microsoft Project for the planning of sprints and Microsoft R Open for statistical tests.

53.6 Operation of the Experiment

53.6.1 Preparation

The steps in the preparation for the conduction of the experiment are described below.

PHASE I – Analysis of Strategic Objectives The main strategic objectives of the company were recorded.

PHASE II – Elaboration of the Macro Business Process In a meeting with the managers of each sector, the macro business process of the food distributor company was elaborated.

PHASE III – Prioritization of Processes Each strategic objective had an assigned weight and processes were prioritized by the partners, according to their strategic contribution. The receive goods process was selected.

The strategic pain points of the Goods Receipt Process were recorded in user stories (Table 53.1).

Table 53.1 User Stories of selected process

Process: receive goods
User story 01: As an entrepreneur, **I want to** reduce the time of the goods receipt process **so that** my customers and suppliers are more satisfied.
User stories 02: As an entrepreneur, **I want to** reduce the number of errors in the goods receipt process, **so that** there is no rework and I can improve the quality of the operation.

53.6.2 Execution

The execution of the *in vivo* experiment started with the diagnostic phase of the AS-IS.

PHASE IV – Elaboration of the AS-IS Diagnostic Through in field interviews and observations, the traditional execution model of the process was designed. The statistics related to the strategic pain points caused by the long receipt time and number of errors began to be formally registered with each received cargo.

PHASE V – Process Transformation After the design of the process and its sub-processes, all the times, costs, errors

and profiles of the professionals allocated to perform each task were recorded. The gaps prioritized by the board and associated with the process were recorded and a summary is shown in Table 53.2.

The digital transformation was applied to the goods receipt process and several changes were implemented (Table 53.3).

PHASE VI – HR Optimization All persons responsible for carrying out the activities of the goods receipt process were interviewed. Their skills and competencies were recorded and the following actions were taken to optimize HR (Table 53.4):

Phase VII – Definition of Process Owners and Indicators of the Process The logistics coordinator was appointed as the owner of the process and the performance indicators were defined.

Phase VIII – Elaboration of the TO-BE Model The new process was designed with changes approved by the board of directors.

Table 53.2 GAPs of process

Priorized GAPS	Proposed solution
G1. Several goods arrive and orders were not previously entered in the ERP system by the buyers to be checked with the invoice during the document validation process.	**S1.** Implementing the ERP integration mode for prior invoice checking through the XML file generated when the vendor invoices the order.
G2. Most of the cargos have their orders in the system, however there are price, quantity or date divergences in relation to the invoice.	**S2.** Taking advantage of the same routine created in S1 to correct the divergent information.
G3. The invoice checking procedure is performed manually, line by line, by the operator.	**S3.** Implementing the ERP integration system with the XML import mode.
G4. The receiving of cargos performed by using the concept of FIFO.	**S4.** Creating weekly schedule for cargo receipt, according to the needs of the warehouse.

Table 53.3 Digital transformations of the process

Digital transformation	Description
TD01. Omnichannel integration with industry.	Development of the integration of the ERP system with the Web Service of a large food company for the automatic receiving of online purchase orders.
DT02. Integration of the ERP system with the XML system of the company.	Automatic pre-validation of the order against the invoice via the XML integration system of the ERP system.
DT03. Implementation of the WMS system.	Implementation of the Warehouse Management System (WMS) to better manage the quantities, volumes, validities and geographic locations of the received goods.
TD04. Process dashboard	Creation of the management dashboard system with monitors in the office of the directors and in the distribution center for real-time monitoring of process indicators.

Table 53.4 HR optimization

HR optimization	Description
HR01. Training sessions for employees.	Training sessions on the system mode for operators that only used basic functions were performed.
HR02. Awareness of the employees.	Employees who did not wear individual protection equipment were aware of its importance.
RH03. Reallocation of activities.	Some activities carried out by the logistics manager were transferred to other employees.
RH04. Implementation of intermittent work.	Due to the seasonality of the number of deliveries on several days of the month, intermittent.

Phase IX – Process Implementation The new process was implemented in the production environment with the training sessions of employees on the new way of execution.

Phase X – Measurement of the Results The statistics related to the strategic pain points were measured again to compare the improvements acquired through the transformation of the process.

53.6.3 Data Collection

With the support of the employee in charge of receiving the goods at the company, the total time, cost and number of goods receipt errors of approximately 200 cargos before the implantation of the new model were registered, and the same data were registered for 200 cargos after the implementation of the TO- BE process. The results of this analysis will be presented in the next session.

53.6.4 Data Validation

Two types of statistical tests, the Shapiro-Wilk Test and the Wilcoxon Test, were used as aid for the analysis, interpretation and validation. The Shapiro-Wilk Test was used for verifying the normality of the samples. The Wilcoxon test was used for comparing the averages of paired samples that did not obtain normality in the data, verifying the difference magnitude [14]. All statistical tests were performed using the Microsoft R Open Tool [17].

53.7 Results and Discussions

53.7.1 Data Analysis and Interpretation

To answer the questions, the following dependent variables were analyzed: (a) the total waiting time of the transportation vehicle and; (b) the errors found in the goods receipt process due to the lack of correction by the purchasing department during the pre-validation of the invoice by the XML file.

Process Execution Time
In order to answer Research Question 01, the results related to the waiting time of the transportation vehicle are presented in Table 53.5. This time reflects the period between the arrival of the transportation vehicle at the gate with the goods and the invoice, and the moment the driver receives the signed invoice stub to leave. The results show that the average time for goods receipt, in the months of April and May, before the implantation of the model, was 636.69 minutes. For the months of June and July, after the implementation of the model, the average time was 483,07 minutes.

Table 53.5 Average waiting time of the transportation vehicle

Month	Received cargos	Average waiting time of the transportation	Averages
April	88	623,52	636,69
May	80	649,86	
June	111	488,76	483,07
July	83	477,39	

The results in Table 53.5 suggest that the goods receipt process, after the implementation of the model has on average, a shorter execution time, when compared to the times obtained before the implementation of the methodology. However, it is not possible to make such assertion without sufficiently conclusive statistical evidence. For this, we first defined a level of significance of 0.05 throughout the experiment and the Shapiro-Wilk test was applied to analyze the normal distribution of the samples. The value of the variable Sig, read p-value, for the cargos received prior to the implementation of the methodology, was 1548e-10, and 1232e-10 after the implementation of the methodology. Such values are smaller than the adopted level of significance. It is assumed that the distribution of data for both samples is not normal. Due to the non-normality of the data, the Wilcoxon Test was applied as a Hypothesis Test. The Hypothesis Test is characterized as non-parametric for paired samples, taking into account that the sample has a continuous and symmetrical behavior. The test, besides comparing the difference among the samples, also verifies the magnitude of such difference. With the application of the Wilcoxon Test for the waiting time of the transportation vehicle, it was possible to verify that the Sig., with a value of 0.01123, is lower than the level of significance of 0.05. It means that the evidence of the difference between the averages of the waiting times of the transportation vehicle was proved. In view of the results, evidence that the time of goods receipt before and after the implementation of the methodology is significantly different is confirmed. That is, the hypothesis (H0), that the traditional management practices and the proposed methodology have the same execution time for the process was rejected.

Process Execution Errors
In order to answer Research Question 02, the results related to the percentage of errors in the goods receipt process are presented in Table 53.6. This percentage reflects the number of cargos received with some errors in the document validation process by the purchasing department. The results show that the average of errors in the months of April and May, before the implementation of the model, was 56%, and for the months of June and July, after the implantation of the model, the average was 38%.

Such results suggest that the goods receipt process, after the implementation of the model, has, on average, a lower

Table 53.6 Percentage of errors in the received cargos

Month	Received cargos	% of errors not corrected by the purchasing department	Averages (%)
April	88	59	56
May	80	53	
June	111	47	38
July	83	28	

number of errors when compared to the results before the implementation of the model. However, it is not possible to make such an assertion without sufficiently conclusive statistical evidence. We performed the normality test again and found that the value of p-value for the cargos received before the implementation of the methodology was 2.2e-16, and 2.2e-16 after the implementation of the methodology. The values are lower than the adopted level of significance. It is assumed that the distribution of data for both samples is not normal. Due to the non-normality of the data, the Wilcoxon Test was applied as a Hypothesis Test. For the number of errors of the received cargos, it was verified that the Sig., with value of 0.001024, is lower than the level of significance of 0.05. It means that the evidence of difference between averages was confirmed. In view of the results, evidence that the time of goods receipt before and after the implementation of the methodology are significantly different is confirmed. That is, the hypothesis (H0), that the traditional management practices and the proposed methodology have the same number of errors for the process was also rejected.

53.7.2 Threats to Validity

Although the results of the experiment were satisfactory, it presents threats to its validity that cannot be disregarded.

Threats to the internal validity: All data used in the *in vivo* experiment were collected manually by the receiving team of the company and recorded in spreadsheets. As the routine in the logistics sector is usually very hectic, some values may not have been correctly registered or entered. To minimize such problem, we collected a large number of samples each month, with records being kept by different people in each shift.

Threats to the external validity: The small number of months observed is a threat, since some particularity of the period may have shifted the averages to more or less. The results are still being monitored and we will be able to carry out a new analysis in the future to guarantee the analysis.

Threats to the construction validity: Since the model was tested in only one process of a company, it is necessary to run the model in other processes of other companies to generalize the found results.

53.8 Conclusion and Future Work

This work presented important contributions to the implementations of BPM projects in organizational environments. The methodology created can be easily replicated in any process in any type of company, increasing the possibilities to reach the strategic objectives. In the analysis of the results, although we monitored only 2 months of operation of the new model, there was evidence of the reduction of time and elimination of errors in the goods receipt process.

Finally, as future works, we intend to extend the model by monitoring the costs associated with the selected process. We know that by reducing the total process time, it is expected that the cost also be reduced proportionally, but a further study must be conducted.

References

1. Carreira, D.: Organização, sistemas e métodos. Editora Saraiva (2017)
2. Cardoso, R.R.: Auditoria como aporte para uma qualidade total. Facit Bus. Technol. J. **1**(4), 47. http://revistas.faculdadefacit.edu.br/index.php/JNT/article/download/220/226 (2017).
3. Pereira, J.R., Regattieri, C.R.: Gestão por processos. Rev. Interface Tecnológica. **15**(1), 409–421 (2018)
4. Razavian, M., Khosravi, R.: Modeling variability in business process models using UML. In: Fifth International Conference on Information Technology: New Generations, pp. 82–87 (2008)
5. De Albuquerque, A.M.M., Rocha, P.S.S.: Sincronismo Organizacional, Editora Saraiva (2017)
6. Feng, L., Zhang, X., Zhou, K.: Current problems in China's manufacturing and countermeasures for industry 4.0. EURASIP J. Wirel. Commun. Netw. **2018**(1), 90 (2018)
7. Neubauer, M., Stary, C., Kannengiesser, U., Heininger, R., Totter, A., Bonaldi, D.: S-BPM's industrial capabilities. In: S-BPM in the Production Industry, pp. 27–67. Springer, Cham (2017)
8. de Souza, K.T.S.: Planejamento Estratégico: uma Análise Estratégica de uma IES Privada de Palhoça/SC, IX SEGet, Simpósio de Excelência em Gestúo e Tecnologia. https://www.aedb.br/seget/arquivos/artigos12/55716704.pdf (2012)
9. Ribeiro, A.L.: Gestão de pessoas. Editora Saraiva (2017)
10. Santana, A.F.L., Alves, C.F.: BPMG–Um Modelo Conceitual para Governança em BPM–Aplicação numa Organização Pública. ISys-Rev. Bras. Sist. Informação. **9**(1), 139–167 (2016)
11. Luz, G.B.: Gestão do processo de vendas: um estudo de caso em uma empresa de tecnologia da informação, UNISINOS. http://www.repositorio.jesuita.org.br/handle/UNISINOS/6161 (2016)

12. Oliveira, J.S.: Análise da Gestão empregada em Universidade Pública sob a perspectiva da Gestão de Processos e seus indicadores, UFRJ. https://pantheon.ufrj.br/handle/11422/345 (2014)
13. Severino, A.J.: Metodologia do trabalho científico. Cortez Editora (2017)
14. de Oliveira, R.A.N., Junior, M.C.: Experimental analysis of stemming on jurisprudential documents retrieval. Information. **9**(2), 28 (2018)
15. Wohlin, C., Runeson, P., Höst, M., Ohlsson, M.C., Regnell, B., Wesslén, A.: Experimentation in Software Engineering. Springer Science & Business Media (2012)
16. Basili, V.R., Weiss, D.M.: A methodology for collecting valid software engineering data. IEEE Trans. Softw. Eng. **SE-10**(6), 728–738 (1984)
17. Naik, N., Jenkins, P., Savage, N., Katos, V.: Big data security analysis approach using computational intelligence techniques in R for desktop users. In: 2016 IEEE Symposium Series on Computational Intelligence (SSCI), pp. 1–8 (2016)

Multi-Agent Based Water Distribution and Underground Pipe Health Monitoring System Using IoT

Lakshmi Kanthan Narayanan and Suresh Sankaranarayanan

54.1 Introduction

In the present industrial era, the agents are focusing on many supplementary fields of Information and Communication Technology (ICT) and Artificial Intelligence (AI). There are various types of agents that are implemented in different applications ranging from small filters for emails to huge and complex critical systems such as missile control, air traffic management etc.

Water is one of the most essential part of human life and the pipeline which is used to transport the water has become an inseparable part of life. The critical part of physical health monitoring of pipeline requires frequent pipeline inspections and online monitoring has become an essential part in a smart water distribution system (SWDS). According to the economic status of developing countries like India, the system should be easily consumable, cost effective, easy to customize and highly scalable in maintenance and deployment of the pipe network [1].

Supervisory Control and Data Acquisition (SCADA) system provide monitoring and control operation for the various distribution systems such as Oil Refineries, Gas Fields, Electrical Energy and Mines etc. SCADA provides the data updating regarding the quantity, pressure and quality of the distributed components to the remote Data Acquisition unit. SCADA also helps in performing the controlling operations during the critical situations. The Water Distribution system SCADA is widely used for monitoring the Quantity, Quality (pH value), Flow and Pressure of water.

There are wired and wireless networks available for the communication of various sensors deployed inside the pipe to the common node. But wide range of real time applications are using wired networks for the communication from devices that are deployed under the ground surface to the control or monitoring station present above the surface of the ground.

The architecture is based on the connectivity and non-breakdown as reliable factor along with maintainability and continuity in power supply in the network is needed [2]. The occurrence of fault in the network or at the node is highly possible at any point of time. So it is highly essential for a SWDS to provide a unique mechanism for the failsafe operation and localization of fault occurrence to the control system or to SCADA engineer. Various problems occur in wired network due to damage in any part of wire and it is hard to provide physical security due to expansion of the pipeline network [2, 3] (Fig. 54.1).

Advanced researches are carried out in the field of Wireless Sensor Network (WSN) and Ad-hoc sensor network resulted in development of advanced sensing technology. These advancements in WSN started attaining attention in various kinds of applications such as agriculture, healthcare, military, ecology and environmental monitoring etc. The rapid growth of advancements in wireless sensor technology facilitates the possibility of implementing real-time automated pipe health monitoring and also detect and report the location of cracks, leakage, corrosion or damage [4–7].

A wide range of implementation of robotics in the water distribution system for an effective water distribution and pipeline health monitoring is available [8, 9].

But in the current modern machinery era we are moving toward building a smart city. Water distribution and management is one crucial area that has to be concentrated and

L. K. Narayanan
Department of Computer Science and Engineering, SRM Institute of Science and Technology, Chennai, Tamilnadu, India

S. Sankaranarayanan (✉)
Department of Information Technology, SRM Institute of Science and Technology, Chennai, Tamilnadu, India
e-mail: suresh.sa@ktr.srmuniv.ac.in

© Springer Nature Switzerland AG 2019
S. Latifi (ed.), *16th International Conference on Information Technology-New Generations (ITNG 2019)*,
Advances in Intelligent Systems and Computing 800,
https://doi.org/10.1007/978-3-030-14070-0_54

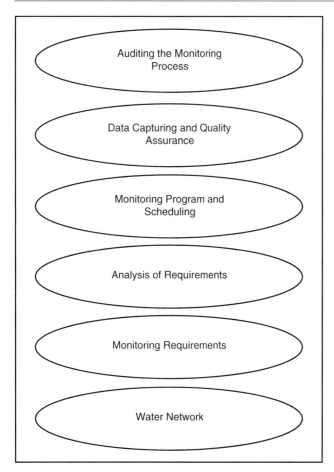

Fig. 54.1 Functional block of WMS functions

with the pipeline monitoring system using multi-agents will be more efficient and effective.

So accordingly, we here propose an multi agent based Smart Water Distribution System employing IoT integrated with Fog computing towards water distribution and pipe health monitoring. The main contribution of the paper is as follows:

- Multi-agent based SWDS architecture using IoT integrated with edge/fog.
- Agents at the edge for pipe health monitoring using rule based model on the basis of fluid dynamics
- Agents in cloud for water demand forecasting and pricing for consumers
- Agent Communication between Aggregator Agent, Node Agent and Cloud Agent towards data transmission, analytics and storage.
- The remaining portion of the paper is organized as follows. Section 54.2 will discuss the state-of-the-art technologies in the development of WSN based water distribution system and pipe health monitoring system, and also the agent based technologies and their limitations in the health monitoring of pipe network. Section 54.3 will provide a detailed description about the proposed multi-agent architecture along with the functionalities and descriptive operations of each agent types in the model. Section 54.4 will discuss about the conclusion and future scope of the system.

54.2 State of the Art Technologies in Water Distribution and Pipe Health Monitoring

There are various agent-based models used for the pipe health monitoring and distribution of water. These are discussed below.

54.2.1 Agent Based Water Distribution and Pipe Health Monitoring

The Software Engineering provides a unique feasibility of developing and designing a software system in terms of software components called as agents. According to Wooldridge, an agent is defined as "an encapsulated computer" that is surrounded by different environment and can act flexibly and autonomously in that environment in order to achieve the design goals [6]. An agent will accumulate four properties such as autonomy, proactivity, reactivity and sociality [10].

According to Jennings and Bussmann, agents are problem-solving entities with well-defined boundaries and interfaces which show a flexible problem-solving behavior

prioritized. But developing a complete system for water distribution, management and pipe health monitoring together along with intelligence is still a huge challenge.

Various types of agents like social, organizational, group and independent agents are in practice. The agents can also be made to interact with the information and knowledge database and can interact with the real time environment. The agents are smart enough which can be directed towards a goal. The agents can interact with the environment and also can interact with the market conditions and pricing policy. The agents are categorized and characterized by their complexity in decision making, operational flow and resources required, rules of behavior and performance attributes etc. There are various types of systems that are implemented using manually operated robots and a few robots with semi-automatic or autonomous control mechanism. The robotic and Ground Penetrating Radar (GPR) based technologies are used to locate the leakages and breakages in the pipeline. Based on the existing models for the pipe health monitoring we can conclude that the pipeline monitoring system should be integrated with wireless sensing technology, which helps in enhancing effective online monitoring. By integrating techniques for locating incidents and physical objects along

[11]. There are basically two types of agent systems namely single agent and multi agent systems. The multi agent system (MAS) is also called as agent-based systems. The MAS system is a software that is made up of multiple number of agents and these systems are proven efficient to control complex environments which consist of interactive parts and also operate in dynamic environment [12].

Water distribution agent implements water distribution automation (WDA) that receives benefits from many areas such as providing a fastest method for improving reliability, efficiency. The water distribution automation is widely accepted and variety of methods are used depending on the past experience. Since the agent-based system are highly complicated, it requires a supervisory function that flows throughout the system. A part of the supervisory system controls the supervising functions of other controllers [13].

In general, various agents like monitoring agents; decision making agents are used in the water distribution system in order to perform effectively and efficiently than the sensor-based distribution system. Simple sensor-based system is deployed and wi-fi communication is used for data transfer and system control. The system is incorporated using Arduino and the distribution regions are classified using postal index codes, forecast of demand and supply [14].

In Valladolid metro city, domestic water management is implemented using a hybrid agent-based model which is used for forecasting of water requirements of the city. ANN is replaced by Agent Based Model (ABM). A general architecture is proposed for water distribution and management [15].

In the agent-based system, one of the key features used is abstraction. Although the agent-based systems may be conceptual, but these systems are implemented without the usage of any software structures. The intelligent agent-based systems do not have overall control over the system. Trust and delegation have become major considerations for its effectiveness approach. The agent types are responsive agents, proactive agents and socio agents which are used in systematic water distribution [16, 17].

54.3 Multi-Agent Based Architecture for Smart Water Distribution Management System Using IoT

The IoT based architecture for the smart water distribution and pipe health monitoring using intelligent agents is proposed. In this architecture the location based nodal agent is deployed in order to perform the online streaming analytics with greater efficiency rate. The agents are deployed at the edge/fog in order to simplify the complexity of computation at cloud level and also to reduce the computation time. The following are the major operations that are performed by the agents.

54.3.1 Pipe Health Monitoring Agents

These agents are integrated with the fog router at each and every location of the pipe network. These agents' helps to perform the pipeline health analytics based on the flow, pressure, velocity and vibration that happens inside the pipe line. The default threshold value is set by the nodal agent based on the hydraulics and fluid mechanics computational models. On the basis of the length, diameter, elevation of the pipe and head loss, the maximum threshold value is computed which the pipe can withstand breakdown, burst and explosion. The following equation is used by the location agent to calculate the total pressure flows in a pipeline.

$$\frac{1}{2}\rho v^2 + \rho gz + P = P_{total}$$

Where P = pressure, v = flow velocity, g = acceleration due to gravity, z = elevation, ρ = density. Ptotal = total pressure.

This total pressure in the pipeline is compared with the maximum pressure the pipe can withstand. Similarly, the loss of water at each junction is determined using the following.

$$L_{total} = Q_{out} - Q_{in}$$

Whereas Qin is the input flow, Qout is the output flow and Ltotal is the Total loss. This computed loss in pipe is compared with the pre predefined head loss computed as per the fluid dynamics. The Hazen-Williams equation of head loss is stated below.

$$h_l = 0.2083\left(\frac{100}{c}\right)^{1.852} q^{1.852}/d_h^{4.8655}$$

Where hl is head loss, q is volume of flow (in gal/min), c is Hazen-Williams roughness constant, dh is the internal diameter of pipe (in inches).

If the actual loss is beyond the computational value, the flow rate analyzed at the previous junctions in order to find the exact point of loss.

$$Q_{jn} = Q_{jn-1}$$

Here Qjn is referred as flow of water in unction "n" and Qjn − 1 is referred as flow of water in previous junction.

The nodal agent with rule based is deployed at the edge towards computing the pressure flow and head loss in pipe based on the equations mentioned above. These rule based agent would take real time value for pipeline in predicting the health of pipeline.

If the flow in present junction is less than the flow in previous junction, then it is a definite water loss occurred at

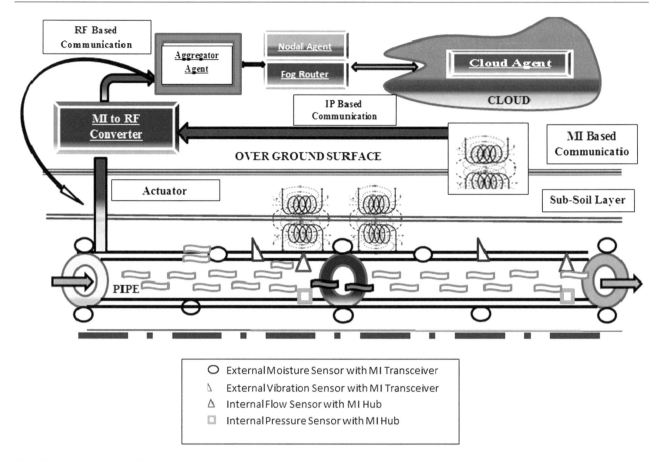

Legend:
○ External Moisture Sensor with MI Transceiver
◊ External Vibration Sensor with MI Transceiver
△ Internal Flow Sensor with MI Hub
☐ Internal Pressure Sensor with MI Hub

Fig. 54.2 Cross-section of SWDS

that particular junction. This is a continuous process happens at each and every junction point in the water distribution network.

The analyzed data will be sent by the nodal agent at the edge/fog which is deployed to collect the information and data from all the aggregator-based agents and filters the unwanted data and also forwards them to the cloud server. The illustration of the pipeline cross section built with IoT based architecture using agents for SWDS is shown in Fig. 54.2.

54.3.2 Demand Forecasting and Pricing Agent

This agent model is useful in the prediction of hourly demand of a particular location and based on the historical data of consumption data which is available in the cloud server. The intelligent prediction analysis of water demand is made in order to provide effective water supply to the consumers demand. The water distribution systems efficiency is said to be very high is it satisfies the following condition

$$T_{demnd} = T_{supply}$$

Here Tdemand refers total demand and Tsupply refers to total supply of water.

In order to achieve this efficiency, the analyzed data is uploaded to cloud server and also communicated to the control engineer assigned for that respective location. Time-series analysis of consumption data of that particular location which is present in the cloud is used for prediction analytics. For effective predictive rate, the past water supply over the years is also taken into considerations. The water consumed by a particular customer in a locality through the water meter which is fixed at the consumer end is estimated on the basis of pricing policy.

$$C_{total} = V_{observed} * C_{perunit}$$

Here Ctotal is total cost (billed amount in Rs), Vobserved is metered reading and Cperunit is cost per unit consumption (as per pricing policy).

54.3.3 Communication Agents

These agents are used to perform communication between the inter agents, consumers and local engineers. In order to communicate, the inter operable signals between the agents located at different geographical locations of a SWDS communicate billed amount for the consumption made by

Fig. 54.3 Control flow of SWDS

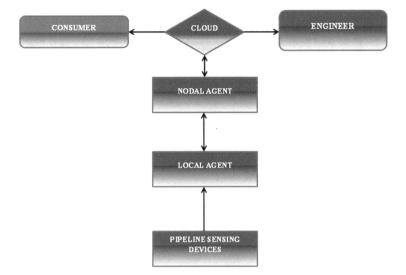

the consumers and also to keep the local engineers posted with the SWDS performance metrices. This communication between the aggregator agent, nodal agents and Cloud agent will happen through Agent Communication language. The Cloud Agent will communicate with the Consumers and SCADA Engineer. The control flow mechanism of SWDS is illustrated in the following (Fig. 54.3).

54.4 Conclusion and Future Work

A novel IOT based architecture for effective and efficient water distribution an pipeline health monitoring is proposed in this paper. This architecture is built with a combination of sensing, intelligent computational agent-based technology for online monitoring. of pipe health of a pipe network and also helps the local engineer to make a rule-based decision for continuous water supply monitoring and the issues that arise due to human errors, hydraulic factors such as ageing of pipe, corrosion, links and joints etc. Moreover, this agent based model is not only useful in monitoring the supply but also very useful in estimation of demand based on the historical data analysis, which is the prime critical section in any water distribution system. In future, this SWDS using agent can be expanded as a visualized localization of fault using mobile agents.

References

1. Jawhar, I., Mohamed, N., Shuaib, K.: A framework for pipeline infrastructure monitoring using wireless sensor networks. In: Wireless Telecommunications Symposium, 2007. WTS 2007, pp. 1–7 (April 2007)
2. Mohamed, N., Jawhar, I.: A fault tolerant wired/wireless sensor network architecture for monitoring pipeline infrastructures. In: SENSORCOMM'08: Proceedings of the 2008 Second International Conference on Sensor Technologies and Applications, pp. 179–184. IEEE Computer Society, Washington, DC (2008)
3. Mohamed, N., Jawhar, I., Shuaib, K.: Reliability challenges and enhancement approaches for pipeline sensor and actor networks. In: ICWN, pp. 46–51 (2008)
4. Murphy, F., Laffey, D., O'Flynn, B., Buckley, J., Barton, J.: Development of a wireless sensor network for collaborative agents to treat scale formation in oil pipes. In: EWSN, pp. 179–194 (2007)
5. Jin, Y., Eydgahi, A.: Monitoring of distributed pipeline systems by wireless sensor networks. In: Proceedings of The 2008 IJAC-IJME International Conference, Nashville, TN (2008)
6. Kim, J., Lim, J.S., Friedman, J., Lee, U., Vieira, L., Rosso, D., Gerla, M., Srivastava, M.B.: Sewersnort: a drifting sensor for in-situ sewer gas monitoring. In: Sixth Annual IEEE Communications Society Conference on Sensor, Mesh and Ad Hoc Communications and Networks (SECON 2009) (2009)
7. Stoianov, I., Nachman, L., Madden, S., Tokmouline, T.: PIPENET: a wireless sensor network for pipeline monitoring. In: IPSN '07: Proceedings of the 6th International Conference on Information Processing in Sensor Networks, pp. 264–273. ACM, New York, NY (2007)
8. Nassiraei, A., Kawamura, Y., Ahrary, A., Mikuriya, Y., Ishii, K.: Concept and design of a fully autonomous sewer pipe inspection mobile robot "KANTARO". In Robotics and Automation, 2007 IEEE International Conference on, pp. 136–143 (April 2007)
9. Scholl, K.-U., Kepplin, V., Berns, K., Dillmann, R.: Controlling a multi-joint robot for autonomous sewer inspection. In: Robotics and Automation, 2000. Proceedings. ICRA '00. IEEE International Conference on, vol. 2, pp. 1701–1706 (2000)
10. Franklin, S., Graesser, A.: Is it an agent, or just a program?: a taxonomy for autonomous agents. In: Proceedings of the Third International Workshop on Agents, Theories, Architectures, and Languages, pp. 21–35 (1996)
11. Jennings, N., Bussmann, S.: Agent-based control systems: why are they suited to engineering complex systems? IEEE Control Syst. Mag. **23**(3), 61–73 (2003)
12. Davis, W.: The distributed intelligent control of complex systems. In: Intelligent Processing and Manufacturing of Materials, 1999. IPMM '99. Proceedings of the Second International Conference on, vol. 1, pp. 615–621 (July 1999)
13. Vale, Z., Morais, H., Silva, M., Ramos, C.: Towards a future SCADA. In: Power Energy Society General Meeting, 2009. PES '09. IEEE, pp. 1–7 (2009)

14. Rudowsky, I.: Intelligent agents. In: Proceedings of the Americas Conference on Information Systems, Communications of the Association for Information System, vol. 14, pp. 275–290 (2004)

15. Milojicic, D.: Mobile agent applications. IEEE Concurr. **7**, 80–90 (September 1999)

16. Jennings, N.R., Wooldridge, M.: Applications of Intelligent Agents, pp. 3–28. Springer Nature, Berlin, Heidelberg (2017)

17. Carrera, A., Iglesias, C.A.: A systematic review of argumentation techniques for multi-agent systems research. Artif. Intell. Rev., Springer. **44**, 509–535 (2015)

Effects of TCP Transfer Buffers and Congestion Avoidance Algorithms on the End-to-End Throughput of TCP-over-TCP Tunnels

Russell Harkanson, Yoohwan Kim, Ju-Yeon Jo, and Khanh Pham

55.1 Introduction

In the context of computer networking, a tunnel is some virtual pathway in which some data stream is encapsulated within some outer data stream, as defined by a tunneling protocol. Oftentimes, one communication protocol will encapsulate another, but could also effectively encapsulate data of the same protocol, such as the case for IP in IP, for Internet Protocol (IP) [1, 2].

Of the many tunneling protocols that exist, none of the common ones opt for a TCP-over-TCP model. This conspicuous absence is indicative of the issues with establishing such a tunnel. In this paper, we investigate the problem with encapsulating a Transmission Control Protocol (TCP) connection within another TCP connection and the performance degradation that such tunnels suffer from. We utilize network simulation to test multiple configurations to determine which parameters could improve the performance of TCP-over-TCP communication for cases where such tunneling would be necessary.

55.2 Relevant Background

In order to understand the problem and the construction of the simulation, we must first briefly disclose some relevant background information. First, we will explain the traffic

R. Harkanson (✉) · Y. Kim · J.-Y. Jo
Department of Computer Science, University of Nevada Las Vegas, Las Vegas, NV, USA
e-mail: harkanso@unlv.nevada.edu; Yoohwan.Kim@unlv.edu; Juyeon.Jo@unlv.edu

K. Pham
Space Vehicles Directorate, Air Force Research Laboratory, Kirtland AFB, Albuquerque, NM, USA
e-mail: khanh.pham.1@us.af.mil

flow of a tunneled connection in Sect. 55.2.1. Then, we will review the relevant aspects of TCP, as originally specified by RFC [3], in the remaining subsections. This overview will explain TCP in the context of a typical untunneled, stateful connection between two peers.

55.2.1 Tunneling Traffic

Traffic tunneling allows for private communication between physically separated networks, enabling them to act as a single, larger network called a Virtual Private Network (VPN). The tunnel allows users of one network to act as virtual members of another, passing information through otherwise restrictive firewalls. After entering the tunnel, from the perspective of any hop along the route, data appears as communication between the two tunnel endpoints, emerging from the entry endpoint and destined for the exit endpoint, potentially granting some additional anonymity to the true source.

We will assume that some sender S, the source of the data, desires to communicate some information to receiver R, the sink for the data, routed over some public, untrustworthy internet network I, which could consist of multiple hops. The entry tunnel node E acts as an encapsulator that starts to tunnel the data it receives, and the exit tunnel node D is the decapsulator that retrieves the original data and continues routing as usual. S, R, E, and D are arbitrarily chosen, as there exists a symmetry in the model dependent upon the direction of data travel. For the remainder of this writing we will assume that data is traveling from left-to-right, S to R, as shown in Fig. 55.1 for untunneled and TCP-over-TCP tunneled connections.

As a packet traverses the networks of nodes, it goes through multiple key states. Through each of the untunneled traffic states, labeled p, $\hat{p}$, and $p\prime$, the packet is largely unchanged as header field values naturally change along

© Springer Nature Switzerland AG 2019
S. Latifi (ed.), *16th International Conference on Information Technology-New Generations (ITNG 2019)*, Advances in Intelligent Systems and Computing 800, https://doi.org/10.1007/978-3-030-14070-0_55

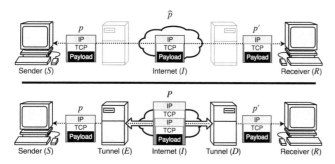

Fig. 55.1 Packet structure between untunneled (top) and TCP-over-TCP tunneled (bottom) connections

the route, such as the Time to Live (TTL) value getting decremented at each hop. For tunneled connections, the data goes through states p, P, and $p\prime$, where P has an additional set of IP and TCP headers between tunnel points, indicative of the new TCP session between the two tunnel endpoints. All of p becomes the payload, including the IP and TCP headers.

If we consider state p at the moment before a packet enters the tunnel, and $p\prime$ to be the state at the moment after it exits, it would be true to say $p = p\prime$ since field values of the inner IP and TCP headers from the original connection are left untouched while being tunneled. Therefore, for the duration of the tunnel, the original packet is essentially frozen in time, retaining the state of field values from before the tunnel was entered.

In addition to providing a virtual route, the contents of the original data stream are often encrypted. Any communication between the private networks of S and R would then be opaque to routers between the tunnel's entry point and exit point at I. Sensitive data could then be shared between the two networks while remaining effectively unreadable by public hops on the greater internet along the way.

55.2.2 TCP Sequence Numbers

For S to initiate a TCP connection with R, a three-way handshake needs to take place between the two. Each endpoint randomly selects an initial sequence number offset, which is shared with the other endpoint during the handshake, and acts as an offset for byte index values. The sequence numbers are used to ensure each byte is received and is in the correct order. This is achieved by TCP header fields that report the sequence number of the first byte of the segment payload as well as the total length of the payload in bytes [4].

After a successful connection is initiated and S starts to send some data segment, R is able to validate the reception of bytes by acknowledgment, requesting the first byte sequence number not yet received. This indicates to S that R has received every byte up until but not including the acknowledged sequence, assuming all of the bytes previously

were received. If some bytes were not received, R could resend an acknowledgment for the sequence number of the first missing byte.

55.2.3 TCP Buffers

In order to ensure byte sequence, each endpoint manages a pair of ring buffers, one responsible for sending and another for receiving bytes. The transfer buffers have some fixed number of bytes they are able to store. Sequence numbers are used to index which position in the buffer the bytes are stored at, wrapping around in the fashion of modular arithmetic when reaching the end. Bytes are added to their correct position in the buffer according to the sequence number indexing and are only able to be dequeued starting from the lowest sequence number the buffer addresses.

After the buffers are filled to capacity, they must deny and drop any incoming bytes until the lowest sequenced bytes are able to be popped off and more space becomes available. At that point, the base sequence number indexing the lowest byte can be increased to the new lowest sequence number allowing for higher sequenced bytes to be stored. This process is outlined in Fig. 55.2, which depicts the events of bytes being acknowledged for S as bytes are about to be delivered to R.

Bytes sent from the upper application layer of S are queued in the send buffer in the order that they are intended to be received by R. S is able to send the sequenced bytes to R, but they are not removed from the buffer until an acknowledgment has been received for them. When S receives an acknowledgment, it is permitted to remove all bytes in its send buffer up until but not including that byte.

Likewise, R is able to enqueue received bytes in its receive buffer. When bytes are received, they are placed in the buffer starting from the sequence number listed by the segment header and then send an acknowledgment for the sequence number of the first byte not stored. Bytes are rolled off from the receive buffer when the upper layer application takes custody of them.

Only contiguous bytes starting from the lowest sequence number are able to be dequeued. If a buffer is at capacity, incoming bytes must be dropped. This end-to-end buffer index sequencing guarantees the reception of each byte in the correct order, allowing TCP to be employed on a lossy channel turning it into a virtually lossless channel.

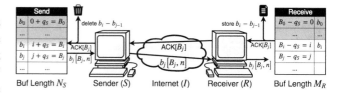

Fig. 55.2 TCP sequences and transfer ring buffers

55.2.4 TCP Congestion Control

Under normal conditions, the congestion control of TCP attempts to maximize the throughput of the connection while minimizing the number of dropped bytes. There are different algorithms for managing congestion, each with slightly different approaches to achieve desirable outcomes. Depending on the operating system and settings of a device, different algorithms may be used. We will be primarily focused on the New Reno and Vegas congestion control schemes.

Congestion control arranges sliding Congestion Windows (cWnds) that dictate how many bytes a node is able to store. This is an imposed limitation as the true maximum is dictated by the size of the buffer, which is often, but not necessarily, fixed per connection [5]. By keeping track of the state of the cWnd, while measuring parameters like Round-Trip Time (RTT) and throughput, the algorithm changes and reports to the other endpoint the size of the window. With this information continually being passed back and forth, the endpoints are able to compute the best number of bytes to send at each moment in order to maximize throughput [6–8].

The congestion control is able to monitor byte loss by monitoring sequence acknowledgment. If duplicate acknowledgments were sent the congestion control can tell that some sequence must have been dropped. If loss was observed the window size will be decreased depending on which congestion state the session is in. If no loss was detected, the receiver could increase the window size to ask for more bytes and continually increase until some loss does occur. The states defined by [9] are slow start, congestion avoidance, fast retransmission, and fast recovery along with treatment for retransmission timeouts. Figure 55.3 portrays a typical evolution of a congestion window's advertised size over time, with transitions into the various states.

TCP New Reno
Congestion avoidance defined by [10] improves upon the older spec TCP Reno. Unsent bytes are allowed to be sent during the fast recovery phase, allowing for more sequence numbers to be filled in.

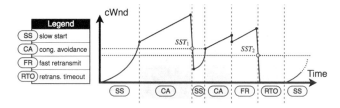

Fig. 55.3 Typical TCP congestion window size over time

TCP Vegas
Tanking a slightly different approach, TCP Vegas measures the RTT for every packet in the transmission buffer, rather than just the last. The window is increased or decreased linearly depending on the expected and actual sending rates [11, 12].

55.3 Problem Description

While investigating solutions for improving the performance of High Assurance Internet Protocol Encryption (HAIPE) encrypted TCP traffic for [13], due to the constraints of the problem necessitating TCP-over-TCP traffic, we quickly discovered the TCP meltdown problem. IPsec encryption applied by the HAIPE obfuscates the underlying TCP header information, and thus congestion control parameters. The Performance-enhancing Proxy (PEP) devices for our situation only recognize this traffic as IP and therefore cannot improve the performance of communication. Our proposal is to rebuild the TCP session on the other side of the HAIPE, after encryption, to allow for the benefits of the PEP. However, wrapping the packet in a rebuilt TCP session is effectively creating a tunnel and hence is prone to TCP meltdown.

TCP meltdown, while not yet thoroughly researched, expresses itself as the degradation of TCP-over-TCP tunneling performance at best, to the complete abortion of the tunneled connection at worst. TCP meltdown, as originally explicitly refered to and explained by [14], is the result of an understandable oversight in the development of TCP, specifically pertaining to the in-built congestion control. Meltdown arises when the sender and receiver have lossless channel between them and TCP is employed upon it.

55.3.1 TCP Meltdown

Consider the following situation: Sender and receiver each have lossless communication with their tunnel endpoint. If the tunnel endpoints are one hop away, such as the first network router, or even zero hops away, running on the same machine, the channel is essentially lossless. That means nearly all of the packet drops occur between tunnel endpoints. Recall, this is a TCP-over-TCP tunnel, so the two tunnel endpoints share a TCP connection used to deliver packets they are tunneling from a TCP session between the sender and receiver. The congestion control from the tunnel's TCP session will manage any loss that occurs, making the channel virtually lossless. With a virtually lossless channel between tunnel endpoints, in addition to true lossless channels in between sender and encapsulator as well as receiver and decapsulator, there now exists a lossless channel, end-to-end, between the sender and receiver that TCP is deployed upon.

Congestion control managing the end-to-end session will not observe any loss since the channel is lossless. This would mean that the receiver will continually increase its window size, asking for more and more bytes from the sender. Without any monitored loss, the window will theoretically increase monotonically. Even when the receiver is demanding more bytes than the physical bandwidth could route, every one of them will eventually arrive since there is no point they could have been dropped, which the congestion control algorithm for the receiver reads as a route that is capable of handling all of those bytes and will then ask for more in its attempt to maximize throughput.

Meanwhile, the outer TCP session between the tunnel endpoints has to manage the increasing demand of inner session. The outer session behaves like a normal TCP session and will attempt to manage congestion to maximize throughput. Practically, routers along the physical path of travel have some bandwidth threshold and will drop any traffic they cannot queue. The tunnel session will drop more packets in this way as the demand from the inner session increases. They will get recovered by congestion control at the cost of remaining in their transfer buffers for a longer time. During this process, the RTT of the inner session will continue to increase. Eventually, the rate at which the buffer is being filled overtakes the rate at which it is being emptied.

At some point, the tunnel buffers will completely fill up and they will be forced to drop any incoming bytes, as shown in Fig. 55.4. It is at this point that the inner session finally observes some packet loss, missing an acknowledgment for the first byte that could not make it into the tunnel's buffer, followed by any other bytes that were unable to be queued. The inner session will now decrease window size and begin to recover as the tunnel session gets to empty its buffer [15].

In general, the problem is that TCP assumes to be deployed upon a lossy channel. When TCP is placed on a lossless channel, congestion control will continue to increase the throughput, thus putting a larger and larger strain on the networks carrying the tunnel. Since the traffic of TCP is ensured through congestion control, an inner TCP session exists within a lossless channel, virtually.

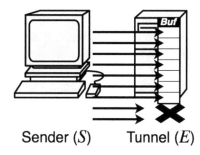

Fig. 55.4 Tunnel buffer cannot fit more traffic and must drop

55.4 Related Work

TCP meltdown was studied next by [16]. It was found that the use of Selective Acknowledgment (SACK), a TCP option initially proposed by [17], lessens the degradation of meltdown. SACK allows for a receiver to enqueue an out-of-order segment to its buffer if the sequence numbers are available to be written within the buffer. The receiver's next acknowledgment to the sender can include the SACK option indicating that acknowledgment of the future byte range. With this mechanism in place, if the sender needs to return to a sequence number that had a duplicate acknowledgment, instead of just starting over from there the sender can just send up until the selectively acknowledged ranges that it knows it can skip.

When the tunnel buffer finally drops a byte, especially considering the prolonged RTT previously mentioned, there will be a longer wait than the usual untunneled case. The delay is due to the fact that the tunnel still needs to empty its send buffer with all of the bytes that made it on after the drop but before the inner connection was notified. With SACK in place, those overshot bytes can still be stored by the receiver and selectively acknowledged. The reduced window size at this point would mean that the RTT is now fast enough again for the sender to recognize those SACKs quickly enough to respond to them. This decreases the overall amount of data needed to be sent and therefore decreases the unnecessary additional demand on the tunnel had the session disabled SACK.

Compeltely avoiding the TCP-over-TCP situation was suggested by [18, 19], opting to tunnel over User Datagram Protocol (UDP). UDP is a simple, stateless protocol with no form of congestion control or retransmission. If a packet gets dropped, it is up to the upper application layer to handle the data loss, making the protocol popular for streaming where the integrity of every byte is not required. If necessary bytes were dropped, the application needs to manually recover the lost data. A TCP-over-UDP tunnel does not face meltdown because any lost UDP packet will be discovered and corrected for by the underlying TCP session. This makes TCP-over-UDP a popular choice for tunnels and VPNs like OpenVPN.

55.5 Simulation

To investigate TCP meltdown in TCP-over-TCP tunnels, we constructed a network simulation to observe the effects that different parameters have on the performance of the end-to-end connection. We have chosen Network Simulator 3 (ns-3), version 3.28, due to its open source codebase which we needed to heavily modify to allow for our experiments [20].

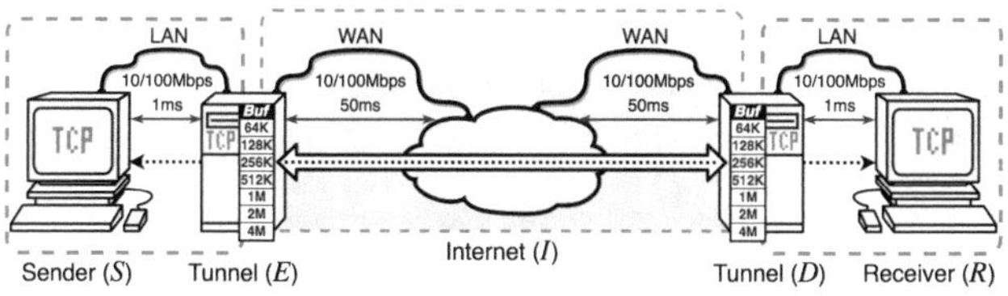

Fig. 55.5 Simulation network node and parameter model

The major modification to ns-3 was the addition of the TCP Tunnel Application class and various companion classes needed for tunneling. A new IP routing protocol class had to be added to inspect traffic being routed through the node and decide to capture it for tunneling. This was necessary to start a new tunnel session for some sender-receiver pair which is uniquely identifiable by the sending IP and port as well as the recipient IP and port. A new TCP option was added to relay metadata from one tunnel endpoint to the other, which necessitated multiple modifications in multiple TCP-related class files to handle the change. When a connection is chosen to be tunneled, the TCP Tunnel Application will create a new TCP Tunnel Session object which will manage the data being encapsulated by keeping track of which byte offsets start newly tunneled packets then relaying that information to the decapsulator via the new TCP option to reconstruct the tunneled traffic.

The network model consists of 5 ns-3 nodes in the configuration of Fig. 55.5, all with an internet stack to allow them to route IP and TCP traffic. The links between each of the nodes are modifiable, allowing for different bandwidths, delays, and drop rates. The sender node S has a slightly modified Bulk Send Application installed to continuously send data to the receiver node R, which has a Packet Sink Application installed to acknowledge reception. The encapsulator E and decapsulator D both have the new TCP Tunnel Application installed to do the actual tunneling. The tunnel endpoints can also be set to UDP mode or completely disabled, acting as normal routing hop, for experimental control cases.

55.5.1 Parameters

Through simulation we test every combination of 6 different parameters for a total of 480 different configurations, 448 test conditions and 32 experimental controls. Each of the 6 parameters can only be in one value state for a single run of the simulation. All of the possible values for each parameter is listed in Table 55.1.

LAN B/W
The bandwidth of both local area networks can be either 10 or 100 Megabits per Second (Mbps). This parameter sets the

Table 55.1 Simulation configurations for experimentation

Parameter	Value						
LAN B/W	10 Mbps			100 Mbps			
WAN B/W	10 Mbps			100 Mbps			
Drop rate	0.1%	0.01%		0.001%		0%	
Internal algo.	New Reno			Vegas			
External algo.	New Reno			Vegas			
Tunnel buffer	64K	128K	256K	512K	1M	2M	4M

Table 55.2 Constant simulation parameters

LAN latency	WAN latency	MTU	Duration
1 ms	50 ms	1500 bytes	300 s

bandwidth for both, the sending network and the receiving network, at the same time.

WAN B/W
The bandwidth of the wide area network can be either 10 or 100 Mbps. This parameter sets the bandwidth of links between the two tunnel endpoints.

Drop Rate
The drop rate is the expected percentage of packets that will randomly be dropped at the internet node. The listed value affects both net device ports of the node, therefore
$$p_{drop-true} = 1 - (1 - p_{drop-listed})^2$$

Internal Algo
Both, the sender and receiver nodes share a TCP connection that can either have a New Reno or Vegas congestion avoidance algorithm.

External Algo
In addition to the congestion avoidance algorithm of the end-to-end connection, the external connection between the two tunnel nodes can be either New Reno or Vegas as well. This parameter is not necessary for the control cases where the tunnel is deactivated, so the nodes just route traffic like normal.

Tunnel Buffer
Tunnel nodes will not accept any data beyond the size of their buffer. When the tunnel is deactivated, this parameter does not apply.

In addition to these variable parameters, a few parameters we kept constant throughout the course of simulation, listed in Table 55.2. Notice, the minimum RTT from sender to receiver and back would be $2 * (1 + 50 + 50 + 1) = 204$ ms.

55.6 Findings

For each configuration of the simulation that we ran, we measured the overall goodput, which is the end-to-end throughput of application data, not including packet headers or bytes added by tunnels. To do this, the receiving node counts all of the bits it receives and divides over the duration of the simulation. To normalize the data collected, the first 20 s were ignored for bit counting and for finding the mean. This way, any chaotic variations associated with initiating both TCP sessions is completely ignored and the goodput is only measured mid-stream, after an established connection. Since

the sender just sends as much non-stop data as it can for the duration of the simulation, no time needs to be cut off of the end.

55.6.1 Congestion Avoidance Algorithm Choice

Four possible combinations of TCP congestion avoidance algorithms have been tested, so that the endpoints and the tunnels are using either TCP New Reno or TCP Vegas. Figure 55.6 contains the performance results for all four algorithm combinations in Mbps. Each sub-figure has 4 graphs,

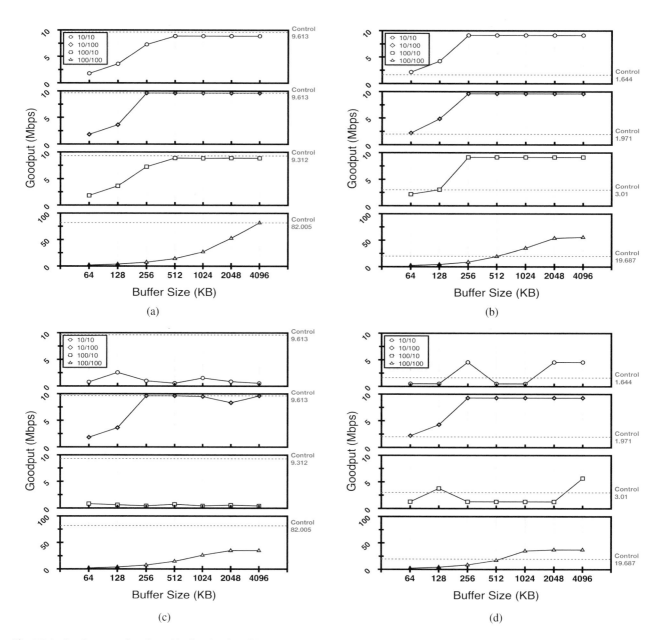

Fig. 55.6 Goodput as a function of buffer size for all bandwidth and TCP congestion algorithm combinations. (**a**) New Reno over New Reno Tunnel. (**b**) Vegas over New Reno Tunnel. (**c**) New Reno over Vegas Tunnel. (**d**) Vegas over Vegas Tunnel

one for each possible combination of network bandwidths, with the LAN or WAN being 10 or 100 Mbps. Lastly, each graph plots the overall goodput of application data in Mbps for each of the 7 possible TCP buffer sizes.

The first group of cases depicted in Fig. 55.6a employs the New Reno algorithm on both end-to-end and tunnel nodes. In these cases the tunnel performed best overall, with the ability to match the control performance of 82 Mbps with a 4 MB buffer for all network bandwidths at 100 Mbps, and match the control for even smaller buffers at lower bandwidths.

The next set of configurations in Fig. 55.6b equip the sender and receiver with TCP Vegas while maintaining New Reno on the tunnel nodes. While this case has the best performances relative to its baseline control case, which is an end-to-end Vegas connection, it did not perform as well overall, as New Reno over New Reno did. The tunnel was able to outperform the control case due to the faster New Reno connection of the tunnel. As Vegas responds to the observed RTT, we would be able to vary the delays of the simulation for future work to see how the overall performance gets altered.

Figure 55.6c, a New Reno endpoint pair through a Vegas tunnel, performed the worst overall. The New Reno connection is hindered by the RTT-following Vegas connection hosting it over the majority of the round-trip delay. When the WAN had a greater bandwidth of 100 Mbps, the connection performed much better, while the LAN bandwidth increase to 100 Mbps from 10 hindered the overall goodput.

The final set of cases in Fig. 55.6d use TCP Vegas for both connections. Again, the tunneled traffic performance was able to outperform the untunneled performance, most evident for 10 Mbps LAN over a 100 Mbps WAN. The performance boost is likely due to the tunnel Vegas connection evaluating its own RTT and optimizing its bandwidth providing a faster tunnel for the inner connection. In addition, the slower LAN implies that the tunnel's buffer would not fill as quickly. A bandwidth of 100 Mbps performed the best overall for this scenario.

55.6.2 Buffer Size Selection

The overall goodput is also affected by the size of the buffer. From Fig. 55.6 we are also able to observe the goodput as the buffer size increases. In general, larger buffer sizes afford greater goodput. In the New Reno tunnel cases where the line bandwidth is capped at 10 Mbps, we see that the small buffer sizes are still the bottleneck of the overall goodput. As soon as the buffer size exceeds 512 MB, it is large enough to handle traffic at those speeds. It is not until the case where the 10 Mbps bottleneck is increased to 100 Mbps that the buffer becomes the bottleneck once more. Only the 4 MB buffer was

able to meet the demands of a 100 Mbps line and put out the theoretical goodput of the control.

55.7 Conclusion

We have studied TCP-over-TCP tunneling for a variety of different cases, spanning over 448 different network configurations. Our focus primarily pertained to the effects of the buffer size and the TCP congestion avoidance algorithm on the overall goodput of said tunnels. We have found that the best results, overall, come from a TCP New Reno connection tunneled over a TCP New Reno tunnel. The bandwidth between tunnel endpoints is of greater priority than the bandwidth of the LAN due to the increased overhead of encapsulation and urgency of maintaining an open buffer. Larger buffer sizes approach the theoretical control goodput of an untunneled connection. Starting from a basis of a 512 KB buffer for a 10 Mbps connection, we recommend that the buffer size should double as the average bandwidth doubles to maintain ideal goodput.

Acknowledgements This research has been supported by 2018 Air Force Research Laboratory Summer Faculty Fellowship Program.

References

1. Abu-Amara, M., Asif, M.A.K., et al.: Resilient internet access using tunnel-based solution for malicious ISP blocking. In: 2011 IEEE 3rd International Conference on Communication Software and Networks, pp. 85–89 (2011)
2. Woo, M., Lee, H., et al.: Evaluation of UDP tunnel for data replication in data centers and cloud environment. In: 2010 6th International Conference on Wireless Communications Networking and Mobile Computing (WiCOM), pp. 1–4 (2010)
3. Information Sciences Institute: Transmission control protocol, RFC 793, Sept 1981. [Online]. Available: https://tools.ietf.org/html/rfc793
4. Kozierok, C.M.: The TCP/IP Guide: A Comprehensive, Illustrated Internet Protocols Reference, 1st edn. No Starch Press, San Francisco (2005)
5. Goel, V., Aggarwal, D.S., Nirwan, A.: System for dynamic configuration of TCP buffers based on operator. In: 2014 International Conference on Advances in Computing, Communications and Informatics (ICACCI), ser. ICACCI'14, pp. 2036–2040. IEEE (2014)
6. Kurose, J.F., Ross, K.W.: Computer Networking: A Top-Down Approach, 6th edn. Pearson, Boston (2012)
7. Jacobson, V.: Congestion avoidance and control. In: Symposium Proceedings on Communications Architectures and Protocols, ser. SIGCOMM'88, pp. 314–329. ACM, New York (1988)
8. Adeel, M., Iqbal, A.A.: TCP congestion window optimization for CDMA2000 packet data networks. In: Fourth International Conference on Information Technology (ITNG'07), ser. ITNG'07, pp. 31–35. IEEE (2007)
9. Allman, M., Paxson, V., Blanton, E.: TCP congestion control. RFC 5681, Sept 2009. [Online]. Available: https://tools.ietf.org/html/rfc5681

10. Henderson, T., Floyd, S., et al.: The newreno modification to TCP's fast recovery algorithm, RFC 6582, Apr 2012. [Online]. Available: https://tools.ietf.org/html/rfc6582

11. Jamal, H., Sultan, K.: Performance analysis of TCP congestion control algorithms. Int. J. Comput. Commun. **2**, 30–38 (2008)

12. Abed, G., Ismail, M., Jumari, K.: Characterization and observation of (transmission control protocol) TCP-vegas performance with different parameters over (long term evolution) LTE networks. Sci. Res. Essays **6**, 2003–2010 (2011)

13. Kim, Y., Jo, J., et al.: TCP-GEN framework to achieve high performance for HAIPE-encrypted TCP traffic in a satellite communication environment. In: 2018 IEEE International Conference on Communications (ICC), pp. 1–7 (2018)

14. Titz, O.: Why TCP over TCP is a bad idea. [Online]. Available: http://sites.inka.de/bigred/devel/tcp-tcp.html (2001)

15. Ali, S.H., Nasir, S.A., Qazi, S.: Impact of router buffer size on TCP/UDP performance. In: 2013 3rd IEEE International Conference on Computer, Control and Communication (IC4), ser. IC4'13. IEEE, pp. 1–6 (2013)

16. Honda, O., Ohsaki, H., et al.: Understanding TCP over TCP: effects of TCP tunneling on end-to-end throughput and latency. In: Proceedings of SPIE, vol. 6011, pp. 138–146 (2005)

17. Mathis, M., Mahdavi, J., et al.: TCP selective acknowledgment options. RFC 2018, Oct 1996. [Online]. Available: https://tools.ietf.org/html/rfc2018

18. Coonjah, I., Catherine, P.C., Soyjaudah, K.M.S.: Experimental performance comparison between TCP vs UDP tunnel using openvpn. In: 2015 International Conference on Computing, Communication and Security (ICCCS), pp. 1–5 (2015)

19. Soyjaudah, K.M.S., Catherine, P.C., Coonjah, I.: Evaluation of UDP tunnel for data replication in data centers and cloud environment. In: 2016 International Conference on Computing, Communication and Automation (ICCCA), pp. 1217–1221 (2016)

20. ns 3: Network simulator 3. [Online]. Available: https://www.nsnam.org/ (2018)

Sung Hoon Park and Yong Cheol Seo

56.1 Introduction

According to the development of sensor technology and ubiquitous networking, autonomous vehicles (AVs) are moving around the road and will be ready for public use in the near future. With the advent of such AVs, the concept of Autonomous Intersection Management (AIM) without traffic light has attracted great attention over the last decade and will be a promising option. In the scheme of AIM, AVs communicate with the intersection controller to exchange the information of their states. Also, AVs are commanded and coordinated by an Intersection Control Unit (ICU) to cross the intersection safely, as the vehicles can control their states accurately by the sensors mounted in them. Live AV data are collected by sensors and sent to the ICU in real time. The right-of-way is assigned to each AV according to its state and the state of the intersection. Only AVs that have the right-of-way can pass through the intersection. AIM aims at calculating the optimized trajectory and determining the intersection passing sequence with the collected information for which the ICU will continue to share information with nearby AVs.

However, to implement the AIM scheme which is based on the ICU as a central control equipment in the intersection, we may face big problems as follows:

First, the scheme requires ICU installation work which is costly and once installed, the design is not flexible (i.e. it is not easy to change or improve).

Second, the serious problem is that if an ICU fails, the entire AIM system will not work and can cause tremendous disruption.

S. H. Park (✉) · Y. C. Seo (✉)
Department of Computer Engineering, Chungbuk National University, Chungju-shi, Chungbuk, South Korea
e-mail: spark@cbnu.ac.kr; ycseo72@naver.com

In order to solve these problems of the existing AIM, the scheme of AIM should be developed based on the distributed computing system. In other words, in order for an AV to safely pass through an intersection without an ICU, it is necessary to thoroughly collaborate through real-time information exchanged between AVs.

However, to the best of our knowledge there have been no studies on the decentralized distributed systems for AIM(D_AIM) except the algorithm of [1–3] and our earlier work of *VTokenIC*, 2017.

In this paper, we design an efficient token-based algorithm for D_AIM that is especially suitable for applications in which group selection probability may be non-uniformly distributed. There is only one token in the intersection and the system proposed consists of continuous flow of sessions.

The rest of this paper is arranged as follows. In Sect. 56.2, we describe the system model assumed and detailed environmental factors of the system. In Sect. 56.3, we present the proposed algorithm for the distributed AIM. In Sect. 56.4, we explain performance analyses of the proposed algorithm. Finally in Sect. 56.5, we give concluding remarks.

56.2 System Model and Definitions

56.2.1 System Model

We assume an asynchronous message-passing distributed system comprising a set of n AVs. Generally, AVs are to continuously arrive at an intersection and leave it and the number of AVs in the system changes all the times. However, at a particular time, there are only fixed number of AVs that are moving in the system. Also, the traffic flow can be truncated into segments by lanes, which simplifies the problem. At that time, there are n AVs, AV $= \{AV_1, AV_2, \ldots, AV_n\}$ in the control range (i.e. system) approaching an intersection from different approaches and passing through the intersection; the essential data of all AVs will be communicated with each other by sending messages over a set of channels.

The mobile ad hoc network model is defined as an undirected graph, G = (V, E). Each vertex of a set of vertices, V = {v_1, v_2, ..., v_n} (n ≥ 1), represents a mobile node.

Information exchange between AVs is done by asynchronous message passing. Further, we assume each vehicle has a unique ID and all the channel are reliable and a network failure does not occur.

56.2.2 Group Mutual Exclusion Problem

In a traditional mutual exclusion theory, only one process can access the critical section at any moment. However, the intersection control problem in our algorithm proposed is based on group mutual exclusion theory. AVs at a same group can access the critical section (i.e. pass the intersection as a group) simultaneously.

The GME problem in the algorithm proposed is to hold the following properties;

1. **Safety**

If two or more AVs are in the critical section at the same time, then they are in a same group.

2. **Liveliness**

AVs that enters the system and requests intersection passing must enter the intersection of the system.

3. **Concurrent entering**

If requests of all AVs belong to the same group, they must not be asked to wait until another AV leaves the intersection.

56.2.3 Intersection

As shown in Fig. 56.1, we assume an intersection with four directions, i.e. north, south, east and west. In each direction, there are two lanes. Typically, an intersection (IC) consists of a number of approaches and a crossing zone (CZ). Each approach may be used by several traffic streams. For example, for the IC in Fig. 56.1, the approach from west to east consist s of two traffic streams (IL 3 and 4). Each stream has its own lane and an independent vehicle queue. The path used by a traffic stream to traverse the IC is called the trajectory. A trajectory connects an approach on which AVs enter the IC to the IC leg on which these AVs leave the IC.

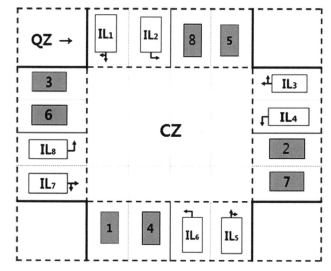

Fig. 56.1 Intersection

The objective of AIM is to transform input traffic flows into output ones while preventing traffic conflicts and satisfying a specific criterion [2–10].

The dashed range represents the IC and all AVs inside the range will be considered as a member of the system to take part in cooperative driving; whereas AVs outside of the circle will not be considered temporarily. The radius of this virtual range should be determined appropriately by inter-vehicle communication protocol that has been selected for this application.

Definition 1: compatible and conflict relationship
When trajectories of two traffic streams do not cross, these streams can simultaneously get the right-of-way and we call these two streams compatible stream. The lanes on which the two streams are moving are called compatible lanes. On the other hand, when trajectories of two traffic streams do cross, the streams are in a conflict relation.

Definition 2: compatible stream group (CSG)
When several stream are compatible with each other, we call the set of these streams a compatible stream groups (CSG). In the system proposed, we can partition the eight streams into four CSGs as below:

(CSG 1 – L1 & L5), (CSG 2 - L2 & L6),
(CSG 3 – L3 & L7), (CSG 4 – L4 & L8)

Definition 3: vehicle passing group (VPG)
A large number of AVs may be waiting in the lane to pass through the CZ. If each AV act and enter the crossing area individually, the performance of the system will be decreased. We therefore define VPG as a set of vehicles from the same stream that pass the CZ without the interruption of

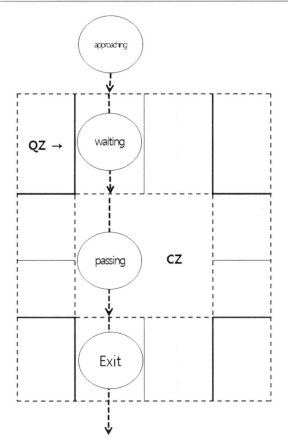

Fig. 56.2 The state of AV_i

the other AVs in the IC. Eight VPGs can be generated in the IC based on each traffic stream. AVs in stream 1 is VPG 1, e.g. AVs in stream i is VPG i. Each VPG consists of AVs in the same lane and the head and tail of the VPG is decided based on the arrival time. Each AV should be in one and only one VPG.

56.2.4 The State of AV_i

The movement of AV_i passing through IC can be described by an automaton with three states, as shown Fig. 56.2.

- *Idle*

At this step, as AV_i is out of the IC, it is not controlled by the system.

- *Waiting* state

The state is that AV_i waits for the CZ passing and AV_i is in the state from when entering the IC to when entering the CZ.

- *Passing* state

The state is that an AV moves to pass the CZ. AV_i will be in the state during the time interval between entering the CZ and exiting it.

56.3 The Algorithm: *TDGim*

In this section, we describe a token-based GME algorithm for the autonomous intersection management (AIM) and we call it *TDGim*. *TDGim* is a distributed AIM in which all AVs in the system collaborate with each other only through message exchanges between vehicles and traverse through the intersection safely without a traffic light.

56.3.1 The Main Idea

TDGim is a token-based GME algorithm in which the token holder manage a session. The token mentioned here means a message with special authority and holding the token means that it is qualified to pass through IC. A single token is used for the system and continuously rotated. The token has a group of the session associated with it and it can only be used to enter the CZ of IC of that type. The session is time interval during the token holder and the all AVs in the same VPG with the token holder enter the CZ and pass through it.

An AV, if it wants to access a session, on receiving the token, initiates the session and also declares the session to all other AVs in the IC. Any AV that listens to this session can enter the CS for the same session concurrently. In other words, an AV, when it wants to access the CZ for a particular session, checks in the entry section whether the session is available. If it is, the AV can join the session concurrently. On the other hand, if an AV wants to access other session, it sends a request to the token holder for the token. After the current session is over, the current token holder sends the token to the requesting AV.

Thus, *TDGim* is the algorithm for a continuous session flow in which the token holder is changed and thereby the group of AVs to be entered the CZ is changed (shown in Fig. 56.3).

56.3.2 Description of TDGim

1. Waiting section
   AV_i approaches to the IC if it is interested in the CZ entering. AV_i is in *Waiting section* when it is located inside the circle of the IC. AV_i in *Waiting section* should wait for the CZ entry in the lane and it may receive the token or the session declare message or a notify message.

Fig. 56.3 Session flow

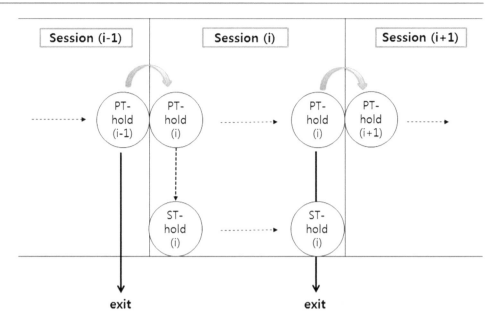

(a) Receiving a session declare

AV$_i$, on receiving a session declare from the *key AV* in *Waiting Section*, updates the variables associated with the session group and ID of the *key AV* and should check the value of the declared session. There are two cases depending on the group of the session as follows.

1. If the value of group of the session declared is same with that of AV$_i$, it can join the session concurrently.
2. If it is not, AV$_i$ should send a *request* message to the *key AV* for the token. The request message contains ID of AV, the value of the session requiring and the current time.

(b) Receiving a *notify* message

AV$_i$, on receiving a *notify* message from the temporal token holder, if it did not receive a *declare* message or another *notify* message yet, sends a *request* message to the temporal token holder. (tph enqueue it to token Q). However if AV$_i$ already received a *declare* message or another *notify* message, it ignores the *notify* message received afterwards.

(c) Generating a token

AV$_i$ generates a token only when two conditions below are met at the same time;

AV$_i$ arrives at the CZ entry border.

AV$_i$ does not receive a declare message or a notify message before the timeout *tmt* occurs.

All AVs that satisfy the conditions above should broadcast a *location* message. If the number of AVs are more than two, the token generator will be decided as below;

AV$_i$ that is located in the stream having the lowest number generates a token and declares the session.

2. Processing Section

Processing Section is the procedure in which *AV$_i$* in *waiting* state enters the CZ and passes through the IC. *AV$_i$* can enter the CZ only if it receives the token or listens to the session declare.

(a) *Receiving the token*

AV$_i$, on receiving the token, starts the procedure of the CZ entering as follows:

1. Session declare

On receiving the token, *AV$_i$* declares the session *session$_i$* by a *declare* message to all other AVs in the IC. We call *AV$_i$* declaring the session *key AV*.

The *declare* message is the form of (ID, PG) where ID is the AV's ID and PG is the passing group of the session.

On the other hand, if the tail of VPG$_i$ receives the token temporarily, broadcasts its position by a *notify* message.

2. Safety check

AV$_i$, on receiving the token from the previous token holder *AV$_j$*, may not be able to use the token immediately. This is because AVs belong to the previous session may not be able to exit the CZ yet. Certainly, *AV$_i$* should check the situation of the CZ before entering the CZ and wait until those AVs belong to the previous session have been released out before it can use the token to enter the CZ.

In *TDGim*, to check the safety of the session start, we utilize a sequence number *session$_i$* that represents the session to which the token belongs.

The sequence number is incremented by one when the token holder transfers the token to the next holder. AVs, on releasing the CZ, piggyback the number of the session it belongs on the *release* message it sends. AV_i records the number of the *release* messages it has received containing the most recent session sequence number (given by variables $noOfRelease_i$ and $session_i$). Further a token contains the number of *<session-declare>* messages that were sent for the previous session (given by variable $oldNoOfDeclare$ in the token). To decide whether the CZ for the new session is safe, AV_i evaluates the following condition [12–15]:

$$(noOfRelease_i = holder_i.oldNoOfDeclare)$$

AV_i, on completing the preparation procedure of the CZ entering, goes into the CZ immediately and switches to *passing* state from *waiting*.

(b) *Receiving a request message*

When the *key AV* or the temporal token holder AV_i receives a *request* message from other AV AV_j, it updates the $tsReq[i]$ in the token and enqueues the request of AV_j into the $tok_i, reqQ$ attached to the token.

(c) Receiving a declare message

If AV_i does not get the token, it may receive a *declare* message from the token holder during the waiting time. On receiving the message, it is able to perceive the VPG of the new session and the ID of the *key AV* and records those information received. Any AVs receive the message can enter the CS for the same session simultaneously.

Note that, only the AV knowing the type of the session can join the session (i.e. can enter the CS). Therefore, other AVs that does not receive the *declare* message cannot join the session at any case and wait.

3. *Exit Section*

Exit Section is the procedure in which AV_i exits the CS after passing through the CS. It is an critical process to decide the new leader of the next session and the performance of the system can be increased by designing the token rotating effectively. There are two cases as below.

(a) *Case 1:* AV_i holds the token.

AV_i, when exiting the CS, if there is a vehicle in VPG_i, transfers the token to the tail of VPG_i, and if not, it transfers the token to the tail of $VPG_{(i+4)}$. On completing the CZ passing, it will be in contact with QBC and receive the boundary information from the sensor. Then, it shall proceed the CZ exit procedure as follows.

1. AV_i, as the leader of the current session $holder_i$, should select an IL in which an AV becomes the

token holder of the next session. The IL for the new session is selected according to the criterion described below and we call the IL selected for th next session NIL.

Criterion for selecting NIL

In *TDGim*, each request is enqueued to the queue $token\text{-}queue_i$ attached to the token, i.e. all the requests are centralized at the token. So we define various NIL selection schemes on the queue of the token.

We can define various selection schemes in the priority $P(IL_i)$ for each IL_i as follows: $NIL(IL_i) = a\ IL(r_i) + b\ IL_{i+1} + c\ IL_{SZ}(r_i) + d\ IL_{AG}(r_i)$, where the following hold:

a, b, c and d are constant parameters.

Order: $IL(r_i)$ is the reserved order of arrival of r_i at each IL counted by $tsReq_i$.

Traffic rule: IL_{i+1} is the reserved order in numerical order of IL.

IL size: $IL_{SZ}(r_i)$ is the number of pending requests that are in the same IL as r_i's.

IL age: $IL_{AG}(r_i)$ is given by the sum of the *ages* of all the requests in each IL. (*age* is the number of sessions that have been initiated since the request was enqueued to $token\text{-}queue_i$.)

Base on this framework, various selection schemes can be obtained by setting four parameters as follows:

$$a = 1, b = 0, \text{ and } c = d = 0.$$

This scheme is the simplest scheme and is the first-come, first-serve scheme in which requests are granted in the order of the enqueue. It yields lower concurrency, but is a non-starving scheme.

$$a = 0, b = 1, \text{ and } c = d = 0.$$

This scheme is the same as the Korea intersection control scheme in which the token is rotated in IL numerical order. It would yield lower concurrency.

$$a = 0, b = 0, \text{ and } c = d = 1.$$

When the QZ of the intersection is heavily loaded, it is desirable that the next session be of the IL for which the number of pending requests in the token queue $token\text{-}queue_i$ is the maximum, However this method would lead to starvation of a request. To avoid starvation, every pending request in $token\text{-}queue_i$ is associated with an attribute called *age*.

The priority of a type of the new session is computed by simply adding two parts such as $IL_{SZ}(r_i)$

and $IL_{AG}(r_i)$. The next session that is initiated of IL excluding the IL it located for which the priority value is the maximum called NIL. Finally, the one of the AVs (which is the oldest in *tsReqIL_i*) in the NIL *NIL-head* is selected to become the next token holder.

2. Then, *AV_i* transfers the token to the *NIL-head* by sending a *<send-pt>* message.

3. Afterwards, *AV_i* sends a release message to the token holder all AVs in the intersection and exits the CZ.

(b) *Case 2*: *AV_i* does not hold the token.

AV_i sends a *release* messages to all AVs in the intersection and exits the CZ.

56.4 Performance Evaluation

TDGim is an algorithm that applies the GME theory to the intersection traffic control system of autonomous vehicles. We analyze the performance in the respect of the measures of a distributed GME algorithms: *message complexity, synchronization delay, concurrency and system throughput*.

56.4.1 Message Complexity

Message complexity is the number of messages needed for AV_i from when entering QZ to when leaving CZ. *Our algorithm, TDGim* exchanges eight kinds of messages and there would be various set of messages based on the situation. Surely the number of messages to enter CZ is the highest when an AV broadcasts *<declare>* message.

As shown in Table 56.1, the message complexity of *TDGim* is better than others considering the specialty for dynamic traffic processing at the intersection.

56.4.2 Synchronization Delay

Synchronization delay is the time between all the vehicles in a session exiting the CZ and a vehicle in next session entering the CZ. In the case of *TDGim*, the *synchronization delay* can

be measured as the delay time due to the PT transmission to the next PT holder selected and is divided into two factors. The first is the network transmission delay (Pt), and the second is the computer processing time delay (Ps). Let the total delay time be P, P = Pt + Ps. The *synchronization delay* of *TDGim* is evaluated as the maximum value of Pt because Ps is negligible time.

56.4.3 Concurrency

Concurrency is defined by the maximum number of vehicles that can enter the CZ simultaneously. Thus, a high degree of concurrency refers to more efficient utilization of resources and the maximum concurrency of GME is **n** where **n** is the total number of processes in the system.

AVs in *TDGim* are divided into 8 ILs and there be an equal amount of AVs in each IL. There are n/8 vehicles per IL. So, the maximum concurrency of *TDGim* is n/4.

56.5 Conclusion

The ongoing researches of traffic signal scheduling and AIM is based on a central control device similar to a client-server in computer system. Since this is controlled by one central device, it has a large influence on the defects of such devices and has a problem of low flexibility for changing devices [15–21]. In order to solve this problem, we design the distributed AIM system that can be controlled only through the communications between AVs. AVs compete for crossing right (or permit) through message exchange between each other. In order to secure safety at an intersection, perfect collaboration must be achieved by exchange of information between AVs.

VTokenIC uses an IL-chain to control traffic flow efficiently and reduce message traffic. The IL-chain is a group of AVs in which a captain of an IL generates the IL-chain. In the system. AV's entry and exit of CZ are done by a chain unit. In other words, by manage ng a chain (a group of AVs) in which all members of one chain will enter CZ together with a single token, message traffic is reduced and traffic flow efficiency is improved.

Table 56.1 Performance

	mc	sd	cr
TDGim	$n + a$	t	$n/8$
Mamun	$n + 2$	$2t$	n
Mittal	$2n-1$	t	n
Weigang	$3n$		$n/8$
Kakugawa	$5Q + 1$	$3-4t$	n

mc message complexity, *sd* synchronization delay, *cr* concurrency
n: number of processes, Q: quorum, t: maximum message delay [8]

References

1. Wu, W., Zhang, J., Luo, A., Cao, J.: A distributed mutual exclusion algorithm for intersection traffic control. IEEE Trans. Parallel Distrib. Syst. **26**(1), 65–74 (2015)
2. Yan, F., Dridi, M., el Moudni, A.: An autonomous vehicle sequencing problem at intersections. Int. J. Appl. Math. Comput. Sci. **23**(1), 183 (2013)
3. Kamal, M.A.S., Imura, J.-i., Hayakawa, T., Ohata, A.: A vehicle-intersection coordination scheme for smooth flows of traffic with-

out using traffic lights. IEEE Trans. Intell. Transp. Syst. **16**(3), 1136–1147 (2015)

4. Li, L., Wang, F.-Y.: Cooperative driving blind crossing using inter-vehicle communication. IEEE Trans. Veh. Technol. **55**(6), 1712–1724 (2006)

5. Qian, X., Gregoire, J., Moutarde, F., De La Fortelle Cañete, A.: Coordination of autonomous and legacy vehicles at intersection (2014)

6. Lee, J., Park, B.: Development and evaluation of cooperative vehicle Intersection control algorithm under the connected vehicles environment. IEEE Trans. Intell. Transp. Syst. **13**(1), 81–90 (2012)

7. Zohdy, I.H., Rakha, H.A.: Intersection management via vehicle connectivity: the intersection cooperative adaptive cruise control system concept. J. Intell. Transp. Syst. **20**(1), 17–32 (2016)

8. Mamun, Q.E.K., Nakazato, H.: A new token based protocol for group mutual exclusion in distributed systems. In: Parallel and Distributed Computing, International Symposium on (2006); Mittal, N., Mohan, P.K.: A priority-based distributed group mutual exclusion algorithm when group access is non-uniform. J. Parallel Distrib. Comput. **67**(7), 797–815 (2007)

9. Kakugama, H., Kemei, S., Masuzawa, T.: A token based distributed group mutual exclusion with quorum. IEEE Trans. Parallel Distrib. Syst. **19**(9), 1153–1166 (2008)

10. Thambu, P., Wong, J.: An efficient token-based mutual exclusion algorithm in a distributed system. J. Syst. Softw. **28**(3), 267–276 (1995)

11. Ricart, G., Agrawala, A.K.: An optimal algorithm for mutual exclusion in computer networks. Commun. ACM. **24**, 9–17 (1981)

12. Joung, Y.-J.: Quorum-based algorithms for group mutual algorithm. IEEE Trans. Parallel Distrib. Syst. **14**(5), 463–476 (2003)

13. Atreya, R., Mittal, N., Peri, S.: A quorum-based group mutual exclusion algorithm for a distributed system with dynamic group set. IEEE Trans. Parallel Distrib. Syst. **8**(10), 1345–1360 (2007)

14. Atreya, R., Mittal, N.: A dynamic group mutual exclusion algorithm using surrogate-quorums (2005)

15. Chandy, K.M., Miara, J.: The drinking philosophers problem (1984)

16. Maekawa, M.: A root N algorithm for mutual exclusion in decentralized systems (1985)

17. Raymond, K.: A tree-based algorithm for distributed mutual exclusion (1989)

18. Swaroop, A., Singh, A.K.: Efficient group mutual exclusion protocols for message passing distributed computing systems (2009)

19. Lamport, L.: Time, clocks, and the ordering of events in a distributed system (1978)

20. Chang, Y.I., Singha, M., Liu, M.T.: A dynamic token-based distributed mutual exclusion algorithm (1991)

21. Bertier, M., Arantes, L., Sens, P.: Distributed mutual exclusion algorithms for grid applications: a hierarchical approach (2006)

Stay Alive: A Mobile Application for the Cardiopulmonary Resuscitation Process According to the Advanced Cardiovascular Life Support Protocol

57

Julio Didier Maciel, Rodrigo Duarte Seabra, and Rodrigo Maximiano Antunes de Almeida

57.1 Introduction

There are more than 135 million cardiovascular-related deaths per year over the world and the prevalence of coronary heart disease has increased. Externally to the hospital environment, the number of visits has a variance of 20–140 per 100,000 people and the survival rate varies from 2% to 11% [1]. For [2], in the same types of care, but in non-hospital settings, the number of people suffering from cardiorespiratory arrest (CRA) who are not promptly rescued is three times as high as those who are rescued by specialized medical services.

In a hospital environment, the professionals have tools to support the cardiopulmonary resuscitation (CPR) process, which provides more positive results [1]. Despite having tools and human resources, as the process requires the extreme attention and physical effort of those who execute it, it is tiring, stressful and, above all, risky.

In the 1990s, the study conducted by [3] concluded that the results of cardiopulmonary resuscitation were not acceptable. The failures resulted in a relatively high number of deaths. The authors state that, in the period of the study, the lack of instruction of the students, or even of the instructors, resulted in the poor training of the professionals meant to apply CPR. As a solution to uncertainty in what must be done in a CPR, the American Heart Association (AHA) created the Advanced Cardiovascular Life Support (ACLS) protocol. Following the creation of the ACLS protocol, several clinics have attempted to adhere to it and to implement new techniques for cardiopulmonary resuscitation processes. Many of these techniques require specialized equipment and human training, and few have been tested in highly selected groups, making current ACLS the most commonly used technique by health professionals [4].

In parallel to this reality the use of cellular devices, especially smartphones, is currently widespread in society in all spheres, be they personal or professional, due to the most varied needs [5]. There are numerous areas of study regarding the use of mobile phones. In health, the area of study directed to the application of cell phones is called mobile health (mHealth). This area has grown significantly in recent years. This is mainly due to the adoption of smartphones by the population and to the increase of the networks speed. In addition, the health area has attracted several studies involving computational solutions [6, 7].

Based on these considerations, this research proposes a solution for the need for an assistance tool during the execution of a CPR in hospital settings. This proposed tool consists of a mobile application, which can be used in mobile phones and tablets, which will act as a decision support tool during the CPR process, following the ACLS guidelines. It is important to highlight that when performing a quick search in the application store, there are still no complete solutions in the area of application proposed by the research in question, that is, there is no tool intended to assist the execution of the ACLS protocol during the occurrence of a CRA.

57.2 Theoretical Foundation

A field survey on cancer was conducted in the United States between 2002 and 2003 by the NCI (National Cancer Institute) and was aimed at analyzing the trends of having cancer according to the behavior of people. In 2004, in the United

J. D. Maciel · R. M. A. de Almeida
Institute of Systems Engineering and Information Technology,
Federal University of Itajubá, Itajubá, Minas Gerais, Brazil

R. D. Seabra (✉)
Institute of Mathematics and Computing, Federal University of Itajubá,
Itajubá, Minas Gerais, Brazil
e-mail: rodrigo@unifei.edu.br

States and Canada, the third most sought-after element on the Internet was about health-related information. Due to the need of research on cancer and the high demand for information, NCI founded the HINTS (Health Information National Trends Survey), an American institute responsible for analyzing health information on the Internet [8].

The creation of HINTS provided an approximation between computer science and medicine. Because of the volume of data available on the network and the unification of nations, it is possible to extract data formerly not possible without the Internet. From this point of view, medical institutions constantly try to computerize their offices by transforming their previously restricted data into digital data so that they can possibly be used or analyzed by other users as well. Today, with the dissemination of cell phones, this proximity of informatics in medicine is even more pronounced. The cell phone is constantly used by health professionals during their work for consulting, acquiring and exchanging information.

57.2.1 Related Researches

Recently, due to the practicality, cost and propagation of mobile phones, there has been a significant growth in the development of health-related applications. Such applications significantly reduce and facilitate the work of their users [9, 10].

In the study conducted by [11], the application presented is related to the detection of falls, focusing on the elderly. By means of a device in the user's pulse (smartwatch), it is possible to detect if a fall has occurred. This device can communicate with health systems, such as a hospital or a rescue. It is thus possible to activate a rescue and prepare the patient's receipt in the hospital a few moments after the occurrence of the fall of the elderly.

There are other ways to use health apps even if indirectly. This is proposed by [12] through an application to be used in health centers. In this study, a prototype of a vaccination portfolio is proposed, since in some countries the traditional model of the paper vaccination portfolio is still used. This system, with physical documentation, provides high levels of inconsistency and loss of information, leaving the patient vulnerable to disease. The proposal of the work is to replace the physical vaccination portfolio with a virtual one, making all the control available in the cloud. Thus, one can access the patient's history, regardless of the health post in which he is consulting, more accurately and easily. This example application is primarily intended for health centers and can be made available to any end user.

57.2.2 Cardiac Arrest

Cardiac arrest, or cardiorespiratory arrest, is the abrupt loss of cardiac function in a person who may or may not have diagnosed heart disease. For cases in which death occurs,

their time and way of occurrence are unexpected, being instantaneous or soon after the appearance of the symptoms.

Even without knowing if the individual is suffering a heart attack, it is advised that the emergency treatment service (firemen, ambulance, hospital, among others) be triggered immediately. This relief aid must follow AHA instructions. In the case of a layperson, the BLS – Basic Life Support protocol, which is a simplified form of first aid, basically consists of manual chest compressions, ventilation (mouth-to-mouth resuscitation), defibrillation and heart rate every 2 minutes.

Unlike regular BLS, the advanced protocol (ACLS) should have more information and be better controlled. Once the patient is in the hospital and being treated by ACLS standards, each action should be monitored based on the dosage of oxygen during CPR, advanced devices of oxygenation, ventilation rate during CPR, among others elements, and time involved in each of the steps performed. It is important to mention that the first step for the ACLS is the BLS, that is, in the absence of the equipment and the necessary structure to execute the ACLS, the professional should choose to execute the BLS protocol [13].

In general, the ACLS protocol involves questions about oxygen dosage during CPR, advanced oxygenation devices, ventilation rate during CPR, exhaled carbon dioxide (CO^2), detection to confirm the insertion of the oxygenation device, physiological monitoring and prognosis during CPR, defibrillation, antiarrhythmic drugs and vasopressors. Cardiac massages should be performed in the same manner as in the BLS protocol, from 100 to 120 times per minute. Airway openings should be conducted by means of a specialized apparatus which, after being installed in the patient, should be activated within 6–8 seconds. The ACLS protocol can be interpreted as an algorithm shown in Fig. 57.1.

57.3 Method

57.3.1 Stay Alive App

In order to raise the requirements of the tool and meet the needs of the protocol, an interview with an experienced ACLS doctor was conducted, with a representative profile of the potential users of the application. The main functions are:

- Supporting screen: Contact and basic information about the developer and AHA guidelines;
- Final Report: In this report, the data and actions taken in the CPR must be present, as well as the time and the time in relation to the beginning of the CPR in which action was taken. This report may be accessed later on if needed;
- Screen for data change during CPR: Some data is subject to change, for example, adrenaline may have a time

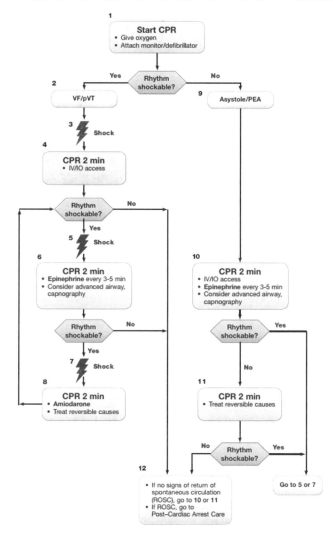

Fig. 57.1 ACLS protocol [13]

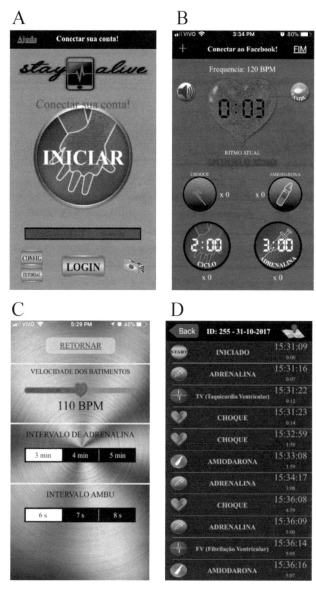

Fig. 57.2 Stay Alive app interface (**a**) The initial screen of the application (**b**) CPR aid screen (**c**) Setup screen (**d**) Application final report

interval of 3–5 minutes. A similar option may be for the heart rate variance and the balloon-mask ventilation (AMBU);

- Usage modes: The application can meet two different types of usage modes. In the first case, the user becomes "freer" to take actions during the CPR. In the second, the application leads the user to take actions according to the algorithm stipulated by the AHA, useful for academics and residents.

For implementing these requirements and their application, the Swift language was chosen for the following reasons: being a native language developed by Apple itself, demonstrating promising growth and using the best hardware performance [14]. Hence, the application runs on the iOS and Android operating systems. The application is illustrated in Fig. 57.2.

The initial screen of the application (Fig. 57.2a) has several buttons for access to more information, tutorial, language change, login and, highlighted in the center of the screen, there is the CPR start button, which aims to take the user to the CPR aid screen (Fig. 57.2b). In this screen, there are all the tools necessary to conduct a CPR following the ACLS protocol. It is possible to change the heart rate, add the number of defibrillations and amiodarone applied, restart the count and add the number of cycles of compressions and adrenaline applied. According to Fig. 57.2c, it is possible to change the compression frequency settings and the adrenaline application intervals and ventilation of the patient through the AMBU. At the end of the CPR, the final report is sent to the user containing the application times of each item (Fig. 57.2d). This report can also be accessed later via the application history.

57.4   Data Analysis

57.4.1 Preliminary App Tests

After the application development (APP), preliminary use tests were performed in an environment as close to reality as possible, in order to provide greater reliability to the tool produced. For reproducing an environment that is faithful to the reality of APP use, interviews were conducted with health professionals with experience in CPR in hospitals to collect information about the most frequent occurrences of CPRs. The main information collected in these interviews was:

- Generally, the first care for the occurrence of a CPR is performed by the nurse, and the physician (professional needed to take the CPR course) is triggered. Upon arriving at the location of the patient, he/she has been in the CPR for some time;
- Some professionals were mentioned to be unprepared to make the chest compressions;
- The absence of a necessary tool or drug for CPR (adrenaline defibrillator and amiodarone ampoules);
- Once the CPR process has started, there is often an uncertainty about whether the drug has been applied in that compression cycle;
- Compressions are not always monitored correctly;
- Professionals in the environment are often exhausted and nervous during the CPR process.

Based on this information, four scenarios were elaborated to simulate the most frequent occurrences of CPR in a hospital. These scenarios are described as:

- Scenario 1: The doctor arrives at the treatment room with a delay of 3 minutes and the cardiac monitor is only brought after some time for the beginning of CPR;
- Scenario 2: The doctor arrives in the waiting room with a delay of 1 minute and the cardiac monitor is brought shortly after the start of CPR;
- Scenario 3: the physician is present as soon as the CPR occurs and the cardiac monitor is already available;
- Scenario 4: The doctor arrives in the treatment room with a delay of 4 minutes and the heart monitor is available.

In scenarios 1, 2 and 3, the patient should have shocking heart rhythms because these heart rhythms are the ones that present the largest line of activities for rescuers. In contrast, in scenario 4, the patient should have non-shockable rhythms

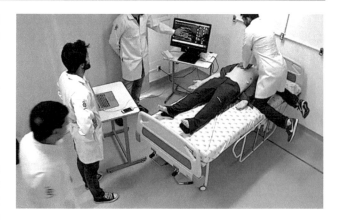

Fig. 57.3 Stay Alive lab test

as a test counterpoint in situations whereby the application prevents certain actions from being taken. The test team consisted of:

- A physician, informing rescuers of the actions to be performed in CPR;
- The first author of this research, administering the application and changing the heart rhythms in the system, following the previously stipulated scenarios;
- Three undergraduates from the Usability Laboratory, who took turns as first responders performing chest compressions and drug applications.

To perform the simulated test, a SimMan3G robotic manikin was used to simulate the patient's vital signs, including the electrical signals of the cardiac arrhythmia, and performed a real monitoring of the quality of the CPR performed, both depth and frequency of achievement. Figure 57.3 illustrates the simulated test of the application in the Usability Laboratory.

57.4.2 Preliminary Results

The preliminary testing of the application was executed as planned, yielding concrete results. Figure 57.4 contains the report issued by the manikin control system. Among the data of the report, the main information for the test of the application related to a CPR are: total time of care; correct percentage of hand positioning on the patient's chest during the execution of the compressions; mean depth of compression; relation between the total of compressions executed and the total of compressions with the correct depth; total return of the chest after the execution of a compression (diastole); compression rate chart (minimum of 100 compressions per minute and maximum of 120 compressions per minute); total and percentage of ventilation performed correctly.

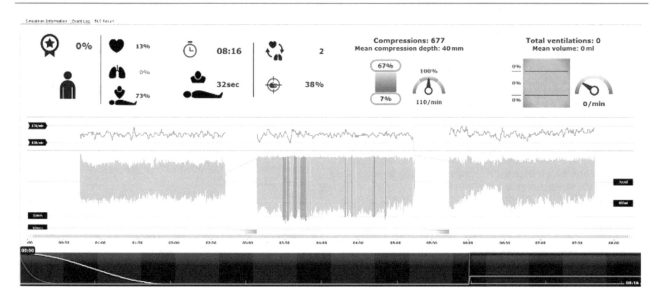

Fig. 57.4 Report issued by the control system

Regarding chest compressions, the correct hand position percentage information during compression shows that the result in this aspect was not good, which is an expected value given the low experience of the first responders. This also applies to the compressions quality, as only 7% of them reached the correct depth interval. Another point is that the application is not intended to fix these parameters.

Prior to the start of the test and to always maintain a proper frequency following the AHA guidelines, the team doctor asked the application configuration related to the frequency of its sound emission to be 110 times per minute. This action brought good results. Figure 57.3 allows observing that the frequency of the compressions performed and, due mainly to the sound emitted by the application, 100% of the compressions were executed within the limit stipulated by the AHA. This is an interesting result because it is one of the variables that the application has and is a feedback structure to guide the rescuer. Although the compressions did not have their correct depth in much of the process, the CPR was executed correctly in the four scenarios following the ACLS protocol, with no decision-making errors. The other scenarios (1, 2 and 4) had similar results to that of scenario 3. In scenario 4, drug administrations occurred in the predicted times according to the ACLS protocol.

Regarding the quality of the compressions, all the scenarios presented unsatisfactory results, probably because of the non-use of trained specialists in the first-aid function and the lack of feedback to the user of the performed task. However, the variables in which the application had some feedback structure (frequency and drug administration) had a very high success rate; in the case of the compression frequency, above 98% were correct. Therefore, it can be concluded from the preliminary tests that the application helped to improve the results of the CPR, contributing to the better execution of the ACLS protocol.

57.5 Final Considerations

This research presented details inherent to the development of a decision support tool during the execution of the ACLS protocol, conducted aiming at successfully completing cardiopulmonary resuscitation in a hospital setting. Based on the literature review and even on the health professionals themselves, it is concluded that there is a lack of a tool to help implement the ACLS protocol. In this context, the development of a mobile application was proposed as a possible solution. The application was developed based on information collected from interviews with experienced professionals in the field.

After the application was developed, it was tested in a laboratory, in an environment simulating a hospital and following the most probable scenarios of occurrence of CPR. The results were concrete and demonstrated that the aid coming from the application was effective. Even using untrained people in lifeguard activity, the variables to which the application provided feedback were executed correctly. Those without feedback presented poor quality results, mainly the small depth of compression. It can therefore be concluded that the use of a suitable tool, which provides feedback to first responders, can aid in a process that is not always followed closely by the interference of the real environment as well as the stress level of the actors. The tool is available for free and the benefits brought by the application help treat one of the leading causes of death in the world.

References

1. Meaney, P., et al.: Cardiopulmonary resuscitation quality: improving cardiac resuscitation outcomes both inside and outside the hospital: a consensus statement from the American Heart Association. Circulation. **128**(4), 417–435 (2013)

2. Ringh, M., et al.: Mobile phone technology identifies and recruits trained citizens to perform CPR on out-of-hospital cardiac arrest victims prior to ambulance arrival. Resuscitation. **82**(12), 1514–1518 (2011)

3. Kaye, W., et al.: The problem of poor retention of cardiopulmonary resuscitation skills may lie with the instructor, not the learner or the curriculum. Resuscitation. **21**(1), 67–87 (1991)

4. Brooks, S.C., et al.: Part 6: Alternative techniques and ancillary devices for cardiopulmonary resuscitation. Circulation. **132**(18 suppl 2), S436–S443 (2015)

5. Dinh, H.T., et al.: A survey of mobile cloud computing: architecture, applications, and approaches. Wirel. Commun. Mob. Comput. **13**(18), 1587–1611 (2013)

6. Gandarilla, H.: Safeguarding personal health information: case study. In: Proceedings of 15th Information Technology – New Generations – ITNG 2018. Las Vegas (2018)

7. Gibbs, M.: Evaluation cyber threats to the United Kingdom's national health service (NHS) spine network. In: Proceedings of 15th Information Technology – New Generations – ITNG 2018. Las Vegas (2018)

8. Nelson, D., et al.: The health information national trends survey (HINTS): development, design, and dissemination. J. Health Commun. **9**(5), 443–460 (2004)

9. Beuscart-Zéphir, M., et al.: Integrating users' activity modeling in the design and assessment of hospital electronic patient records: the example of anesthesia. Int. J. Med. Inform. **64**(2), 157–171 (2001)

10. Hamborg, K., Vehse, B., Bludau, H.: Questionnaire based usability evaluation of hospital information systems. Electron. J. Inf. Syst. Eval. **7**(1), 21–30 (2004)

11. Maurício, C.P.P., et al.: Home E-Care: monitoramento de quedas em idosos através de smartwatches. In: Anais do XV Congresso Brasileiro de Informática em Saúde. Goiânia (2016)

12. Muratt, P., et al.: Carteira de vacinação virtual. In: Anais do XV Congresso Brasileiro de Informática em Saúde. Goiânia (2016)

13. Hazinski, M.F., et al.: Highlights of the 2015 American Heart Association Guidelines Update for CPR and ECC. American Heart Association, Dallas (2015)

14. Housden, B.E., et al.: Loss-of-function genetic tools for animal models: cross-species and cross-platform differences. Nat. Rev. Genet. **18**(1), 24 (2017)

Towards a Reusable Framework for Generating Health Information Systems

58

André Magno Costa de Araújo, Valéria Cesário Times, and Marcus Urbano Silva

58.1 Introduction

As recent studies indicate, the software industry and academics have put much effort in finding solutions to improve the lifecycle of Health Information Systems (HIS) [1–3]. HIS have the important function of improving the quality of services provided by health organizations, facilitating the access to patients' medical records and the Electronic Health Record (EHR) storage management.

Common problems faced by an HIS are the lack of uniformity in EHR data and high maintenance costs due to the natural evolution of clinical concepts and advances in medical technology. They are not designed to allow dynamic changes in order to adapt to a given problem domain, nor do they allow end users to create new functionalities, thus reinforcing their reliance on a software programmer.

Several studies show that openEHR specifications can be used as an alternative to support the development of flexible and interoperable EHR systems [4–7]. The specification of archetype-based EHR consists in organizing its components through a two-level modeling approach [8].

An archetype consists of a computational expression based on a reference model and represented by domain constraints and terminologies [8]. A template is a structure used to group archetypes and allow their use in a particular context of application. It is often associated with a graphical user interface (GUI). In the current approach, the components responsible for modeling the clinical and demographic data of the EHR are specified through generic data structures, which are composed of data types, constraints and terminologies.

Such data structures allow the representation of heterogeneous EHR data through the following types: ITEM_SINGLE, ITEM_LIST, ITEM_TREE and ITEM_TABLE. In the openEHR community, current practices recommend the use of ITEM_TREE, which specifies a hierarchical data structure that is logically represented as a tree. It can be used, for instance, to model a patient's physical or neurological evaluations.

OpenEHR archetypes have been used to develop GUIs [9], to model clinical concepts found in legacy applications [10], to create ontologies for web data processing [11] and to evaluate database performance [12]. Additionally, research studies used archetypes to evaluate the maintainability and usability of health applications [13, 14]. Although the above related works represent an advance in the state of the art, it is notable that the software industry needs tools that aid the development process of HIS using archetypes. The creation of an application framework based on openEHR specifications is meant as a solution to improve the flexibility and reusability of EHR systems.

This article extends the cloud service proposed in [15] and specifies a framework to assist in the development of health applications using archetypes. For this, a software architecture was developed to create relational and NoSQL data schemas, independent of storage technology, as well as generating GUIs with data persistence capabilities. A health application was developed using the proposed framework and archetypes available in the openEHR clinical knowledge manager (CKM). The evaluation results indicate a 72% reduction in coding efforts.

The other sections of this article are organized as follows: Sect. 58.2 brings an analysis of the main related works,

A. M. C. de Araújo (✉)
Department of Information Systems, Federal University of Alagoas, Penedo, Brazil
e-mail: andre.araujo@penedo.ufal.br

V. C. Times · M. U. Silva
Center for Informatics, Federal University of Pernambuco, Recife, Brazil
e-mail: vct@cin.ufpe.br; mus@cin.ufpe.br

© Springer Nature Switzerland AG 2019
S. Latifi (ed.), *16th International Conference on Information Technology-New Generations (ITNG 2019)*, Advances in Intelligent Systems and Computing 800,
https://doi.org/10.1007/978-3-030-14070-0_58

while Sect. 58.3 presents the software architecture and main functionalities developed. Section 58.4 describes a study performed to evaluate the framework and discusses the main results obtained. Finally, Sect. 58.5 presents final considerations and future work.

58.2   Related Works

In this section, we conduct an analysis of the main related works in the domain of archetype mapping in databases and GUI generation.

One of the pioneering works in the field of archetype [16] mapped a set of archetypes for a legacy database and exposed the lack of tools and methodologies that would have helped in modeling archetypes in a database. In [9], a set of rules were specified to map archetype data attributes in a relational database, generating templates in a specific problem domain. However, the proposed solution could not map archetypes with hierarchical data structures (i.e., ITEM_TREE).

Node+Path is an openEHR solution in which the data attributes of an XML archetype are serialized in a database management system (DBMS) [17]. This approach uses the Entity-attribute-value (EAV) approach to store the path (i.e., address) of an attribute in the first column, while the value of such attribute is stored in the second column. Yet, this solution requires the creation of complex logical expressions for data manipulation. The persistence solution proposed in [18] specifies a set of mapping rules, extracts the archetypes attributes and stores them in tables of a relational data schema. However, the terminologies and constraints specified in those archetypes are not considered.

In [12], the authors evaluated the performance of the NoSQL database in managing archetypes. Two main aspects were evaluated: the space allocated for data storage and the processing time of a set of queries. For the study, four legacy relational databases were used to generate clinical data documents in JSON and XML format. The results show that the JSON storage space is smaller than for XML. In addition, the performance of XML was lower compared to JSON when processing a set of queries for data recovery.

The EHRScape [19], EHRServer [20] and EtherCIS [21] frameworks support health application development using openEHR specifications. EHRScape is a solution from Marand which provides a set of application programming interfaces (APIs) to query terminologies, integrate legacy EHR data and consult clinical decision support systems. Cloud EHRServer is maintained by CaboLabs and features cloud services for EHR storage and dynamic building of GUIs that are compatible with the openEHR standard. Finally, the solution developed by EtherCIS offers a RESTful API compatible with archetypes, templates and archetype query language (AQL).

Recent studies based on openEHR specifications include the implementation of an information model for medical data visualization [22] and a medical data visualization method [23]. However, the use openEHR archetypes to build heterogeneous data schema, store and standardize EHR data is an open issue found in the state of the art.

58.3   Architecture and Framework Overview

The framework proposed herein consists of a computational environment focused on the development of health applications using openEHR specifications. As shown in Fig. 58.1, the framework generates GUIs and data schemas using archetypes and templates created by the openEHR community with tools such as Archetype editor and Template designer.

The software architecture uses classes, services and interfaces based on the reference model to create relational and NoSQL data schemas and GUIs with data persistence capabilities. To achieve this, the framework classes retrieve data attributes that define the EHR, the terminologies that give a semantic meaning to clinical data and the specified constraints on data attributes. Using the extracted elements, a REST API creates a GUI by performing the following tasks: it maps data attributes to data entry components in the GUI, uses the constraints as a GUI data entry validation mechanism (e.g., range of value, data type constraint) and makes the terminologies available in their respective GUI components. In addition, data attributes, terminologies and constraints are used to create tables, columns, and referential integrity constraints in relational data schemas and hierarchical structures in NoSQL documents.

An organization can offer different types of health services like laboratory testing services, emergency care, hospitalization, etc. The framework therefore supports the creation and configuration of domains and subdomains that represent the services offered, where every GUI generated is linked to a subdomain. To ensure the functioning of the data schemas generated by the framework, independently of the technology used, an architectural standard allows the business logic to access data objects regardless of the data storage technology. Called a repository, this pattern encapsulates the set of objects persisted in the data layer and the operations performed on them. In addition, mediation between the components of the presentation layers, business logic, repository and data access is performed through interfaces.

Figure 58.2 shows the software architecture specified for the framework using unified modeling language (UML) package diagram notation and repository pattern formalism. It is possible to observe that the components communicate through interfaces. The main advantage of using interfaces

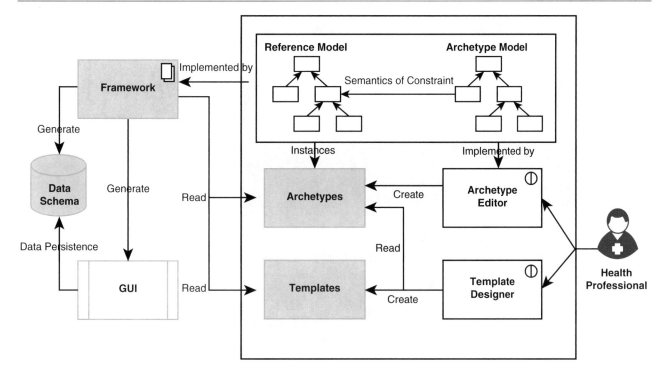

Fig. 58.1 Conceptual vision of the framework

is that they establish the form of communication between the components and allow modifications to be made in the business logic without affecting the operation of external applications that may interact with the framework. For example, it is possible that an external component is consuming the relational data schema generation methods of the DataSchema component (Fig. 58.2), while new methods are being implemented in the Generator component to create NoSQL data schemas.

The IRepository interface contains the signatures of the persistence methods to be implemented in the different data storage technologies (e.g., Oracle, SQL Server, Cassandra, Neo4J, etc.). The IMapper interface provides resources for business logic to map archetypes into data schemas. Finally, the IDataSource interface provides the connection parameters for various types of database management system. Every GUI dynamically created by the framework has data persistence capabilities (i.e., insert, update, select and delete). As the framework deals with storage in different database paradigms, a REST API captures the data entered in the GUI and stores it in the different databases created by the framework.

All features in the framework were developed using the C# programming language and ASP.NET technology.

58.4   Framework Evaluation

One of the major challenges in evaluating a framework is measuring how fast and at what cost a software project can be built, and how much development effort can be reduced. Therefore, the two main points investigated in this evaluation were the cost of developing a health application and the effort reduction in the coding phase. For this evaluation, a health application was developed based on the care given to public healthcare patients from a hospital located in the northern region of Brazil.

The functionalities to be developed in this environment were: hospitalization, patient family history, patient evaluation, physical examination archiving and vital signs collection.

Two types of professionals took part in our assessment. First, two health professionals were asked to choose from the openEHR CKM, archetypes and their respective templates to represent the functionalities. The following archetypes were chosen: Admission, family_history, blood_pressure, exam and respiration. Second, three programmers with over 10 years of experience in the healthcare sector were invited to build the features using the framework and the archetype-

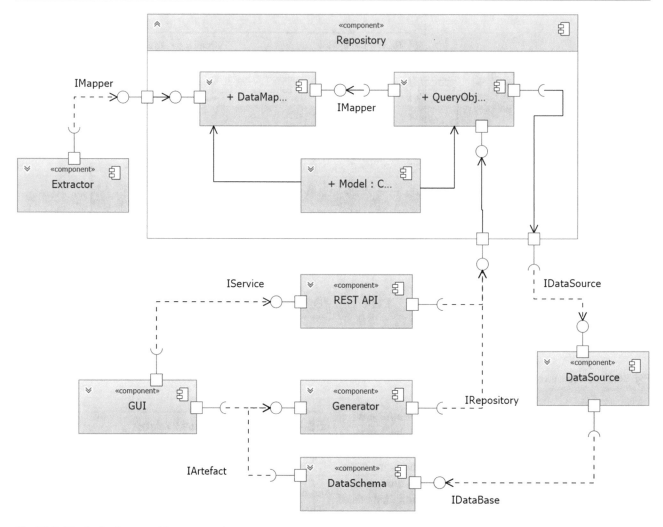

Fig. 58.2 Standard software architecture repository

s/templates selected. Before the evaluation, each programmer was given the framework documentation describing its functionalities, the software architecture with its respective components and instructions on how to generate the software artifacts. For this evaluation, each programmer was asked to:

- Install the framework on the workstation
- Generate application features using the framework and archetypes/templates
- Answer the evaluation questionnaire

In this study, we used the framework evaluation proposed in [24], which consists in measuring the software design costs by observing the following items:

- System cost: This metric measures the cost of software licensing and the effort spent in configuring and installing the technologies to be used before the development.

- Project cost: Evaluates classes, interfaces, design patterns and data schemas.
- Learning cost: Investigates the time required to understand the programming language syntax and the features offered by the framework.
- Implementation cost: This metric evaluates how much effort the framework reduces in implementing the functional and non-functional requirements of an application.
- Testing cost: This metric measures whether the developers have tools to test, analyze and debug the source code of the framework.

The purpose of the evaluation is to calculate the total cost of developing an application and the benefit of using a framework. For each question in the questionnaire, the answer should be a value from 0 to 5. The total cost is calculated by multiplying the total of questions by 5, that is, the maximum value that a question can receive. The cost of the framework is calculated using the sum of the value

Table 58.1 Programmer 1 evaluation results

Programmer 1	Total cost	Framework cost	Benefit
System	15	6	9
Project	25	2	23
Learning	20	7	13
Implementation	25	6	19
Tests	10	2	8
Total	95	23	72

Table 58.2 Programmer 2 evaluation results

Programmer 2	Total cost	Framework cost	Benefit
System	15	7	8
Project	25	1	24
Learning	20	10	10
Implementation	25	8	17
Tests	10	2	8
Total	95	28	67

Table 58.3 Programmer 3 evaluation results

Programmer 3	Total cost	Framework cost	Benefit
System	15	5	10
Project	25	4	21
Learning	20	9	11
Implementation	25	7	18
Tests	10	1	9
Total	95	26	69

assigned to each response (i.e., 0 to 5), while the benefit is calculated by subtracting the total cost from the framework cost. Finally, to measure the cost reduction rate, the benefit of the framework is divided by the total cost.

Following the evaluation model described in [24], the project total cost, the framework cost and the benefit of its use were initially calculated. Since all programmers used the same scenario for the application development, the total cost is the same for the three evaluations, that is, 95.

For programmer 1, the framework cost is 23 and the benefit of its use, 72 (Table 58.1). The calculated cost reduction rate is 0.75 (72 ÷ 95), thus, 75% of efforts and costs were reduced by using the framework.

The evaluation results from programmer 2 show that the framework cost is 28 with a benefit of 67 (Table 58.2). The cost reduction rate of the framework is 0.7 (67 ÷ 95), which represents 70% of efforts and costs.

Finally, evaluation results from programmer 3 show a framework cost of 26 and a benefit of use of 69 (Table 58.3). The cost reduction rate of using the framework is 0.72, thus 72% of efforts and costs.

Analyzing the results of the three participants, a 72% reduction in the costs and efforts of developing a health

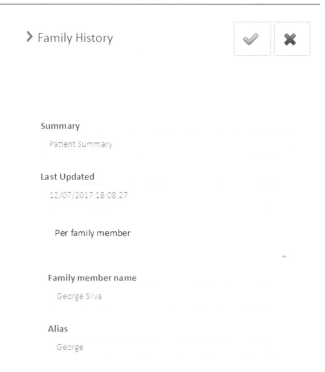

Fig. 58.3 GUI generated from the family history archetype

application using archetypes was observed. At the end of the evaluation, the participants were able to present their opinions and improvement suggestions. Usability and the way the GUI is made available to end users were suggested improvements for the framework.

Figure 58.3 shows the GUI generated by reading the family history archetype.

58.5   Conclusion

This article presents a framework to support the development of health applications using openEHR specifications. Key features include a software architecture with a set of classes, services and interfaces that generate graphical user interfaces as well as relational and non-relational data schemas, regardless of the data storage technology used. To validate the proposed framework, a health application was developed using archetypes available in the openEHR clinical knowledge manager. The results indicated a 72% reduction in coding effort to develop health applications.

Software quality assessment using ISO/IEC 9126 and an EHR storage performance evaluation are suggestions of future work.

The results presented in this study are restricted to a local context with few participants and few functionalities evaluated. Thus, the results presented here cannot be generalized.

References

1. Karamjit, K., Rinkle, R.: Managing data in healthcare information systems: many models, one solution. Computer. **48**, 52–59 (2015). https://doi.org/10.1109/MC.2015.77

2. Noor, A.M., Nur, F.A.A., Nurul, A.M.Z.: A novel conceptual framework of Health information systems (HIS) sustainability. In: International Conference on Research and Innovation in Information Systems (ICRIIS) 2017, pp. 1–6. IEEE, Langkawi, Kedah (2017). https://doi.org/10.1109/ICRIIS.2017.8002458

3. Thuemmler, C., Paulin, A., Jell, T., Lim, A.K.: Information technology – next generation: the impact of 5G on the evolution of health and care services. In: Latifi, S. (ed.) Information Technology – New Generations. Advances in Intelligent Systems and Computing, vol. 558. Springer, Cham (2018). 978-3-319-54978-1

4. Linder, J.A., Schnipper, J.L., Middleton, B.: Method of electronic health record documentation and quality of primary care. J. Am. Med. Inform. Assoc. **19**(6), 1019–1024 (2012). https://doi.org/10.1136/amiajnl-2011-000788

5. Martínez, C.C., Menárguez, T.M., Fernández, B.J.T., Maldonado, J.A.: A model-driven approach for representing clinical archetypes for Semantic Web environments. J. Biomed. Inform. **42**(1), 150–164 (2009). https://doi.org/10.1016/j.jbi.2008.05.005

6. Bernd, B.: Advances and secure architectural EHR approaches. Int. J. Med. Inform. **75**(3–4), 185–190 (2006). https://doi.org/10.1016/j.ijmedinf.2005.07.017

7. Buck, J., Garde, S., Kohl, C.D., Knaup-Gregori, P.: Towards a comprehensive electronic patient record to support an innovative individual care concept for premature infants using the openEHR approach. Int. J. Med. Inform. **78**(8), 521–531 (2009). https://doi.org/10.1016/j.ijmedinf.2009.03.001

8. Beale, T.: Archetypes: constraint-based domain models for future-proof information systems. In: Eleventh OOPSLA Workshop on Behavioral Semantics: Serving the Customer, pp. 16–32 (2002)

9. Georg, D., Judith, C., Christoph, R.: Towards plug-and-play integration of archetypes into legacy electronic health record systems: the ArchiMed experience. BMC Med. Inform. Decis. Mak. **13**(1), 1–12 (2013). https://doi.org/10.1186/1472-6947-13-11

10. Bernstein, K., Bruun, R.M., Vingtoft, S., Andersen, S.K., Nøhr, C.: Modelling and implementing electronic health records in Denmark. Int. J. Med. Inform. **74**(2–4), 213–220 (2005). https://doi.org/10.1016/j.ijmedinf.2004.07.007

11. Lezcano, L., Miguel, A.S., Rodríguez, S.: Integrating reasoning and clinical archetypes using OWL ontologies and SWRL rules. J. Biomed. Inform. **44**(2), 343–353 (2010). https://doi.org/10.1016/j.jbi.2010.11.005

12. Freire, M.S., Teodoro, D., Wei-kleiner, F., Sundvall, E., Karlsson, D., Lambrix, P.: Comparing the performance of NoSQL approaches for managing archetype-based electronic health record data. PLoS One. **11**(3), e0150069 (2016). https://doi.org/10.1371/journal.pone.0150069

13. Atalag, K., Yang, H.Y., Tempero, E., Warren, J.R.: Evaluation of software maintainability with openEHR – a comparison of architectures. Int. J. Med. Inform. **83**(11), 849–859 (2014). https://doi.org/10.1016/j.ijmedinf.2014.07.006

14. Araujo, A.M.C., Times, V.C., Silva, M.U.: Template4EHR: building dynamically GUIs for the electronic health records using archetypes. In: 16th IEEE International Conference on Computer and Information Technology, pp. 26–33. IEEE, Nadi (2016). https://doi.org/10.1109/CIT.2016.43

15. de Araújo, A.M.C., Times, V.C., da Silva, M.U.: A Cloud service for graphical user interfaces generation and electronic health record storage. In: Latifi, S. (ed.) Information Technology – New Generations. In Advances in Intelligent Systems and Computing, vol. 558, pp. 257–263. Springer, Cham (2018). 978-3-319-54978-1

16. Späth, M.B., Grimson, J.: Applying the archetype approach to the database of a biobank information management system. Int. J. Med. Inform. **80**(3), 205–226 (2010). https://doi.org/10.1016/j.ijmedinf.2010.11.002

17. Node + Path Persistence. Available online: https://openehr.atlassian.net. Accessed on 26 Mar 2018

18. Wang, L., Min, L., Lu, X., Duan, H.: Archetype relational mapping – a practical openEHR persistence solution. BMC Med. Inform. Decis. Mak. **16**(21), 1–18 (2015). https://doi.org/10.1186/s12911-015-0212-0

19. API Explorer – EHRScape. Available online: https://www.ehrscape.com. Accessed on 26 Mar 2018

20. EHRServer – CaboLabs. Available online: https://cabolabs.com. Accessed on 26 Mar 2018

21. EtherCIS – Enterprise Clinical Data Repository. Available online: https://ethercis.org. Accessed on 26 Mar 2018

22. Kopanitsa, G., Veseli, H., Yampolsky, V.: Development, implementation and evaluation of an information model for archetype based user responsive medical data visualization. J. Biomed. Inform. **55**, 196–205 (2015). https://doi.org/10.1016/j.jbi.2015.04.009

23. Kopanitsa, G.: Evaluation study for an ISO 13606 archetype based medical data visualization method. J. Med. Syst. **39**, 82 (2015). https://doi.org/10.1007/s10916-015-0270-y

24. Lee, C.: An Evaluation Model for Application Development Frameworks for Web Applications. Thesis, Degree Master of Science. The Ohio State University, Ohio (2012)

Use of Eigenvector Centrality to Rank the Vertices in a Disease-Disease Network

Md Atiqur Rahman, Mahzabin Akhter, and Natarajan Meghanathan

59.1 Introduction

Centrality metrics are useful to assess the topological importance of a vertex in complex real-world networks [1]. The centrality metrics for complex network analysis could be broadly classified [2] as either neighborhood-based or shortest path-based. Among the various centrality metrics that fall into one of these two categories, the eigenvector centrality (EVC) metric is a widely studied neighborhood-based centrality metric that has been used to rank the vertices in several real-world networks that are modeled as undirected weighted graphs (e.g., [3]). The EVC of a vertex (ranging from 0 to 1) comprehensively captures the importance of the vertex with respect to the following [1]: the number of neighbors of the vertex (also referred to as the degree of the vertex), the weights of the edges incident on the vertex, the degree of its neighbors as well as the weights of the edges incident on the neighbors. Though the EVC metric has been used to analyze several weighted real-world network graphs, to the best of our knowledge, it has not been used to assess the topological importance of the vertices in a disease-disease network graph in which the vertices are the diseases and the edge weights represent the number of shared genes between the end vertices. In this paper, we use the EVC metric to rank the diseases in a disease-disease network graph for humans and we construct such a graph based on the

results of the disease-gene association studies reported in the NIH GWAS (Genome-Wide Association Studies) catalog [4] and OMIM (Online Mendelian Inheritance in Man) database [5].

The Human Genome Project (HGP) [6] contributed to the identification of the genes (nearly 30,000) that are part of the human DNA and the sequences of chemical bases that constitute the human DNA. A prominent offshoot of the HGP research project has been the several disease-gene association studies (e.g., [7]) conducted by researchers to explore the complex relationship between the human diseases (as well as the diseases traits) and the genes. Though gene mutations are not always harmful, genetic disorders (resulting from gene mutations) are considered to be the principal cause of the diseases that could impact the normal functioning of one or more organs of the body [8, 9]. The mutations in a particular gene could lead to a person procure one or more diseases. If a disease is associated with several such genes, then the mutations in even one of these genes could lead to the person procuring the disease. Moreover, if the gene is associated with several diseases, a person is likely to procure one or more of these diseases that share the gene. Several studies (e.g., [7, 10]) in the literature have reported the association between the diseases and the genes in the human genome, all of which could be compiled together to build a complex disease-gene association network.

From a graph theoretic perspective, the disease-gene association network could be modeled as a bipartite graph such that a disease (in one partition: the set of diseases) could be linked to one or more genes (in the other partition: the set of genes) that are associated with the disease. Based on such a bipartite graph, we could build a disease-disease network wherein the vertices are the diseases and two vertices are connected with an edge if the corresponding diseases share at least a gene. The number of shared genes between two diseases is the weight of the edge connecting the correspond-

Md. A. Rahman · M. Akhter
Computational and Data-Enabled Science and Engineering, Jackson State University, Jackson, MS, USA
e-mail: md.a.rahman@students.jsums.edu;
mahzabin.akhter@students.jsums.edu

N. Meghanathan (✉)
Department of Computer Science & Engineering, Jackson State University, Jackson, MS, USA
e-mail: natarajan.meghanathan@jsums.edu

© Springer Nature Switzerland AG 2019
S. Latifi (ed.), *16th International Conference on Information Technology-New Generations (ITNG 2019)*,
Advances in Intelligent Systems and Computing 800,
https://doi.org/10.1007/978-3-030-14070-0_59

ing vertices. We compute the EVC values of the vertices (diseases) in the disease-disease network and observe the distribution of the EVC values to be Pareto in nature (80-20 rule) [11]. Only about 18% of the diseases had higher EVC values that are appreciably different from each other; the remaining 82% of the diseases had lower EVC values that are not much different from each other. A vertex/disease with a higher EVC is more likely to be an influential disease contracting which would put a person vulnerable to other related diseases that are incident on the vertex in the disease-disease network graph. A person procuring a higher EVC disease is thus at a greater risk of procuring other related diseases (that share a larger number of genes with the higher EVC disease) compared to person procuring a lower EVC disease.

The rest of the paper is organized as follows: Sect. 59.2 presents the procedure employed to build a disease-disease network from a disease-gene association network and compute the EVC values of the vertices in the disease-disease network. Section 59.3 presents the real-world dataset of the disease-gene association network and presents the EVC values of diseases computed on the corresponding disease-disease network. Section 59.4 discusses related work on the use of EVC for assessing the topological importance of vertices in real-world networks. Section 59.5 concludes the paper and outlines plans for future work. Throughout the paper, the terms 'link' and 'edge', 'node' and 'vertex', 'network' and 'graph', 'disease' and 'disease trait' are used interchangeably. They mean the same.

59.2   Procedure

In this section, we present in detail the procedure followed to build a weighted disease-disease network graph based on a disease-gene association network and compute the eigenvector centrality (EVC) of the vertices in the disease-disease network graph. We use a running example to illustrate the different stages of our procedure.

59.2.1 Construction of Disease-Disease Network

The input to our procedure is the raw data (referred to as the disease-gene association network) extracted from the results of the disease-gene association studies. The disease-gene association network (shortly referred to as the DG graph) is a bipartite graph wherein the diseases form a partition (referred to as the set D) and the associated genes form the other partition (referred to as the set G). For every disease (say, d_i; where $1 \leq i \leq |D|$), there could be one or more associated genes (mutations in which could lead to the disease). A gene (say, g_j; where $1 \leq j \leq |G|$) could be shared by two or more diseases: in such a case, mutations in the gene could lead

to the occurrence of one or more of the diseases that share the gene. The adjacency matrix for a DG graph comprises of the diseases as rows and the genes as columns: the entry for a cell (i, j) is 1 if the gene corresponding to column j is an associated gene for the disease corresponding to row i; otherwise, the entry is a 0.

We obtain a disease-disease network (shortly referred to as the DD graph) by performing a "group projection" [12] of the DG adjacency matrix. The Math behind the group projection is illustrated below. An entry (i, k) in the DD adjacency matrix indicates the number of shared genes between the two diseases d_i and d_k. If two diseases d_i and d_k share at least one gene, we link them in the DD graph and the weight of the edge is the number of shared genes.

$$DD_{ik} = \sum_{j=1}^{|g|} DG_{ij} DG_{kj} \Rightarrow DD_{ik}$$

$$= \sum_{j=1}^{|g|} DG_{ij} DG_{jk}^{T} \Rightarrow DD = DG * DG^{T}$$

A sample disease-gene (DG) association network (that will also be used throughout this section as part of the running example) and its adjacency matrix are shown in Fig. 59.1 wherein there are five diseases (d1, d2, ..., d5) and fifteen genes (g1, g2, ..., g15). The disease-disease (DD) network for this sample DG network is built via group projection, as illustrated in Fig. 59.2. The entries in the adjacency matrix for the DD network indicate the number of shared genes between the corresponding two diseases. The size of a node in the DD network is proportional to its EVC value (the procedure is explained in Sect. 2.2).

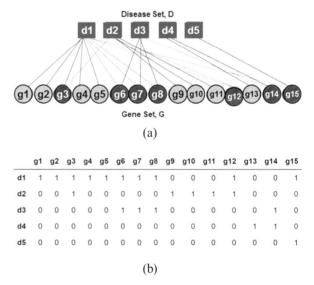

	g1	g2	g3	g4	g5	g6	g7	g8	g9	g10	g11	g12	g13	g14	g15
d1	1	1	1	1	1	1	1	1	0	0	0	1	0	0	1
d2	0	0	1	0	0	0	0	0	1	1	1	1	0	0	0
d3	0	0	0	0	0	1	1	1	0	0	0	0	0	1	0
d4	0	0	0	0	0	0	0	0	0	0	0	0	1	1	0
d5	0	0	0	0	0	0	0	0	0	0	0	0	0	0	1

(b)

Fig. 59.1 Bipartite graph representation and the adjacency matrix for a disease-gene network. (**a**) Disease-gene bipartite graph representation, (**b**) Adjacency matrix of the disease-gene bipartite graph

	g1	g2	g3	g4	g5	g6	g7	g8	g9	g10	g11	g12	g13	g14	g15
d1	1	1	1	1	1	1	1	1	1	0	0	0	1	0	0
d2	0	0	1	0	0	0	0	0	1	1	1	1	0	0	0
d3	0	0	0	0	0	1	1	1	0	0	0	0	0	1	0
d4	0	0	0	0	0	0	0	0	0	0	0	0	0	1	1
d5	0	0	0	0	0	0	0	0	0	0	0	0	0	0	1

Disease-Gene Adjacency Matrix (DG)

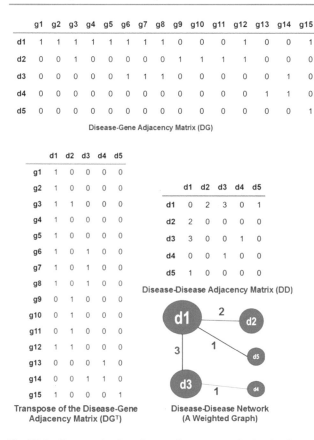

Transpose of the Disease-Gene Adjacency Matrix (DGᵀ)

Disease-Disease Adjacency Matrix (DD)

Disease-Disease Network (A Weighted Graph)

Fig. 59.2 Group projection: disease-disease network (for the disease-gene network of Fig. 59.1)

59.2.2 Computation of Eigenvector Centrality

We implement the well-known power-iteration algorithm [13] to determine the eigenvector centrality (EVC) values of the vertices in the undirected weighted disease-disease (*DD*) network graph. We start with a unit vector of all 1 s as the tentative principal eigenvector of the graph and go through a series of iterations to arrive at the final principal eigenvector, whose entries correspond to the EVC values of the vertices. In each iteration, we multiply the *DD* adjacency matrix with the tentative principal eigenvector for that iteration and normalize the resulting product vector (using the square root of the sum of the squares of the entries in the product vector). The tentative principal eigenvector for the next iteration is the normalized product vector at the end of current iteration. We repeat this process until the entries in the tentative principal eigenvector between two successive iterations do not change beyond a certain level of decimal precision.

The EVC values (in the decreasing order) and the rank of the five vertices in the disease-disease network of Fig. 59.2 computed using the power-iteration algorithm are as follows: Rank-1: d1, EVC = 0.69; Rank-2: d3, EVC = 0.58; Rank-3: d2, EVC = 0.36; Rank-4: d5; EVC = 0.18 and Rank-5: d4, EVC = 0.15. The EVC value of a vertex (ranging from 0 to 1) is a complex function of the following: the number of neighbors (degree) of the vertex, the weights of the edges incident on the vertex, the degrees of the neighbors of the vertex as well as the weights of the edges incident on the neighbors of the vertex. The vertices in the *DD* graph of Fig. 59.2 are sized proportional to their EVC values. Vertex d1 incurs the largest EVC value owing to its larger degree as well as larger weights for the incident edges. Vertex d1 is the most influential disease among the five diseases, procuring which could lead to procuring one or more of its neighboring diseases (due to the presence of a larger number of shared genes). On the other hand, vertex d4 has the lowest EVC (0.15) and is the least influential of all the five diseases. Note that, even though, both d4 and d5 have the same degree, vertex d4 has an edge with d3; whereas, vertex d5 has an edge with d1 (that has a larger EVC than d3). Hence d5 has a relatively larger EVC than d4.

59.3   Dataset and Results

Our dataset comprises of 277 diseases/traits and their associated 9445 genes, as reported in the NIH GWAS (Genome-Wide Association Studies) catalog [4] and OMIM (Online Mendelian Inheritance in Man) database [5]. The number of genes associated with a disease ranges from 5 to 254. Out of the 9445 genes, 3545 genes are shared by two or more diseases. Note that some of the nodes in the disease set could be 'traits' that represent the characteristics of a person or a specific disease (like height, tumor biomarker, etc). For simplicity, we refer to all the nodes as 'disease nodes' and treat them equally. More information about the dataset can be found in [8].

We first construct a bipartite graph of the disease-gene (*DG*) association network by identifying the diseases and the genes associated with each of them. We then perform group projection of the *DG* bipartite graph to generate the disease-disease (*DD*) network. We observe four of the 277 diseases/traits (Blood trace element, Pancreatitis, Parent of origin effect on language impairment: paternal, and Social communication problems) to be isolated (i.e., do not share genes with any other node). Hence, we do not consider them for further analysis, and the *DD* network is treated to be a graph of 273 nodes. A node could share the same gene with two or more neighbors. In order to handle such scenarios for quantitative analysis, we consider two variants for the number of genes shared by a disease/node with its neighbors: the number of unique shared genes (a shared gene is counted only once even if shared with two or more neighbors) and the fraction of raw genes shared (the ratio of the number of unique shared genes and the raw gene count associated with the disease).

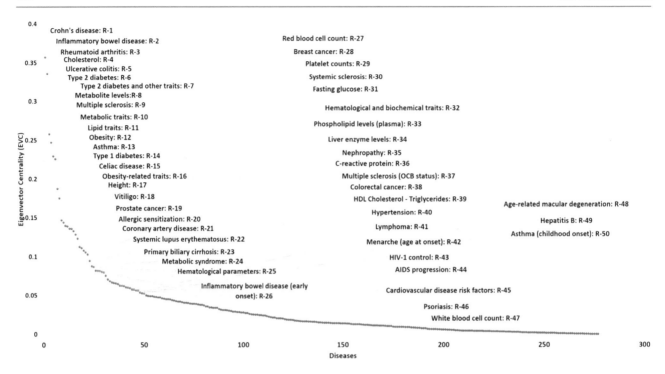

Fig. 59.3 Distribution of the EVC values of the diseases/disease traits (incl. names and ranks of top 50)

The diseases with higher EVC scores share a significant number of genes with the neighboring diseases, but such diseases count less than 20% of the total number of diseases. Whereas, lower and comparable EVC scores for more than 80% of the diseases is an indication that most of the diseases share fewer genes (but comparable in number) and have only few neighbors in the disease-disease network. A closer look at the distribution of the EVC values of the top 50 diseases (i.e., 18% of the diseases with higher EVC scores) in Fig. 59.3 indicates that it is self-similar in nature to that of the overall distribution of the EVC values of the diseases: the top 10 EVC values (among the top 50 diseases) range from 0.14 to 0.36, whereas, the bottom 10 EVC values (among the top 50 diseases) range from 0.05 to 0.06.

Figure 59.4c, d illustrate an interesting and contrasting pattern. In Fig. 59.4c, we observe the distribution of the EVC values vs. the fraction of raw genes to be half-normal in nature (with a concave down pattern of increase). The EVC value for a vertex need not be high even if the vertex has a larger fraction of raw genes. Whereas, from Fig. 59.4d, we observe the distribution of the EVC values vs. the number of unique shared genes to exhibit a concave up pattern of increase: i.e., a disease with a larger value for the number of unique shared genes is more likely to incur a larger EVC value.

The top ten among the 50 diseases/traits (the most influential) in the decreasing order of their rank are as follows: R-1: Chron's disease [EVC 0.36], R-2: Inflammatory bowel disease [EVC 0.34], R-3: Rheumatoid arthritis [EVC 0.26],

R-4: Cholesterol [EVC 0.25], R-5: Ulcerative colitis [EVC 0.229], R-6: Type 2 diabetes [EVC 0.225], R-7: Type 2 diabetes and other traits [EVC 0.19], R-8: Metabolite levels [EVC 0.18], R-9: Multiple sclerosis [EVC 0.15], R-10: Metabolic traits [EVC 0.14].

59.4 Related Work

The eigenvector centrality (EVC) metric has been widely applied for research problems in several network domains, including biological networks. For undirected biological networks, EVC has been primarily used to identify the dominant genes or proteins in gene-gene interaction networks [14] or protein-protein interaction networks [15] respectively. For directed biological networks, in a recent work [16], the EVC metric has been used for quantifying the topological importance of nodes based on their occurrence in motifs of different sizes. Though a neighborhood-based metric, EVC has been observed to exhibit a strong correlation with the shortest path-based closeness centrality metric for transcription regulation networks [17]. In [10], the authors proposed an automatic approach based on text mining (dependency parsing and support vector machines) and network analysis (use of centrality metrics) to predict disease-gene associations that could not be experimentally determined. The EVC and degree centrality metrics were observed to be accurate in highly ranking the genes that were related to the disease of interest (Prostate Cancer)

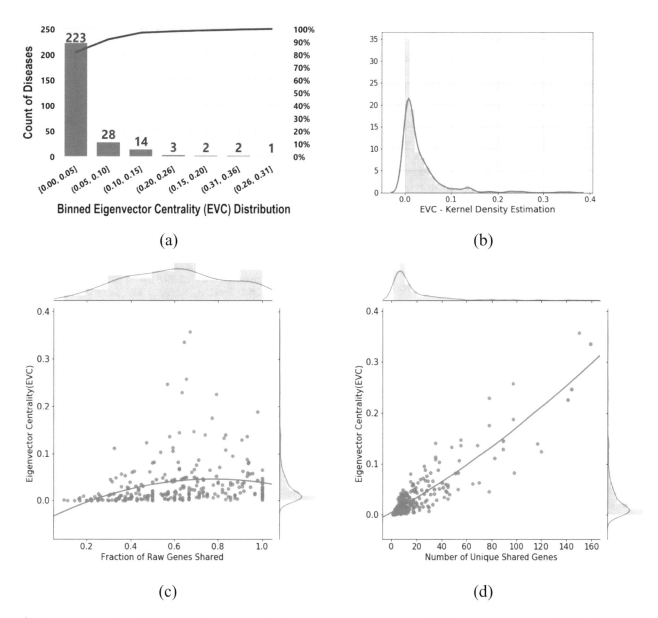

Fig. 59.4 Analysis of the EVC values of the vertices in the disease-disease network graph. (**a**) EVC: Binned distribution, (**b**) EVC: Kernel density estimation, (**c**) Fraction of raw genes shared vs. EVC, (**d**) Number of unique shared genes vs. EVC

in that paper. To the best of our knowledge, we have not come across any work that has used the EVC metric to identify the most influential diseases in a weighted disease-disease network graph, wherein the weight of an edge is the number of shared genes between the constituent end vertices (diseases). The approach presented in this paper can be applied for analyzing the disease-gene associations for any organism as well as prescribe personalized medication (as in [18]).

Some of the applications of the EVC metric for non-biological networks are as follows: In [19], it was shown that for two graphs to be isomorphic, it is necessary for the sorted order of the EVC values of the vertices in the sorted order of the EVC values of the vertices could be an effective pre-requisite check that be conducted to test for graph isomorphism. In [3], the EVC metric was effectively used to trace the trajectory of a radioactive (RDD) device in a sensor network: the weight of an edge represents the strength of the radioactive signals detected by the end vertices of the edge, and the RDD is predicted to be in the vicinity of the node with the largest EVC. In [20], the EVC metric has been used to quantify the stability of links in cognitive radio sensor networks wherein the weight of an edge is the number of common primary channels that are available in the neighborhood of two secondary nodes.

59.5   Conclusions and Future Work

We have essentially proposed an eigenvector centrality-based framework to study the disease-disease network for any organism on the basis of the number of shared genes associated with the diseases. Though it is common knowledge that only very few diseases are highly influential and could lead to other diseases, to the best of our knowledge, we have not come across any prior work that presents a ranking of these highly influential diseases on the basis of the number of shared genes. In this paper, we present a ranking of the diseases based on the EVC values of the vertices and show that the distribution of the EVC values of the vertices exhibits a Pareto pattern (80-20 rule): only about 18% of the diseases have significantly larger EVC values (that are also appreciably different from each other and share a significant number of genes with other diseases) and are the most influential diseases in the network; the remaining 82% of the diseases have low and comparable EVC values (with fewer shared genes and fewer neighbors). We observe disease vertices whose incident edges have a larger number of unique shared genes to typically incur a larger EVC value; on the other hand, disease vertices with a larger fraction of the raw genes need not necessarily have a larger EVC value.

For future work, we plan to build a gene-gene network based on the disease-gene associations such that two genes are connected with an edge if they are associated with at least one disease (the weight of a gene-gene edge would be the number of such associated diseases) and rank the genes on the basis of the EVC values. We are interested in exploring whether the distribution of EVC values of the genes would also be Pareto in nature. We also plan to run community detection algorithms on the disease-disease network to identify disease clusters based on the number of shared genes.

Acknowledgements   This research was funded by the NASA EPSCoR subaward #: NNX14AN38A from the University of Mississippi; the NSF MRI Grant 13-38192 and NSF CNS Grant 14-56638.

References

1. Newman, M.: Networks: An Introduction, 1st edn. Oxford University Press, Oxford, UK (2010)
2. Meghanathan, N.: Correlation coefficient analysis of centrality metrics for complex network graphs. In: Proceedings of the 4th Computer Science Online Conference, (CSOC-2015), Intelligent Systems in Cybernetics and Automation Theory: Advances in Intelligent Systems and Computing, April 27–30, vol. 348, pp. 11–20 (2015)
3. Meghanathan, N.: An eigenvector centrality-based mobile target tracking algorithm for wireless sensor networks. Int. J. Mob. Netw. Des. Innov. **6**(4), 202–211 (2016)
4. GWAS Catalog. https://www.ebi.ac.uk/gwas/. Last accessed: 9 Nov 2018
5. OMIM Database. https://www.omim.org/. Last accessed: 9 Nov 2018
6. Human Genome Project. https://www.genome.gov/12011238/an-overview-of-the-human-genome-project/. Last accessed: 9 Nov 2018
7. Zhang, Y., Shen, F., Mojarad, M.R., Li, D., Liu, S., Tao, C., Yu, Y., Liu, H.: Systematic identification of latent disease-gene associations from PubMed articles. PLoS One. **13**(1: e0191568), 1–23 (2018)
8. Yang, J., Huang, T., Song, W., Petralia, F., Mobbs, C.V., Zhang, B., Zhao, Y., Schadt, E.E., Zhu, J., Tu, Z.: Discover the network mechanisms underlying the connections between aging and age-related diseases. Sci. Rep. **6**(32566), 1–12 (2016)
9. Patnala, R., Clements, J., Batra, J.: Candidate gene association studies: a comprehensive guide to useful in silico tools. BMC Genet. **14**(39), 1–11 (2013)
10. Özgür, A., Vu, T., Erkan, G., Radev, D.R.: Identifying disease-gene associations using centrality on a literature mined gene-interaction network. Bioinformatics. **24**(13), 277–285 (2008)
11. Arnold, B.C.: Pareto Distributions, 1st edn. International Cooperative Publishing House, Fairland (1983)
12. Banerjee, S., Jenamani, M., Pratihar, D.K.: Properties of a projected network of a bipartite network. In: Proceedings of the 2017 International Conference on Communication and Signal Processing, pp. 143–147, Chennai (2017)
13. Panju, M.: Iterative methods for computing eigenvalues and eigenvectors. Waterloo Math. Rev. **1**(1), 9–18 (2011)
14. Liseron-Monfils, C., Ware, D.: Revealing gene regulation and associations through biological networks. Curr. Plant Biol. **3–4**, 30–39 (2015)
15. Jalili, M., Salehzadeh-Yazdi, A., Gupta, S., Wolkenhauer, O., Yaghmaie, M., Resendis-Antonio, O., Alimoghaddam, K.: Evolution of centrality measurements for the detection of essential proteins in biological networks. Front. Physiol. **7**(375), 1–4 (2016)
16. Wang, P., Lu, J., Yu, X.: Identification of important nodes in directed biological networks: a network motif approach. PLoS One. **9**(8: e106132), 1–15 (2014)
17. Koschutzki, D., Schreiber, F.: Comparison of centrality for biological networks. In: Proceedings of the German Conference on Bioinformatics, October 4–6, pp. 199–206, Bielefeld (2014)
18. Siddiqi, J., Akhgar, B., Gruzdz, A., Zaefarian, G., Ihnatowicz, A.: Automated diagnosis system to support colon cancer treatment: MATCH. In: Proceedings of the Fifth International Conference on Information Technology: New Generations, pp. 201–205, Las Vegas (2008)
19. Meghanathan, N.: Exploiting the discriminating power of the eigenvector centrality measure to detect graph isomorphism. Int. J. Found. Comput. Sci. Technol. **5**(6), 1–13 (2015)
20. Meghanathan, N.: Eigenvector centrality-based stable path routing for cognitive radio ad hoc networks. Int. J. Netw. Sci. **1**(2), 117–133 (2016)

Thiago Silva Chiaradia, Rodrigo Duarte Seabra, and Adriana Prest Mattedi

60.1 Introduction

The demographic transition can be observed at different stages round the world. In conjunction with this transition, the main demographic phenomenon of the twenty-first century is verified, known as population aging [1]. National demographic studies carried out by the Brazilian Institute of Geography and Statistics (IBGE) show that the aging of the population is not only a characteristic of developed countries. According to the data of the Demographic Census of 2010, released by the IBGE, an increase was observed in the participation of the population aged 60 years or more in Brazil, rising from 5.9% in 2000 to 7.4% in 2010, and the population of the Brazilian elderly represents a contingent of almost 15 million people, which represents 8.6% of the Brazilian population. The absolute growth of the Brazilian population from 2000 to 2010 occurred especially in the light of the growth of the adult population [2]. This growth has put the elderly as an emerging age group [3].

One of the most obvious assets of this technological era is related to mobile issues. According to data collected by [4], smartphones are being adopted at an accelerated rate, with evident benefits and scenarios in which the use of these devices are essential. The increase in the elderly population has also led to their growing introduction in the labor market, increasing the interest of this public in new technologies, among which, mobile devices.

In the face of this scenario, both accessibility and usability must prevail in the products and systems developed. The technology should be designed to meet diverse needs, with emphasis on support to older users who seek, by means of technological resources, to have an independent and interactive life [5]. However, the way the user interfaces are designed today does not favor the interaction of this population in general, because the different needs of users are not considered, especially of those who are not digitally literate [6]. Even among senior individuals, there are differences in terms of experience with technologies. The user interface designers should consider that as people get older, their sensory and cognitive skills present different types of responses [5].

One of the greatest challenges of the researchers in the area of Human-Computer Interaction (HCI) is to provide interfaces that meet to the largest possible number of users regardless of their sensory, physical, cognitive and emotional abilities. According to [7], during the design of the development of a project of interfaces, it is necessary to take into consideration the aspects of usability for all stakeholders in the system, such as the elderly user in the interaction with cell phones software.

With the advancement of artificial intelligence used in mobile devices, applications such as virtual assistants on smartphones are available to facilitate the user tasks. A more specific example is the Siri virtual assistant, present in the iOS devices of the iPhone smartphone and used as an object of study herein. This assistant has the ability to understand the context in which users are inserted and help them perform various types of tasks via voice commands [8].

From the scenario above, this research aims to evaluate the usability of the Siri virtual assistant focusing on elderly users. In this context, the research aims to analyze the impact of age

T. S. Chiaradia · R. D. Seabra (✉) · A. P. Mattedi
Institute of Mathematics and Computing, Federal University of Itajubá,
Itajubá, Minas Gerais, Brazil
e-mail: rodrigo@unifei.edu.br

© Springer Nature Switzerland AG 2019
S. Latifi (ed.), *16th International Conference on Information Technology-New Generations (ITNG 2019)*,
Advances in Intelligent Systems and Computing 800,
https://doi.org/10.1007/978-3-030-14070-0_60

on users' performance in tasks using the virtual assistant, on the basis of the usability criteria defined by software engineering. In addition, the elderly people have attracted several studies involving their behavior [9, 10].

60.2 Theoretical Foundation

Usability of mobile device interfaces is differentiated because, when it comes to these devices, the physical limitations should be considered. For example, smartphones screens are considerably smaller than those of personal computers [11]. In their studies, [12] found that, in the mobile platform, the successful applications focused design not only on the screen physical limitation, but also on other limitations, such as smaller access bands, processing power and reduced memory, besides differences between methods of data entry, when compared to personal computers. [13] conducted a review of the usability models for smartphones and showed that usability can be measured by three attributes: effectiveness, efficiency and satisfaction. However, when it comes to the elderly, [14] states that a different treatment should be given to the usability issues in applications targeted to that audience.

To reduce access barriers to technology, solutions are needed to deal with usability and accessibility, because the proximity of technology, more precisely the use of mobile devices, can reduce the limitations and difficulties provided by age and bring many benefits to the lives of the elderly [15]. Due to the benefits informatics offers, an increasing number of seniors who are increasingly interested by the cyber world has been verified both at the national and world level [16]. If, on the one hand, the new generations are familiar with the use of technological innovations that arise rapidly, the elderly are against it, feeling lost amid rapid technological advancement which causes awkwardness and fear [17]. Some individuals of this generation feel illiterate before new technologies, revealing difficulties in understanding the new language and in dealing with technological advances and can find complications in even basic questions, for example, simply making a call on a smartphone. Along aging, several competencies are developed, while others are lost, above all, the ability to adapt to new technologies.

60.3 Method

The research involved the participation of young and old people, who acted as volunteers. Three distinct tasks were proposed (T1, T2 and T3) to participants using the Siri virtual assistant. Each completed task in the assistant was followed by a questionnaire, so that three questionnaires were applied to each volunteer. The objective was to in-

vestigate the perceptions of those volunteers with regard to the tasks proposed using the virtual assistant. A fourth questionnaire was applied with for profiling individuals. To ensure the highest possible quality of responses, the tasks and the questionnaires were always applied in the presence of the researchers involved in the study. In this way, the necessary explanations for the proper conduct of the experiment were provided, ensuring that the questions were always solved. The experiment was performed using the same appliance model smartphone iPhone 5s 16GB with the same 3G Internet connection, allowing to eliminate any type of variation in the test environment. The tests were conducted in the participants' homes or in their work environments and all agreed to participate in the survey by signing an informed consent form.

For completing the research, the basic principles of usability and mobile usability aspects relevant to elderly people [5, 18] were selected in the literature [19, 20], which served as the basis for preparing the questionnaires. These aspects cover both usability and hedonic issues. Thus, it is necessary to take into account the progressive limitations that occur in elderly individuals. The research was restricted to considering the limitations related to the performance: loss of sight, hearing, motor and cognitive function. As regards hedonic aspects, the study also cared to evaluate the satisfaction users of the Siri virtual assistant had to complete tasks. Siri was selected as the object of this research due to not yet being a virtual assistant popularized in Brazil, so that most of the participants of the study had no prior knowledge of their use.

Whereas reductions of cognition, vision and hearing occur with the passage of time, we observe the time of learning, performance and the rate of errors made by the user in the tasks performed in the virtual assistant. Other factors related to the reduction of cognition also influence the sedimentation of knowledge because with the advancement of age, the memory performance is also affected [21]. The subjective satisfaction of each participant in accomplishing the task was also measured, in order to evaluate how users felt satisfied when using the system interface. The objective was to evaluate the following usability criteria [19, 20]: learning time, performance, rate of mistakes made by the user, sedimentation of knowledge by experience and subjective satisfaction.

The set of questionnaires was composed of an initial group of issues, for tracing the profile of each volunteer. The contents of this group involved aspects such as: age, knowledge and smartphone use time, purpose of use and if the participant had already been in contact with some virtual assistant, in addition to their educational level and occupation. This initial step intended to verify if the cultural or customary aspects of the volunteers would influence the performance of tasks using the virtual assistant.

Table 60.1 Issues related to usability criteria

Learning time	1. Was performing the tasks easy?
	2. Were you able to understand what was happening during the execution of the tasks?
	3. Are the pieces of information presented on the screen easily understood?
	4. Can I quickly understand most of the information presented on the screen?
	5. Did you have to learn a lot of things to accomplish the tasks?
Performance	6. How would you rate the time spent to accomplish the tasks?
	7. How do you evaluate the simplicity to accomplish the task?
	8. Does the Siri virtual assistant make executing the tasks more agile?
Rate of mistakes made by the user	9. Were you able to perform the tasks without making mistakes?
	10. If you made any mistakes during the execution of the tasks, was it easy to fix it?
	11. If you made a mistake in some task, would you be able to detect the cause?
	12. If you made a mistake, did the Siri virtual assistant help you to solve the mistake?
Sedimentation of knowledge by experience	13. Would you be able to perform the same tasks again?
	14. Is it easy to remember how to perform the tasks?
	15. Was the path taken to complete the tasks intuitive?
Subjective satisfaction	16. Did you feel satisfied when performing tasks on your smartphone?
	17. Is the interface of the Siri virtual assistant attractive?
	18. Did you feel comfortable using the Siri virtual assistant?
	19. Would you use the Siri virtual assistant again to repeat the task?

After the completion of each task, a questionnaire containing 19 questions was applied, with answers according to the five-point Likert scale, which encompassed the usability criteria considered in the study. One question about the completion of the tasks was also employed as an instrument of evaluation along with the analysis of the time taken in it. The issues related to the usability criteria set are listed in Table 60.1.

60.3.1 Participants and Description of Method

The research involves individuals of two different ages groups, randomly selected, covering young people and adults, in the range of 17–29 years of age, as well as older people with ages varying between 60 and 85 years of age. Two groups were then composed: one consisting of young people and the other of elderly people (over 60). Both groups were composed of 30 volunteers each, totaling 60 participants. The average age of the group of young participants was 23.6 years of age, with a standard deviation of 3.6 years. Among the elderly, the average age was 71 years, with a standard deviation of 8.2 years.

The profile of the young participants was diverse, consisting of high school students, academics and workers. Among these, 97% responded that they used smartphones daily, with an average of daily use time of 3 hours and 57 minutes. The interests in use ranged from accessing social networks, using various applications, conducting research on the Internet, making calls, exchanging messages, sending e-mails and playing. The profile of the elderly regarding the use of smartphones was 33% use or have used it and 67% had never had contact with the device. Among those who use or have used the device, the average daily use was 1 hour and 53 minutes. As for education, 46.7% of elderly

participants have incomplete elementary school, 16.7% have incomplete high school, 13.3% have complete elementary school, 10% have incomplete higher education incomplete, 6.7% completed high school and 6.6% of the participants had completed higher education and graduate studies. The interests in the use of smartphones for this audience was the same as for the non-elderly participants.

The layout of the tasks performed by using the virtual assistant occurred as follows: the first task (T1) requested that the participant made a call to the fictional contact "Thiago Silva", which was previously registered on their smartphone. The second task (T2) involved sending a message to the same fictional contact "Thiago Silva" with the following content: "Test message to Thiago". The third task (T3) requested that the participant created a reminder concerning the scheduling of an appointment, for example, "meeting next Wednesday". The volunteers received the guidelines containing the steps required for executing the tasks. Before beginning each task, each volunteer read all the guidelines and subsequently followed its implementation. Researchers counted the time of execution of each task of the study, registering the time in seconds. Then, the participants answered a questionnaire concerning that task. The researchers did not exert any influence on the participants with regard to the execution of tasks, or any assistance on questions of how to use the smartphone or the Siri virtual assistant, find information or perform steps.

60.4 Discussion of Results

Among the 60 participants, 30 elderly volunteers and 30 non-elderly, is relevant to mention that 10% were unable to complete any task using the Siri virtual assistant, giving up the participation in the research. Therefore, the responses of these participants were not considered during the survey data. It was found that participants who did not complete any task were elderly, with incomplete basic education, that had never used the Internet or had contact with any kind of smartphone. The average age of this portion of the elderly group was 79 years of age. Two factors were key to the withdrawal of the completion of the task, being the degree of education and loss of hearing due to age. In relation to the level of education, illiteracy interfered directly, so that those volunteers were unable to read or identify what was displayed on the screen during the interaction with the smartphone. In the case of a participant with hearing loss at a very advanced stage, the difficulty in hearing the feedback given by the virtual assistant prevented the task continuation.

The conclusions regarding the results from the experiment begin with evident differences in the performance by the two groups. The elderly volunteers concluded the three tasks with an average of 552 seconds; the group of young people, with an average of just 38 seconds. In this context, the young people completed the same tasks as the elderly at a time almost 15 times smaller.

Analyzing Task 1, it a relation was verified between the learning time for accomplishing the task using the Siri virtual assistant and the use of the smartphone. Users who already used smartphones were able to accomplish the task aided by the virtual assistant 75% faster as compared to users who had never used smartphones, taking 160 seconds on average to complete T1. With regard to individuals who use smartphones, none of them considered the task execution with the virtual assistant difficult or very difficult. Note, therefore, that users' experience influences usability. In relation to the run time of T1, among elderly individuals who had difficulty in understanding the information presented on the screen, an average of 832 seconds to complete the task was established. In turn, those that had no trouble took an average of 95 seconds. Similarly, the volunteers who judged it necessary to learn a lot of things to accomplish the task using the Siri virtual assistant consumed more time to complete the task taking 782 seconds on average. Those who expressed the need to learn almost nothing or nothing obtained an average of 186 seconds. Still in the question concerning the time of learning in relation to T2, the time taken to perform the task with the virtual assistant was longer for the elderly, even considering the participants used with smartphones. These consumed 237 seconds, on average, to complete the task, i.e. 48% more than the time spent for T1. The increase in time was also observed for elderly users not using smartphones, registering an average of 1077 seconds, i.e. nearly 30% more than the time spent to complete T1. This increase, comparing the runtimes between T1 and T2 in the elderly group, is directly connected to the criterion evaluating the simplicity of the task. In T1, the percentage of volunteers that deemed the task very simple or simple was 41%. On

the other hand, the percentage that evaluated T2 very simple or simple was only 12%. Therefore, it was found that the level of difficulty of the task and the number of interactions directly influenced the time necessary to complete it. About the learning time criterion, it is necessary to emphasize that the limitations arising from age also were directly related to the time used by the participants. Elderly participants who claimed not being able to quickly understand most of the information presented on the screen took, on average, 775 seconds to complete the task; in turn those that claimed to be able to quickly understand it took 136 seconds on average to conclude the task about 80% faster.

Observing the evaluation of users as regards the performance provided by the Siri virtual assistant during the execution of T2, it was found that only 10% of the elderly fully agreed that the virtual assistant made the task execution faster, which represents a drop compared to the percentage of T1, which was 20%. On the same criterion, T3 obtained the best evaluation indexes for elderly users, and 37% fully agreed that the virtual assistant became faster task execution. The low index established for T2 after the evaluation by the elderly resulted from the complexity and the longer time taken to complete the task. Among young users, the percentages obtained for T1, T2 and T3 were 60%, 73% and 96%, respectively, suggesting that the complexity of the task did not affect this group of volunteers.

On the rate of errors made by the user, it is relevant to mention that during the execution of T1, T2 and T3, 78% of the elderly made some kind of mistake. When checking the same criterion for non-elderly participants, a percentage of 26% was established. T2 was the task with the highest occurrence of errors between the two groups, because it demanded more interactions from participants, which directly resulted in a great number of errors. Participants who made some sort of mistake negatively evaluated the issues concerning the ease of correcting the error and identifying what had been done wrong. They also negatively evaluated the question regarding if, in the event of errors, the virtual assistant helped solve them. It is important to check that this negative assessment directly relates to the feedback from the Siri virtual assistant on the occurrence of the error. To better serve not only older adults who have limitations inherent to age, but the general users, more specific feedback about what went wrong during the execution of a task is necessary.

In questions aimed at assessing the sedimentation of knowledge by experience, a positive assessment was verified regarding the ease of rerunning the tasks proposed. For the two groups of participants, none of them entirely disagreed they would be able to accomplish the same task again. Still, 74% was the average of the elderly that agreed partly or wholly on its being easy to remember how to perform tasks. The third task was considered more intuitive, and the learning factor influenced how much the participant considered the task intuitive or not; since T3 was the last task performed, the prior knowledge generated by running T1 and T2 resulted in an increase in the facility for executing the T3.

Relating issues *ability to accomplish the same task again* (Q13) with the *intuitiveness of the path taken to complete the task* (Q15) present in the questionnaires of T1, T2 and T3, a high degree of correlation between the two variables was observed, showing that the ability to perform the same task again is linked to the intuitiveness of the path taken by the user while performing the task. The correlation coefficient found between these variables was 0.68.

The time spent in users' interaction with the virtual assistant also influences the subjective satisfaction of the elderly. Individuals who were dissatisfied with T1, on average, were found to take 906 seconds to complete the task, unlike individuals who were satisfied, and took an average of 199 seconds. On the other hand, when the elderly volunteers were asked if they would use the Siri virtual assistant to accomplish the tasks, 73% responded positively. Note, also, that even the elderly who committed some type of error would use the virtual assistant, pointing out that the subjective satisfaction may be perceived differently for each user.

60.5   Final Considerations

This study aimed to analyze the performance impact on the elderly using the Siri virtual assistant in comparison with young people. The performance of people aged over 60 was verified to be affected. Older adults spent more time to complete the same tasks performed by young people. Thus, from the example of other studies, age clearly influences the usability of the Siri virtual assistant, probably due to a gradual reduction of cognitive and motor skills.

In future research, it is relevant to assess the usability of other virtual assistants such as Google Now and Cortana, involving senior users. Another possibility in the continuity of this work lies in the scope of using the Siri virtual assistant, demonstrating the performance of its interface and how the application can contribute to improving the performance of the tasks, especially by elderly users.

References

1. Nasri, F.: O envelhecimento populacional no Brasil. Einstein. **6**(Supl 1), S4–S6 (2008)
2. IBGE: Instituto Brasileiro de Geografia e Estatística. Available in: http://www.ibge.gov.br. Accessed: Aug 2017
3. Da Silveira, M.M., et al.: Educação e inclusão digital para idosos. RENOTE. **8**(2), 4 (2010)
4. Falaki, H., et al.: Diversity in smartphone usage. In: Proceedings of the 8th International Conference on Mobile Systems, Applications, and Services, pp. 179–194. ACM (2010)
5. Carneiro, R.V., Ishitani, L.: Aspectos de usabilidade de *mobile learning* voltado para usuários com restrições decorrentes da idade. Rev. Bras. Comput. Apl. **6**(1), 81–94 (2014)
6. Pattison, M., Stedmon, A.W.: Inclusive design and human factors: designing mobile phones for older users. Psychnol. J. **4**(3), 267–284 (2006)
7. Gonçalves, V.P.: Um estudo sobre o design, a implementação e a avaliação de interfaces flexíveis para idosos em telefones celulares. Tese de Doutorado. Universidade de São Paulo (2010)
8. iOS 9: Siri. Available in: http://www.apple.com/br/ios/siri/. Accessed: Sept 2017
9. Mello, J.L.C., et al.: Application of an effective methodology for analysis of fragility and its components in the elderly. In: Proceedings of 15th Information Technology – New Generations – ITNG 2018, Las Vegas (2018)
10. Dinh, A., et al.: Implementation of a physical activity monitoring system for the elderly people with built-in vital sign and fall detection. In: Proceedings of 6th Information Technology – New Generations – ITNG 2009, Las Vegas (2009)
11. Rauch, M.: Mobile documentation: usability guidelines, and considerations for providing documentation on Kindle, tablets, and smartphones. In: Professional Communication Conference (IPCC), 2011 IEEE International, pp. 1–13. IEEE (2011)
12. Oinas-Kukkonen, H., et al.: Developing successful mobile applications. In: International Conference on Computer Science and Technology (IASTED), pp. 50–54, Cancun (2003)
13. Harrison, R., Flood, D., Duce, D.: Usability of mobile applications: literature review and rationale for a new usability model. J. Interact. Sci. **1**(1), 1–16 (2013)
14. Mol, M.A.: Recomendações de usabilidade para interface de aplicativos para smartphones com foco na terceira idade. Minas gerais, 2011. 81f. Dissertação de Mestrado. Pontifícia Universidade Católica de Minas Gerais, Belo Horizonte (2011)
15. Wagner, N., Hassanein, K., Head, M.: The impact of age on website usability. Comput. Hum. Behav. **37**, 270–282 (2014)
16. Sharples, M., Taylor, J., Vavoula, G.: A theory of learning for the mobile age. In: Medienbildung in neuen Kulturräumen, pp. 87–99. VS Verlag für Sozialwissenschaften, Wiesbaden (2010)
17. Ribeiro, S.C., Mattedi, A.P., Seabra, R.D.: Avaliando a usabilidade de *websites* com ênfase em usuários idosos: um estudo de caso. RENOTE. **13**(2), 1–10 (2015)
18. Díaz-Bossini, J.M., Moreno, L.: Accessibility to mobile interfaces for older people. Procedia Comput. Sci. **27**, 57–66 (2014)
19. Nielsen, J., Budiu, R.: Mobile Usability (MITP-Verlags GmbH & Co. KG., Pearson Education, San Francisco, 2013)
20. Shneiderman, B.: Designing the User Interface: Strategies for Effective Human-Computer Interaction (Pearson Education, San Francisco, 2010)
21. Bonardi, G., et al.: Incapacidade funcional e idosos: um desafio para os profissionais. Sci. Med. **17**(3), 138–144 (2007)

A Study on the Usability of Facebook on Mobile Devices with Emphasis on Brazilian Elderly Users

Anna Beatriz Dias Morais, Rodrigo Duarte Seabra, and Adriana Prest Mattedi

61.1 Introduction

Following a worldwide trend, in Brazil, the demographic data demonstrate an increase in the proportion of the number of elderly people in relation to the total population [1]. According to the Synthesis of Social Indicators of the Brazilian Institute of Geography and Statistics (IBGE), an increase was observed in the participation of the population aged 60 years or more in Brazil, rising from 5.9% in 2000 to 7.4% in 2010, and the population of the elderly represents a contingent of almost 15 million people in Brazil, representing 8.6% of the Brazilian population. At the same time, an increase of elderly users who use the Internet is observed. According to the 2016 demographic data, disclosed by IBGE, between 2008 and 2013, the percentage of elderly Internet users more than doubled. The Brazilian index of 60 year-olds or older with access to the network increased from 5.7% to 12.6%, corresponding to an increase of 121% [2].

In parallel to this context of population aging and the elderly interaction with virtual environments, the spread of Information and Communication Technologies (ICT) can be observed. One of the most obvious assets of this technological era is related to mobile issues. Nowadays, the use of smartphones has become current in the everyday setting of the third age. However, the technology can also be a barrier in the lives of elderly people who have no familiarity with it, including those who are not able to keep updated with the frequent innovations [3]. For this, there is a need for adaptation and learning for adequately using these devices [4].

With the advancement of technologies, social networks integrated with smartphones make more daily users, including the elderly. The evolution in the number of connected elderly is high in social networks: between 2005 and 2011, there was a growth of 222.3%, from 2.514 billion for 8.101 billion, totaling 18.4% of all the Internet users in the country. Note that the population of 60 year-olds or older was 39 million, allowing a much larger growth [2]. Social networks are spaces for socialization, with resources for sharing information, photos, videos, and, above all, communication between its users. The social network Facebook, present in mobile devices, is one of them, which will be used for research purposes herein.

The facilities generated by Facebook to adopt features with titles and content in Portuguese and the ability to post and to communicate with friends and family, including the popular "like" in the contents and comments, are important factors when the target audience in question is composed of elderly users. "Facebook is a communication media that enables you to communicate directly with friends and family, is fast and always available. It's like having a family member on your side" [5].

From the scenario above, this research aims to evaluate the usability of social network Facebook, on the mobile device interface, emphasizing users of the third age, comparing their performances to that of young users. A field research was carried out with elderly and youth to evaluate if the usability of this social network is suitable for the target audience. In order to achieve these results, certain tasks were proposed for users to perform and, subsequently, a questionnaire was applied to serve as a basis for evaluating the usability of the social network when used for this audience. In addition, the elderly people have attracted several studies involving their behavior [6, 7].

A. B. D. Morais · R. D. Seabra (✉) · A. P. Mattedi
Institute of Mathematics and Computing, Federal University of Itajubá, Itajubá, Minas Gerais, Brazil
e-mail: rodrigo@unifei.edu.br

S. Latifi (ed.), *16th International Conference on Information Technology-New Generations (ITNG 2019)*,
Advances in Intelligent Systems and Computing 800,
https://doi.org/10.1007/978-3-030-14070-0_61

61.2 Theoretical Foundation

Usability is a field of research in the area of Human-Computer Interaction (HCI) that studies a better interaction between man and machine, i.e. a better development of interfaces in which the end user feels comfortable with the design of the applications. These studies allow the custom development of systems for a specific audience, improving not only the users' relationship with technology but also how technology can adjust the restrictions present in their life [8]. The concept of usability was defined by [9] as the ability of something to be used by humans with ease and effectiveness and which can be used as a metric to define how a product can be used by multiple users, meeting goals such as efficiency, effectiveness or satisfaction. To [10], the usability is generally accepted as the guarantee that the interactive products are easy and pleasant to use by the user. [11] defines usability as a technical term used to describe the quality of use of an interface.

Currently, given the different types of devices and the new perspectives of users in relation to the consumption of content, the development of interfaces presents enormous challenges in terms of design and evaluation methodologies [12]. With this, it is necessary to take into account the specificity of the smartphone systems to evaluate the usability of their interfaces. The interaction of mobile devices is notedly more dynamic and developing applications is associated with mobility because people can carry all the computer services, and the content available can be accessed from any location.

In this context, research involving the usability of social networks, especially Facebook on mobile devices for the elderly has increased over the years. However, most of the research related to social networks found in the literature is focused on the reason for the increasing number of elderly using social networks and the reasons for this growth. In order to evaluate social networking sites in general, [13] conducted a systematic search in the scientific literature looking for several articles that report the experience of the elderly in social networks. Since Facebook is one of the largest social networks, several articles found are on this social network. The results from the research were separated into topics, such as articles included in review articles reporting tests of social networking sites with the elderly, and articles that assess the attitudes of the elderly in social networking sites. Finally, they discuss what makes the elderly use social networks, and if the interfaces of the social networking sites are adaptable to the elderly. As conclusions and future research, the authors deem interesting to conduct research into the use of social networks for smartphones.

In order to develop a new interface for the Facebook application in tablets with an emphasis on the elderly audience, [14] conducted a survey in the field to assess the use of the Facebook application on these devices for the elderly. In the first phase of the research, the authors raised what the activities carried out by the target audience in the application were. In the second phase, the elderly carried out some tasks in the Facebook application so that authors could check their difficulties. Then, the problems found in Facebook application interface in tablets were reported, and the authors developed a new Facebook application by correcting the errors found. To test the application developed, another field research was conducted and elderly users performed the same tasks in two different applications. It was possible to conclude that the application is very easy for the elderly to understand as compared to the current Facebook app.

61.3 Method

The research involved the participation of young and old people, who acted as volunteers. Three distinct tasks were proposed (T1, T2 and T3) to participants using the Facebook app. The tasks were selected based on the potential interest of social network usage for both profiles, i.e. tasks traditionally performed by users of social networks. Each completed task in the app was followed by a questionnaire, so that three questionnaires were applied to each volunteer. The objective was to investigate the perceptions of those volunteers with regard to the tasks proposed using Facebook. A fourth questionnaire was applied for profiling individuals. To ensure the highest possible quality of responses, the tasks and the questionnaires were always applied in the presence of the researchers involved in the study. Thus, the necessary explanations for properly conducting the experiment were provided, ensuring that the questions were always solved. The experiment was performed using a smartphone with an iOS operating system with the same 3G Internet connection, allowing eliminating any type of variation in the test environment. The tests were conducted in the participants' homes or in their work environments and all agreed to participate in the survey by signing an informed consent form.

For completing the research, the basic principles of usability and mobile usability aspects relevant to elderly people [10] were selected from the literature [15, 16], serving as the basis for preparing the questionnaires. These aspects cover both usability and hedonic issues. It is thus necessary to take into account the progressive limitations that occur to elderly individuals. The research was restricted to considering the limitations related to the performance: loss of sight, hearing, motor and cognitive function. As regards hedonic aspects, the study also concerned evaluating the satisfaction users of the Facebook had to complete tasks.

Other factors related to the reduction of cognition also influence the sedimentation of knowledge because, with the advancement of age, the memory performance is also affected. The subjective satisfaction of each participant in accomplishing the task was also measured, in order to evaluate how satisfied users felt when using the system interface. The objective was to evaluate the following usability criteria [15, 16]: learning time, performance, rate of mistakes made by the user, sedimentation of knowledge by experience and subjective satisfaction.

The set of questionnaires was composed of an initial group of issues, for tracing the profile of each volunteer. The contents of this group involved aspects such as: age, knowledge and smartphone use time, purpose of use and if the participant had already been in contact with a social network, in addition to their educational level. This initial step intended to verify if the cultural or customary aspects of the volunteers would influence the performance of tasks using the Facebook app.

After the completion of each task, a questionnaire containing 19 questions was applied, with answers according to the five-point Likert scale, which encompassed the usability criteria considered in the study. One question about the completion of the tasks was also employed as an instrument of evaluation along with the analysis of the time taken in it. The issues related to the usability criteria set are listed in Table 61.1.

61.3.1 Participants and Description of Method

The research involves individuals of two different ages groups, randomly selected, covering young people and adults, in the range of 15–29 years of age, as well as older people with ages varying between 60 and 83 years of age. Two groups were then composed: one consisting of young people and the other of elderly people (over 60). Both groups were composed of 30 volunteers each, totaling 60 participants. The average age of the group of young participants was 23.3 years of age, with a standard deviation of 4.5 years. Among the elderly, the average age was 68.4 years, with a standard deviation of 6.8 years.

The profile of the young participants was diverse, consisting of high school students, college students, graduate students and postgraduates. Among these, 100% responded that they used smartphones daily, with an average of daily use time of 6 hours and 34 minutes. The interests in use ranged from accessing social networks, using various applications, conducting research on the Internet, making calls, exchanging messages, sending e-mails, playing and using it as an alarm clock. The profile of the elderly regarding the use of smartphones was 66.6% use or have used it and 33.4% had never had contact with the device. Among those who use or have used the device, the average daily use was 3 hours and 30 minutes. As for education, 16.7% have incomplete elementary school, 30% have complete elementary school, 16.7% have incomplete high school, 13.3% have complete high school, 3.3% have incomplete higher education and 20% of the participants had completed higher education. The interests in the use of smartphones for this audience were the same as for the non-elderly participants.

The tasks performed in the Facebook application were conducted from a generic account created for conducting research, as follows: the first task (T1) requested the participants to add a friend to their profile, being that they could choose the friend. The second task (T2) involved posting a

Table 61.1 Issues related to usability criteria

Learning time	1. Was performing the tasks easy? 2. Were you able to understand what was happening during the execution of the tasks? 3. Are the pieces of information presented on the screen easily understood? 4. Can I quickly understand most of the information presented on the screen? 5. Did you have to learn a lot of things to accomplish the tasks? 6. Are the pieces of information presented on the screen excessive? 7. Does the interface of Facebook application helped to easily recall the steps to accomplish the same task again?
Performance	8. How would you rate the time spent to accomplish the tasks? 9. How do you evaluate the simplicity to accomplish the task? 10. Does the Facebook application interface make executing the tasks more quickly?
Rate of mistakes made by the user	11. Were you able to perform the tasks without making mistakes? 12. If you made any mistakes during the execution of the tasks, was it easy to fix them? 13. If you made a mistake in a task, would you be able to undo the action and correct the mistake?
Sedimentation of knowledge by experience	14. Would you be able to perform the same tasks again? 15. Is it easy to remember how to perform the task? 16. Was the path taken to complete the tasks intuitive?
Subjective satisfaction	17. Did you feel satisfied when performing task in the Facebook application? 18. Is the interface of the Facebook application attractive? 19. Was the language used in the Facebook application easy to understand?

message in their profile with the following content: "Today I am happy". The third task (T3) requested the participant to check if he/she had any notification and answer it in an affirmative case. The volunteers received the guidelines containing the steps required for executing the tasks. Before beginning each task, each volunteer read all the guidelines and subsequently followed the implementation. Researchers counted the time of execution of each task of the study, registering the time in seconds. Then, the participants answered a questionnaire concerning that task. The researchers did not any influence the participants with regard to the execution of tasks, or provide any assistance regarding how to use the smartphone or the Facebook application, find information or perform steps.

61.4   Discussion of Results

The conclusions regarding the results from the experiment begin with evident differences in the performance by the two groups (Fig. 61.1). The elderly volunteers concluded the three tasks with an average of 683 seconds; the group of young people, with an average of just 25 seconds. The young participants therefore completed the same tasks as the elderly at a time almost 27 times smaller. Comparing the tasks individually, the average time for the group of young participants generated the following results: 21 seconds for T1, 26 seconds for T2 and 28 seconds for T3. The average time to carry out the same tasks by the elderly was: 595 seconds for T1, 642 seconds for T2 and 813 seconds for T3.

Among the elderly users, a noticeable performance difference with regard to participants owning smartphones was observed. After collecting the data, it was found that the group of seniors who owned a smartphone showed better performance during the tasks, taking on average only 25% of the time spent by the other elders. Among the young

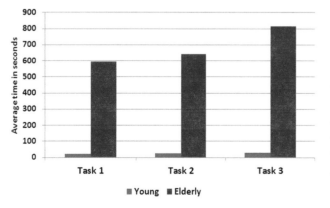

Fig. 61.1 Average time to carrying out of tasks among the elderly and young participants

participants, there were no relevant differences in performance considering this variable.

Analyzing Task 1, completed by the group of elderly participants, a relation was verified between the learning time for accomplishing the task using the Facebook app and the use of smartphone. Users who already used smartphones daily were able to accomplish the task aided by the Facebook app four times faster as compared to users who did not use smartphones daily, taking 286 seconds on average to complete T1. With regard to the daily use of smartphone for those individuals, 10% considered carrying out the task difficult and these individuals did not use any type of social network. Note, therefore, that users' experience on smartphones and social networking influences usability. In relation to the run time of T1, among elderly individuals who had difficulty in understanding the information presented on the screen, an average of 1415 seconds to complete the task was established. In turn, those that had no trouble took an average of 205 seconds. Similarly, the volunteers who judged it necessary to learn a lot of things to accomplish the task using the Facebook app consumed more time to complete the task taking 1255 seconds on average. Those who expressed the need to learn almost nothing, nothing or were neutral obtained an average of 232 seconds.

Still concerning the time of learning in relation to T2, the time taken to perform the task with the Facebook app was longer for the elderly, even considering the participants who used smartphones. These consumed 642 seconds, on average, to complete the task, i.e. a time greater than the time spent for T1, which was 595 seconds. The increase in time was also observed for elderly users not using smartphones, registering an average of 1206 seconds, i.e. 5% more than the time spent to complete T1. This increase, comparing the runtimes between T1 and T2 in the elderly group, is directly connected to the criterion evaluating the simplicity of the task. In T2, no volunteer judged the performance very easy. Conversely, in T1, the percentage of volunteers that deemed the task very simple was 10%. Therefore, it was found that the level of difficulty of the task and the number of interactions directly influenced the time necessary to complete it. About the learning time criterion, it is necessary to emphasize that the limitations arising from age were also directly related to the time used by the participants. In the question about if the interface of the application helps performing the task more quickly, the volunteers who agreed with this question totally or partially have taken, on average, 205 seconds to complete the task; those who disagreed with this statement took, on average, 1040 seconds.

Observing the evaluation of users as regards the performance provided by the Facebook app during the execution of T1, it was found that 87% of the elderly positively responded that it was simple to perform the task in the Facebook app, and no volunteer said to totally disagree with that statement.

In the same criterion, during the execution of T3, 7% of the elderly differed totally from the simplicity of the task and we conclude that this number of users evaluated the task as complicated to be performed. In relation to the time taken to complete tasks T1, T2 and T3, 33% of elderly agreed partly or totally in both tasks.

On the rate of errors made by the user, it is relevant to mention that during the execution of T1, T2 and T3, 77% of the elderly made some kind of mistake. When checking the same criterion for non-elderly participants, it can be observed that any volunteer of this group made some kind of mistake. T1 and T2 were the tasks with the highest occurrence of errors between the two groups. Some users who were not used to the interface of the Facebook app had some difficulty at the beginning, until getting used. Both tasks had the same percentage of error; 26% of users made some kind of mistake. Participants who made some sort of mistake positively evaluated the issues concerning the ease of correcting the error. They also positively evaluated the question regarding if, in the event of errors, the Facebook app helped solve them. Note that this positive assessment directly relates to the feedback from the Facebook app.

In questions aimed at assessing the sedimentation of knowledge by experience, a positive assessment was verified regarding the ease of rerunning the tasks proposed. In the elderly group, only 6% was the average of participants that disagreed partly or totally on its being easy to remember how to perform tasks again. Still, 70% was the average of elderly volunteers who agreed totally or partially to its being easy to remember how to perform tasks again. In both tasks, the elderly volunteers positively judged the Facebook interface in terms of intuitiveness to help complete the task. On average, 77% of the participants agreed totally or partially with this aspect.

The time spent in users' interaction with the Facebook app also influences the subjective satisfaction of the elderly. Individuals who were dissatisfied or very dissatisfied with T3, on average, were found to take 1529 seconds to complete the task, unlike individuals who were satisfied or very satisfied, and took an average of 706 seconds. Note, also, that no elderly person totally disagreed that the Facebook interface is attractive, that is, everyone considered the interface attractive.

61.5 Final Considerations

This study aimed to analyze the performance impact on the elderly using the Facebook app in comparison with young people. The performance of people aged over 60 was verified to be affected. Older adults spent more time to complete the tasks and took longer to find the commands in the application interface when compared to the non-elderly. Thus, from the example from other studies, age clearly influences the usability of the Facebook app, probably due to a gradual reduction of cognitive and motor skills.

In future research, it is relevant to assess the usability of other virtual assistants, such as Instagram and Twitter, involving senior users, seeking to promote a comparison between the usability of the different interfaces offered by these social networks.

References

1. Mol, M.A.: Recomendações de usabilidade para interface de aplicativos para smartphones com foco na terceira idade. Minas Gerais, 2011. 81f. Dissertação de Mestrado. Pontifícia Universidade Católica de Minas Gerais, Belo Horizonte (2011)
2. IBGE: Instituto Brasileiro de Geografia e Estatística. Available in: http://www.ibge.gov.br. Accessed: Aug 2017
3. European Communities: The key competences for lifelong learning: a European framework. Available in: http://ec.europa.eu/dgs/education_culture/publ/pdf/-lllearning/keycomp_en.pdf. Accessed: Mar 2017
4. Matos, E.M.L.: Idosos e os dispositivos móveis: novas abordagens de interação. Lisboa, 2014. 142f. Dissertação de Mestrado – Universidade de Lisboa (2014)
5. Patrício, M.R., Osório, A.: Como os adultos idosos usam o Facebook para literacia digital e aprendizagem ao longo da vida: um estudo de caso de aprendizagem intergeracional. In: III Congreso Ibérico de Innovación en Educación com las TIC (2014)
6. Mello, J.L.C., et al.: Application of an effective methodology for analysis of fragility and its components in the elderly. In: Proceedings of 15th Information Technology – New Generations – ITNG 2018, Las Vegas (2018)
7. Dinh, A., et al.: Implementation of a physical activity monitoring system for the elderly people with built-in vital sign and fall detection. In: Proceedings of 6th Information Technology – New Generations – ITNG 2009, Las Vegas (2009)
8. Hong, Y.W.B.: Matters of design. Commun. ACM. **54**(2), 10–11 (2011)
9. Shackel, B.: Usability – context, framework, definition, design and evaluation. Interact. Comput. **21**(5–6), 339–346 (2009)
10. Preece, J., Rogers, Y., Sharp, H.: Interaction Design: Beyond Human-Computer Interaction. Wiley, New York (2002)
11. Bevan, N.: Usability is quality of use. Adv. Hum. Factors Ergon. **20**, 349–354 (1995)
12. Bertini, E., et al.: Appropriating heuristic evaluation for mobile computing. Int. J. Mobile Hum. Comput. Interact. **1**(1), 20–41 (2009)
13. Nef, T., et al.: Social networking sites and older users – a systematic review. Int. Psychogeriatr. **25**(7), 1041–1053 (2013)
14. Gomes, G., Matos, J.C.E., Duarte, C.: Estudo de uma nova interface para o Facebook centrada em utilizadores idosos. In: 5th Conferência Nacional em Interacção Pessoa-Máquina. Education India (2013)
15. Nielsen, J., Budiu, R.: Mobile Usability. MITP-Verlags GmbH & Co. KG. Pearson Education, San Francisco (2013)
16. Shneiderman, B.: Designing the User Interface: Strategies for Effective Human-Computer Interaction. Pearson Education, San Francisco (2010)

A Survey on the Needs of Visually Impaired Users and Requirements for a Virtual Assistant in Ambient Assisted Living

62

Juliana Damasio Oliveira and Rafael H. Bordini

62.1 Introduction

Ambient Assisted Living (AAL) technologies are an excellent opportunity for improving people's lives, especially for those who live with disability, illness, or aging [1]. This is an emerging area that brings intelligence to our everyday environments and makes those environments sensitive to us [2]. It is expected that a suitable AAL system should allow people to interact with it in a natural and personalised way, taking into consideration their specific interests, needs, demands, requirements, and abilities as well as disabilities [3].

According to the Brazilian Institute of Geography and Statistics (IBGE) [4], there are approximately 6.5 million people who are visually impaired in Brazil. These people face many barriers in their daily lives to carry out activities that are simple for most people, such as identifying objects, finding objects, predicting obstacles, locomotion, and orientation and mobility. Several systems have been introduced in recent years to address some of the related issues, such as locomotion assistance [5], identification of cash [6], and identification of product labels [7]. The available applications become great allies to their users, specially when providing different features in a single device, reducing costs and providing portability [8]. However, despite efforts being made, there are still few applications to help people who are visually impaired become more independent in their homes.

Much work has also been developed for assistance through robots and Ambient Assisted Living technologies but the majority focusing on elderly people instead [9, 10], who have different needs from people who are visually impaired. Moreover, several solutions lack features and are difficult to use, because they are not oriented particularly to the specific requirements of blind users and are therefore not accepted by them [11, 12].

This work is part of a larger work that aims at the creation of a multiagent architecture for environmental control including a virtual assistant for people who are visually impaired. A virtual assistant is a software agent that can perform tasks or services for an individual [13]. This virtual assistant may be included in a companion robot in the future.

In this first part that composes our work, we adopted a human-centered approach to investigate the needs and requirements for a virtual assistant in AAL for people who are visually impaired. We created a qualitative/quantitative survey questionnaire for two main research questions: (1) What are the needs and barriers faced by people who are visually impaired in their homes? and (2) Are people receptive to the idea of companion robots or a virtual assistant in their home?

The main contribution of this paper is to pave the way for an innovative assistant or companion robot for people who are visually impaired that are able to assist the user with daily tasks such as locating objects and preventing accidents due to things being left in the wrong place. Before any research group can embark in such endeavours, it is necessary to make sure the right set of features is known, so that the resulting products can be of the best possible assistance and also effectively adopted by the users.

This paper is organised as follows. Section 62.2 reviews prior related papers. Section 62.3 describes the method used. Section 62.4 shows the results obtained. Section 62.5 presents the our discussion about the results. Section 62.6 presents final remarks and future work.

J. D. Oliveira (✉) · R. H. Bordini
School of Technology, Pontifical Catholic University of Rio Grande do Sul, Porto Alegre, Brazil
e-mail: juliana.damasio@acad.pucrs.br; rafael.bordini@pucrs.br

© Springer Nature Switzerland AG 2019
S. Latifi (ed.), *16th International Conference on Information Technology-New Generations (ITNG 2019)*,
Advances in Intelligent Systems and Computing 800,
https://doi.org/10.1007/978-3-030-14070-0_62

62.2 Related Work

We concentrate this section on reporting about previous work that has surveyed the needs of the use of assistive technology. Below, we use n to mean the number of respondents of the questionnaires or participants of experiments.

Heek et al. [14] investigate benefits and barriers of AAL technologies, which were contrasted in four user groups: healthy "not-experienced" people, disabled, their relatives, and professional care givers. They applied a qualitative interview pre-study ($n = 9$) and a validating questionnaire study ($n = 279$). Results indicate that disabled and people in need of care show a higher acceptance and intention to use an AAL system than "not-experienced" people or care givers. Also, the results show the importance to integrate diverse user groups (age, disabilities) into the design and evaluation process of AAL technologies.

Maan and Gunawardana [15] investigate the barriers in acceptance of Ambient Assisted Living Technologies among elder Australians. For this purpose, they used mixed approaches through a combination of written questionnaire and qualitative methods such as focus groups ($n = 25$). The results show that there are different factors that restrict the use of these technologies and elderly people have certain preferences when using the technology. An understanding of these barriers is gained to provide solutions according to user needs.

Dautenhahn et al. [16] explored people's perceptions and attitudes towards the idea of a future robot companion for the home. A human-centered approach was used with questionnaires and human-robot interaction ($n = 28$ adults). A series of questionnaires were collected before and after an interaction session with a PeopleBot robot. Results indicate that 40% of the participants in that study were in favour of the idea of having a companion robot in the home. Most subjects saw the potential role of a companion robot in the home as being an assistant, machine or servant. In terms of specific tasks for a robot companion, 90% stated that it would be useful for the robot to do the vacuuming.

62.3 Survey

We created a qualitative and quantitative online questionnaire study to assess the needs and barriers faced by people who are visually impaired in their homes. The aim was to investigate the necessary characteristics in ambient assisted living for these people. We chose this approach to reach more people, as well as being affordable and effective [17,18]. The questionnaire was made available from March to September 2018, and was divided into three parts:

- Term of consent to participate in the survey with information about the project and questionnaire.

Table 62.1 The content of the questionnaire

Questions	Response type
Which of these technologies have you heard about? (virtual assistant, computer, companion robot, smartphone, tablet, smart tv, home automation, robot)	Multiple choice
Which of these technologies have you had contact with? (virtual assistant, computer, companion robot, smartphone, tablet, smart tv, home automation, robot)	Multiple choice
What is your level of knowledge about companion robots?	5-point likert scale
What is your level of knowledge about virtual assistants?	5-point likert scale
Do you like the idea of having a virtual assistant at home?	5-point likert scale
What technologies would you like for helping with daily activities at home?	Open question
What difficulties do you face in daily activities while being alone at home?	Open question
What are the technologies that most help you (when alone) in your daily activities at home?	Open question
What tasks would you like a virtual assistant to be able to carry out? The tasks are listed in Table 62.3	5-point likert scale

- This part addressed demographic aspects, such as age, gender, visual acuity, state of Brazil where they live, and whether living alone.
- 9 questions about prior experience with technology, the level of knowledge about robot companion and virtual assistants, receptivity to the idea of a virtual assistant, technologies to help with daily activities at home, difficulties and easiness faced in daily activities whilst alone at home, and important tasks a virtual assistant could potentially carry out (Table 62.1). The questions had explanations about what was being asked (e.g., what is a virtual assistant, companion robot, home automation, among others).

The same questionnaire was also applied in person, as interviews, in the Association of Blind People of Rio Grande do Sul (ACERGS). This strategy was used to obtain more respondents.

The participants were recruited by personal contact and posted online in social network groups. Table 62.2 gives the demographics of the participants. All the participants are from Brazil, living in different states of Brazil ($n = 27$); 16 of them answered the questionnaire online, while 11 people answered in person. The data analysis shows that the majority of the participants were male (55,6%). Most of the participants were blind (81,5%). More than half of the respondents (59,3%) stated being visually impaired from birth. Most of the participants live with other people (81,5%). The age of the participants was distributed over 18–73, with the age group of 18–28 being the largest with 48,1%.

Table 62.2 Sample

Sample demographics ($n = 27$)	
Gender	%
Male	55,6%
Female	44,4%
Age	%
18–28	48,1%
29–39	18,5%
40–50	11,1%
51–61	11,1%
62–72	11,1%
73>	0%
Visual acuity	%
Low vision	18,5%
Blindness	81,5%
Living alone	%
Living with other people	81,5%
Living alone	18,5%

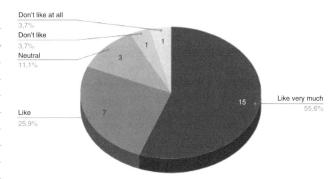

Fig. 62.1 Receptivity to a virtual assistant at home

62.4 Results

Prior experience with technology: when asked "What technologies have you heard about?", the participants answered computer (96,2%), smart TV (96,3%), smartphone (92,5%), tablet (85,1%), virtual assistant (51,8%), robot (55,5%), home automation (48,1%), and companion robot (29,6%). While when asked if they had contact with these technologies the distribution was computer (96,2%), smart TV (62,9%), smartphone (96,3%), tablet (66,6%), virtual assistant (33,3%), robot (0%), home automation (3,7%), and companion robot (0%).

Level of knowledge about companion robots and virtual assistants: when asked "What is your level of knowledge about companion robot?", most of the participants had no knowledge (66,7%). When asked the same question about the virtual assistant, less than half the participants (40,7%) had any knowledge. 11,1% of participants have knowledge about companion robot, and 33,3% have knowledge about the virtual assistant.

Receptivity to the idea to having a virtual assistant at home: 81,5% of participants liked or liked very much the idea of having a virtual assistant at home (Fig. 62.1). The reasons for not liking the idea were: "Because it does the same things I did before, but in a different way, with more care" and "Because I learned by myself how to do things, so it is not needed".

Technologies to help: when asked "What technologies would you like for helping with daily activities at home?" the technologies that appeared the most in the responses of the participants were: mobile apps (48,1%), robots (25,9%),

virtual assistant (18,5%), and sensors (14,8%). Some responses by participants are given below:

- "A virtual assistant, to inform something relevant in my case about the weather or something, a mobile application that could send messages every hour with the information I need and a robot that could use sensors to check for brief changes in the environment, and informing me. Anything to give me more independence".
- "I often need to use various types of applications on my cell phone or computer to perform simple tasks, like checking the colour of something or reading a correspondence and having a robot say so I do all this (all in one) as if I used the robot's eyes. And of course, sweeping the floors because it is annoying to do because I have to keep running my hand on the floor to see if the floor is clean".

Barriers in daily activities: when asked "What difficulties do you face in daily activities while being alone at home?" the participants answered: house cleaning in general (22,2%), reading (e.g., correspondences, newspapers, and medicines) (22,2%), non-accessible household appliances (18,5%), identify clothing colours (11,1%), finding items (e.g., clothes, medicines, and food) (11,1%), check if lights are on/off (7,4%), and identify obstacles (7,4%). Some responses from the participants are given below:

- "Some difficulties are to know if the light is on or off, to check the colours of the clothes, to know if it is dirty or smudged to find something like remote control, to phone recharge, things that we use in daily tasks, to identify the bills. Find some clothes like some clothes of my favourite colour, find some medicine or a specific food. Warn some danger or something that is like an emergency warning".
- "The only difficulty is solitude".

Technology facilities: when asked "What are the technologies that most help you (when alone) in your daily activities at home?" the participants answered: mobile apps

(33,3%), smart TV (18,5%), cellphone (25,9%), computer (18,5%) and screen readers (14,8%). Some responses from the participants are given below:

- "Mobile apps help identify the colours and some products. But still needs improvement. There are a number of accessible, leisurely, and helpful applications that help you keep in touch with others".
- "Technologies that help me are the screen readers that I use on my computer and on the phone".

Tasks for a virtual assistant: we asked "What tasks would you like a virtual assistant to be able to carry out?" and we listed options to indicate the level of importance (0: not important – 4: very important). The tasks presented to the respondents are listed in Table 62.3. Considering the combination of responses important and very important, more than half of the respondents wanted the virtual assistant to find objects (85,2%), notify if objects are out of place (81,5%), warn if there is another person in the environment (77,8%), prevent obstacles (74,1%), identify whether the light is on/off (70,4%), turn on/off devices (70,4%), reminders to take medicine (59,2%), warn how many meters away objects are (59,2 %), warn about appointments as a virtual calendar (55,5%),request help from any registered contact (55,5%), inform the weather (51,9%), and notify location (e.g., degrees and hours) of the objects (51,80%). Less than half of participants wanted it to identify whether the door is open or closed (40,7%), inform the time (40,7%), and identify whether the window is open or closed (37,0%). Figure 62.2 shows the tasks ordered by the total of votes in the important and very important scale.

Table 62.3 Tasks for a virtual assistant

Id	Tasks
1	Find objects
2	Detect obstacles
3	Identify whether the light is on or off
4	Identify whether the window is open or closed
5	Identify whether the door is open or closed
6	Inform the time of the day
7	Inform the weather
8	Warn about appointments as a virtual calendar
9	Reminders to take medicine
10	Turn devices on and off
11	Notify when objects are out of the usual places
12	Warn how many meters away objects are
13	Warn if there is another person in the environment
14	Notify location (e.g., in degrees) of objects
15	Request help from any registered contact

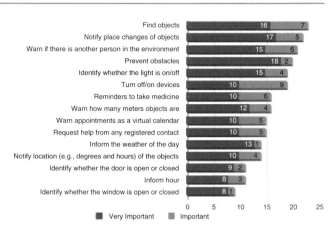

Fig. 62.2 Top tasks for a virtual assistant

62.5   Discussion

Through the questionnaire survey, we found that the respondents accept the idea the having a virtual assistant in their home, answering our research question 2. Less than half of the participants to had heard about (Computer – 96,2%, Smart TV – 96,3%, Smartphone – 92.5%) and had contact (Computer – 96,2%, Smart TV – 62,9%, Smartphone – 96,3%) with technologies. Incoherent responses also reflect in the percentages, for example, 92.5% said to had heard about Smartphone, but 96,3% said they have contact with Smartphone. This is a frequent problem when using the questionnaire method because it is not known if there was a lack of understanding about the questions or if the users did not pay enough attention to the questions. Nevertheless, the results clearly indicate that much help can be provided to such users through ambient assisted living and a custom-made virtual assistant for blind users. When we asked about the technologies that most help them at home, they answer mobile apps (33,3%), smart TV (18,5%), and computer (18,5%).

The majority of barriers/difficulties faced by participants in their home are related to house cleaning in general, reading (e.g., correspondences, newspapers, and medicines), non-accessible household appliances, identifying clothing colours, finding items (e.g., clothes, medicines, and food), checking if light are on/off, and avoiding obstacles. Most of the participants would like some technologies such as mobile apps, robots, virtual assistant, and sensors in order to support their daily activities at home. It was a much encouraging finding for our ongoing work. This findings answer our research question 1.

In terms of specific tasks for a virtual assistant, most of the respondents wanted to find objects, to be notified of objects being out of place, begin warned if there is another person in the environment, avoiding obstacles, checking whether the light is on/off, turn devices on/off, reminders to take medicine, information about how far away objects are, reminders about appointments as a virtual calendar, and request help from any registered contact (answers our research question 1). Some of these tasks are also related to the needs of the elderly, for example, reminders to take medicine, reminders of appointments as a virtual calendar, and requesting help from any registered contact, while others are very specific to blind users, such as for example colours.

We focused this study to the needs of blind users in our country (Brazil). However, this exploratory study has revealed important findings that could be relevant for future research ideas for AAL technologies around the world.

62.6 Conclusion

The current study explored the needs of people who are visually impaired and requirements for the design of a virtual assistant in their home. Despite the limitations of the method used, positive results have emerged, indicating that most people have interest in the idea of a virtual assistant. Results have highlighted the specific tasks that people who are visually impaired need in a virtual assistant, and which are their main difficulties/barriers while being alone at home. These findings allow studies to consider such user's needs and requirements when designing AAL technologies. Due to the level of detail of the questionnaire survey applied, we believe that this study can be replicated in other countries.

There are several machine learning, computer vision, and Internet of things techniques that can be used and adapted to the development of a virtual assistant that includes the features mentioned in our survey. Our future work is to identify these techniques and incorporate them into a virtual assistant. Moreover, we aim to integrate this assistant into a multiagent system for ambient assisted living.

Acknowledgements This study was financed in part by the Coordenação de Aperfeiçoamento de Pessoal de Nivel Superior – Brasil (CAPES) – Finance Code 001.

References

1. Pinnelli, S., Fiorucci, A.: Giving voice to images: audio description and visual impairment: technological solutions and methodological choices. In: Ambient Assisted Living, pp. 347–355 (2015)
2. Cook, D.J., Augusto, J.C., Jakkula, V.R.: Ambient intelligence: technologies, applications, and opportunities. Pervasive Mob. Comput **5**(4), 277–298 (2009)
3. Mukasa, K.S., Holzinger, A., Karshmer, A.I.: Workshop on intelligent user interfaces for ambient assisted living. In: Proceedings of the 13th International Conference on Intelligent User Interfaces, pp. 436–436 (2008)
4. Demográfico, C.: Características gerais da população, religião e pessoas com deficiência. Brasil Ministério do Planejamento, Rio de Janeiro, vol. 29 (2010)
5. Ganz, A., Schafer, J.M., Tao, Y., Wilson, C., Robertson, M.: Percept-ii: smartphone based indoor navigation system for the blind. In: Engineering in Medicine and Biology Society, pp. 3662–3665 (2014)
6. Radványi, M., Solymár, Z., Stubendek, A., Karacs, K.: Mobile banknote recognition: topological models in scene understanding. In: Proceedings of the 4th International Symposium on Applied Sciences in Biomedical and Communication Technologies, p. 185 (2011)
7. Oliveira, J.D., Borges, O.T., Paixão-Cortes, V.S.M., de Borba Campos, M., Damasceno, R.M.: Lêrótulos: a mobile application based on text recognition in images to assist visually impaired people. In: International Conference on Universal Access in Human-Computer Interaction, pp. 337–354 (2018)
8. Damaceno, R.J.P., Braga, J.C., Chalco, J.P.M.: Mobile device accessibility for the visually impaired: problems mapping and empirical study of touch screen gestures. In: Proceedings of the 15th Brazilian Symposium on Human Factors in Computing Systems, p. 2 (2016)
9. Ferreira, G., Penicheiro, R., Bernardo, P., Mendes, L., Barroso, J., Pereira, A.: Low cost smart homes for elders. In: Antona, M., Stephanidis, C. (eds.) Universal Access in Human–Computer Interaction. Human and Technological Environments. UAHCI 2017. Lecture Notes in Computer Science, vol. 10279. Springer, Cham (2017)
10. Połap, D., Woźniak, M.: Introduction to the model of the active assistance system for elder and disabled people. In: Dregvaite, G., Damasevicius, R. (eds.) Information and Software Technologies. ICIST 2016. Communications in Computer and Information Science, vol. 639. Springer, Cham (2016)
11. Choraś, M., D'Antonio, S., Iannello, G., Jedlitschka, A., Kozik, R., Miesenberger, K., Vollero, L., Wołoszczuk, A.: Innovative solutions for totally blind people inclusion. In: Garcia, N.M., Rodrigues, J.J.P. (eds.) Ambient Assisted Living. CRC Press (2015)
12. Chan, A.T., Gamino, A., Harris, F.C., Dascalu, S.: Integration of assistive technologies into 3D simulations: an exploratory study. In: Latifi S. (eds.) Information Technology: New Generations. Advances in Intelligent Systems and Computing, vol. 448. Springer, Cham (2016)
13. Kerekešová, V., Babič, F., Gašpar, V.: Using the virtual assistant alexa as a communication channel for road traffic situation. In: International Conference on Multimedia and Network Information System, pp. 35–44 (2018)
14. van Heek, J., Himmel, S., Ziefle, M.: Helpful but spooky? acceptance of AAL-systems contrasting user groups with focus on disabilities and care needs. In: Proceedings of the 3rd International Conference on Information and Communication Technologies for Ageing Well and e-Health, pp. 78–90 (2017)
15. Maan, C., Gunawardana, U.: Barriers in acceptance of ambient assisted living technologies among older australians. In: Life Sciences Conference, pp. 222–225 (2017)
16. Dautenhahn, K., Woods, S., Kaouri, C., Walters, M.L., Koay, K.L., Werry, I.: What is a robot companion – friend, assistant or butler? In: IEEE/RSJ International Conference on Intelligent Robots and Systems, pp. 1192–1197 (2005)
17. Preece, J., Rogers, Y., Sharp, H.: Interaction Design: Beyond Human-Computer Interaction. Wiley, Chichester (2015)
18. Barbosa, S., Silva, B.: Interação humano-computador. Elsevier, Brasil (2010)

Yousef Alsahafi, Daniel Lemmond, Jonathan Ventura, and Terrance Boult

63.1 Introduction

Research on fine-grained object classification on videos has been limited in part due to the lack of datasets. Still-image classification produces exciting results but completely ignores the temporal aspect of available data and reduces the dimensions for classification. Datasets do exist for traditional classification in videos but lack the granularity and labeling needed for fine-grained object classification.

The goal is to identify objects as belonging to a subclass, e.g., classifying different bird types [1], car models [2], dog breeds [3], or flower species [4]. Fine-grained object classification is generally considered more difficult than typical object classification tasks [5] because of the high degree of similarity between classes [6]. Several years of work on fine-grained classification has led to significant progress [1, 7–9]. Early methods for fine-grained object classification combined hand-crafted features with machine learning, e.g., using a histogram of oriented gradients (HOG) and support vector machines (SVMs) [1]. More recently, deep convolutional neural networks [10–13] have been shown to exhibit significant improvements over traditional approaches [8, 14]. Most previous works in this domain have treated the fine-grained object classification task as a still-image classification problem [2, 15].

Still-image classification represents a narrow view of object classification and is especially limiting for fine-grained categorization where many of the object's views are insufficient for classification. Videos remedy this problem through the volume of data they provide and the complexity at which this data is available, providing dozens of views and varying angles in rapid succession. Previously, the limiting factor in the creation of video classification datasets was the availability of data but with several billion videos now available on YouTube and an additional 24 h of video uploaded every minute, video classification has become a much more approachable problem [16]. Many object classification methods can use videos, and transitioning to video based classification provides new opportunities for fine-grained object detection/classification. Due to the prevalence of video data and the variability in its posing and lighting, video data is more likely to include the features for proper classification.

This paper introduces a new dataset, CarVideos, for fine-grained car classification in videos. The most common dataset for fine-grained car classification in still-images is the *Cars* dataset [15] which contains 16,185 annotated images and spans 196 classes of cars. This dataset is still widely used, but by now many models and methodologies have achieved near ceiling performance [15]. The CarVideos dataset contains more than one million frames obtained from 124 videos, with each video labeled with one of ten different makes, models, and years as shown in Table 63.2. The chosen classes cover different types of cars: three SUVs, four sedans, one hatchback, one compact sedan and one station wagon. We divided each video into a number of clips, and annotated each clip with meta information for training and evaluation. An additional contribution of this work is the expansion of data labels provided for fine-grained video data, as "present/non-present" is insufficient for labeling the target object. We use our new dataset to evaluate current state-of-the-art video classification systems such as Temporal Segment Networks (TSN) [17] and 3D ConvNets [18], both of which rely on both spatial and temporal information. Our approach outperforms both of the previous state-of-the-art methods.

Our novel approach uses a Single Shot Multibox Detector (SSD) [19] to extract the most important regions from each frame and sends them to a convolutional neural network

Y. Alsahafi (✉) · D. Lemmond · J. Ventura · T. Boult
Department of Computer Science, University of Colorado at Colorado Springs, Colorado Springs, CO, USA
e-mail: yalsahaf@uccs.edu; dlemmond@uccs.edu; jventura@uccs.edu; tboult@uccs.edu

© Springer Nature Switzerland AG 2019
S. Latifi (ed.), *16th International Conference on Information Technology-New Generations (ITNG 2019)*,
Advances in Intelligent Systems and Computing 800,
https://doi.org/10.1007/978-3-030-14070-0_63

(CNN) that is a ResNet architecture [20] for classification. The ResNet is pre-trained on ImageNet [21] and fine-tuned on our dataset. One of the most important steps in our process is the data augmentation technique from [20] in which each bounding box is cropped into four corner crops and a center crop. These crops relay additional information to the ResNet and assist us in reducing the chance for overfitting during training.

The contributions of this paper are twofold:

(1) We introduce CarVideos, a new video dataset for fine-grained car classification. The dataset poses several challenges for state-of-the-art methods, such as clutter, large variations in viewpoint, scale and camera movement. The images are annotated with bounding boxes, ten car classes, and multiple visibility attribute labels. CarVideos provides a test-bed for future research on fine-grained object detection and classification in video. We plan to make this dataset publicly available.

(2) We provide a baseline performance evaluation of state-of-the-art video classification methods on our dataset. Our evaluation shows that, on this dataset, a fine-grained classification approach that combines an object detector with a classifier outperforms both temporal segment networks and 3D convolutional networks, which are state-of-the-art methods for video classification.

63.2 Background

Fine-grained object classification is a highly studied area [8, 14, 22–25]. Each fine-grained object classification task has a unique dataset such as the Caltech-UCSD Birds-200-2011 dataset for birds [22] and the Dog Breed dataset for dogs [26]. Fine-grained object classification for cars has two prominent datasets, the *Cars* dataset [15] mentioned previously and the *CompCars* dataset [2]. Each of the aforementioned datasets is annotated with bounding boxes, part locations, and attribute labels such as the type of car and number of doors. These datasets have images that are hard even for humans to classify due to the minute visual differences that exist between subclasses. The task is made even harder by the varied poses and lighting that can mask these critical visual differences.

To tackle these tasks, many researchers localize various parts of the object they want to classify and train the model on these detected locations [14, 23]. These techniques originated on hand-crafted features [8, 14] but were rapidly outpaced by CNNs. One of the clear disadvantages of localizing various parts of the object is the difficulty to annotate these parts.

Recently, some researchers in [24, 25, 27] have considered fine-grained object classification in video. Videos offer a unique challenge by presenting the object and associated ob-

Table 63.1 Comparison of datasets for fine-grained video classification in video

Dataset	Classes	Videos	Frames
CRP (pedestrians)	14	7	261,645
VB100 (birds)	100	1,416	14 video clips on each class. Length of a video is 32 s
IBC127 (birds)	127	8,014	Numbers of frames per video between 18 and 159
CarVideos (cars)	10	129	1,529,846

ject parts in dozens of viewpoints and with various scalings in a short amount of time. Ge [24] and Saito [27] created a dataset of videos of birds datasets while Hall [25] presented a pedestrian videos dataset. In this paper we introduce a new dataset which, to the best of our knowledge, is the first public dataset for evaluating fine-grained car classification in videos. Our CarVideos dataset consists of videos with camera motion and cluttered backgrounds making for a difficult classification task. Table 63.1 compares previous fine-grained video classification datasets to our novel dataset.

Video classification has seen significant improvements in recent years with the introduction of two new architectures. Ge [24] adapted two-steam networks [21] with bilinear CNN features [10]. Their approach combines the information of the convolutional layers of two-stream networks. Saito [27] used a similar approach. They utilized a CNN to extract features from frames and used dense trajectories (iDT) to extract mention features. Ge [24] and Saito [27] focused their work on bird object classification while Hall [25] focused on human object classification. Hall work is comprised of two tasks: fine-grained categorization and pose estimation. Video action recognition is closely related to object classification in a video, and has gained a significant amount of popularity in recent years [21, 28, 29]. Recent work on video action recognition uses two-stream CNNs [21] to recognize actions in the video, where one stream uses temporal information and the other uses spatial information. We implement both two-stream networks [21] and 3D ConvNets [18, 30] to establish an appropriate baseline for fine-grained object classification in video on our dataset.

63.3 CarVideos Dataset

63.3.1 Dataset Collection

The goal of our dataset is to provide the community with a challenging, large-scale dataset that contains numerous different car models, car manufacturers, and car styles to allow for a complex classification task.

In this section we describe our strategy to effectively assemble a large, fine-grained car classification dataset. Our dataset contains over a million of frames and ten different

Fig. 63.1 A series of examples from our dataset highlighting the diversity and quality of the data

Table 63.2 CarVideos dataset details. The first three columns present (make, model, year) for each class. The fourth column shows types of cars. The fifth and sixth columns show number of easy and hard videos on each class. The last one presents the number of frames on each class

Make	Model	Year	Type	Easy	Hard	Frames
Honda	CR-V Touring	2017	SUV	10	10	304,577
Honda	Accord V6 Touring	2017	Sedan	7	3	88,355
Mazda	CX-5	2017	SUV	10	10	277,185
Mazda	Mazda3 i Touring	2014	Hatchbacks	5	5	95,461
Toyota	Avensis Touring Sports	2015	Station Wagon	9	0	73,145
Toyota	RAV4	2017	SUV	4	5	96,423
Toyota	Camry XSE	2018	Sedan	9	0	108,423
BMW	7 Series 750i	2016	Sedan	9	3	168,684
Kia	Forte EX	2014	Compact Sedan	8	7	139,890
Volvo	S90	2017	Sedan	10	5	177,703
Total				81	48	1,529,846

classes corresponding to specific makes and models of car, including sedans, SUVs, convertibles, hatchbacks, and station wagons as shown in Fig. 63.1.

To build our videos dataset for fine-grained car classification we searched YouTube for professional and amateur review videos about recent-model cars. We identified ten different models for which we could find many videos. Table 63.2 presents the number of videos and frames for each class.

We separated the videos in our dataset into two separate categories: easy and hard. Easy videos contain a single car belonging to one of our chosen car models. Hard videos contain multiple cars in the same frame, some of which might not belong to our chosen set of car models.

We pruned our dataset by removing videos that do not have a good variety of views of the target car. For example, we pruned videos which provide views of only one side of the car or do not actually contain the target car at all. We also ensured that video clips do not overlap between the training and testing sets, since some of the videos we found share source material.

63.3.2 Annotations

We divided each video into multiple clips with an open source scene detection library.[1] The software divides the video into clips by detecting large changes in the scene. We manually checked and corrected each clip to ensure they did not cross scene changes.

For each video, we know the exact year, make, and model of the car being reviewed. We call this car the "target." There may be other cars present in the video for which we do not know the exact year, make, and model.

We used the Google Object Detection API [31] to find a bounding box for each car in each frame. We manually inspected each frame, fixed incorrect bounding boxes, and

---

[1] https://pyscenedetect.readthedocs.io/en/latest/

added bounding boxes for cars missed by the automatic car detector. Each bounding box is labeled with one of our ten classes or "other" if it does not belong to one of the ten classes.

We also annotated each frame with one of the following attributes: car is obvious, car is not obvious, part of car is obvious, car is not our target, more than one car model, no car in the frame, and car is occluded.

63.4 Fine-Grained Video Classification

We start with a brief review of our chosen baseline methods that make use of spatial and temporal information: a two-stream network [17] and a 3D convolutional network [18]. These two methodologies are currently considered state-of-the-art on spatial-temporal feature learning for action recognition and represent a suitable baseline for our new dataset. We then describe our approach which combines an SSD [19] and a CNN [32].

63.4.1 Temporal Segment Networks & 3D ConvNets

Video-based recognition represents an interesting classification problem since it includes both temporal and spatial features. Several network architectures have been proposed for fine-grained object detection in videos such as TSNs [17, 21], 3D ConvNets [18], and others. The primary goal of these networks is to utilize both the temporal and spatial information available in videos, including motion between frames, to classify the object.

Temporal Segment Networks

TSNs are built to capture long-range temporal structure [17]. TSNs are an extension of the two-stream network architecture [21]. Two-stream networks have achieved great success in video-based recognition [17, 20, 21] and incorporate optical flow (temporal) and RGB color (spatial) information. The TSN approach divides a video into K-segments and then randomly selects one short snippet from each segment. TSNs take these short snippets as input and produce class scores for all the classes from each individual snippet. All of these

results from the TSN are passed to a segment consensus function to obtain a video-level prediction.

3D Convolutional Networks

3D ConvNets add an additional time dimension to a ConvNet to allow for both spatial and temporal information to be used for classification. 3D ConvNets use 3D convolutional filters instead of 2D filters and take a segment of a video as input. 3D ConvNets have been shown to outperform 2D ConvNets for video classification [18, 30, 33].

SSD-CNN

Our novel fine-grained video object classification method uses a CNN [32] for classification but introduces an SSD object detector [19] as a pre-processing step. Using an object detector allows us to be much more precise when training and testing our network as we are able to ignore much of the non-essential information while still capturing and training on the most valuable parts of the image. The CNN can take either an RGB image or optical flow image as input. In addition, to capture multi-frame information, we compute an average prediction over a sequence of frames from the video.

Figures 63.2 and 63.3 give an overview of our SSD-CNN architecture. Objects or object part bounding boxes are detected by the SSD and then cropped and resized for input to the CNN. Each bounding box is separately classified by the CNN. Our architecture uses a ResNet-152 [32] for both the spatial and temporal (optical flow) streams.

63.4.2 Implementation Details

In this section we explain the implementation details for the baseline methods (TSN, 3D ConvNets and SSD-CNN) we evaluated with the CarVideos dataset.

TSN

Our first baseline divides the clip into three segments and a short snippet is randomly chosen from each segment. Each snippet is sent to a two-stream ConvNet, both streams of which use the BN-Inception architecture [34]. The BN-Inception architecture weights are initialized from models pre-trained on ImageNet [35]. The spatial stream utilizes a single RGB image while the temporal stream uses five

Fig. 63.2 Each image shows bounding boxes predicted by the SSD network

Fig. 63.3 The entirety of our pipeline, including the output from our SSD and then the following resize and pass to our CNN

consecutive optical flow images as input. To extract the optical flow for our CarVideos dataset frames we use the TVL1 optical flow algorithm from [17]. The TSN approach uses the mini-batch stochastic gradient descent algorithm to learn the BN-Inception network parameters with a batch size of 256 and momentum of 0.9. We use the same TSN learning rate for spatial and temporal networks. For spatial networks, the learning rate starts at 0.001 and is multiplied by 0.1 every 2000 iterations to a maximum of 4,500 iterations. For the temporal network, the learning network starts at 0.005 and is multiplied by 0.1 after 12,000 and 18,000 iterations before training ceases at 20,000 iterations. The TSN model is tested by taking 25 RGB images or optical flow stacks from the test video. Each frame is cropped to four corners and one center and flipped horizontally as well. Each clip is classified by taking the average score over the cropped and flipped frames.

3D ConvNets

The 3D ConvNet has eight convolutional layers, five pooling layers, two fully connected layers, and a softmax output layer. The network uses convolutional kernels of $3 \times 3 \times 3$ to model (16) frames of information simultaneously. We trained a 3D ConvNet fine-tuned from the model pre-trained on the Sports-1M. We follow the original paper's approach [18] which takes five two-second long clips from each training video. Then we crop each two-second long clip into $16 \times 112 \times 112$ crops for spatial and temporal images. To obtain a video prediction, we randomly extract ten sub-clips from the video clip and then take the average score over the ten clips.

SSD-CNN Training

Our implementation consists of two networks: an SSD [19] and a CNN [20]. The SSD network is used to locate objects in each frame and segment them using a bounding box and a label. We only use the bounding boxes labeled "car." The SSD network is an SSD300 that is pretrained and validated on the Visual Object Classes (VOC) train and validation datasets [20] provided by https://github.com/amdegroot/ssd. pytorch. The SSD300 architecture [19] takes an input image

Fig. 63.4 An example of the crops for each possible classification image which consists of four corner crops and a single center crop. The crops are then passed from this output into our CNN

size of size 300×300 and outputs a set of bounding boxes with confidence values in the range [0 1]. Figure 63.2 shows examples of the SSD output.

Following the SSD detector stage, cars are cropped from the frame and converted into individual images before passing them to the CNN. Due to the variance in the size of each bounding box, we scale each individual box to 224×224 before passing it to the CNN, as shown in Fig. 63.3.

We prune and label the bounding boxes before providing them to the network as training data. We reject any bounding boxes with confidence less than 0.6. If a bounding box overlaps the target vehicle with intersection-over-union (IOU) greater than 0.7, it is labeled as the target vehicle. Otherwise, it is labeled as not the target vehicle. If the image has no predicted bounding boxes, we pass the entire image to the network and label it as the target vehicle.

We consider data augmentation to be one of the important parts of training the SSD-CNN combination. Our data augmentation approach is adapted from [20]. After resizing each bounding box to 224×224, we apply four corner crops and a center crop as shown in Fig. 63.4. This allows the network to train on parts of cars and thus be robust to partially occluded cars.

Following [20], we fine-tune a ResNet-152 [32] that has been pre-trained on ImageNet [35]. We fine-tune the network with stochastic gradient descent (SGD) momentum of 0.9, and a batch size of 25. We start with a learning rate of 0.001, divide it by 10 after every 4,000 iterations, and stop after 10,000 iterations.

SSD-CNN Testing

Our testing is similar to [20] and [21]. For testing we extract a series of 25 frames and apply our data augmentation approach to each frame. Each clip is classified by taking the average score over the cropped and flipped frames.

We tested three different configurations of the SSD-CNN: RGB input from a single frame, RGB input with 25 frames, and optical flow input with 25 frames.

63.4.3 Results

We trained and tested each model on our CarVideos dataset. We divided our dataset into three sets: 60% for training, 20% for validation, and 20% for testing. We used mean classification accuracy to measure the performance of each method on the test set. A clip is counted as a true positive if the classifier correctly predicts the target car, if the target car is in the clip, or the classifier correctly predicts "no car." Table 63.3 gives the results.

Table 63.3 Comparison of classification accuracy on the CarVideos dataset. The first four raws presents TSN [17] and 3D ConvNets[18] that are our baselines. Last two rows show our methods results. Clearly, SSD-CNN using (RGB) provides the best result

Method	Accuracy
TSN (RGB)[17]	62.39%
TSN (FLOW)[17]	37.88%
TSN (RGB + FLOW)[17]	56.44%
3D ConvNets[18]	52.6%
SSD-CNN Our (RGB – single frames)	73.53%
SSD-CNN Our (RGB – 25 frames)	**76.18%**
SSD-CNN Our (FLOW – 25 frames)	42.3%

Our evaluation shows that our SSD-CNN outperforms our two baselines from the video domain for object classification. Our method achieves an accuracy of 76.18% on the test set when using 25 RGB frames; the best baseline method we evaluated achieved an accuracy of 62.39%.

The comparison between using a single frame and 25 frames with SSD-CNN shows that using more frames from each video increases the accuracy. However, all methods did worse when using optical flow information; this suggests that optical flow information might be useful for coarse object categories, e.g. bikes, but is unlikely to be useful for identifying fine-grained bike categories, e.g. mountain bike vs. road bike vs. electric bike (Fig. 63.5).

63.5   Conclusions

In this paper we have proposed two contributions. First, we have created a large-scale dataset for fine-grained car classification in videos. Second, we present an evaluation of the performance of baseline methods for detection and recognition of fine-grained object categories in videos. Our evaluation includes a new detection and recognition approach uses SSD object detection to identify objects or object parts on frames and a deep CNN with appropriate training to classify these inputs. Our SSD-CNN outperforms Temporal Segment Networks (TSN) and 3D Convolutional Networks, which are state-of-the-art on human action recognition in videos. Likewise, we showed that using optical flow information did not improve the accuracy of fine-grained object classification on our dataset.

Future work includes evaluating methods designed specifically for fine-grained object classification, such as bilinear

Volvo 2017 S90 classified correctly

Volvo 2017 S90 classified as BMW 2016 7 Series 750i

BMW 2016 7 Series classified as no car in the image

SSD missing bounding box

Toyota Avensis classified as no car in the image

Fig. 63.5 Four test samples of classification and their prediction results. The first row present the correct classification while the rest shows the failure cases. Rows from 2 to 3 present SSD network provides the correct object (car) to CNN. However, CNN missed classification. The last row shows SSD network could not detect the object in the frame and then CNN miss classification the car

pooling [10]. We also intend to add more videos and classes to the dataset in the future.

References

1. Berg, T., Belhumeur, N.P.: Poof: part-based one-vs.-one features for fine-grained categorization, face verification, and attribute estimation. In: Computer Vision and Pattern Recognition (CVPR), 2013 IEEE Conference on. IEEE, pp. 955–962 (2013)
2. Yang, L., Luo, P., Change Loy, C., Tang, X.: A large-scale car dataset for fine-grained categorization and verification. In: Proceedings of the IEEE Conference on Computer Vision and Pattern Recognition, pp. 3973–3981 (2015)
3. Sermanet, P., Frome, A., Real, E.: Attention for fine-grained categorization. arXiv preprint arXiv:1412.7054 (2014)
4. Angelova, A., Zhu, S.: Efficient object detection and segmentation for fine-grained recognition. In: Computer Vision and Pattern Recognition (CVPR), 2013 IEEE Conference on. IEEE, pp. 811–818 (2013)
5. Macanhã, P.A., Eler, M.D., Garcia, E.R., Junior, W.E.M.: Handwritten feature descriptor methods applied to fruit classification. In: Information Technology-New Generations. Springer, pp. 699–705 (2018)
6. Xiao, T., Xu, Y., Yang, K., Zhang, J., Peng, Y., Zhang, Z.: The application of two-level attention models in deep convolutional neural network for fine-grained image classification. In: Computer Vision and Pattern Recognition (CVPR), 2015 IEEE Conference on. IEEE, pp. 842–850 (2015)
7. Chai, Y., Lempitsky, V., Zisserman, A.: Symbiotic segmentation and part localization for fine-grained categorization. In: Computer Vision (ICCV), 2013 IEEE International Conference on. IEEE, pp. 321–328 (2013)
8. Farrell, R., Oza, O., Zhang, N., Morariu, I.V., Darrell, T., Davis, S.L.: Birdlets: subordinate categorization using volumetric primitives and pose-normalized appearance. In: Computer Vision (ICCV), 2011 IEEE International Conference on. IEEE, pp. 161–168 (2011)
9. Gavves, E., Fernando, B., Snoek, G.C., Smeulders, W.A., Tuytelaars, T.: Fine-grained categorization by alignments. In: Computer Vision (ICCV), 2013 IEEE International Conference on. IEEE, pp. 1713–1720 (2013)
10. Lin, T.-Y., RoyChowdhury, A., Maji, S.: Bilinear CNN models for fine-grained visual recognition. In: Proceedings of the IEEE International Conference on Computer Vision, pp. 1449–1457 (2015)
11. Zhang, N., Donahue, J., Girshick, R., Darrell, T.: Part-based R-CNNS for fine-grained category detection. In: European Conference on Computer Vision. Springer, pp. 834–849 (2014)
12. Rezende, E., Ruppert, G., Carvalho, T., Theophilo, A., Ramos, F., de Geus, P.: Malicious software classification using VGG16 deep neural network's bottleneck features. In: Information Technology-New Generations. Springer, pp. 51–59 (2018)
13. Santos, A.F., do Nascimento, F.B., Santos, S.M., Macedo, T.H.: Training neural tensor networks with the never ending language learner. In: Information Technology-New Generations. Springer, pp. 19–23 (2018)
14. Bourdev, L., Maji, S., Malik, J.: Describing people: a poselet-based approach to attribute classification. In: Computer Vision (ICCV), 2011 IEEE International Conference on. IEEE, pp. 1543–1550 (2011)
15. Krause, J., Stark, M., Deng, J., Fei-Fei, L.: 3D object representations for fine-grained categorization. In: Computer Vision Workshops (ICCVW), 2013 IEEE International Conference on. IEEE, pp. 554–561 (2013)
16. Kuehne, H., Jhuang, H., Garrote, E., Poggio, T., Serre, T.: HMDB: a large video database for human motion recognition. In: Computer Vision (ICCV), 2011 IEEE International Conference on. IEEE, pp. 2556–2563 (2011)
17. Wang, L., Xiong, Y., Wang, Z., Qiao, Y., Lin, D., Tang, X., Van Gool, L.: Temporal segment networks: towards good practices for deep action recognition. In: European Conference on Computer Vision. Springer, pp. 20–36 (2016)
18. Tran, D., Bourdev, L., Fergus, R., Torresani, L., Paluri, M.: Learning spatiotemporal features with 3D convolutional networks. In: Computer Vision (ICCV), 2015 IEEE International Conference on. IEEE, pp. 4489–4497 (2015)
19. Liu, W., Anguelov, D., Erhan, D., Szegedy, C., Reed, S., Fu, C.-Y., Berg, C.A.: SSD: single shot multibox detector. In: European Conference on Computer Vision. Springer, pp. 21–37 (2016)
20. Wang, L., Xiong, Y., Wang, Z., Qiao, Y.: Towards good practices for very deep two-stream convnets. arXiv preprint arXiv:1507.02159 (2015)
21. Simonyan, K., Zisserman, A.: Two-stream convolutional networks for action recognition in videos. In: Advances in Neural Information Processing Systems, pp. 568–576 (2014)
22. Wah, C., Branson, S., Welinder, P., Perona, P., Belongie, S.: The caltech-UCSD birds-200–2011 dataset (2011)
23. Branson, S., Van Horn, G., Belongie, S., Perona, P.: Bird species categorization using pose normalized deep convolutional nets. arXiv preprint arXiv:1406.2952 (2014)
24. Ge, Z., McCool, C., Sanderson, C., Wang, P., Liu, L., Reid, I., Corke, P.: Exploiting temporal information for dcnn-based fine-grained object classification. In: Digital Image Computing: Techniques and Applications (DICTA), 2016 International Conference on. IEEE, pp. 1–6 (2016)
25. Hall, D., Perona, P.: Fine-grained classification of pedestrians in video: benchmark and state of the art. In: Computer Vision and Pattern Recognition (CVPR), 2015 IEEE Conference on. IEEE, pp. 5482–5491 (2015)
26. Liu, J., Kanazawa, A., Jacobs, D., Belhumeur, P.: Dog breed classification using part localization. In: European Conference on Computer Vision. Springer, pp. 172–185 (2012)
27. Saito, T., Kanezaki, A., Harada, T.: IBC127: video dataset for fine-grained bird classification. In: Multimedia and Expo (ICME), 2016 IEEE International Conference on. IEEE, pp. 1–6 (2016)
28. Gan, C., Yao, T., Yang, K., Yang, Y., Mei, T.: You lead, we exceed: labor-free video concept learning by jointly exploiting web videos and images. In: Computer Vision and Pattern Recognition (CVPR), 2016 IEEE Conference on. IEEE, pp. 923–932 (2016)
29. Peng, X., Wang, L., Wang, X., Qiao, Y.: Bag of visual words and fusion methods for action recognition: comprehensive study and good practice. Comput. Vis. Image Underst. **150**, 109–125 (2016)
30. Ji, S., Xu, W., Yang, M., Yu, K.: 3D convolutional neural networks for human action recognition. IEEE Trans. Pattern Anal. Mach. Intell. **35**(1), 221–231 (2013)
31. Huang, J., Rathod, V., Sun, C., Zhu, M., Korattikara, A., Fathi, A., Fischer, I., Wojna, Z., Song, Y., Guadarrama, S., et al.: Speed/accuracy trade-offs for modern convolutional object detectors. In: IEEE CVPR (2017)
32. He, K., Zhang, X., Ren, S., Sun, J.: Deep residual learning for image recognition. CoRR, vol. abs/1512.03385 (2015). [Online]. Available: http://arxiv.org/abs/1512.03385

33. Soomro, K., Zamir, R.A., Shah, M.: Ucf101: a dataset of 101 human actions classes from videos in the wild. arXiv preprint arXiv:1212.0402 (2012)

34. Ioffe, S., Szegedy, C.: Batch normalization: accelerating deep network training by reducing internal covariate shift. arXiv preprint arXiv:1502.03167 (2015)

35. Deng, J., Dong, W., Socher, R., Li, L.-J., Li, K., Fei-Fei, L.: Imagenet: a large-scale hierarchical image database. In: Computer Vision and Pattern Recognition, 2009. CVPR 2009. IEEE Conference on. IEEE, pp. 248–255 (2009)

64

A Technique About Neural Network for Passageway Detection

Pedro Lucas de Brito, Félix Mora-Camino, Luiz Gustavo Miranda Pinto,
José Renato Garcia Braga, Alexandre C. Brandão Ramos,
and Hildebrando F. Castro Filho

64.1 Introduction

In recent times there has been a breakthrough in the area of automotive winged micro vehicles especially in the area of drones. At the same time, with the advancement of the internet, the efficiency of gathering and sort a large number of information became much more agile, which helped the development of artificial intelligence techniques, using machine learning.

Given the current scenario of model airplanes a challenge that stands out in the accomplishment of autonomous tasks is in the identification of passages like doors and windows in constructions.

In this document we bring an approach that wrap some programming techniques, such as: Neural Network of Deep Learning, Filters and Image Processing. Both applied together for detection of paths through the drone's camera.

The technique is not fully implemented yet but has already achieved satisfactory results on the problems faced.

P. L. de Brito · L. G. M. Pinto · A. C. B. Ramos (✉)
Institute of Mathematics and Computing, Federal University of Itajubá,
Itajubá, Brazil
e-mail: ramos@unifei.edu.br

F. Mora-Camino
Federal Fluminense University, Brazil Computing, Rio De Janeiro,
Brazil

J. R. G. Braga
Nacional Research Institute (INPE), São José dos Campos, Brazil
e-mail: jose.braga@inpe.br

H. F. C. Filho
Aeronautical Institute of Technology, Brazil Computing, Rio De
Janeiro, Brazil

64.2 Related Works

The techniques on detection of objects through the neural network are increasingly present, both for business technologies and for the personal field of the user (in cases of WEB Big Data).

In the area of drones, one can quote the object tracking algorithms, which work with pattern recognition through a deep learning network. As example the study done by the students at Charles University in Czech Republic implemented by Roman Barták and Adam Vyskovský, "Any Object Tracking and Following by a Flying Drone" [1].

In the image processing application there is the case of obstacle avoidance, which uses image filter applications to obtain a map of objects present on the drone path. The work of the Brazilian student at Federal University from Itajubá, Wander Mendes, "The Computer Vision Based Algorithm for Obstacle Avoidance" employs this type of processing [2, 3].

64.3 Resources and Equipments

In this project was used the Tello model drone of Ryze Tech startup, which uses Intel and DJI technology. And Tello has an enviable stability, thanks to its high-definition global shutter camera, which in conjunction with its hardware stabilizes the drone's flight by comparing captured frames. Despising the use of GPS.

It also has a presence sensor located beneath its casing, which assists in the landing and the recognition of the edges of a passage, where it is desired to apply the presented technique.

The model has low autonomy, but its size compensates for the case of indoors flights. Figure 64.1 below shows the drone given by the university for the study.

S. Latifi (ed.), *16th International Conference on Information Technology-New Generations (ITNG 2019)*,
Advances in Intelligent Systems and Computing 800,
https://doi.org/10.1007/978-3-030-14070-0_64

Tello creates a Wi-Fi network to connect with smartphones and microcomputers. Which are responsible for their control of hard data processing.

The implementation of the technique was based on the library developed by Hanyazou: Tellopy, a collection presented in the python language that implements several basic functions of control and it also converts the images captured by the drone to a version recognizable by the OpenCV [4].

64.4 Techinique

In the beginning the proposed algorithm uses a neural network to detect the passages between the locations of a building. This network must be pre-trained in a supervised way, with manually defined characteristics by the user. When identifying a known passage, it starts recognizing the path through implemented PDIs to detect contours that forms the passage, this outline is compared to the value stored in the database, if it is correct, the drone follows the path inside it.

Fig. 64.1 Tello Model used [10]

64.4.1 Classification Deep Learning

According to the author A. Vargas, "a Convolutional Neural Network is a variation of the networks of Perceptrons of Multiple Layers, having been inspired by the biological process of visual data processing" [5]. The CNN network can apply filters on visual data, maintaining the neighborhood relation, just like a matrix convolution graphical processing.

A CNN is divided into multiple layers with certain functions. The first layer usually has a convolution function. In convolution layer type, there is a map of neurons where each one is responsible of performing an operation on a group of pixels within the image. Then, the image is filtered, and it classifies characteristics that are possibly important for the processing of the current set. CNN may have another very relevant layer type, the pooling layer. This type of layer works grouping and dividing data to reduce the size of information and increase network agility, the reduction is applied at the height and width of a map of neurons. The architecture of a CNN can have several different combinations of layers of pooling and convolution [5–7]. Figure 64.2 above shows the structure example of a CNN.

This approach will use the Single Shot MultiBox Detector (SSD) network, which is a CNN real-time object detection network [8]. This network works with bounding box hypotheses, creating a label in the form of a frame around the detected object, allowing multiple simultaneous detections, it also implements several improvements such as the single shot detector that leaves the network with higher accuracy and lower execution cost, these improvements that named the network [9].

SSD works with few image inputs, an advantage for a quicker pre-training, but it does use multiple bounding boxes for the same image. With this bounding box set through a block mapping of the image itself, it can predict the real po-

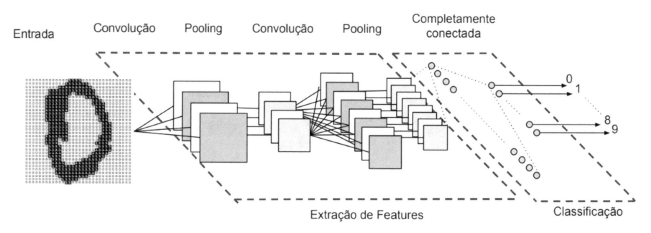

Fig. 64.2 CNN structure [5]

sition of the object, with greater accuracy and performance, this is called the single shot technique [9, 10]. Figure 64.3 shows an example.

The implementation was done on the Google Machine Learning library, TensorFlow [11], through the programming language Python.

64.4.2 Supervised Training

Prior to the use of this model, a pre-training for the proposed environment is required. In this pre-training, the image must be passed to the neural network with the approximate demarcation of the passage made by a colored box as shown in Fig. 64.4 below. At this point the training will be done for the obstructed path, avoiding that the robot tries to fly across a closed window. In this figure it is possible to see one green box for obstructed path, one blue box for unobstructed path and red box for passageway inside of the window. The

network will have a standard model defined by the algorithm with a reasonable window and door dataset, but whenever the drone changes the environment it is necessary to perform a new training to specify the classification of the network, due to the low performance of the hardware used.

For identification of the interior passageway and combination with the information obtained by the image processing, it will be necessary to make use of the multiple bounding box selection of the model. Where for each element detected as an unobstructed path, there will also be a sub detection of the interior passageway. To ease the implementation will be used two network models, one that will detect the unobstructed or obstructed path and another to detect the passageway in the obtained object path. The use of the two networks avoids the detection of a passageway without a window or door for example. As soon as the model encounters a window, it is cropped from the image and sent to a sub netting detection passages within the window, so it guarantees the passage search only within the window threshold. Upon finding

Fig. 64.3 Single shot improvement [9]

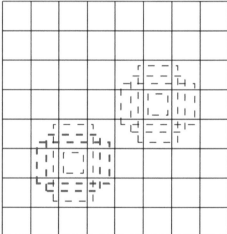

Fig. 64.4 Bounding boxes

the passageway it makes the comparison with the image processing data to pass the execution responsibility to its.

64.4.3 Box Detector by Image Processing

When the network model detects passage within a window object for example, the time comes to deal with the problem through image processing. This processing in this case becomes more agile than the network, but its accuracy may change according to the environment, light state, state of the object to be detected, etc. This can make detection impossible. To create a general case of detection the code wants to do the interpolation of its processing parameters for each case which will be discussed in the next section.

The image processing detection code works with the OpenCV framework implemented in the programming language Python. OpenCV has a contour grouping function to find polygons on the images. To optimize the detection of the contours, a pre-treatment is done through the Canny and Threshold segmentation filters.

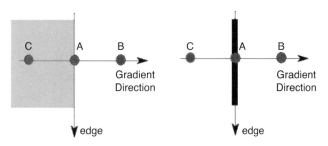

Fig. 64.5 Edge of Canny filter [12]

Fig. 64.6 Threshold examples [15]

The Canny filter takes a grayscale image and transforms it into a binary image, where the edges identified by pixel neighborhood analysis are highlighted in white. This segmentation initially works with noise reduction through the Gaussian filter, then walks over the pixels of the image with a gradient vector that calculates the direction of the pixel's color intensity, thus finding the thresholds faster, Fig. 64.5 then denote this action, where point A marks the horizontal threshold of the image to be detected [13].

The Threshold filter works by generating an image where all the pixels will be divided into two groups, with different pixel shades [14].

Figure 64.6 above shows some examples of thresholds, applied to image segmentation the defined threshold that defines the separation of the image.

When a frame arrives to be identified, the code generates several images from a range of values defined by the parameters passed by the user. This value oscillates in the range, dividing the obtained frame in the threshold filter images. When the threshold returns values without significance, the binary image is passed to be handled by a Canny border filter. At the end of the treatment, the OpenCV's *approxPolyDP* function scans all the generated images from the frame and contour search that forms a four-sided polygon [16]. When satisfying some conditions, the code saves these polygons in vector which is ordered by size of area. So, at the end of the execution it returns the median of these values, to avoid the great rate of input and output of data, it works only in the case of the middle one. After the identification, the drone detects the passage inside the frame it begins to cross the path.

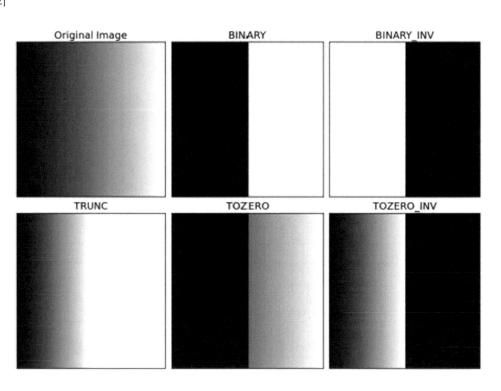

64.4.4 Parameters Interpotalion

As previously mentioned, the image processing algorithm depends heavily on the variations of the environment, since it is a generic detection procedure. To correct this problem, acting on specific cases, variation of the parameters of the code is made.

At that moment the drone encounters a pre-trained pass through the network, it searches for the image that marks the passage of the unobstructed path in the dataset, to compare with the found data by image processing. It is important to note that until the defined moment, the drone has not yet made any movement which ensure more time to process these previous steps. The code will make use of a base of test parameters, pre-defined empirically. And it does the interpolation in the surroundings of this base, avoiding visiting a great number of possibilities values. After finding a set of acceptable parameters, it updates the database with the new environment values, and then the robot starts moving toward crossing the gate.

64.4.5 Path Decision and Movement

The TelloPy library used in the project has precise movement functions for the Tello. Even more taking into consideration that the drone will perform in indoor environment, the precision of the drone's movement becomes enviable.

To work with the location of a passage, the drone rotates in its own axis seeking to recognize a pre-trained passage, recognizing it, the drone starts to analyze the environment to find the parameters necessary to image processing, then draw the picture of the path to be crossed.

Upon finding the parameters, it begins to align with the center of the polygon, analyzing the parallelogram found, when the parallelogram becomes a rectangle and aligns with the center point of the frame, the drone starts to move in a straight line [17]. While moving, it makes use of the presence sensor located beneath the body, that makes possible to check if it had already crossed the passageway.

64.4.6 The Algorithm

The algorithm has been divided into three parts. The first part contains the main code call, the Algorithm 64.1 the following shows its structure.

In the first step the algorithm makes a call of the turn_arround function, which searches for a pre-trained passageway through the network. Then the best parameters

for image processing, are interpolated to the environment via the calibration function. With the use of these parameters, it is possible to align the drone with the passage, and finally when crossing, through the following functions.

Algorithm 64.1

main loop

```
1. main():
    2. LOOP:
        3. turn_arroud (passage)
        5. calibration (passage, parameters)
        6. adjust_angle (parameters)
        7. go_away (parameters)
    8. END_LOOP
9. end
```

The Algorithm 64.2 implements the turn_arround function, at each step of the loop, the drone rotates on its own axis through the function move_yaw. Then the SSD network starts to identify pre-trained passages, through the function find_path_SSD.

Algorithm 64.2

turn_arroud function

```
1. turn_arroud ():
    2. LOOP:
        3. move_yaw ()
        4.find_path_SSD (passage)
        5. IF is_valid (passage): RETURN
    6. END_LOOP
7. end
```

The Algorithm 64.3 is the calibration function, which works with the passage value returned by the neural network. It does the interpolation of the parameters and returns the values to main.

Algorithm 64.3

calibration function

```
1. calibration (passage, parameters[]):
    2. LOOP:
        3. interpolation (passage, parameters)
        4. IF is_good (passage, parameters):
            RETURN
    5. END_LOOP
6. end
```

It is important to notice that only the adjust_angle and go_away functions works with image processing. They receive only the parameters and will detect the square of the passage in real time.

64.5 Applications

There are two major application niches for the technique presented. One is in the field of security and autonomous supervision of closed environments, and another in the branch of rescue in risky environments.

In both cases it is possible to specify the use of the network. For a supervisory environment the technique can be implemented with a good supervised network training, which guarantees a great efficiency, and a possible independence of the image processing code. In the rescue environment, such as in a burning building for example, it would be interesting to use windows with marked edges, in the public buildings such as hospitals and schools, improving edge detection by image processing, which can give the technique an independence of the neural network.

64.6 Conclusion

In this article a passageway detection technique was presented, which gives the ability of an autonomous drone to perform an indoor flight through buildings.

In this implementation the difficulty is in the accuracy of the model and in its performance, to increase the accuracy the implementation of a neural network pre-trained for each specific case of environment. And to increase the processing speed during its navigation, image processing was used in the detection of contours which has a more agile execution than the network.

It is observed that the hardware limitation is still a problem for applications such as these, the modeling becomes more agile and precise when some improvements in this aspect is made.

In the future, it is expected that more hardware will be available to raise the area's research. It would also be interesting that the community creates standardized window models with details recognizable by an AI.

64.7 Future Works

One possible improvement would be creating a large dataset to perform a test with a generic network for pass-through detection.

Another proposal is the use of a more agile technology that implements the technique, as in the case of the PixHawk control board with Robotic Operational System (ROS) [18, 19]. For an indoor flight with this system would require a stabilization by camera similar as Tello. PX4FLOW is an ideal camera model for this job [20]. So, its stability would be linked to the frame being processed in the installed auxiliary camera, and not to a GPS like outdoor flight.

Acknowledgment The authors would like to gratefully thank the founding institution CAPES.

References

1. Barták, R., Vyskovský, A.: Any Object Tracking and Following by a Flying Drone. Charles University in Prague, Faculty of Mathematics and Physics, Prague (2015)
2. Martins, W.M., Braga, R.G., Ramos, A.C.B., Camino, F.M.: A computer vision based algorithm for obstacle avoidance. In: Springer International Publishing AG. (Org.) (ed.) Advances in Intelligent Systems and Computing, vol. 738, 1st edn, pp. 569–575. Springer International Publishing AG, Las Vegas (2018)
3. Ramos, A.C.B., Shiguemori, E.H., Serokhvostov, S.V., Gupta, P.K., Zhong, L., Hu, X.B.: Solar-powered UAV platform system: a case study for ground change detection in BRIC countries. In: Advances in Intelligent Systems and Computing, vol. 738, 1st edn, pp. 613–618. Springer International Publishing AG, Las Vegas (2018)
4. TelloPy Library. Available in: https://github.com/hanyazou/TelloPy
5. Vargas, A., Paes, A., Vasconcelos N.: Um Estudo sobre Redes Neurais Convolucionais e sua Aplicação em Detecção de Pedestres. Institute of Computing from Federal University Fluminense Niterói (2016)
6. Girshick, R.: Fast R-CNN. In: The IEEE International Conference on Computer Vision ICCV, Las Condes, Chile (2015)
7. He, K., Zhang, X., Ren, S., Sun, J.: Spatial pyramid pooling in deep convolutional networks for visual recognition. In: European Conference on Computer Vision ECCV, Zurich, Switzerland (2014)
8. Ren, S., He, K., Girshick, R., Sun, J.: Faster R-CNN: towards real-time object detection with region proposal networks. In: Neural Information Processing Systems NIPS, Montréal Canada (2015)
9. Liu, W., Anguelov, D., Erhan, D., Szegedy, C.: SSD: Single Shot MultiBox Detector. From University of Michigan In: European Conference on Computer Vision ECCV, Amsterdam, The Netherlands (2016)
10. Ning, C.; Zhou, H.; Song, Y.; Tang, J.: Inception single shot multibox detector for object detection. In: IEEE International Conference on Multimedia & Expo Workshops ICMEW, Hong Kong (2017)
11. TensorFlow. Available in: https://www.tensorflow.org
12. Canny OpenCV Function. Available in: https://docs.opencv.org/3.4.3/da/d22/tutorial_py_canny.html
13. Venugopala, P.S., Sarojadevi, H., Ankitha, A., Niranjan, N.: An Approach to Improvise Canny Edge Detection using Morphological Filters
14. Lee, K., Lee, Y.: Threshold boolean filter. IEEE T. SIGNAL PROCES. **42**(8), (1994)
15. Threshold OpenCV Function. Available in: https://docs.opencv.org/3.4/d7/d4d/tutorial_py_thresholding.html
16. approxpolyDP OpenCV Function. Available in: https://docs.opencv.org/2.4/modules/imgproc/doc/structural_analysis_and_shape_descriptors.html
17. Braga, R.G., da Silva, R.C., Ramos, A.C.B., Mora-Camino, F.: Collision avoidance based on reynolds rules: a case study using quadrotors. In: Advances in Intelligent Systems and Computing, 1st edn. Springer International Publishing, Las Vegas (2018)
18. PixHawk. Available in: http://pixhawk.org/
19. ROS. Available in: http://www.ros.org/
20. PX4FLOW. Available in: https://docs.px4.io/en/sensor/px4flow.html

A SSD – OCR Approach for Real-Time Active Car Tracking on Quadrotors

Luiz Gustavo Miranda Pinto, Félix Mora-Camino, Pedro Lucas de Brito, Alexandre C. Brandão Ramos, and Hildebrando F. Castro Filho

65.1 Introduction

Drones have grown in popularity during the last years, principally quadrotors, due to the development of new models with inexpensive material and easy commands such as the Tello DJI drone, which a child may be able to use it [1]. This kind of drone is flexible and enables coded applications which may be useful for both military and civil necessities. A great example of its utility, and object of study of this paper, is the active car track mission, where the drone needs to identify a car, check its plate to make sure it is the right target and follow it.

Since the command control of a drone, in its essence, is generally guided by a pilot with the assistance of a Radio Control, there is the need for a person to stay in charge. More than this, the max distance a drone can perform is bound to the max wave length of a Radio Control. Having this in mind, the active car track is a challenge where the machine operates everything by itself.

From this point of view, the goal of this paper is to create an algorithm which grants total control to a drone during an active car tracking mission. To better simulate the human vision during the track process, the authors decided to use an SSD deep neural network to identify an object instead of

using other techniques, such as geometry approximation or tag detection. At the same time, the OCR were chosen to validate the target found.

The structure of this paper starts by presenting the background and related work in the Sect. 65.2. Next, in Sect. 65.3, there are described the materials and methods used during the experiments. In Sect. 65.4, the authors present the experiment results and some discussions about it. The last section presents the conclusions about the paper and the future works, being followed by references and acknowledgment.

65.2 Background and Related Work

The main concepts for building an algorithm like the one described in this paper comes from the application of object detection alongside with digital image processing. The object detection is characterized as a source of identification of instances or classes for an object inside an image and the main goal is to find all these classes and instances [2]. This may be used to innumerous cases, such as a car detection or any other vehicle.

The digital image processing is a method where operations are performed on a digital image to extract information or get a new image with enhanced features [3]. Also, there are innumerous applications, such as finding the boundaries between types of tissue in medicine [4] or detecting the contour of characters and numbers, in the case of an OCR system.

The combination of these techniques makes almost all kinds of pattern recognition possible inside an image. In that way, they may be responsible to guide an autonomous quadrotor during an active car tracking or any kind of tracking/detection mission.

In the field of object tracking, there are innumerous papers and researches that have inspired the development of the

L. G. M. Pinto · P. L. de Brito · A. C. B. Ramos (✉)
Institute of Mathematics and Computing, Federal University of Itajubá, Itajubá, Brazil
e-mail: ramos@unifei.edu.br

F. Mora-Camino
Federal Fluminense University, Brazil Computing, Rio de Janeiro, Brazil

H. F. Castro Filho
Aeronautical Institute of Technology, Brazil Computing, São José dos Campos, Brazil

S. Latifi (ed.), *16th International Conference on Information Technology-New Generations (ITNG 2019)*,
Advances in Intelligent Systems and Computing 800,
https://doi.org/10.1007/978-3-030-14070-0_65

Fig. 65.1 VGG16 structure

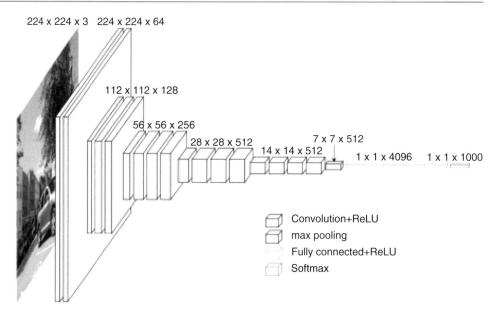

algorithm presented in this paper. Barták and Vyškovský, presented a Track Learn and Detect - TLD solution for object tracking and following implemented on a Parrot AR Drone, that uses pixels comparison for each frame [5]. Patil implemented a version of the CamShift algorithm where it uses angles formed in the camera as the source of the centroid coordinates of an object, alongside with a NavStik autopilot board [6].

At the same time, among the OCR researches, Tavares, Causin and Gonzaga [7] presented an algorithm for Brazilian car plates recognition using Google's Tesseract platform [7]. Alongside with them, Qadri and Asif used the Automatic Number Plate Recognition – ANPR to extract information from car plates and turn them into characters and numbers [8]. Without the ideas presented in those researches, it wouldn't be possible to create the algorithm present in this paper.

convolution and ReLU layers to convolution and fully connected layers, in addition to max pooling layers and a final softmax layer [11], as shown in Fig. 65.1.

The second part is composed by 6 more auxiliary convolution layers after the VGG16. In five of those layers 6 predictions are make, where in total, 8732 predictions occur during the algorithm's execution. In that process, default boxes are selected manually, where a scale value for each feature map are chosen. For comparison, YOLO uses k-means clustering to define those default boxes [10], causing it to run a lot slower. The whole structure for SSD may be seen in Fig. 65.2.

The last part is composed by enumerating the positive matches, used to calculate cost in bounding box mismatch, and evaluate the loss function where the difference between the real boundary box and the predicted boundary box [10]. The model keeps running until a desired loss cost is achieved.

65.3   Materials and Methods

65.3.1 Single Shot MultiBox Detector – SSD

In 2016, the paper "SSD: Single Shot MultiBox Detector" [9] presented a new concept for an object detection algorithm using deep neural networks. It was designed for real-time applications, where it speeds the process for not resampling pixels or features for inferring new bounding boxes, while uses improvements as default boxes and multi-scale features to prevent accuracy loss [10].

The SSD is composed of extraction of feature maps and the application of convolution filters to detect objects. The extraction is executed using a convolutional neural network called VGG16, where it is composed of combinations from

65.3.2 Optical Character Recognition – OCR

Optical Character Recognition is a technique that deals with recognition of optically drawn characters and is needed when information must be read by both humans and computers when another kind of input is not given [12]. OCR is a complex problem because of the variety of languages, fonts and styles in which texts can be written, and the complex rules of languages [13] and needs a sort of steps to achieve reasonable results.

Inside the major steps for an OCR algorithm, we have image acquisition, where the data is feed; preprocessing, where the quality of the image is improved, and character segmentation, where characters are separated, where they compose the first stage of the algorithm.

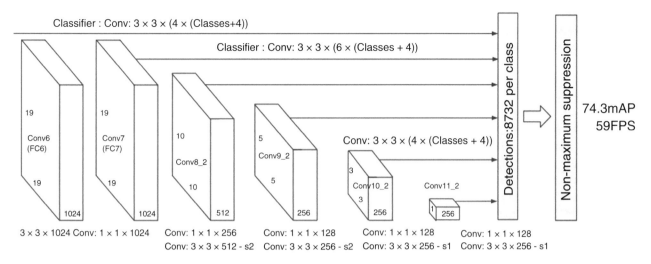

Fig. 65.2 SSD structure

Fig. 65.3 Steps of OCR. (**a**) Image acquisition. (**b**) Preprocessing. (**c**) Character segmentation. (**d**) Feature extraction. (**e**) Classification and post-processing

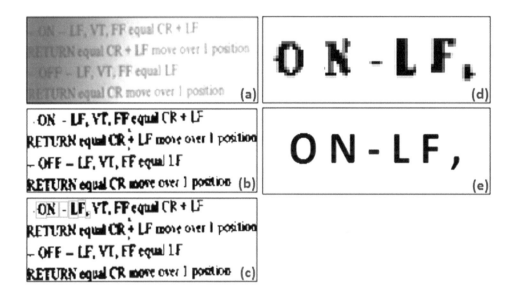

The second stage are composed by feature extraction, where each character segment is processed to obtain unique features; classification, where the mapping between segmented images to classes are made, and post-processing, that is the refine stage to guarantee the right result [13, 14]. The steps for OCR may be seen in Fig. 65.3.

65.3.3 Quadrotor

The quadrotor is an aircraft made up of four engines and holds the electronic board in the middle and the engines at four extremities [15]. A quadrotor is controlled by adjusting the angular velocities of the rotors which are spun by electric motors, where you have 6 degrees of freedom, being x or roll, y or pitch, and z or yaw axis [16, 17, 22], as seen in Fig. 65.4.

The altitude and position of the quadrotor can be controlled with desired values by changing the speeds of the four motors [15]. For example, by increasing the speed in

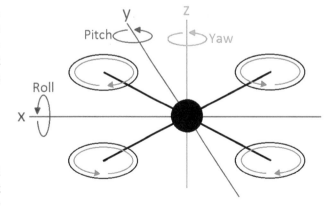

Fig. 65.4 Degrees of freedom on a quadrotor

the motors on the right and decreasing it in the motors on the left, the quadrotor rolls in a counterclockwise direction. The same applies to the pitch but controlling the front or back motors [17].

In this way, when you decide to pass total control to an algorithm, every speed variable needs to reflect the reality of each motor, to achieve a smooth movement.

65.3.4 Tello

The Tello DJI is a very easy-to-use quadrotor drone, with simple controls that allows children to use it, as seen in Fig. 65.5. It comes with a 720p HD transmission camera used both for taking pictures and streaming [1]. The camera type is global-shutter where it records the whole frame on each iteration, instead of scanning rows of pixels which leads to loss of precision, as seen in rolling-shutter models.

It doesn't have a GPS, so it uses the frontal camera and a distance sensor underneath for a stabilized flight. Both are essential components for its collision avoidance system, since it doesn't have approximation sensor on every side.

A great advantage of having Tello DJI is its SDK [18] that allows the development of different applications without the need of a communication interface such as ROS. The SDK is written in Python language, and enables remote control having a computer or smartphone as a source of transmission.

Since Python is a robust programming language, almost any kind of application may be created. That allows the SDK to make good use of deep learning libraries for object recognition, as well as image processing libraries.

Fig. 65.5 Tello DJI

The Tello SDK connects to the aircraft through a Wi-Fi UDP port, allowing users to control the drone with text commands [18]. After the connection is stablished, a text-command 'command' is needed to check if the connection was successful, returning an 'OK' statement if that was true, and 'ERROR' in any other case.

After validating the statement, the aircraft is ready to receive action commands, such as 'takeoff', 'land', 'flip x', and many others. Some commands are used to check the flight current state, such as 'battery?', 'height?' and 'speed?', returning information that may be useful to a control algorithm.

Despite the variety of commands already implemented in the SDK, there isn't an official command to access the feed of the aircraft camera on a computer. Having this in mind, the TelloPy package was created to supply this necessity [19].

This package includes a byte stream translator that capture the stream packages through Wi-Fi while the camera is on, seen in Fig. 65.6. When launching TelloPy, the SDK works as a background assistant, as seen in Fig. 65.6, supporting the already developed commands. With both APIs the implementation of an active car tracking is possible with Tello.

65.4　Experiments and Results

65.4.1 Algorithm Implementation

The algorithm starts with a loop control, where the main goal is to capture each frame received from the camera, apply the SSD object recognition model to identify a car's front or back sides as a sub-image, feed this sub-image to OCR, return the string identified and its bounding box position in the sub-image, and then calculates the drone position. The main loop control may be seen in Algorithm 65.1.

The first step is to make sure there's a valid new frame. In other words, this means that the algorithm will keep processing until the end of the quadrotor's camera stream. If a frame could not be read or cannot be processed, the algorithm discards that frame and acquire another one.

The sequence advances with the application of the SSD model for object recognition, to identify if the image is

Fig. 65.6 Tello system overview

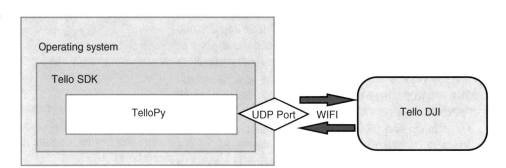

Algorithm 65.1 Main loop of active car tracking

1: procedure control
2:   while frame= true:
3:     sub_image = SSD(frame)
4:     bbox, info = OCR(sub_image)
5:     setPosition(bbox, info)
6:   end while
7: end procedure

Algorithm 65.2 SSD object recognition

1: procedure SSD (frame)
2:   inf = load_inference_graph(model)
3:   objects = calculate_objects(frame, inf)
4:   selected = get_higher_accuracy(objects)
5:   bbox = bbox_selection(selected)
6:   sub_image = crop_image(bbox)
7: end procedure

Algorithm 65.3 OCR

1: procedure OCR (sub_image)
2:   new_image = preprocess(image)
3:   seg, bbox = char_segment(new_image)
4:   features = extract_features(seg)
5:   info = classify(features)
6: end procedure

a car, as seen in Algorithm 65.2. The first step for SSD is to load the inference model, where the inference rules of a trained model is stored. In this case, the model was trained with a car dataset. Having the inference model, the sequence infers possible candidates from the selected frame to be the desired object. This step is covered by calculate_objects() and returns a list of objects with highest matching.

After the objects are gathered, the next step is to select the one with highest score. In this context, it means the one with the highest probability of being a car. After receiving the selected object, is necessary to calculate its bounding box. A bounding box is a minimum delimiter box that surrounds the contours of an object [20]. In other words, it's the object's exact location inside the image. The last process is to crop a sub image using the bounding box as a region delimiter and return it to the control loop.

Back to the control loop, now it's time to recognize if the car plate it's the correct one. In this step the sub image is sent to the OCR model, being presented in Algorithm 65.3. The first step of this algorithm is to apply a preprocessing procedure, where filters are used to highlight the desired characteristics of the image and a new improved image is generated. From there, the algorithm selects the position where a sequence of characters is probably located. Having

this position, it calculates the bounding box, in this case, of the car plate and returns it alongside with a vector of segmented character images.

The next procedure is to extract the features from each segmented image. Having the features, the algorithm now classifies every character recognized and returns it to the control loop. The last step involves calculating the new position for the quadrotor based on the car plate bounding box and checking if it is the right plate to follow. This last part is still in development.

65.4.2 Experiments

The main goal of the experiments was to determine the object detection and the character recognition is already working, since the drone positioning is still in development. For the tests, an Ubuntu v16.04_64 OS were used in a laptop with Intel Core i5 with 1,7 GHz and 8GB of RAM memory.

For the SSD model, the Stanford's Car Dataset were used for its training, containing 16,185 images of 196 classes of different cars [21]. It was built using Google's TensorFlow platform and uses the SSD Mobilenet V1 Pets configuration file for its training.

The first test realized was the recognition of a car from an image It was used to verify the consistence of the model. As seen in Fig. 65.7 (top), the car was identified with 89% during the processing. The generated image was then fed to the OCR algorithm to find the car plate. After identifying the right position, the algorithm generated a string information from the image. As seen in Fig. 65.7 (bottom), the text 'NKC 6590' was created and matched the original information from the plate.

65.5   Conclusions and Future Works

The main goal of this paper is still in development. However, major steps were already finished. Having both the SSD and the OCR working is one of the main sub goals to achieve the result proposed, and their results were consistent.

In SSD the car was detected with 89% of accuracy, which is great in comparison with some other techniques. At the same time, the OCR managed to recover the right information from the original source. Using both algorithms the problem of identifying the plate from the right car was solved.

In the future, the next step is to quantify the drone distance in relation to the car plate, its position and speed. For this step, an interaction with Fuzzy Logic may be possible, to be easier to identify how far, close, fast or slow the drone is in relation to the target. Finally, the position may be adapted according to the center of the plate's bounding box.

Fig. 65.7 Results collected. (top) The result from SSD. (bottom) The result from OCR

Aknowledgement The authors would like to thank the funding institution FAPEMIG, where without its resources this project wouldn't be possible.

References

1. Tello: Ryze Robotics. Available in https://www.ryzerobotics.com/tello
2. Amit, Y., Felzenszwalb, P.: Object detection. In: Computer Vision, A Reference Guide. Springer, New York (2014)
3. Anbarjafari, G.: Introduction to image processing. In: Digital Image Processing. University of Tartu, Tartu (2014)
4. Namee, B.M.: Digital Image Processing. Introdution. Dublin Institute of Technology, School of Computer Science, Dublin (2018)
5. Barták, R., Vyškovský, A.: Any object tracking and following by a flying drone. 2015 Fourteenth Mexican International Conference on Artificial Intelligence (2015)
6. Patil, K.: Object Tracking Using Quadcopter. Department of Electrical Engineering, Indian Institute of Technology, Mumbai (2017)
7. Tavarez, D.T., Caurin, G.A.P., Gonzaga, A.: Tesseract OCR: A Case Study for License Plate Recognition in Brazil. Laboratório de Mecatrônica, Departamento de Engenharia Mecânica, EESC-USP (2003)
8. Qadri, M.T., Asif, M.: Automatic number plate recognition system for vehicle identification using optical character recognition. 2009 International Conference on Education Technology and Computer (2009)
9. Liu, W., et al.: SSD: Single Shot MultiBox Detector. The 14th European Conference on Computer Vision, Amsterdam, 29 Dec 2016
10. Hui, J.: Choosing default boundary boxex. In: SSD object detection - single shot MultiBox detector for real-time processing. Deep Learning. Available at: https://medium.com/@jonathan_hui/ssd-object-detection-single-shot-multibox-detector-for-real-time-processing-9bd8deac0e06 (2018)
11. Frossard, D.: VGG in TensorFlow: Model and Pre-trained Parameters for VGG16 in TensorFlow. Department of Computer Science, University of Toronto, Toronto, ON (2016)
12. Eikvil, L.: OCR: Optical Character Recognition. Department of Statistical Analysis, Image Analysis and Pattern Recognition (SAMBA). Norsk Regnesentral Norwegian Computing Center, Oslo (1993)
13. Islam, N., Islam, Z., Noor, N.: A survey on optical character recognition system. J. Inf. Commun. Technol. **10**(2), (2016)
14. Srinivasan, S., et al.: Performance Characterization and Acceleration of Optical Character Recognition on Handheld Platforms. 978-1-4244-9296-1/10/$26.00 ©2010 IEEE
15. Sabatino, F.: Quadrotor control: modeling, nonlinear control design, and simulation. Master's Degree Project, KTH Electrical Engineering, Stockholm, Sweden June 2015, XR-EE-RT 2015:XXX
16. Luukkonen, T.: Modelling and control of quadcopter. Independent research project in applied mathematics. Aalto University School of Science, Espoo, August 22, 2011
17. Braga, R.G., da Silva, R.C., Ramos, A.C.B.: Collision avoidance based on Reynolds rules: a case study using quadrotors. 14th International Conference on Information Technology. ITNG, (2008)
18. Rize Robotics, Tello SDK 1.3.0.0.: Available in: http://www.ryzerobotics.com, 2018
19. Hanyazou: TelloPy: DJI Tello drone controller python package. Available in: https://github.com/hanyazou/TelloPy, (2018)
20. Hsu, K., Lin, Y., Chuang, Y.: Augmented multiple instance regression for inferring object contours in bounding boxes. IEEE Trans. Image Process. **23**(4), (2014)
21. Krause, J., Stark, M., Deng, J., Fei-Fei, L. 3D object representations for fine-grained categorization. 4th IEEE Workshop on 3D Representation and Recognition, at ICCV 2013 (3dRR) (2013)
22. Martins, WM., Braga, R.G., Ramos, Alexandre, C.B., Camino, F.M.: A computer vision based algorithm for obstacle avoidance. In: Springer International Publishing AG (Org.). Advances in Intelligent Systems and Computing, 1st edn, pp. 569–575.-13. Springer, Las Vegas (2018), v. 738. Sydney. 8 Dec 2013

Learning Through Simulations: The Ship Simulator for Learning the Rules of the Road

Sonu Jose, Siming Liu, Sushil J. Louis, and Sergiu M. Dascalu

66.1 Introduction

Simulations are instructional scenarios that could be integrated into education, particularly because through them students experience a stronger sense of reality. In general, traditional instructional approaches such as classroom-based learning or reading from textbooks can be challenging. Through traditional education the students learn a diversity of subjects but they may not know how to apply their gained knowledge to real-world problems. An alternative to the traditional learning techniques is the introduction of computer simulations for instructional purposes. In [1], the authors determined that when simulations are incorporated into the education they have a positive influence on the learning outcomes. They facilitate students to acquire knowledge through direct experience, especially if the simulations are utilized as attractive, easy to understand interactive media. Simulation-based learning has the capability to incorporate real world training scenarios that otherwise are strenuous to simulate in a traditional teaching environment. Thus, it can help students sustain their attention throughout extended learning periods.

The main objective of simulation-based learning is that students learn while doing or experiencing things in the simulated world, which resembles the real world. The integration of effective features of simulations into the education can enhance learning. More effective training systems are vital to the survival of education because they can help students learn a variety of subjects actively rather than passively.

Many real world situations and problems could be represented in simulation-based systems, which can be useful for the students to get a better feel of practical experiences. Software tools for creating simulation environments are becoming more advanced and easier to employ [2].

Although simulations can provide enhanced learning environments, some disadvantages include: (a) students need training before using simulation tools; (b) teaching methods and the students' learning may not match; (c) there is no instant feedback provided to the user; (d) before the students use the simulation environment, it is necessary that they have basic computer skills and also know the various components of the simulation [2].

66.1.1 Rules of the Road Ship Simulator

The Rules of the Road Ship Simulator (RoRSS) we created as part of our research work provides a platform for students to learn about ship navigation rules by facilitating interactive real-world experiences. Figure 66.1 shows a screenshot taken during a practice section with RoRSS. It is worth noting that RoRSS has two main components or sections: a *practice section* and a *quiz section*.

In the practice section, the students learn about the ship's type, target angle, light configurations, and locations of ships, thereby acquiring the concepts and information needed for safe-navigation on the open ocean. Navigation is the maneuvering of a ship from one point to another on the open ocean, similar to driving the car on the road [3]. Safe ship navigation requires knowledge about ship types, shapes, lights, light configurations, target angles, and the associated rules of the road to control the vessel and avoid collisions between ships. So far we have incorporated in RoRSS information about 10 ships used by the US Navy.

S. Jose · S. Liu · S. J. Louis · S. M. Dascalu (✉)
Department of Computer Science and Engineering, University of Nevada, Reno, NV, USA
e-mail: sjose@nevada.unr.edu; simingl@unr.edu; sushil@cse.unr.edu; dascalus@unr.edu

© Springer Nature Switzerland AG 2019
S. Latifi (ed.), *16th International Conference on Information Technology-New Generations (ITNG 2019)*,
Advances in Intelligent Systems and Computing 800,
https://doi.org/10.1007/978-3-030-14070-0_66

Fig. 66.1 Screenshot from a practice session with RoR

In the practice section, the students can use the right and left arrow keys to rotate and maneuver the ship around. The target angle and view of the ship given at the top-center of the screen enable the users to know about the direction to which the target ship is headed. Users can use the buttons *Night*, *Dusk*, and *Day* to switch between the night, dusk, and daytime views of the ship. With the system's *night option* available in the practice section, the students can understand different light configurations: *Underway, Not Under Command, Making Way, Not Making Way, Restricted in Ability to Maneuver, Constrained by Draft, Pushing Ahead, Towing Astern, Engaged in Fishing, Engaged in Sailing*, and *At Anchor*. For example, *Underway* means not attached to the shore but not necessarily moving through the water while *Making Way* means moving always through the water [3]. The users can also view other ships at different distances by using the *Near*, *Medium*, and *Far* buttons that are provided in the simulator's Graphical User Interface (GUI). To avoid collisions, one must know the target angles and positions of the ships as well as all possible light configurations during the night.

66.1.2 Problem Statement

When traditional learning methods are used, students extract knowledge from textbooks, classroom lectures, labs and tutorials, or from material available online. However, it has been shown that this way of learning has several limitations, especially in terms of students applying their newly acquired knowledge to actual real-world scenarios [4]. This lag between the learning process and the application of the learned knowledge manifests in students having difficulty using new principles and concepts in practice. Therefore, this problem merits investigation to determine whether other learning methods, such as simulation-based learning, could be more effective in this respect.

66.1.3 Purpose of the Research

The purpose of the study is to compare the effects of student performance of simulation-based learning versus traditional

studying of online material. Our research study employs RoRSS as an environment for students to self-study the rules of ship navigation. In particular, the research explores the effectiveness of simulation in learning concepts and information related to safe navigation, including ship types, shapes, lights, light configurations, and target angles. The design method applied involves qualitative and quantitative analysis of data. The following is the research question we are trying to address in this paper:

Can we improve student learning by using a simulation-based approach over employing a traditional learning method such as the studying of online material?

The sub-questions are:

(i) Do users perform better (as reflected in quiz scores) when learning using a simulation environment rather than when using a traditional learning method?
(ii) Do users learn a given set of concepts in less time when using simulation versus when studying online material?

The *null hypothesis* is:

H0: The performance achieved and time taken by students to learn ship navigation concepts and information using a simulation environment do not significantly differ from the performance achieved and time taken by students who use online study material.

The *alternate hypothesis* is:

H1: The performance achieved and time taken by students to learn ship navigation concepts and information using a simulation environment are different from the performance achieved and time taken by students who use online study material.

The remaining of this paper is organized as follows: Sect. 66.2 covers the related work, followed by Sect. 66.3 which describes the methodology applied. Results are presented and discussed in Sect. 66.4 while Sect. 66.5 presents the conclusions of the paper and outlines directions of future work.

66.2 Related Work

Integrating technology into education has become popular nowadays. In particular, simulation has been employed as a teaching tool to provide students with a realistic experience. In [2], the authors discussed the impact of simulation-based teaching on students' learning outcomes compared with the hands-on approach in the lab-based environment. The results of their study revealed that simulation is effective in promoting learning when it is combined with hands-on activities. In [5], Sulaiman et al. address the effectiveness of simulation on students' academic performance by comparing a simulation-based technique with a lecture based method of teaching. The study found that the students taught with the simulation game technique performed better than the students who used the lecture method for learning. According to Veenman, Elshout and Busato [6], simulations that address problem-oriented concepts improve cognitive abilities, creativity, and problem-solving capabilities of the students. Simulation-based learning helps students develop critical and strategic thinking skills as well as enables them to learn interactively various theories and concepts. Thus, simulations promote active learning. The main advantage of simulation-based training systems is that they provide a strong tool for students to develop strategic planning and thinking.

Simulations are not only used in education but also in other areas such as business, medicine, transportation, and search and rescue operations. In [7], Siddiqui et al. address the potential advantage of simulation for disaster operations. They investigated the effect of a simulation environment for Unmanned Aerial Vehicles (UAVs) on training personnel who can experience real-world control of UAVs for rescue operations. In [8], Shin examined the effectiveness of patient simulation in nursing education through meta-analysis of several primary studies. These studies focused on the effects of simulation on patients in nursing. The authors identified significant post-intervention betterment among participants who received simulation education compared to the control group, thereby concluding that simulation was a more potent educational tool than the traditional learning method [8].

In [9], the authors investigated the effect of an online simulation-based learning system on students' performance in learning the concepts of linked-list structures. Their results exhibited that the participants considered simulation as a useful learning environment. Furthermore, simulations can help learners enhance their skills through practice and prepare them to more easily understand complex concepts. In particular, simulation enhances learning by inclusion of tasks, problems, or cases [10]. A study was conducted in Osun-State, Nigeria to identify how simulations and game environments aid in increasing student motivation and improving their attitude towards learning mathematics [11]. The results of their study showed that the students were positively affected by the use of games and simulations. In [12], Roger and Michelene investigated the difference between learning using computer simulations and reading from text. They used the domain topic *project management* for their study. The students in the experimental group learned from simulation about project management activities such as interpreting the relationship between project management decisions and measure project outcomes. The control group focused on learning different sections about project management by reading related texts. Their results showed that computer simulations enhanced the learning of implicit domain knowledge from pre-test to post-test.

In [13], the author examined the impact of an interactive computer simulation to foster student learning outcomes.

The results indicated that both observation and experimental interactivity have an equal effect on facilitating students' performance. Moreover, the results also showed that when compared with structured prompt scaffolding, the driving questions scaffolding helped the students to learn the concepts of sinking and floating better. In [14], the author investigated the use of computer simulations as a training tool to reinforce model-based reasoning in students. Cristian et al. in [15] proposed simulation for training in cyber space. Additionally, the authors provided an interface prototype for the training in digital battlefield. In [16], Chan et al. presented an exploratory study of integrating assistive technologies into 3D simulations, pertaining to Brain-Computer Interface (BCI) technology.

66.3 Methodology

We studied the effectiveness of using RoRSS in learning the concepts of ship navigation rules. To this end, the control group of students learned these concepts through related online material while the experimental (intervention) group of students learned using the ship simulator. The results of the study were based on the students' scores and quiz completion times, for both the simulation and the online study material approaches. Part of this user study, the responses from 44 participants were examined quantitatively and qualitatively.

66.3.1 Participants

The participants for this user study were university graduate and undergraduate students with very limited knowledge of the domain (ship navigation rules). 44 participants with no prior experience in our simulation-based learning system were recruited by word of mouth. Among 44 participants, there were 22 participants in the intervention group and the other 22 participants were considered to be the control group. The participants ages ranged from 18 to 35 years. Users included both genders, and there was an equal percentage of males and females in the experimental group.

66.3.2 Apparatus

The platform used for simulation was Windows. The hardware used consisted of an Intel core i7 processor and the total memory of 16 GB. The application was developed using the Unity game engine with C# as the programming language. The data from the user study was stored in a SQLite database which has been integrated into the Unity game engine.

66.3.3 Procedure

The study was approved by our university's IRB (Institutional Review Board) office and the experiment was conducted in the university library. Prior to the experiment, the participants were provided with a consent form that explained the objective of the research and the procedure of the experiment. The participation in this study was on a voluntary basis. Participants were directed to the further steps only if they agreed to be in the study. The experimental group used the RoRSS, whereas the control group used online study material.

The online study material provided to the control group was the abbreviated guide to navigation rules of the road available from [17]. The control group as well as the intervention group first completed a pre-test survey which included demographic data, their current knowledge about ship navigation rules, and their most preferred learning method. Due to the complexity of the subject matter, both groups were allowed to familiarize themselves about ship navigation by reading a related online study material for 5 min.

Afterwards, they were asked to take a pre-test. The pre-test given was a quiz that contained 20 questions. After the quiz, users underwent a self-study session. The intervention group used RoRSS' practice section, whereas the control group was presented with the online material "Abbreviated guide to navigation rules of the road" [17] for their self-study session. The time allotted for both groups during their self-study was 20 min. The self-study session was followed by another quiz (post-test) consisting of 20 questions with the same level of difficulty as that of the pre-test. For consistency, every quiz provided to the users was generated automatically. Snapshots taken during the RoRSS quiz session are shown in Figs. 66.2, 66.3, and 66.4.

Figure 66.2 displays one of the ships included in the simulator. The students had to identify the type of ship displayed by choosing one of three possible answers. Once they finished answering the question, the students were able to proceed to the following question by clicking on the *Next Question* button.

Fig. 66.3 Simulator quiz question regarding the light configuration of a ship

Fig. 66.4 Simulator quiz question regarding the target angle of a ship

Figure 66.3 shows a ship's light configuration during the night. At night, lights are the only option to identify a ship and its target angle. During the quiz, the students were asked to identify the particular light configuration of the ship.

Figure 66.4 displays the ship heading towards a specific direction. Students had to select the target angle of the displayed ship. To do that, they had to move the needle of the compass to select the estimated target angle.

The quiz score and the time took for each question were recorded in the SQLite database. Finally, a post-test survey was given to the participants to help analyze whether they liked the system and whether they prefer to use the simulation feature or other type of learning. In addition, their comments about the simulation were also collected. For each participant, the whole experiment lasted approximately 40 min.

66.3.4 Design

The experiment was a 2×2 in-between subjects design. There were two independent variables and two dependent variables.

Independent variables: Simulation, online study material

Dependent variables: Score, quiz completion time

For the design, we used four methods of evaluation. In the first method, we assessed the pre-test and the post-test scores of both the intervention and control groups. We analyzed the

Fig. 66.2 Simulator quiz question regarding the type of a ship

scores as a criterion to evaluate the students' improvement in learning. In the second method, pre-test and post-test quiz completion times were analyzed for both the intervention and control groups.

In the third method, we compared the delta, or change, in the pre-test and post-test quiz scores of the students who used RoRSS' practice section with that of the students who learned using online study material. Finally, in the fourth method, we compared the delta in the pre-test and post-test quiz completion times between the two groups.

66.4   Results and Discussion

The first phase of the analysis consisted of collecting data based on the quiz attempted by the participants. The IBM SPSS Statistics Data Editor tool was used for analyzing data. The Paired-Samples T-test was used for testing within-subject factors and the Mixed-Design ANOVA (Split-Plot ANOVA) was used for testing between subject factors.

The quantitative analysis of the data collected from the user study is presented next.

Table 66.1 shows the analysis of quiz scores for the within subjects design. For the experimental group, there was a significant improvement from pre-test (Mean = 33.41%, Std. Deviation = 0.0931) to post-test (Mean = 53.86%, Std. deviation = 0.1185), with a p-value less than 0.05. This result thus provided strong evidence that the simulation-based learning helped the students to improve their scores.

However, for the control group who used online study material for their self-study session, there was no significant improvements from pre-test (Mean = 31.82%, Std. deviation = 0.1006) to post-test (Mean = 35.45%, Std. deviation = 0.0987), with a p-value = 0.73. The result thus indicates that the students were not able to improve their scores when they used online study material for their self-study session. However, not statistically significant because p > 0.05.

Table 66.2 shows the analysis of quiz times for the within subjects design. From the details given in Table 66.2, it can be seen that for the experimental group there was a slight decrease in mean time from pre-test (Mean = 11.97 min, Std. deviation = 2.8421) to post-test (Mean = 10.93 min, Std. deviation = 3.2428). However, the p-value obtained was p = .118 (p >.05) indicating that there was no statistical significance difference even though they used simulation for learning.

Table 66.2 also shows the times taken by students who used online study material for their self-study session. In this case, the mean value obtained for the post-test was 10.65 min which is slightly greater than the mean value (10.31 min) obtained for the pre-test. The p-value was 0.549 (p >0.05). These results indicate that there was not a statistically significant difference between the times taken for the quiz before and after self-study using online material.

Based on the delta mean values presented in Tables 66.1 and 66.2, Table 66.3 shows the differences in scores between subjects: the experimental group and the control group. Results from the mixed-design ANOVA test indicate that there was a significant difference in the pre-test and post-test scores of both the control and the experimental groups. This is based on the fact that the calculated p-value was less than 0.05 and the F-value is 822.079. Therefore, the null hypothesis was rejected. Similarly, the difference in times taken for the quiz between subjects was also statistically significant because the p-value was less than .05. From Table 66.3, it can be implied that the simulation-based approach improved the outcomes of the students' learning when compared with outcomes of the online material study approach.

Figure 66.5 shows the graph of the scores between subjects. The x-axis represents when the quiz was taken (1 = pre-test and 2 = post-test) and the y-axis represents the estimated marginal means of the scores. The graph indicates that there was only a small difference (0.32 vs 0.35) in scores between pre-test and post-test for the control group. In contrast, the experimental group's scores increased substantially from pre-test to the post-test (0.33 vs 0.54). This indicates that the

Table 66.1 Analysis of quiz scores within subjects

Description	Mean [%]	Delta Mean [%]	Std. deviation	P-value
Quiz scores before training using simulation	33.41	20.5	0.0931	0.000
Quiz scores after training using simulation	53.86		0.1185	
Quiz scores before training using online study material	31.82	3.6	0.1006	0.73
Quiz scores after training using online study material	35.45		0.0987	

Table 66.2 Analysis of quiz times within subjects

Description	Mean [minutes]	Delta Mean [minutes]	Std. deviation	P-value
Time taken for the quiz before training using simulation	11.97	−1.03	2.8421	0.118
Time taken for the quiz after training using simulation	10.93		3.2428	
Time taken for the quiz before training using online study material	10.31	0.34	2.9193	0.549
Time taken for the quiz after training using online study material	10.65		2.9213	

Table 66.3 Analysis between subjects

Description	F-value	P-value
Scores of the quiz between subject (Simulation vs Online)	822.079	.000
Time taken for the quiz between subjects (Simulation vs Online)	762.709	.000

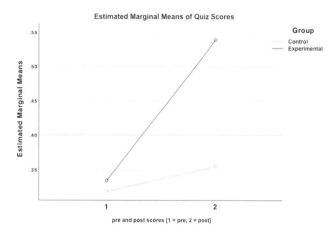

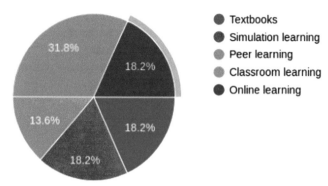

Fig. 66.7 Students' most preferred learning method before self-study using RoRSS

Fig. 66.5 Online vs Simulation in terms of pre-test and post-test quiz scores

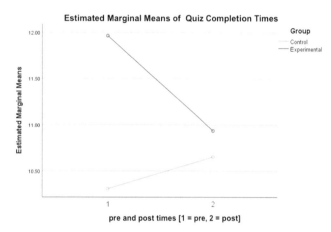

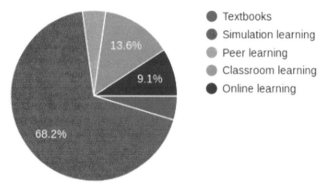

Fig. 66.8 Students' most preferred learning method after learning using RoRSS

Fig. 66.6 Online vs Simulation in terms of pre-test and post-test quiz completion times

performance of the students who used simulation for learning increased substantially.

Figure 66.6 shows the graph of the time taken for the quiz between subjects. The x-axis represents when the quiz was taken (1 = pre-test and 2 = post-test) and the y-axis represents the estimated marginal means of the times taken for the quiz. The graph indicates that the time taken for the quiz by the experimental group decreased significantly (the marginal mean decreased from 11.96 to 10.93) whereas the control group had a slight increase in the time taken from pre-test to post-test (the marginal mean increased from 10.30 to 10.65).

Overall, the results from the quantitative analysis demonstrated that the simulation was a more effective tool for learning than the online study material.

In addition to the analysis of scores and time taken for pre-test and post-test quizzes, we also collected the students' responses and feedback about RoRSS. This survey was used to evaluate the students' satisfaction in using simulation for learning and determine how comfortable they were while using the system. The feedback received indicated that many students liked the simulation-based approach and considered it an effective tool for understanding complex concepts. The results of the survey are described next. The pie chart presented in Fig. 66.7 shows that only 18.2% of students were interested in simulation before they used the RoRSS, whereas as shown in Fig. 66.8, the percentage increased to 68.2% after they interacted with the simulation we developed.

Figure 66.9 shows the number of experimental and control group participants' responses regarding the level of difficulty of learning during their self-study sessions. It can be seen that the highest percentage (55%) of the experimental group characterized their difficulty level of learning using

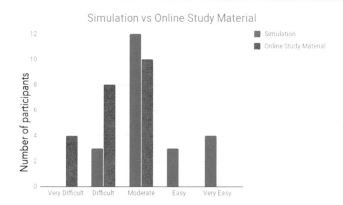

Fig. 66.9 The difficulty level of learning using simulation vs online study material

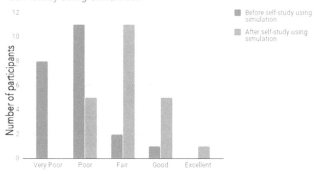

Fig. 66.10 Level of knowledge about ship navigation before and after self-study using simulation

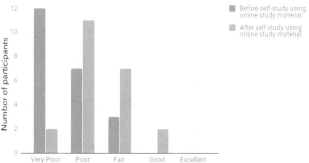

Fig. 66.11 Level of knowledge about ship navigation before and after self-study using online study material

learning about ship navigation, most of the control group participants (59%) were not satisfied with this learning technique.

66.5   Conclusion and Future Work

This paper has presented a user study that investigated the effects of a simulation-based approach on the students' learning of the concepts pertaining to ship navigation. Based on the findings of the study, it is concluded that the users who used simulation achieved significantly better results on the post-test quiz, whereas the users who used online study material did not perform well.

The intended users of this simulation are midshipmen where they can experience a realistic environment for learning about ship types, shapes, lights, light configurations, and target angles. The domain of this particular simulation (ship navigation) is complex for regular students, who largely are not familiar with it. The important finding is that the simulation-based learning significantly improved the experimental group's performance over time, despite the fact that the domain is complex and the participants were not the intended users of the simulation. The performance-based evaluation provided evidence that with help from simulation, students could improve their post-test quiz scores. This work thus demonstrated that a simulation-based approach could enhance the learning outcomes when compared with a traditional educational method. Furthermore, the results of the satisfaction analysis showed that the students were interested and willing to use the simulation as a learning tool in the future.

For future work, we are interested in investigating the students' performance using a Bayesian network approach. To this end RoRSS' Feedback GUI will be further modified so that the students could have a better understanding of their ongoing performance during the quiz. We also plan to

simulation as moderate. On the other hand, only 45% of the control group indicated their response as moderate. Out of 22 experimental group participants, 19 favored the use of simulation whereas 3 found the simulation difficult to use. In the case of self-study using online material, 12 of the 22 control group participants described their level of learning as difficult or very difficult. Moreover, none of the control group participants found the online study material easy or very easy for learning the concepts of ship navigation.

Figure 66.10 shows the students' responses regarding their level of knowledge about ship navigation before and after self-study using simulation. A glance at the graph enlightens us of the fact that the students were able to learn more by using simulation-based learning. More than 70% of the students gave positive feedback about the simulation-based learning method. Only 22% of the students in the experimental group did not find the simulation as useful.

Figure 66.11 shows the students' responses regarding their level of knowledge about ship navigation before and after self-study using online study material. It is evident from the chart that although the online study material helped in

integrate a game-based approach to the simulation so that learning could also be fun.

Acknowledgements This material is based upon work supported in part by the Office of Naval Research under grant No. N00014-15-1-2015 and by the National Science Foundation under grant No. IIA-1301726.

References

1. Vlachopoulos, D., Makri, A.: The effect of games and simulations on higher education: a systematic literature review Int. J. Educ. Technol. High. Educ. **14**, 1–33 (2017)
2. Taher, M.T., Khan, A.S.: Impact of simulation-based and hands-on teaching methodologies on students learning in an engineering technology program. In: Proceedings of 121st ASEE Annual Conference and Exposition, Indianapolis (2014)
3. Michael, R.B., Joseph, A.C.: COLREGS – based navigation of autonomous marine vehicles. In: Proceedings of the 2004 IEEE/OES Autonomous Underwater Vehicles, Sebasco, 17–18 June 2004, pp. 32–39 (2004)
4. Raymond, C.: Do role-playing simulations generate measurable and meaningful outcomes? a simulation's effect on exam scores and teaching evaluations. Int. Stud. Perspect. **11**, 51–60 (2010)
5. Sulaiman, B., Mustapha, I.B., Ibrahim, B.B.: Effect of simulation techniques and lecture method on students academic performance in Mafoni Day Secondary School Maiduguri, Borno State. J. Educ. Pract. **7**, 113–117 (2016)
6. Veenman, M.V., Elshout, J., Busato, V.: Metacognitive mediation in learning with computer-based simulations. J. Comput. Hum. Behav. **10**, 93–106 (1994)
7. Siddiqui, K.T.A., Feil-Seifer, D., Jiang, T., Jose, S., Liu, S., Louis, S.: Development of a Swarm UAV simulator integrating realistic motion control models for disaster operations. In: Proceedings of the ASME Dynamic Systems and Controls Conference, Tysons Corner, Oct 2017, pp. 1–10 (2017)
8. Shin, S., Park, J.H., Kim, J.H.: Effectiveness of patient simulation in nursing education: meta-analysis. Nurse Educ. Today **35**(1), 176–182 (2015)
9. Lai, A., Wu, T., Lee, G., Lai, H.: Developing a web-based simulation-based learning system for enhancing concepts of linked-list structures in data structures curriculum. In: Proceedings of the 3rd International Conference on Artificial Intelligence, Modelling and Simulation, pp. 185–188 (2015)
10. Teresa, C.: Using simulations to enhance teaching and learning: encouraging the creative process. Educ. Res. VSTE Va. Soc. Technol. Educ. J. **21**, 1–7 (2006)
11. Akinsola, M.K.: The effect of simulation-games environment on students achievement and attitudes to mathematics in secondary schools. Turk. Online J. Educ. Technol. TOJET **6**, 1–7 (2007)
12. Taylor, R.S., Chi, M.T.H.: Simulation versus text: acquisition of implicit and explicit information. J. Educ. Comput. Res. **35**, 289–313 (2006)
13. Chang, H.: How to augment the learning impact of computer simulation? The designs and effects of interactivity and scaffolding. J. Interact. Learn. Environ. **25**, 1083–1097 (2017)
14. Develaki, M.: Using computer simulations for promoting model-based reasoning. Sci. Educ. **26**, 1001–1027 (2017)
15. Barria, C., Rusu, C., Cubillos, C., Diaz, J.: Training through simulation for digital battlefield. In: Proceedings of the 12th International Conference on Information Technology – New Generations, pp. 540–545 (2015)
16. Chan, A., Gamino, A., Harris, F.C. Jr., Dascalu, S.: Integration of assistive technologies into 3D simulations: an exploratory study. In: Proceedings of the 13th International Conference on Information Technology: New Generations, Advances in Intelligent Systems and Computing, Springer International, vol. 448, pp. 425–437 (2016)
17. QUICK REFERENCE abbreviated guide to navigation rules of the road based on the navigation rules international – Inland (Commandant Instruction M16672.2D). (1999) [Online]. Available: https://www.uscg.mil/hq/cgcvc/cvc3/references/Rules_of_Road_Quick_Reference.pdf

Image Classification Using TensorFlow

Kiran Seetala, William Birdsong, and Yenumula B. Reddy

67.1 Methods and Controls

67.1.1 Python

The programming language used is Python an object-based programming language. Python objects may include numeric types, Boolean expressions, as well as lists and arrays. Using a read-eval-print loop architecture, Python allows for a user to easily interact with and use nearly any program. Since Python is so popular, Python can be used as a language to connect to most libraries to complete more difficult tasks. Python 2.7.15rc was used to connect the libraries as it was already installed on the Linux machine.

67.1.2 TensorFlow

TensorFlow is an open source library that allows for high performance computing and machine learning [3, 5]. TensorFlow is used throughout many companies ranging from Nvidia to Intel, and many more. TensorFlow operates using multidimensional arrays called tensors. These tensors allow Python to complete more complicated computations that are needed when working with machine learning. This format of holding information is used to save much more complicated information in a tensor than in a typical one dimensional array. The version of TensorFlow utilized is TensorFlow 1.9.

K. Seetala
Department of Electrical Engineering, Louisiana Tech University, Ruston, LA, USA

W. Birdsong · Y. B. Reddy (✉)
Department of Computer Science, Grambling State University, Grambling, LA, USA
e-mail: ybreddy@gram.edu

TensorFlow 1.9 has updated documentation, improved data loading methods, and has an expanded Python interface compared to older versions. The expanded interface was very useful as it allows Python to use the command line and its pip installation. As seen in Fig. 67.1, TensorFlow 1.9 can be easily downloaded from the official Google TensorFlow website using simple pip installation to the Linux-based system. TensorFlow Hub also had to be downloaded using the pip command, as seen in Fig. 67.2, to ensure all packages that were needed to train the neural network were present within the system.

67.1.3 Machine Learning

Machine learning (ML) is the study of algorithms and mathematical models that computer systems use to progressively improve their performance on a specific task. Machine learning algorithms build a mathematical model of sample data, known as "training data," in order to make predictions or decisions without being explicitly programmed to perform the task. Machine learning is becoming so advanced that programs can be used for speech and image recognition. In this research, a specific type of Machine Learning called an Artificial Neural Network was used.

67.1.4 Artificial Neural Networks

An Artificial Neural Network is a type of machine learning system based on the connections of neurons in a human brain. In an Artificial Neural Network neurons act as containers that can hold values. As seen in Fig. 67.3a, these neurons are organized into three different types of layers: an input layer, an output layer, and a number of hidden layers that can change from network to network [6]. Each neuron has a specific activation function that can be used to tell whether or not a neuron is "firing." Moreover, a neuron is connected to

S. Latifi (ed.), *16th International Conference on Information Technology-New Generations (ITNG 2019)*,
Advances in Intelligent Systems and Computing 800,
https://doi.org/10.1007/978-3-030-14070-0_67

```
$ pip install -U tensorflow
```

Source: TensorFlow, 2018
<https://www.tensorflow.org/install/install_linux>.

Fig. 67.1 TensorFlow install command

```
$ pip install tensorflow-hub
```

Source: TensorFlow, 2018
<https://www.tensorflow.org/hub/installation>.

Fig. 67.2 Pip Install command to install TensorFlow

a

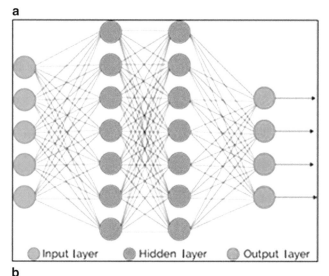

b

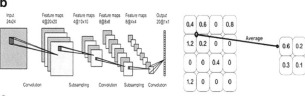

c

INPUT -> FC Implements a linear classifier
INPUT -> CONV -> ReLU -> FC

INPUT -> [CONV -> ReLU -> POOL] * 2 -> FC -> ReLU -< FC
 (Single CONV layer between every pool)

INPUT -> [CONV -> ReLU -> CONV -> ReLU -> Pool] * 3 -> [FC -< ReLU] * 2 -> FC
 (Here there are two CONV layers stacked before every pool)

Fig. 67.3 (**a**) Artificial Neural Networks with two hidden layers. (**b**) CNN architecture. (**c**) CNN architecture Convolutional layers stacked before each pool

all the neurons in the preceding layers as well as all neurons in the layer directly after it. These connections have a specific weight, which will be manipulated later on in the training process, attributed to them [2]. The value in a neuron will be the sum of all of the values in the signal multiplied by their individual weights, all summed together.

Information flows from the input layers towards the output layers. At first, random weights are used but errors are minimized using a process called backpropagation. In backpropagation, the error is computed at the output and then the neural network's weights in the connections are manipulated to minimize the error. This process starts from the outputs, moves backwards through the hidden layers, and ends with the input layer [4]. This process minimizes error and successfully trains the system.

67.1.5 Convolutional Neural Networks

A fully connected neural network transforms inputs to outputs (Fig. 67.3a). We call it fully connected because the input affects the output. The layers possess many learnable parameters and assume no structure in the input. In convolutional neural networks (CNN), the inputs are closer and semantically related. Therefore, CNN is more suitable for image classification. The deep neural network (DNN) architecture (a CNN model) requires a significant number of hidden layers, input parameters, and a sufficient number of images for training. DNN provides the breakthrough architecture in image classification, speech recognition, and natural language processing. Fundamentally, DNN possesses series of layers between input and output for feature identification, similar to our brain (process in series of stages). The existing algorithms are good for learning weights in neural networks with one or two layers. Current algorithms are not suitable for more hidden layers since they need to make thousands of adjustments to make the network slightly better. CNN works in such situations. In the image analysis, the relative position is significant.

In CNN, each image is represented as numbers (each number is a pixel). We apply a series of operations to conclude the probability of closeness to original object. The network contains multiple layers, including: convolution, rectified linear units (ReLU), pooling, fully connected and loss. Suppose the input is a 28x28 area. The input area and other areas consist of grids and resulting grids. The grid has several neurons and takes input from all grids of the previous layer. The weights for each neuron in the current section are the same. After each CNN layer, there is a pooling layer (filter) to produce a resolution of the future map. In pooling, each grid creates a value. The value may be average or maximum or a linear combination of grid values. Figure 67.3b shows the CNN architecture and the example of maximum pooling.

The pooling layer controls overfitting of the number of parameters and the amount of computation. The pooling layer operates independently on every depth slice of input and sizes spatially. Recent trends tried to discard the pooling filter due to the control of the size of the CNN architecture and used many other filters. The Rectified Linear Units (ReLU) increase the nonlinear properties of the decision function and overall network, without affecting the receptive fields of the convolution layer. The process of input to fully connected (FC) is shown in Fig. 67.3c.

67.2 Image Classification

Two Python files retrain.py and Label_image.py were down loaded in the proposed implementation.. Retrain.py was the program that actually created a neural network and went through the process of training the system. Label_image.py was used to label an image using a fully trained neural network. Both were downloaded from https://github.com/googlecodelabs/tensorflow-for-poets-2/tree/master/scripts as both were needed for the next step [3, 5].

Using a Google Chrome extension called Fatkun Batch Download Image, which can be found at [7], images can be quickly downloaded for use as training images. These images, totaling over a hundred images for each subject, were placed in folders corresponding to their subject. The folders were placed in a folder, named "tf_files" in the local home directory.

Using "retrain.py" file from the command line, bottle-necks were created and saved in the "tf_files" folder. Bot-tlenecks are vectors that represent each image that acts as a summary so that the classifier can work on new classes. These bottlenecks are also saved in the "tf_files" folder along with a "retrained_labels.txt" that contained the names of the training folders to be used in classification. Moreover, a "retrained_graph.pb" file was created and saved as well. The "retrained_graph.pb" file contains the newly trained neural network.

Using the "label_image.py" Python file, a new image can be classified using the trained neural network. Once called, "label_image.py" will evaluate the image and send back a list of objects that the image could be. The names, are also printed with a score ranging from 0 to 1. A higher score shows that the neural network is more confident in the classification of the image.

67.3 Results

After testing the program many times with various images, the results were positive. The program works correctly and shows 99% accuracy. It can successfully determine what an image is. The main problem that was encountered was installing some the TensorFlow packages onto the terminal. After the program was properly tested, such as the tiger image in Fig. 67.4, the neural network shows the possibilities of what the image could have been. It also shows us the how long it took to determine the image or the evaluation time. For the tiger test image, the score of it being a tiger

Fig. 67.4 The Tiger Test Image to be classified

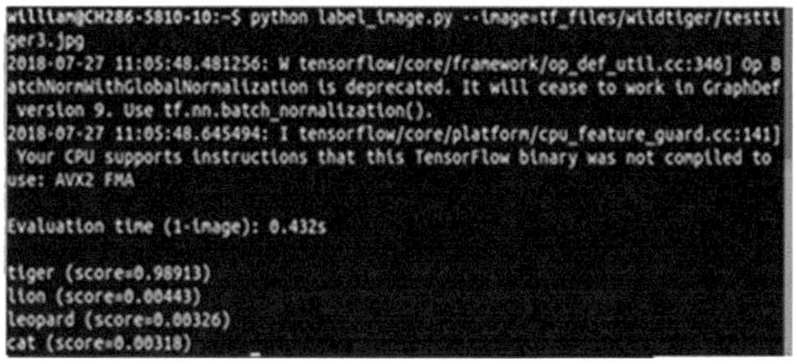

Fig. 67.5 The results which show that the neural network defines the picture as a tiger with 98.913% confidence

was 0.98913. In Fig. 67.5, the other categories include lions, leopards and cats which scores were 0.00443, 0.00326 and 0.00318 respectively.

67.4 Recent Developments

Deep learning models are used in face recognition par-ticularly cropped face [8]. This shows that computer can reconstruct the face image from cropped one. The image recognition was also used in recognition of hand written characters [9]. Further, DL models used for Iris Recognition and the effectiveness can reach up to 99.35% of accuracy [10]. Recently, the DL models were used as a tool for digital image processing [11]. Vaezipour et al., used DL models for face recognition [12]. The identification extended to automatic recognition by establishing age and gender as initial parameters. Bannister provided the faster algorithms for evaluation of face change and recognition [13]. Moor developed a robust system to accurately determine the age and gender of a person from a single image by the application of a deep multi-task learning architecture [14].

Dandan presents a survey that DL systems that improved the accuracy of speech recognition [15]. Xavier et al., dis-cussed visual speech information from the speaker's that has been successfully shown to improve noise robustness of automatic speech recognizers [16]. This survey discussed the human-computer relation extensively. Mihalj et al. discussed the DL models for medical diagnosis [17]. In this review the application of DL model for medical diagnosis is addressed.

A thorough analysis of various scientific articles in the domain of deep neural networks application in the medical field has been conducted.

67.5 Conclusions

The proposed research uses TensorFlow to train the set of images and test the specific image to identify closely. We used Google images of dogs, cats, tigers, and lions and tested the specific image to recognize. The tested image was identified closely with 98.9% confidence. To do this, TensorFlow is used, an open source software library that allows for high performance numerical computation and machine learning. Almost everything that is downloaded in this project is Python based. This was done using the operating system Ubuntu 16.04. The main goal is to classify the presented image accurately. Tiger was the image tested and recognized 98.9% accuracy.

Acknowledgement This work was supported by the AFRL Minority Leaders Research Collaboration Program, contract FA8650-13-C-5800. The authors greatly acknowledge AFRL/RY for their assistance in this work.

References

1. Fatkun Batch Download Image: *Google*, Google, chrome.google.com/webstore/detail/fatkun-batch-download-ima/nnjjahlikiabnchcpehcpkdeckfgnohf
2. Neural Networks: Artificial Intelligence: A Modern Approach, by Stuart Jonathan. Russell and Peter Norvig, pp. 736–748. Prentice Hall Pearson Education International (2003)
3. TensorFlow: *TensorFlow*, www.tensorflow.org/
4. Wasserman, P.D.: Neural Computing: Theory and Practice. Van Nostrand Reinhold, New York (1989)
5. MarkDaoust: Tensorflow-flow-poets-2, (2018). https://github.com/googlecodelabs/tensorflow-for-poets-2
6. Kang, N.: Multi-Layer Neural Networks with Sigmoid Function – Deep Learning for Rookies. https://towardsdatascience.com/multi-layer-neural-networks-with-sigmoid-function-deep-learning-for-rookies-2-bf464f09eb7f
7. https://chrome.google.com/webstore/detail/fatkun-batch-download-ima/nnjjahlikiabnchcpehcpkdeckfgnohf
8. Mosavi, A., Vaezipour, A.: Reactive search optimization; application to multiobjective optimization problems. Appl. Math. **3**, 1572–1582 (2012)
9. Goel, B.: Developments in the field of natural language processing. Int. J. Adv. Res. Comput. Sci. **8**, (2017)
10. Lee, T., David, M.: Hierarchical Bayesian inference in the visual cortex. JOSA. **20**, 1434–1448 (2003)
11. Mosavi, A., Rituraj, R., Varkonyi-Koczy, A.R.: Reviewing the multiobjective optimization package of mode frontier in energy sector. Adv. Intel. Syst. Comput. **519**, 349–355 (2017)
12. Vaezipour, A., Mosavi, A., Seigerroth, U.: Machine learning integrated optimization for decision making. 26th Europian Conference on Operational Research, Rome (2013)
13. Bannister, A.: Biometrics and AI: how face sentinel evolves 13 times faster thanks to deep learning, Editor, IFSEC Global (September 13, 2016)
14. Moor, J.: The Turing Test: The Elusive Standard of Artificial Intelligence. Springer, Dordrecht (2003)
15. Mo, D.: A survey on deep learning: one small step toward AI, https://daisypodcast.files.wordpress.com/2014/01/mo.pdf (2012), pp. 1–16
16. Glorot, X., Bordes, A. et. al.: Domain adaptation for large-scale sentiment classification: a deep learning approach, Proceedings of the 28th International conference on international conference on machine learning (ICML), (July 2, 2011), pp. 513–520
17. Mihalj, B., Dragica, R.: Deep learning and medical diagnosis: a review of literature. Multimodal Technol. Interact. **2**(3), 47 (2018)

Effectiveness of Social Media Sentiment Analysis Tools with the Support of Emoticon/Emoji

68

Duncan C. Peacock and Habib Ullah Khan

68.1 Introduction

Opinions are a central driver of human behaviour. Individuals naturally seek the opinions of others before making decisions, such as buying products and services, investing, and voting in elections. This consultation is being increasingly done using microblogging platforms such as Twitter, posts on social media, discussion forums or reviews on sites like TripAdvisor [1, 2]. Organisations also need feedback on their products and services so that resources can be allocated efficiently to find new investment opportunities, to publicise and improve products, and to anticipate problems. Consequently, interest has grown in a field of study called sentiment analysis to extract meaning from the vast amounts of digital opinion data available. One key feature of a post (or group of posts) that is frequently desired is whether its sentiment polarity is positive, neutral, or negative about a subject. This can be used to give a single sentiment signal, or be aggregated to give an opinion over time [3]. It is vital, therefore, that the increasing number of sentiment analysis tools developed for this purpose classify posts as accurately as possible.

The main approaches used in sentiment analysis are lexicon-based, data- or corpus-based, or a combination of the two. Depending on the algorithm used and the training data, there can potentially be wide differences in the results. For example, unsupervised (lexicon-based) methods can perform better across different subject domains, whereas supervised methods (trained, e.g., on product data), may be better in specialist areas. Analysis of posts made by the wider public must deal with slang, sarcasm, abbreviations, misspellings, grammatical aspects (e.g., multiple exclamation marks), demographics, and technology changes. For example, emojis and emoticons, which are increasingly used on smartphones, can be used to clarify, enhance, or sometimes reverse the sentiment of a post.

Sentiment analysis tools are offered as stand-alone products, but increasingly through APIs as web services. This could potentially offer organisations the chance to compare products, select specialist tools depending on requirements, and benefit from online lexicons and ongoing algorithm development.

Sentiment analysis of short social media messages on microblogging platforms such as Twitter or Instagram is of high interest to organisations that increasingly want to use social media to study the public mood in addition to or in place of traditional methods of obtaining feedback, such as surveys and opinion polls. An increasing number of specialist tools, that can rate the sentiment of a post in a microblog, are being offered to organisations as web services to cater for this need.

Analysis of microblogging messages must be able to handle short messages, varied language use, and specifics such as emoticons, emojis, and hashtags. Emoticons and emojis are increasingly being used in short social media messages and appear to have a significant effect on the sentiment of a tweet and the accuracy of classification. For example, one study [4] suggested that using only the emoticon to rate sentiments could achieve accuracy rates of above 80% [5]. further suggested that emoticon sentiment is likely to be more important than text sentiment and may increase accuracy across subject domains. However, [6, 7], in a limited test, cautiously suggested that there may be classification errors with some sentiment analysis tools in the case in which the emoticon sentiment disagrees with the text sentiment.

D. C. Peacock
University of Liverpool, Liverpool, UK

H. U. Khan (✉)
College of Business and Economics, Qatar University, Doha, Qatar
e-mail: habib.khan@qu.edu.qa

© Springer Nature Switzerland AG 2019
S. Latifi (ed.), *16th International Conference on Information Technology-New Generations (ITNG 2019)*,
Advances in Intelligent Systems and Computing 800,
https://doi.org/10.1007/978-3-030-14070-0_68

Details of the approach used in developing commercial web services for sentiment analysis are not always available; and therefore comparing them is difficult. It seems that the effect of emoticons and emojis should be considered.

68.1.1 Main Aims of the Project

1. To develop a prototype application that can be used to compare web-service-based sentiment analysis tools and prove or disprove the hypotheses.
2. Through this artefact, study inconsistencies in the treatment of emoticons, emojis, and subject area by different sentiment analysis tools.
3. Evaluate whether one tool or method of analysis is more accurate than another by comparing tools against a manually labelled data set.
4. Provide an application that could be expanded in the future into a platform for testing, comparing, and benchmarking sentiment analysis tools and generate test sets for wider study.

68.1.2 Subsidiary Aims

5. Demonstrate a technique for organisations to find the best sentiment analysis tools for their needs.
6. Demonstrate a tool for organisations offering sentiment analysis web services the ability to benchmark their tool against others on the market.

68.2 Literature Review

Sentiment analysis or opinion mining – a subtopic of natural language processing (NLP) – is the study of public opinion, emotions, and attitudes through the analysis of written language [1, 8]. It is a popular area of research with 7000 research articles already written by [9]. Interest from business and other organisations has grown as the amount of digital data, and the use of social media and smart devices has increased [10–12]. There are currently around 319 million monthly active Twitter users, compared with 1.817 billion users of Facebook, 1 billion users of WhatsApp, 600 million users of Instagram, and 877 million users of QQ – a Chinese microblogging platform [13].

The ability of microblogs to give instant feedback is valuable in many domains. Applications have been written that give organisations the ability to, for example, assess whether a target audience is happy (a positive sentiment polarity) or unhappy (a negative sentiment polarity) using a sample of tweets from the live Twitter stream on a desired subject. This could be used to track the sentiment of a brand or feeling about a product. Other uses include politics (judging reactions to policies or predicting election results), financial markets (tracking sentiment on stocks) and tracing the spread of a disease [3, 14].

Given the amount source material, a solid foundation in the subject was provided by sources such as a widely-cited book [1] and survey papers [9]. These followed the evolution of the subject since it was identified around the year 2000, including the strengths and weaknesses of the different techniques – supervised, unsupervised or hybrid – used in sentiment analysis research. A series of conferences called SemEval [15] tracks ongoing developments of computation techniques in semantic evaluation and has a competition to improve techniques in sentiment analysis applications, such as the 'support vector machine classifier and hashtag' used successfully by [16]. Twitter is of particular interest because of the availability of data and the ease-of-use of the public API. At the recent 2016 SemEval conference, Twitter research was the most popular [15], but techniques refined on Twitter could be applied to other platforms. For these reasons, Twitter was chosen as the data source in this project.

The first recorded emotional icon or emoticon in digital communication was a smiley ':-)' used in 1982 at Carnegie Mellon University to indicate that a piece of text was a joke [17]. They began (and are still commonly used) as text characters indicating facial expressions. More recently, icons such as have been increasingly used in place of the text – either through substitution or by allowing the user to select one from a list of icons. Emojis (meaning picture character in Japanese) are a step further, allowing short messages to be sent with pictograms showing concepts such as celebrations, weather, vehicles, thumbs up, and so on [15].

There is some confusion about the terms emoticon, emoji, and smiley, and they are often used interchangeably. However, there are differences in the history, usage, and technical implementations of the text and pictorial variants. Hence, following [15], they are defined in this paper as follows:

- *Emoticons* are pictures made up from the standard ASCII character set used to indicate a facial expression. For example, the smiley emoticons ':)' (read sideways) and ' (^_^) ', and the sad emoticon ':('.
- *Emojis* are pictorial evolutions of emoticons that allow a wider range of ideas (such as weather, directions, and vehicles) as well as emotions. They are stored as Unicode characters with the first set introduced in 1995 [18].

Since the introduction of smartphones (and their addition to popular apps), emojis have become increasingly popular [15] found 4% of tweets and up to 50% of Instagram messages contained emoticons or emojis, whereas [19], found a 20% occurrence in a database of Japanese tweets [20]. noted that the Sina microblog contains a larger number of emoticons and emojis than Twitter. Academic studies such as those by [4, 5, 15] have investigated how emoticons and emojis could be used to improve the accuracy of sentiment analysis classification tools. One challenging area is where emoji or emoticon sentiment is different from the text, perhaps indicating sarcasm.

Several downloadable sentiment analysis tools, such as SentiStrength (2016), have been developed and methods have been created to compare their performance [15] and set benchmarks [21]. However, sentiment analysis is also increasingly being offered in the form of commercial web service APIs. many of which are aimed at Twitter. Factors such as increasing emoticons/emoji use, online dictionaries, machine classification, and ongoing development may make the performance of such tools differ from generalised tools, and vary in relation to each other over time [3, 22, 23]. included some web services in their benchmarking tests; however, to the author's knowledge, no comparison framework specifically aimed at Twitter-based web services exists. There is a need for a specialist service so that organisations with specific social sentiment analysis requirements can find the best tool for their needs.

68.2.1 Sentiment Analysis Research

Liu [1] described how sentiment analysis research chiefly consists of breaking a piece of text down into its constituent parts at three levels:

- Document level – A document is assumed to refer to one entity or subject (like a product), and a positive or negative sentiment is calculated for the whole document.
- Sentence (or even clause) level – where sentences within a document are classified first as subjective or objective and then as positive, negative, or neutral.
- Entity/aspect level – attempts to analyse exactly which aspects of the entity (price, size, etc.) are being rated.

As sentiment analysis concerns natural language, many difficult research problems have had to be addressed, including:

- Comparative opinions, for example, 'iPhones are better than Samsungs'.
- Sentences that mean different things in different subject domains, for instance, 'this vacuum cleaner sucks' [1].
- Sarcasm (particularly in the political sphere).

- Sentences with sentiment words that express no feelings (i.e., factual) and sentences with no sentiment words that express an opinion, for example, 'Can you recommend a good restaurant?'.

Both supervised and unsupervised approaches are used for sentiment analysis, and there are challenges at each level. For example, at the level of extracting aspects from a body of text, [1] noted how a supervised model trained on a test set from one domain (e.g., product reviews) might not perform as well in a different domain. Unsupervised (lexicon or dictionary based) methods can perform better across different domains. This was supported in research by, among [3, 5].

68.2.2 Sentiment Analysis on Microblogging Platforms

Twitter-like microblog posts differ from sources traditionally used in sentiment analysis in several ways:

- Tweets are limited to 140 characters, meaning that they are usually short and to the point. Other platforms may not be as limited, but there is more of a focus on short messages.
- Emoticons and emojis are used both to enhance the sentiment of a tweet [5] and to indicate a joke or sarcasm [24].
- Language use is more casual, less composed, uses slang, and can vary by subject [25]. noted that other emotional signals, such as certain word pairings, exist. However, [26] found a low correlation between the emotional words used in social media and the emotional state of the user, suggesting that using words alone is not sufficient to identify sentiment.
- Volume, speed, variation, and noisiness of data.
- The use of hashtags, both for subject identification and for sentiment annotation [16].
- A group view rather than Individual views on a topic is the target of research [15]
- Other features such retweets, follows, and mentions [27].

Giachanou and Crestani [27] added that there are also specific issues in processing microblogging messages in areas like topic identification, tokenization, and data sparsity (incorrect language and misspellings). This has led to two further approaches to sentiment analysis: hybrid models that combine lexical and machine-learning methods, and graph-based models that include social networking features.

Note: Research is still in progress.

References

1. Liu, B.: Sentiment Analysis and Opinion Mining. Morgan & Claypool (Synthesis lectures on human language technologies, 16), San Rafael (2012)
2. Khan, H.U., Gadhoum, Y.: Measuring internet addiction in Arab based knowledge societies: a case study of Saudi Arabia. J. Theor. Appl. Inf. Technol. **96**, (2018)
3. Abbasi, A., Hassan, A., Dhar, M.: Benchmarking twitter sentiment analysis tools. In: ResearchGate. 9th Language Resources and Evaluation Conference. Available at: https://www.researchgate.net/publication/273000042_Benchmarking_Twitter_Sentiment_Analysis_Tools (2014). Accessed 15 Dec 2018
4. Go, A., Bhayani, R., Huang, L.: Twitter sentiment classification using distant supervision. CS224N Proj. Rep. Stanford. **1**, 12 (2009)
5. Hogenboom, A., Bal, D., Frasincar, F., Bal, M., de Jong, F., Kaymak, U.: Exploiting emoticons in sentiment analysis. In: Proceedings of the 28th Annual ACM Symposium on Applied Computing, pp. 703–710. ACM (2013)
6. Teh, P.L., Rayson, P.E., Pak, I., Piao, S.S., Yeng, S.M.: Reversing the polarity with emoticons. In 21st International Conference on Applications of Natural Language to Information Systems. Available at: http://eprints.lancs.ac.uk/79422/1/nldb_2016.doc (2016). Accessed 20 Dec 2018
7. Khan, H.U., Awan, M.A.: Possible factors affecting internet addiction: a case study of higher education students of Qatar. Int. J. Bus. Inf. Syst. **26**(2), 261–276 (2017)
8. Brock, V.F., Khan, H.U.: Big data analytics: does organizational factor matters impact technology acceptance? J. Big Data. **4**(1), 1–28 (2017)
9. Feldman, R.: Techniques and applications for sentiment analysis. Commun. ACM. **56**(4), 82–89 (2013). https://doi.org/10.1145/2436256.2436274
10. Urabe, Y., Rzepka, R., Araki, K.: Comparison of emoticon recommendation methods to improve computer-mediated communication. In: Ulusoy, Ö., Tansel, A.U., Arkun, E. (eds.) Recommendation and Search in Social Networks, pp. 23–39. Springer International Publishing (Lecture Notes in Social Networks) (2015). https://doi.org/10.1007/978-3-319-14379-8_2
11. Heang, J.F., Khan, H.U.: The role of internet marketing in the development of agricultural industry: a case study of China. J. Internet Commer. **14**(1), 1–49 (2015)
12. Bashir, G.M., Khan, H.U.: Factors affecting learning capacity of information technology concepts in a classroom environment of adult learner. 15th International Conference on Information Technology Based Higher Education and Training (IEEE Conference), Istanbul, Turkey, September 8th – September 10, 2016. (Conference Proceeding) (2016)
13. Statista: Twitter MAU worldwide 2016/Statistic, Statista. Available at: https://www.statista.com/statistics/282087/number-of-monthly-active-twitter-users/ (2017). Accessed 10 Jan 2019
14. Najmi, E., Hashmi, K., Malik, Z., Rezgui, A., Khan, H.U.: ConceptOnto: an upper ontology based on Conceptnet. 11th ACS/IEEE International Conference on Computer Systems and Applications (AICCSA' 2014), November 10–13, 2014, Doha, Qatar, pp. 366–372. (Conference Proceeding) (2014)
15. Nakov, P., Ritter, A., Rosenthal, S., Sebastiani, F., Stoyanov, V.: SemEval-2016 task 4: sentiment analysis in Twitter. In: Proceedings of the 10th International Workshop on Semantic Evaluation (SemEval 2016), San Diego, USA.

Available at: http://alt.qcri.org/semeval2016/task4/data/uploads/semeval2016_task4_report.pdf (2016). Accessed 20 Dec 2018
16. Mohammad, S.M., Kiritchenko, S., Zhu, X.: NRC-Canada: building the state-of-the-art in sentiment analysis of tweets. arXiv preprint arXiv:1308.6242. Available at: http://arxiv.org/abs/1308.6242 (2013). Accessed 27 Dec 2018
17. Dresner, E., Herring, S.C.: Functions of the nonverbal in CMC: emoticons and illocutionary force. Commun. Theory. **20**(3), 249–268 (2010). https://doi.org/10.1111/j.1468-2885.2010.01362.x
18. Unicode.org: Emoji Versions, v3.0. Available at: http://www.unicode.org/emoji/charts/emoji-versions.html (2016). Accessed 4 Jan 2019
19. Yamamoto, Y., Kumamoto, T., Nadamoto, A.: Role of emoticons for multidimensional sentiment analysis of Twitter. In: Proceedings of the 16th International Conference on Information Integration and Web-based Applications & Services. ACM, pp. 107–115. Available at: http://dl.acm.org/citation.cfm?id=2684283 (2014). Accessed 27 Dec 2018
20. Zhang, L., Pei, S., Deng, L., Han, Y., Zhao, J., Hong, F.: Microblog sentiment analysis based on emoticon networks model. In: Proceedings of the Fifth International Conference on Internet Multimedia Computing and Service. ACM, pp. 134–138. Available at: http://dl.acm.org/citation.cfm?id=2499832 (2013). Accessed 27 Dec 2018
21. Ribeiro, F.N., Araújo, M., Gonçalves, P., André Gonçalves, M., Benevenuto, F.: SentiBench – a benchmark comparison of state-of-the-practice sentiment analysis methods. EPJ Data Sci. **5**(1), (2016). https://doi.org/10.1140/epjds/s13688-016-0085-1
22. Bankole, O.A., Lalitha, M., Khan, H.U., Jinugu, A.: Information technology in the maritime industry past, present and future: focus on Lng carriers. 7th IEEE International Advance Computing Conference, Hyderabad, India, 5–7 Jan 2017. (Conference Proceeding) (2017)
23. Smuts, R.G., Lalitha, M., Khan, H.U.: Change management guidelines that adress barriers to technology adoption in an HEI context. 7th IEEE International Advance Computing Conference, Hyderabad, India, 5–7 Jan 2017. (Conference Proceeding) (2017)
24. Rajadesingan, A., Zafarani, R., Liu, H. Sarcasm detection on Twitter: a behavioral modeling approach. In: Proceedings of the Eighth ACM International Conference on Web Search and Data Mining, pp. 97–106. ACM (WSDM '15), New York, NY (2015). https://doi.org/10.1145/2684822.2685316
25. Hu, X., Tang, J., Gao, H., Liu, H.: Unsupervised sentiment analysis with emotional signals. In: Proceedings of the 22nd International Conference on World Wide Web, pp. 607–618. ACM (WWW '13), New York, NY (2013). https://doi.org/10.1145/2488388.2488442
26. Beasley, A., Mason, W.: Emotional states vs. emotional words in social media. In: Proceedings of the ACM Web Science Conference, pp. 31:1–31:10. ACM (WebSci '15), New York, NY (2015). https://doi.org/10.1145/2786451.2786473
27. Giachanou, A., Crestani, F.: Like it or not: a survey of twitter sentiment analysis methods. ACM Comput. Surv. **49**(2), 28:1–28:41 (2016). https://doi.org/10.1145/2938640
28. Khan, H.U., Uwemi, S.: Possible impact of E-commerce strategies on the utilization of E-commerce in Nigeria. Int. J. Bus. Innov. Res. **15**(2), 231–246 (2018)
29. Khan, H.U., Ejike, A.C.: An assessment of the impact of mobile banking on traditional banking in Nigeria. Int. J. Bus. Excell. **11**(4), 446:463 (2017)
30. Khan, H.U., Alhusseini, A.: Optimized web design in the Saudi culture. IEEE Science and Information Conference 2015, pp. 906–915. London, UK, 28–30 July 2015 (2015)

Deep CNN-LSTM with Word Embeddings for News Headline Sarcasm Detection

69

Paul K. Mandal and Rakeshkumar Mahto

69.1 Introduction

Sarcasm detection has being a difficult problem in traditional Natural Language Processing (NLP)/Artificial Intelligence (AI). As described by Pozzi and colleagues [1], *"The difficulty in recognition of sarcasm causes misunderstanding in everyday communication and poses problems to many NLP systems."* Due to the importance and complexity of the problem, many automatic sarcasm detection techniques were reviewed in [2]. The approaches presented for sarcasm detection in [2] were rule-based AI, statistical based AI, and machine learning based AI. However, the rule-based AI in sarcasm detection shows an inability in understanding the context or meaning of words [1–2]. Additionally, the rule-based AI is quite onerous to program. The rule-based modeling for NLP presented in [3] was only able to understand active voice sentences. In this paper, we present a unique deep neural network-based sarcasm detection technique in the news headlines that weight in the advantages of convolution neural network (CNN) and long short-term memory layer (LSTM).

The rest of this paper is organized as follows. Section 69.2 describes the dataset used in this paper. Section 69.3 presents detail CNN-LSTM architecture we used for sarcasm

detection. Section 69.4 highlights the results obtained after applying CNN-LSTM architecture. Finally, Sect. 69.5 concludes the paper.

69.2 Description of Dataset

The dataset consists of 26,709 news headlines. Of these headlines, 43.9% are satire, and 56.1% are real as shown in Table 69.1. Each record consists of three attributes. The first was a Boolean variable indicating whether the headline is sarcastic or not. The second was the news headline itself. The third was the URL of the article. Since the goal was to determine whether a news headline was sarcastic or not, we omitted the URL from our model (Figs. 69.1 and 69.2).

The Natural Language Tool Kit (NLTK) library has a variety of methods that greatly simplify data preprocessing. First, the tokenizer function was used to split each headline into a vector of words. Then, NLTK's PorterStemmer method was used to stem the words. Finally, a frequency list of the entire corpus was created. Later, a dictionary of the 10,000 most common words and assigned each word a number corresponding to its index in the dictionary. Figure 69.3, illustrates the word cloud for the sarcastic and non-sarcastic words. Each headline was then zero padded or truncated so that it was 20 words long in order to feed them into the Neural Network.

69.3 Proposed CNN-LSTM Architecture

The architecture that proposed in this paper consists of an embedding layer, a CNN, and a bidirectional LSTM on word-level vector encodings of news headlines. This network architecture leverages advantages of the LSTM proposed in [4] and the CNN described by [5].

P. K. Mandal
Department of Computer Engineering, California State University, Fullerton, CA, USA
e-mail: pmandal@csu.fullerton.edu

R. Mahto (✉)
Department of Computer Engineering, California State University Fullerton, Fullerton, CA, USA
e-mail: ramahto@fullerton.edu

© Springer Nature Switzerland AG 2019
S. Latifi (ed.), *16th International Conference on Information Technology-New Generations (ITNG 2019)*,
Advances in Intelligent Systems and Computing 800,
https://doi.org/10.1007/978-3-030-14070-0_69

69.3.1 Embedding

The first layer of our network architecture accepts the news headlines as a sequence of word indices corresponding to the dictionary outlined in Sect. 69.3. The embedding layer accepts the inputted headlines and encodes each word into a vector of size e. In the proposed network, sequences are 20 words long, and the embedding size is 128. Thus, this layer will output a matrix of size 20×128.

69.3.2 Convolution

The output of the embedding layer is then fed into a 1-dimensional (1-D) convolution layer. For the 1-D convolutional layer, 32 filters with a kernel size of 7 were used. This layer will perform 1-D convolution on words rather than the 2-D convolutional windows commonly applied on image data which would cut our embedding vectors (and thus our words) into pieces.

The CNN layer allows viewing word combinations equivalent to the kernel size. This permits the neural network to have a sense of context when words are used with other words. In the proposed case, the kernel size is 7. Therefore the filter will establish 7-word combinations. Each neuron uses a rectified linear unit (ReLU) in order to learn non-linear features. It is worth noting that if a linear activation function is used, then the subsequent convolutional layers would be redundant.

69.3.3 Max Pooling

The output from the proposed convolutional layer then undergoes 1-dimensional max pooling. This layer converts each kernel size of the input into a single output by selecting the maximum value observed in each kernel. There are other variants of pooling such as min pooling or average pooling, but most often the most substantial value in the kernel tells

Table 69.1 Type of datasets used

Statistics/dataset	Headlines
Total number of records	26,709
Number of sarcastic records	11,725
Number of non-sarcastic records	14,984

the most about the data. Pooling is used to reduce overfitting; this allows to add more layers to proposed architecture and in turn allows the neural network to extract higher-level features.

69.3.4 Convolution

The output from the max pooling layer is now fed into another 1-dimensional convolutional layer. For this layer, 32 filters are again used and chose a kernel size of 7. To feed in the data, the output of the max pooling layer is padded in order to apply the filer of size 32 (note that zero padding does not affect the data in any other way).

It is more difficult to describe what associations this second convolutional layer will build. However, one can intuitively conceptualize that if the outputs of the first convolutional layer are groupings of words, then this second convolutional layer will output groupings of phrases. The activation function for this layer was also a ReLU.

69.3.5 Bidirectional LSTM

Technically, this layer can be viewed as two layers. First, the output of our convolution is duplicated. One of these sets is reversed. We now have one set in "chronological" order and another in "reverse" order. One LSTM is trained on the chronologically ordered set and another on the reversed set. The output of these two LSTM's is then merged together.

It is worth noting that if an RNN is trained on a chronological set of language data, it can also extract a set of useful features from the reverse order. The idea of a Bidirectional LSTM is to learn features from both the chronological and reverse orderings of language data. Although the data has already passed through two CNN's and isn't "chronological" in the conventional sense, the LSTM can still take advantage of both orderings of the output. A recurrent dropout of 0.5 is used.

69.3.6 Output, Loss Function, and Hyperparameters

Since the news headlines used in this work, ultimately fall under the categories of sarcastic or not sarcastic, our output

Fig. 69.1 CNN-LSTM based architecture

layer is a single sigmoid neuron trained with loss function binary cross-entropy. To train the neural network, a sarcastic headline is represented as a 1 and a real headline as a 0. The output of the sigmoid corresponds to its confidence in deciding whether the headline is sarcastic or not.

The neural network is trained using the optimizer Adam. RMSprop and stochastic gradient descent also used but did not perform as well. The optimal batch size was 128.

69.4 Results

This deep CNN-LSTM with word embedding was able to achieve an accuracy of 86.16% as shown in Fig. 69.4. Many models were tested before developing this "optimal" model. The primary challenge was to design a larger model that would not overfit. More data would have allowed us to design more complex models.

69.4.1 Comparison of Other Models

Currently, no other works have trained neural networks on this dataset. As a baseline, a feedforward neural network with 32 ReLU neurons in the first hidden layer, 4 ReLU neurons in the second neurons in the second hidden layer, and a single sigmoid neuron in the output layer was trained.

```
Real Example:

        Former Versace Store Clerk Sues Over Secret
'Black Code' For Minority Shoppers

Satirical Example:

        Mom Starting To Fear Son's Web Series
Closest Thing She Will Have To Grandchild
```

Fig. 69.2 Example of real and satirical text analyzed

Fig. 69.3 Word cloud for (**a**) non-sarcastic text (**b**) sarcastic text

This yielded an accuracy of 84.56%. Later, another feedforward neural network is trained with three hidden units with ReLU neurons. The first two hidden layers consisted of 32 neurons with a dropout of 0.5; The third hidden layer was composed of four neurons. The output neuron was a single sigmoid neuron. This network architecture had an accuracy of 84.92%.

Various LSTM architectures without an embedding layer were tested. None of them achieved an accuracy above 75%. It should be noted that models with recurrent dropout performed much better than those without it. A network architecture with an embedding layer of 32 features and an LSTM with 32 units had an 85.26% accuracy. A similar architecture with an embedding layer of 256 achieved an accuracy of 85.52%.

Using 1-dimensional convolution yielded further gains. Architecture with an embedding layer of 128 features, a bidirectional LSTM with 32 neurons, and a convolutional layer with 32 filters and a kernel size of 7 with global max pooling achieved an accuracy of 86.04%. The next architecture that performed better is the one proposed in this paper.

69.4.2 Layer Performance

A few things were consistent regardless of the network architecture that we designed. The first is that weight regularization (whether it be the L1 norm, L2 norm, or any other variant) did not help for any of our network architectures. Secondly, all of our models served to benefit from dropout if they were big enough. For LSTM and CNN layers, embedding proved to be invaluable in boosting accuracy. Furthermore, using convolution before our bidirectional LSTM yielded further improvements.

a)   b)

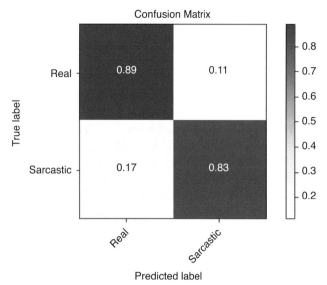

Fig. 69.4 Confusion matrix to determine the accuracy of our model

69.5 Conclusion

In this paper, we presented a CNN-LSTM based sarcasm detection from the news headlines with 86.16% accuracy.

A severe limitation in our experiments was the lack of data at hand. More data would not only yield a higher accuracy on the current model, but also lead to the development of more advanced architectures. Furthermore, we did not experiment with pretrained word embeddings. Significant practical gains could be made with word2vec or other pretrained models.

References

1. Pozzi, F.A., Fersini, E., Messina, E., Liu, B.: Sentiment Analysis in Social Networks. Morgan Kaufmann, Burlington, MA (2016)
2. Joshi, A., Bhattacharyya, P., Carman, M.J.: Automatic sarcasm detection: a survey. ACM Comput. Surv. **50**(5), 73
3. Bajwa, I., Choudhary, M.: A rule based system for speech language context understanding. J. Donghua Univ. **23**(6), 39–42 (2006)
4. Rahman, L., Mohammed, N., Azad, A.K.A.. A new LSTM model by introducing biological cell state. In: 2016 3rd International Conference on Electrical Engineering and Information Communication Technology (ICEEICT), pp. 1–6. (2016)
5. Kim, Y.: Convolutional neural networks for sentence classification. In: Proceedings of the 2014 Conference on Empirical Methods in Natural Language Processing (EMNLP), pp. 1746–1751. Doha, Qatar (2014)

Prioritizing Capabilities of Blockchain Technology in Telecommunication for Promoting Customer Satisfaction

70

Fatemeh Saghafi, Maryam Pakyari, and Masoud Rezaei

70.1   Introduction

Information technology is the foundation of today's communications industry. The telecommunications industry has one of the highest growth rates among other industries. In various industries, the service sector of each industry has the fastest growth rate. This rapid growth stems from the dramatic competition of telecom operators for providing optimum services for customers and achieving customer satisfaction. Since the acquisition of a new customer is much more difficult and more expensive than keeping the existing.

Customer, business owners in the industry intend to understand customer preferences and behaviors more rigorously [1]. Maintaining a balance between the growing need for complex IT services and costs is a major challenge for the telecommunications industry. To keep customers loyal and rise profits these needs must be met along with information security and privacy, optimization of operations and process speeds. The growth of every technology, in conjunction with its benefits, is also causing challenges that convince us to use the technologies that address the weaknesses of the old services. Nowadays, blockchain technology is able to address the weaknesses of past technology. This technology can prevent errors in the telecommunications industry and provide an integrated user experience as well as mechanisms

for protection of sensitive information from being hacked in telephone networks; security of the third and fourth generation services as well as wireless and other communication services based on cloud networks; management of consumer packages, network speed and value services [1].

Currently, countries such as Saudi Arabia, UK, Kazakhstan, South Korea and India use this technology in different ways [1]. Because of the rapid pace of globalization, communication development is essential for developing countries and specially Iran.

In this study we will look into telecommunication service methods and challenges. Then we will define blockchain technology potentials to overcome these challenges and will rank them according to quality aspects which satisfy customers the most. Results of this study will be beneficial for overall planning of telecommunication industry to optimize provision of services to customers.

70.2   Introducing Blockchain Technology

Blockchain has been in the technology field for 10 years [1]. The word Blockchain is made up of blocks and chains. In fact, this technology is a chain of blocks. Blockchain is a database that is not located on one or more specific servers, instead; it is distributed on all of the computers that are connected to a network. This distributed database; which is one of the distinctive features of this technology; basically creates a completely different digital structure. In blockchain, each block in the network records an independent report and each independent record comes together with the other records and a collection of official reports is manufactured. When a transaction is announced, each block will create its own version of those updated events. In fact, this technology is an interesting innovation in the recording and distribution of information that makes it unnecessary to

F. Saghafi (✉)
Faculty of Management, University of Tehran, Tehran, Iran
e-mail: fsaghafi@ut.ac.ir

M. Pakyari
Department of Industrial Engineering, Payame Noor University, Tehran, Iran

M. Rezaei
Alborz Campus, Department of Management, University of Tehran, Tehran, Iran
e-mail: Rezaei2017@ut.ac.ir

have a third party in order to facilitate digital relationships. In fact, blockchain is a general ledger for records and reports, and because of the type of encryption and its registration on all network computers, registered reports cannot be hacked or deleted. By using blockchain, many people can import different reports into an archive, and users can also control the process of registering and updating the information [1].

70.3 Blockchain Technology

The main technologies in blockchain are fundamental trust mechanism and synchronization process between nodes. Li said: "Blockchain systems guarantee the reliability and consistency of the data and transactions by adopting the decentralized consensus mechanism. In the existing blockchain systems, there are four major consensus mechanisms: PoW (Proof of Work), PoS (Proof of Stake), PBFT (Practical Byzantine Fault Tolerance), and DPoS (Delegated Proof of Stake). PoW mechanism uses the solution of puzzles to prove the credibility of the data. PoS mechanism uses the proof of ownership of crypto currency to prove the credibility of the data [2].

Block propagation and synchronization is another mechanism. In the blockchain, each full node stores the information of all blocks. Being the foundation to building consensus and trust for blockchain, the block propagation mechanisms can be divided into the following categories: Advertisement-based propagation, Sendheaders propagation, Unsolicited push propagation, Relay network propagation, and Push/Advertisement hybrid propagation." [2]

70.4 Blockchain Features

This technology has some key features, namely: consensus, authenticity, immutability and certainty. Consensuses necessitate that in order to validate a transaction, all parties must agree on the validity of the transaction. Authenticity means the participants know where the asset comes from and how its ownership changes over time. Immutability insures that no participant can manipulate a transaction after registering it in the general ledger (blockchain). Certainty means the general ledger is where we are referred to in order to determine the ownership of an asset or the completion of a transaction [3].

Blockchain technology with all benefits is not a new technology by itself though, rather a combination of empirical technologies that work in a new way.

70.5 Telecommunications Challenges

Telecom challenges such as high costs, time-consuming operations, attacks on telecommunication networks, anti-spam necessity, unsolicited calls and privacy violations are the major issues of telecoms recipients. Blockchain technology with its own features has the ability to fix these problems, which will be discussed in next subsections [2].

70.5.1 Telecom's Time-Consuming Operations

Insufficient speed of service delivery has led many customers to switch their service provider hoping for more appropriate service. In this situation the technology that can reduce operating time will be able to satisfy customers and gain more market share. Blockchain reduces transaction time for complex and multi-party interactions from a few days to a few minutes. Transaction billing is faster in this system because it does not require a central authority to approve the bill. Some features have been introduced in a licensed blockchain that play a significant role in reducing the time of telecom operations. The following are some of these features.

Improved audit: a shared ledger, used as a single source of information, improves the ability to track and audit transactions.

Increased operational efficiency: net asset digitization makes property transfer more efficient, and as a result, transactions can run at a pace that is more relevant to business [3].

70.5.2 High Telecoms Costs

Price is one of the most important elements of marketing mix; paying attention to cost-cutting strategies will encourage industry owners to use a variety of cost-effective methods (new technologies like blockchain). The cost cutting mechanisms that exist in blockchain technology reduce costs in several ways, including:

1. Monitoring necessity: blockchain requires less monitoring, since all partners in a network are known, which makes it a self-policing net [3].
2. Intermediary necessity: in this technology intermediaries are reduced because, participants can exchange valuable items directly [3].
3. Duplicate work: since all contributors have access to a common ledger duplicate duties have been removed [3].

70.5.3 Attacks on a Telecommunication Network

Majority of Attacks[1] (51% of attacks) are the attacks that compromise digital security. In these attacks, a network member with an aggressive force which accounts for more than 50% of the total network power tries to control a blockchain. When such an attack is detected, the blockchain network interrupts the execution of transactions automatically and prevents the malicious effects of a majority attack. Blockchain is also able to control Distributed Denial of Services (DDOS) attacks [4].

70.5.4 Spam

Getting unsolicited advertisement is one of the problems in today's telecommunication world. Contact information of millions of users can be encrypted and secured using blockchain technology. In this case, contact information of the users will be provided to the registered markets only in special circumstances. In this system, the approval of the subscriber is required for any SMS advertisement. Also, users are allowed to revise their consent at any time and cancel the relevant request. At the moment, a request to cancel the application of goods and information or advertisements takes a week, while using the blockchain technology, the order to cancel the application of such calls and SMSs will be immediately processed and approved [5].

70.5.5 Unsolicited Calls

The increase in social media use, which requires the subscribers to share their contact numbers or emails while signing up, the number of unsolicited calls or messages has also increased. This trend has gone so far that in many cases it has forced users to find a way to get rid of this nuisance. So far, the proposed solutions have been mostly at the user level of software and cellphone apps, or have been limited to warning users for caution in sharing their information on social media. Blockchain is a communicational solution designed to achieve customer satisfaction by recording all contacts between telemarketing companies and subscribers in the blockchain's ledger. This is useful for protecting the customer from abuses of unsolicited and frequent contacts [5].

70.5.6 Insecurity of Data and Privacy

The increase in supervisory and security breaches have resulted in the lack of confidentiality and breach of privacy of the users and has become one of the most important problems of the telecommunication systems. This issue is related to the current model of telecommunication systems in which; a third party collects and controls a large amount of personal information. Blockchain technology provides an infrastructure that adopts a decentralized personal information management. In this system, users control their data. Blockchain runs a protocol that turns it into a control manager with automated access that does not trust a third party. In these circumstances, transactions are used to carry instructions such as storing, querying and sharing information.

The security of data transfers in this system, comes with a major blockchain security feature, namely; protection against tampering, fraud and cybercrime. Permission blockchain networks have a special network that only includes specific documented members. This feature ensures that, the identity of the members is exactly the same as the have declared before and the goods or assets which will be traded are exactly the same as the items that were presented before. The privacy of these blockchain permission networks has improved [3].

- Enhancement of privacy: in this system users, with the help of their IDs and permissions, can choose which transactions other contributors are allowed to see. Permissions can be developed for specific users such as auditors who may need access to more transaction details [3].

70.6 Customer Satisfaction

Satisfaction is each person's good or bad feeling about the services they have experienced, compared to their initial perception of the service quality. Today, technology services which set high expectations and deliver an equivalent decent performance will be successful. The purpose of these services is to achieve total customer satisfaction [6].

Quality of service delivery is one of the determinants of satisfaction which includes a set of characteristics and features of services that includes the ability to meet the implicit and specific (stated) needs. In order to achieve customer satisfaction, the features of the provided services should be prioritized and presented in the framework of those needs.

The specific requirements which are specified in the contract are called constraints. These are the requirements that are mentioned, but the customer does not expect them in the product or service, however, if they are met, there will be more customer satisfaction. Implicit needs are market

---

[1]A majority attack (usually labeled 51% attack or >50% attack) is an attack on the network. This attack has a chance to work even if the merchant waits for some confirmations, but requires extremely high relative hash rate.

dependent. These are the requirements that, the customer expects from the product, even if they are not mentioned. The existence of implicit needs for a product or service is deemed necessary. If they are not met, there will be customer dissatisfaction.

Quality has 9 dimensions, namely: performance, features, compliance, reliability, durability, service, accountability, aesthetics and reputation. These dimensions are somewhat independent and a service can be excellent in one of them, while moderate or weak in others. Few products or services can satisfy all of these dimensions simultaneously. The responsibility of a service or a product marketing team, is to identify and rank the indicators that are consistent with these dimensions, and put them on the agenda to create new services or improve previous ones [7].

70.7 Research Methodology

In this research, 9 dimensions of service quality were extracted from the research literature. This dimensions were shared by 15 industry experts, of whom 5 were professors. According to experts, 3 of these dimensions have an importance of 80% and the rest are less than 60%. So, these three dimensions were introduced in this article as a benchmark for quality assessment of blockchain based services in the telecommunications industry. The dimensions selected by experts are presented in Table 70.1 and Fig. 70.1 The inconsistency rate is 0.05%, so, the results of this research are acceptable.

The hierarchy plan of the research is shown in Fig. 70.2.

Subsequently, a panel of 20 clients was prepared and the 6 applications were prioritized using hierarchical analysis and the three mentioned criteria. The applications weights are computed and ranked by the output of Expert Choice software assigning a number from 1 to 9.

Table 70.2 and Fig. 70.3 show the ranking and estimated weighting of the blockchain applications.

The results showed that the option data security and privacy has the highest weight and the option curbing spam has the least weight. The inconsistency rate is 0.08, making the results acceptable.

The 6 blockchain applications that were considered in this research are all part of customers implicit needs, which, if not met, will cause dissatisfaction.

70.8 Conclusion

The present research shows that blockchain technology has the capabilities to upgrade and optimize the services provided in telecommunication industry. This technology can overcome the shortcomings of previous technologies in a safe and low-cost manner as quickly as possible. In this research, data security and privacy were considered as the most important shortcomings of previous technologies. Since reliability is one of the most basic features a customer is concerned about in choosing any product or service; it's clear that the use of this technology can help the telecom operators to gain more market share. Blockchain technology, can achieve a larger share of the market for telecommunication services by meeting quality dimensions such as reliability, functionality and features. In this study, six applications for blockchain technologies were introduced in detail and were ranked by the customers who receive telecommunication services. The results showed that data security and privacy are the most important factors for customers, and curbing spam is the least important. Then these applications were examined regarding the coverage of the quality dimensions of services that are effective in customer satisfaction and market share. As blockchain applications are important at all personal, organizational and national levels, the results of this research can help policy makers, investors and industry players to upgrade and optimize the performance of the industry using this information in national development plans.

Importance of Data security and privacy is proved by giving the added value of telecommunication services to customers. These services have both advantages and disadvantages. In both cases, if they don't guarantee Security and privacy, they will threaten the assets and dignity of the customers.

Table 70.1 Prioritizing the dimensions of service quality in the telecommunications industry

	Dimensions of quality	weights
1	Performance	0.594
2	Reliability	0.249
3	Features	0.157

Inconsistency = 0.05

Performance 0.595

Reliability   0.249

Features   0.157

Fig. 70.1 Prioritization of service quality dimensions in the telecommunications industry

Fig. 70.2 Hierarchy plan identifying the most important blockchain application in telecommunication according to customers

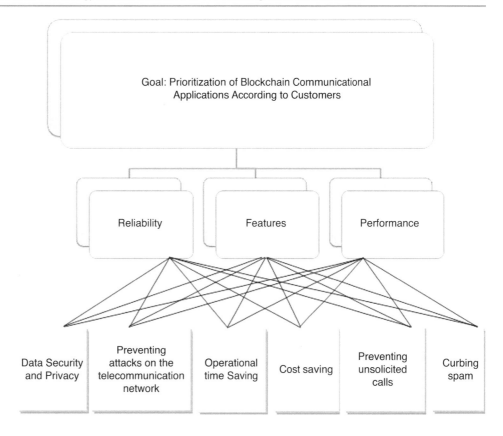

Table 70.2 Prioritization of blockchain communicational applications according to customers

Determining the weights of blockchain communicational applications		
Options	Weight	Ranks
Data security and privacy	0.345	1
Preventing attacks on the telecommunication network	0.327	2
Operational time saving	0.139	3
Cost saving	0.095	4
Preventing unsolicited calls	0.058	5
Curbing spam	0.036	6

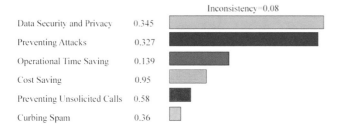

Fig. 70.3 prioritization of blockchain communicational applications according to customers

Therefore, Data security and privacy are more important in comparison to other factors.

References

1. Akhavan, P.: Digital Currency, vol. 272, 1st edn, pp. 62–63. Tehran (2018). [In Persian]
2. Li, X., et al.: A survey on the security of blockchain systems. Futur. Gener. Comput. Syst. 2 (2017). https://doi.org/10.1016/j.future.2017.08.020
3. Gupta, M.: Blockchain for Dummies, vol. 51, 2nd edn, pp. 3–10. Wiley, Hoboken, NY (2018)
4. De, N.: British Telecom Awarded Patent for Blockchain Security Method. [online]. Available: https://www.coindesk.com/british-telecommunications-receives-patent-for-blockchain-protection/ (2017)
5. CNN news: India's Telecom Regulator Taps Blockchain to Curb Spam Calls, SMSes. [online]. Available: https://www.ccn.com/indias-telecom-regulator-taps-blockchain-to-curb-spam-calls-smses/ (2018)
6. Kotler, P., et al.: Marketing Management: Analysis, Planning, Implementation and control, vol. 756, 7th edn, pp. 70–80. Prentice-Hall, Scarborough (1991)
7. Jafari, M., et al.: Total Management, vol. 206, pp. 17–18. Tehran (2000). [in Persian]

Jorge Ramón Fonseca Cacho, Kazem Taghva, and Daniel Alvarez

71.1 Introduction

Accuracy and effectiveness in Optical Character Recognition (OCR) technology continues to improve due to the increase in quality and resolution of the images scanned and processed; however, even in the best OCR software, OCR generated errors in our final output (which we will refer from now on as OCR'd text) still exist. To solve this, many OCR Post Processing Tools have been developed and continue to be used today. An example of this is OCRSpell that takes the errors and attempts to correct them using a confusion matrix and training algorithms. Yet no software is perfect or at least as perfect as a human reading and correcting OCR'd text can be. This is because humans can not only identify candidates based on their memory (something easily mimicked with a dictionary and edit distances along with the frequency/usage of a word), but more importantly, they can also use context (surrounding words) to decide which of the candidates to ultimately select.

In this paper we attempt to use the concept of context to correct OCR generated errors using the Google Web 1T Database as our context. We also use the freely available TREC-5 Data set to test our implementation and as a benchmark of our accuracy and precision. Because one of our main goals is to produce reproducible research, all of the implementation code along with results will be available on multiple repositories including docker and git. The paper is split into several sections, first we discuss the concept of context and its importance, then we introduce the Google Web-1T corpus [1] in detail and the TREC-5 Data set [2], then we explain our workflow and algorithm including our

usage of the tool *OCRSpell*, and then discuss the results and propose improvements for future works.

71.2 Why Using Context is So Important

How can we mimick this behavior of understanding context with a machine? Natural language processing is a complicated field, but one relatively simple approach is to look at the frequency that phrases occur in text. We call these phrases n-grams depending on the amount of words they contain. For example, a 3-gram is a 3 word phrase such as "were read wrong". Suppose that the word *read* was misrecognized by OCR as *reaal* due to the letter *d* being misread as *al*. If we tried to correct this word using a dictionary using conventional methods we could just as easily accept the original word was *real* by simply removing the extra a. Using Levenshtein distance [3], where each character insertion, deletion, or replacement counts as +1 edit distance we can see that $reaal \rightarrow real$ or $reaal \rightarrow regal$ both have an edit distance of 1. If we wanted to transform the word into *read* this would be an edit distance of 2 ($reaal \rightarrow read$). Notice that $reaal \rightarrow ream$, $reaal \rightarrow reap$, $reaal \rightarrow reel$, $reaal \rightarrow meal$, and $reaal \rightarrow veal$ all also have an edit distance of 2. In most cases we want to create candidates that have a very small edit distance to avoid completely changing words since in this case having an edit distance of 5 would mean that the word could be replaced by any word in the dictionary that is 5 or less letters long. However, in this case the correct word has an edit distance of (2) from the erroneous word. This is greater than the minimum edit distance of some of the other candidates (1).

This is where context would help us to decide to accept the word with the bigger edit distance by seeing that some of these candidates would not make sense in the context of the sentence. Our 3-gram becomes useful in deciding which of these candidates fit within this sentence. What's further is

J. R. Fonseca Cacho (✉) · K. Taghva · D. Alvarez
Department of Computer Science, University of Nevada, Las Vegas, NV, USA
e-mail: Jorge.FonsecaCacho@unlv.edu; kazem.taghva@unlv.edu; Alvarez5@unlv.nevada.edu

© Springer Nature Switzerland AG 2019
S. Latifi (ed.), *16th International Conference on Information Technology-New Generations (ITNG 2019)*,
Advances in Intelligent Systems and Computing 800,
https://doi.org/10.1007/978-3-030-14070-0_71

if we use the concept of frequency we can then pick which is the more likely candidate. While both "were real wrong" and "were read wrong" could be possible answers. It is more likely that "were read wrong" is the correct phrase as "were real wrong" is very rare. This however is flawed that while the frequency of some phrases is higher than others, it is still possible that the correct candidate was a low frequency 3-gram. This is a limitation to this approach that could only be solved by looking at a greater context (larger n in n-gram).

There is also the issue of phrases involving special words where ideally rather than representing something like "born in 1993" we could represent it as "born in *YEAR*" as otherwise correcting text with a 3-gram such as "born in 1034" where born has an error could prove hard as the frequency of that specific 3-gram with that year may be low or non-existent. A similar idea was attempted at the University of Ottawa with promising results [4]. This same concept could be extended to other special words such as proper nouns like cities and names; however, a line must be drawn to how generalized these special phrases can be to the point that context is lost and all that remains are grammatical rules.

Even with these limitations context can help us improve existing OCR Post Processing tools to help select a candidate, or even correct from scratch, in a smarter, more efficient way than by simple Levenshtein distance [5]. However, to do this we need a data set of 3-grams to 'teach' context to a machine. This is where Google's Web-1T Corpus comes in.

71.3   Google Web-1T Corpus

The Google Web 1T 5-gram Version 1 corpus (Google-1T) is a data set "contributed by Google Inc., contains English word n-grams and their observed frequency counts. The length of the n-grams ranges from unigrams (single words) to five-grams" [1]. For the purposes of this paper we will be using the unigrams (1-grams) and trigrams (3-grams). The unigrams serve as a dictionary of all words contained in the data set. The trigrams will server as our context. The source of the data, "The n-gram counts were generated from approximately 1 trillion word tokens of text from publicly accessible Web pages" [1].

What makes Google-1T great is that having it based on Web pages allows it to have a broader variety of subjects that enable it to understand more context than other corpus. The variety does come at a price as n-grams with fewer than 40 occurrences are omitted [6]. This means that many correct 3-grams that could be very specialized to a topic are missing. A possible solution to this would be to add subject-specific 3-grams to the data set for the documents being

corrected. There are 13,588,391 unigrams and 977,069,902 trigrams in the corpus contained in approximately 20.8 GB of uncompressed text files.

71.4   The TREC-5 Data Set

As part of our ongoing research with OCR Post Processing and Error correction [7, 8] we were searching for a data set that had a large corpus and included the original 'source' text and the OCR'd version so we could easily test our corrections and compare the accuracy of them. The U.S. Department of Commerce's National Institute of Standards and Technology (NIST) Text REtrieval Conference (TREC) TREC-5 Confusion Track [2] fit these needs. The TREC-5 Confusion Track file *confusion_track.tar.gz* [9] is freely available to be downloaded from their website. Among visiting the TREC-5 Website, please see our previous work with the data set for more details on how we use it and its benefits and limitations [7].

71.5   OCR Spell

Initially released in 1994, OCRSpell is a tool designed to correct OCR generated errors that continues to be relevant today [10, 11]. OCRSpell is written in a mix of Linux Bash and C. It works by generating candidate words through different techniques including the use of ispell. In a previous work on reproducible research [12], we were able to take the old source code, compile it after installing the necessary dependencies, and sucessfully ran it in a Docker container we later uploaded for everyone to easily download and use. Docker greatly enables and encourages reproducible research through the use of container technology [13]. See *docs.docker.com* for more information.

For this project we use OCRSpell to detect OCR generated errors along with a way to benchmark our correction against OCRSpell and to then try and complement it as a way to select the best candidate suggested by OCRSpell. The ease in which we were able to use OCRSpell in this project is a good example on why reproducible research allows building on top of previous work very easily and increases its relevance to scientists. During our work with OCRSpell, we detected a bug when using a single quote (') surrounded by white space or by other quotes that were surrounded by whitespace. This is most likely a parsing error that we intend on fixing for the benefit of the community. This shows that as Peng points out, "reproducibility is critical to tracking down the bugs of computational science" [14].

71.6 The Algorithm and Our Implementation Workflow

The workflow of our algorithm and `run-script` is as follows. Start with OCR'd text and identifying OCR errors, we then generate 3-grams for each error, search in Google-1T for 3-grams where the first and third word match the given 3-gram and then pick the one with the highest frequency as our top candidate if any. Finally, we verify the output using an alignment algorithm we developed [7] that will align the original text with the OCR'd text to generate statistics on the accuracy of our solution. The implementation is split into several modules that perform the aforementioned tasks

71.6.1 Preparation and Cleanup

This pre-processing step transforms the TREC-5 files into clean, human-readable documents. Initially the TREC-5 Data set contains several tags like <DOC>, <DOCNO>, <SUMMARY> that are markers for programs and not actual content to be used for 3-gram collection and thus can be safely removed. We do take advantage of the <DOCNO> to denote where each Document begins and ends since each text file in TREC-5 contains 100 individual documents. Separating them makes it easier to split and transfer into the OCRSpell container we will use in the following steps. These tags are in both the degrade5 and original copies of the files. In the original TREC-5 files there are additional HTML-style character references such as & and &hyph;. These character references can be handled in a variety of ways depending on what it represents. For this step, we take the common ones and transform them into the ASCII character they represent. It is important to point out that our results could be improved by handling these character references with more care than eliminating them or replacing them, but doing so would bias our results to look better with this data set than any other data set that could be thrown into our program. Therefore it is best to remain as neutral as possible. At the end of this step our tidy data [15] is ready to be transfered into an OCRSpell container.

71.6.2 Identifying OCR Generated Errors Using an OCRSpell Container

In order to identify what words are errors we send our split files into a Docker container running an instance of OCRSpell. OCRSpell then generates an output file where each line represents the next word, separated by whitespace, in the given documents. If the word is correct it marks an

asterisk (*), if the word is incorrect and has no suggestions it marks a number sign (#), and if the word is incorrect and has suggestions/candidates it marks an ampersand (&) followed by the candidates. The first line of OCRSpell has information about the program and is preceded by an at sign (@).

```
@(#) Ispell Output Emulator Version 1
* ← Correct Word
& ← Incorrect Word with Suggestions
# ← Incorrect Word with No Suggestions
```

For example, here are the first few lines of the first OCRSpell output file from the first document, `degrade.FR940104-0-00001.000.output`:

```
@(#) Ispell Output Emulator Version 1
*
& hegister 2 0: register, heister
*
# 391 0
*
# 2 0
*
& anuary 1 1: January
# 41 0
```

The matching text for this file is found in `degrade.FR940104-0-00001.000`:

```
Federal hegister Vol. 391 No. 2 )
Tuesday1 'anuary 41 1994 hules and
```

The word *hegister* is identified as misspelled and OCR-Spell has two candidates. It orders them by likeliness that they are the correct answer. In our experiment we only care about identifying the word as incorrect and not using the candidates offered. However in future experiments we will be using this data along with this experiment's output to increase efficiency. Furthermore, numbers are identified as misspellings with no candidates. In our next section we must take this into consideration and skip numerical words as they are assumed to be correct. Also notice that the word *anuary* should match *'anuary* but it does not because OCRSpell has done pre-processing on this word and removed punctuation. This is an issue when trying to align our words that we deal with in a later section. Finally, notice that the single character ')' should be classified as a word but is instead ignored and discarded by OCRSpell. This causes alignment problems with the amount of lines in OCRSpell not matching the number of words in the documents fed to OCR Spell. This is dealt with in the next section.

71.6.3 Get 3-Grams

Once we generate the OCRSpell output and copy it back from the container to the host machine we use this to locate the mispelled words in the cleaned texts by using the line as the location in the document. We then collect the neighboring words to form the 3-grams whose middle word is the incorrectly spelled word and continue on to the next misspelled word. For now, this process is sensitive to the choices made by OCRSpell. If OCRSpell chooses to ignore a word, then the discrepancy between the cleaned form of the source document and the OCRSpell output will increase the error rate by producing erroneous 3-grams. We ignore any 3-grams with a number as the second center word because a number is value-specific and we cannot know which value it is supposed to take. As mentioned earlier, this could be improved by replacing and merging frequencies in the Google-1T data set that involve special phrases. In the second version of our code we implement a crude version of our alignment algorithm [7] to overcome the alignment problem and properly generate 3-grams without errors. In this relatively simple case we were able to achieve 100% accuracy on re-aligning OCRSpell output with the given clean text files as long as the Levenshtein edit distance was smaller than the threshold. Because most OCR words have a small number of errors this covers most cases. In our third version of the code we wanted to make sure that, alignment or not, we could guarantee all matches so we implemented our own Spell Checker that uses the Google Web 1T 1-gram vocabulary which has all words that appear in the corpus, sorted alphabetically, to identify if a word has an error and ensure that for every word in the text there is a corresponding line generated in our output that mimics the OCRSpell formatting to ensure 100% alignment:

```
@(#) Google 1T OCRSpell Output Emulator
*
# hegister 0
*
# 391 0
*
# 2 0
*
*
# 'anuary 0
# 41 0
```

Notice that here we have the additional line with asterisk to indicate that the single parenthesis is considered a word, since it is separated with whitespace, that is correct. Also note that we do not generate candidates so we mark all errors with the number sign. Everything is done with the same format as OCRSpell.

Here is an example of an output file after we have generated the corresponding 3-grams for all Errors. The following is from Version 3 using 1-grams to mimic OCRSpell for file `degrade.FR940104-0-00001.000.3gms`:

```
Federal hegister Vol. 1
Tuesday1 'anuary 41 8
and hegulations Vol. 13
Tuesday1 'anuary 41 19
```

After we have generated all 3-grams we can then move on to searching the Google1T data set.

71.6.4 Search

For this step, we consult Google Web 1T to produce a list of candidate 3-grams for each erroneous word. Since the Google1T database is stored as a list of indexed files, we utilize the index to quickly locate which file(s) our 3-grams may be in. Note that it is possible that the 3-grams candidates may be in multiple Google1T files due to them being near the end of one file or in the beginning of another. Realistically in the worse case they are in 2 files at the most and even so this is very rare. Once we've collected the names of the files, to we utilize bash's *grep* tool to search each file for all the 3-grams whose first and last word match those of our 3-gram. We then save these candidates as potential corrections for our misspelled word. The output file is separated into sections whose beginning is marked by a line with a sequence of characters "%%%" followed by the 3-gram of an incorrectly spelled word. Subsequent lines before the next section are the Google1T 3-grams with the candidate correction at the center. The following is an excerpt of the results from `degrade.FR940104-0-00001.000.candidates`:

```
%%% Federal hegister Vol. 1
Federal Domestic Volunteer 76
Federal Employee Volunteer 42
Federal Loan Volume 47
Federal Register Vol 8908
Federal Register Vol.59 41
Federal Register Vol.60 75
Federal Register Vol.61 81
Federal Register Vol.62 83
Federal Register Vol.63 69
Federal Register Vol.65 82
Federal Register Vol.66 57
Federal Register Vol.70 63
Federal Register Volume 1833
Federal Regulations Volume 68
Federal Reserve Volume 8013
Federal Surplus Volunteer 380
```

```
%%% Tuesday1 'anuary 41 8
%%% and hegulations Vol. 13
and destroy Voldemort 67
and development Vol 58
*rest of output truncated*
```

As we can see the first 3-gram generated 16 candidates of which the most frequent is `Federal Register Vol` with 8908 frequency. On the opposite end the next 3-gram generated no candidates due to having an error in a neighboring word. This could be fixed with multiple passes and allowing edit distance in the 3-gram generation but would greatly increase the number of candidates generated and the runtime. Something for the future. The next 3-gram after that generated over 2000 candidates due to having such common words as the first and third word `and hegulations Vol.`. To solve this issue we reduce the number of candidates each 3-gram generates in the next step.

71.6.5 Refine

Currently any word, regardless of edit distance, appears in our candidate list of 3-grams. However, it is unlikely that the correct word has a large edit distance compared to the candidates so in the refine step we eliminate candidates with large Levenshtein edit distances even if they have high frequency. Test indicated this to be the case. Doing so reduced our candidate list to a more manageable size at the cost of some 3-grams losing all available candidates the more the edit distance is restricted. Here is the output on the same file we have shown `degrade.FR940104-0-00001.000.clean`:

```
%%% Federal hegister Vol. 1
Federal Register Vol 8908
Federal Register Vol.59 41
Federal Register Vol.60 75
Federal Register Vol.61 81
Federal Register Vol.62 83
Federal Register Vol.63 69
Federal Register Vol.65 82
Federal Register Vol.66 57
Federal Register Vol.70 63
Federal Register Volume 1833
&&& Register
%%% Tuesday1 'anuary 41 8
&&&
%%% and hegulations Vol. 13
and Regulations Volume 146
&&& Regulations
%%% Tuesday1 'anuary 41 19
```

Another addition to refine is a list of candidates beneath all the remaining 3-grams sorted by highest frequency appearing first for multiple candidates. We can see that the first 3-gram has only 1 candidate left which happens to be the correct answer. Our testing indicated that an edit distance of 4–5 is the ideal values in terms of having about 70–80% of all the 3-grams that originally had candidates keep at least 1 candidate. This was taking into consideration 3-grams with candidates that were all false positives or incorrect candidates. As mentioned before it is far easier to correct large words as smaller words can be greatly mutated with a small edit distance. Ideally we could have a variable edit distance depending on the length of the word.

71.6.6 Verify

Finally for the last step we want to take the original file and compare it to our candidates and corrections in order to produce statistics to verify our correction accuracy. This is accomplished by consulting a document with each incorrect word mapped to its correct version. This document was generated as part of our alignment algorithm project for the same data and comes from a MySQL database with all of the documents. The Alignment allows for text to be out of order and not aligned due to extra words added the OCR'd text or words missing in the OCR'd text [7]. Using this data we count how many words were corrected properly and how many at least had the correct word in its candidate list. This information is saved to a file with the filename, count corrected, and count of words with correct word in suggestions on a single line for further analysis.

71.7 The Results

For the three runs performed. More than triple the amount of words were corrected in the OCR Aligned run versus the Non-Aligned run. This may seem surprising seeing how little the percentage of OCR Errors detected changed, but is not because we no longer have invalid 3-grams that never existed in the original text or the OCR'd Text.

Surprisingly the 1-gram Aligned method performed less than the OCR Aligned method and is more due to the limited vocabulary from which it can detect errors. However, this version is the one that has the greatest room for improvement as it is a mere comparison to the vocabulary list. For words that the correct version exists in the Google1T 1-gram vocabulary, the main reasons why they were unable to generate the correct candidate were:

- Mistakes in neighboring words to the 3-gram: If we have two or three words that have OCR Generated Errors in them then matching to valid 3-grams is very unlikely. A solution to this as mentioned would be to match to 3-grams with a short edit distance whenever there are no candidates. Also doing several passes of the algorithm to slowly fix words one at a time. This would require to do 3-grams where the misspelled word could appear as the first or third word and to use the other two as context.

- Numbers in Neighbor words: OCRSpell removes numbers and punctuation in a word before checking if it is spelled correctly. Doing this is good; however unless we also do that in our code then we face trying to find 3-grams with words that have numbers as part of the word.

- Not a word: Numbers or combination of numbers and letters like, *SFV93-901-1An* or US.C.001014 are practically impossible to find even if we had Levenshtein. These sets of letters could be valid, or invalid, in the original file and may create false positives.

- Hyphens between words: For example, "California-Arikona" has issues because the Google1T data set separated those words with hyphens into separate words so not only does that word have a spelling mistake, but it would be much scarce to find such 3-grams.

- Rare words: Especially for 1-gram search. Google1T has a limit of not showing words with frequencies lower than 40. Therefore a lot of valid but rare or specialized words will not appear in the corpus and in that case.

- False Negatives are also present when a word that has an OCR Generated Error is marked as correct when the word itself is valid but not what was originally intended. An example of this was the word "hules" in the first document. The correct word is *rules* but hules is a valid word and is marked as OK by 1-gram search but not by OCRSpell. A solution to this would be to use a confusion matrix to check if a word is misspelled.

- Acronymns in both neighboring words when generating the 3-grams or as the word being identified if correct or not pose a problem in detecting. One solution would be to implement the old Acronymn finder to recognize these as acronymns and leave them alone [16] (Figs. 71.1 and 71.2).

A: Non-Aligned OCRSpell
B: Aligned OCRSpell
C: Aligned 1-gram Mimicking OCRSpell

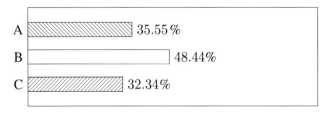

Fig. 71.1 Percentage of identified OCR generated errors with candidates (after refine)

A: Non-Aligned OCRSpell
B: Aligned OCRSpell
C: Aligned 1-gram Mimicking OCRSpell

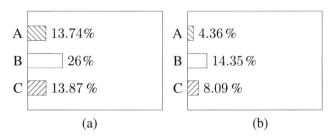

Fig. 71.2 Percentage of identified OCR generated errors with candidates that include the correct word (after refine) (**a**). Percentage of those that include the correct word as the first suggestions (**b**)

As Gritta et al. state, "If research is optimised towards, rewarded for and benchmarked on a possibly unreliable or unrepresentative dataset, does it matter how it performs in the real world of Pragmatics?" [17]. Several papers have reportedly shown higher correction success rates using Google1T and n-gram corrections [18,19] than our results. Unlike those cases we publicly provide our data and code.

71.8 Challenges, Improvements, and Future Work

In addition to the issues we faced and resolved with OCR-Spell, another major workflow concern, one that is always present with Big Data, is Space and Time complexity. Deciding what to read and have in main memory versus reading from a file was a great challenge. For the 1-gram search we found it to be much more efficient to hold this in main memory, but even then a better searching algorithm could enable this method to run at least as fast as OCRSpell. On the other hand the Google1T database was too large to run on main memory; however we take advantage of the index file, stored in memory, to only search relevant files in the data set. An improvement we could do to increase search speed would be to sort our 3-grams and then search for them this way so that caching could help us search through the database much quicker. Memory was always a concern and even when collecting statistics from our output and log files we had to be careful on how this was done to maintain efficiency. On the other side time complexity could be improved at the cost of more space if we implemented better sort and search algorithm.

The idea of a confusion matrix as a way to generate and/or decide between candidates is commonly used in OCR error correction. Usually such confusion matrix are generated through previous training, but have limitations in that each OCR'd document has its own unique errors. In our previous work on global editing, [20], we found that correcting longer words was easier due to having less number of candidates regardless of method. This could then be used to generate a confusion matrix based on what the error correction did. This self-training data could then be used to help in correcting shorter words that can be hard to generate valid candidates since a large edit distance allows for far too much variation.

One way to speed up the query process is to insert the Google Web 1T corpus into a database such as MySQL or SQLite [6].

71.9   Conclusion

As we continue to tackle ways to improve OCR Post Processing, we find that there is no single solution, but at the same time the solution entails many different pieces. Context-based candidate generation and candidate selection are key pieces. In this paper we took the TREC-5 Confusion Track's Data Set and used Google1T to correct OCR Generated Errors using contex given by 3-grams. We explained our implementation and workflow as part of ensuring our work is Reproducible [21] and discussed on ways to improve the results. The implementation of this paper and the entire OCR Post Processing Workflow for running the Google-1T experiment will be available on multiple repositories including docker, zenodo & git (Search **unlvcs** or see DOI: 10.5281/zenodo.2536408).

References

1. Brants, T., Franz, A.: Web 1t 5-gram version 1 (2006)
2. Kantor, P.B., Voorhees, E.M.: The trec-5 confusion track: comparing retrieval methods for scanned text. Inf. Retr. **2**(2–3), 165–176 (2000)
3. Levenshtein, V.I.: Binary codes capable of correcting deletions, insertions, and reversals. Sov. Phys. Dokl. **10**(8), pp. 707–710 (1966)
4. Islam, A., Inkpen, D.: Real-word spelling correction using Google web it 3-grams. In: Proceedings of the 2009 Conference on Empirical Methods in Natural Language Processing: Volume 3-Volume 3. Association for Computational Linguistics, pp. 1241–1249 (2009)
5. Guyon, I., Pereira, F.: Design of a linguistic postprocessor using variable memory length Markov models. In: Document Analysis and Recognition, 1995 Proceedings of the Third International Conference on, vol. 1, pp. 454–457. IEEE (1995)
6. Evert, S.: Google web 1t 5-grams made easy (but not for the computer). In: Proceedings of the NAACL HLT 2010 Sixth Web as Corpus Workshop. Association for Computational Linguistics, pp. 32–40 (2010)
7. Fonseca Cacho, J.R., Taghva, K.: Aligning ground truth text with OCR degraded text. Paper presented at Computing Conference. London, UK (2019)
8. Fonseca Cacho, J.R., Taghva, K.: Using linear regression and MySQL for OCR post processing (To Appear)
9. Trec-5 confusion track. https://trec.nist.gov/data/t5_confusion.html, Accessed: 2017-10-10
10. Taghva, K., Stofsky, E.: Ocrspell: an interactive spelling correction system for OCR errors in text. Int. J. Doc. Anal. Recognit. **3**(3), 125–137 (2001)
11. Taghva, K., Nartker, T., Borsack, J.: Information access in the presence of OCR errors. In: Proceedings of the 1st ACM Workshop on Hardcopy Document Processing. ACM, pp. 1–8 (2004)
12. Fonseca Cacho, J.R., Taghva, K.: Reproducible research in document analysis and recognition. In: Information Technology-New Generations. Springer, pp. 389–395 (2018)
13. Boettiger, C.: An introduction to docker for reproducible research. ACM SIGOPS Oper. Syst. Rev. **49**(1), 71–79 (2015)
14. Peng, R.D.: Reproducible research in computational science. Science **334**(6060), 1226–1227 (2011)
15. Wickham, H., et al.: Tidy data. J. Stat. Softw. **59**(10), 1–23 (2014)
16. Taghva, K., Gilbreth, J.: Recognizing acronyms and their definitions. Int. J. Doc. Anal. Recognit. **1**(4), 191–198 (1999)
17. Gritta, M., Pilehvar, M.T., Collier, N.: A pragmatic guide to geoparsing evaluation. arXiv preprint arXiv:1810.12368 (2018)
18. Bassil, Y., Alwani, M.: OCR context-sensitive error correction based on google web 1t 5-gram data set. arXiv arXiv:1204.0188 (2012)
19. Mei, J., Islam, A., Wu, Y., Moh'd, A., Milios, E.E.: Statistical learning for OCR text correction. arXiv preprint arXiv:1611.06950 (2016)
20. Taghva, K., Borsack, J., Bullard, B., Condit, A.: Post-editing through approximation and global correction. Int. J. Pattern Recognit. Artif. Intell. **9**(06), 911–923 (1995)
21. Fonseca Cacho, J.R., Taghva, K.: The state of reproducible research in computer science (to appear)

Ruth Obidah and Doina Bein

72.1 Introduction

The limited platform in traditional education systems cannot serve the diverse needs of students. Classes are often comprised of students with diverse learning styles and differing proficiency backgrounds and as a result, the traditional education system often under-serves many students. With current technologies, there is no reason for this condition in education to perpetuate. For students to be better served; a system that can be designed to adapt education delivery to students' needs as well as give students an immersive education experience is needed. Researchers and some educators looked into Game Based learning (GBL) to address the limitations of the traditional education systems.

Game based learning is the concept and practice of using games and simulated environments to teach students. The idea behind game based learning is that with the incorporation of a gaming environment in learning, students will be forced to have more engagement in the learning process resulting in deeper learning and mastery of the subject taught. GBL includes applications and software used to train, teach, or facilitate the learning of a subject. In addition, GBL has been shown to provide better mastery of subjects due to the ability to deliver a customizable experience to users. The premise for this project is that, GBL can go further from being an education facilitator in courses, to being a standalone course medium, if gaming technology is well integrated into good GBL design.

This paper presents an Unreal Engine-based game called Ecomerica, to show how a GBL system can lead to better learning of introductory topics on economics. The game design, and other aspects of game functionality requirements including the emphasis on teaching, interactivity, assessment, and feedback to foster deeper learning and mastery of introductory concepts in Economics, are explored. The benefit and uniqueness of a GBL system as implemented in Ecomerica are explained, and the market outlook for such a product is established.

The paper is organized as follows. In Sect. 72.2 we present traditional learning model, pros and cons, and the need for GBL. A description of the project functionality is given in Sect. 72.3. Concluding remarks and future work are presented in Sect. 72.4.

72.2 Traditional Learning Model Versus GBL

As society advances, more people begin to understand human psychology and human behavior. It has become apparent that acquiring knowledge is simply not a one size fits all affair. As a result, tutoring programs geared to complement traditional training or learning, train their employees or tutors on the various learning styles and how to help learners with different learning styles [1]. Through many years of research five major learning styles have been established. They include the Visual learners, the tactile/kinesthetic learners, the auditory learners, the logical learners, and the verbal learners. Depending on what literature you read or the culture of an organization, the learning styles may be seven and would include social learners and solitary learners [1]. In any given class in the traditional model, at least four to five of these categories of learners are represented among the students. In most cases, the traditional model of lecturing the students will fail in providing the diversity needed for students to get deep understanding of subjects. This occurs because the traditional model is mostly unidimensional [2].

R. Obidah · D. Bein (✉)
Department of Computer Science, California State University, Fullerton, Fullerton, CA, USA
e-mail: ruthobidah@csu.fullerton.edu; dbein@fullerton.edu

© Springer Nature Switzerland AG 2019
S. Latifi (ed.), *16th International Conference on Information Technology-New Generations (ITNG 2019)*,
Advances in Intelligent Systems and Computing 800,
https://doi.org/10.1007/978-3-030-14070-0_72

Motivating learners and delivering knowledge in a way that would facilitate understanding, is the heart of education and many companies and researchers recognize that. It is obvious that within traditional education, effort has been made to enhance student learning outcomes, as well as to optimize the depth of student understanding. To this end, it is not uncommon for lectures to include visual guides, PowerPoint presentations, audio presentations, videos, and even repeatable online quizzes to encourage student learning. Moreover, effort has been made to incorporate some digital learning games in the traditional model to maximize student learning. For example, with young children, edutainment is often used to maximize their learning within the traditional model [3]. Some companies such as MobLab create learning games for students of Economics and other Social Sciences to experiment on concepts in their field of study. These games are typically sold to individual students or to instructors as part of their courses. However, these solutions still have their inadequacies. While integration of presentation technologies is great and have proven useful and helpful to students, there is still the limitation of the traditional classroom format which will inadvertently result in undeserving the diverse needs of students. Tutoring is helpful to students, but can be very expensive, and often dependent on the level of mastery the tutor has attained as well. In addition, digital learning games that teach concepts within a topic are undeniably very good when incorporated into a very good classroom experience, but there is still the danger of students not having a well-rounded approach since these games can be concept-focused.

Researchers, especially in the fields of psychology and computer science, have been looking into ways to deliver education to learners in such a way that would result in deeper understanding. So far, the area of Game Based Learning (GBL) has been promising to this end. Game based learning includes applications/software that employ interactive environments that motivate and actively engage students in the learning process [2]. GBL has shown great promise in learning outcomes. For example, pilots use GBL to simulate the flying experience and increase learning. In addition, some GBL experts can design teaching games to create a compelling environment for surgical students to practice or hone their skills for laparoscopic procedures [2].

The ability for GBL to create an interactive environment is at the core of its effectiveness. Motivation is a huge component of successful learning and a big issue that the traditional model has limited ability to address. The advantage of a game includes immediate feedback, graduated learning to complete otherwise difficult challenges, quick application of new concepts, and consequences or gains in a simulated world that give players somewhat of an instantaneous purpose to their learning pursuit, and as a result, the motivation to keep learning. As can be observed, the psychology that drives gameplay, is often impacted by the design, including the constraints of the virtual world, which can motivate gamers to learn new tricks to gain advantage in a game. This can be gleaned from Pang's analysis on gamer psychology [4]. According to her analysis, the sense of achievement, socialization (in some online games), and the immersion component of games motivate players to learn gameplay rules to advance. In addition to the aforementioned factors, Pang also pointed to studies on the self-determination theory and motivation to play games. Based on the theory of self-determination, behavior in humans is driven by a need for competence, autonomy, and relatedness. As a result, a well-designed game with appropriate flow meets this human need and motivate players to play [4]. Hence, depending on the quality of design, GBL can create and sustain the motivation to learn.

The entire learning environment created in GBL with the high level of interactivity it requires, can successfully maintain learners' attention, another issue the traditional model increasingly finds difficult to mitigate. Attention is distinctly different from motivation in that attention does not just deal with sustaining an interest in continuing the activities leading to learning but maintaining the learner's engagement while in the moment of learning. A motivated student may lose attention as she become tired, or the presentation method is unsuitable. With a well-developed learning game environment, the learners can be well paced based on good and professionally designed game play and they can have the ability to not just have a highly interactive learning process in a virtual world, but also the ability to customize their learning process through self-pacing given the ability to save progress and continue at a different time. In addition, a well-designed virtual world can be very immersive that it also captures the attention of learners for longer periods of time relative to other mediums of teaching.

Additionally, a well designed GBL can facilitate mastery of a subject through graduated learning, interactivity, and learning by doing. The traditional model is mostly limited with this since it relies heavily on memorization and the accumulation of knowledge without practice [5]. To demonstrate this, assuming a player fails to achieve a passing score in a quiz, traditional learning is limited in this because it is not practically feasible to provide customized care for learning needs in this scenario. While an instructor can allow multiple trials on an online quiz before a closing period, there are just not enough hints, game scenarios and other resources to package for focused mastery on unique needs of the student given the deficiencies exhibited on that quiz scenario. A GBL system can account for this, through requiring the student to play another learning scenario, do another challenge, or suggesting more focused or detailed resources needed for the deficiencies exhibited in the quiz.

GBL can be designed to be very user specific. That means that psychological needs, human learning behavior, and cognitive development can be addressed with GBL expertly designed to meet those needs [6]. In the traditional model, it is hardly feasible to meet these needs especially with a diverse audience, however, in a game options can be created based on a user's self-selection or a more intricately design system can implement the analysis of the player's learning patterns based on data collected during game play to customize the learning track for players.

72.3 The Case and Feasibility of GBL

In contrast to seeing GBL as mostly a learning aid, GBL has great promise as a standalone medium for education. This is because of the following reasons:

- *Game Design*: Technology in gaming has exploded throughout the years. From World of Warcraft, to the Sims, to Grand Theft Auto, Entertainment gaming has shown that an entire world can be designed with great designing principles. These designing principles are the same principles necessary for creating the optimal learning environment. For example, to create an effective learning environment some key elements would include, user centricity, usability, graduated learning, and testability. All these elements already exist in games for entertainment and can be honed for GBL. For example, in games, players get hints or tutorials to prepare them for the next challenge to come. Furthermore, during gameplay, game developers often integrate hints and routines after some indications that the player may be having a hard time getting over the obstacle. With this, it is clear to see that with a well-designed GBL system, assuming a calculus topic is being taught as part of the game, a game can have side challenges, references to resources, tutorials, etc. incorporated into an interactive lecture. Particularly, after several failures in response, the game can hint to a topic in algebra where the student might be experiencing a learning gap, causing failure in the ability to solve the calculus problem.
- *Game Multidimensionality*: GBL currently in use allows a multidimensional approach to learning and holds even greater promises as the field is explored. In the traditional model, differences in past experiences and deficiencies are extremely difficult to address. With GBL, a design that gives immediate feedback and incorporates graduated learning can address this issue. Also, a GBL system can be designed to address gaps in past knowledge with routines, hints, suggestions, and/or tutorials to address specific prior knowledge needed to successfully complete a challenge. This will assist students with knowledge gaps

that often occur because of the impossibility of uniform learning in tradition education. Hence instead of learners resigning to the notion of being just bad at it, focused training in a specific need for a specific challenge can be designed in the game.
- *Cost Effectiveness*: With current advancements in gaming and computer technology, and with people owning multiple devices for gaming in their homes. GBL is no longer an expensive feat for the consumer. People will be able to purchase GBL applications/software at very affordable prices.
- *A Ready Market*: According to [2], the workforce is rapidly changing as the older generation retires and 18 to 40-year old's take over. This new generation of workers are very tech savvy and have been greatly exposed to computer technology and gaming. With this new workforce, GBL is a feasible future for education [6].

72.4 Proposal and Requirements

The following requirements (more like guiding principles), are necessary for the type of GBL proposed:

- *GBL shall be user centric in approach*
- *GBL shall emphasize usability*
- *GBL shall have clear goals and/or tasks*
- *GBL shall provide frequent feedback*
- *GBL shall have comprehensive evaluation criteria*

In the requirements above, user centricity refers to proper design of a GBL application to approach delivery of material based on the user's cognitive, developmental, learning, and other psychological needs.

We propose a standalone GBL system that can teach a given subject. The idea behind this is that given a selected subject area, a game can be developed that will teach and facilitate mastery of that subject area. Hence the goal of this project was to develop a game software in a way that teaches a defined subject area and facilitates understanding of that subject area. To do so, the employment of knowledge and research on learning styles and current ideas based on current teaching practices to design the game was necessary. Some key elements of traditional education that were included are: lectures, quizzes, homework or projects, and test or exams.

The general concept is shown in Fig. 72.1. The system in the model acts as the teacher and is responsible for lecturing, assessing, and giving feedback to the student until the game is successfully completed by calling various subroutines (ex. Tutorials) based on students' response. The subject taught in this project is an introduction to Macroeconomics concepts. Topics on the fundamentals of economics, budget constraints, and supply and demand were implemented. The

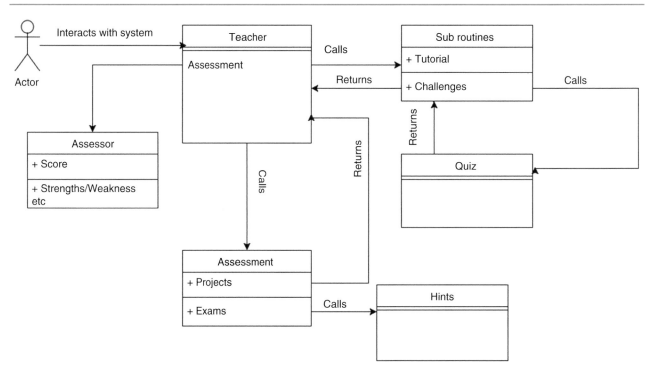

Fig. 72.1 GBL model of Ecomerica

goal was that with the right ordering of lectures, quizzes, and tests, learners playing the game will be able to master these core concepts. With regards to serving students with different learning needs and facilitating mastery of subjects, the software, while borrowing some elements from traditional education and other learning settings, will take a different approach. The game software has been designed to have an open world to practice concepts learned in lecture presentations with risk-free quizzing to master concepts. Student mastery has been measured by the ability of the student to score at least a minimum score set by the system. Should the student score less than the minimum score, the system will select the next set of interactions.

The concept borrowed from these key elements from traditional education involves human psychology of learning that needs information to be given, retained, and tested. Lectures provide a means to give students information on a subject, quizzes, homework or projects, and tests, act to gauge understanding of the subject. Taking these ideas, the software's goal will be teaching and assessing interactively by using an increasing level of quizzing the students on the topics taught, the system will output some set of interactions and ask for action to be taken by displaying options to the player. The student, player, will make a selection, and the system, will either continue with its predefined routine or call a subroutine based on the selection of the student. After the subroutine completes, the student is returned to the main routine. The reason for approaching teaching is this manner, is to maximize the interactivity

allowable in this medium in such a way that events occur in logical sequences while taking into account the need to customize the user's experience to achieve optimal understanding of the subject being taught. The emphasis on interactivity including the integration of effective assessment with instantaneous feedback, facilitates active learning unlike in a traditional lecture setting, which leads to better learning outcomes.

72.5   Project Description

The game is set in a fictitious colony named Ecomerica. The land is an open natural land and the settlers must settle the land. The player is the governor and must embark on quests to earn money, as well as develop the colony's economy to generate more resources for advancement. The player's role is to manipulate outcome. The quests are in learning principles of economics that will enable the player to understand economic decision making, get quizzed and tested, in order to acquire large sums of money for colony's development. The governor also needs to encourage economic growth for a constant growth of resources. Ecomerica was implemented as a real time strategy game.

Ecomerica is a resource management game with learning integrations. The player, who is the manager of the economy is tasked with learning the principles of economics for the purpose of promoting economic growth in the colony. The game is an open world game and to build out a massive city, the player must participate in lessons, quizzes, and tests to gain resources for gameplay. As a GBL system, the purpose of this game is to create an integrated experience of gameplay with learning. Unlike other systems in the market, the focus of this game is not to teach a focused concept and/or practice. Rather the game teaches a course sequence, much like an online course but with less rigidity, more flexibility, more interactivity, and a continuous gameplay experience where the player's decisions and decision-making process can stay persistent and evolve.

Hence, the purpose of Ecomerica, is not just to create a game, or concept learning game to assist in the learning process, but to create a unique learning experience in a full gameplay. Factors in gameplay that were implemented include, a practice space for ideas behind the lessons which also included the ability to construct 3 styles of residences and 3 styles of businesses. Other elements include the implementation of a goods market and a service market that automatically spawn at the beginning of gameplay, and the ability to generate AI characters as members of the population.

Given the constraints, the part of teaching was implemented as succinct PowerPoint presentations. The presentations were made short and concise dealing with a few topics each time to allow for better absorption of material and maintenance of attention. For player assessment, quizzes and tests were implemented. For the purposes of the game, three lessons were implemented, and each lesson was broken down into two parts. In addition, six quizzes were implemented and three tests per lesson were implemented as well. A standardized number of quiz questions and tests were also implemented to keep the project within a practical scope. Feedback was limited to giving the options to return to lecture and to go to a textbook for review in the case that the player did not attain the minimum score for a quiz. In the case that the player does not attain the minimum score for a test, feedback included directions to view some helpful videos.

The game design was based on the circular flow model of economics (Fig. 72.2).

The game was designed to have an economy to motivate players to learn and practice the concepts taught in the game through the gaming experience. The virtual economy in the game allows the player to place one goods market and one service market at no cost. Firms and homes are the key resources that the player can purchase to encourage the growth of the economy. Lectures, quizzes, and tests provide avenues for the player to earn large sums of game money to grow the economy. Other than the lectures, tests, and quizzes, a percentage of the transactions in the economy is added to the game money as well, to motivate the player to grow the economy. Transactions in the economy can only occur when the population has wealth, which is different from the

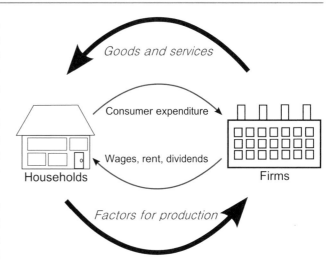

Fig. 72.2 Circular flow model of economics

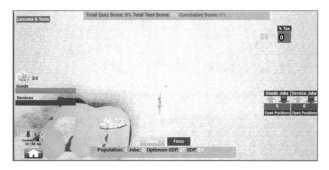

Fig. 72.3 Starting the game

game money, so the player must keep a balance of population and firms to generate this wealth to enable transactions. As is noticeable in the circular flow model of a realistic economy, the virtual economy described above bears a close resemblance. The reason for the design is for the creation of a simplified, yet more realistically modelled virtual market economy for students to practice concepts learned as well as have an immersive gaming experience.

Game money is the resource for buying all buildings. The player can earn money via going through lessons, tests, and quizzes. The player can also earn money by growing the virtual economy's GDP through population transactions in both the goods and the services markets. Initially the player starts with $0 (Fig. 72.3), then clicks on Lesson tab (Fig. 72.4), and can open all tests, quizzes and lectures (Figs. 72.5, 72.6, 72.7, 72.7, 72.8, 72.9, 72.10, 72.11).

72.6 Conclusion and Future Work

The potential for education by the development of GBL systems is immense with a strong business case for it. With the inclusion of critical learning factors such as teaching, assessment, feedback, and interactivity, highly effective GBL

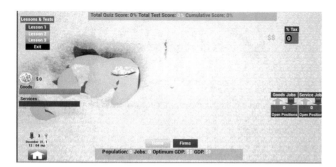

Fig. 72.4 Starting the lesson

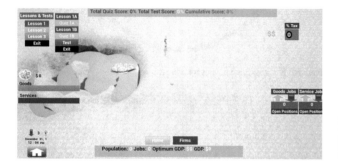

Fig. 72.5 Navigate and select lesson or quiz or test

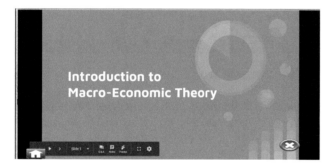

Fig. 72.6 Sample lesson in PowerPoint

Fig. 72.7 Sample quiz

systems can be designed/created as standalone teaching systems [7]. The game developed in the project, Ecomerica has successfully modeled the benefits and potential of a stan-

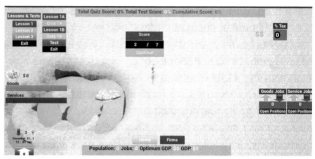

Fig. 72.8 Score of the quiz is displayed

Fig. 72.9 Score of the quiz is too low so the student must retake the lesson

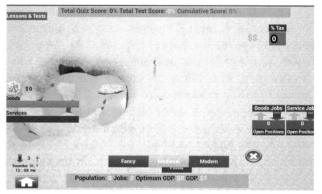

Fig. 72.10 Score of the quiz is high enough to gain $

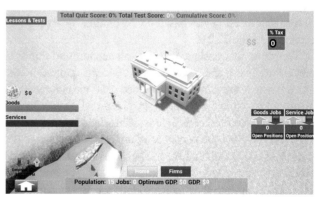

Fig. 72.11 Add firms and grow population needed for services

dalone GBL system. Although it has its current limitations, Ecomerica has future potential to become a larger and more immersive learning and gameplay experience. On the long-run, AI characters, with differing needs, incomes etc. can be implemented. More diversified homes and firms, can also be implemented to create a more sophisticated gameplay. With regards to the learning part of the system, more lectures, quizzes and tests can be implemented on the short run. In addition, lectures that are not outsourced to the web, but integrated within the system can be developed in the long run. Such lectures can be designed with more interactivity with a more focused hint/suggestion system. The current system utilizes short PowerPoint presentations to maintain the attention of the user. That can come at the expense of utilizing the power of the platform to approach topics to increase the breadth of knowledge that can be difficult to accomplish with traditional learning styles. An alternative approach of making Ecomerica into a role-playing game can provide a more directed user experience [8–10].

References

1. OLS16, Overview of Learning Styles: [Online]. Available: http://www.learning-styles-online.com/overview/. Accessed 26 Oct 2018
2. Trybus, J.: Game-Based Learning: What it is, Why it Works, and Where it's Going. New Media Institute, 2014, [Online]. Available: http://www.newmedia.org/game-based-learning%2D%2Dwhat-it-is-why-it-works-and-where-its-going.html (2014). Accessed 26 Oct 2018
3. I Liao, Y.H., Shen, C.-Y.: Heuristic evaluation of digital game based learning: a case study. Proceedings of the 2012 IEEE Fourth International Conference On Digital Game And Intelligent Toy Enhanced Learning IEEE (2012). doi:https://doi.org/10.1109/DIGITEL.2012.54
4. Pang, C.: Understanding Gamer Psychology: Why Do People Play Games? [Online]. Available: https://www.sekg.net/gamer-psychology-people-play-games/ (2017). Accessed 26 Oct 2018
5. Monsalve, E.S., Do Prado Leite, J.C.S., Werneck, V.M.B.: Transparently teaching in the context of game-based learning: the case of simulES-W. In: 2015 IEEE/ACM 37th IEEE International Conference on Software Engineering (ICSE), vol. 2, pp. 343–352. IEEE Press, Piscataway (2015)
6. Tan, P.-H., Ling, S.-W., Ting, C.-Y.: Adaptive digital game-based learning framework. In: Proceedings of the 2nd International Conference on Digital Interactive Media in Entertainment and Arts (DIMEA 2007), pp. 142–146. ACM, New York, NY (2007)
7. Janarthanan, V.: Serious Video Games: Games for Education and Health, pp. 875–878 (2012). https://doi.org/10.1109/ITNG.2012.79
8. Pellegrini, R.M., da Silva, C.E.S., de Souza, A.D.: Teaching communication management in software projects through serious educational games. In: Latifi, S. (ed.) Information Technology – New Generations. Advances in Intelligent Systems and Computing, vol. 738. Springer, Cham (2018)
9. de Souza, A.D., Seabra, R.D., Ribeiro, J.M., da Silva Rodrigues, L.E.: An experience of using a board serious virtual game for teaching the SCRUM framework. In: Latifi, S. (ed.) Information Technology – New Generations. Advances in Intelligent Systems and Computing, vol. 738. Springer, Cham (2018)
10. de Vasconcelos, L.E.G., de Vasconcelos, L.G., Oliveira, L.B., Guimarães, G., Ayres, F.: Gamification applied in the teaching of agile scrum methodology. In: Latifi, S. (ed.) Information Technology – New Generations. Advances in Intelligent Systems and Computing, vol. 738. Springer, Cham (2018)

System Architecture for an In-House Developed Admission System Intended for Higher Education Institution in Kazakhstan

73

Askar Boranbayev, Ruslan Baidyussenov, and Mikhail Mazhitov

73.1 Purpose of the System

The "Admission" system is an automated information system designed to automate the processes of the Admissions Department for the reception and processing of applicants' data, namely, on-line reception and processing of data, structuring received data, viewing application forms and related documents of applicants, sending notifications to groups of applicants, preparing lists of applicants for taking exams and making a decision about enrollment [1].

Stakeholders and the list of objects of automation of the Information system are [2]:

– Admission Department;
– School of Business admission.

The Admission System is part of a university-wife Student Information System Environment, which implements the following functions [2]:

– Candidates enrollment;
– Filling, editing and sending forms;
– Entering and reviewing exam results;

– Maintaining students database;
– registration of applications of applicants, the formation of lists of applicants, control of input information;
– payment and formation of questionnaire forms by applicants;
– accounting of the results of entrance exams, preparation of documents;
– formation of a rating of applicants;
– automated preliminary enrollment procedure;
– automated preliminary enrollment procedure.

The university-wide Integrated Student Information System Environment consists of multiple software systems, modules and applications integrated with each other. Core automation objects of the System are candidates, students and graduates related business-processes. Fig. 73.1 below shows a simplified scheme of business process flow that are being automated in our university.

Goals of the integrated Student Information System Environment [3–7]:

– Developing single information space – Allows for sharing data and dramatically reduce time spent for processing, information analysis, and making decisions.

A. Boranbayev (✉)
Department of Computer Science, Nazarbayev University, Astana, Kazakhstan
e-mail: aboranbayev@nu.edu.kz

R. Baidyussenov · M. Mazhitov
Nazarbayev Univesity Library and IT Services, Nazarbayev University, Astana, Kazakhstan
e-mail: rbaidyussenov@nu.edu.kz; mmazhitov@nu.edu.kz

© Springer Nature Switzerland AG 2019
S. Latifi (ed.), *16th International Conference on Information Technology-New Generations (ITNG 2019)*,
Advances in Intelligent Systems and Computing 800,
https://doi.org/10.1007/978-3-030-14070-0_73

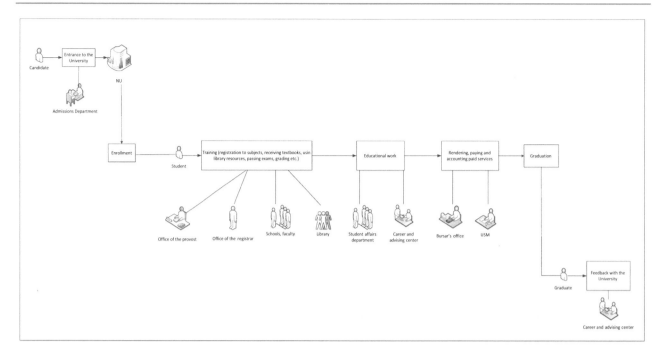

Fig. 73.1 The scheme of business process flow

– Single technical platform for System development with single interface – Searching and justification of a single technical platform for System realization with respect to cost, quality, University needs and opportunities.
– Single data repository – Creating single architecture of DB system, which will as an information source for users.

Stakeholders of the integrated Student Information System Environment are as follows [8] (Table 73.1):

– University leadership,

For "candidate" object:

– Admissions Department,

For "student" object:

– Bursars Office,
– Office of Provost,
– Office of the Registrar,
– University schools,

– Faculty,
– Department of Students Affairs,
– USM,
– Library staff.

For "graduate" object:

– Career and Advising Center.

73.2 Structure of the Admissions Information System

The Fig. 73.2 bellow shows a simplified scheme of a business process flow for the "candidate" object.

The Admission system consists of the client part, the server part and the database (DB). The client part connects to the server, in turn the server connects to the database and transfers the processed data to the client part.

The developed application is integrated with other applications of the University [9]. Figure 73.3 shows the structure of the Admission System and Fig. 73.4 show input and output data from the University's integrated systems.

Table 73.1 Stakeholders of the Student Information System and their functions

Users	Functions
University leadership	Reviewing personal data (candidate, student, graduate, and faculty); Drafting consolidated statements and surveys.
Admissions Department	Registering candidates' applications, listing candidates, and input information control; Recording admission exams results, and preparing documents. Automated formation of exam records; Forming candidates ratings; Automated procedure of preliminary enrollment; Automated selection to vacant positions in University Schools for candidates failed to enter the School; Candidates enrollment; Drafting consolidated statements and surveys.
Bursars Office	Concluding contracts for paid education; Issuing invoices to students; Control over education contracts execution; Reviewing students' personal data; Drafting consolidated statements and surveys.
Office of Provost	Approving curriculums; Approving lists of students for dismissal, academic leave etc.; Reviewing students', graduates' personal data; Drafting consolidated statements and surveys.
Office of the Registrar	Maintaining students data base; Creating courses; Approving students' and teachers' schedule; Opening/closing registration for a course; Issuing references, transcripts; Reviewing students' personal data and academic progress; Drafting consolidated statements and surveys.
School moderator	Coordinating students' and teachers' schedules; Reviewing students' personal data and academic progress; Drafting lists for dismissal, academic leave etc.; Drafting consolidated statements and surveys.
School administrator	Creating curriculums and disciplines; Drafting schedules; Reviewing students' personal data and academic progress; Drafting lists for dismissal, academic leave etc.; Maintaining data base of potential students/attendees; Drafting consolidated statements and surveys.
Faculty	Drafting a list of training and tutorial literature; Scoring; Reviewing students' personal data and academic progress; Reviewing schedules; Drafting consolidated statements and surveys.
Department of Students Affairs	Maintaining students databases (vulnerable students, those students who appealed to psychological help etc.); Supervising and moderating student clubs; Informing students about extracurricular events; Allocating place in students dormitory; Carrying out psychological trainings; Drafting consolidated statements and surveys.
USM manager	Residing and dispossessing students from dormitories; Issuing invoices; Providing paid services; Drafting consolidated statements and surveys.
Accounting office	Maintaining tariffs for provided services; Issuing invoices; Accounting provided paid services; Confirming material funds writing off (lost and faulty books etc.); Drafting consolidated statements and surveys.

(continued)

Table 73.1 (continued)

Users	Functions
Library	Delivering and receiving training materials etc.; Cataloging; Drafting a list of necessary literature and procuring these books; Digitizing printed publications; Drafting consolidated statements and surveys.
Career and Advising Center	Maintaining data base of students who passed internship; Maintaining graduates database; Maintaining potential employers database; Carrying out carrier guiding for students; Drafting consolidated statements and surveys.
Candidate	Filling registration form for Personal cabinet creation; Filling, editing and sending forms; Payment for questionnaire form and guarantee deposit; Reviewing exam results.
Student	Reviewing personal information; Reviewing grades, transcripts and schedule; Registration/deregistration to a subject; Request for issuing references and other documents; Involvement into University clubs activity; Using library resources; Requests for accommodation in a dormitory, inclusion into discount lists etc.; Reviewing internships and vacancies; Paying for services, rendered by the University or private entities.
Graduate	Reviewing, and editing personal data; Reviewing transcript; Request for issuing references and other documents; Using library resources; Reviewing vacant positions from employers; Paying for services, rendered by the University or private entities.

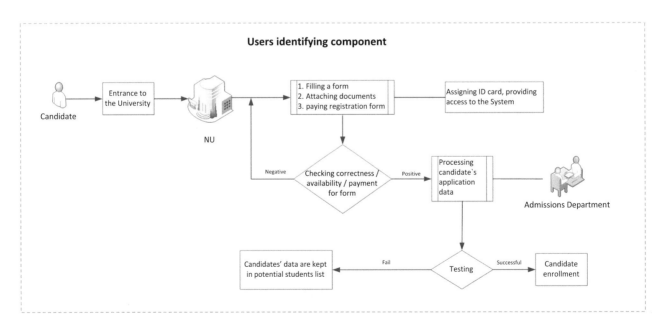

Fig. 73.2 The business process flow for the "candidate" object during enrollment

Fig. 73.3 The structural scheme of the Admission System

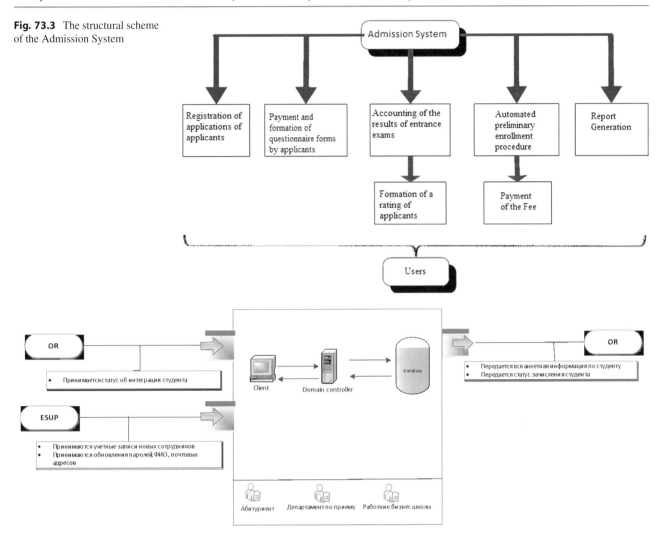

Fig. 73.4 Scheme of integration of the Admission System

Table 73.2 Specifications of the hardware platform

Oracle Database Enterprise Edition 11.2.0.4 – 64bit
IBM WebSphere Portal Server 7.0.0.2
HDD 400 GB
RAM 16 GB
CPU 4
Red Hat Enterprise Linux Server 6.7 (64-bit)
IBM Tivoli Directory Server 6.2.0.0
IBM HTTP Server 7.0.0.31
6 GB
4
IBM Forms 8.0.1
Throughput is 100 Mbit/second

The technical specifications of the hardware used in the project are shown in Table 73.2.

Table 73.3 describes the roles and the functionality.

"Red Hat Enterprise Linux Server release 6.7" is operating system, which is installed on servers, where the Admission System is running. LDAP and the Dynamic cache service is used [10]. Oracle Database 11 g is the main databased used in the system.

73.2.1 Major Components of the Developed Software Application

The entry point of the system is the IBM HTTP Server Web Server 7.0.0.31. which serves as a proxy server to protect against direct access to the portal server and as a server for static files. The logic and data processing of the Admission system is implemented on the IBM WebSpere Portal 7 portal solution that interacts with the Oracle 11 g DBMS for storing data on applicants and with the IBM DB 2 DBMS for storing the system data. To work with user data, the Portal server interacts with IBM Tivoli Directory Server 6.2.0.0. The Admission system implements a flexible system for developing and processing application forms using IBM Forms 8.0.1 technologies (Fig. 73.5).

Table 73.3 Description of system roles

Names of roles	Description of functional abilities of roles
Main Admin	Full Rights to customize various system parameters; Full rights to provide access to users
System Admin	Setting up various system parameters; Management of access rights to functions and objects of the system
Employee of the Admission Department	Processing of applicants' information through the "Application Forms" module; Notification of applicants about the change in the status of the application through the module "Notifications"; Formation of the schedule of entrance exams, the distribution of applicants for exam flows, loading the exam results Through the module "Exams"; Keeping a record of decisions on enrolled applicants through the "Decisions" module; Determining the funding of training through the "Funding" module; Program change through the module "Change Program"; Generation of reports and their uploading to various formats through the module "Reports"; Search and view introductory documents for previous years through the module "Archive".
Employee of the School of Business Admission	Processing of applicants' information through the "Application Forms" module; Notification of applicants about the change in the status of the application through the module "Notifications"; Determining the funding of training through the "Funding" module; Program change through the module "Change Program"; Appointment of applicants checking the written work through the "Assign Reader" module; Generation of reports and their uploading to various formats through the module "Reports"; Search and view introductory documents for previous years through the module "Archive".
Applicant	Online submission of application forms, uploading documents, online payment, viewing and changing personal data in the system through the Applicants module.

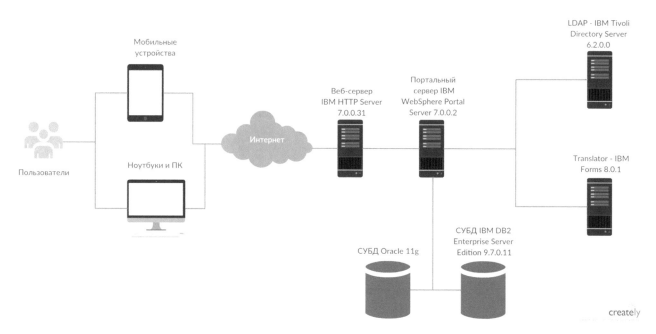

Fig. 73.5 The architecture of the system

73.3 Description of Integration with Other Information Systems

The developed Admission System has been integrated with the following existing information systems:

1. Microsoft Forefront Identity Management System.
2. Registrar System.
3. Internal Student Portal.

Figure 73.6 shows the general scheme of interaction between the information systems of the University and the developed Admission System [2].

Information exchange between the Admission System and other systems is carried out according to the request-response principle [10].

Figure 73.7 shows the scheme of information interaction [2].

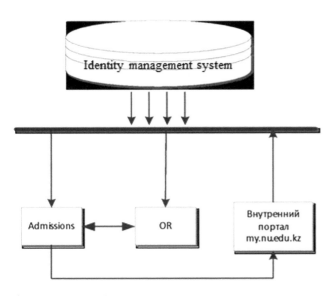

Fig. 73.6 Scheme of integration

73.4 Description of Modules and Interface Components (Table 73.4)

73.5 Conclusion

The process of designing, implementing and enhancement of the Admission System is an ongoing process [11]. In this paper we described the proposed design and implementation architecture of the Admission System that was developed in-house. It has become vital for both the Admission Department and the Information Technology Department to work closely together to make use of the system. Based on this information, the researcher will be able to know what kinds of design and implementation techniques were used for the development of an Admission System [12].

The Admission System is part of a university-wide integrated student information system, which is a full complex environment, where the System components are built in accordance with single ideology, methodology and technology and work in alignment. The System realization allows for integrating administrative, financial and training-educational components of the System that in its turn will ensure quality increase in University activity organization and man-

Table 73.4 Description of modules/information pages

Modules and functionality of the admission system	
Registration/Authorization of and applicant	The module is intended for: Initial registration in the system; Access to the system; Change the current password.
Profile of an applicant	The module is intended for: Changes to the admission program; Changes in the applicant's email address; Change the current password.
Application Forms	The module is intended for filling, editing, sending a questionnaire form.
Ad & announcements	The module is intended for ad and announcements placement.
Instructions	The module is intended for posting reference information.
Authorization of the Employee of the Admission Department	The module is intended for: View application forms; Setting status; The appointment of the applicants checking the essay; Definition of financing; Checking the payment status of the security deposit; Send a notification when the status of an applicant changes; Review and supplement the necessary documents for admission; Change the applicant's program.
Exams	The module is intended for: Determining the dates of participants in the exam; Determining the admission of applicants to the exam; Importing results
Reports	Generation of Reports
Orders	The module is intended for formation of lists of enrolled applicants.
Archive	The module is intended for storage, viewing and processing of data of the applicant.

Fig. 73.7 The scheme of information interaction

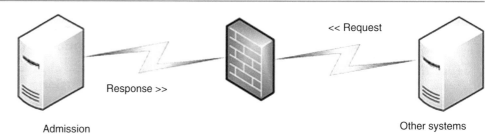

Response >>

<< Request

Admission

Other systems

agement in order to achieve strategical goals set to the University [8].

Overall, we are satisfied with the results of implementation of the Admission System. In a short period of time, we have been able to assemble a skilled and talented technical staff that has delivered numerous information systems to our university constituents [2].

As the university continues to grow and expand, we will consider conducting internal assessments with our stakeholders to determine whether, or not, existing solutions are continuing to meet their growing business needs [2].

References

1. Boranbayev, A., Shuitenov, G., Boranbayev, S.: The method of data analysis from social networks using apache hadoop. In: 14th International Conference on Information Technology - New Generations, Springer Verlag, Advances in Intelligent Systems and Computing, vol. 558, pp. 281–288, Las Vegas, NV (2017)
2. Boranbayev, A., Baidyussenov, R., Mazhitov, M.: Software Architecture for In-House Development of a Student Web Portal for Higher Education Institution in Kazakhstan. In: 15th International Conference on Information Technology - New Generations, Springer Verlag, Advances in Intelligent Systems and Computing, vol. 738, pp. 759–760, Las Vegas, NV (2018)
3. Boranbayev, A., Boranbayev, S., Nurusheva, A., Yersakhanov, K.: The Modern State and the Further Development Prospects of Information Security in the Republic of Kazakhstan. In: 15th International Conference on Information Technology - New Generations, Springer Verlag, Advances in Intelligent Systems and Computing, vol. 738, pp. 33–38, Las Vegas, NV (2018)
4. Scott Austin, A., Zeder, P.A.: Fifteen Things You Need to Know to Advise Your Clients about Websites http://cyber.law.harvard.edu/ecommerce/austin.ppt
5. Boranbayev A.S. Defining methodologies for developing J2EE web-based information systems. J. Nonlinear Anal. Theory Methods Appl., Volume 71, Issue 12, 15 December 2009, p.1633–1637
6. Li-Rui, W., Zhi-Fu, W.: Implementation on Ecommerce Network marketing based on web Mining technology IEEE-978-1-4673-7143-8/15 (2015)
7. Boranbayev, A., Boranbayev, S., Yersakhanov, K., Nurusheva, A., Taberkhan, R.: Methods of Ensuring the Reliability and Fault Tolerance of Information Systems. In: Proceedings of the 15th International Conference on Information Technology: New Generations, pp. 729–730, Las Vegas, NV (2018)
8. Keynan, I.: Knowledge as responsibility: Universities and society. J. Higher Educ. Outreach Engag. **18**(2), 179 (2014)
9. Boranbayev, A., Boranbayev, S.: Assel Nurusheva. Analyzing Methods of Recognition, Classification and Development of a Software System. In: Proceedings of Intelligent Systems Conference (IntelliSys) 2018, pp. 1055–1061, London (2018)
10. Vinitha Stephie, V., Lakshmi, M.: Design and implementation of e-commerce web application. ARPN J. Eng. Appl. Sci.. **12**(16), 4769–4772 (2017)
11. Boranbayev, A.S., Boranbayev, S.N., Khassanova, A.A.: Comparative Analysis of Methods of Face Detection and Classification of Images, pp. 71–89. Bulletin of L.N. Gumilyov Eurasian National University №2 (2017)
12. Tunardi, Y., Layona, R., Yulianto, B.: Gallery portal web for promoting students' & lecturers' masterpieces(Conference Paper). In: Proceedings of 2016 International Conference on Information Management and Technology, ICIMTech 2016, pp. 257–260. Aston TropicanaBandung; Indonesia; 16 November 2016 до 18 November 2016, Paper number 7930340 (2016)

A Survey on Algorithmic Approaches on Electric Vehicle Adaptation in a Smart Grid: An Introduction to Battery Consolidation Systems

74

Dara Nyknahad, Wolfgang Bein, and Rojin Aslani

74.1 Introduction

The trend for reducing the carbon emission effect on the environment is globally appreciated, and is being progressively adopted by many countries [1]. One of the most significant aspects of this trend is related to the positive effects of Electric Vehicle (EV) incorporation into the electrical gird. According to the Annual Energy Outlook 2016 [1], the perspective of US energy trends is to focus on the analysis of potential changes in US rules, regulations, and policies related to energy through 2040 [1]. As a result, California has been a pioneer in preparing The California Low-Emission Vehicle Regulations, which define the emission regulations required for vehicles from 2018 through 2025 [2]. Among the components in a Smart Grid (SG), EVs have interesting features. As the Global EV Outlook in [3] emphasized that the number of new EV registrations passed 750K in 2016, and a total of 2 million EVs are in the transportation system. China has the most significant market for EVs. According to the EV Outlook 2017 in [4], EVs will be more than 54% of new car sales by 2040.

In order to accomplish effective CO_2 emission reduction, renewable energy (RE) resources like wind, solar energy, or hydro-power must be incorporated into the SG,. However, some of these resources, by their nature, have daily fluctuations, e.g. solar panels, or seasonally, e.g. wind turbines [5]. Having fluctuation makes the grid more vulnerable if timing cannot be managed. The storage capability of EVs

can be considered as a balancing tool for these fluctuations, and can lead to a better performance of the SG in regard to the valley effect of high demand time. During the commute of an EV, there is a possibility of recuper- ating energy (while going down hill), which could also be considered in the formula. From the SG point of view, an EV fleet could be considered as reliable storage if it is effectively managed [6]. This storage may make it possible to purchase electricity, Grid to Vehicle (G2V), or sell electricity to the gird at some point, Vehicle to Grid (V2G). Therefore, having reliable methods for incorporating these vehicles as energy storage objects could assure the effective usage of these RE resources in SGs. Moreover, based on an independent report from the Reinventing Parking Organization [7], cars are parked for more than 95% of their daily use. This gives another reliable feature to a fleet of EVs to be considered as storage. Most of the motor vehicle manufacturers are in the process of designing EVs with different capacities and ranges. The time required to charge them can vary from 30 min to 12 h. The process of charging an EV is a time-consuming procedure, which decreases the satisfaction level of users, who are used to the quick refueling of gas-operable cars. It is worth mentioning that these data are for personal cars, which use considerably smaller battery packs compared to commercial vehicles. Therefore, when considering EVs as commercial vehicles, an exchange battery becomes a must.

With the advancement in battery installation, Battery Exchange Stations (BES) have reasonable features and can provide comfort for drivers, while being incorporated in the SG in the coming years. One of the features of a BES is its scalability on the demand side. According to current studies [8], in most EVs, an average of 6–8 h is required to charge the battery for an average of 200 Km for personal cars. A BES can have a variety of batteries in use to trade out. Additionally, it can manage the scheduling of its batteries based on the electricity market in order to maximize its profits, while shaving the peak load demand in the grid in

D. Nyknahad · W. Bein (✉)
Department of Computer Science, University of Nevada Las Vegas, Las Vegas, NV, USA
e-mail: nyknahad@unlv.nevada.edu; wolfgang.bein@unlv.edu

R. Aslani
Electrical and Computer Engineering Department, University of Nevada Las Vegas, Las Vegas, NV, USA
e-mail: rojin.aslani@unlv.edu

© Springer Nature Switzerland AG 2019
S. Latifi (ed.), *16th International Conference on Information Technology-New Generations (ITNG 2019)*,
Advances in Intelligent Systems and Computing 800,
https://doi.org/10.1007/978-3-030-14070-0_74

order to smooth the performance of the SG. Hypothetically, through the contribution of BESs in the peak load time of the smart grid, the peak-to-average (PAR) of the grid should be reduced. Reducing the PAR is one of the objectives of the optimization problem in the SG. According to [9], there could be billions of dollars of savings achieved by shifting the charging time of one million smart appliances from on-peak to off-peak. However, this shift will challenge the electric gird. Another benefit of BESs is serving the drivers demands at their peak demand time, which is the same as the grid peak load time, while reducing the load on the grid. The drivers habits for charging their vehicles will not change significantly in this manner.

There will be a gradual procedure in replacing conventional cars with Plugin Electric Vehicles (PEV). According to [9], if 30% of the conventional cars in US were exchanged to PEVs, there would be an 18% decrease of US summer peak load. Therefore, more flexible and distributed approaches are needed to get better acceptance of EV adaptation. Decentralized algorithms that focus on managing the demand side from vehicles are interesting, since they exchange the least (limited) amount of information, which is more appealing from the drivers point of view. Moreover, optimal BES placement plays a significant role in the SG, since extensive battery charging time and limited energy storage makes each mistake of battery outage in a vehicle an expensive one. Another benefit of BESs in the SG are the facilities that BESs can provide for public vehicles like taxis. Taxis are inherently prone to need more energy during a 24-h life cycle. Therefore, having reliable network of BESs in the SG makes the idea of electric taxis more attainable.

Focusing on the electricity market, in [8], a mathematical market strategy is formulated based on the short term management, and the proposed model is tested using the revenue of the station. The authors of [10] presented a strategy mathematically to show that BESs are financially justified for the owners. Additionally, they showed that having BESs in the grid is more environmentally friendly, compared to not having them. However, they stated that optimizing the capacities of components in a BES is complex and still under research. Their proposed objective function was to minimize the annual cost of the BES while constraining the construction costs [10].

The authors of [11] discussed that a business and operating model and operating model are useful for successful deployment of stations. The study's objectives were to maximize profits, while offering alternative means of charging batteries. They had the electricity price market one day ahead to afford them the capability to combine facility configurations and operational policies [11]. However, the topology and reinforcement of the SG must be incorporated during the model definition and design. In [12], a methodology was proposed based on queuing theory for electric taxi fleets.

The authors of [13] also used queuing theory to facilitate their idea of mobile charging methods for vans. However, the queuing theory has a weakness in extreme cases for solving this class of problems. Therefore, it is a better choice to design some multi-objective optimization functions to optimize grid component features in the system. As a response to immediate demand from car drivers, BESs can play a satisfaction role in the SG. In [14], it is discussed that BESs can be a reliable way of frequency control units in the SG. As a result, the electricity market can rely on BESs in their peak operations if their incorporation is managed accordingly. More specifically, BESs have characteristics that will serve all engaged parties suitably: drivers can manage their arrival times to BESs to minimize wait time; while the BESs can plan to transact with the grid to maximize their financial benefit; and the SG can maximize its reliability by purchasing electricity from BESs at their chosen time in order to minimize the possibility of instabilities and smooth the grid performance.

In this paper, the authors focus on BES optimization, which facilitates the adaptation of EVs into the SG. The researchers first propose the problem statement of BES in the SG in Sect. 74.2. Then in Sect. 74.3, the authors present a survey on BES optimization methods in the literature, based on the proposed algorithms and considered uncertainties. Next, in Sect. 74.4, the authors introduce the concept of battery consolidation, which focuses on optimizing the incorporation and transaction of all of the components in the SG. The authors propose a system model for a battery consolidation system, and formulate the stated BES problem in the SG by defining the objective function and variables. They then present different scenarios for some theoretical situations for EV adaptation in the SG. Finally, Sect. 74.5 contains the conclusion.

74.2 Problem Statement

Incorporation of BESs in a SG is a complement for the use of RE resources like solar systems and wind turbines. The nature of these two systems, BES and RE, is such that they can complete each other. While the RE systems cannot store the generated electricity, they can rely on the BESs for their ability to store electricity. A PEV can have the by-product role of an off-line uninterruptible power supply when it is parked in the local (micro-island) grid, not only for the owners house, but also for nearby neighbors. The BESs have byproduct benefits also, which might be invisible. The BESs can be used to control the charging time more precisely, while minimizing waiting time for users. Another advantage is the possibility of offering a price plan based on the usage of the different users. Comfortable online ordering and trade off with the SG are also appealing.

The problem of finding and optimizing the place-ment and capacity of BESs is inherently stochastic. The sources of uncertainty that make the problem inherently stochastic are: the (1) unknown yet justifiable PEVs charging/discharging schedules, (2) stochastic wind speed, (3) seasonal and daily variation in received solar power and intensity, (4) fluctuation of the energy market, (5) stochastic demand of the PEV users, and (6) randomness notion of energy loss in the grid. These uncertainties are important in the design of algorithms and frameworks to define the problem. In the next section, the authors categorize the proposed methods in the literature for finding optimum BES placing, and their capacities in regards to optimiz- ing some of the features of the system, such as objective functions based on proposed algorithms and considered uncertainties.

74.3 Literature Review

In this section, the authors present a survey on BES opti-mization methods in the literature based on the proposed algorithms and considered uncertainties. First, the authors categorize the related works based on uncertainties consid-ered in their stated problem. Then, they classify the proposed algorithms for addressing the problem of BES optimization in the literature, based on centralized and decentralized manners.

74.3.1 Uncertainties

A critical part of studying a SG is related to modeling the uncertainties of the system in a stochastic yet predictive way that simulates reality as much as possible. Following is a classification of some studies in the literature based on uncertainties.

Customer Demand Forecasting

Driver (user) demand is one of the uncertain resources in the SG. Part of this uncertainty is rooted in the forecasting of demands, like vehicle charging demand and scheduling of consumers in the SG. There is research that has studied the prediction of energy demand in SGs in an algorithmic manner [15–17].

Electricity Load Forecasting

(1) There is a direct relation between the PEV demand and the electricity demand in the system. However, with the predicted electricity market, a BES can sell and buy electric- ity in the SG, which could be considered as a source of uncertainty. There have been a wide range of studies on electricity load forecasting. Each re- searcher used a different methodology to predict the short-term or long-

term load forecasting with the aim of precise prediction. These methodologies are com- putationally expensive and are usually solved in a centralized manner. However, a centralized approach is heavily based on much real time data ex-change, which is undesirable not only for the owner of the computing system and grid management, but also for the drivers. Researchers approached this load forecasting with different methodologies, of which some were studied for this paper [18–22].

Solar Energy Forecasting

The renewable supply energy resources in the system are inherently stochastic. As the authors of [23] discussed that the use of low voltage level rooftop solar generators has been growing rapidly during previous years. Despite having positive effects on reducing CO_2 emissions, this growth introduces more complexities to the management of the grid distribution operation. This forecasting could be spatial and/or temporal. How- ever, it is worth mentioning that forecasting precision is based on the accuracy of the col-lected data from solar devices. These data might have noise; however, some researchers like [24] proposed probabilistic approaches to minimize the effect of it [20, 23, 24].

Wind Power Forecasting

Wind power fore- casting is the prediction of the wind speed. Therefore, this problem could be classified into three categories: the digital weather forecasting method, stochastic (statistical) method, and learning algorithm method. The learning algo- rithms, specifically artificial neural networks, have the weakness of a tendency for over-fitting, which leads to falling into a local minimum. Among these methods, a support vector machine or an enhanced least square support vector machine showed that it can overcome the problem of the tendency into local minimum. Some researchers tried to use Adaboost to create an integrated model to predict short-term wind speed and find the significant features, which might not be possible with a single support vector machine [25–40].

Electricity Price Forecasting

Electricity price market variation can initiate fluctuation in the demand, and uncertainties in the response of the gird. Different researchers approached this problem with different methods. Generally, there are two approaches to this prob-lem. One is based on parametric statistical modeling. The most popular methods are regression models and time series prediction. The other approach is non-parametric methodolo-gies like a support vector machine, a neural network, fuzzy logic, and new big data analytic. Non-parametric approaches are more compatible with sensor-collected data from the devices. These machine-learning algorithms create the best match between the observed data and the output [9, 41–46].

74.3.2 Proposed Algorithms in the Literature

Energy management in any grid has many dimensions, and each researcher is exploring these dimensions with different goals. However, the basis for all research is to focus on finding new methodologies for more efficient energy consumption in a certain grid. There are several approaches to classify algorithms, but here the focus is on centralized and decentralized algorithm classification. Energy management and the search for utilizing new methods of battery charging in girds is not just limited to the SG, but other grids like wireless and communications [47]. To determine the placement of BESs and their capacities, one of approach is to use the graph theory, which can be solved in centralized or decentralized manners. In this regard, the objective functions and their optimality could be a non-convex problem, which makes them more difficult. Selecting optimal BES placement and capacity under a multi-objective function is a computationally expensive problem.

Centralized Approaches

In the centralized approaches, the authors assume that a centralized system will coordinate all of the transactions and resource allocation, mainly favoring the grid itself. This class of algorithms has some features that make them interesting for the researchers. Centralized algorithms are easier to implement and cheaper for maintenance. The authors of [10] used a version of a non-dominated sorting genetic algorithm in their research. They had a certain district for the placement of battery stations and optimized their multi-objective optimization problem to find optimum capacity. In [11], a centralized model for capacity optimality was developed by including uncertainty to inventory robust optimization for the battery de- mand. On the other hand, to consider electricity price uncertainties, the authors deployed a multi-band robust optimization with the focus of financial profitability justification. However, they did not consider the battery degradation cost, which still has a con- sequential effect. In [12], the authors used the Monte- Carlo method and Dijkstra in a centralized way for taxi operations, and used the queuing theory in taxi waiting modeling. In [13], a centralized way was used to minimize the waiting time of PEVs for the battery exchange and delivery scheduling when the PEV was immobilized and out of battery. The authors of [14] used a Monte-Carlo stochastic simulation method to predict the capacity of the BESs. Additionally, they used an automatic generation control strategy to allocate power among generators and BESs. These types of problems are complex and challenging. Integrating RE in the SG, the evolution of the SG itself according to urban demand, and incorporating the uncertainties give this class of problems more layers of complexity.

74.4 Battery Consolidation Systems

As mentioned before, the existing approaches in the literature for the BES optimization problem as a solution for EV adaptation in the SG have proposed algorithms in a solus manner towards improving the incorporation of existing components into the SG. However, none of these studies have proposed a comprehensive model that considers the incorporation and transaction of all of the components in the SG. The battery consolidation system is a class of problems and solutions which focuses on optimizing the incorporation and transaction of all of the components in the SG. In the following, the researchers first propose a system model for a battery consolidation system, and then formulate the problem of finding the optimum BES placement and capacity in regard to optimizing some of the features of the system. Finally, the authors present different scenarios for some theoretical situations for EV adaptation in the SG.

74.4.1 System Model

By using the functional flow block diagram method, the researchers propose a system model as shown in Fig. 74.1. The system model includes:

1- Current grid
2- Renewable energy resource uncertainty
3- Electricity market
4- EV demand and scheduling
5- BES

The components that are used in previous work like [48–51] mostly combined BESs and the electricity market, as well

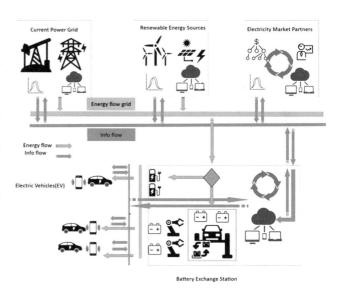

Fig. 74.1 Battery consolidation system model

as some RE resources discussed previously. In this research, the authors would like to use a more comprehensive and general system model as seen in Fig. 74.1.

74.4.2 Problem Formulation

In the previous sections, the researchers stated the problem of finding the optimum BES placement and capacity in regard to optimizing some of the features of the system and subject to satisfying some constraints raised by uncertainties of the system. Therefore, the formulation of the stated BES problem for the adaptation EVs in the SG consists of variables, objective function, and constraints. The variables include placement and capacity of the BES in the SG. The objective function for this class of problems can be evaluated from the perspective of different parties involved in the system. One approach could be the minimization of the cost of the operation, battery quantity, and other facilities in a BES. On the other hand, the objective function can be designed in a way that the performance of the gird should be optimized. For this reason, the performance should be defined and designed. For instance, valley filling of the SG is a good measurement for enhancing the performance of the grid. For this purpose, the design should be constructed in a way to encourage energy use during the lowest system demand periods. This valley filling is acquired by load shifting. Usually, this load shifting is with the aim of shifting the peak demand usage to the periods that have the lowest demand. As one of the hypotheses, this design makes the grid more reliable, while maximizing the profitability of the battery exchange stations. Therefore, the system model and formulation could combine these ideas together and reach an optimal solution. The optimal solution should configure the BES components to minimize the costs and schedule operation of the BES to recharge the batteries in the lowest demand period of the grid. The sale of batteries are usually made at the peak demand period on the usual days. Another objective function that might be interesting is maximizing the throughput of the vehicles in all BESs available on the grid. Part of the design of the objective function and the optimality question here is the modeling of vehicles arrival patterns.

The objective function for the price includes the following independent variables for each BES.

(1) BES installment Cost, i.e., BES_{inst}
(2) BES operations Cost, i.e., BES_{op}
(3) BES maintenance Cost, i.e., BES_{main}
(4) BES loss of electricity Cost, i.e., BES_{loss}
(5) BES Battery Cost, i.e., BAT (capacity to meet the demand)
(6) RE resources
(7) Electricity cost

The dependent variables are:

(1) Placement of the BES in the SG
(2) Capacity of the BES in the SG

The Constraints are:

(1) Max/Min number of battery charge/discharge
(2) Available energy for users
(3) Amount of energy provided by RE resources
(4) Fluctuation of the energy market
(5) Stochastic demand of the PEV users

74.4.3 Methodologies for Addressing the Formulated Problem

As the stated BES problem for adapting EVs in a SG is a NP hard problem due to the multi- objective nature and constraints raised by different uncertainties of system, proposing an algorithmic solution to address it is a challenging issue.

74.4.4 Scenarios

In this section, the researchers first define some indexes for evaluation of the approaches and algorithms to address the BES problem. For instance, assuming that a vehicle decided to exchange its battery and requested it, the closest BES will be found on the map and the driver can submit the request. Based on the level of the EV battery (percentage) and possibility to reach the BES, the researchers can define a new index, hypothetically an Anxiety Level Index (ALI), that shows how user friendly and practical the BES is. Besides, the researchers can use the new online data streaming to update (online) the scheduling methods in order to lower the anxiety level index. Parameters that can be measured to evaluate the performance of the applied methodologies are:

(1) Power supply reliability
(2) Network loss
(3) User satisfaction

Then the researchers define a few scenarios, such that in each case the objective function can be seen from a different point of view. More specifically, a scenario means the combination of the demands and components in a graph with a specific goal.

Some of the scenarios will be maximizing the BES financial benefits from the BES point of view, some will minimize the probability of power loss and unreliability of

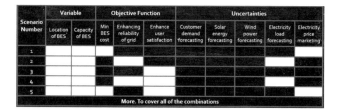

Fig. 74.2 Scenarios of EV adaptation in the SG

the grid; while other objective functions will maximize the PEV user satisfaction with the BES. For each sub-problem the criteria for performance evaluation of the algorithms should be developed. In Fig. 74.2, some initial scenarios are shown.

74.5 Conclusions

In this paper, the researchers have studied the BES optimization problem as a promising approach for EV adaptation in a SG. The authors have presented a survey on existing algorithms in the literature to address the BES optimization problem in a SG. However, to the best of the authors knowledge, none of these studies have proposed a comprehensive model that considers the incorporation and transactions of all of the components such as EV, RE, and BES on the basis of a SG. The researchers have introduced the concept of a battery consolidation system, which focuses on optimizing the incorpora- tion and transaction of all of the components in a SG. The authors have proposed a comprehensive system model for a battery consolidation system, in which all of the transactions between the components of a SG have been considered. Then, the authors have formulated a BES optimization problem based on the proposed system model. The researchers have proposed algorithmic approaches to address the formulated problems. Finally, some indexes and theoretical scenarios have been proposed to evaluate the efficiency of algorithmic approaches to address the BES optimization problem for EV adaptation in a SG on the basis of a battery consolidation system.

Acknowledgements The authors gratefully acknowledge that this research was supported in part by "NV Energy Renewable Energy Fellowship".

References

1. Annual Energy 2016. John J. Conti (john.conti@eia.gov) Natural Gas, and Biofuels Analysis; James T. Turnure (james.turnure@eia.gov), no. 202 (2016)
2. Zero-Emission Vehicle Standards for 2018 and Subsequent Model Year Passenger Cars, Light-Duty Trucks, and Medium-Duty Vehicles (2018)
3. International Energy Agency, I.: Global EV Outlook 2017 Together Secure Sustainable Global EV outlook 2017 (2017)
4. Electric Vehicle Outlook 2017|Bloomberg New Energy Finance|Bloomberg Finance LP (2017)
5. Gonzalez Vaya, M., Andersson, G.: Self scheduling of plug-in electric vehicle aggregator to provide balancing services for wind power. IEEE Trans. Sustainable Energy **7**(2), 886–899 (2016)
6. Al-Hallaj, S., Wilke, S., Schweitzer, B.: Energy storage systems for smart grid applications. In: Badran, A., Murad, S., Baydoun, E., Daghir, N. (eds.) Water, Energy & Food Sustainability in the Middle East. Springer, Cham (2017)
7. Barter, P.: Cars are parked 95% of the time. Let's check! ̃ Reinventing Parking. Technical Report (2013)
8. Yang, S., Yao, J., Kang, T., Zhu, X.: Dynamic operation model of the battery swapping station for EV (electric vehicle) in electricity market. Energy **65**, 544–549 (2014)
9. Fan, Z.: A distributed demand response algorithm and its application to PHEV charging in smart grids. IEEE Trans. Smart Grid **3**, 1280–1290 (2012)
10. Liu, N., Chen, Z., Liu, J., Tang, X., Xiao, X., Zhang, J.: Multi-objective optimization for component capacity of the photovoltaic-based battery switch stations: towards benefits of economy and environment. Energy **64**, 779–792 (2014)
11. Sarker, M.R., Pandzic, H., Ortega-Vazquez, M.A.: Optimal operation and services scheduling for an electric vehicle battery swapping station. IEEE Trans. Power Syst. **30**(2), 901–910 (2015)
12. Jing, Z., Fang, L., Lin, S., Shao, W.: Modeling for electric taxi load and optimization model for charging/swapping facilities of electric taxi. In: 2014 IEEE Conference and Expo Transportation Electrification Asia-Pacific (ITEC Asia-Pacific), pp. 1–5. IEEE, Aug (2014)
13. Shao, S., Guo, S., Qiu, X.: A mobile battery swapping service for electric vehicles based on a battery swapping van. Energies **10**(10), 1667 (2017)
14. Xie, P., Li, Y., Zhu, L., Shi, D., Duan, X.: Supplementary automatic generation control using controllable energy storage in electric vehicle battery swapping stations. IET Gener. Transm. Distrib. **10**(4), 1107–1116 (2016)
15. Andro-Vasko, J., Bein, W., Nyknhad, D., Ito, H.: Evaluation of online power-down algorithms. In: 2015 12th International Conference on Information Technology – New Generations, pp. 473–478. IEEE (2015)
16. Bein, W., Madan, B.B., Bein, D., Nyknhad, D.: Algorithmic approaches for a dependable smart grid, pp. 677–687. Springer, Cham
17. Tushar, M.H.K., Zeineddine, A.W., Assi, C.: Demand-side management by regulating charging and discharging of the EV, ESS, and utilizing renewable energy. IEEE Trans. Ind. Inf. **14**(1), 117–126 (2018)
18. Joshi, P.A., Patel, J.J.: Computational analysis and intelligent control of load forecasting using time series method, pp. 297–306. Springer, Singapore
19. Kuo, P.-H., Huang, C.-J.: A high precision artificial neural networks model for short-term energy load forecasting. Energies **11**(1), 213 (2018)
20. Yao, E., Samadi, P., Wong, V.W.S., Schober, R.: Residential demand side management under high penetration of rooftop photovoltaic units. IEEE Trans. Smart Grid **7**(3), 1597–1608 (2016)
21. Muralitharan, K., Sakthivel, R., Vishnuvarthan, R.: Neural network based optimization approach for energy demand prediction in smart grid. Neurocomputing **273**, 199–208 (2018)
22. Goswami, D.Y., Kreith, F.: Energy efficiency and renewable energy handbook, 2nd edn. In: Goswami, Y., Kreith, F. (eds.). CRC Press, Flordida
23. Bessa, R.J., Trindade, A., Miranda, V.: Spatial-temporal solar power forecasting for smart grids. IEEE Trans. Ind. Inf. **11**(1), 232–241 (2015)

24. Sheng, H., Xiao, J., Cheng, Y., Ni, Q., Wang, S.: Short-term solar power forecasting based on weighted Gaussian process regression. IEEE Trans. Ind. Electron. **65**(1), 300–308 (2018)

25. Li, Y., Yang, P., Wang, H.: Short-term wind speed forecasting based on improved ant colony algorithm for LSSVM. Clust. Comput. 1–7 (2018). https://doi.org/10.1007/s10586-017-1422-2

26. Mana, M., Burlando, M., Meissner, C.: Evaluation of two ANN approaches for the wind power forecast in a mountainous site. Int. J. Renew. Energy Res. (IJRER) **7**(4), 1629–1638 (2017)

27. Georgilakis, P.S.: Technical challenges associated with the integration of wind power into power systems. Renew. Sustain. Energy Rev. **12**(3), 852–863 (2008)

28. Lahouar, A., Ben Hadj Slama, J.: Hour-ahead wind power forecast based on random forests. Renew. Energy **109**, 529–541 (2017)

29. Zhao, X., Wang, S., Li, T.: Review of evaluation criteria and main methods of wind power forecasting. Energy Procedia **12**, 761–769 (2011)

30. Tascikaraoglu, A., Uzunoglu, M.: A review of combined approaches for prediction of short-term wind speed and power. Renew. Sustain. Energy Rev. **34**, 243–254 (2014)

31. Alam, M.M., Rehman, S., Al-Hadhrami, L., Meyer, J.: Extraction of the inherent nature of wind speed using wavelets and FFT. Energy Sustain. Dev. **22**, 34–47 (2014)

32. Colak, S., Sagiroglu, M., Yesilbudak: Data mining and wind power prediction: a literature review. Renew. Energy. **46**, 241–247 (2012). https://doi.org/10.1016/j.renene.2012.02.015

33. Jung, J., Broadwater, R.P.: Current status and future advances for wind speed and power forecasting. Renew. Sustain. Energy Rev. **31**, 762–777 (2014)

34. Zhang, Y., Wang, J., Wang, X.: Review on probabilistic forecasting of wind power generation. Renew. Sustain. Energy Rev. **32**, 255–270 (2014)

35. Cassola, F., Burlando, M.: Wind speed and wind energy forecast through Kalman filtering of numerical weather prediction model output. Appl. Energy **99**, 154–166 (2012)

36. Louka, P., Galanis, G., Siebert, N., Kariniotakis, G., Katsafados, P., Pytharoulis, I., Kallos, G.: Improvements in wind speed forecasts for wind power prediction purposes using Kalman filtering. J. Wind Eng. Ind. Aerodyn. **96**(12), 2348–2362 (2008)

37. Foley, A.M., Leahy, P.G., Marvuglia, A., McKeogh, E.J.: Current methods and advances in forecasting of wind power generation. Renew. Energy **37**(1), 1–8 (2012)

38. Wang, C., Liang, Z., Liang, J., Teng, Q., Dong, X., Wang, Z.: Modeling the temporal correlation of hourly day-ahead short-term wind power forecast error for optimal sizing energy storage system. Int. J. Electr. Power Energy Syst. **98**, 373–381 (2018)

39. Bludszuweit, H., Dominguez-Navarro, J.A.: A probabilistic method for energy storage sizing based on wind power forecast uncertainty. IEEE Trans. Power Syst. **26**(3), 1651–1658 (2011)

40. Ma, X.-Y., Sun, Y.-Z., Fang, H.-L.: Scenario generation of wind power based on statistical uncertainty and variability. IEEE Trans. Sustain. Energy **4**(4), 894–904 (2013)

41. Lahmiri, S.: Comparing variational and empirical mode decomposition in forecasting day-ahead energy prices. IEEE Syst. J. **11**(3), 1907–1910 (2017)

42. Wang, K., Xu, C., Zhang, Y., Guo, S., Zomaya, A.Y.: Robust big data analytics for electricity price forecasting in the smart grid. IEEE Trans. Big Data. **5**(1), 34–45 (2017). 1 March 2019. https://doi.org/10.1109/TBDATA.2017.2723563

43. Hosking, J., Natarajan, R., Ghosh, S., Subramanian, S., Zhang, X.: Short-term forecasting of the daily load curve for residential electricity usage in the smart grid. Appl. Stoch. Model. Bus. Ind. **29**(6), 604–620 (2013)

44. Hong, T., Fan, S.: Probabilistic electric load forecasting: a tutorial review. Int. J. Forecast. **32**(3), 914–938 (2016)

45. Bassamzadeh, N., Ghanem, R.: Multiscale stochastic prediction of electricity demand in smart grids using Bayesian networks. Appl. Energy **193**, 369–380 (2017)

46. Edwards, R.E., New, J., Parker, L.E.: Predicting future hourly residential electrical consumption: a machine learning case study. Energ. Buildings **49**, 591–603 (2012)

47. Aslani, R., Rasti, M.: Distributed power control schemes in in-band full-duplex energy harvesting wireless networks. IEEE Trans. Wirel. Commun. **16**(8), 5233–5243 (2017)

48. Adler, J.D., Mirchandani, P.B., Xue, G., Xia, M.: The electric vehicle shortest-walk problem with battery exchanges. Netw. Spat. Econ. **16**(1), 155–173 (2016)

49. Mak, H.-Y., Rong, Y., Shen, Z.-J.M.: Infrastructure planning for electric vehicles with battery swapping. Manag. Sci. **59**(7), 1557–1575 (2013)

50. Cheng, Y., Zhang, C.: Configuration and operation combined optimization for EV battery swapping station considering PV consumption bundling. Prot. Control Mod. Pow. Syst. **2**(1), 26 (2017)

51. Yao, L., Damiran, Z., Lim, W.H.: A fuzzy logic based charging scheme for electric vechicle parking station. In: 2016 IEEE 16th International Conference on Environment and Electrical Engineering (EEEIC), pp. 1–6. IEEE, June (2016)

A New Efficient Power Management Interface for Hybrid Electromagnetic-Piezoelectric Energy Harvesting System

Sara Zolfaghar Tehrani, Hossein Ranjbar, Peter Vial, and Prashan Premaratne

75.1   Introduction

Vibration energy is one of the free available and wasted energy sources in the environment, which originated from the flow of water and air, body movement, or even breathing, heartbeat, and so on [1]. The transformation of any vibration energy utilizes individual electrostatic, piezoelectric, electromagnetic devices or their hybrid systems (HVEH) to generate enough power for electronic devices such as wireless sensor networks (WSN) [2]. Therefore, these potential energy harvesting systems could be adopted as an alternative to the batteries. Since batteries' lifespan and their replacement accessibility confined, developing an entirely self-powered circuitry is an objective consideration in a dynamic energy harvesting system [3].

The piezoelectric vibration energy harvesting (PZVEH) systems work with the piezoelectric effect of materials. In particular, the benefits are the ease transformation and integration of bulk to Micro/Nanosystems, having an unlimited lifetime, a tendency of direct energy conversion and producing an extended voltage for small deflections [4]. On the other hand, the electromagnetic vibration energy harvesting (EMVEH) systems are inductive generators generally retain a strong magnetic field with a permanent magnet (PM), so they are preferred in many applications [5].

The efficiency of vibration energy harvesting systems (VEHS), which refers to the ratio of the output power to the total input electrical power, is usually low. Thus, there are several separate power control circuits from the perspective of circuit optimization to adjust the transferred power to the electrical load or the energy storage media. The energy extraction interfaces (EEI) could be an AC-DC converter to modify the alternating vibration energy to extract the dc power, which is suitable for WSN and mobile applications [6]. The other EEI would be a DC-DC converter with the function of either the impedance adaptation or the voltage regulation [5].

The most basic interface circuit of a VEHS, defined as standard energy harvesting (SEH) circuit, is an AC-DC converter, which is composed of a traditional full bridge diode rectifier (FBR) or a voltage-doubling rectifier (VDR) and a DC filter capacitor [1]. This capacitor is expected to be large enough to generate the constant output voltage essentially [1]. The output results of the VEHS is mostly limited and depending on the load impedance or the storage voltage in the DC filter capacitor [1]. The optimization of SEH technique during a certain period in each cycle cannot guarantee the blocking of returned energy from the electrical part to the mechanical part, which is called the energy return phenomenon (ER) [6].

The efficiency estimation of the other interface circuits usually are compared to the output result of SEH. Several researches adopted a type of switching interface to enhance the capability of the VEHS. These switches are managed with a controller and they are noted for their high efficiency.

The researchers have suggested various nonlinear EEIs in comparison to SEH in a PZVEH. For instance, the synchronized switch harvesting on inductor (SSHI) circuit attached in parallel or series to the piezoelectric elements called (P-SSHI) [7–9] and (S-SSHI) [10, 11] respectively. The SSHI is a controlled electromechanical interface. There are also some of the upgraded circuits which use double synchronous switching called (DSSH) circuit [12] or the optimized syn-

S. Z. Tehrani (✉) · P. Vial (✉) · P. Premaratne (✉)
Department of Electrical, Computer and Telecommunications
Engineering, Wollongong University, Wollongong, Australia
e-mail: szt685@uowmail.edu.au; peter_vial@uow.edu.au;
prashan@uow.edu.au

H. Ranjbar (✉)
Department of Electrical Engineering, Sharif University of
Technology, Tehran, Iran
e-mail: h_ranjbar@ee.sharif.edu

© Springer Nature Switzerland AG 2019
S. Latifi (ed.), *16th International Conference on Information Technology-New Generations (ITNG 2019)*,
Advances in Intelligent Systems and Computing 800,
https://doi.org/10.1007/978-3-030-14070-0_75

chronous electric charge extraction (OSECE) interface [13]. These adjustments are made to enhance the output power. However, there are others, who with the aid of a self-powered SSHI (SP-SSHI) found the output power is significant when the excitation level is high enough [14].

Lallart et al. in [11] firstly, implement the classical circuit of S-SSHI, but uses two switches S_1 and S_2 in parallel to detect and segregate the maximum and minimum of switches. Besides, they are in series toward an inductor. Then, they superseded the switches by two diodes of FBR in their recommended design. The benefits of modifications are (i) the reduction of the size of the circuit and its cost because of fewer components, also (ii) overcoming the voltage gaps due to fewer losses and higher storage energy.

Ramadass et al. [12] take advantage of DSSH technique as same as in [11]. There are switches S_1 and S_2, which are in series and their shared inductor L_{share}. This interface is parallel to FBR called bias-flip rectifier circuit. The L_{share} in this interface has the benefit of decreasing the off time of switches although causes delay in acknowledgment of signal request. Moreover, the extra switch in series causes resistance in the inductor sharing named R_{BF} [12], but The extracted energy is four times of SHE. However, if the switch only rectifier (SOR) interface is attached to the energy harvester, it could generate double power in comparison to a FBR or a VDR [12].

The modified interface circuits for an EMVEH, which compare to the SHE, are various. Wang et al. introduced an interface design for wearable applications using an active rectifier with a comparator as a controller of conducting paths and a switched-inductor DC-DC boost converter [15]. An active load using the two-stage charge-pump circuit uses a comparator in synchronous rectifier and generates 35% power efficiency [16].

For an electromagnetic generator with ultra-low power a nonsynchronous, full-wave boost rectifier introduced in [17] which increases the output result to 70%. They exploit the parasitic coil inductance as the boost inductor where the coil resistance is negligible for the energy harvester [17]. Moreover, the two FBR diodes replaced with switches to do rectification and voltage boosting at the single stage [17]. There are other improved rectifiers such as gate cross-coupled rectifier (GCCR) [18, 19] and bootstrap rectifier (BSR) which upsurge the efficiency of the circuit [18].

The characterization results of EMVEH system in terms of generated DC output voltage and power from two different passive interface electronic rectifiers GCCR and BSR in [18] are compared with FBR output results. The result reveals that BSR configuration could significantly decrease the effect of the threshold voltage of diodes in FBR.

Using of multimodal coupling factor or hybrid harvesters such as a hybrid piezoelectric and electromagnetic system (HPES) could enhance the output power along with design-

ing new power management interfaces [20, 21]. Sriramdas et al. proposed a new prototype model to optimize the output power using the load optimization and achieve the efficiency of 55% [20]. The other MEMS piezoelectric cantilever array utilized to design the hybrid system with low operation frequency and the power management circuit including an AC-DC rectifier and DC-DC buck converter to generate low power consumption [21]. They indicated many other methods to boost the output result using triple hybrid VEHS with adding a thermoelectric harvester to an HPES [22] or a frequency up-conversion method under low-frequency wide-spectrum vibration [23].

Based on reviewed researches, a few works have performed on interface circuits of HPES. Thus, this paper contributes a novel interface circuit for the HPES to enhance the output power and the efficiency. This system can supply enough energy, which is available for many applications. The stated interface consists of two independent circuits for the piezoelectric and the electromagnetic sectors. The interface of the piezoelectric part employs a typical P-SSHI based on [7], and the electromagnetic interface circuit combines a common FBR and a DC-DC boost converter.

75.2 Principles

75.2.1 Electromechanical Model and Equivalent Circuit

A typical HPES device consists of the cantilever-beam structure made of steel substrate and d_{31} mode piezoelectric material layers. There is one NdFeB (N35) magnet as a tip mass, and a copper coil fixed under it. The Euler–Bernoulli beam theory is an accurate model to derive the governing equations of motion for the electromechanical model of the HPES device, which is given in Fig. 75.1. There are damping losses associated with electromagnetic, piezoelectric and

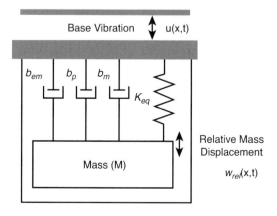

Fig. 75.1 The electromechanical model of a typical hybrid piezoelectric and electromagnetic energy harvesting system

mechanical vibration including b_{em}, b_p and b_m respectively. There are K_{eq} the spring constant and M the mass of magnet.

The total motion of the beam w(x, t) is at point x and time t. Firstly, this is the function of relative transverse displacement of mass, $w_{rel}(x, t)$ with boundary $x_0 = 0$. Secondly, it relates to the axial deflection of structure at the base u(x, t) while the small rotation neglected ($t_0 = 0$). The relative displacement in a continuous system, $w_{rel}(x, t)$, according to the Rayleigh-Ritz procedure could be defined as,

$$w_{rel}(x, t) = \sum_{r=1}^{\infty} \phi_r(x)\, \eta_r(t) \qquad (75.1)$$

Where $\phi_r(x)$ and $\eta_r(t)$ represent the mechanical mode shape eigenfunction and the mechanical temporal modal coordinates, respectively.

Here we follow the procedure outlined in [20]. The governing equations at low excitation frequency in the HPES could be written as,

$$\ddot{\eta}_r + 2\zeta_r\omega_r\dot{\eta}_r + \omega_r^2\eta_r - f_{pr} - f_{er} = f_{inr}(t) \qquad (75.2)$$

$$\sum_{r=1}^{\infty} \chi_r\dot{\eta}_r - C_p^S\dot{\upsilon}(t) - \frac{1}{R_{Lp}}\upsilon(t) = 0 \qquad (75.3)$$

$$\sum_{r=1}^{\infty} \beta_{er}\dot{\eta}_r + L_c\dot{i}_c + (R_c + R_{Le})\,i_c = 0 \qquad (75.4)$$

Where R_{Lp} and R_{Le} indicate the load resistances of piezoelectric and electromagnetic separately in the order.

According to (75.2)–(75.4), the equivalent circuit of a classic HPES hybrid can be shown as Fig.75.2 which consists of a mechanical and two electrical sections related to piezoelectric and electromagnetic domains [20]. As can be seen, the mechanical part of the system modeled by one RLC circuit and an alternating vibration source. The piezoelectric domain governed by a piezoelectric capacitance C_p with the coupling force $f_{pr} = \chi_r\upsilon(t)$ (the modal electromechanical coupling coefficient χ_r) and the electromagnetic domain using the coil resistance and inductance R_c, L_c respectively. In addition, a gyrator used in the circuit to model the coupling force $f_{er} = \beta_{er}i_c$ (the modal electrodynamic coupling coefficient β_{er}) where each domain is connected through an AC-DC converter to the load part.

75.3 Circuit Design

The proposed interface for the HPES circuit follows the purpose of the output and efficiency optimization. The coupling interface between the mechanical and electrical domain in the piezoelectric system introduced by a 1: K step-up transformer where the variable is χ_r. Moreover, this variable in the electromagnetic part β_{er} modeled in the circuit by an ideal gyrator, which neither stores nor dissipates energy [24].

As an effective technique the AC-DC converter of the SHE has improved in the suggested design with an standard synchronized switch called P-SSHI [8, 10]. Besides, the electrical part of the electromagnetic section with the micro-level output power from the coil's inductance needs a DC-DC boost converter just after the FBR. The suggested interface makes a pick-up to the output power while it regulates the extracted electric charge with high ripples. The considered circuit design for the HPES with a specific power management circuit is shown in Fig. 75.3.

75.4 The Simulation Result

Here we present the simulation results for the standard PZVEH, the PZVEH with P_SSHI interface, the standard HPES, and the HPES with proposed interface, which we

Fig. 75.2 Equivalent electromechanical circuit of the standard hybrid piezoelectric and electromagnetic energy harvesting

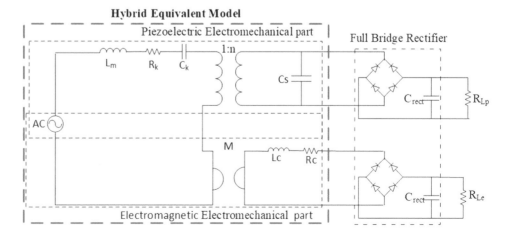

Hybrid Harvester Equivalent circuit

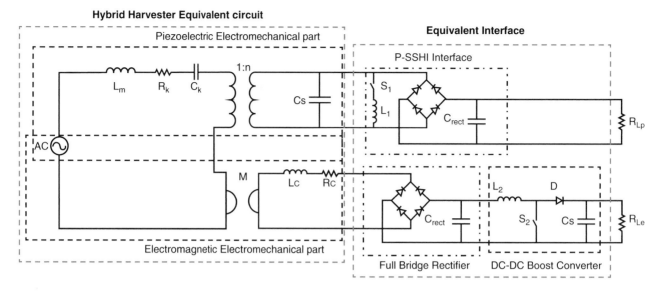

Fig. 75.3 The proposed energy harvesting interface circuit structure for the hybrid system

Table 75.1 Component models and parameters

Circuit components	Name	Values of models
All diodes	D	$V_f = 0.6$ V, $R_{on} = 0.3$ Ω
Buffer capacitor	C_{rect}	2 μF
P-SSHI inductor	L_1	1 mH
Boost inductor	L_2	23 mH
Boost capacitor	C_s	120 μF
Piezoelectric load	R_{LP}	50 kΩ
Electromagnetic load	R_{Le}	50 kΩ

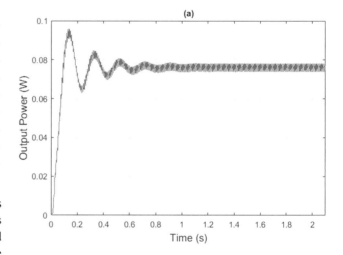

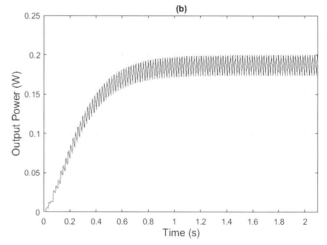

Fig. 75.4 The piezoelectric energy harvester (**a**) with Standard interface (**b**) with P_SSHI interface

described in Sect. 75.3. The electrodynamic parameters such as piezoelectric, substrate, magnet and coil materials and their geometric, and physical properties for the hybrid system given in [25, 26]. The circuit parameters of the proposed HPES are listed in Table 75.1. The Simulink environment in the software [27] was used to simulate the mentioned models. Figures 75.4 and 75.5 show the output power of mentioned designs. In addition, the efficiency of interfaces are presented and compared in Table 75.2.

According to the results, the efficiency of the proposed interface for HPES is around 80%. However, the standard HPES just has efficiency 54%. In fact, the output power from the suggested HPES is 250 mW as shown in Fig. 75.4b although the standard HPES only generates 90 mW the output power as indicated in Fig. 75.4a. Moreover, to verify the accuracy of the results from the HVEH simulation the output power result of the standard PZVEH and the PZVEH with P_SSHI interface is given in Fig. 75.5a, b, which are 80 and 200 mW respectively.

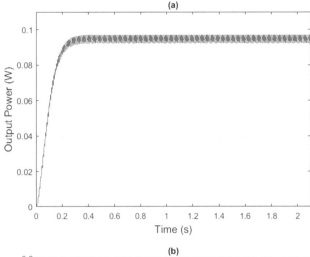

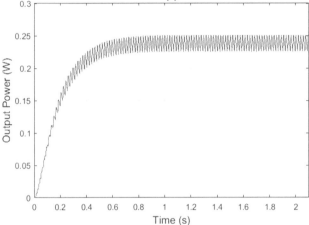

Fig. 75.5 The output power of hybrid piezoelectric-electromagnetic energy harvester (**a**) with Standard interface (**b**) with the proposed interface

Table 75.2 The efficiency of all the model harvesters

Efficiency (%)	VEHS type
41.72	Standard piezoelectric
48.20	Piezoelectric with P-SSHI
54.30	Standard hybrid
80.17	Proposed hybrid

References

1. Ge Shi, Y.X., Ye, Y., Qian, L., Li, Q.: An efficient self-powered synchronous electric charge extraction interface circuit for piezoelectric energy harvesting systems. J. Intell. Mater. Syst. Struct. **27**(16), 2160–2178 (2016)
2. Anton, S.R., Sodano, H.A.: A review of power harvesting using piezoelectric materials (2003–2006). Smart Mater. Struct. **16**(3), R1–R21 (2007)
3. Dallago, E., Danioni, A., Marchesi, M., Nucita, V., Venchi, G.: A self-powered electronic interface for electromagnetic energy harvester. IEEE Trans. Power Electron. **26**(11), 3174–3182 (2011)., Art. no. 5756243
4. Tzeno Galchev, E.E.A., Najaf, K.: A piezoelectric parametric frequency increased generator for harvesting low-frequency vibrations. Microelectromech. Syst. **21**(6), 1311–1320 (2012)
5. Cao, X., Chiang, W.-J., King, Y.-C., Lee, Y.-K.: Electromagnetic energy harvesting circuit with feedforward and feedback DC–DC PWM boost converter for vibration power generator system. IEEE Trans. Power Electron. **22**(2), 679–685 (2007)
6. Liang, J.: A systematic investigation on piezoelectric energy harvesting with emphasis on interface circuits. Chinese University of Hong Kong (2010)
7. Badel, A., Guyomar, D., Lefeuvre, E., Richard, C.: Piezoelectric energy harvesting using a synchronized switch technique. J. Intell. Mater. Syst. Struct. **17**(8–9), 831–839 (2006b)
8. Guyomar, D., Badel, A., Lefeuvre, E., Richard, C.: Toward energy harvesting using active materials and conversion improvement by nonlinear processing. IEEE Trans. Ultrason. Ferroelectr. Freq. Control. **52**(4), 584–595 (2005)
9. Kumari, S., Sahu, S.S., Gupta, B.: Efficient SSHI circuit for piezoelectric energy harvester uses one shot pulse boost converter. Analog Integr. Circ. Sig. Process. **97**, 545–555 (2018)
10. Badel, A., Benayad, A., Lefeuvre, E., Lebrun, L., Richard, C., Guyomar, D.: Single crystals and nonlinear process for outstanding vibration-powered electrical generators. IEEE Trans. Ultrason. Ferroelectr. Freq. Control. **53**(4), 673–684 (2006)
11. Lallart, M., Guyomar, D.: An optimized self-powered switching circuit for non-linear energy harvesting with low voltage output. Smart Mater. Struct. **17**(3), 035030 (2008)
12. Ramadass, Y.K., Chandrakasan, A.P.: An efficient piezoelectric energy harvesting interface circuit using a bias-flip rectifier and shared inductor. IEEE J. Solid State Circuits. **45**(1), 189–204 (2010)
13. Wu, Y., Badel, A., Formosa, F., Liu, W., Agbossou, A.: Self-powered optimized synchronous electric charge extraction circuit for piezoelectric energy harvesting. J. Intell. Mater. Syst. Struct. **25**(17), 2165–2176 (2014)

75.5   Conclusion

This study presents a new interface for hybrid piezoelectric-electromagnetic energy harvesting system and a comprehensive analysis of its performance compared to the standard HPES, PZVEH and the piezoelectric energy harvesting system with P-SSHI interface. The effect of the new interface on the harvesting energy is analyzed by calculating efficiency and the output power. The proposed interface can provide high-level efficiency of 80.2% while the output power increases to 250 mW. As a result, the new interface can harvest up to three times more output power than the standard HVEH interface circuit.

14. Liang, J., Liao, W.-H.: Improved design and analysis of self-powered synchronized switch interface circuit for piezoelectric energy harvesting systems. IEEE Trans. Ind. Electron. **59**(4), 1950–1960 (2012)

15. Wang, S.-W., Ke, Y.-W., Huang, P.-C., Hsieh, P.-H.: Electromagnetic energy harvester interface design for wearable applications. IEEE Trans. Circuits Syst. II Express Briefs. **65**(5), 667–671 (2018)

16. Rahimi, A., Zorlu, O., Kulah, H., Muhtaroglu, A.: An interface circuit prototype for a vibration-based electromagnetic energy harvester. In: Energy Aware Computing (ICEAC), 2010 International Conference on, IEEE, pp. 1–4 (2010)

17. Szarka, G.D., Burrow, S.G., Proynov, P.P., Stark, B.H.: Maximum power transfer tracking for ultralow-power electromagnetic energy harvesters. IEEE Trans. Power Electron. **29**(1), 201–212 (2014)

18. Rahimi, A., Zorlu, Ö., Muhtaroğlu, A., Külah, H.: An electromagnetic energy harvesting system for low frequency applications with a passive interface ASIC in standard CMOS. Sensors Actuators A Phys. **188**, 158–166 (2012)

19. Rahimi, A., Zorlu, Ö., Muhtaroğlu, A., Külah, H.: A compact electromagnetic vibration harvesting system with high performance interface electronics. Procedia Eng. **25**, 215–218 (2011)

20. Sriramdas, R., Pratap, R.: An experimentally validated lumped circuit model for piezoelectric and electrodynamic hybrid harvesters. IEEE Sensors J. **18**(6), 2377–2384 (2018)

21. Yu, H., Zhou, J., Yi, X., Wu, H., Wang, W.: A hybrid micro vibration energy harvester with power management circuit. Microelectron. Eng. **131**, 36–42 (2015)

22. Uluşan, H., Chamanian, S., Pathirana, W., Zorlu, Ö., Muhtaroğlu, A., Külah, H.: A triple hybrid micropower generator with simultaneous multi-mode energy harvesting. Smart Mater. Struct. **27**(1), 014002 (2017)

23. Edwards, B., Aw, K.C., Hu, A.P., Tang, L.: Hybrid electromagnetic-piezoelectric transduction for a frequency up-converted energy harvester. In: Advanced Intelligent Mechatronics (AIM), 2015 IEEE International Conference on, IEEE, pp. 1149–1154 (2015)

24. Cheng, S., Wang, N., Arnold, D.P.: Modeling of magnetic vibrational energy harvesters using equivalent circuit representations. J. Micromech. Microeng. **17**(11), 2328–2335 (2007)

25. Yang, Y., Tang, L.: Equivalent circuit modeling of piezoelectric energy harvesters. J. Intell. Mater. Syst. Struct. **20**(18), 2223–2235 (2009)

26. Park, J.C., Bang, D.H., Park, J.Y.: Micro-fabricated electromagnetic power generator to scavenge low ambient vibration. IEEE Trans. Magn. **46**(6), 1937–1942 (2010)

27. Moore, H.: MATLAB for Engineers. Pearson (2017)

Online Competitive Control of Power-Down Systems with Adaptation

James Andro-Vasko and Wolfgang Bein

76.1 Introduction

In Information Technology energy consumption is an issue in terms of availability as well as terms of cost. According to Google [6] energy costs are often larger than hardware costs. Ways to minimize energy consumption are crucial and power usage has increasingly become a first order constraint for data centers. A growing body of work on algorithmic approaches for energy efficiency exists, see Albers et al. for a general survey [3].

To manage power usage, power-down mechanisms are widely used: for background on algorithmic approaches to power down see [1, 2, 8, 9]. In power down problems a machine needs to be in an on-state in order to handle requests but over time it can be wasteful if a machine is on while idling. To increase efficiency devices are often designed with power saving states such as a "hibernate state", a "suspend state" or various other hybrid states. Power down algorithms exists to control single machines or systems with multiple machines, such as in distributed machine environments.

Power-down is studied for hand-held devices, laptop computers, work stations and data centers. However, recent attention has been on power-down in the context of the smart grid [7]: Electrical energy supplied by sustainable energy sources is more unpredictable due to its dependence on the weather, for example. When renewables produce a surplus of energy, such surplus generally does not affect the operation of traditional power plants. Instead, renewables are throttled down or the surplus is simply ignored. But in the future

where a majority of domestic power would be generated by renewables this is not tenable. Instead it may be the traditional power plant that will need to be throttled down.

Power-down problems are studied in the framework of online competitive analysis, see [4,5]. In online computation, an algorithm must make decisions without knowledge of future inputs. Online algorithms can be analyzed in terms of competitiveness, a measure of performance that compares the solution obtained online with the optimal offline solution for the same problem, where the lowest possible competitiveness is best. Online competitive models the advantage that no statistical insights are needed, instead a worst case view is taken: this is appropriate as request in data centers, or short term gaps in renewable energy supply are hard to predict.

76.1.1 Our Contribution

In this paper we study power-down problems with three states using a budget, that is computed from a set of delay times between requests, to compute an updated delay time for future requests. We begin in Sect. 76.2, where we introduce the three state machine, in Sect. 76.3 we introduce the taper down mechanism, in Sect. 76.4, we show simulations using experimental parameters, and in Sect. 76.5 we conclude our work along with future work.

76.2 Three State Machine

In this section, we give a characterization of the three state system. In the three state problem, the states are ON, OFF, and INT where INT is some lower power intermediate state. The online algorithm has two switching times, one for when there is a switch from ON to INT which we denote as x_1 and one for when there is a switch from INT to OFF which we denote as x_2 (Table 76.1).

J. Andro-Vasko · W. Bein (✉)
Department of Computer Science, University of Nevada, Las Vegas, NV, USA
e-mail: androvas@unlv.nevada.edu; wolfgang.bein@unlv.edu

© Springer Nature Switzerland AG 2019
S. Latifi (ed.), *16th International Conference on Information Technology-New Generations (ITNG 2019)*,
Advances in Intelligent Systems and Computing 800,
https://doi.org/10.1007/978-3-030-14070-0_76

Table 76.1 Three state costs

State	Idle cost	Power up cost
ON	1	0
INT	$a \in (0, 1)$	$d \in (0, 1)$
OFF	0	1

The machine can handle the request only if it is in the highest power state, the ON state. If a request arrives while the machine is in the ON state, the machine will simply handle the request. If the machine is idle long enough where it switches to the INT state, where less power is consumed during that time, but then there is a small power up cost in order to switch to the ON state to handle the request. If the machine is idle long enough, idle for time longer than x_2, then the machine powers up from OFF to ON to handle the request, where powering from OFF to ON requires the highest power up cost. This is the online model used for the three state problem, and the worst case cost occurs when a request arrives right at the moment when the machine switches to a lower power state.

When measuring the competitive ratio of an online algorithm, we also need to consider the optimal offline algorithm. In an online algorithm, the input is not known in advance and thus needs to use a strategy to lower its cost. An offline algorithm knows the entire input in advance and thus can always schedule itself to determine the optimal cost. For example, if the next request arrives after an arbitrarily long amount of time, the machine can power down immediately. However, if it is not known when the next request arrives, then powering down too soon might yield a larger cost than it would if the machine rather had been idle for a brief amount of time. We then compare an online algorithm to this optimal cost offline algorithm, which we call the competitive ratio. Given the cost of an online algorithm $A(\sigma)$ and optimal offline algorithm $OPT(\sigma)$, where σ is the input sequence, we say A is c-competitive if the following holds $A(\sigma) \leq cOPT(\sigma)$.

The goal is to find an online algorithm that minimizes the competitive ratio c for the input sequence. In order to accomplish this, we try different strategies and compare them to the optimal offline cost. For the power down problem, we have various idle and power up costs which may have various competitive ratios, and we adjust the times in which we switch to lower power states. For our experimentation, we use various a and d values. From [4], it is proven that $a = 0.6$ and $d = 0.4$ yields the optimal online algorithm for the case when $a + d = 1$. For the case where we have $a = 0.6$ and $d = 0.4$, the following figures show the competitive ratios for various x_1 times.

We see that in Fig. 76.1, that the competitive ratio is minimized when $x_1 = 0.5$, when $x_1 \neq 0.5$ we see we obtain a larger competitive ratio shown in Figs. 76.2 and 76.3. The

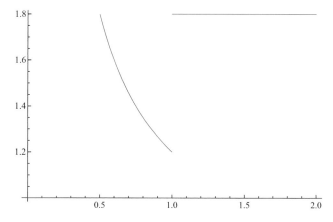

Fig. 76.1 Competitive ratio $x_1 = 0.5$

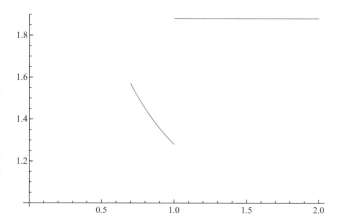

Fig. 76.2 Competitive ratio when $x_1 > 0.5$

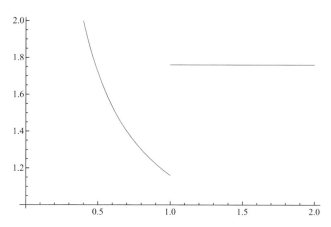

Fig. 76.3 Competitive ratio $x_1 < 0.5$

time at which we switch OFF is $x_2 = 1$, if $x_2 \neq 1$ then the competitive ratio would also increase, since the offline algorithm uses the OFF state if the request arrives at time x_2 or after, the proof can be seen in [4]. Using these optimal switch times for x_1 and x_2, we will introduce a tapering down approach that does adjust the switch times that will increase the competitive ratio slightly but does show better results in practice using input sequences generated by a normal random distribution rather than using worst case scenario for input. This concept holds for any value of a and d where there is only one possible value for x_1 and x_2 such that the competitive ratio is minimized.

76.3 Adjusted Delay Times

From the previous section, we determined that given a three state system where $a = 0.6$ and $d = 0.4$, we obtain the optimal online algorithm when $x_1 = 0.5$ and the competitive ratio is 1.8. For this apparatus, we increase the competitive ratio by a small amount, by $1.8(1 + \epsilon)$, where ϵ is an arbitrarily small constant. By adjusting the competitive ratio, we can obtain a different values for x_1 and x_2. We decrease the value of x_1 and x_2 by a small amount which will cause the competitive ratio to be $1.8(1+\epsilon)$, and using experimental request delay times, we determine if we are able to save energy as long as the request does not arrive right when we switch to the INT state or OFF state, using this updated x_1 and x_2 value which we will call y_1 and y_2. We have the cost function C

$$C(X_1, X_2, r) = \begin{cases} X_1 + d & \text{if } r \leq X_1 \\ X_1 + a(r - X_1) + d & \text{if } X_1 < r < X_2 \\ X_1 + a(X_2 - X_1) + 1 & \text{if } r \geq X_2 \end{cases}$$

Where r denotes the delay time between requests, we have x_1, x_2, y_1, and y_2 which we use to compute the cost and determine the amount of loss or gain we obtain using the updated switch times. The gain can be computed by

$$\text{gain} = C(x_1, x_2, r) - C(y_1, y_2, r)$$

We use the gain to update the budget amount, as the budget increases we adjust the values of y_1 and y_2. The budget increases or decreases based on the value of the gain after the most recent request. If the budget becomes negative, we reset it back to the value of 0 and thus the value of y_1 and y_2 reset back to their original value.

$$y_1 = \max\left\{ \frac{d - b}{CR(1 + \epsilon) - 1}, 0 \right\} \tag{76.1}$$

$$y_2' = \max\left\{ \frac{1 - b - dCR(1 + \epsilon) + y_1(1 - a)}{aCR(1 + \epsilon) - a}, 0 \right\} \tag{76.2}$$

$$y_2'' = \max\left\{ \frac{1 - b + y_1(1 - a)}{CR(1 + \epsilon) - a}, 0 \right\} \tag{76.3}$$

From Eqs. 76.2 and 76.3, the value of $y_2 = \max\{y_2', y_2''\}$.

Figure 76.4 shows the how the delay times decrease when $a + d = 1$ as the budget increases. The horizontal line shows the value the offline algorithm uses when deciding an initial

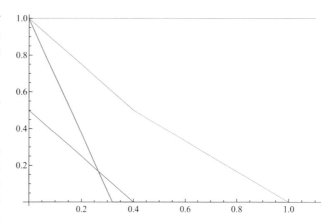

Fig. 76.4 y_1 and y_2 values $a = 0.6\ d = 0.4\ \epsilon = 0.001$

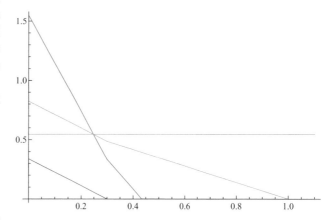

Fig. 76.5 y_1 and y_2 values $a = 0.45\ d = 0.3\ \epsilon = 0.001$

state based on the delay time for a request, this time allows the optimal algorithm to decide whether to use the ON state or INT state initially when the system becomes idle. There is also a decide time for whether the offline algorithm initially starts in the OFF state right when the machine begins idling which is the value of x_2. When $a + d = 1$ the y_2' and y_2'' curves always lie below the x_1 value, this is due to the fact that when $a + d = 1$, the offline decision time to use INT and OFF state are the same and thus the optimal cost does not use the INT state.

Figure 76.5, we have $a + d < 1$ and thus the y_2' and y_2'' curves both lie above the horizontal line. We choose the larger of the two curves and once the budget increases enough, both curves fall below the threshold and we switch to the other curve since the max values change. The reason for that behavior is when the y_2' and y_2'' curves fall below the horizontal line, the optimal cost changes which which causes the curve that was smaller to increase since the optimal cost curve goes through a sudden decrease.

Figure 76.6 uses the same parameters as from Fig. 76.5 except we use a larger ϵ value. This larger ϵ value causes a larger overall competitive ratio and also sets the adjust delay

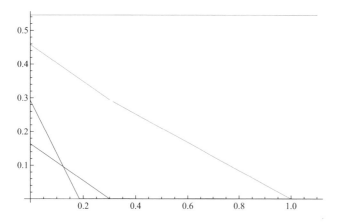

Fig. 76.6 y_1 and y_2 values $a = 0.45$ $d = 0.3$ $\epsilon = 0.1$

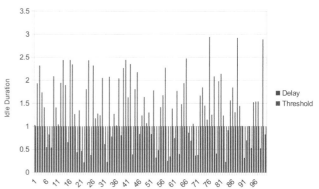

Fig. 76.7 Slack system 1

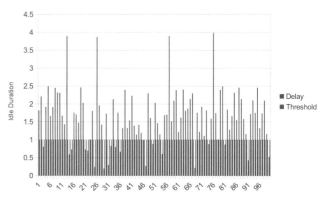

Fig. 76.8 Slack system 2

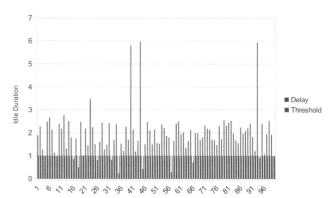

Fig. 76.9 Slack system 3

times to a smaller initial value. Once again since both y_2' and y_2'' curves both lie below the horizontal threshold curve. The next section will use this concept of using a budget to adjust the delay times using experimental data.

76.4 Experimental Results

For our experimentation, we use three states where $a = 0.6$ and $d = 0.4$ since it yields an optimal competitive ratio. For our simulation, we have 100 requests each with delay time between each using a normal distribution. Each request arrives either when the machine is in ON or INT, or when the machine is OFF. If a request occurs after the machine powers down we denote this as a slack request and if the request arrives before that then we denote this as a busy request.

76.4.1 Slack Systems

We experiment with inputs where we have more slack requests than busy request and then when we have more busy requests than slack requests. The figures below are a few input sets for various systems with more slack requests than busy requests.

In Fig. 76.7 we have our threshold, that distinguishes a slack request or a busy request, any request below the threshold is considered a busy request and anything above is considered a slack request. For this figure, we have twice as many slack requests than busy requests.

Figure 76.8, we have four times as many slack requests than we have busy requests. We increase the number of slack requests for analysis purposes, it is known if many requests arrive after the threshold, we have 1.8-competitive online algorithm, however we evaluate how the budget based tapering technique is effected if we have more slack requests than busy requests.

Figure 76.9, we have eight times as many slack requests than busy requests. We further increase the ratio of slack requests to busy requests to see if we obtain more favorable results if we further increase the amount of busy requests. Now we compare the costs of the 1.8-competitive strategy and the budget based delay time strategy.

Even though we increase the competitive ratio when applying the adjust wait times with the budget, we see that when we have random inputs, we obtain better costs than the 1.8-competitive strategy which is shown in Table 76.2. This is due to the fact that many requests arrive after the machine powers down, and powering down sooner when the request

Table 76.2 Costs using Fig. 76.7 input

ϵ	Budget based algorithm	1.8-competitive strategy
0.0005	107.88	151.81
0.005	113.52	
0.05	102.77	

Table 76.3 Costs using Fig. 76.8 input

ϵ	Budget based algorithm	1.8-competitive strategy
0.0005	108.79	172.27
0.005	105.58	
0.05	105.18	

Table 76.4 Costs using Fig. 76.9 input

ϵ	Budget based algorithm	1.8-competitive strategy
0.0005	116.83	185.55
0.005	105.77	
0.05	102.77	

is slack allows energy to be saved. We can also notice that when ϵ becomes larger, the energy consumption decreases since the delay times decrease at a more rapid rate.

Table 76.3 shows the costs when we have yet more slack requests. We obtain better results using the budget based algorithm. Once again, since most of the requests arrive after the machine is in the OFF state, switching to lower power states yields better results. The pattern remains the same, a higher ϵ value gives slightly minimal cost compared to a lower ϵ value.

Table 76.4, when the ratio of slack requests to busy requests increases even further, the budget based technique sees significant results. Overall, the budget based technique does not improve however the cost of the 1.8-competitive strategy increased considerably, which implies that the adjusting of the wait times using a budget is an improvement from the 1.8-competitive strategy. However, we only see when the inputs are mostly slack requests. The next section covers a system when the ratio between slack to busy is significantly lower.

76.4.2 Busy Systems

Let us now consider input where there are more busy requests than slack requests.

Figure 76.10 shows the requests where the majority of the requests lie below the threshold value which makes them busy requests. In this case we have four times as many busy requests over slack requests.

Figure 76.11 decreases the ratio of busy requests over slack requests, the amount of busy requests is twice as many slack requests, however, the ratio of busy to slack decreases

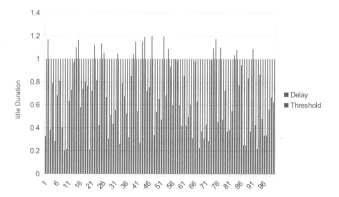

Fig. 76.10 Busy system 1

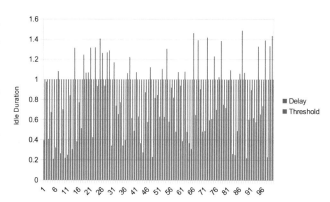

Fig. 76.11 Busy system 2

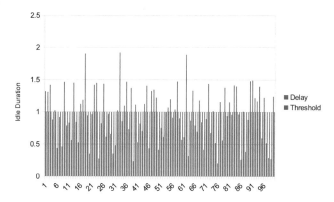

Fig. 76.12 Busy system 3

and we analyze if there is a pattern if the number of busy requests decrease to the overall power consumption using our strategy.

Figure 76.12 shows the input sequence where the ratio between slack and busy requests are relatively even. And we analyze the cost when we have a relatively even distribution of the different request types.

Table 76.5 shows that when the input sequence contains significantly higher number of busy requests over slack requests, the adjusted delay times produce a worse result compared to the 1.8-competitive strategy. Although the budget based technique is slightly worse, the 1.8-competitive

Table 76.5 Costs using Fig. 76.10 input

ϵ	Budget based algorithm	1.8-competitive strategy
0.0005	85.98	84.67
0.005	88.96	
0.05	94.08	

Table 76.6 Costs using Fig. 76.11 input

ϵ	Budget based algorithm	1.8-competitive strategy
0.0005	108.49	105.50
0.005	109.94	
0.05	97.84	

Table 76.7 Costs using Fig. 76.12 input

ϵ	Budget based algorithm	1.8-competitive strategy
0.0005	128.40	125.58
0.005	131.46	
0.05	117.75	

strategy is minimal. Also, we can notice that when ϵ is arbitrarily higher, the budget technique is actually worse than when ϵ is arbitrarily lower which contrasts when the input sequence contains more slack input, from the previous section. This behavior is due to the fact that since the requests are arriving when the machine is in either ON or INT state, powering down too soon actually increases the overall power consumption. In other words, when ϵ is arbitrarily lower, the budget based algorithm does not decrease the wait times too rapidly, and thus does not power down too soon which causes an extra power consumption since remaining in a higher power state for small amount of time saves more energy then to power down and power up right after each other.

Table 76.6 shows the cost when the amount of busy requests is twice the number of slack requests. The cost for the budget algorithm does improve slightly from the input sequence where there are four times as many busy requests over slack requests, from the previous figure. In this case, we see that when ϵ is increased, the budget based technique sees an improvement when the ϵ value is decreased. Once again, this is due to the fact that we have more slack requests and the budget technique powers down sooner which saves power consumption since powering down sooner for a slack request produces favorable results as shown in the previous section.

Table 76.7 shows an improvement of the budget based technique over the 1.8-competitive strategy when ϵ is arbitrarily smaller or larger. Similar to the slack input sets, a larger ϵ value yields to a decrease in the power consumption. As the number of slack requests increase, our technique yields favorable results.

76.5 Conclusions

We introduced our budget based algorithm and ran simulations on random input. Even though the competitive ratio is increased by a small constant, which yields worse results in the worst case, yield better results than the optimal algorithm in certain practical situations. We noticed that when the small constant ϵ is larger, then the initial delay times are decreased by a larger amount which implies that the machine would switch to lower power states earlier. When we have a set of inputs where the majority of the inputs arrive after the machine is in the OFF state, our algorithm yields a better cost than the optimal 1.8-competitive algorithm, which does not adjust its wait times. Since the inputs where mostly slack requests, powering down sooner caused energy to be conserved, however we only save if the requests arrive in that fashion, when the requests were mostly busy we did not see any power saved. Which implies that this technique is better used if we have a more slack system, and of course the budget based technique can adapt when the request types change and improve itself however is not the optimal algorithm over the 1.8-competitive strategy. Future work can be done to evaluate this budget technique on systems with more power states, say five power states. We can also apply this technique on a continuous state problem.

References

1. Agarwal, Y., Hodges, S., Chandra, R., Scott, J., Bahl, P., Gupta, R.: Somniloquy: augmenting network interfaces to reduce pc energy usage. In: Proceedings of the 6th USENIX Symposium on Networked Systems Design and Implementation, NSDI'09, Berkeley, pp. 365–380 (2009). USENIX Association
2. Agarwal, Y., Savage, S., Gupta, R.: Sleepserver: a software-only approach for reducing the energy consumption of pcs within enterprise environments. In: Proceedings of the 2010 USENIX Conference on USENIX Annual Technical Conference, USENIXATC'10, Berkeley, pp. 22–22 (2010). USENIX Association
3. Albers, S.: Energy-efficient algorithms. Commun. ACM **53**, 86–96 (2010)
4. Andro-Vasko, J., Bein, W., Nyknahad, D., Ito, H.: Evaluation of online power-down algorithms. In: Proceedings of the 12th International Conference on Information Technology – New Generations, pp. 473–478. IEEE Conference Publications (2015)
5. Augustine, J., Irani, S., Swamy, C.: Optimal power-down strategies. In: IEEE Symposium on Foundations of Computer Science, pp. 530–539. Cambridge University Press (2004)
6. Barroso, L.A.: The price of performance. ACM Queue **3**, 48 (2005)
7. Bein, W., Madan, B.B., Bein, D., Nyknhad, D.: Algorithmic approaches for a dependable smart grid. In: Latifi, S. (ed.), Information Technology: New Generations: 13th International Conference on Information Technology, pp. 677–687. Springer International Publishing (2016)

8. Nedevschi, S., Chandrashekar, J., Liu, J., Nordman, B., Ratnasamy, S., Taft, N.: Skilled in the art of being idle: reducing energy waste in networked systems. In: Proceedings of the 6th USENIX Symposium on Networked Systems Design and Implementation, NSDI'09, Berkeley, pp. 381–394 (2009). USENIX Association

9. Reich, J., Aman Kansal, M.G., Padhye, J., Padhye, J.: Sleepless in seattle no longer. Technical report, Microsoft (2010)

Laxmi Gewali and Binay Dahal

77.1 Introduction

Problems dealing with visibility on the surface of terrain have attracted the interest of many researchers in diverse scientific areas that include (i) geographic information systems, (ii) path planning for aerial vehicles, (iii) transportation networks, (iv) emergency response planning, and (v) wireless communications. In geographic information system (GIS) the topography of the terrain is modified by discretizing the surface by placing nodal points. These nodal points are selectively connected to obtain covering faces consisting of triangles and quadrilaterals. In a telecommunication network it is required to construct towers on the surface of terrain to cover a given region. Just placing towers on triangles having a high visibility index may need a prohibitively large number of towers. This issue has attracted the interest of many researchers from the algorithm community to develop efficient algorithms for covering a given terrain region of TIN with only a few numbers of towers. Most researchers have adopted the convention of line-of-sight communication for developing tower placement algorithms. In line-of-sight communications, two towers can directly communicate with each other if they are in each other's line of sight, i.e. the line segment connecting the top of the towers does not intersect with the terrain. In rare cases, towers not in line-of-sight, may be able to communicate by exchanging feeble signals. However, most researchers have adopted the line-of-

sight model as fairly good, adequate, and intuitive for many applications and we stick with this model.

In this paper, we examine algorithmic approaches for the placement of towers in terrain. Some versions of tower placement problems are known to be intractable [1] and consequently our motivation is in the development of tower placement approximation algorithms that are efficient in practice and easy to implement. In Sect. 77.2, we present a critical review of important algorithms reported in publication avenues. In Sect. 77.3, we present the main contribution of our work. We first formulate the Tower Placement Problem (TPP) and present an $O(n^2)$ algorithm that finds the location for placing a tower of given height to maximize the visible region in given 1.5D terrain. In Sect. 77.4, we discuss approaches for making the proposed algorithms efficient. In particular we discuss the need in characterizing transition points where tower placement are actually needed.

77.2 Review of Tower Placement Algorithms

The terrain surface satisfies an interesting structural property called 'the projection containment property' which can be elaborated as follows. The term **h-cross-section** is used to indicate the intersection between the terrain and a horizontal plane. The area of h-cross-section $I(h_i)$ at height h_i decreases monotonically as height h_i increases. Specifically, consider two h-sections $I(h_1)$ and $I(h_2)$ at height h_1 and h_2 (h_1 above h_2). The projection of $I(h_1)$ on the horizontal plane at h_2 is contained inside $I(h_2)$. This projection containment property has been used extensively to develop efficient algorithms for solving geometric problems on terrain. Due to the validity of the projection inclusion property, computational geometry investigators often refer to terrain as a geometric shape in two and half dimension or simply 2.5D-terrain: the dimensional-

L. Gewali (✉) · B. Dahal
Department of Computer Science, University of Nevada, Las Vegas, Las Vegas, NV, USA
e-mail: laxmi.gewali@unlv.edu; dahalb1@unlv.nevada.edu

© Springer Nature Switzerland AG 2019
S. Latifi (ed.), *16th International Conference on Information Technology-New Generations (ITNG 2019)*,
Advances in Intelligent Systems and Computing 800,
https://doi.org/10.1007/978-3-030-14070-0_77

Fig. 77.1 A 1.5D terrain

ity of a terrain is viewed between two dimensions (2D) and three dimensions (3D).

When the terrain is restricted to two dimensions, the surface becomes a monotone polygonal chain, monotone along the x-axis. It is noted that in a x-monotone chain Ch, any vertical line intersects with Ch in at most one point. Consequently, a terrain in two dimensions is viewed as a 1.5D-terrain. A 1.5D terrain is illustrated in Fig. 77.1.

77.2.1 Placement of Single Tower

One of the extensively investigated problems on terrain visibility is the placement of shortest tower(s) on the surface of 2.5D terrain so that all points on the surface are visible from the top of the tower. The problem can be formally stated as follows:

Shortest Tower Problem (STP)
Given: A 2.5D terrain L.

Question: Find a position $p_0(x,y)$ to place a shortest vertical tower on L so that all points on L are visible from the top of the tower.

It is noted that a point $p_i(x,y)$ on the surface of L is visible from the top point t_p of tower if the line segment connecting t_p to p_i does not intersect with the surface of L. Details about the concept of visibility can be found in O'Rourke's book [2]. One of the first algorithms for computing the shortest tower was reported by Sharir [3]. Sharir's paper establishes that STP reduces to the problem of computing the shortest distance between two polyhedrons L and S. Polyhedron S is formed by the intersection of half planes formed by the 2D faces of L. It turns out that while L is not a convex polyhedron, S is convex. For the purpose of clarity of presentation, we can illustrate the formation of S in 1.5D as shown in Fig. 77.2.

In the figure, we illustrate Sharir's idea in 1.5D terrain. The area below the terrain can be represented by a simple polygon (not necessarily convex) and the intersections of half planes is represented by a convex polygon which we call the reference polygon. In Fig. 77.2, the polygon representing terrain is filled with a darker shade and the convex region

Fig. 77.2 Illustrating S and L used in Sharir's algorithm

is filled with a lighter shade. To construct the reference polygon, Sharir [3] used the idea of intersecting rays that originate from segments of the terrain and extend above. Each of these rays defines a half plane (either to the left or to the right as appropriate). Specifically, for a ray proceeding to the north-east direction, the half plane is to the left of the ray. Similarly, for the rays proceeding to the north-west direction the half plane is to the right of the ray. The intersection of these half planes precisely forms the reference convex polygon. It is remarked that the reference convex polygon is unbounded. Sharir proved by geometric analysis that the point on the terrain that minimizes the distance to the reference polygon is the point where the shortest watch tower should be located. To sketch the resulting algorithm, three cases are distinguished. The first case is to find the distance between a vertex of the reference polygon and a line segment of the terrain. The second case is to find the distance between a vertex of the terrain and the reference polygon. Finally, the third case is to find the distance between a line segment of the reference polygon and a line segment of the terrain. While the first two cases can be solved easily in O($n \log n$) time by using a standard technique in computational geometry [4], the third case is slightly complicated and intricate point location techniques are used in [3] to obtain the distance in O($n \log^2 n$) time. It took another 9 years to obtain a faster algorithm for solving the shortest watch tower problem. Binhai Zhu [5] reported a faster algorithm. Binhai used Dobkin-Kirkpatrik's [6] hierarchical representation of convex polyhedron to store additional information on the polyhedron. This approach resulted in a faster algorithm which executes in O($n \log n$) time.

77.2.2 Placement of Two Towers

Illuminating terrain by the placement of two towers has been investigated. The problem can be formally stated as follows.

Two Tower Placement Problem (TTPP)

Given: A 2.5D terrain L.

Question: Find the placement of two towers of common smallest height to cover L.

This problem can be further stated in two version. In the first version (the discrete version), the base of the tower is restricted to be among the vertices of L. In the second version (called the continuous version) the base of the tower could be anywhere on the surface of the tower. As observed in Agarwal et al. [7], the optimal solution for the TTPP could be either on vertices or on interior points.

By using a parametric search technique, it is established in [7] that the discrete two-watchtower position can be determined in $O(n^2 \log^4 n)$ time, where n is the number of edges in 1.5D terrain. It is further shown in [7] that within the same time complexity, the semi-continuous version of the two-watchtower problem can be solved. It is remarked that in the semi-continuous version, one of the towers can be anywhere while the other is required to be placed at one of the vertices. For the continuous version of the two-watchtower problem, it is proved in [7] that the optimum placement points can be computed in $O(n^3 \alpha(n) \log^3 n)$ time, where $\alpha(n)$ is the inverse of the Ackermann function.

77.2.3 Intractability and Approximation

The Tower placement problem is closely related to the well-known art gallery problem [2] of computational geometry. In the art gallery problem, it is asked to find the minimum number of point guards inside a simple polygon so that any point in the interior of the polygon is visible to some point guard. It is noted that a point g_i sees a point p_j inside the polygon if the line segment (g_i, p_j) does not intersect with the exterior of the polygon. The standard art gallery problem is known to be NP-Hard [2]. A 1.5D terrain can be viewed as a part of a monotone polygon. The complexity of finding the minimum number of point guards to illuminate a 1.5D terrain, often called the Terrain Illumination Problem (TIP) was not settled for quite some time. Finally, in 2010, King and Krohn [1] were able to establish that TIP is NP-Hard. Some interesting approximation algorithms for solving TIP have been proposed. One of the first such algorithms was reported by Stephen Eidenbenz in [8]. The approach taken in this paper is the development of a relationship between the minimum set-cover (**minSet**) problem and TIP. The minimum set cover problem (minSet) is a well-known intractable problem [9] and a few approximation algorithms have been reported [9]. Specifically, in the minSet problem,

two sets (i) $E = \{e_1, e_2, \ldots, e_n\}$ and (ii) $S = \{s_1, s_2, s_3, s_m\}$ are given, where each s_i is a subset of E. The minSet problem asks to find a minimum subset S_0 of S such that every element of E is in at least one member of S_0. This problem is known to be NP-Hard [9] and an approximation algorithm with approximation ratio $\log n + 1$ is known [9]. In [8] the space above the 2.5D terrain is partitioned into 3D convex cells.

These cells can be viewed as set S in the minSet problem. Analyzing this approach, it is established in [8] that an approximation algorithm for solving TIP can be developed. The approximation ratio is also $O(\log n)$. The time complexity of the algorithm is $O(n^6)$. This algorithm is of theoretical interest and not efficient enough for practical application. Another approximation algorithm for covering 1.5D terrain is reported in [10]. This algorithm is based on placing point guards (watchtowers of zero height) at (i) the vertices of the convex hull CH(T) of terrain T and (ii) at the carefully selected vertices on sub-terrains defined by consecutive vertices of CH(T). A complicated case analysis is done in [10] to select desired vertices for placement in sub-terrains. It is reported in this paper that the resulting algorithm yields a constant factor approximation for placing guards on terrain T. It is however not clear about the value of the constant factor.

77.3 Placement Algorithms

77.3.1 Problem Formulation

We are given a 1.5D terrain T1 and a watch tower R1 of height h1. Find a placement of R1 so that the portion of T1 visible from the top of the tower is maximized. The problem is relevant when the height of the tower is not long enough to visibly cover the whole terrain. It was observed in Sect. 77.2 that the solution to a single tower placement problem need not be in one of the vertices of the terrain. When the solution is one of the interior points on the edge of the terrain it is not clear how to locate such a point. The placement problem can be formally stated as follows:

Tower Placement Problem (TPP)

Given: (i) A 1.5D terrain T1, (ii) A tower R1 of height h1.

Question: Find the location on the terrain to place tower R1 such that the portion of T1 visible from the top of the tower is maximized.

If we move the tower from the leftmost points in T1 to the right then the length of T1 visible from the top of R1 changes. At some intervals the change in visible length is gradual (increasing or decreasing), while at some intervals the change is abrupt and discontinuous.

Definition 1 (Transition Point): The placement point on the terrain that corresponds to a discontinuity in visible length is called transition point. In the terrain of Fig. 77.3, there are 12 transition points.

Definition 2 (Critical Points): The set of points on the terrain consisting of transition points and terrain vertices are called critical points.

Definition 3 (Basic Interval): The interval on the terrain between two consecutive critical points is referred to as a basic interval.

A transition point could be any point on the terrain. For a given terrain, the position of a transition point depends on (i) the structural shape of the terrain, and (ii) the height of the tower. Consider the change in visibility (portions of terrain) from the top of the tower as its placement point moves along the basic internal segment. Visible portions of terrain consist of several sub-segments (we refer to them **v-edges**). As the tower moves, some v-edges shrink and other v-edges expand, increasing or decreasing their lengths monotonically. This can be established as stated in the following lemma.

Lemma 1 *The change in the length of v-edges as the tower moves along the basic interval is increasing or decreasing monotonically.*

77.3.2 Computing Transition Points

Consider the image $\text{Im}(T_1, h_1)$ of terrain T_1, formed by lifting it by height h_1 of the tower. The image is shown in Fig. 77.3b, drawn in thin segments. From each peak points z_i of the terrain we can construct two grazing rays r_{left} and r_{right} that originate at the peak point and extend upward along the terrain edges incident on z_i. In Fig. 77.3c, grazing rays are drawn as dashed edges. The points of intersections between grazing rays and terrain image $\text{Im}(T_1, h_1)$ are referred to as guiding points. Guiding points are illustrated in Fig. 77.3d drawn as small red circles. We can project guiding points vertically downward on the terrain to obtain transition points, drawn as blue dots. A straightforward algorithm for computing transition points is to directly use their constructive definition. Such an algorithm can be described as follows:

The image chain $\text{Im}(T_1\ h_1)$ can be constructed by adding height h_1 of tower to the y-coordinates of terrain chain T_1. Specifically, if (x_i, y_i) is the co-ordinate of vertex v_i of Ti then the coordinates of the corresponding image $\text{Im}(T_1, h_1)$ is $(x_i,$

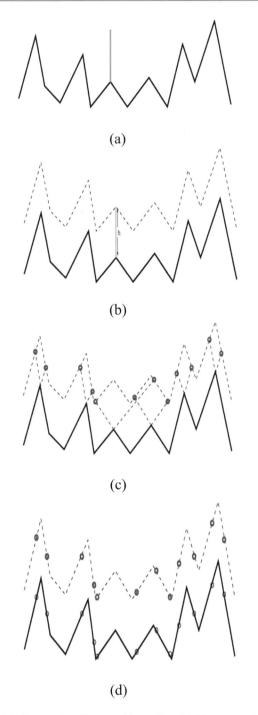

(a)

(b)

(c)

(d)

Fig. 77.3 Stages of finding transition points. (**a**) Placement of tower. (**b**) Lifting-up terrain image. (**c**) Rays from top-vertices. (**d**) Formation of marker points

$y_i + h_1$). Grazing rays from each vertex v_i of terrain T_1 can be constructed by using the slope of segments incident on v_i in constant time. We can then check for intersection between grazing rays and segments of image chain $\text{Im}(T_1, h_1)$. A formal sketch of our algorithm based on this straightforward approach is listed as Algorithm 77.1.

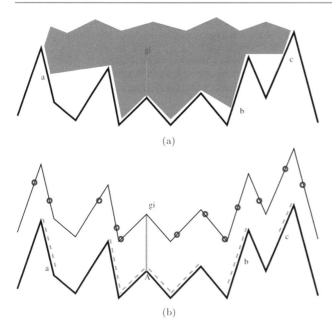

Fig. 77.4 Visibility polygon and parts of visible edges. (**a**) Visibility polygon VP(g_i) from point g_i. (**b**) Illustrating visible portions from position A

Algorithm 77.1

Straightforward Intersection Algorithm for Computing Transition Points

---

```
1:  Input: (i) Terrain T₁, (ii) Tower height h₁
2:  Output: Transition points U={u₁, u₂,...,uₖ}
3:  Construct Im(T₁,h₁) by lifting T₁ by h₁
4:  V = ∅
5:  for each n in N do
6:    Construct grazing rays r_left and r_right
         for z_i
7:    Let W_i be the intersection points between
         Im(T₁,h₁) and grazing rays r_left and r_right
8:    Add W_i to V
9:    Project points in V vertically downward to
         T₁ to obtain U
10: Output U
```

This happens when a grazing ray originating from a peak point intersects with almost all edges of the terrain's image. Such an input instance is depicted in the full version of the paper. Computing guiding points by using Algorithm 77.1 is rather slow. Observation 1 leads to the conclusion that Step 7 in Algorithm 77.1 can take O(n) time for computing intersection points corresponding to one pair of grazing segments. Since there are O(n) peaks, a straightforward approach for computing guiding points can take O(n^2) time.

By using the plane sweep technique of computational geometry [4], all guiding points can be computed more efficiently. Detail of this approach is in the full version of the paper.

77.3.3 Maximizing Tower Coverage

Once we have the critical points, we are ready to describe an algorithm for placing a tower T_l of given length h_l to maximize the coverage. Consider the visibility polygon.

VP(g_i) from a critical point gi as shown in Fig. 77.4a. The interior of VP(g_i) is shaded in the figure. The visibility polygon from a point inside a simple polygon can be computed in O(n) time [11]. The portion of T1 visible from g_i, denoted by L(T_l, g_i), can be extracted from VP(g_i) is straightforward manner. L(T_l, g_i) is indicated in Fig. 77.4b where the visible portions of terrain edges are indicated by dashed edges. The Visible Portion L(T_l, g_i)'s from all critical vertices are computed and we select the one that maximizes the length of the visible portions. A formal sketch of the algorithm is listed as Algorithm 77.2.

Algorithm 77.2

Placement to Maximize Coverage

```
Input: (i) Terrain T₁, (ii) Height h₁
Output: Placement point t' on T₁ that maximizes
         coverage
1:  Compute critical points g₁,g₂, ... ,gₘ
  using Algorithm 77.1
2:  for each point g_i do
3:      Compute Visibility polygon V P(g_i)
4:      Extract L(T₁, g_i) from V P(g_i)
5:  Set g' to g_i that maximizes L(T₁, g_i)
6:  Project down g' to T₁ to obtain t'
7:  Output t'
```

77.4 Conclusion

We presented a cursory review of existing algorithms for placing towers in 1.5D and 2.5D terrain. We proposed a novel method for discretizing the problem domain that resulted in the development of efficient algorithm for placing a tower of given length to maximize the coverage.

It was observed in Sect. 77.3 that the number of transition points could be quadratic in the number of vertices in the terrain. Not all transition points, as mentioned in Sect. 77.3, are necessary to search for the optimum placement. So, it would be interesting to reduce the number of transition points to make the proposed algorithms efficient. It would also be interesting to characterize 1.5D terrain for which the number of transition points is linear in the number of vertices of the terrain. The proposed algorithms are for 1.5D terrain. It would be a valuable exercise to extend our proposed algorithms to 2.5D terrain.

Another variation of the tower placement problem is the positioning of two towers of common height to maximize the coverage. We can call this problem 2 T-Max. A solution for 2 T-Max can be obtained by exploiting the structure of transition points formulated in Sect. 77.3. What we need is to check the coverage for all pairs of transition points and pick the one that maximizes the cover. It would be interesting to solve 2 T-Max efficiently without using all pairs of transition points.

References

1. King, J., Krohn, E.: Terrain guarding is np-hard. SIAM J. Comput. **40**(5), 1316–1339 (2011)
2. O'Rourke, J.: Art Gallery Theorems and Algorithms, vol. 57. Oxford University Press, Oxford (1987)
3. Sharir, M.: The shortest watchtower and related problems for polyhedral terrains. Inf. Process. Lett. **29**(5), 265–270 (1988)
4. O'Rourke, J.: Computational Geometry in C. Cambridge University Press, Cambridge (1998)
5. Zhu, B.: Computing the shortest watchtower of a polyhedral terrain in O($n \log n$) time. Comput. Geom. **8**(4), 181–193 (1997)
6. Dobkin, D.P., Kirkpatrick, D.G.: A linear algorithm for determining the separation of convex polyhedra. J. Algorithms. **6**(3), 381–392 (1985)
7. Agarwal, P.K., Bereg, S., Daescu, O., Kaplan, H., Ntafos, S., Zhu, B.: Guarding a terrain by two watchtowers. In: Proceedings of the Twenty-First Annual Symposium on Computational Geometry, ACM, pp. 346–355 (2005)
8. Eidenbenz, S.: Approximation algorithms for terrain guarding. Inf. Process. Lett. **82**(2), 99–105 (2002)
9. Garey, M.R., Johnson, D.S.: Computers and Intractability, vol. 29. W. H. Freeman, New York (2002)
10. Ben-Moshe, B., Katz, M.J., Mitchell, J.S.: A constant-factor approximation algorithm for optimal 1.5 d terrain guarding. SIAM J. Comput. **36**(6), 1631–1647 (2007)
11. De Berg, M., Van Kreveld, M., Overmars, M., Schwarzkopf, O.C.: Computational Geometry: Theory and Applications. Springer, Berlin Heidelberg (2000)

Prabhdeep Kaur Bhullar, Chary Vielma, Doina Bein, and Vlad Popa

78.1   Introduction

"What other people think" was always a better way to make informed decisions regarding purchasing, seeking services, or voting. Before Internet it included friends, family and neighbors, now due to Internet, it includes people we have never heard of. And conversely, more and more people are making their opinions available to strangers via the Internet Sentiment analysis is the task of identifying whether the opinion expressed in a text is positive or negative in general, or about a given topic. For example, "Django is such a good movie, highly recommends 10/10", expresses positive sentiment towards the movie, named Django, which is considered as the topic of this text [1]. Sometimes, the task of identifying the exact sentiment is not clear even in humans, for example in the text: "I'm surprised so many people put Django in their favorite films ever list, I felt it was a good watch but definitely not that good", the sentiment expressed by the author toward the movie is probably positive, but not as good as in the message that was mentioned above.

Sentiment analysis is a series of methods, techniques, and tools about detecting and extracting subjective information, such as opinion and attitudes, from language. Traditionally, sentiment analysis has been about opinion polarity, i.e., whether someone has positive, neutral, or negative opinion towards something. The object of sentiment analysis has typically been a product or a service whose review has been made public on the Internet. This might explain why sentiment analysis and opinion mining are often used as synonyms, although, we think it is more accurate to view sentiments as emotionally loaded opinions.

The goal of the project is to classify movies' reviews, by analyzing the polarity (positive or negative) of each paragraph in a review [2, 3]. We experimented with various RNN models on the Neon deep learning framework, an open-source framework developed by Nervana Systems [4], in order to improve accuracy in training and validation data. We experimented with network architecture, hyper parameters (batch size, number of epochs, learning rate, batch normalization, depth, vocabulary size) in order to find out which model works best for sentiment classification.

The paper is organized as follows. In Sect. 78.2 we present basic notions of sentiment analysis and how to setup Neon Nervana. A description of the project functionality is given in Sect. 78.3 together with analysis of the results. Concluding remarks and future work are presented in Sect. 78.4.

78.2   Related Work

The interest on other's opinion is probably almost as old as verbal communication itself [5]. Historically, leaders have been intrigued with the opinions of their subordinates to either prepare for opposition or to increase their popularity. Examples of trying to detect internal dissent can be found already at Ancient Greece's times. Ancient works in East and West mingle with these subjects. "The Art of War" has a chapter on espionage that deals with spy recruiting and betrayal, while in the beginning of "Iliad" the leader of Greeks Agamemnon tries to gauge the fighting spirit of his men. The first papers that used sentiment analysis among their keywords were published about a decade ago, but the field can trace its roots back to the middle of the nineteenth century. One of the pioneering resources for sentiment anal-

P. K. Bhullar · C. Vielma · D. Bein (✉)
Department of Computer Science, California State University, Fullerton, Fullerton, CA, USA
e-mail: prabhbhullar06@csu.fullerton.edu;
chary.vielma@csu.fullerton.edu; dbein@fullerton.edu

V. Popa
Liceul Tehnologic Petru Poni, Iași, Romania

© Springer Nature Switzerland AG 2019
S. Latifi (ed.), *16th International Conference on Information Technology-New Generations (ITNG 2019)*,
Advances in Intelligent Systems and Computing 800,
https://doi.org/10.1007/978-3-030-14070-0_78

ysis is the General Inquirer [5]. Although it was launched already in the 1960s, it is still being maintained.

Sentiment identification is a very complex problem, and thus much effort has been put into analyzing and trying to understand movies and product reviews, blogs and Twitter posts. As news stories have traditionally been considered neutral and free from sentiments, little focus has been on them. However, the interest in this domain is growing, as automated trading algorithms account for an ever-increasing part of the trade. A fast and simple method for determining the sentiment of a text is using a pre-defined collection of sentiment-bearing words and simply aggregating the sentiments found more advanced methods do not treat all words equally but assign more weight to important words depending on their position in the sentence.

Much work on analyzing sentiment and opinions in politically-oriented text focuses on general attitudes expressed through texts that are not necessarily targeted at a particular issue or narrow subject. For instance, the authors of [6] experimented with determining the political orientation of websites essentially by classifying the concatenation of all the documents found on that site. We group this type of work under the heading of "viewpoints and perspectives", and include under this rubric work on classifying texts as liberal, conservative, libertarian, etc., placing texts along an ideological scale, or representing Israeli versus Palestinian viewpoints. Although binary classification may be used, here, the classes typically correspond not to opinions on a single, narrowly defined topic, but to a collection of bundled attitudes and beliefs. This could potentially enable different approaches from polarity classification.

On the other hand, if we treat the set of documents as a meta-document, and the different issues being discussed as meta-features, then this problem still shares some common ground with polarity classification or its multi-class, regression, and ranking variants [5]. Indeed, some of the approaches explored in the literature for these two problems individually could very well be adapted to work for either one of them. In polarity classification, labels being considered are more about attitudes that do not naturally correspond with degree of positivity.

While assigning simple labels remains a classification problem, if we move farther away and aim at serving more expressive and open-ended opinions to the user, we need to solve extraction problems [5]. For instance, one may be interested in obtaining descriptions of opinions of a greater complexity than simple labels drawn from a very small set, i.e. one might be seeking something more like "achieving world peace is difficult" than like "mildly positive" [5]. In fact, much of the prior work on perspectives and viewpoints seeks to extract more perspective-related information (e.g., opinion holders). The motivation was to enable multi-perspective question answering, where the user could ask

questions such as "what is Miss America's perspective on world peace?", rather than a fact-based question (e.g., "who is the new Miss America?").

There has been a lot of work done in the field of sentiment analysis in natural language and movie reviews post. The research ranges from document level classification, contextual polarity disambiguation to topic based on sentiment classification. Initially, classification of sentiments has been done with traditional linear classification methods, named, Supper Vector Machines (SVM) and logistic regression [5] Also, there have been other classification techniques which have also been applied for Sentiment Analysis, which are, Maximum Entropy and Naive Bayes techniques. Since that time, there have been tremendous growth and improvement in the field of deep learning, which opened new ways to classify sentiments on text data.

Deep convolutional neural networks have also been used for sentiment of text data. Deep CNN has an advantage over the traditional techniques, because Deep CNN do not depend on extensive manual feature engineering and it extract the features automatically [1]. [S75] introduced a Recursive Neural Network which maps phrases through word embedding and a parse tree. CNNs have shown good results for the sentiment classification of movie reviews which outperforms all the above mentioned techniques.

In our project, we have worked on models based on RNN architecture using LSTM, BILSTM, RNN, BIRNN implemented using the Neon Nervana deep learning framework. These models, and different experimentations we performed on our model, gave us better results in comparison to other methods used for sentiment classification. Prior work on perspectives and viewpoints seeks to extract more perspective-related information (e.g., opinion holders). The motivation was to enable multi-perspective question answering, where the user could ask questions such as "what is Miss America's perspective on world peace?", rather than a fact-based question (e.g., "who is the new Miss America?").

There has been a lot of work done in the field of sentiment analysis in natural language and movie reviews post. The research ranges from document level classification, contextual polarity disambiguation to topic based on sentiment classification. Initially, classification of sentiments has been done with traditional linear classification methods, named, Supper Vector Machines (SVM) and logistic regression [5]. Other classification techniques have been applied for sentiment analysis, e.g. Maximum Entropy and Naive Bayes techniques. Since that time, there have been tremendous growth and improvement in the field of deep learning, which opened new ways to classify sentiments on text data.

Deep convolutional neural networks have also been used for sentiment of text data. Deep CNN has an advantage over the traditional techniques, because Deep CNN do not depend on extensive manual feature engineering and it extract the features automatically [1]. The authors of [7] introduced a Recursive Neural Network which maps phrases through word embedding and a parse tree. CNNs have shown good results for the sentiment classification of movie reviews which outperforms all the above mentioned techniques.

In our project, we have worked on models based on RNN architecture using LSTM, BILSTM, RNN, BIRNN implemented using the Neon Nervana deep learning framework. These models, and different experimentations we performed on our model, gave us better results in comparison to other methods used for sentiment classification.

78.3   Project Description

Our project uses the Large Movie Review Dataset collected and organized by Stanford University researchers for their research paper Learning Word Vectors for Sentiment Analysis [8]. The dataset consists entirely of movies reviews by users of the website Internet Movie Database (IMDB) [9]. It contains 12,500 negative samples for validation set, 12,500 positive samples for validation, 12,500 negative samples for training, 12,500 positive samples for training, 25,000 unlabeled samples for unsupervised training, a vocab file containing approximately 89,500 vocabularies, a bag-of-words for the labelled training set, and another bag-of-words for the unlabeled set for unsupervised learning.

The dataset and our models use the polar classification of negative and positive. IMDB uses a rating scale from 1 to 10 stars. All positive examples consist of ratings between 7 and 10, while all negative examples consist of ratings between between 1 and 4. Neutral ratings of 5 and 6 are excluded.

For our project, we only use the vocab file and the labelled datasets for training and validation. Each movie review in the dataset are variable in length – that is some reviews may be a short paragraph while others may consist of multiple paragraphs. These reviews by many different users and for many different movies have an extensive vocabulary [9].

To reduce the complexity of the datasets, all reviews are preprocessed before they enter the neural networks. For our models we limited vocabulary to 20,000 which is the size chosen by the RNN examples by Nervana Systems. Less frequent words are replaced by an Out-Of-Vocab (OOV) character. If a review was larger than 128 words, we truncated to 128 words. If a review was smaller than 128 words, we padded it with whitespace.

The steps of operating a deep learning model implemented using Neon Nervana are [10]:

1. Generate a backend – The backend defines where computations are executed in neon. Nervana supports both CPU and GPU (Pascal, Maxwell or Kepler architectures) backend. The backend we used for our project is NervanaCPU.

2. Load data – Neon supports loading of both common and custom datasets. Data should be loaded as a python iterator, providing one mini-batch of data at a time during training. In our models, we have tried loading data in two ways, manually using ArrayIterator (method of loading data in Nervana) and with in-built data loading API for IMDB movies dataset.

3. Specify model architecture – Create your model by providing a list of layers. For layers with weights, provide a function to initialize the weights prior to training.

4. Train Model – to train a model, provide the training data (as an iterator or through in-built API), cost function, and an optimization algorithm for updating the model's weights. To modify the learning rate over the training time, provide a learning schedule.

5. Evaluate – evaluate a trained model based on a validation dataset and a provided metric. In our case, we evaluate our model by performing inference on our model by giving a movie review one by one or by providing the entire test dataset to our inference model in the form of a loop.

For our dataset which is IMDB movie reviews dataset, Neon provides an object classes for loading, and sometimes pre-processing the data. The online source is stored in the "_ _init_ _" method. Initializers we used in our project includes Uniform and Glorot Uniform. Uniform initializer is used for uniform distribution from low to high. Glorot Uniform is used for uniform distribution from −k to k, where k is scaled by the input dimensions k = SQRT(6/(din + dout)), where din and dout refer to the input and output dimensions of the input tensor. We used Adagrad (Adaptive gradient) algorithm that adapts the learning rate individually for each parameter by dividing L2-norm of all previous gradients. Given the parameters θ, gradient ∇J, accumulating norm G, and smoothing factor ϵ, we use the update equations:

$$G' = G + (\nabla J)^2$$
$$\theta' = \theta - \frac{\alpha}{\sqrt{G'+\varepsilon}} \nabla J$$

where the smoothing factor epsilon prevents from dividing by zero. By adjusting the learning rate individually for each parameter, Adagrad adapts to the geometry of the error surface. Differently scaled weights have appropriately scaled update steps.

Our network models consisted of a word Embedding Layer, Long Short-Term Memory layers, bidirectional Long Short-Term Memory layers, Recurrent Layers, bidirectional Recurrent Layers, Dropout Layer, and Affine Layer. A brief description of the layers is given below:

1. LookupTable is a word embedding that maps from a sparse one-hot representation to dense word vectors. This embedding is learned from data.

2. rlayer could basically be any layer of our choice, for example, Long Short-Term Memory layer, Recurrent layer, bidirectional Long Short-Term Memory layer, bidirectional Recurrent layer.
3. RecurrentSum is a recurrent output layer that collapses over the time dimension of the LSTM by summing outputs from individual steps.
4. Dropout performs regularization by silencing a random subset of the units during training.
5. Affine is a fully connected layer for the binary classification of the outputs.

OpenBLAS and Numpy is not required for Neon to work, but it is a recommended installation to make use of the multithreaded operations [11].

VirtualEnv is a tool used to isolate our Python development environments. It is not required but it is recommended. It makes it easier to setup Neon and its dependencies in an isolated environment. This prevents problems involving dependency conflicts. Instructions are provided in the Python Guide.

We specifically used Python 2.7 for our Neon setup. Additionally, we recommend using PIP to install any Python libraries that Neon may ask for. Depending on the operating system configuration, some Python libraries may not be installed by default. Neon with notify the user of the missing module that need to be installed. We used "pip search <module-name>" and "pip install <module-name>" to search for and install modules. Note that the it is recommended to search the Python module before installing it, to make sure you install the correct module.

Once training is done, the system has saved your file with .o extension. We use is .o extension file to produce the error graphs for our models. We used the command "nvis -i mymodel101.o -o myplots" to store in neon/myplots the rror graph "mymodel101", with the name "cost-hist.html" file. This .html error graph can be viewed using web browser. The error graphs for various models are shown in Fig. 78.1.

We use three general methods to perform data loading. The first method loads data from a pickle file. The second method loads data from the file format tab-separated values. The third method loads data with the help of the IMDB class that Nervana provided for convenience.

The training parameters are presented in Table 78.1. Neon evaluation function shows that the LSTM network with embedding layer has the best training and validation accuracy.

We modified the RNN inference example to load any saved RNN model and perform inference on individual movie reviews. We ran some models through the program to see if the models effectively classified movie reviews on IMDB.

The performance of each model is shown in Table 78.2.

We noticed that some models frequently underperformed their predictions even though Neon's eval() function indicated validation accuracy rates around 85%. Some models predicted the wrong classes when given inputs from either the validation set or recent movie reviews on IMDB. Furthermore, many of the incorrect predictions had a probability for the wrong class. Many of the correct predictions had less than 60% probability for the correct class. We were unsure if our outputs were outliers as a result of a ~15% inaccuracy given a ~85% accuracy, so we decided to run our own evaluation on these models to confirm that the accuracy rates were below 85%.

We made another modification of the RNN inference example to load any saved RNN models and evaluate that model's accuracy. The Neon script allowed us to re-evaluate the accuracy of all trained models and compare the results with Neon's built-in evaluation function. The script can run inference on the entire validation set containing all 25,000 examples split evenly into two folders labelled "neg" and "pos". The script also provides the user the option to specify how many examples per set to run through our model evaluation. When we ran the LSTM models through our evaluation program, the accuracy dropped by more than 30% for the models that used the IMDB class to perform the data loading (see Table 78.3). Models that trained without the IMDB class did not have this problem. We have not discovered yet the cause of this drop.

78.4 Conclusion and Future Work

We trained various RNN models, including LSTM, multi-layer BIRNN, multi-layer BILSTM, single layer BILSTM, RNN with embedding layer, and LSTM with embedding layer. Following the training process, we used the Neon evaluation function to evaluate model accuracy of the training set and the validation set. The results showed that LSTM with embedding layer had the best accuracy for both categories.

However, when we passed new inputs into the LSTM w/ embedding network, we noticed that the predictions were wrong more often than not. We suspected that the model evaluation was inaccurate. So we wrote our own program to re-evaluate the model and confirm our suspicion. This program loads the saved model with its associated weights and performs inference on the entire data set. We logged the results and found that models that used the IMDB class had training and validation accuracies of approximately 52% which is far below 85%. However, trained models that performed data loading without the IMDB class had had high accuracy upon re-evaluation with our program.

We suspect that the cause could be either incorrectly loading saved models, using a faulty IMDB helper function, or an error in the Neon eval() function. We have not found

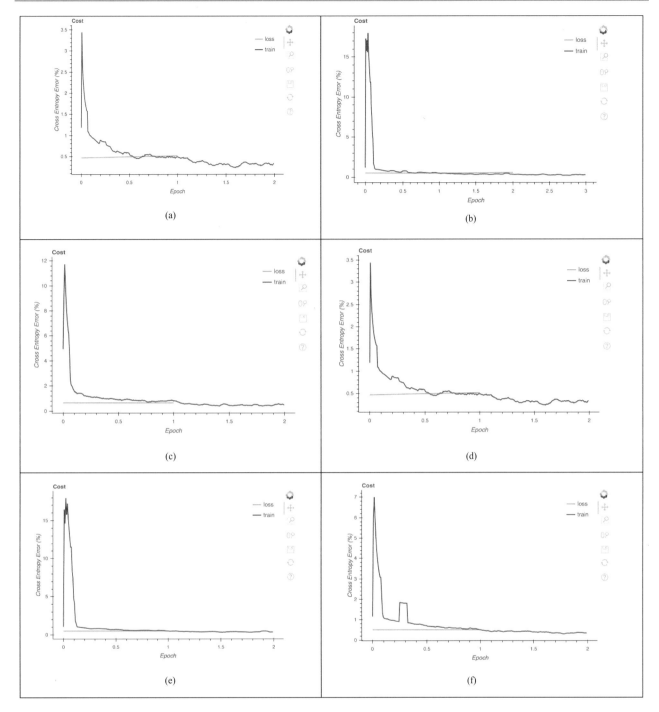

Fig. 78.1 Error graphs for various models. (**a**) Screenshot of interactive .html NVIS error graph. (**b**) Screenshot of interactive .html NVIS error graph for 2layer_bilstm. (**c**) Screenshot of interactive .html NVIS error graph for 2layer_birnn. (**d**) Screenshot of interactive .html NVIS error graph for 3layer_birnn. (**e**) Screenshot of interactive .html NVIS error graph for imdb_lstm. (**f**) Screenshot of interactive .html NVIS error graph for LSTM_embedding

Table 78.1 Training parameters

Model	Batch nor-malization	Learning rate	Epochs	Batch size
Imdb_lstm model	N/A	0.01	2	150
2-Layer BIRNN	False	0.01	2	200
3-Layer BIRNN	True	0.01	2	200
3-Layer BIRNN	True	0.05	5	200
LSTM w/ embedding layer	N/A	0.04	5	150
2-Layer BILSTM	N/A	0.02	2	150
RNN w/ embedding layer	N/A	0.01	2	128
1-Layer BILSTM	N/A	0.03	2	150
1-Layer BILSTM	N/A	0.01	2	150
imdb.p	N/A	0.01	2	32
jupyter-model.pkl	N/A	0.01	2	128
tutorial-model.pickle	N/A	0.01	2	1
mymodel16.pickle	N/A			

Table 78.2 Performance of each model

Network	Batch normalization	Learning rate	Train accuracy	Test accuracy
Imdb_lstm model	**N/A**	**0.01**	**94.03**	**84.03**
2 Layer birnn network	False	0.01	95.17	81.86
3 Layer birnn network	True	0.01	94.87	81.41
3 Layer birnn network	True	0.05	51.61	52.39
LSTM network with embedding layer	**N/A**	**0.04**	**95.6**	**85.6**
2 Layer bilstm	N/A	0.02	87.19	81.77
RNN network with embedding layer	N/A	0.01	94.88	85.13
1 Layer bilstm Network	N/A	0.03	95.93	84.93
1 Layer bilstm Network	N/A	0.01	95.73	85.36

the cause for this discrepancy, but we plan to select a data loading method that does not use the IMDB class, re-train all the models with the selected data loading method (method 1, 2), log the model accuracy, run our evaluation program csufnet-check-inference.py, and re-analyze our results. We

Table 78.3 Model accuracy: manual data loading vs IMDB class

	Evaluation method		
Trained models	Built-in evaluation function (%)	Re-evaluation (1000 × 2 files) (%)	Re-evaluation (12,500 × 2 files) (%)
imdb.p	85.1	84.1	84.8
jupyter-model	85.6	83.8	84.9
tutorial-model	~85	50.7	51.5
mymodel16 (lstm w/ embedding)	85.6	51.9	52.1

also plan to figure out if we are loading some saved models incorrectly or if the problem is a Neon issue. If our inference and evaluation program does not load these models correctly, we plan on writing a separate program to load only the models that are trained using the IMDB class.

References

1. Kowalska, K., Cai, D., Wade, S.: Sentiment analysis of polish texts. Int. J. Comput. Commun. Eng. **1**(1), 39–42 (2012)
2. Chowdhuri, I.K., Latif, S., Hossain, M.S.: Sentiment intensity analysis of informal texts. Int. J. Comput. Appl. **147**(10), 24–31 (2016)
3. Cui, Z., Shi, X., Chen, Y.: Sentiment analysis via integrating distributed representations of variable-length word sequence. Neurocomputing. **187**, 126–132 (2016)
4. Neon: Retrieved Oct. 28, 2018 [Online]. http://neon.nervanasys.com/docs/latest/ (n.d.)
5. Kabir, I., Latif, S., Saddam, M.: Sentiment intensity analysis of informal texts. Int. J. Comput. Appl. **147**(10), 24–31 (2016)
6. Grefenstette, G.: Explorations in Automatic Thesaurus Discovery. The Springer International Series in Engineering and Computer Science. Springer, New York (2012)
7. Socher, R., Lin, C.C.-Y., Ng, A.Y., Manning, C.D.: Parsing natural scenes and natural language with recursive neural networks. In: Proceedings of the 28th International Conference on Machine Learning
8. Maas, A.: Large movie review dataset. Retrieved May 18, 2017, from http://ai.stanford.edu/~amaas/data/sentiment (n.d.)
9. Harer, S., Kadam, S.: Sentiment classification and feature based summarization of movie reviews in mobile environment. Int. J. Comput. Appl. **100**(1), 30–35 (2014)
10. Neon IMDB sentiment classification implementation. Retrieved Oct. 26, 2018, from https://gist.github.com/nervanazoo/976ec931bb4549131ae0
11. https://hunseblog.wordpress.com/2014/09/15/installing-numpy-and-openblas/

Suyash Vardhan Singh and Rakeshkumar Mahto

79.1 Introduction

As technology is moving forward, feature size continues to shrink, and additional transistors are added following Moore's law. In order to continue scaling to reduce power consumption, the voltage also needs to be scaled by technology [1]. In recent times, the demand for low power computing in portable electronics, mobile devices, and wearable device etc. is on the rise. For any processor, an adder is one of the main building blocks. It participates in arithmetic logic unit, floating point unit, cryptography and many more. Most of the processors which are in use today, extensively use pipeline technique to improve the throughput. The implementation of a low power adder with low latency in a pipeline architecture is still a challenging problem. Improving the performance of pipelined adder circuits will improve the overall system performance.

One of the most critical aspects of designing a low power processor is dynamic power. Various techniques are used to reduce power consumption. One such technique is clock gating [2, 3] where a portion of the D-flip flop is shut off thereby not switching states. This results in a reduction of dynamic power. Another approach of reducing dynamic power is by using pre-defined dual voltage source (Vdd) and dual threshold voltage (Vt) [4]. By using clock gating and dual Vt technique in a 4-bit BCD adder, average power in a 4-bit BCD adder was reduced by 62.8% [5]. Additionally, dynamic scaling of voltage and frequency [6] with changing load conditions are among a few other techniques used in portable embedded systems that can increase throughput and

lower the dynamic power as per load requirement. Dynamic scaling of the frequency (DSF) is done through PLL which multiply the low frequency clock generated from the crystal oscillator. Dynamic scaling of the frequency and Dynamic Scaling Voltage (DSV) techniques were applied [7] on the sparse module 2^{n+1} Adder. Experimental data revealed that using DVS and DFS techniques lowered the power by 11.4% in sparse module 2^{n+1} Adder.

The throughput and latency are the other important factors to consider while designing a low power processor. The throughput can be improved by utilizing a pipelined architecture. For implementing a pipeline, the task is divided into subtasks. Pipeline registers are added between each subtask. This addition of registers at each stage increases the latency since most of these registers are single edge triggered (SET) FF. Latency issue can be improved by using dual edged triggered (DET) flip-flop (FF) based register. The DET flip-flop processes data at both rising and falling edge as compared to regular SET flip-flop. It is shown that DET FF is more energy efficient as compared to the SET FF [8]. Implementing a pipelined architecture on an adder circuit with DET FF based registers can mitigate pipeline latency and power consumption issues to some extent when compared to a conventional design.

Recently, many novel design methods have been proposed for DET FF to speed up the system with capturing data at both the clock edges. In work [9–11] DET FF is built using a lesser number of transistors to save power and area. In another technique [12–14], the authors generated a pulse at both edges for creating a DET FF. Both the techniques have an area overhead. In the latter technique, an increase in switching activity increases dynamic power consumption. It is also presented in [15] a high input activity to the DET FF may lead to higher power dissipation than single edged one. Instead of just DET FF, it is desirable to build a technique that can dynamically switch between DET to Single Edge

S. V. Singh · R. Mahto (✉)
Computer Engineering Program, California State University, Fullerton, Fullerton, CA, USA
e-mail: ramahto@fullerton.edu

© Springer Nature Switzerland AG 2019
S. Latifi (ed.), *16th International Conference on Information Technology-New Generations (ITNG 2019)*,
Advances in Intelligent Systems and Computing 800,
https://doi.org/10.1007/978-3-030-14070-0_79

Triggered (SET) Flip Flop depending on workload. In this work, we present a technique that can be used in an existing pipelined adder architecture with minimum area overhead. The technique allows dynamically switching between SET FF and DET FF without the need of a special dual-edged FF or a pulse generating circuit. The proposed technique can be easily implemented for reducing the latency without the requirement of replacing all the D-Flip Flop with DET FF.

The organization of rest of the paper is as follows. Section 79.2 describes the proposed shift register circuit and its comparative simulation with a conventional shift register. In Sect. 79.3 describe the implementation of the shifter circuit on a 4-bit pipelined parallel adder circuit. The conclusion of this paper is presented in Sect. 79.4.

79.2   Single/Dual Edged Shift Register

Inter-modular communication design is part of the overall System on Chip (SOC) design. There are two techniques used for data transfer between two modules, parallel data transmission and bit serial data transmission. In parallel data transmission technique data is sent through a bus. Whereas in serial data transmission technique the data is sent one bit at a time through a single wire. Although the bit serial data transmission is better compared to the parallel data transmission in terms of area, less leakage, fewer driver and better routability [16, 17], they are bounded by the clock frequency. Since the data moves from input to output terminal synchronously at one of the clock edges only this reduces the throughput. To solve this problem, the DET flip-flop can be used since they transfer data from input to output at both edges thereby increasing the throughput for the same frequency. For implementing DET flip-flop different techniques have been used in past that can be classified in three different categories (1) Conventional dual edge triggered flip-flops, (2) explicit pulsed DET flip-flop and (3) implicit pulsed flip-flops [8]. The conventional DET flip-flop utilizes negative and positive edge flip-flop using a multiplexer that let these FFs operate at adjacent clock edges. This technique is non-ideal since it requires more area and increase in switching activity that results in higher power consumption. The explicit pulsed DET flip-flop uses the pulse generating circuit to generate a pulse at both the rising and falling edges. Nevertheless, this technique imposes higher power consumption. Whereas, the implicit pulsed FF utilizes clock and delayed clock into two series device embedded in the latching part. However, this technique has lower performance due to deeper N-Type Metal Oxide Semiconductor (NMOS) stack and power overhead due to the pulse generating circuit.

However, none of the techniques in all the three categories can be switch between SET FF and DET FF based on load requirement and input switching activity to save power.

Moreover, for reducing the latency of pipeline DET FF will be required to replace all the D-Flip Flops in the design which has huge area overhead. Whereas the techniques presented in this paper uses the existing D-Flip Flop based register to mimic DET FF based register.

79.2.1 Proposed Switchable Register

The proposed technique uses a pair of D-flip flops as a register where the first one work as single edged (rising edge) flip-flop (SET FF) whereas the other acts as a switchable single edged flip-flop as shown in Fig. 79.1. When the signal M is set to low, then both flip-flops work on the same edge (rising edge). Whereas, when the signal M is set to a high, this results in clock inversion to the second FF which makes it capture data at the falling edge of the original clock signal. In this condition the flip-flop pair transfer data from input to output at both edges.

When M = '1', the first D-Flip flop holds the data until the falling edge is detected. As soon as falling edge is detected by the next switchable flip-flop, it reads the data and hold it until the arrival of next falling edge.

These pair of adjacent edge triggered flip-flop imitate a dual edge triggered flip-flop. The basic operation and truth table is shown in Table 79.1.

The minimum clock period for this configuration can be calculated by determining worst case logical delay between two adjacent registers. This is called critical delay. Let's assume the critical delay to be equal to T_{Max_logic}. The setup and hold time for the registers are T_{setup} and T_{hold} respectively. The minimum clock period required for the sequential circuit is given by

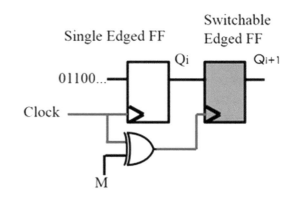

Fig. 79.1 Switchable DET and SET FF register pair

Table 79.1 Truth table

M	Edge switchable D-flip flop	Flip flop pair
0	Rising edge	Single edge shift register
1	Falling edge	Dual edge shift register

$$\frac{T}{2} \geq T_{C2Q} + T_{\text{Max\_logic}} + T_{setup} - T_{Xor\_gate}$$

For satisfying the hold time condition the contamination delay (T_{Min_logic}) or the minimum propagation delay through logic network plays a very important role. The hold time is given by,

$$T_{hold} \leq T_{C2Q} + T_{\text{Min\_logic}} - T_{Xor\_gate}$$

Where, T_{Xor_gate} is propagation delay across the XOR gate.

79.2.2 SPICE Simulation Result

The SPICE simulation for both the cases, SET FF and DET are shown in Fig. 79.2 is performed using the Berkeley Predictive Technology (BPTM) in a 45 nm CMOS technology [18]. The Clock to Q (C2Q) delay from clock at the normal D Flip Flop to the output terminal of the switchable flip-flop is improved by 95.1%.

The average power consumption of the modified shifter was measured to be 40.3 µW for clock operating at 250 MHz whereas for the regular shifter circuit it is decreased by small margin due to the presence of extra XOR gate for switching

the switchable FF. For simulation, the supply voltage (Vdd) for both the cases was considered 1.1 V.

79.3 Pipeline Adders

Pipelining is one of the technique which can be used to increase the throughput of a sequential set of distinct data inputs. For implementing pipelining requires including registers at each stage of the design. However, adding of the extra register for creating pipeline consumes more area and increases the latency. Different techniques were used to reduce the area and latency in the pipelined adder architecture. The overlapping clock is used on a conventional parallel pipelined adder [19] to reduce the sources of overhead [20]. For achieving high throughput, Lin et al. [21] uses proper scheduling a cascaded pipeline to eliminate data hazards. Carry propagation delay is one of the leading speed limiting factor in an adder design. For reducing the carry propagation delay across the pipelined adder, various techniques have being used [22, 23].

79.3.1 Modified 4-Bit Pipelined Parallel Adder

To speed up the execution time, we implemented switchable DET register in a 4-bit pipelined parallel adder [19] and compared the results obtained in both the cases.

A 4-bit pipelined parallel adder is shown in Fig. 79.3. As can be seen in the block diagram in Fig. 79.3 the register pair was replaced with the switchable SET FF to DET FF. As it is shown in Fig. 79.4, latency for M = '1' is one clock period when the shifter circuit is shifting the capture data at both the clock edges. Whereas, for M = '0' it is equal to 2 clock period which is same as the conventional 4-bit adder. This clearly present a decrease in the latency for M = '1'.

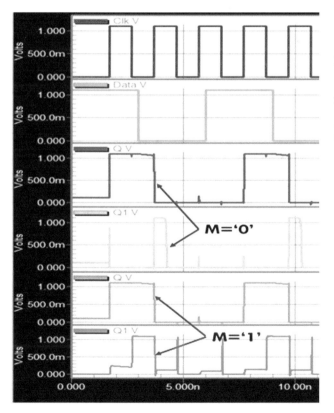

Fig. 79.2 The SPICE simulation of shifter circuit

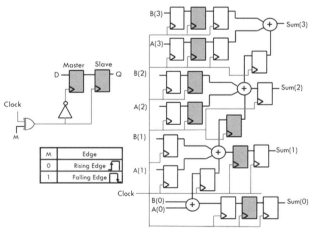

Fig. 79.3 Modified 4-bit pipelined parallel adder

Fig. 79.4 Circuit simulation result of modified 4-bit pipelined adder

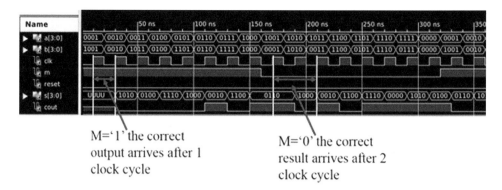

M='1' the correct output arrives after 1 clock cycle

M='0' the correct result arrives after 2 clock cycle

Table 79.2 Simulation result

	Area (μm^2)	Delay (ns)	Power (μW)	Power delay product (fs-W)
Pipeline adder [19]	293	2.0069	24.26	48.7
M = '0'	297	2.0069	26.471	53.1
M = '1'	297	1.0286	26.471	27.2

The latency is reduced by 50%. We implemented the design using 45 nm PD-SOI standard cell library [21] to compare the area overhead, power consumption, and delay. Table 79.2 summarizes the numerical result for the 4-bit adder with the regular register and modified register for M = '0' and M = '1'.

79.3.2 Simulation Result

The propagation delay presented in Table 79.2 is from clock to the output terminal of the switchable register pair. So the propagation delay or C2Q delay for M = '0' and conventional pipeline adder are same. The propagation delay for M = '1' is reduced by 95.1% compared to conventional 4-bit parallel pipelined adder.

The power consumption of modified 4-bit pipeline adder is increased by 9% due to the addition of XOR gates for making register switchable. Whereas, the area for implementing 4-bit adder is increased by 1.365% due to additional XOR gates. The power delay product for M = '1' is decreased by 44.16% compared to the convention 4-bit adder.

79.4 Conclusion

A switchable register proposed in this paper, can be switched between DET FF and SET FF depending on the load condition. The proposed register has an ability to reduce the latency by half with a minor expense in terms of area and power overhead. The proposed register reduces the latency in 4-bit pipeline adder circuit by 50% compared with conventional pipeline adder. This technique also reduced the power delay product by 44.16%.

References

1. Rabaey, J.M., Chandrakasan, A., Nikolic, B.: Digital Integrated Circuits, 2nd edn. Prentice Hall, Upper Saddle River, NJ (2003)
2. Donno, M., Ivaldi, A., Benini, L., Macii, E.: Clock-tree power optimization based on RTL clock-gating. In: Design Automation Conference, 2003. Proceedings, pp. 622–627 (2003)
3. Oliver, J.P., Curto, J., Bouvier, D., Ramos, M., Boemo, E.: Clock gating and clock enable for FPGA power reduction. In: 2012 VIII Southern Conference on Programmable Logic (SPL), pp. 1–5 (2012)
4. Li, F., Lin, Y., He, L., Cong, J.: Low-power FPGA using pre-defined dual-Vdd/dual-Vt fabrics. In: Proceedings of the 2004 ACM/SIGDA 12th International Symposium on Field Programmable Gate Arrays, New York, NY, USA, pp. 42–50 (2004)
5. Saha, D., Basak, S., Mukherjee, S., Sarkar, C.K.: A low-voltage, low-power 4-bit BCD adder, designed using the clock gated power gating, and the DVT scheme, 2013 IEEE Int. Conf. Signal Process. Comput. Control ISPCC, pp. 1–6 (September 2013)
6. Choi, K., Lee, W., Soma, R., Pedram, M.: Dynamic voltage and frequency scaling under a precise energy model considering variable and fixed components of the system power dissipation. In: IEEE/ACM International Conference on Computer Aided Design, 2004. ICCAD-2004, pp. 29–34 (2004)
7. Jagadeeswari, M., Surabhi, S.: A robust power downgrading technique using sparse modulo 2n+1 adder. Int. J. Adv. Res. Comput. Commun. Eng. **2**(3), (2013)
8. Tschanz, J., Narendra, S., Chen, Z., Borkar, S., Sachdev, M., De, V.: Comparative delay and energy of single edge-triggered and dual edge-triggered pulsed flip-flops for high-performance microprocessors. In: Low Power Electronics and Design, International Symposium on, 2001, pp. 147–152 (2001)
9. Lu, S.-L., Ercegovac, M.: A novel CMOS implementation of double-edge-triggered flip-flops. IEEE J. Solid State Circuits. **25**(4), 1008–1010 (1990)
10. Afghahi, M., Yuan, J.: Double-edge-triggered D-flip-flops for high-speed CMOS circuits. IEEE J. Solid State Circuits. **26**(8), 1168–1170 (1991)
11. Devarapalli, S.V., Zarkesh-Ha, P., Suddarth, S.C.: A robust and low power dual data rate (DET) flip-flop using c-elements. In: 2010 11th International Symposium on Quality Electronic Design (ISQED), pp. 147–150 (2010)

12. Nedovic, N., Walker, W.W., Oklobdzija, V.G., Aleksic, M.: A low power symmetrically pulsed dual edge-triggered flip-flop. In: Solid-State Circuits Conference, 2002. ESSCIRC 2002. Proceedings of the 28th European, pp. 399–402 (2002)

13. Young, S.P., Menon, S.M., Sodha, K., Carberry, R.A., Hassoun, J. H.: Double data rate flip-flop. US6525565 B2 (25 Feb. 2003)

14. Ghadiri, A., Mahmoodi, H.: Dual-edge triggered static pulsed flip-flops. In: 18th International Conference on VLSI Design, 2005, pp. 846–849 (2005)

15. Strollo, A.G.M., Napoli, E., Cimino, C.: Analysis of power dissipation in double edge-triggered flip-flops. IEEE Trans. Very Large Scale Integr. VLSI Syst. **8**(5), 624–629 (2000)

16. Lee, K., Lee, S.-J., Yoo, H.-J.: SILENT: serialized low energy transmission coding for on-chip interconnection networks. In: IEEE/ACM International Conference on Computer Aided Design, 2004. ICCAD-2004, pp. 448–451 (2004)

17. Dobkin, R.R., Morgenshtein, A., Kolodny, A., Ginosar, R.: Parallel vs. serial on-chip communication. In: Proceedings of the 2008 International Workshop on System Level Interconnect Prediction, New York, NY, USA, pp. 43–50 (2008)

18. Predictive Technology Model (PTM) [Online]. Available: http://ptm.asu.edu/. Accessed 14 Oct 2014

19. Ercegovac, M.D., Lang, T.: Digital Arithmetic, 1st edn. Morgan Kaufmann, San Francisco, CA (2003)

20. Sukumar, V., Pan, D., Buck, K., Hess, H., Li, H., Cox, D., Mojarradi, M.M.: Design of a pipelined adder using skew tolerant domino logic in a 0.35 mu;m TSMC process. In: 2004 IEEE Workshop on Microelectronics and Electron Devices, pp. 55–59 (2004)

21. Lin, M., Cheng, S., Wawrzynek, J.: Cascading deep pipelines to achieve high throughput in numerical reduction operations. In: Proceedings of the 2010 International Conference on Reconfigurable Computing and FPGAs, pp. 103–108 (2010)

22. Chan, P.K., Schlag, M.D.F., Thomborson, C.D., Oklobdzija, V.G.: Delay optimization of carry-skip adders and block carry-lookahead adders. In: 10th IEEE Symposium on Computer Arithmetic, 1991. Proceedings, pp. 154–164 (1991)

23. Huang, C.-H., Wang, J.-S., Yeh, C., Fang, C.-J.: The CMOS carry-forward adders. IEEE J. Solid State Circuits. **39**(2), 327–336 (2004)

A Genetic Algorithm for the Maximum Clique Problem

Rebecca Moussa, Romario Akiki, and Haidar Harmanani

80.1 Introduction

The *Maximum Clique Problem* is a combinatorial optimization problem that has been studied by various researchers due to its importance and wide application in the areas of social network analysis, telecommunication networks, bioinformatics, information retrieval, and computer vision.

Given an arbitrary graph $G = (V, E)$ where V is the set of vertices and E is the set of edges, a clique is a subset of vertices, $V' \in V$ such that $\forall (v_i, v_j) \in V' \times V'$ and $(v_i, v_j) \in E$ or $v_i = v_j$. The clique number of G, $\omega(G)$, is the size of maximal clique in G. The *maximum clique problem* is to find in an arbitrary graph a clique partition $|S|$ with maximal cardinality:

$$\omega(G) = \max\{|S| : S \text{ is a clique in G}\}.$$

The maximum weight clique problem is a variation of the maximum clique problem that finds cliques of maximum weight. The *Maximum Clique Problem* is formally defined as follows [1]:

Instance: Graph $G = (V, E)$
Solution: A clique in G, i.e., a subset of $V' \in V$ such that every two vertices in V' are joined by an edge in E.
Measure: Cardinality of the clique, i.e., $|V'|$

The maximum clique problem has been shown to be $\mathcal{NP}$-hard and the associated decision problem to be NP-complete [1]. Although many approximation algorithms were proposed, no algorithm with constant performance ratio has been found. Ausiello et al. [2] provide a proof of the problem's non-approximability. The maximum clique is not ap-proximable in the sense that no deterministic polynomial-time algorithm can find cliques of size $|V|^{\frac{1}{4}-\epsilon}$ for any $\epsilon > 0$. The problem is reducible to the maximum *3-satisfiability problem*, and has been shown to be equivalent to finding the *maximum independent set* on the complementary graph $\overline{G}$, as well as to the *maximum vertex packing*, and the *minimum vertex cover* [2]. The maximum clique problem can be found in linear time in *Chordal* graphs [3].

80.1.1 Related Work

Harary et al. [4] proposed the first serial algorithm for solving the maximum clique problem. The inductive algorithm worked by enumerating all cliques in an arbitrary graph. Bron et al. [5] proposed a greedy approach that finds maximal cliques through a depth-first search. The algorithm branches are formed based on candidate keys, and backtracks once a maximal clique is found. Later, Tomita et al. [6] proposed a modified *Bron-Kerbosch* algorithm with a worst case time complexity of $O(3^{\frac{n}{3}})$ for an $n-$vertex graph. Alusaifeer et al. [7] proposed a parallel method that finds the maximal clique in a graph based on enumeration. The algorithm was parallelized on GPUs using *CUDA*. Balas [8] proposed a branch and bound algorithm for finding a maximum clique in arbitrary undirected graphs. Feo et al. [9] proposed a parallel randomized heuristic for maximum independent set. The approach was effective in finding large cliques in randomly generated graphs. Homer et al. [10] experimented with various heuristic approaches and concluded that simulated annealing outperformed other heuristics for finding maximal cliques in a graph. Jogota [11] proposed several approaches for solving the maximum clique using the Hopfield Neural Network model. Ouyang et al. [12] proposed a molecular biology technique based on DNA computing to solve the maximum clique problem. Ji et al. [13] proposed

R. Moussa · R. Akiki · H. Harmanani (✉)
Department of Computer Science and Mathematics, Lebanese American University, Byblos, Lebanon
e-mail: haidar@lau.edu.lb

© Springer Nature Switzerland AG 2019
S. Latifi (ed.), *16th International Conference on Information Technology-New Generations (ITNG 2019)*,
Advances in Intelligent Systems and Computing 800,
https://doi.org/10.1007/978-3-030-14070-0_80

an algorithm for solving the maximum clique problem in *k-partite* undirected weighted graph. The algorithm finds significant stems conserved across at least *K* sequences in RNA secondary structure motifs. Bui et al. [14] proposed a genetic algorithm for solving the maximum clique problem based on a hybrid strategy. The results were promising. Fleurent et al. [15] proposed a hybrid genetic algorithm and tabu search. The results were encouraging, but the running time was quite high. Marchiori et al. [16] proposed a hybrid approach that combines the simple genetic algorithm and a naive greedy heuristic procedure. The algorithm outperforms previous genetic-based clique finding procedures over various DIMACS graphs, both in terms of quality of solutions and speed. Trefftz et al. [17] proposed an algorithm to find the maximal clique on interval graphs. The algorithm is based on channel assignment [18], where the maximum number of required channels is the same as the size of the maximum clique in the internal graph. The algorithm was parallelized on GPUs using *CUDA*.

This paper proposes a constructive genetic algorithm for solving the *maximum clique problem*. The algorithm is based on a combination of deterministic and stochastic moves. We report experimental results from the DIMACS problem instances whose size goes up to 1,500 vertices, and compare our results with the best known so far. The remainder of the paper is organized as follows. Section 80.2 formulates the genetic maximum clique problem and presents the chromosomal representation, genetic operators, and cost function. Section 80.3 presents the maximum clique genetic algorithm and discuss experimental results in Sect. 80.4. We conclude with remarks in Sect. 80.5.

80.2 Genetic Maximum Clique Problem

Genetic Algorithms are search techniques that are based on the principles of natural selection [19]. Genetic algorithms work by selecting during each generation individuals from the current population and applying transformations based on a certain probability in order to produce more fit off-springs. Just like the natural process, genetic algorithms subject the individual chromosomes to reproduction using operators such as *crossover* and *mutation*. At the end of each reproduction cycle, individuals are evaluated based on the fitness function, and chromosomes are copied to the next generation according to a probability proportional to their fitness.

In order to formulate the *maximum clique problem* using genetic algorithms, one needs to the specify the following components: (a) problem representation, (b) generation of an initial population, (c) fitness evaluation, (d) variation operators, and (e) parents selection mechanism. In what follows, we describe our formulation for the problem.

80.2.1 Chromosomal Representation

We solve the maximum clique problem using a vector-based chromosomal encoding. The length of the vector is equal to the number of vertices in the graph. A gene can take on either a 0 or a 1, depending on whether the vertex belongs to the clique. Figure 80.1a shows a simple graph along with the corresponding chromosome encoding in Fig. 80.1b. The shown solution represents clique $V_2 V_4 V_5 V_6$, a maximum clique in this case.

80.2.2 Initial Population

The solution quality is highly affected by the construction of the initial solution. The initial population is randomly generated as follows. The algorithm iterates over each chromosome and for each vertex V_i it randomly sets the corresponding gene to either 0 or 1. The algorithm next ensures that the solution is feasible using the *ExpandClique* operator, which we will explain next.

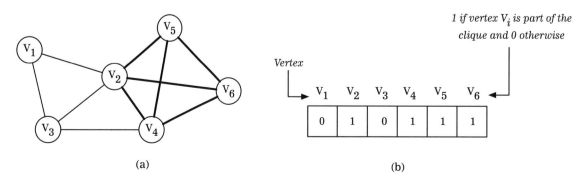

(a)   (b)

Fig. 80.1 (a) Example graph, (b) Maximal clique chromosomal encoding

80.2.3 Cost Function

The cost function is crucial for chromosomes survival from one generation to the next. Given a graph $G = (V, E)$, the cost of a solution is the size of the largest clique, $\omega(G)$:

$$\omega(G) = \sum_{i=0}^{n} w_i * V_i \qquad (80.1)$$

The algorithm accommodates weighted maximum cliques by simply adjusting the cost function as shown in Eq. 80.1.

80.2.4 Selection

The selection process is essential in order to increase the mean quality of solutions in the population. During every generation, the algorithm selects individuals for reproduction. We use the classical *roulette wheel* approach in order to ensure that we have a mating pool that is sampled based on a fitness proportionate probability distribution. The approach will guarantee that "healthy" individuals are selected in a higher probability than "weak" individuals. Thus, a random value $r \in [0, 1]$ is selected and compared with the respective cumulative probability q for the individual under consideration. If the random number falls in the interval $q_{i-1} < r \le q_i$, or if its less than the cumulative probability of the first individual when considering it, then the individual is selected.

Algorithm 80.1 Crossover($Parent_1$, $Parent_2$)

Input: $Parent_1$ and $Parent_2$
Output: *offspring*
 1: **for** $i = 0$ to n **do**
 2:     Offspring[i] = $Parent_1[i]$ & $Parent_2[i]$
 3: **end for**

80.2.5 Genetic Operators

We explore the design space using two genetic operators, *mutation* and *crossover*, which we apply iteratively with their corresponding probabilities. Both operators ensure that the generated solutions are feasible using the *ExpandClique* operator.

Mutation

The *mutation* operator is used in order to create a random, unbiased change while inhibiting premature convergence. The operator, Algorithm 80.3, randomly selects a gene in a chromosome and inverts its value based on a certain probability. The mutation operator is followed by the *deterministic* operator *ExpandClique* that will ensure the chromosome feasibility.

Crossover

The crossover operator is an essential operator in genetic algorithms that ensures necessary diversity within the population. The crossover operator selects two chromosomes and apply the bitwise and operator. Thus, only common edges are maintained in the solution. The algorithm is followed by the *ExpandClique* operator.

80.2.6 ExpandClique

The *ExpandClique* is an operator that aims at deterministically expanding the size of the clique that was found so far. The operator selects a random vertex V_k and iterates over the genes from vertices $k+1$ till n. At each iteration, the operator randomly sets gene i to 1 if the corresponding vertex V_i is pairwise connected to all vertices in the current clique. The final outcome is a larger clique.

Algorithm 80.2 ExpandClique(Offspring, V_k)

Input: $G = (V, E)$, Vertex V_k
Output: $V' = $ Expanded clique starting from V_k
 1: $V' = \emptyset$
 2: $V' = V' \cup V_k$
 3: **for** $i = k + 1$ to n **do**
 4:     **if** V_i is pairwise connected to every vertex in V' **then**
 5:         $V' = V' \cup V_i$
 6:         $V = V - V_i$
 7:     **end if**
 8: **end for**

Algorithm 80.3 Mutate($Parent$)

Input: $Parent$
Output: *Offspring*
 1: i = Random()
 2: *offspring[i]* = $\sim Parent[i]$

Algorithm 80.4 GA_MaxClique()

Input: $G = (V, E)$
Output: A Clique of size $\omega(G)$
 1: Create an initial population
 2: **for** i = 0 to N_g **do**
 3:     **for** each chromosome **do**
 4:         Calculate fitness value = $\frac{1}{\omega(g)}$
 5:     **end for**
 6:     **while** Not all chromosomes have been visited **do**
 7:         $parent\,1$ = RouletteWheel_Selection(population)
 8:         $parent\,2$ = RouletteWheel_Selection(population)
 9:         $r_1 \leftarrow random()$
10:         **if** ($r_1 < \alpha$) **then**
11:             $Offspring$ = crossover($parent\,1$, $parent\,2$)
12:             mark $parent\,1$ and $parent\,2$ as visited
13:             Select random vertex V_k
14:             ExpandClique(Offspring, V_k)
15:         **end if**
16:         $r_2 \leftarrow random()$
17:         **if** $r_2 > \beta$ **then**
18:             $offspring$ = mutate($Offspring$)
19:             Select random vertex V_k
20:             ExpandClique(Offspring, V_k)
21:             **if** $Offspring$ is better than either parent **then**
22:                 Replace the worst parent by the $offspring$
23:             **else**
24:                 Accept $offspring$ with a certain probability
25:             **end if**
26:         **else**
27:             Mutate $parent_1$ with probability β
28:             Mutate $parent_2$ with probability β
29:         **end if**
30:     **end while**
31: **end for**

80.3 Genetic Maximum Clique Algorithm

The proposed genetic maximum clique algorithm is shown in Algorithm 80.4. The algorithm starts by randomly creating an initial feasible population of chromosomes. During each generation, the cliques within the chromosomes are successively evolved using the crossover and mutation operators.

The operators are always followed by the the *ExpandClique* operator that randomly adds vertices to the clique in order to expand its size. The best chromosomes are maintained for the next generation. The algorithm repeats for N_g generations.

80.4 Experimentation Results

The proposed algorithm was implemented in *Java* and evaluated on the DIMACS clique benchmark graphs [20] which arise from various applications and problems. Table 80.1 illustrate all the benchmarks. We report the instance name, the best known solution, and the number of vertices and nodes in the graph. For reference purposes, we also report the median as well as the interquartile range (*iqr*) related to the graph degrees as well as the median and interquartile range of the degrees of the vertices lying in the best known solution. We have empirically determined that a population size of 20 and a generation number of 10,000 are sufficient to achieve good results. We have also experimentally determined the crossover probability to be 80% and the mutation probability to be 40%. We compare our results with the most optimal results obtained so far for the maximal clique problem. As it can be seen, our algorithm was able to find the optimal answer in most of the attempted cases (Table 80.1).

80.5 Conclusion

This paper presented a genetic algorithm for finding a maximal clique in an arbitrary graph, an $\mathcal{NP}$-hard problem. The proposed algorithm found optimal results in 21 of the attempted DIMACS benchmarks. The algorithm was also very competitive for the other results producing suboptimal answers in a very reasonable time.

Table 80.1 Experimental results

Instance	$\omega(g)$	Best known	Vertices	Edges	Graph degrees		Best degrees		Our solution	Error
					median	iqr	median	iqr		
brock200_2	12	12	200	9,876	99.0	(10.00)	101.0	(11.00)	12	0.00%
brock200_4	17	17	200	13,089	131.0	(8.00)	134.0	(6.00)	17	0.00%
brock400_2	29	29	400	59,786	299.0	(10.00)	299.0	(9.00)	25	13.79%
brock400_4	33	33	400	59,765	299.0	(11.00)	299.0	(9.00)	33	0.00%
brock800_2	24	24	800	208,166	521.0	(18.00)	516.5	(20.25)	20	16.67%
brock800_4	26	26	800	207,643	519.0	(18.25)	512.0	(20.25)	21	19.23%
C125.9	34*	34	125	6,963	112.0	(5.00)	114.5	(4.75)	34	0.00%
C250.9	44*	44	250	27,984	224.0	(6.00)	227.0	(5.00)	44	0.00%
C500.9	≥57	57	500	112,332	449.0	(9.00)	455.0	(9.00)	56	1.75%
C1000.9	≥68	68	1,000	450,079	900.0	(13.00)	907.0	(11.25)	64	5.88%
C2000.9	≥80	80	2,000	1,799,532	1800.0	(18.00)	1,803.0	(15.25)	70	12.50%
gen200_p0.9_44	44	44	200	17,910	180.0	(8.00)	179.5	(4.25)	44	0.00%
gen200_p0.9_55	55	55	200	17,910	179.0	(7.25)	179.0	(5.50)	55	0.00%
gen400_p0.9_55	55	55	400	71,820	360.0	(13.25)	359.0	(6.00)	52	5.45%
gen400_p0.9_65	65	65	400	71,820	361.0	(14.00)	359.0	(9.00)	65	0.00%
gen400_p0.9_75	75	75	400	71,820	359.0	(13.00)	359.0	(8.00)	75	0.00%
hamming10-4	40	40	1,024	434,176	848.0	(0.00)	848.0	(0.00)	16	0.00%
hamming8-4	16	16	256	20,864	163.0	(0.00)	163.0	(0.00)	40	0.00%
keller4	11	11	171	9,435	110.0	(8.00)	112.0	(17.00)	11	0.00%
keller5	27	27	776	225,990	578.0	(38.00)	578.0	(33.00)	27	0.00%
keller6	≤59	59	3,361	4,619,898	2,724.0	(50.00)	2,724.0	(50.00)	54	6.90%
p_hat300-1	8	8	300	10,933	73.0	(39.00)	103.0	(20.00)	8	0.00%
p_hat300-2	25	25	300	21,928	146.5	(73.00)	213.0	(18.00)	25	0.00%
p_hat300-3	36	36	300	33,390	224.0	(38.00)	251.0	(15.25)	36	0.00%
p_hat700-1	11	11	700	60,999	174.5	(87.00)	250.0	(22.50)	11	0.00%
p_hat700-2	44	44	700	121,728	353.0	(177.50)	508.0	(31.50)	44	0.00%
p_hat700-3	62*	62	700	183,010	526.0	(89.00)	602.0	(14.00)	62	0.00%
p_hat1500-1	12	12	1,500	284,923	383.0	(197.00)	509.0	(82.00)	12	0.00%
p_hat1500-2	65*	65	1,500	568,960	763.0	(387.00)	1,100.0	(37.00)	65	0.00%
p_hat1500-3	94*	94	1,500	847,244	1,132.5	(192.00)	1,297.5	(25.75)	93	1.06%

Table 80.2 Problem parameters

Parameters	Values
Number of generations	10,000
Crossover	0.8%
Mutation	0.4%

References

1. Garey, M.R., Johnson, D.S.: Computers, Complexity, and In-tractability. W. H. Freeman and Co, San Francisco (1979)
2. Ausiello, G., Crescenzi, P., Gambosi, G., Kann, V., Marchetti-Spaccamela, A., Protasi, M.: Complexity and approximation: combinatorial optimization problems and their approximability properties. Springer, Berlin (2000)
3. Gavril, F.: Algorithms for minimum coloring, maximum clique, minimum covering by cliques, and maximum independent set of a chordal graph. SIAM J. Comput. **1**, 180–187 (1972)
4. Harary, F., Ross, I.: A procedure for clique detection using the group matrix. Sociometry **20**(3), 205–215 (1957)
5. Bron, C., Kerbosch, J.: Algorithm 457: finding all cliques of an undirected graph. Commun. ACM **16**(9), 575–577 (1973)
6. Tomita, E., Tanaka, A., Takahashi, H.: The worst-case time complexity for generating all maximal cliques and computational experiments. Theor. Comput. Sci. **363**, 28–42 (2006)
7. Alusaifeer, T., Ramanna, S., Henry, C.J., Peters, J.: GPU implementation of MCE approach to finding near neighbourhoods. In: Lecture Notes in Computer Science (including subseries Lecture Notes in Artificial Intelligence and Lecture Notes in Bioinformatics) (2013)
8. Balas, E., Yu, C.S.: Finding a maximum clique in an arbitrary graph. SIAM J. Comput. **15**(4), 1054–1068 (1986)
9. Feo, T.A., Resende, M.G.C.: A greedy randomized adaptive search procedure for maximum independent set. Oper. Res. **42**, 860–878 (1994)

10. Homer, S., Peinado, M.: Experiments with polynomial-time CLIQUE approximation algorithms on very large graphs. In: Johnson, D., Trick, M. (eds.), Cliques, Coloring, and Satisfiability: Second DIMACS Implementation Challenge, pp. 147–167. American Mathematical Society, Providence (1996)

11. Jagota, A.: Adaptive, restart, randomized greedy heuristics for maximum clique. J. Heuristics **7**(21), 565–585 (2001)

12. Ouyang, Q., Kaplan, P.D., Liu, S., Libchaber, A.: DNA solution of the maximal clique problem. Science **278**, 446–449 (1997)

13. Ji, Y., Xu, X., Stormo, G.D.: A graph theoretical approach for predicting common RNA secondary structure motifs including pseudoknots in unaligned sequences. Bioinformatics **20**, 1591–602 (2004)

14. Bui, T.N., Eppley, P.H.: A hybrid genetic algorithm for the maximum clique problem. In: Proceedings of the 6th International Conference on Genetic Algorithms, pp. 478–484 (1995)

15. Fleurent, C., Ferland, J.: Object-oriented implementation of heuristic search methods for graph coloring, maximum clique, and satisfiability. In: Johnson, D.S., Trick, M. (eds.) Cliques, Coloring, and Satisfiability: Second DIMACS Implementation Challenge, DIMACS Series in Discrete Mathematics and Theoretical Computer Science. American Mathematical Society, Providence (1996)

16. Marchiori, E.: A simple heuristic based genetic algorithm for the maximum clique problem. In: Proceedings of the 1998 ACM Symposium on Applied Computing – SAC'98 (1998)

17. Trefftz, C., Santamaria-Galvis, A., Cruz, R.: Parallelizing an algorithm to find the maximal clique on interval graphs on graphical processing units. In: IEEE International Conference on Electro Information Technology (2014)

18. Dekel, E., Sahni, S.: Parallel scheduling algorithms. Oper. Res. **31**(1), 24–49 (1983)

19. Goldberg, D.E., Holland, J.H.: Machine Learning **3**, 95 (1988)

20. Second DIMACS Challenge on Cliques, Coloring and Satisfiability: http://iridia.ulb.ac.be/~scia/maximum_clique/DIMACSbenchmark (1993)

Strategies Reported in the Literature to Migrate to Microservices Based Architecture

81

Heleno Cardoso da Silva Filho and Glauco de Figueiredo Carneiro

81.1 Introduction

Microservices are a suite of usually small autonomous deployable and modular services running an unique process. They are network-accessible and communicate through well-defined, lightweight mechanisms to serve a business goal. Microservices can be effective to build complex software solutions in less time when compared to traditional software architectural solutions [1]. In fact, not only microservices but also the container-based approaches are associated with the boom of the so called cloud-native applications [2]. Microservices are a promising target to encourage the modernization of monolithic legacy applications to allow to take advantage of the benefits provided by cloud computing [3].

The aim of this work is to provide an updated overview regarding the lessons learned reported in the literature related to the migration and associated difficulties/challenges of this process. The migration from a monolithic architecture into microservices is not trivial due to decisions such as how to distribute the legacy functionalities into microservices and establish the dependencies among them in order to preserve their originality [4].

The remainder of this paper is organized as follows. The background and related work are presented in Sect. 81.2. Section 81.3 describes the steps we followed to find relevant studies in the literature. In Sect. 81.4, we discuss lessons

H. C. da Silva Filho
UnniRuy (Wyden Área 1), Salvador, Brazil
e-mail: heleno.filho@area1.edu.br

G. de Figueiredo Carneiro (✉)
Universidade Salvador (UNIFACS), Salvador, Brazil
e-mail: glauco.carneiro@unifacs.br

learned, difficulties, benefits and challenges of the migration process based on evidence obtained from the selected studies. Finally, in Sect. 81.5, we present conclusions and opportunities for future research.

81.2 Background

The migration to a microservice based architecture relies on the principle that the main focus must be on the services instead of infrastructure owner-ship [5]. Microservice, the software equivalent of toy bricks, have been enjoyed increasing popularity and diffusion in industrial environments to build complex solutions [1, 6]. The microservice architecture is the result of applying the single responsibility principle at the architectural level [4]. The high level of independence of microservices allows them to be separately deployable from each other. This enables that parts of the application can be changed and updated without affecting other parts [7]. Considering that the number of services involved in applications based on the microservices architecture can increase, manual deployment processes is no more effective in such architectures due to the frequency of new deployments. For this reason, automated deployment solutions such as continuous delivery pipelines can be an effective solution for this situation [7]. As a result of the services independence, the underlying technologies and adopted programming languages can be diversified. If one service can be implemented using Java EE, another one could be implemented using .Net, Ruby, or Node.js [7]. Microservice architectures illustrate the principle of *smart endpoints* and *dumb pipes* [8], where the lightweight and minimal middleware components such as messaging systems are *the dumb pipes* and the intelligence of each service is *the smart endpoint* [7].

S. Latifi (ed.), *16th International Conference on Information Technology-New Generations (ITNG 2019)*,
Advances in Intelligent Systems and Computing 800,
https://doi.org/10.1007/978-3-030-14070-0_81

81.3   Steps to Find Relevant Studies in the Literature

This section describes the steps to find relevant studies in the literature. We do not intend to perform a precise and rigorous literature review, based on well-defined and structured review protocols to extract, analyze, and document results [9]. Our goal is to perform a non-structured review process to gather information to answer the research questions of this paper as follows: **Research Question 1 (RQ1)**: *Which strategies have been reported in the literature to support the migration of legacy software systems to microservices-based architecture?* The knowledge of strategies applied both by researchers in the academia and practitioners in the industry to support the migration of legacy systems to microservices can be an opportunity to encourage them to embrace this challenge. **Research Question 2 (RQ2)**: *Which lessons learned have been reported in the literature regarding challenges and advantages perceived as a consequence of the aforementioned migration?* The reported challenges and advantages are key to improve mentioned strategies for the migration.

To answer the research questions, we performed a search in ACM Digital Library, Science Directory, IEEE Xplore and Springer scientific repositories to identify relevant studies published in top software engineering venues. A selection of keywords was made to perform the search in the mentioned repositories. We used an adjusted version of the PICOC method proposed by Petticrew and Roberts [10] as presented in Table 81.1.

We defined the following search string to identify the primary studies in the target scientific repositories to select studies published from 2008 to 2018: *((software or application or monolithic) and migration and microservice)*. Considering differences in the syntax of the target search engines, we adjusted the search string in each repository to fulfill their respective search requirements as presented in Table 81.2. All searches were conducted in May 4th, 2018. The result of the automated search returned a set of 95 studies (14 IEEE, 13 ACM, 25 SCD and 43 SPRINGER), from which we selected five studies that met the inclusion, exclusion and quality criteria as described below (Tables 81.3, 81.4 and 81.5). Moreover, we included seven studies based on suggestions of the authors (Tables 81.6).

Table 81.1 Defining the search string

Component	Definition
Population (P)	Monolithic, software, application
Intervention (I)	Migration
Outcomes (O)	Microservices

Table 81.2 Adjusted search strings for each repository

Repository	Adjusted search strings
ACM Digital Library	"query": ((software or application) and monolithic and microservice) "filter": "publicationYear": "gte":2008, "lte":2018 , owners.owner=HOSTED, acmPubGroups.acmPubGroup=Journal & Proceeding
IEEE Xplore	((software or application) and monolithic and microservice) and refined by Year: 2008–2018
Springer	((software or application) and monolithic and microservice) and refined by Year: 2008-2018
Science Directory	2008 and ((software or application) and monolithic and microservice) [All Sources(Computer Science)]

Table 81.3 Inclusion criteria

Inclusion criteria	Criteria
IC1	The articles should address difficulties in migrating legacy systems to cloud-based architecture in the presence of cloud computing, software architecture, legacy system, migration, change, evolution, strategies, approaches, techniques, type of change, change category, support, analyze AND
IC2	The papers are reported in peer reviewed conference or Journal AND
IC3	The papers are reported in peer reviewed conference or Journal AND
IC4	The publication date of the article should be between 2008 and 2018

Table 81.4 Exclusion criteria

Exclusion criteria	Criteria
EC1	Articles that do not address difficulties in migrating legacy systems to cloud-based architecture in the presence of cloud computing, software architecture, legacy system, migration, change, evolution, strategies, approaches, techniques, type of change, change category, support, analyze OR
EC2	The papers are not published in a peer reviewed conference or journal OR
EC3	The papers are not described in English OR
EC4	Date of publication of the article outside the period 2008 e 2018 OR
EC5	Duplicated reports of the same study available in different sources, consider the most complete version of the study

81.4   Results and Discussion

We based on the results of the selection process presented in Sect. 81.3 to list and classify the selected primary studies in Table 81.6 aiming at enabling a clear understanding of the lessons learned and findings reported to answer research questions **RQ1** and **RQ2**. Results obtained from both the automatic string search in the selected repositories and manual inclusion were maintained in a dataset. Studies with the same title, author(s), year of publication and abstract were considered duplicated and thus discarded. We organized the selected studies based on fields as follows: (i) identification number; (ii) year; (iii) title; (iv) objectives or aims; (v) strategies to support the migration of legacy software systems to microservices-based architecture (**RQ1**); and (vi) lessons learned regarding challenges and advantages perceived as a consequence of the aforementioned migration (**RQ2**). In the following section, we discuss findings to answer **RQ1** and **RQ2**. These findings are presented in Fig. 81.1.

81.4.1 Evidence to Answer RQ1

According to Fig. 81.1, 12 studies (S05, S06, S08, S79, S88, S96, S97, S98, S99, S100, S101, S102) proposed different strategies to support the migration of legacy software systems to microservices-based architecture. These strategies are presented in the following paragraphs.

S05 proposed formal coupling strategies and the clustering algorithm to support the migration. The strategy consisted in transforming the monolith application into the graph

Table 81.5 Quality criteria

Quality criteria	Criteria
QC1	Is the paper a primary study (or is it a review, secondary study or "lessons learned" document based on an expert point of view)?
QC2	Is there a clear statement of the goals of the research?
QC3	Is there an appropriate description of the context in which the research was performed?
QC4	Was the research design appropriate to address the aims of the research?
QC5	Was the data collected in a way that addressed the research issue?
QC6	Is there a clear statement of findings?

representation, while the clustering step proposes a new version of the graph representation of the monolith into microservice candidates [11]. The authors also proposed a quality evaluation that can support software architects to execute the approach according to their specific needs, making viable the reduction of the team size and lowering the domain redundancy of extracted services [11].

S06 discussed the use of a dataflow-driven mechanism as a systematic methodology for microservice-oriented decomposition. It is an algorithm based on a semi-automated process intended to reduce the complexity during the decomposition practices [12].

S08 proposed a methodology to convert a monolithic system into an architecture based in microservices. The methodology consists in a sequence defined by the phases: analysis and design, implementation, testing and continuous integration within an evolutionary life cycle [13].

S79 proposed a solution based on the semantic similarity of functionalities related to the OpenAPI specifications. Through the use of a reference vocabulary, the approach seeks for potential candidates for microservices, as fine-grained groups of cohesive operations (and their associated resources) [14]. The approach has as input an OpenAPI specification of the application that describes its different interfaces, operations, and resources and Schema.org is given as reference vocabulary [14].

S88 discussed the use of relevant requirements for the decomposition of services through the *Service Cutter*, a knowledge management method and supporting tool framework for microservice decomposition that requires as input a set of specification documents and a set of weighted coupling criteria [15]. The output is a graph representing candidate microservices as nodes, and how cohesive and/or coupled two candidates are through the use of weighted arcs. The authors emphasized the intention to support the decision making process instead of automating it completely [15]. Despite been proposed for generic services applications, the framework proposed in [15] can be an useful support the migration from legacy software monolithic systems towards microservices.

In S96, the authors described a manual migration based on the identification of Domains, Non-functional Require-

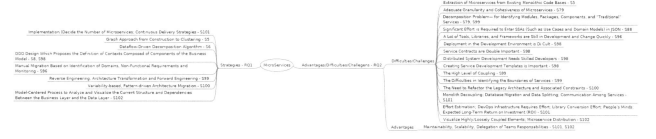

Fig. 81.1 Findings from the selected studies

ments and Monitoring, [16]. The authors of S97 used a panoramic view of the grey and informal literature to contextualize scenarios of the migration with a special emphasis on the concern of microservices' granularity. They selected three decision problem scenarios and proposed a solution based on the MAPE-K loop [17] regarding the migration [18].

S98 used the approaches of Domain Driven Design and Bounded Context to support the identification of potential functionalities to be converted into microservices [19]. The authors described the steps they used to the migration, when they also emphasized the use of continuous integration pipeline and continuous delivering [19].

The S99 paper presented the results of a survey with 18 practitioners from which the authors obtained data related to the migration to the microservices architecture. The authors recognized three main phases during the migration: (i) architecture recovery of the legacy system, (ii) architecture transformation from the legacy system to the new architecture, and (iii) the implementation of the new microservice-based architecture system [20].

The authors of study S100 proposed an approach called *Variability-based, Pattern-driven Architecture Migration (V-PAM)*, to support the migration method based on three items; (i) a catalogue of fine-grained service-based cloud architecture migration patterns focusing on multi-cloud scenarios, (ii) a framework to support the pattern selection and composition, and (iii) a variability model to guide the system migration towards a coherent framework [21]. The V-PAM approach considered empirical evidence and data from different migration projects, best practice from cloud architectures and a systematic literature review focusing on the theme [21].

In S101, the authors proposed a framework as a result of the analysis of three different migration processes reported by the interviewed practitioners and respective motivations and challenges faced throughout the migration process [22]. According to the authors, two of the analyzed processes targeted the migration of a legacy monolithic system to a microservice-based through the implementation of the new system from scratch. The third process aims at implementing new features as microservices to replace external services provided by third parties or develop specific features due to new changes that must be implemented to progressively replace the legacy system [22].

The authors of S102 proposed a model-centered process to analyze and visualize the current structure of a legacy software system and dependencies between their components or layers. The goal is to group functionalities into clusters and microservices supported by four different diagrams [4].

81.4.2 Answers to RQ2

The selected studies also provided evidence regarding advantages and challenges related to the migration to microservices (right side of Fig. 81.1).

A non exhaustive list of expected benefits over a traditional monolithic architecture are independence of deployability, language, platform and technology adoption to implement the microservices to accomplish scalability and flexibility from the architectural point of view [23]. The study S101 points out traceability, accountability and auditing as advantages that can be achieved through the use of microservices. It takes into account that the migration accomplished the isolation of business functionalities into microservices that interact among themselves through standardized interfaces [18]. According to S08 [13], the architecture of microservices facilitates the refinement of the limits of business logic, allowing the isolation of units to be tested, making them simpler and easier to understand and maintain.

However, according to S101 many practitioners are not confident to migrate due to the perception that microservice can be a hype and due to the lack of a well-know migration process [22]. In fact, during the migration process, practitioners often deal with common challenges and issues, mainly as a result of their lack of knowledge of best practices and patterns [22]. For example, S101 reported that the main issues associated to the migration are decoupling from the monolithic system, database migration, data splitting, and communication among services.

For medium-sized systems, the adoption of microservices can result in agility, quality improvement, cost reduction, and less time to market. For large cloud systems, they can represent a relevant change in terms of scalability, integration and release frequency [1]. Although microservices can provide substantial benefits, their implementation requires extra machinery and resources, which can impose substantial costs [1]. It is worth to mention that some aspects of this migration are still blurred for practitioners. For example, establishing the appropriate level of granularity and implementing an effective trade-off between size and number of microservices requires in fact expertise to accomplish them [18].

According to [24], the main issue regarding the migration is related to the separation of functionalities intro microservices, in other words, the extraction of microservices from existing monolithic code bases [11], especially in the cases in which the modules are tightly coupled [25]. To accomplish these goals, authors argue that there is the need for tools to automatically to deploy, scale and manage microservices, as well as to log and monitor them. Authors have also mentioned organizational challenges related to the migration as the need for more freedom for teams to implement DevOps tasks [24].

The authors of S96 argued that the migration to a microservice based architecture requires high effort due to the need to analyze every small part of the whole system individually and hence decide what should be converted to microservice [16].

S79 argued that the use of Service Cutter [15], a state-of-the-art tool for microservice decomposition, requires a set of specification artifacts and other specification artifacts containing coupling criteria [15]. In most of the cases, the availability of these documents and respective data is arguable [14].

In S98 [19], the authors argued that the migration to the microservices architecture is not a trivial task. For example, the deployment of the new implemented microservices in the development environment can be difficult, especially for novices. Moreover, the authors explain that service contracts are of vital importance and small changes in the contracts can impact part or the whole system. They suggest the use of service versioning to deal with this problem. The authors conclude the paper stating that microservices is not a silver bullet [19].

The authors of S99 [20] reported that the main challenge was the high level of coupling among the modules and/or components of the legacy system. This implies that the more the modules are coupled among themselves, the more difficult it is to extract functionalities from the legacy system [20]. The identification of service boundaries is another challenge mentioned by the same authors. The need for a *different* mindset for developer was reported by the participants of the study. For example, developers were used to get everything in one single database. In case they needed to get data, they would just query it from the appropriate table. With the distribution of the persistence layer through the microservices, they need to do an HTTP call, including authentication and identification to get this data [20].

In S101 [22], the authors present a list of issues and challenges regarding the migration to microservices based on data obtained from a survey with practitioners. The list is as follows: monolith decoupling, database migration and data splitting, communication among services, effort estimation, DevOps infrastructure requires effort, library conversion effort, peoples minds, expected long-term return on investment (ROI) [22]. As can be seen, many of these items were already mentioned in previous paragraphs, revealing that the results presented in [22] are in line with the other studies.

81.5 Conclusions and Future Work

In this paper, we presented a structured body of knowledge to characterize lessons learned, as well as difficulties and challenges related to the migration from a monolithic software application to a microservice based architecture.

The goal of this study is to identify main successful strategies and corresponding challenges reported in the literature during the migration of legacy software systems to microservices.

As future work, we intend to conduct a mapping study and a survey in the industry to identify which guidelines have been adopted by practitioners and compare them with guidelines reported in the literature.

Appendix

Table 81.6 Selected primary studies

Selected studies	Repository
S5 – Extraction of microservices from monolithic software architectures	IEEE Xplorer
S6 – From monolith to microservices: a dataflow-driven approach	IEEE Xplorer
S8 – Methodology to transform a monolithic software into a microservice architecture	IEEE Xplorer
S79 – Microservices identification through interface analysis	Springer
S88 – Service cutter: a systematic approach to service decomposition	Springer
S96 – Highly-available applicationss on unreliable infrastructure - a microservice architectures in practice	Manual
S97 – Microservices and their design trade-offs a self-adaptive roadmap	Manual
S98 – Migrating to cloud-native architectures using microservices: an experience report	Manual
S99 – Migrating towards microservice architectures: an industrial survey	Manual
S100 – Pattern-based multi-cloud architecture migration	Manual
S101 – Processes, motivations, and issues for migrating to microservices architectures: an empirical investigation	Manual
S102 – Towards the understanding and evolution of monolithic applications as microservices	Manual

References

1. Singleton, A.: The economics of microservices. IEEE Cloud Comput. **3**(5), 16–20 (2016)
2. Kratzke, N., Quint, P.-C.: Understanding cloud-native applications after 10 years of cloud computing-a systematic mapping study. J. Syst. Softw. **126**, 1–16 (2017)
3. Furda, A., et al.: Migrating enterprise legacy source code to microservices: on multitenancy, statefulness, and data consistency. IEEE Soft. **35**(3), 63–72 (2018)
4. Escobar, D., Cárdenas, D., Amarillo, R., Castro, E., Garcés, K., Parra, C., Casallas, R.: Towards the understanding and evolution of

monolithic applications as microservices. In: Computing Conference (CLEI), 2016 XLII Latin American, pp. 1–11. IEEE (2016)

5. Reza Bazi, H., Hassanzadeh, A., Moeini, A.: A comprehensive framework for cloud computing migration using meta-synthesis approach. J. Syst. Softw. **128**, 87–105 (2017)

6. Taibi, D., Lenarduzzi, V.: On the definition of microservice bad smells. IEEE Softw. **35**(3), 56–62 (2018)

7. Leymann, F., Breitenbücher, U., Wagner, S., Wettinger, J.: Native cloud applications: why monolithic virtualization is not their foundation. In: Cloud Computing and Services Science, pp. 16–40. Springer, Cham (2016)

8. Fowler, M.: Microservices resource guide. Martinfowler. com. Web 1 (2015)

9. Wohlin, C., Runeson, P., Höst, M., Ohlsson, M.C., Regnell, B., Wesslén, A.: Experimentation in Software Engineering. Springer Science & Business Media, Berlin/Heidelberg (2012)

10. Petticrew, M., Roberts, H.: Systematic Reviews in the Social Sciences: A Practical Guide. Blackwell Publishing CrossRef Google Scholar, Malden (2006)

11. Mazlami, G., Cito, J., Leitner, P.: Extraction of microservices from monolithic software architectures. In: 2017 IEEE International Conference on Web Services (ICWS), pp. 524–531 (2017)

12. Chen, R., Li, S., Li, Z.: From monolith to microservices: a dataflow-driven approach. In: 2017 24th Asia-Pacific Software Engineering Conference (APSEC), pp. 466–475 (2017)

13. Acevedo, C.A.J., Y Jorge, J.P.G., Patio, I.R.: Methodology to transform a monolithic software into a microservice architecture. In: 2017 6th International Conference on Software Process Improvement (CIMPS), pp. 1–6 (2017)

14. Baresi, L., Garriga, M., Renzis, A.D.: Microservices identification through interface analysis. In: Service-Oriented and Cloud Computing, pp. 19–33. Springer, Cham (2017)

15. Gysel, M., Klbener, L., Giersche, W., Zimmermann, O.: Service cutter: a systematic approach to service decomposition. In: Service-Oriented and Cloud Computing, pp. 185–200. Springer, Cham (2016)

16. Richter, D., Konrad, M., Utecht, K., Polze, A.: Highly-available applications on unreliable infrastructure: microservice architectures in practice. In: Software Quality, Reliability and Security Companion (QRS-C), 2017 IEEE International Conference on, pp. 130–137 (2017)

17. De Lemos, R., Giese, H., Müller, H.A., Shaw, M., Andersson, J., Litoiu, M., Schmerl, B., et al.: Software engineering for self-adaptive systems: a second research roadmap. In: Software Engineering for Self-Adaptive Systems II, pp. 1–32. Springer, Berlin/Heidelberg (2013)

18. Hassan, S., Bahsoon, R.: Microservices and their design trade-offs: a self-adaptive roadmap. In: Services Computing (SCC), 2016 IEEE International Conference on, pp. 813–818. IEEE (2016)

19. Balalaie, A., Heydarnoori, A., Jamshidi, P.: Migrating to cloud-native architectures using microservices: an experience report. In: European Conference on Service-Oriented and Cloud Computing, pp. 201–215 (2015)

20. Paolo Di Francesco, I.M., Lago, P.: Migrating towards microservice architectures: an industrial survey. IEEE Cloud Computing bibtex em 12052018 ainda nao disponivel (2018)

21. Jamshidi, P., Pahl, C., Mendona, N.C.: Pattern-based multi-cloud architecture migration. Softw. Pract. Exp. **47**(9), 1159–1184 (2017)

22. Taibi, D., Lenarduzzi, V., Pahl, C.: Processes, motivations, and issues for migrating to microservices architectures: an empirical investigation. IEEE Cloud Comput. **4**(5), 22–32 (2017)

23. Lewis, J., Fowler, M.: Microservices: a definition of this new architectural term (2014). http://martinfowler.com/articles/ microservices. html (cit. on p. 26) (2017)

24. Kalske, M., Mkitalo, N., Mikkonen, T.: Challenges when moving from monolith to microservice architecture. In: Current Trends in Web Engineering, pp. 32–47. Springer, Cham (2017)

25. DAgostino, D., Danovaro, E., Clematis, A., Roverelli, L., Zereik, G., Galizia, A.: From lesson learned to the refactoring of the DRIHM science gateway for hydro-meteorological research. J. Grid Comput. **14**(4), 575–588 (2016)

George Samuels, Debarshi Dutta, Paul Mahon, and Sheetal Vasant Nikam

82.1 Introduction

Quantum Computing is a method that not many people can articulate or even imagine. What is Quantum Computing? What is it used for? How do developers interact with these computers? During the course of our studies these questions and more will be addressed. Quantum Computing is known as the process of assembling instructions called quantum programs which are capable of running on a quantum computer. These instructions can be written in either Python or JavaScript,both have a variety of Quantum Computation libraries [1]. For this project in particular we will use Python to write programs that will be complied on a local machine, and ran on IBM's cloud quantum computer which is for public use. A vital asset in understanding Quantum Computing is knowing about classical computation. Classical computation are illustrated by classical bits, which are seen as singular strings of 1 or 0. The bits of 1 or 0 are representing Boolean values of true or false.

The basic entity of quantum information is a qubit or a quantum bit. Qubits can represent a 1, a 0 or both at once, in contrast to the binary digits used in classical computing [2]. Consider the electron in a hydrogen atom. It can be in its ground state (i.e. an s orbital) or in an excited state. If this were a classical system, we could store a bit of information in the state of the electron: ground = 0, excited = 1 (Fig. 82.1).

The qubits are usually thought to be spatially separated. The greater the distance apart each are, the more each exhibit perfect correlation even though there is no way to tell which

state each qubit is in a situation that cannot be explained without quantum mechanics. Qubits represent atoms, ions, photons or electrons and their respective control devices that are working together to act as computer memory and a processor. Because a quantum computer can contain these multiple states simultaneously, it has the potential to be millions of times more powerful than today's most powerful supercomputers [3].

Understanding what qubits are and how they are represented now helps us, as we look to understand Bell states. One Bell state can be defined as a maximally entangled quantum state of two qubits. The qubits are seen as a spatially separated . Quantum entanglement can be understood by breaking it down in simply terms. It can be imagined as if a person's hair is tangled, the two strains of hair are now connected as one; however, these strains can be separated once more creating two parts. In the quantum space these strains are not simply taken apart once they are entangled. Due to the entanglement, measurement of one qubit will assign one of two possible values to the other qubit instantly, where the values are assigned depends on which Bell state the two qubits are in.

Bell states can also be measured, the Bell measurement is an important concept in quantum information science: It is a joint quantum-mechanical measurement of two qubits that determines which of the four Bell states the two qubits are in. The four Bell states are

$$|\Phi^+\rangle = \frac{1}{\sqrt{2}} (|00\rangle + |11\rangle) = \frac{1}{\sqrt{2}} \begin{bmatrix} 1 \\ 0 \\ 0 \\ 1 \end{bmatrix}$$

$$|\Phi^-\rangle = \frac{1}{\sqrt{2}} (|00\rangle - |11\rangle) = \begin{bmatrix} 1 \\ 0 \\ 0 \\ -1 \end{bmatrix}$$

Thanks to the IBM Faculty Award that made this research possible.

G. Samuels · D. Dutta (✉) · P. Mahon · S. Vasant Nikam
Pace University, Pleasantville, NY, USA
e-mail: gsamuels@pace.edu; dd50506n@pace.edu;
pm07433n@pace.edu; sn07217n@pace.edu

S. Latifi (ed.), *16th International Conference on Information Technology-New Generations (ITNG 2019)*,
Advances in Intelligent Systems and Computing 800,
https://doi.org/10.1007/978-3-030-14070-0_82

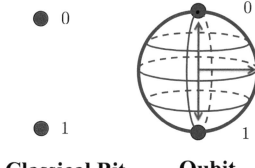

Classical Bit   Qubit

Fig. 82.1 Qubit verse classical bit

$$|\Psi^+\rangle = \frac{1}{\sqrt{2}}\left(|01\rangle + |10\rangle\right) = \begin{bmatrix} 0 \\ 1 \\ 1 \\ 0 \end{bmatrix}$$

$$|\Psi^-\rangle = \frac{1}{\sqrt{2}}\left(|01\rangle - |10\rangle\right) = \begin{bmatrix} 0 \\ 1 \\ -1 \\ 0 \end{bmatrix}$$

Each state can be achieved based on the polarization of a single photon (spin up, spin down). The states (Φ) and (Ψ) are represented by the polarization of the photon, either being horizontal or vertical. The polarization being the same are represented by the Greek symbol (Φ), while the polarization being opposites are shown as the Greek symbol (Ψ) [4]. These two divisions are subdivided by the rotational state. The rotations include the positive horizontal state, negative horizontal state, positive vertical state, and negative vertical state. The public computers at IBM now allow full exploration into all four Bell states, where previously researchers and computer scientist were limited to one Bell state.

82.2   Project Requirements

To setup our Quantum Information System Kit (Qiskit) development environment we used both Microsoft Windows and Apple Mac configurations.

Our team selected Jupyter Notebook for the Integrated Development Environment (IDE) platform. Jupyter Notebook is an cross-platform, open source application that is based on a server-client structure to enable an interactive programming experience for Python [5]. Jupyter Notebook is highly suited to data science and also works well as a presentation tool.

Our team used the Anaconda distribution of Python, which includes the Jupyter Notebook IDE, for our programming interpreter. Anaconda is a popular distribution of

Python for data science and suited this project well [6]. The main advantage of using the Anaconda distribution is that it comes with a package manager (conda) that seamlessly installs key packages "out of the box" that were needed for our Qiskit project. The libraries needed for this project are as shown:

1. IBM IBMQuantumExperience
2. Numpy
3. Scipy
4. Matplotlib

Anaconda also has a graphical user interface, "Anaconda Navigator", that enables our team to launch applications and manage these conda packages.

To setup our Qiskit Development Environment (QDE) we first downloaded the Anaconda distribution of Python at anaconda.com. After verifying the installation, we created our Qiskit virtual environment with the Anaconda tool ("conda create"). The virtual environment enabled us to isolate the Python libraries and package installations from the local computer systems.

Once we created the virtual environment, installing the Qiskit libraries with the Python package management system (PIP) followed. We now had our programming environment setup with the necessary libraries installed.

The next step was to clone the open source code project from Github, a code sourcing repository (repo). This code is a combination of efforts from IBM and other research institutions. Microsoft Windows systems required the installation of the Github client prior to pulling the code from GitHub. Github was already installed on our Mac systems. The team also cloned the Qiskit tutorial repo from Github that provided us with help files and configuration templates (Qconfig.py.template).

IBM has taken the first initiative to build a quantum computer for research [7] and they have provided access to this computer through an application programming interface (API). For the Qiskit project, our team needed a token (code) to access this API for our code testing. To get the code, we logged in to IBM Q experience site [8] and generated an API token. This token was subsequently entered in to the Qiskit configuration file for access to the IBM quantum computer via the API.

Finally, Jupyter Notebook was launched through the Anaconda Navigator and loaded the Qiskit home page (index.ipynb).

- IBM's QISKit
- Q-Experience API key
- Anaconda
- Python
- Jupyter Notebook

- Slack for QISKit
- Understanding of entangled states [9]

82.3   Literature Review

A review of the literature available on quantum computing reveals that the focus of the research thus far has been on the first bell state (Φ +). This can be attributed to the fact that the technology which was used to analyze the Bell states has been made available only relatively recently. In fact, prior to 2017 the research on Bell states was largely hypothetical. Liao [10] has investigated entanglement generated from polar molecules of two-dimensional rotation in a static electric field. The concurrence is used to estimate the degree of entanglement. Parallel and perpendicular application of the electric field to the inter-molecular direction reveals two overlapping features, which corresponds to the existence of Bell-like states. The characteristics of Bell-like states and overlapping concurrences are kept independent of the modulation of dipole–field and dipole–dipole interactions. The Bell-like states however do not coexist in other field directions, which signifies non-overlapping concurrences. Dissimilar suppressed concurrences occur due to different energy structures for the two specific field directions. Friis, Marty et al. [11] has characterized entangled states of a registry of twenty individually controlled qubits. Each qubit was encoded into the electronic state of a trapped atomic ion. Entanglement is generated during the out-of-equilibrium dynamics of an Ising-type Hamiltonian, which was built through laser fields. The qubit-qubit interactions decay with distance and entanglement is generated early between neighboring qubits.

According to Hu et al. [12], the search for the n-variable Boolean functions fulfilling global cryptographic constraints is computationally hard due to the sheer exponential size $O(2^{2n})$ of the space. So a codification of the relevant constraints in the ground state of an Ising Hamiltonian, has been introduced which provided a quantum speedup. Adding to this, small n cases in a D-Wave machine has been set as a point of reference which demonstrates its capacity of devising bent functions, the most relevant set of cryptographic Boolean functions. Belte, Hacker [13] suggests that neutral atoms trapped inside an optical cavity provide an ideal platform for the implementation of quantum networks. A quantum network has nodes containing multiple atomic qubits, which are essential for the construction of a quantum repeater as they allow for entanglement swapping and thus the generation of entanglement between qubits over long distances. The realization of such a multi-qubit network node containing two Rubidium (atomic number- 87) atoms in an optical cavity. Local entanglement between the two atoms is created with an experimental technique called quantum state carving.

The entanglement properties of the grid states form a discrete set of mixed quantum states. These states have been graphically represented by Lockhart et al. [14]. More precisely the entire entanglement properties have been defined, and evaluation methods for the entanglement criteria for the grid states have been computed graphically. The experiment shows that entanglement theory for grid states; although being a discrete set, grid states have a complexity similar to that for general states. Krenn et al. [15] demonstrate a link between high-dimensional multipartite quantum states and graph theory. The paths of photons are identified in such a fashion that the photon-source information is never created. In fact, each specific setup corresponds to an undirected graph on its own, which in turn points to a separate experimental setup. In order to correlate graph theory and quantum states, Krenn et al. have rephrased theorems from graph theory like Hall's marriage problem in the language of pair creation. The issue, however is in calculating the final quantum state which is in the P-complete complexity class. Hence the evaluation could not be done efficiently.

82.4   Methodology

Certain terms need to explained before delving into the implementation of the bell states-

- Hadamard (H) gate- The Hadamard gate acts on a single qubit. It is one qubit version of the quantum fourier transform. It maps the basis state

$$|0\rangle \, to \frac{1}{\sqrt{2}} \left(|0\rangle + |1\rangle \right)$$

and

$$|1\rangle \, to \frac{1}{\sqrt{2}} \left(|0\rangle - |1\rangle \right),$$

which means that a measurement will have equal probabilities to become 1 or 0 (i.e. creates a superposition). It represents a rotation of π about the $(\hat{x} + \hat{z}) \, / \, \sqrt{2}$. Equivalently, it is the combination of two rotations, π about the X-axis

$$H = \frac{1}{\sqrt{2}} \begin{bmatrix} 1 & 1 \\ 1 & -1 \end{bmatrix}$$

- Controlled NOT gate (CNOT)- The CNOT is a quantum gate that is an essential component in the construction of a quantum computer. The CNOT gate operates on a quantum register consisting of 2 qubits. The CNOT gate flips the second qubit (the target qubit) if and only if the first qubit (the control qubit) is $|1\rangle$. The CNOT gate can be represented by the matrix (permutation matrix form):

$$\begin{bmatrix} 1 & 0 & 0 & 0 \\ 0 & 1 & 0 & 0 \\ 0 & 0 & 0 & 1 \\ 0 & 0 & 1 & 0 \end{bmatrix}$$

- The IBM Q Experience is a cloud based quantum computing platform, that gives users in the general public access to a set of IBM's prototype quantum processors via the Cloud. There are two variations of the IBM Q 5 available –
  - ibmqx2 in Yorktown
  - ibmqx4 in Tenerife

82.5　Preliminary Results

Our goal was to look at each gate through IBM Q and become familiar with the graphical user interface(GUI). While testing in this environment we would need to know exactly how to create each Bell state using the quantum gates and also how to measure what was being produced after simulation. First we explored the first Bell state which is the most documented bell state out of the four (Fig. 82.2). Creating the first Bell state involved using two quantum gates and placing each on qubits, q[0] and q[1]. The first gate being an CNOT, which was placed on q[0] with a measure gate. During the process it was found that if each line must have a measuring gate along with the conventional flipping gates in order to see the results of the test. Previously we only measured one line with the CNOT gate and received graphs and other data based on one measurement. Placing both an Hadamard gate and an measurement gate to the next line q[1], we were able to see the amount of bit flips (0 to 1) would take place in each gate, the blue line from q[0] to q[1] represents entanglement of the two qubits as shown in Fig. 82.3. Next was to explore the second Bell state through simulation, this would be done similar to the first Bell state by placing a Hadamard gate on q[1]; however, q[0] does not receive a gate. Instead a CNOT

Fig. 82.2 Quantum gates

Fig. 82.3 Results of the first Bell state (Φ+)

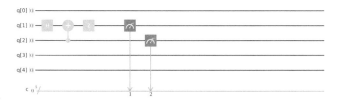

Fig. 82.4 Results of the second Bell state (Φ−)

Fig. 82.5 Results of the third Bell state (Ψ+)

gate is place on the same line as the Hadamard gate, and new gate would then be needed. The Pauli Z gate as shown in Fig. 82.2, is a gate that has a property of X to −X , Z to Z. The Z gate is known as the phase-flip gate and will do such while being place on q[1] to flip the outcome from CNOT and the Hadamard gate creating (Φ−). The results are as shown in Fig. 82.4.

Simulations continue with the third Bell state, at this point it is a lucid notion that each Bell state will be the same with q[1] having both the CNOT and the Hadamard gates respectively. Similar to the gate before it, have some newly introduced gate to flip the bits on each qubit to receive the outcome needed to enter each Bell state. As we seen in the last example the Z gate was introduced at q[1] to flip the result of the Hadamard gate; for the third Bell state another flip gate is needed, the Pauli X gate. The X gate is known as the bit-flip and will flip bits based on the outcome of the CNOT gate and the Hadamard gate, while being entangled.The important difference to notice from the second Bell state to the third is the removal of the Z gate (phase-flip) from q[0] allowing the Bell state to remain positive (Ψ+) (Fig. 82.5).

Testing the final Bell state does follow the rule in which we discover while on the third Bell state, that being similar to it's predecessor. Entering this state required the addition of the Z gate(phase-flip),and also leaving the X gate for the bit-flipping which took place in the third Bell state. The last

Fig. 82.6 Results of the fourth Bell state ($\Psi-$)

Bell state helped us as we were able to compare what makes each Bell state so different and how these states are achieved through bit manipulation using the bit-flipping gate of X and the phase flipping gate of Z.

The more interesting part of our findings began when testing different qubits by placing the quantum gates on q[2], q[3] and so on then, analyzing the results. There is no one or two methods of entering the 4 Bell states, there are several other implementations not explained in this section. This was great for us to see, prior to these results the thought was the contrary. Quantum Computing as we explore it continues to deepen our understanding of the growing technology as well as widen the possibilities of to what it may be used for in years to come (Fig. 82.6).

82.6   Conclusion

Using the resources given to the public by IBM, we were able to test the 4 Bell states by actually creating them and simulating each, producing graphed results of the qubit flipping. Reviewing what has already been achieved through the first Bell state will helped us look past and explore the other three Bell states; writing code for quantum machines has only helped in learning about each Bell state and it's polarization. Through our research it is our goal to learn more about what we can use this technology for not being limited to one Bell state. We hope that our research educates

and intrigues many to help grow this very new technology further.

References

1. X-Team: Quantum Computation Python JavaScript (2018). Available at https://x-team.com/blog/quantum-computation-python-java script/
2. Maheswaran, M.: QP-difference (2018). https://www.quora.com/What-is-the-difference-between-a-regular-computer-and-a-quantum-computer
3. BONSOR, K.: Qbits (2018). https://computer.howstuffworks.com/quantum-computer1.htm
4. Vijayan, T.: Youtube-video (2018). https://www.youtube.com/watch?v=lLbMvHRQ6IY
5. Vasconcellos, P.H.: python-ide (2018). https://www.datacamp.com/community/tutorials/data-science-python-ide
6. Anaconda: ANA-IDE (2018). https://www.anaconda.com/what-is-anaconda/
7. IBM: IBM-IDE (2018). https://www.research.ibm.com/ibm-q/
8. IBM: IBM-quantum (2018). https://quantumexperience.ng.bluemix.net/qx/experience
9. Thanasekaran, V.: Quantum spin, entanglement and teleportation (2014). Available at https://youtu.be/-Tw-GzAWvNI
10. Liao, Y.-Y.: Bell states and entanglement of two-dimensional polar molecules in electric fields. Eur. Phys. J. D **71**, 277 (2017)
11. Friis, N., Marty, O., Maier, C., Hempel, C., Holzäpfel, M., Jurcevic, P., Plenio, M.B., Huber, M., Roos, C., Blatt, R., et al.: Observation of entangled states of a fully controlled 20-qubit system. Phys. Rev. X **8**(2), 021012 (2018)
12. Hu, F., Lamata, L., Sanz, M., Chen, X., Chen, X., Wang, C., Solano, E.: Quantum computing cryptography: unveiling cryptographic boolean functions with quantum annealing. arXiv preprint arXiv:1806.08706 (2018)
13. Welte, S., Hacker, B., Daiss, S., Li, L., Ritter, S., Rempe, G.: Quantum state carving of two atomic qubits in an optical cavity. In: 2017 Conference on Lasers and Electro-Optics Europe & European Quantum Electronics Conference (CLEO/Europe-EQEC), paper EB_7_3 (2017)
14. Lockhart, J., Gühne, O., Severini, S.: Entanglement properties of quantum grid states. Phys. Rev. A **97**, 062340 (2018)
15. Krenn, M., Gu, X., Zeilinger, A.: Quantum experiments and graphs: multiparty states as coherent superpositions of perfect matchings. Phys. Rev. Lett. **119**, 240403 (2017)

Using the Random Tree Classifier to Improve the Project's Cost Predictability in the Earned Value Management: An Empirical Study

Ana C. da S. Fernandes and Adler Diniz de Souza

83.1 Problem and Motivation

To assess whether or not a project will achieve its cost and time objectives, various measures are collected during its execution, and various performance indexes are produced and periodically analyzed. When deviations above tolerable are found in some performance index, corrective actions are implemented in order to improve them. Among the main techniques to analyze cost and time performance, the EVM – is considered the most reliable [8].

Several formulas derived from EVM measurements are available and have been studied in the last 15 years. However, studies intended to improve the predictability of the results of time and cost have remained stagnant over the last decade and still require further studies [9].

The application and reliability of the CPI for the realization of projections have been widely discussed as shown by the works of [3, 8] and [6].

The discussions focus on the CPI's stability. The statement of the CPI's stability is important, as it is used to make cost projections. According to [3], the CPI is considered stable if there is a variation of plus or minus 10% of the value reached when the project has been 20% executed.

A study performed in [3] evaluated the stability of the CPI of various projects of the U. S. DoD, and found that there was stability of the indicator, after 20% of project execution. This study generalized the result, stating that any project could use the technique reliably after 20% of execution.

However, several other studies have questioned the generalization of these results in different contexts that showed different results, i.e., the CPI showed instability during much

of the project, [3, 7, 8] and [13]. One of the motivations for performing this work is the fact that much of the work of software development be such projectized. Another motivation is the possibility of using the technique proposed as a model of cost performance for projects with high levels of maturity, such as CMMI-Dev [10].

83.2 Backgroud and Related Works

The method of EVM allows the calculation of variances and performance indices of cost and time, which generate forecasts for the project, given its performance so far, allowing the implementation of actions aimed at correcting any deviations [1]. This allows the project's manager and your team to adjust their strategies, make trade-offs based on the goals, on the project's the current performance, on trends, and on the environment in which the project is being conducted [9].

The method of EVM is based on three basic measures (Planned Value – PV, Earned Value – EV and Actual Cost – AC), which are derived to generate other measures and performance indicators, i.e. Cost Performance Index – CPI_{Accum}.

Related works have been presented in [5, 12] and [4], using statistical methods to predict the CPI.

83.3 Problem Description

The traditional EVM, use data of the current project (EV_{Accum} and AC_{Accum}) to calculate the CPI_{Accum} to: (i) to measure the cost efficiency (only analyzing the index), or (ii) to make a cost projections, thought EAC (BAC/CPI_{Accum}).

Techniques like COCOMO [2] and Use Case Points [11] use the historical data to provide a final estimate of size,

A. C. da S. Fernandes (✉) · A. D. de Souza
Universidade Federal de Itajuba, Itajuba, Minas Gerais, Brazil

S. Latifi (ed.), *16th International Conference on Information Technology-New Generations (ITNG 2019)*,
Advances in Intelligent Systems and Computing 800,
https://doi.org/10.1007/978-3-030-14070-0_83

effort and cost. However this technique don't provide a way to monitoring and controlling the projects, based in the performance of the current project.

This paper proposes the integration of the EVM with CPI_{Accum}. historical data of processes selected by the classifier Naive Bayes. In this work, this classifier was chosen because this algorithm is ideal for small data sets. The Naive Bayes is a classification algorithm implemented in the scikit learn library which is a binomial or multinomial probabilistic classification model.

This integration consists in gathering and using the CPI_{Accum}. Historical data of each process of the software lifecycle, with traditional measures of the EVM technique, calculated separately to each process, as in Eq. (1). Proposed by [4, 5].

The CPI_{Accum} historical performance of the processes is important because it will be used to predict the future behavior of the project CPI_{Accum}, which consists of the individual performance of each process.

Thus, given its performance and each EVM individual measure of processes, it is possible the equation (83.3) to calculate the CPI_{Accum} projected to the final of project execution:

$$CPI_{Exp}$$
$$= \frac{EV_{AccumProject} + \sum_1^n (BAC_{PN} - EV_{AccumPN}) + \sum_1^n BAC_{PN}}{AC_{AccumProject} + \sum_1^n AC_{ExpectedPN} - AC_{AccumPN} + \sum_1^n AC_{ExpectedPN}},$$
$$(83.3)$$

Where:

- $EV_{AccumProject}$ and AC_{Accum}Project: are respectively the traditional EV_{Accum} and AC_{Accum} of the project, that can be calculated using the traditional EVM equations;
- Bac_{PN}: can be calculated adding every PV activities of the process. It can be calculated using Eq. (83.4):

$$Bac_{PN} = \sum_1^n PVActivityNofProcessPN; \qquad (83.4)$$

- $EV_{AccumPN}$ and $AC_{AccumPN}$: are respectively the EV_{Accum} and AC_{Accum} of each process. It can be calculated adding every EV and AC of executed activities of the process, like Eq. (83.5) presented previously;
- $AC_{Expected}$ PN: is the AC_{Accum} expected by each process after it be executed. The $AC_{Expected}$ PN use the historical CPI_{Accum} of each process and can be calculated using Eq. (83.5):

$$AC_{Expected}PN = \frac{BAC_{PN}}{HistoricCPIofPN} \qquad (83.5)$$

83.3.1 APP Validation

Since the largest cost component in a software project is the hours required for product development, all the basic measures and indicators of the traditional EVM were calculated, based on estimated hours and actual hours, both multiplied for a fixed cost to generated the EV_{Accum} and AC_{Accum}. The calculations were made through an application created in Python. The methodology followed in this paper is presented in the flow chart of Fig. 83.1.

For each activity planned and executed in the projects, EV (through the estimated effort for the execution of the activity) and Actual Costs (through real effort calculated after performing the activity) were calculated. Based on this information and on the project progress, CPI_{Accum} for processes and for the project were calculated.

For each activity planned and executed in the projects, EV (through the estimated effort for the execution of the activity) and Actual Costs (through real effort calculated after performing the activity) were calculated. Based on this information and on the project progress, CPI_{Accum} for processes and for the project were calculated.

The projects that participated in the study were executed on different dates, and therefore, different periods were designed for an application of the Naive Bayes technique. For this purpose, a function was performed that evaluated the starting date of the current project with the end date of the previous projects. If the final date was less than the start date, the final date project was included for processing. Table 83.1 shows the project CPI_{Accum} separated by phase and the period during which each project was carried out.

The input data to generate the probabilities were: the estimated effort of each process accumulated per project or per process, the number of project activities and classes of CPI.

Fig. 83.1 Flowchart of the Naive Bayes to improve the extension of the earned value management technique

Table 83.1 Separating the CPI_{Accum} into classes

CPI_{Accum}	Classes
$CPI_{Accum} > 1.5$	Class1
$1.5 < CPI_{Accum} =< 2$	Class2
$2 < CPI_{Accum} =< 2.5$	Class3
$2.5 < CPI_{Accum} =< 3$	Class4
$3 < CPI_{Accum} =< 3.5$	Class5
$3.5 < CPI_{Accum} =< 4$	Class6
$4 < CPI_{Accum}$	Class7

Table 83.2 Accuracy (error between EAC estimates techniques)

		25%		50%		75%	
% Exec							
Projects		EAC HIST + Naïve Bayes	EAC TRAD	EAC HIST + Naïve Bayes	EAC TRAD	EAC HIST + Naïve Bayes	EAC TRAD
P. 10		12.5	38.7	12.1	35.5	11.2	29.3
P. 11		5.4	25.7	5.5	22.1	5.6	20.4
P. 12		11.6	19.6	10.7	17.4	8.3	17.4
P. 13		29.3	16.8	28.8	17.9	24.1	15.4
P. 14		38.5	37.2	31.2	34.8	23.6	28.4
P. 15		6.6	18.2	7.2	16.1	9.6	13.6
P. 16		84.8	130.2	68.4	94.6	66.8	84.9
P. 17		29.8	50.6	27.2	49.4	19.1	12.9
P. 18		3.6	16.9	3.6	16.9	1.0	0.8
P. 19		43.8	52.8	42.5	50	30.4	1.9
P. 20		5.5	91.4	5.5	76.4	9.6	58
P. 21		25.2	38.1	23.1	29.8	20.7	6.5
P. 22		20.6	37.4	19.9	32.8	11.4	46.3

Table 83.3 Accuracy hypothesis test

Hypothesis	Tests	T	P	Conclusion
HO $_{Accuracy}$	Error EAC.Trad − Error EAC.Hist + Naïve Bayes = 0	2.01	0.03	Refute HO
HO $_{Variation}$	Variation CPI.Trad − Variation CPI.Hist + Naïve Bayes = 0	3.913	1.7×10^{-3}	Refute HO

The Python programming language was used to implement the algorithm. According to the category of the project being evaluated, the projects belonging to a class for the implementation of the data technique and calculated by the CPI are selected.

83.4   Planning of the Study

One of the aims of this study was to answer the following question: "Does the traditional EVM technique have higher accuracy than the EVM with historical performance technique near the beginning, middle and end of the project execution?" Thus, the following hypotheses were set up to evaluate the accuracy of the techniques:

- $H0_{Accuracy}$: the traditional EVM technique is as accurate as the EVM technique with the history of performance (Error EAC_{EVM} − Error EAC_{Hist} = 0)
- $H1_{Accuracy}$: the traditional EVM technique is less accurate than the EVM technique with the history of performance (Error EAC_{EVM} − Error EAC_{Hist} > 0).

It was established that the techniques would be compared when projects had 25% of execution (near the beginning), 50% of execution (middle), and 75% of execution (near the end).

83.5   Results and Contributions

Measures from 23 software development projects were collected between 2009 and 2010. The proposed technique requires that the processes used to perform the CPIEst

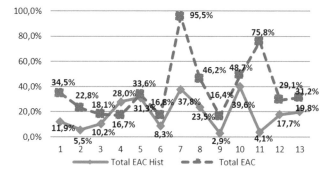

Fig. 83.2 Mean error (total accuracy of EAC techniques)

projection are under statistical control, and therefore stable. The IM/R control chart was selected because, according to [8], this graph is used in situations in which the sample size consists of a single unit, as in the case of the problem addressed in this paper. All processes showed stable behavior after being plotted on the chart I/MR.

Table 83.2 shows the average error (accuracy) of each technique for each technique for 13 projects, in different times. No errors from the 9 first projects were collected because they formed the basis of historical data.

The analysis of data in Tables 83.2 and 83.3, and Fig. 83.2 allows inferring that the proposed technique provides greater accuracy in cost estimations, considering the average error of EAC, and lower variation in CPI_{Accum}. Although the technique shows a positive result, we must take into account that the database is small and of the same type of project. A very recurrent problem in software engineering, so before using the technique it is recommended to do a similar study, with the intention of determining if the proposed technique will give a result for this project [13].

The empirical study showed that the proposed technique is more accurate and more stable (less variation) than the traditional technique. The most of the hypothesis tests conducted with a select historical database composed 10 projects showed significant results, at 95% significance level.

References

1. Anbari, F.T.: (2003) Earned value project management method and extensions. Project Manag. J. **34**(4), 12–23 (2003)
2. Boehm, B.W., Madachy, A.C., Brown, A.W., Chulani, S., Bradford, K.C., Horowitz, E., Madachy, R., Reifer, D.J., Steece, B.: Software Cost Estimation with COCOMO II, 1st edn. Prentice Hall, New Jersey (2000)
3. Christensen, D.S., Heise. S.R.: Cost performance index stability. Natl. Contract Manag. J. **25**(1), 7 (1992)
4. de Souza, A.D.: Uma proposta para melhoria da previsibilidade de custo de projetos, utilizando a técnica de gerenciamento de valor agregado e dados históricos de custo e qualidade. Ph.D. Dissertation. Universidade Federal do Rio de Janeiro (2014)
5. de Souza, A.D., Rocha, A.R.C.: A proposal for the improvement predictability of cost using earned value management and quality data. In: European Conference on Software Process Improvement. Springer, pp. 190–201 (2013)
6. Fenton, N., Marsh, W., Neil, M., Cates, P., Forey, S., Tailor, M.: Making resource decisions for software projects. In: 26th International Conference on Software Engineering, ICSE 2004. Proceedings. IEEE, pp. 397–406 (2004)
7. Henderson, K., Zwikael, O.: Does project performance stability exist? A re-examination of CPI and evaluation of SPI(t) stability. Crosstalk. **21**, 1–13 (2008)
8. Lipke, W.: Statistical methods applied to EVM. . . the next frontier. Meas. News **19**, 18–30 (2006)
9. Lipke, W., Zwikael, O., Henderson, K., Anbari, F.: Prediction of project outcome: the application of statistical methods to earned value management and earned schedule performance indexes. Int. J. Proj. Manag. **27**(4), 400–407 (2009)
10. Sei, S.E.I.: CMMI® for Development (CMMIDEV), V1. 2, CMU/SEI-2006-TR-008. Software Engineering Institute (2006)
11. Smith, J.: The Estimation of Effort Based on Use Cases, Rational Software, Cupertino, TP-171 (1999)
12. Souza, A.D., Rocha, A.R.C.: A proposal for the improvement the predictability of project cost using EVM and Historical Data of Cost. In: 35th International Conference of Software Engineering-ICSE. ACM SRC, San Francisco. Accepted Feb 2013
13. Zwikael, O., Globerson, S., Raz, T.: Evaluation of models for forecasting the final cost of a project. Project Manag. J. **31**(1), 53–57 (2000)

A Proposal to Improve the Earned Value Management Technique Using Quality Data in Software Projects

84

Christopher de Souza Lima Francisco and Adler Diniz de Souza

84.1 Introduction

To assess whether or not a project will meet its deadline and cost objectives, a number of measures are collected during its execution, generating performance indicators that should be reviewed periodically. When deviations greater than the tolerable are found in some performance indicator, corrective actions are performed in order to improve them. Among the main techniques used to analyze cost, time and scope performance, the Earned Value Management (EVM) technique is considered the most reliable [1]. The EVM technique integrates scope, time, and cost data to measure project performance and predict its cost and deadline based on current team performance. The technique gained great importance when, in 1967, the United States Department of Defense (DoD) began to demand its use as a means to control the costs of contracted projects [2].

Several formulas derived from the EVM technique measurements are available and have been studied over the past 15 years. However, research aimed at improving the predictability of time and cost results has remained stagnant during the last decade and still needs further study [1].

Because this technique works with measures related to the schedule and cost of projects and these measures are important for the achievement of the business objectives of the organizations, the technique may be particularly important in the context of software companies that seek high maturity, as reported by [3] and [4].

Organizations also seek a better relationship with their customers through improvements in the planning and management of their development activities and the decrease in the number of defects in the delivered products. The demand for software quality has brought about the need to develop quality models for the software community. Quality software is easy to use, works properly, is easy to maintain and provides data integrity to prevent possible failures outside the control zone. To the despair of its users, flaws present themselves without warning, generating an economic and social impact many times irreparable. Costs resulting from software failures or errors for both business and consumer could be catastrophic: banks could lose millions of dollars and customers would see their money disappear, justifying the importance of software quality and the need for indicators to ensure its efficiency [5] and [6].

84.2 Problem Description and Related Work

The problem addressed by this research is the lack of the quality component in the cost and time results of the software projects that use the EVM technique. Although this technique has been used by several companies for more than 35 years to predict time and cost results, many studies such as [7] and [8] found vulnerabilities, among which it is possible to mention that there is no integration of quality data into the EVM technique [4, 9, 10] and [11].

In order to contribute to the inclusion of quality data in the EVM technique, such as [10] and [11], several articles were published aiming at the vulnerability indicated in [12–14]. Solomon [4] suggests the use of Capability Maturity Model Integration to strengthen the adhesion of EVM, especially related to quality assurance. Yerabolu [13] proposed the integration of critical quality metrics to EVM. Solomon [9] and Solomon [10] shows a set of principles and guidelines that specify effective technical performance measures to be used integrated with EVM. The lack of quality data in the EVM

C. de Souza Lima Francisco (✉) · A. D. de Souza
Federal University of Itajubá, Itajubá, Minas Gerais, Brazil
e-mail: christopher@inatel.br

© Springer Nature Switzerland AG 2019
S. Latifi (ed.), *16th International Conference on Information Technology-New Generations (ITNG 2019)*,
Advances in Intelligent Systems and Computing 800,
https://doi.org/10.1007/978-3-030-14070-0_84

technique can cause erroneous projections and contribute to the delivery of software projects out of time, over budget and without complying with customer needs. Wrong projections can pass optimistic results and postpone the execution of corrective or preventive actions that would improve the final performance of the projects, avoiding delays and costs higher than the estimated ones [3].

Thus, the main research question of this work is: "In what way can new quality measures be incorporated into the EVM technique to improve the performance of its traditional indicators: Cost Performance Index (CPI) and Schedule Performance Index (SPI) in the context of software projects?" knowing that Schedule Performance Index (SPI) indicates how efficiently the project is actually progressing compared to the planned project schedule and Cost Performance Index (CPI) helps to analyze the efficiency of the cost utilized by the project. It measures the value of the work completed compared to the actual cost spent on the project.

This paper will discuss the results of this research, presenting the extension of the EVM technique developed to support the monitoring and control of projects and will show how this elaborated extension was based on the results of secondary studies.

84.3 Methodology

In order to support the definition and improvement of the new proposal for the EVM technique that will be developed in this work, the evidence-based approach presented in [15] will be used. According to [16], this approach is an extension of the methodology proposed in [17] for the introduction of software technologies in industry, which is based on experimental studies as a way of determining what works or not in the application of the proposed technology.

According to [15] there is a need to perform secondary studies before the primary studies suggested in [17], so that the definition of the new technology could be based on evidence from the literature. Thus, they propose the adoption of initial activities to conduct secondary studies, more precisely systematic reviews. Secondary study means the execution of a study that aims to identify, evaluate and interpret all relevant results in a particular topic of research, phenomenon of interest or research question. A systematic review is a type of secondary study [18].

As suggested by [16], in the extension proposed by [15], the methodology is divided into two parts, the first part containing the proposed extension and the second containing the original methodology defined by [17]. In the first part the initial definition of the technology is made, with the execution of two activities:

(1) Performing secondary studies to identify, evaluate and interpret all relevant results according to the desired research question.
(2) Creation of the initial version of the technology based on the results collected from the secondary studies.

This paper used an adaptation of the methodology proposed by [17] and the extension proposed by [15]. The observation study phase will be suppressed, as the result of the present work does not propose a different set of steps for the application of the new techniques.

The activities of the methodology will be used to develop the new extensions of the EVM technique and the resources to support the monitoring and control of projects, and are described below:

(1) **Performing secondary studies:** a study was carried out based on a systematic review on "Earned Value Management", aiming to characterize problems and proposals to improve the EVM technique integrating the quality component into the traditional technique.
(2) **Initial Proposal**: Based on the definition of the main problems resulting from the systematic review carried out, an initial version of the proposal was developed.
(3) **Analyzing Feasibility**: In order to characterize the proposed technique and verify its feasibility, feasibility studies were conducted using simulated project data.

84.4 Proposal of Quality EVM

It is clear that producing quality software is an essential and basic goal of Software Engineering, which offers methods, techniques and tools to achieve this goal. What is needed is that the software be reliable, effective and follow the standards required by the context. Crosby [5] presents a distinction between basic quality and extra quality. In basic quality the author lists: functionality, reliability, ease of use, economy and safety of use. In extra quality: flexibility, ease of repair, adaptability, ease of understanding, good documentation and ease of adding improvements. These priorities will depend a lot on each case and the cost of each of these qualities.

According to [5] "Quality is conformity to requirements". This definition is very interesting because it shows the way forward to judge the quality of software. Based on this assumption it is necessary to consider three factors for the correct verification of conformity to the requirements proposed by [5]:

(1) **Conformity Definition**: For the correct verification of the quality it is necessary a prior definition of the preci-

sion margins of the expected results. Making it possible to measure the quality of the final product.

(2) **Results Observation Methodology**: It is necessary to keep in mind that the observed measurement can contain error margins. There are several factors that can corrupt the data used in the observation.

(3) **Reconciling the Interests of the Various Stakeholders**: The quality is closely linked to the requirements and these requirements are defined by someone, so the quality depends on the choices that are made.

The proposal of integrating quality data into EVM technique consists of collecting quality data, based on the contributions of [5], such as the number of tasks performed and their respective quality requirements.

Quality requirement in this proposal refers to the quality value reached by a given task, ranging from 0% to 100%. Zero percentage means that the activity was not completed with the expected quality and 100% means that the activity was fully completed within the expected quality. The quality indicators for the implementation of the proposed technique will be presented and discussed below.

84.4.1 Quality Requirements

The proposed technique consists of identifying the actual quality of a particular project activity and its subsequent comparison with the desired quality. The two measures collected in this activity will be used to calculate the activity quality indicator in the project phases. As these measures will be calculated by phases and by activity, they must also be collected in phases and by activity. In this proposal, desired quality requirement and real quality requirement are represented by DqR and AqR, respectively:

(1) **DqR – Desired Quality Requirement**: Represents the desired quality value for any activity. DqR is always represented by the value 1 which corresponds to 100% of quality, i.e. the activity has been completed within the desired quality requirements. DqR is the value to be achieved of quality in the activities of a project.

(2) **AqR – Achieved Quality Requirement**: This indicator represents the actual measured value of quality of any activity in the phases of a project. It is represented with values between 0 and 1 to indicate from 0% to 100% quality. AqR is compared to DqR to generate the quality reached for a given activity. AqR equal to 0.5 means that the activity has reached 50% of the desired quality.

The AqR indicator may raise the question: "How can the quality be measured in order to have AqR defined?". Studies such as [11, 19–21] and [22] have contributed to this matter.

It can be very difficult to obtain Quality measures during projects execution, but very easy once the final release is distributed to customers. Defect Counts are a very common measure at that final stage. Measuring something with unit counts of flaws and discovered problems without clear and positive measures can mean very little to the project. Real quality indicators are needed in order to better show the real situation of the project. Even projects that have delivered zero defects, when closely analyzed, can be perceived by developers, project managers and clients, to lack quality. To effectively measure the Project's Quality, as explained by [22], two aspects must be considered:

(1) Indicators, Defect Counts or Positive Counts that measure Technical Quality.

(2) Customer Involvement and Satisfaction that indicates the subjective factor of Perception of Quality.

It is also possible to use Quality Indicators. These can be very useful in any stage of the project, especially very early on when the development team can make adjustments with little to no impact and it might not be possible to acquire Defect Counts. So, Quality Indicators are proof that quality aspects of the Project are met. These Indicators can refer to the whole project or individual tasks. Project Managers can monitor these Indicators for improvement, strategic decisions and adjustments [20].

As explained by [20], Indicators can be used in small to multi-billion dollar projects. There even are subjective Indicators that are attached to the idea of Perception of Quality:

- Engagement Measures: Customer involvement in important project activities
- Planned vs. Actual Cumulative Review Count
- Assessment Measures: Customer satisfaction surveys; stakeholder expectations evaluation

Customer's level of acceptance can be indicated early on though Engagement Measures. Customer participation in projects activities such as Documentation, Training, Requirements Definition and Design decisions can have a huge impact on project success.

To use the technique proposed in this paper, project managers can chose freely how to measure the quality of the projects they manage and then establish the value fit for AqR.

84.4.2 Quality Performance Index

After identifying the values of DqR and AqR it is possible to generate the quality performance index (QPI) of an activity in a project phase by the following equation:

$$QPI = \frac{AqR}{DqR} \qquad (84.1)$$

QPI indicates how efficiently an activity is conducted throughout a project to meet quality requirements. QPI can range from 0 to 1, where:

- QPI = 1, means that 100% of the activity was performed within the desired quality;
- QPI = 0, means that 0% of the activity was performed within the desired quality;
- QPI = 0.5 means that 50% of the activity was performed within the desired quality.

84.4.3 Quality Correction Cost (QCC)

After collecting the data of Desired Quality Requirement (DqR) and Achieved Quality Requirement (AqR) and calculate the Quality Performance Index (QPI), it is possible to calculate the cost to correct an activity finalized outside the expected quality. Any activity that does not achieve the expected quality generates an additional cost to the project, this cost is calculated and added to the cost of the project in the execution phase of the next activity following the project activities. Thus, if the first activity is being executed – activity 1 – and this activity generated an additional cost due to lack of quality, this additional cost will be added to the project during the execution of activity 2. In the equations discussed below, we adopted a simplified notation to reference the project's activities. The first activity performed will be called "a", the next activity will be referenced as "a + 1", in the same manner "a − 1" represents the previous activity.

The following equation shows how to calculate the cost of correcting an activity (QCC) taking into account the quality achieved (AqR) and its actual cost (AC).

$$QCC(a) = \frac{AC(a)}{QPI(a)} - AC(a) \qquad (84.2)$$

where:

- QCC(a) (Quality Correction Cost): Represents the cost to correct an activity that is not within the quality requirements, i.e. the value of AqR is lower than DqR;
- AC(a) (Actual Cost): represents the actual cost of a project activity without taking into account the quality of this activity;

84.4.4 Quality Actual Cost (QACAcum)

From the Quality Correction Cost (QCC) of an activity that did not meet the quality requirements, one can calculate the Quality Actual Cost (QACAcum). Note that there is a difference between the Actual Cost (AC) and the Quality Actual Cost (QACAcum) of the project, Actual Cost (AC) does not take into account the quality with which the activities were developed, Quality Actual Cost (QACAcum) does. The following equation shows this difference and how to calculate the QACACum.

$$QACAcum(a) = QCC(a-1) + AC(a) + QAC(a-1) \qquad (84.3)$$

where:

- QACAcum(a) (Quality Actual Cost): Represents the actual cost of the project taking into account the Quality Correction Cost (QCC) of an activity that is not within the quality requirements;
- QCC(a − 1) (Quality Correction Cost): Represents the cost of correcting the previous activity to the analyzed activity, if the analyzed activity is the first activity executed in the project, then QCC (a − 1) = 0;
- AC(a) (Actual Cost): Represents the actual cost of a project activity without taking into account the quality of this activity;
- QAC(a − 1) (Quality Actual Cost): Represents the actual quality cost calculated for the activity prior to the activity analyzed, if the activity analyzed is the first activity performed on the project, then QAC (a − 1) = 0.

Performing the calculations of QACAcum the result is the value of the real cost of quality of the project, cumulatively. This value can be compared to the value obtained for the Actual Cost (AC) using the traditional EVM technique. If the values are equal, it means that the project is completing the activities with the ideal quality, that is, according to the quality requirements imposed by the client. If QACAcum is greater than AC, it means that activities are being delivered out of expected quality, generating additional project costs (QCC).

84.4.5 Quality Planned Value (QPVAcum)

The goal of Quality Planned Value (QPVAcum) is to measure if the project can deliver planned quality requirements during project execution. This measure is based on the quality of the activities studied in the previous sections and on the planned value (PV) of the project. QPVAcum shows whether or not the delivery of a task has met the planned value (PV). To perform this calculation, it is necessary that the planned project value (PV) is already available to use. The PV is calculated using the conventional EVM technique. The following equation shows how the QPVAcum is calculated:

$$QPVAcum(a) = (QPI(a) * PV(a)) + QEV(a-1) \quad (84.4)$$

where:

- *QPVAcum(a) (Quality Planned Value)*: Represents the cumulative Quality Planned Value of the project;
- *QPI(a) (Quality Performance Index)*: Represents the quality indicator of an activity;
- *PV(a) (Planned Value)*: represents the planned value for an activity, calculated by the traditional EVM technique;
- *QPV(a − 1) (Quality Planned Value)*: Represents the Quality Planned Value of the previous activity, if the activity under analysis is the first activity of the project, QPV (a − 1) = 0 .

After the calculation of QPVAcum, it is possible to compare this value with the planned value (PV) of the project. This comparison shows whether project activities are being delivered in the planned way:

- If QPVAcum equals PV, it means that the project activities are being delivered within the expected quality. This is the best scenario possible.
- If QPVAcum is less than PV, it means that project activities are being delivered out of expected quality.
- If QPVAcum is greater than PV, it means that the quality with which an activity was developed is greater than the desired quality, i.e. AqR > DqR. This means that what was delivered to the customer is better than what the customer asked for. This scenario does not seem to fit the real world, but it is possible mathematically.

84.4.6 Quality Cost Performance Index (QCPIAcum) and Quality Schedule Performance Index (QSPIAcum)

The two main indicators of the EVM technique are SPI (Schedule Performance Index) and CPI (Cost Performance Index) as have stated [6, 21, 23, 24] and many other articles.

The SPI indicates the efficiency with which the project is progressing compared to the planned schedule. According to [25], "The Schedule Performance Index (SPI) is a measure of schedule efficiency, expressed as the ratio of value to planned value." The Schedule Performance Index provides

information about project schedule performance. It is the efficiency of the time used in the project.

The CPI helps to analyze the cost efficiency used by the project. It measures the value of completed work compared to the actual cost spent on the project. According to [25], "The Cost Performance Index (CPI) is a measure of the cost efficiency of budgeted resources, expressed as a ratio of the amount earned to the actual cost." The Cost Performance Index specifies how much the project is earning for each unit of expenditure. The Cost Performance Index is an indication of how well the project stays within budget.

In this proposal, after generating all the quality indicators shown in Sects. 84.4.1, 84.4.2, 84.4.3, 84.4.4 and 84.4.5, it is possible to compare the performance of the traditional EVM (SPI and CPI) indicators with two new cost and schedule indicators related to the quality of the project executed:

$$QSPI(acum) = \frac{QPV(acum)}{PV(acum)} \quad (84.5)$$

$$QCPI(acum) = \frac{QPV(acum)}{QAC(acum)} \quad (84.6)$$

QSPI (Quality Schedule Performance Index) and QCPI (Quality Cost Performance Index) relate to SPI and CPI, respectively, of the traditional EVM technique. QSPI and QCPI indicates the values of SPI and CPI taking into account the quality of the project's activities, the values are compared to demonstrate the error of the traditional technique in relation to the proposed technique. As the traditional indicators (SPI and CPI) do not take into account the quality of the activities performed, these indicators can lead project managers to believe in an unreal scenario where quality is not taken into consideration. QSPI and QCPI take into account quality indicators and update the SPI and CPI values to demonstrate the reality of the project. If the calculated values for QSPI and QCPI demonstrate a worse scenario than the traditional EVM technique scenario, the project manager can use the proposed technique to help identify: (i) where the problem is occurring; (ii) whether the problem is critical or not; and (iii) which action will bring the project back to the planned track.

84.5 Planning of the Feasibility Study

The study's objective was to answer the following question: "Is the traditional EVM technique more accurate than the EVM technique with quality?". Thus, the following hypotheses were set up to evaluate the accuracy of the technique:

- **H0Accuracy:** the traditional EVM technique is as accurate as the EVM technique with quality. H0Accuracy = (Error TraditionalEVM – Error QualityEVM = 0).

- **H1Accuracy:** the traditional EVM technique is less accurate than the EVM technique with quality. H1Accuracy = (Error TraditionalEVM – Error QualityEVM > 0).

The technique presented in Sect. 84.4 was evaluated through a feasibility study, in which the objective was to measure the accuracy of the proposed technique and compare it with the traditional EVM technique.

The accuracy of the proposed technique can be calculated for any of the indicators (CPI, QCPI, SPI or QSPI), using a variation of the equation of accuracy proposed by [26] and [21]:

$$Error(Indicator(a)) = \left\| 1 - \frac{Indicator(a)}{IndicatorwithQualityAcum.} \right\| \tag{84.7}$$

For example, if Eq. 84.7 was applied to calculate the accuracy of the CPI indicator it would look like:

$$Error(CPI(a)) = \left\| 1 - \frac{CPI(a)}{QCPI(Acum.)} \right\| \tag{84.8}$$

One of the main difficulties presented by [21, 27], and [26] in studies related to this one, was the lack of project performance data, available for studies. Thus, it was decided to validate the proposed technique through the performance of project simulations, similarly to the studies conducted by [21] and [26]. For the simulation, the tool Microsoft Excel was used to generate and store data of effort, quality and consequently cost, as well as information about the existence of defects and the respective effort to correct them. Oracle Crystal Ball tool was used to compile macros and generate several simulations results.

The simulation model for this proposal consists in generating a database of software projects based on an already executed project and to use it as a starting point for the simulation. The simulation requires input values which are the data of effort stored and quality information (DqR, AqR and QPI) explained in Sects. 84.4.1 and 84.4.2. The output values are the new indicators introduced by this proposal and shown in Sect. 84.4 (QCC, QACAcum, QPVAcum, QCPIAcum and QSPIAcum).

84.6   Feasibility Study

In order to evaluate the accuracy, or error, of the proposed technique in relation to the traditional EVM technique, Eq. 84.7 presented in Sect. 84.5 was simulated in the available database. As stated by [21] and [26], the largest cost component in a software project are the man-hours necessary for product development, all the necessary simulations required for the calculation of the base measures and indicators

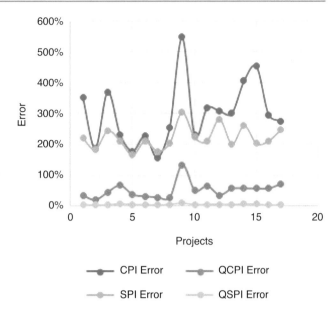

Fig. 84.1 CPI, SPI, QCPI and QSPI error throughout the project's execution (95% significance level)

of traditional EVM were based on the planned effort and actual effort for a set of activities of possible processes of any given project. These activities were calculated using the available database on the MSExcel tool.

Figure 84.1 shows that the Cost Performance Index (CPI) of the traditional EVM technique has a greater error than the Quality Cost Performance Index (QCPI) proposed in this work. The CPI error has an average of 299% and varies between 155% and 547% throughout the project's execution. The QCPI indicator proposed in this work and illustrated in Fig. 84.1, presents less variation than the traditional EVM technique, since the values forecast for QCPI are between 11% and 134% with an average of 52%.

Similarly, the forecast of the Schedule Performance Index (SPI) of the traditional EVM technique and the Quality Schedule Performance Index (QSPI) of the proposed technique, also shown in Fig. 84.1, demonstrates that the proposed technique also has the smallest error throughout the project's execution for the schedule indicators. SPI has an average of 221% error and ranges between 166% and 305% while QSPI has an average error of 6% ranging from 3% to 10%.

In attempt to evaluate the hypotheses shown in Sect. 84.5, statistical tests based in the data of Fig. 84.1 were performed to confirm that the differences in accuracy found in applying the proposed techniques were significant. The MSExcel was used to perform the hypotheses tests of T paired samples, with significance level of 95%.

The analysis of data in Table 84.1 and Fig. 84.1 allows inferring that the proposed technique provides greater accuracy in cost and schedule estimations.

Table 84.1 Accuracy hypothesis test

Hypothesis	Indicator	Tests	T	P	Result
H0Accuracy	CPI	Error traditional EVM −	2,119	2,960E-09	**Refute H0**
	SPI	Error quality EVM = 0	2,119	2,989E-14	

With these results in place, it is possible to answer both hypotheses presented in Sect. 84.5 and calculated utilizing Eq. 84.7 for CPI, QCPI, SPI and QSPI. **H0Accuracy** is false and **H1Accuracy** is true, thus the technique presented in this work, that integrates quality data into the EVM, is more accurate than the traditional EVM technique.

84.7 Validity Threats

According to [28] the internal validity verifies if the treatments really cause the expected results. In this study, the expected results is to decrease the CPIAcum and SPIAcum variability. The proposed technique achieved the expected result. However, it's necessary to consider that, the technique was validated through an empirical study using simulated data. The simulated data was generated from one real software project from the industry. It is important that a more widely study is conducted with more organizations, in different domain of applications.

The main problem in this study is the number of available projects to conduct the hypothesis test to evaluate the statistical significance of the proposed technique. This is a known problem in Software Engineering. Thus, the result cannot be considered conclusive, but just a clue that the technique works. Before using the technique the company is recommended to make a similar study, intending to determine if the proposed technique provides better results to its projects [28].

84.8 Conclusion

This work presented the EVM technique with quality data (DqR, AqR, QPI, QCC, QACAcum, QPVAcum, QCPIAcum and QSPIAcum), which is the proposal to improve the EVM technique to carry out cost and schedule forecasts when there is quality data available. An empirical study was carried out based on data from 1 real executed project and 17 generated projects through simulation.

The proposed technique was evaluated and compared to the traditional technique through different hypothesis tests, utilizing data from simulated projects. Hypotheses tests with 95% significance level were performed, and the technique

was more accurate than the traditional EVM for the calculation of the Cost Performance Index – CPI and the Schedule Performance Index – SPI.

References

1. Lipke, W.: Statistical methods applied to EVM: the next frontier. CrossTalk **19**, 20–23 (2006)
2. Vandevoorde, S., Vanhoucke, M.: A comparison of different project duration forecasting methods using earned value metrics. Int. J. Proj. Manag. **24**(4), 289–302 (2006)
3. Lipke, W.: Schedule is different. PMI CPM J. Meas. News **1**, 31–34 (2003)
4. Solomon, P.: Using CMMI to improve earned value management. Technical report CMU/SEI-2002-TN-016, Software Engineering Institute, Carnegie Mellon University, Pittsburgh (2002)
5. Suarez, J., Crosby, P., Total Quality Leadership Office Arlington VA, Deming, W., Juran, J., United States. Navy Department, Total Quality Leadership Office: Three Experts on Quality Management: Philip B. Crosby, W. Edwards Deming, Joseph M. Juran. TQLO publication, Department of the Navy TQL Office (1992)
6. Dodson, M., Defavari, G., de Carvalho, V.: Quality: the third element of earned value management. Proc. Comput. Sci. **64**, 932–939 (2015). Conference on ENTERprise Information Systems/International Conference on Project MANagement/Conference on Health and Social Care Information Systems and Technologies, CENTERIS/ProjMAN/HCist 2015, 7–9 Oct 2015
7. Lipke, W., Zwikael, O., Henderson, K., Anbari, F.: Prediction of project outcome. Int. J. Proj. Manag. **27**(4), 400–407 (2009)
8. Henderson, K., Zwikael, D.O.: Does project performance stability exist? A re-examination of CPI and evaluation of SPI(t) stability. CrossTalk **21** (2008)
9. Solomon, P.J.: Practical performance-based earned value. In: Systems and Software Technology Conference (2006)
10. Solomon, P.J.: Performance-based earned value. In: INCOSE International Symposium, vol. 15 (2007)
11. Solomon1, P.J.: Basing earned value on technical performance. CrossTalk **26**, 25–28 (2013)
12. Leu, S.-S., Lin, Y.-C., Chen, T.-A., Ho, Y.-Y.: Improving traditional earned value management by incorporating statistical process charts, In: 23rd International Symposium on Automation and Robotics in Construction (2006)
13. Yerabolu, R., P.M. Institute: Framework for Integrating Project Quality, Risk Management, and Integration Management Disciplines into Earned Value Management (EVM) for Deriving Performance Based Earned Value (PBEV) (2010)
14. Ma, X., Yang, B.: Optimization study of earned value method in construction project management. In: 2012 International Conference on Information Management, Innovation Management and Industrial Engineering, vol. 2, pp. 201–204 (2012)
15. Mafra, S.N., Barcelos, R.F., Travassos, G.H.: Aplicando uma Metodologia Baseada em Evidência na Definição de Novas Tecnologias de Software. In: XX Simpósio Brasileiro de Engenharia de Software, Florianópolis, pp. 239–254, Oct 2006

16. Conte, T., Massolar, J., Mendes, E., Travassos, G.H.: Web usability inspection technique based on design perspectives. IET Softw. **3**, 106–123 (2009)
17. Shull, F., Carver, J., Travassos, G.H.: An empirical methodology for introducing software processes. SIGSOFT Softw. Eng. Notes **26**, 288–296 (2001)
18. Kitchenham, B.: Procedures for performing systematic reviews. Keele University, Technical report tr/se-0401, Department of Computer Science, Keele University (2004)
19. Pohl, M.J.: A total quality management (Tqm) strategic measurement perspective with specific reference to the software industry. PhD thesis (1997)
20. Rever, H.: Quality in project management – a practical look, chapter 8. In: Paper presented at PMI® Global Congress 2007– Latin America, Cancún, Mexico. Project Management Institute, Newtown Square (2007)
21. de Souza, A.D., Rocha, A.R.C., Cristina, D., Constantino, B.A.: A Proposal for the Improvement of Project's Cost Predictability Using Earned Value Management and Quality Data – An Empirical Study, pp. 170–181. Springer, Berlin/Heidelberg (2014)
22. Peccoud, J.: If you can't measure it, you can't manage it. PLoS Comput. Biol. **10**(3), e1003462. https://doi.org/10.1371/journal. pcbi.1003462
23. Lipke, W.: Is something missing from project management? CrossTalk **26**, 16–20 (2013)
24. Khalid, T.A.: Controlling software cost using fuzzy quality based EVM. In: International Conference on Computing, Control, Networking, Electronics and Embedded Systems Engineering (2015)
25. Project Management Institute: A Guide to the Project Management Body of Knowledge (PMBOK Guides). Project Management Institute (2004)
26. de Souza, A.D., Rocha, A.R.C.: A Proposal for the Improvement Predictability of Cost Using Earned Value Management and Quality Data, pp. 190–201. Springer, Berlin/Heidelberg (2013)
27. Iranmanesh, S.H., Hojati, Z.T.: Intelligent Systems in Project Performance Measurement and Evaluation, pp. 581–619. Springer International Publishing, Cham (2015)
28. Wohlin, C., Runeson, P., Höst, M., Ohlsson, M.C., Regnell, B., Wesslén, A.: Experimentation in Software Engineering: An Introduction. Kluwer Academic Publishers, Norwell (2000)

Market Prediction in Criptocurrency: A Systematic Literature Mapping

André Henrique de Oliveira Monteiro, Adler Diniz de Souza, Bruno Guazzelli Batista, and Mauricio Zaparoli

85.1 Introduction

When Nakamoto [1] introduced Bitcoin in 2009, there was a major revolution in how to handle the economy, and a new market model emerged alongside bitcoin with its innovative design and ensuring security without the need for an agent external, thanks to Blockchain technology, thus becoming one of the most popular research topics in the field of economics and machine learning.

85.2 Methodology

Given the importance of Bitcoin as well as the other crypto-currencies, the need to validate and classify which research topics, technologies, methodologies and tools are being studied and addressed in crypto-coins pricing and which are currently the greatest challenges and limitations that need further study. To answer these questions, it was decided to use a systematic mapping process to identify relevant research related to crypto-coin pricing. In the systematic mapping study, a research protocol was designed based on the researches [2, 3] and [4] to search for articles in scientific databases [5, 6] and as a result a map was produced containing the current articles on algorithms and pricing techniques in crypto-currencies.

The execution of the research can be summarized in Fig. 85.1, it was used a search string: ("Technical analysis" OR "Fundamental analysis" OR "Artificial Intelligence" OR "Text Mining" OR "Data Mining" OR "Sentiment Analysis"

OR "Machine learning" OR algorithms OR regression) AND (prediction OR "Pricing Prediction" OR "price fluctuation" OR forecasting) AND (cryptocurrency OR bitcoin OR "Virtual Money" OR litecoin OR micropayment OR "Virtual Currency "OR blockchain).

After the final selection of the articles, a data extraction procedure was performed from the complete reading of each article. The main data extracted were: data samples used in the studies, how the training and test sets were selected, how the data was obtained and what types of data, which technique or techniques were used, how the technique was validated and how the results were analyzed, classification rate and forecast. It was possible to create a Table 85.1 which shows the summarized demographic characteristics of the analyzed studies and Fig. 85.2 is a conceptual map that presents the techniques used by the articles.

85.3 Results

The methods, algorithms, and tools that have been developed for market predictability in cryptocurrencies are presented as a map described in Fig. 85.2 and a classification in the Table 85.1 of all technical characteristics and algorithms found in the studies, aiming at two main strands identified in the studies.

On the price prediction in crypto-coins research gaps, Kristjanpoller [24] presents a wide review of the areas related to market prediction. Based on this study we identified some gaps on the found studies:

Análise de sentimento e análise emocional: only one article [17] addressed this subject.

Análise de bolhas em mercados: only one article [23] addressed this criptocurrency subject.

Negociação algorítmica: no article addressed this subject.

A. H. de Oliveira Monteiro (✉) · A. D. de Souza · B. G. Batista
M. Zaparoli
Institute of Mathematics and Computation, Federal University of Itajubá, Itajubá, MG, Brazil
e-mail: brunoguazzelli@unifei.edu.br

© Springer Nature Switzerland AG 2019
S. Latifi (ed.), *16th International Conference on Information Technology-New Generations (ITNG 2019)*,
Advances in Intelligent Systems and Computing 800,
https://doi.org/10.1007/978-3-030-14070-0_85

Fig. 85.1 Research
methodology. (Adapted from [3])

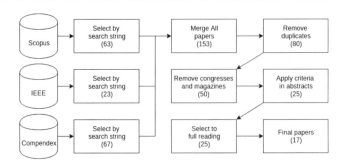

Table 85.1 Demographic characteristics of the included studies

N^o	Data	Period	Algorithms	Validation criteria	Analyze
1 [7]	Serie prices Bitcoin (BTC)	10/13/2011 to 8/26/2017	GARCH and ANN-GARCH	MSE, MAPE and MCS	Technical
2 [8]	Daily quotes BTC, ETH, LTC, XEM, XRP, XLM (OHLC)	28/03/2013 to 1/16/2018	SVM, ANN, deep learning and BoostedNN	MAPE	Technical
3 [9]	BTC price series and 15 bitcoin blockchain features	04/2016 to 12/2017	Linear regression (LR), random florest regression and descending gradient	RMSE and MAE	Fundamentalist
4 [10]	Daily quotes BTC (close and volume)	2012 to 2018	LR, linear SVM and polinomial SVM	MSE	Technical
5 [11]	Purchase price of BTC	02/01/2017 to 02/02/2018	ANN	MSE	Technical
6 [12]	BTC price series and 19 characteristics of Bitcoin's blockchain	02/05/2015 to 06/20/2017	ANN	MSE	Fundamentalist
7 [13]	BTC daily quotes (OHLC) and 2 blockchain variables (difficulty rate and hash)	19/08/2013 to 07/19/2016	RNN, LSTM and ARIMA	RMSE	Fundamentalist
8 [14]	Daily quotes BTC (OHLC) and volume BTC	10/06/2013 to 04/02/2017	BPNN, GANN, GABPNN, NEAT	MAPE, MSE and MAE	Technical
9 [15]	Price series BTC, ETH, DASH and Price Series Euro, Pound Sterling and Japanese Yen	04/01/2016 to 07/31/2017	GARCH, EGARCH, SVM-GARCH	MAE, RMSE and Diebold-Mariano	Technical
10 [16]	BTC price series and 5 bitcoin blockchain features	2012 to 2017	Bayesian regression and GLM/random florest	Do not depict	Fundamentalist
11 [17]	7,000,000 tweets, daily quotation BTC (Open)	2013	LSTM and 6 classification models (Logistic regression, ADL, KNN, random tree, Naive Bayes and SVM)	MAE	Fundamentalist
12 [18]	BTC price series and 4 bitcoin blockchain features	25/01/2017 to 22/01/2018	Multiple linear regression, random florest, LSTM	MSE	Fundamentalist
13 [19]	BTC price series, number of transactions, Google treands and Wikipedia usage	01/12/2013 to 9/21/2016	NLP and deep learning	Matthews correlation coefficient and F-test	Fundamentalist
14 [20]	BTC price series and 336,859 blocks of BTC blockchain with 55 million transactions	01/01/2011 to 12/31/2014	VAR and models of analysis of flow in graphs	Granger causality, F1-Score and p-value	Fundamentalist
15 [21]	Book price and order BTC	06/03/2014 to 07/24/2014	Bayesian regression	Own analysis	Technical

(continued)

Table 85.1 (continued)

Nº	Data	Period	Algorithms	Validation criteria	Analyze
16 [22]	BTC price series and 9 characteristics of the blockcoin bitcoin, 11 macroeconomic indices and 5 currency exchange rate fit	11/09/2011 to 08/22/2017	BNN	RMSE and MAPE	Fundamentalist
17 [23]	Daily quotation BTC, ETH, XMR and LTC (Closed and Volume) and Posts of Twitter and Reddit	2015 to 2016	Hidden Markov Model	own analysis	Fundamentalist

Fig. 85.2 Conceptual map of pricing techniques based on the numbering of the articles referring to the Table 85.1

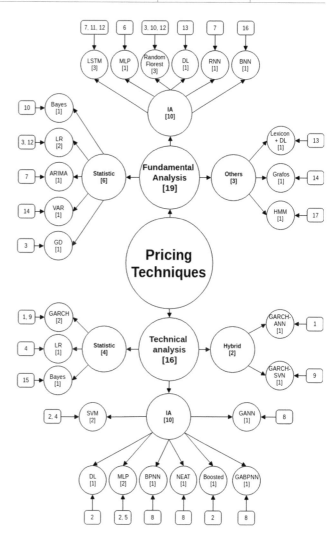

Criptomoedas: apenas três artigos [13,15,23] abordaram outras criptomoedas além do bitcoin.

85.4 Conclusion

To achieve this goal a mapping of all relevant research was done using the systematic mapping process. One hundred and fifty three articles were extracted and analyzed from scientific databases and a total of 17

articles were obtained. The main market prediction systems for crypto-coins were identified and classified, and it is possible to identify some of the gaps in the research area.

This work is believed to be the first effort to provide a broad review with a comprehensive and interdisciplinary perspective on crypto-currency pricing by giving an overview of different approaches, algorithms and predictability techniques, as well as identifying opportunities for research in the area.

References

1. Nakamoto, S.: Bitcoin: a peer-to-peer electronic cash system, p. 9. Www.Bitcoin.Org (2008)
2. Kitchenham, B., Charters, S.: Guidelines for performing systematic literature reviews in software engineering. EBSE Technical report, EBSE-2007-01, Software Engineering Group, School of Computer Science and Mathematics, Keele University, Keele, Staffs, ST5 5BG, UK and Department of Computer Science, University of Durham, Durham, UK (2007)
3. Petersen, K., Feldt, R., Mujtaba, S., Mattsson, M.: Systematic mapping studies in software engineering (2008)
4. Petersen, K., Vakkalanka, S., Kuzniarz, L.: Guidelines for conducting systematic mapping studies in software engineering: an update. Inf. Softw. Technol. **64**, 1–18 (2015)
5. Dyba, T., Kitchenham, B.A., Jorgensen, M.: Evidence-based software engineering for practitioners. IEEE Softw. **22**(1), 58–65 (2005)
6. Sjoberg, D.I.K., Dyba, T., Jorgensen, M.: The future of empirical methods in software engineering research. In: Future of Software Engineering (FOSE'07), pp. 358–378 (2007)
7. Kristjanpoller, W., Minutolo, M.: A hybrid volatility forecasting framework integrating GARCH, artificial neural network, technical analysis and principal components analysis. Expert Syst. Appl. **109**, 1–11 (2018)
8. Hitam, N., Ismail, A.: Comparative performance of machine learning algorithms for cryptocurrency forecasting. Ind. J. Electr. Eng. Comput. Sci. **11**(3), 1121–1128 (2018)
9. Saad, M., Mohaisen, A.: Towards characterizing blockchain-based cryptocurrencies for highly-accurate predictions. In: INFOCOM 2018 – IEEE Conference on Computer Communications Workshops, pp. 704–709 (2018)
10. Karasu, S., Altan, A., Sarac, Z., Hacioglu, R.: Zaman serisi verilerini kullanarak makine orenmesi yontemleri ile bitcoin fiyat tahmini. In: 26th IEEE Signal Processing and Communications Applications Conference, SIU 2018, Izmir, pp. 1–4 (2018) [Online]. Available: http://dx.doi.org/10.1109/SIU.2018.8404760
11. Sakiz, B., Kutlugün, E.: Bitcoin price forecast via blockchain technology and artificial intelligence algorithms. In: 2018 26th Signal Processing and Communications Applications Conference (SIU), pp. 1–4 (2018)
12. Sin, E., Wang, L.: Bitcoin price prediction using ensembles of neural networks. In: 2017 13th International Conference on Natural Computation, Fuzzy Systems and Knowledge Discovery (ICNC-FSKD), pp. 666–671 (2017)
13. McNally, S., Roche, J., Caton, S.: Predicting the price of bitcoin using machine learning. In: Proceedings – 26th Euromicro International Conference on Parallel, Distributed, and Network-Based Processing, PDP 2018, pp. 339–343 (2018)
14. Radityo, A., Munajat, Q., Budi, I.: Prediction of Bitcoin exchange rate to American dollar using artificial neural network methods. In: 2017 International Conference on Advanced Computer Science and Information Systems (ICACSIS), pp. 433–438 (2017)
15. Peng, Y., Albuquerque, P., Camboim de Sá, J., Padula, A., Montenegro, M.: The best of two worlds: forecasting high frequency volatility for cryptocurrencies and traditional currencies with Support Vector Regression. Expert Syst. Appl. **97**, 177–192 (2018)
16. Velankar, S., Valecha, S., Maji, S.: Bitcoin price prediction using machine learning. In: 2018 20th International Conference on Advanced Communication Technology (ICACT), p. 1 (2018)
17. Kinderis, M., Bezbradica, M., Crane, M.: Bitcoin currency fluctuation. In: COMPLEXIS 2018 – Proceedings of the 3rd International Conference on Complexity, Future Information Systems and Risk, vol. 2018, pp. 31–41 (2018)
18. Snihovyi, O., Ivanov, O., Kobets, V.: Cryptocurrencies prices forecasting with anaconda tool using machine learning techniques In: CEUR Workshop Proceedings, vol. 2105, pp. 453–456 (2018)
19. Kim, Y., Lee, J., Park, N., Choo, J., Kim, J.-H., Kim, C.: When Bitcoin encounters information in an online forum: using text mining to analyse user opinions and predict value fluctuation. PLoS ONE **12**(5), e0177630 (2017)
20. Yang, S., Kim, J.: Bitcoin market return and volatility forecasting using transaction network flow properties. In: Proceedings – 2015 IEEE Symposium Series on Computational Intelligence, SSCI 2015, pp. 1778–1785 (2015)
21. Shah, D., Zhang, K.: Bayesian regression and Bitcoin. In: 2014 52nd Annual Allerton Conference on Communication, Control, and Computing, Allerton 2014, pp. 409–414
22. Jang, H., Lee, J.: An empirical study on modeling and prediction of bitcoin prices with Bayesian neural networks based on blockchain information. IEEE Access **6**, 5427–5437 (2018) [Online]. Available: http://ieeexplore.ieee.org/document/8125674/
23. Phillips, R., Gorse, D.: Predicting cryptocurrency price bubbles using social media data and epidemic modelling. In: 2017 IEEE Symposium Series on Computational Intelligence, SSCI 2017 – Proceedings, vol. 2018, pp. 1–7 (2018)
24. Khadjeh Nassirtoussi, A., Aghabozorgi, S., Ying Wah, T., Ngo, D.C.L.: Text mining for market prediction: a systematic review. Expert Syst. Appl. **41**(16), 7653–7670 (2014)

Using Agile Testing in an Academic Health System Case Study

Daniela America da Silva, Samara Cardoso dos Santos, Rodrigo Monteiro de Barros Santana, Filipe Santiago Queiroz, Gildarcio Sousa Goncalves, Victor Ulisses Pugliese, Alexandre Nascimento, Luiz Alberto Vieira Dias, Adilson Marques da Cunha, Johnny Marques, and Paulo Marcelo Tasinaffo

86.1 Introduction

This paper presents the practical application of interdisciplinary concepts using Interdisciplinary Problem Based Learning (IPBL) and the agile Scrum development method and its best practices [1,2] in the development of an academic project with students from three different courses taught at the Brazilian Aeronautics Institute of Technology (*Instituto Tecnologico de Aeronautica* – ITA). The list of courses is presented in Table 86.1.

This academic project was driven by an aeronautics institute of technology to generate expertise in bio-engineering, as a continuation of previous studies performed in the area [3].

The STAMPS Academic Project, in Portuguese *Solucoes Tecnologicas Aplicaveis a Midias e Produtos em Saude*, means in English "Technological Solutions Applicable to Medias and Products in Health Care" has considered the development of a Computer System based on Big data, Internet of Things (IoT), and other cutting-edge technologies [22].

It applies the agile method to ensure the application of quality, testability, reliability, and safety assurance. In addition, the following aspects of certification were used and also the International Classification of Diseases (ICD) [11] and The Health Level Seven International (HL7) [4].

To teach how to develop a computer system to meet these requirements in a period of only 17 academic weeks, the work was distributed amongst 4 teams composed of students from the 3 different courses and specified 4 subsystems with high cohesion and low coupling characteristics: PATIENT, PHYSICIAN, HOSPITAL AND SUPPLIER.

At each Sprint the user histories of each subsystem were prioritized and after this prioritization the development was performed using the agile Scrum and Accepted Testing (ATDD) [5] method with Python, Java, PHP, Spark, databases NoSQL, Kafka and other technologies. At the end of each Sprint, revisions and learnings were conducted and the resulting set of artifacts was posted on the project site (https://sites.google.com/site/stampsacademico/) [6].

Also, more than one test case and acceptance test were developed for each user story. Cucumber Tool for test automation were used to improve the Quality of Software (QoS).

The STAMPS application was developed in a cloud system and each of its subsystem was hosted in the Heroku. Each Development Team was responsible for modeling and building its own application in the Cloud.

They simulated a healthcare system architecture and built the software, working in collaboration with developers and testers remotely-based.

The integration strategy between groups of subsystem components used a subsystem called STAMPSNet, an intelligent system that defines the macro strategy of integration between teams and used the events submitted by the subsystems for decision making.

At the end of the Project, a Proof of Concept was presented at the ITA Bioengineering Laboratory to some entrepreneurs, doctors and local authorities, demonstrating that the STAMPS Project provided appropriate management of health crises by fulfilling User Stories specified in each Sprint project development.

D. A. da Silva (✉) · S. C. dos Santos · R. M. de Barros Santana
F. S. Queiroz · G. S. Goncalves · V. U. Pugliese · A. Nascimento
L. A. V. Dias · A. M. da Cunha · J. Marques · P. M. Tasinaffo
Brazilian Aeronautics Institute of Technology, Sao Jose dos Campos, Sao Paulo, Brazil

© Springer Nature Switzerland AG 2019
S. Latifi (ed.), *16th International Conference on Information Technology-New Generations (ITNG 2019)*,
Advances in Intelligent Systems and Computing 800,
https://doi.org/10.1007/978-3-030-14070-0_86

Table 86.1 The Graduate Courses

Course	Goal
Software testing $CE - 229$	Main techniques involved in the software testing
Database systems project $CE - 240$	Main techniques involved in the development of a database systems
Information technology $CE - 245$	Main information technologies (IT) used in the process of developing computer systems

Table 86.2 The STAMPS academic project – roles description

Roles	Description
Product owner	As a stakeholder representative and with the goal of delivering a valuable product at the end of the process, the Product Owner defined the vision, product requirements, and team goals [8,9]
Development team	It is the self-organized and self-managed team, responsible for the product development and delivery [10]
Stakeholders	They are the interest involved or are somehow affected by the product outcome
Scrum master	He/She is responsible to help the team in the execution process, in order to maximize efficiency and deliver the goal of the Sprint

86.2 Project Overview

86.2.1 The Scrum Agile Method and Its Best Practices

By using Scrum it is possible to develop applications iteratively and incrementally in work cycles called Sprints, which are performed consecutively one after the other and usually have a fixed duration of 2–4 weeks [1,7].

At each Sprint start, a planning meeting is held where the Product Owner and Team Scrum review the Product Backlog and discuss the user stories that will be developed. After this selection the user stories will be divided into a set of individual tasks and will be developed and completed by the end of each Sprint.

The description of the Scrum papers is shown in Table 86.2.

In each Scrum team an experienced student in the process was selected as Scrum Master and the Professors took on the role of stakeholders. One student from each team (preferably from the CE-229 course) was selected by each development team to be its Product Owner.

During this STAMPS Academic Project development, all students have participated in the Development Team. Both CE-240 and CE-245 students were responsible for modeling and programming and also for evaluating compliance with the ICD-10 and the HL7 protocols. The CE-229 students were responsible for the definition and execution of acceptance testing and its oracles.

The STAMPS Academic Project was developed in 4 Sprints. Sprint 0 has focused on some technical training, warming up with GitHub, and also in some multiple courses from the IBM Cognitive Class (https://cognitiveclass.ai/).

The project product backlog had 106 previously defined User Stories (US). However, only 36 were prioritized and implemented, during the 17 weeks: 16 USs in Sprint 1, 13 USs in Sprint 2 and 12 USs in Sprint 3.

Planning Poker, Sprint Review and Sprint Retrospective are Scrum techniques also applied in this project conducted asynchronously and remotely by students. The Sprint Reviews were formalized by the Product Owners, assessing whether the development of the User Stories (US Stories) was successful. The Sprint Retrospectives were also held and captured the contributions of all team members, indicating what went right or wrong and also some potential improvements.

86.2.2 Using the ICD

The International Classification of Diseases (ICD) is the international "standard diagnostic tool for epidemiology, health management, and clinical purposes" [11]. It is maintained by the World Health Organization (WHO) [12].

The ICD data was loaded into the MongoDB NoSQL database and used to map symptoms and suggest potential diagnosis of patients from the STAMPS Academic Project.

The map was performed by using a statistic method called Jaccard Index [13] that helped to identify a list of diseases that could match a symptom. The Jaccard Index is a measure of similarity for the two sets of data (ICD and Patient symptoms), with a range from 0% to 100%, as shown in Fig. 86.1.

86.2.3 The HL7 Standard Protocol

In order to use a comprehensive framework and standard for the exchange, sharing and retrieval of electronic health information, the project chose to use the Health Level Seven International protocol (HL7), enabling the STAMPS project to support clinical practice, management, delivery and evaluation of health services [4].

In the STAMPS Academic Project, it was considered the Level 4 of the HL7 that defines standards for Data Exchange. Figure 86.2 shows the HL7 Level 4 Data exchange standard protocol.

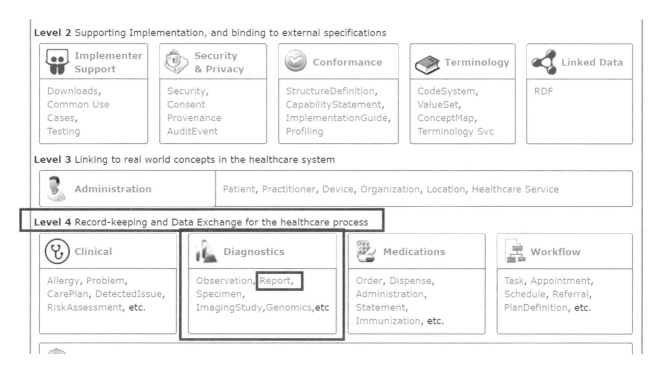

Fig. 86.1 Using Jaccard Index to match potential diagnosis

Fig. 86.2 The HL7 data exchange standard pROTOCOL

86.2.4 The Architecture Overview

For the development of the STAMPS application was used the programming languages Python, PHP or Java. MySQL was used as relational database and MongoDB for unstructured data storage (NoSQL).

JSON and Apache KAKFA were adopted for the exchange of messages. The development was streamlined using Python Web Framework. GitHub was used as a repository of the development platform, allowing teams to review the code, manage the project in a collaborative way with developers and testers working remotely.

Each team identified the most appropriate technology in accordance with the team's internal competence and thus defined its own architecture according to the team members'

best knowledge of technologies. The subsystem of each team was integrated through a common data bus operating according to the STAMPSNet architecture definitions.

The STAMPS application was hosted as a service on the Heroku Platform and in the cloud.

86.2.5 The STAMPSNet

The STAMPSNet is an intelligent system running on top of PATIENT, PHYSICIAN, HOSPITAL, and SUPPLIER subsystems. It is responsible to ensure the interoperability among modules through a data bus for data exchanging, by using the JSON and HL7 format protocol. The STAMPSNet operates according to the stages listed in Table 86.3.

Table 86.3 The STAMPS stages

Stage	Description
Detection	Get data from social media (e.g. Twitter, Google, and Facebook), city hall, hospitals, and others
Screening	Organize data captured by using an intelligent system, to enable the events localization, the identification the common symptoms, and the identification of the potential diseases, according to the ICD and authorities involved
Treatment	According to information identified and organized in the screening stage, it processes data and sends alerts. Also, It treats false/positive alerts, before send it
Response	This subsystem will notify the other subsystems: PATIENT, PHYSICIAN, HOSPITAL and SUPPLIER about what should be done in a crisis situation. For example, preventing traffic at critical locations, and what should be provided to manage the crisis. Notifications will also be sent to authorities such as city hall and army. When the crisis is under control, the alert will be closed

86.3 The Project Development

86.3.1 The STAMPS Development Environment

Heroku is a service platform that supports multiple programming languages and is designed to host application systems using cloud computing capabilities and runs on Amazon EC2 which is IaaS or Infrastructure as a Service.

Heroku supports multiple programming languages, is practical in application management and abstracts service configuration, providing continuous and free implementation for testing cloud applications [14–16].

The creation of a new virtual machine (dyno) as well as the configuration of the development environment is automated.

The development teams of each subsystem were able to share the same Heroku resources as the MLab (MongoDB) and ClearDB (MySQL) database systems.

The MLab (MongoDB) was populated with a large set of ICD-10 data (with more than 100,000 MongoDB documents) and also the symptoms of the disease were added in English (DETAIL collection).

The ClearDB relational database (MySQL) has stored some tables and master data.

Development teams stored their source code in GitHub – Git Control Version.

86.3.2 The STAMPS Academic Groups of Subsystems Structure

The STAMPS Academic Project had 4 development teams consisting of 7 plus or minus 4 students from 3 different disciplines, and each group was responsible for the development of a specific group of subsystem components.

The STAMPS Academic Project was also divided into 4 groups of components of the application subsystem, classified according to their high cohesion and low coupling characteristics, such as: PATIENT, MEDICAL, HOSPITAL (4) and SUPPLIER, as shown in Table 86.4.

Table 86.4 Subsystems

Subsystem	Description
Patient	Provides Patients with information about health provided by the STAMPS Academic Project. Key features were developed integrated with Facebook, reaching about 1.7 billion users, also making it possible to extract usage statistics and geo-location; among others
Physicians	It has digital components that assist doctors in PATIENT consultations, such as the identification of symptoms, diagnosis and number of visits in a geo-location. This subsystem utilized the International Classification of Diseases (ICD), which is the international standard diagnostic tool for epidemiology, health management and clinical purposes. And for data exchange, HL7 was used as the standard for the exchange, integration, sharing and retrieval of electronic health information that supports clinical practice and the management, delivery and evaluation of health services. The use of HL7 provided interoperability with other STAMPS subsystems
Hospitals	It allows hospitals to manage the progress of hospital treatments and manage resources per patient and uses the ICD to develop their main functionalities. For example, identify the patient's location, medications, prescriptions, vital signs inside the hospital, among others
Suppliers	This subsystem helps suppliers to register as providers of various products and services required for health care solutions managed by the STAMPS Academic Project. Among the benefits are identifying necessary resources and buying others according to a geographical location, among other characteristics
StampsNet	Subsystem developed to integrate each PATIENT, MEDICAL, HOSPITAL and SUPPLIER data subsystem and send alerts to each subsystem to take action to respond to health crisis management using the STAMPS application

Based on the Model View Controller (MVC) strategy, using Django, the STAMPS Academic Project was developed over a period of only 17 academic weeks [17–19].

86.4 The Proof of Concept Presentation as an Assigned Mission

Inspired by the needs of public and/or private organizations dealing with crisis management involving health events (eg epidemics), which need to manage data and information for decision-making, the STAMPS application was developed in 17 weeks and used as proof of concept a fictional Ebola epidemic, described below.

"During a national holiday, a Person faints in the middle of a Shopping Mall. The safety team will direct the person to the local ambulatory where a doctor will examine symptoms such as very high fever, tremors, vomiting with blood, red eyes, mental confusion."

During the care, the Patient reports recent work in areas affected by Ebola, and in the face of its serious symptoms, the medical staff reports that it may be an Ebola infection. This patient can be considered as the first PATIENT to start an epidemic in a region if it is readily identified.

From there on, a set of epidemiological events started to be managed in such way that:

- *Promptly identify and/or treat who has had contact with the infection of patients with Ebola*
- *Since this case could evolve into an epidemic, it is recommended that society has a computational tool capable of providing appropriate records, management, and controls; and*
- *From a confirmed case and to avoid unnecessary panic and despair of society, the tool should be able to provide preventive management and control of the diagnosis of patients in hospitals.*

Also, from there on, it is assumed that a set of epidemiological events are managed in such way that:

- *Through appropriate technologies and efficient screening methods, hospitals can collaborate in the process of managing large flows of medical and patient care*
- *Drug and device suppliers also need adequate tools to participate in a material and drug inventory management process as there are several materials involved in care to ensure the safety of health care professionals (doctors) and the well-being of patients*
- *The inventory control of a hospital will be updated in each case confirmed due to the use of care kits containing ma-*

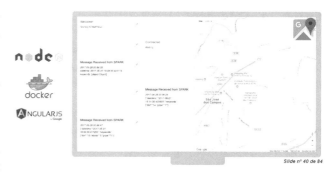

Fig. 86.3 The STAMPSNet timeline

terials and medicines to start the treatments. And besides medicines, protective equipment will also be considered, given the rapid rate of disease contagion, such as Ebola; and
- *The public administration has reliable data for decision-making in crisis situations.*

From this point on, it is also assumed that all events in the assigned mission is presented in a dashboard, as a time line, listing all actions taken in every phase of the STAMPSNet and describing all actions taken into account, since the first patient was identified in an epidemic, as shown in Fig. 86.3.

86.5 Tests

86.5.1 Testing Methods

The STAMPS Academic Project was developed by using the TDD (Test Development Driven) methodology. Therefore, the requirements were converted into very specific test cases. They were prepared right on the first week of the Sprint.

In this context, the development took into consideration test case specifications. The product was validated against those tests cases every week of each Sprint. An iterative development process was used, where requirements have evolved through collaboration among self-organized teams, development team members, product owner, and stakeholders.

Issues were determined in advance.While in conventional methods, tests use to be performed only after implementations, on this current PoC scenario, testing was done during the implementation.

The Behaviour Driven Development (BDD) technique was implemented by some teams in the 3rd Sprint. The Cucumber tool was used to automate tests related to the behavior of the application. It saved time from the team to invest in exploratory tests.

86.5.2 Test Evidences

The developed software for the PATIENT subsystem was successfully modeled, generated, tested, and verified. Figure 86.4 illustrates the "Finding Physicians" functionality on a map.

The developed software for the PHYSICIANS subsystem was also successfully modeled, generated, tested, and verified, as it is illustrated on Fig. 86.5.

The developed software for the HOSPITAL subsystem was successfully modeled, generated, tested, and verified too, as Fig. 86.6 illustrates.

86.5.3 The Learning Curve

During the academic development of the STAMPS Academic Project, different design, development and test activities were carried out. All the participants had the opportunity to carry out their tasks using agile development concepts.Some students had their first contact with the Agile Scrum method in this experience. It was also their first hands-on experience with some of the used technologies, concepts and standards.

The Agile Method helped to reduce the development time. Probably its activity-centric approach was one of the greatest

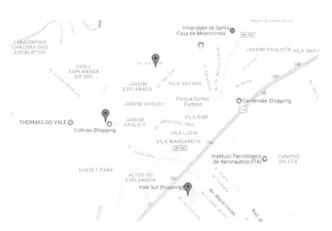

Fig. 86.4 PATIENTS finding PHYSICIANS, according to their locations

Fig. 86.6 Managing HOSPITAL information per patient

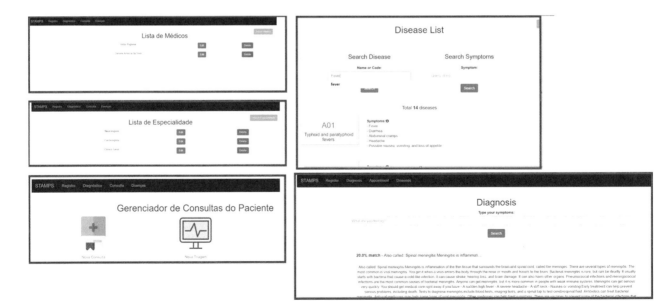

Fig. 86.5 PATIENTS appointment and potential diagnosis

contributors. In this context, the focus was on the activities, and each team member had a small activity that could be executed quickly and accurately. That is different from other traditional methods, in which one member does not execute other tasks that do not belong to it [20].

Although there was no quantitative measurements, a qualitative evaluation showed the learning curve was accentuated and performance was superior than a scenario where students would use other traditional development methods. Frequent communication enhanced the feedback-loop frequency. As a result, it accelerated the learning process and increased productivity.

The degree of complexity of the project and the coordination among the teams were also evaluated. All groups had a real and true insight into the complexity of this process and the high level of difficulty in dealing with integrated tasks.

The definition of the integration strategy between subsystems in the initial phase of the project, allowed to build a product oriented for continuous integration. This avoided the common trap of most traditional software development approaches, where integration is performed late in the development schedule.

As a result, the project was successfully completed on time, reinforcing the qualitative evaluation of a more steeper learning curve and a superior performance.

86.6   Conclusion

This paper aimed to describe the development of an academic interdisciplinary project using the Scrum agile method and its best practices. It described an Interdisciplinary Problem-Based Learning (IPBL) project named STAMPS Academic Project.

The educational goal of the project was the development of Proof of Concept (PoC) prototype using agile testing methods to improve software quality [21, 22]. This PoC was built on the top of many utting-edge technological architectures, frameworks and concepts, such as Big Data, Internet of Things, among many others. Cloud-computing resources were used to support the distributed collaborative team efforts without any disruption on the communication among team members and cross-teams. Cloud based tools also ensured the synchronism inside team and among teams.

For the system purpose selection, the potential social impact was taken into account. Therefore, the selected purpose of this system was to aggregate data and integrate actors such as PATIENT, HOSPITAL, PHYSICIAN, and SUPPLIER for the decision making process related to health crisis management, such as epidemics.

The STAMPS Academic Project was developed by students from 3 different courses taught at the Brazilian Aero-

nautics Institute of Technology (Instituto Tecnologico de Aeronautica – ITA), on the 1st Semester of 2017.

86.6.1   Specific Conclusions

The STAMPS project provided the application of cloud computing and agile test methods into a interdisciplinar context successfully, as presented in the section 'The STAMPS Academic Project'. Through interdisciplinarity in three Computer Science courses, students were able to work as a team to successfully develop a complex system, using cloud computing extensively for collaborative distance work with online meetings, and sharing subject reports on personal website and the project artifacts on an official project website.

In order for a team of around 30 students to deliver value to stakeholders, at the end of each Sprint and also at the end of this project, the Scrum framework was adapted to the reality of interdisciplinarity in the academic environment of ITA.

In order to guarantee the quality of the product and the services, the Agile Methods of testing were used from the concept of the system to its delivery (such as TDD, BDD and ATDD). The use of these techniques was closely related to the use of interdisiplinarity, since the acceptance tests were created by the students of course CE-229, at the same time that the software was being implemented by the students of the other two courses CE-240 and CE-245 .

The project closure milestone was the presentation performed by the students. The target audience was teachers, businessmen and also for some guests from industry and academia. In this presentation, the students involved in the project had the opportunity to present the STAMPS Final Academic Project PoC successfully.

86.6.2   General Conclusions

Interdisciplinary context, agile approach, cloud computing tools, and hand-on experience on cutting-edge technologies were fundamental to keep students motivated and involved. Also, that supported the teams members to work at the right pace to achieve the goals in 17 weeks. Although this experience was applied to the health care area, the presented results of this learning process can be extended to other knowledge domains.

During this academic project, students have acquired skills and experience that can be applied to solve real-world problems. The experience was rewarding and could also be replicated in different undergraduate and postgraduate courses.

86.6.3 Recommendations

The authors recommend that the assigned mission to be clearly defined in the initial stages of the project. This made the User Stories specification and Product Backlog much faster. As a result, is allowed the creation of the PoC quickly.

They also recommend the use an integration-oriented architecture since the early stages of the project. This allows different teams to work on communication protocols, refinements of layouts and integrated tests throughout the project development.

Finally, they recommend the use automated testing tools. They accelerate the development process and increase the level of quality reached by the final product. However in this project there was not enough time for students to benefit significantly from this advantage.

86.6.4 Future Works

The authors suggest that the process used in this STAMPS Academic Project be extended to other systems projects, such as in the health care field, to collect and track personal health conditions. And they also suggest extending academic cooperation with industry, providing projects, partnerships and products aligned with market needs and real products to manage crises in the field of health care knowledge.

Acknowledgements The authors thank the Brazilian Aeronautics Institute of Technology (ITA), the Ecossistema Negocios Digitais Ltda, and the Casimiro Montenegro Filho Foundation (FCMF) for their general and finantial support, during the development of this STAMPS Academic Project.

References

1. Sutherland, J.: Scrum Handbook. Scrum Training Institute Press, Somerville (2010)
2. Cohen, D., et al.: An introduction to agile methods. In: Fraunhofer Center for Experimental Software Engineering Advances in Computers, vol. 62. Elsevier (2004). http://www.cse.chalmers.se/~feldt/courses/agile/cohen_2004_intro_to_agile_methods.pdf
3. ITA: Integrated real-time management system for alerts and crises (2016). https://sites.google.com/site/projetosigtrac2016/
4. HL7: Health Level Seven International (2017). http://www.hl7.org/documentcenter/public/training/IntroToHL7/player.html
5. Pugh, K.: Lean-Agile Acceptance Test-Driven Development: Better Software Through Collaboration. Addison-Wesley. ISBN:978-0321714084
6. STAMPS: Technological solutions applicable to medias and products in health (2017). https://sites.google.com/site/stampsacademico/
7. Maria, R., et al.: Applying scrum in an interdisciplinary project using big data, Internet of Things, and credit cards. In: 2015 12th International Conference on Information Technology – New Generations (2015)
8. Cohn, M.: Succeeding with Agile: Software Development Using Scrum, 1st edn. Pearson Education, Inc., Boston (2010)
9. Leandog: Agile Discussion Guide, Version 3.1, Cleaveland, Ohio (2012)
10. Schwaber, K., Sutherland, J.: The Scrum Guide (2013). Available: https://www.Scrum.org/Scrumguide. Access in 07th Mar 2014
11. ICD: International Classification of Diseases (2017). http://www.who.int/classifications/ICD/en/
12. WHO: World Health Organization (2017). http://www.who.int/about/en/
13. Niwattanakul, S., Singthongchai, J., Naenudorn, E.: Using of Jaccard-coefficient for keywords similarity. In: Proceedings of the International Multi Conference of Engineers and Computer Scientists, IMECS2013, Hong Kong, vol. I, 13–15 Mar 2013
14. IMasters: Development, creation and innovation (2017). https://imasters.com.br/box/ferramenta/heroku. Access in 11th June 2017
15. DevMedia: Codes for who develops codes (2017). http://www.devmedia.com.br/primeiros-passos-em-paas-com-heroku/29465. Access in 11th June 2017
16. Heroku: Heroku Dev Center (2017). https://devcenter.heroku.com. Access in 11th June 2017
17. Freeman, E., Bates, B., Sierra, K., Robson, E.: Head First Design Patterns. O'Reilly, Sebastopol (2004)
18. Model view controller. Available: http://c2.com/cgi/wiki?ModelViewController. Access in 16th Aug 2013
19. MVC pattern and Django. Available: https://overiq.com/django/1.10/mvc-pattern-and-django/ in 15th June 2017
20. Silva, G., et al.: Integrating amazonic heterogeneous hydrometeorological databases. In: 2009 Sixth International Conference on Information Technology: New Generations (2009)
21. Goncalves, G., et al.: An interdisciplinary academic project for spatial critical embedded system agile development. In: 2015 IEEE/AIAA 34th Digital Avionics Systems Conference (DASC) (2015)
22. Silva, D., et al.: Health Care Information Systems: A Crisis Approach (Chapter 34). In: Part of the Advances in Intelligent Systems and Computing book series (AISC, volume 738). Springer Nature America, Inc (2018). https://link.springer.com/chapter/10.1007/978-3-319-77028-4_34

Leadership Ethics Conduct: A Viable View on Project Management in Defense and Aerospace Industry

Joseph A. Ojo

87.1 Introduction

Kant's categorical imperative can be described as a moral law and principle that every person must live and abide by. It is called categorical imperative because it is a command and not a choice. Grosch, Large, Wenham, Appleton, Astley, Fiddes, and Hick [1] explained that the philosophy of Kant could be related to deontological ethics. The idea of the philosophy being based on deontological principles means that it is based on obligation and duty. The principle of Kant is based on the idea that people must execute their affairs centered on a strict moral law [1]. Kant explained the relationship between the willful act of human and the qualification of good will. The morality of humans is relatively related to the commitment that they have in a universal law, which binds each individual in existence [2]. Kant explained that morality and doing the right thing is imperative, which means a must in every situation.

87.2 Categorical Imperative

Kant's categorical imperative can be described as the moral philosophy that determines the will and rationale in a moral action [3, 4]. This theory is also known as the centerpiece of Kant's ethics theory [4]. The categorical imperative can be described as the universal code and rational that cannot be disputed. Schulzke [4] referred categorical imperative as a universal command that must be followed. Sensen [5] explained that these laws must be followed as orders or commands, and the fact that reasons dictate why they must be followed. The categorical in the Kant's categorical imperative means that there are no exceptions to the laws, and they apply in all cases [4]. Guseinov [6] explained that the

motive of duty is the most important aspect that categorical imperative recognizes. The categorical imperative addresses the behavior of people that are guided exclusively by moral laws.

Grosch et al. [1] explained categorical imperative as an unconditional and absolute command that must be consistently followed and obeyed. It is called imperative because the laws are absolute and must be consistently obeyed. The value behind these laws can be described as intrinsic or internal due to the fact that it is a universal law [7]. Kant's categorical imperative form of ethics does not leave room for rationale or reasoning behind a morality law that should be followed. It can be said that people behave in a specific ways due to constraints or consequences, but Kant's categorical imperative claims that morality and values should be adhered to because they are rational, universal laws.

The laws, values, and morality standard should not be followed due to personal gain or recognition, but should be followed because it is logical and not contradictory [1]. One of the important aspects behind Kant's categorical imperative is the notion of acting in a way that will be accepted universally. For example, an individual that steals from another person would not want the whole universe to steal other's property, including his or her own. After reflecting on their decision, they must decide if it is a moral standard that can be adopted by the whole word. Kant explained that if a moral law is imperative and it is good, the principle of this law should be categorical because it conforms to reasons [1]. Categorical imperative is ethics that are necessary and must be obeyed not because it is convenient, but because they are universal law.

Kant's Categorical Imperative in Project Management
Kant's categorical imperative is engraved with an ethical guideline to avoid doing bad and consistently doing the right thing [8]. This categorical imperative applies to the field of

J. A. Ojo (✉)
Doctorate Student, Chicago, IL, USA

© Springer Nature Switzerland AG 2019
S. Latifi (ed.), *16th International Conference on Information Technology-New Generations (ITNG 2019)*,
Advances in Intelligent Systems and Computing 800,
https://doi.org/10.1007/978-3-030-14070-0_87

project management. Based on the aggressive expectation of shareholders and highly paid executives, it can be said that ethical practice has suffered due to personal gain.

Barsh and Lisewski [9] described business ethics as the process of organization to evaluate and determine actions that are right and wrong. Leaders are essential in balancing the organization's ability to respond to ethical principles and norms [10]. Ethical concerns in the field of project management include honesty, financial risk and reporting, and consumer safety. This ethical dilemma is common in this field either to hide a technical error from the customer or withhold valuable information in respect to the status of the program [11].

Kant's categorical imperative is a moral law that explained how people should act in the maxim if the law can be universal [1]. If Kant's ethical theory is followed in the field of project management, there will be a sense of reflection that project managers will have before intentionally providing inaccurate information to their stakeholders. There is a selfish interest that some project managers have in giving false information to cover potential issues in a program in order to save face in the organization.

87.3 Conclusion

Kant's categorical imperative appears to be appropriate framework from which to view how project leaders should manage and lead organization. What makes Kant's categorical imperative the ideal ethics methodology is the fact that it does not leave room for exception and guessing. It enforces and encourages everybody in the organization to follow the same ethical principle. Finally, this type of ethical structure is what is needed in the aerospace and defense industry and should be used during business transaction and in corporate industry due to the fact that it enables people to reflect on the ethical decision that they are about to make.

References

1. Grosch, P., Large, W., Wenham, D., Appleton, J., Astley, J., Fiddes, P., Hick, J.: Kant's categorical imperative. Retrieved from https://curriculum.leadinglearning.org.uk/phil/Shared%20Documents/A%20Level%20Dialogues%20Issues/03%20Cat%20I%20LAD%20Euth%20Rel%20ex.pdf (1994)
2. Bagnoli, C.: Morality as practical knowledge. Anal. Philos. **53**(1), 61–70 (2012). https://doi.org/10.1111/j.2153-960X.2012.00549.x
3. Geiger, I.: What is the use of the universal law formula of the categorical imperative? Br. J. Hist. Philos. **18**(2), 271–295 (2010). https://doi.org/10.1080/09608781003643568
4. Schulzke, M.: Kant's categorical imperative, the value of respect, and the treatment of women. J. Mil. Ethics. **11**(1), 26–41 (2012). https://doi.org/10.1080/15027570.2012.674241
5. Sensen, O.: Kant's conception of inner value. Eur. J. Philos. **19**(2), 262–280 (2011). https://doi.org/10.1111/j.1468-0378.2009.00385.x
6. Guseinov, A.A.: The golden rule of morality. Russ. Soc. Sci. Rev. **55**(6), 84–100 (2014)
7. Kitcher, P.: Kant's argument for the categorical imperative. Noûs. **38**(4), 555–584 (2004)
8. Succi, G.: Agile methods: between categorical imperatives and lean production. Commun. ACM. **49**(10), 31–32 (2006)
9. Barsh, A., Lisewski, A.: Library managers and ethical leadership: a survey of current practices from the perspective of business ethics. J. Libr. Adm. **47**(3–4), 27–67 (2008)
10. Dierksmeier, C.: Kant on Virtue. J. Bus. Ethics. **113**(4), 597–609 (2013). https://doi.org/10.1007/s10551-013-1683-5
11. Johnson, C.E.: Ethical challenges of leadership: Casting light or shadow (Kindle version), 4th edn. SAGE Publications, Thousand Oaks, CA (2012). ISBN 978-1-4129-822-1

Greater Autonomy for RPAs Using Solar Panels and Taking Advantage of Rising Winds Through the Algorithm

88

Leandro Diniz de Jesus, Felix Mora-Camino, Luciano V. Ribeiro, Hildebrando Ferreira de Castro Filho, Alexandre C. B. Ramos, and José Renato Garcia Braga

88.1 Introduction

Unmanned aerial aircraft are increasingly being used [1] for various purposes such as forest monitoring, locating endangered people or objects in a large area.

For countries where there is a vast border airport where it is difficult to monitor even manned aircraft where it ends up being a very high price, due to the fact of the fuel, maintenance of the aircraft, then there are emerging alternatives of using unmanned electric aircraft for being a cheaper solution and letting manned aircraft be used in more emergency locations.

However it still has a very big problem because the autonomy of the batteries used by these RPA are of very short duration, having to be forced at some time to land them to replace these batteries and continue their mission. There are many works being done to lower the energy consumption of the aircraft's embedded system [2], as well as altering its design [3], but the focus of this work is to use the researched ones carried out on flights with electric aircraft using solar cells in order to recharge the battery, where for Brazil that is a tropical country we can have many benefits of this technology and use the research done on upwind that can be much used to take the expense of the engines, letting only the air current keep the aircraft flying.

L. D. de Jesus · L. V. Ribeiro · A. C. B. Ramos (✉)
Federal University of Itajuba (UNIFEI), Itajubá, Brazil
e-mail: lucianoribeiro@unifei.edu.br; ramos@unifei.edu.br

F. Mora-Camino
Federal Fluminense University (UFF), Rio das Ostras, Brazil

H. F. de Castro Filho
Aeronautical Institute of Technology, São José dos Campos, Brazil

J. R. G. Braga
Nacional Research Institute (INPE), São José dos Campos, Brazil
e-mail: jose.braga@inpe.br

In the next chapter will be approached about photoelectric energy, in sequence I will talk about thermic winds, to the present the experiment and thus to be able to enter in the final chapters that I will talk about the future works and possible conclusions.

88.2 Background and Related Work

88.2.1 Solar Powered Charging

There are many works being carried out on the use of photo voltaic plates in RPAs, where they are used to increase the duration of the flight, being used to only recharge the battery, as well as to recharge the battery, this already be consumed by eletronic parts [4].

What matters most to us is the fact that whis helps significantly increase the duration of an RPA in flight by thinking of flights that need to cover a long stretch of land or forest.

For the climate of Brazil where most of the time we have a high amount of sun, this type of energy generation would be ideal.

In the research carried out by Prof Dr. Sergey [5] we can see that for high altitudes the yield of photo voltaic plate is much higher than at low altitudes.

Since the idea is to carry out monitoring in regions that require a low altitude flight, it was necessary to implement something more than just photo voltaic plates to increase autonomy.

88.2.2 Updraft Winds

Because of the need to perform low flights by reducing the efficiency of the photo voltaic plate, it was necessary to look

© Springer Nature Switzerland AG 2019
S. Latifi (ed.), *16th International Conference on Information Technology-New Generations (ITNG 2019)*,
Advances in Intelligent Systems and Computing 800,
https://doi.org/10.1007/978-3-030-14070-0_88

fo alternatives, só we reached the upwinds that is very easy to be found in Brazil because of the tropical climate.

Winds up of thermals are those where they are generated by heating the surface, where it generates a flow of hot air going from the bottom up.

There are many jobs being done to track the hot springs, such as [6], but the idea of this project is only to identify if the RPA is entering a thermal to decrease the use of the motors that is one of the components that most consumes battery and then turn it off while in the thermal.

With earlier experiments, we can see that it is very advantageous to use hot springs, which manned aircraft use to improve their autonomy.

88.3 Experiment

We will use for this experiment a model airplane known as wing-zags with a wingspan of 2.25 meters, has a angle of 30 degrees, ZAGUI12 wing profile, done on styrofoam type 5 (P3). Figure 88.1 shows the UAV previously referenced already with the embedded system, but without the solar plates.

An amount of 20 solar cells will be allocated in the wing where on the one hand will have on average 10 cells connected in series and each wing will be connected in parallel. The solar panels are made of mono-crystalline silicon material, measuring 63×125 mm, which generates 0.574 volts, generate a chain 2915 amperes, with a 21.8% efficiency.

For the autonomous flight will be used a pixhawk controller board where with it will be a zybo zynq 7000 type computer that has 650 MHz dual-core ARM Cortex-A9 processor, DDR3 memory and programmable logic equivalent to FPGA Artix-7, where it will be running Ubuntu 16.04 with the ROS program to do optimization of performance in a thermal. The algorithm for optimization will be implemented in python.

A MPPT (Maximum power point tracking) circuit is implemented to maximize the solar irradiation [4].

The algorithm being presented is very similar to that of [6], but its focus in to find thermals, I will adapt it so that instead of the RPA being running until it reaches the desired height and going to another thermal, it will just shut down the motor and coninue the autonomous path until the next point of the mission.

The purpose is to generate a mission autonomously for aircraft and to measure the time that it has been accomplished this mission to obtain results in the part of autonomy.

88.4 Conclusion and Future works

With the observed experience of other works, the few tests carried out, we can observe that the gain of autonomy will be very satisfactory.

For future work, more data will be collected to verify the work. Improvement of the implemented algorithm for the use of thermal springs.

Acknowledgment This work is supported by CAPES (Coordenação de Aperfeiçoamento de Pessoal de Nível Superior).

References

1. Braga, R.G., da Silva, R.C., Ramos, A.C.B.: Collision avoidance based on Reynolds Rules: a case study using quadrotors. In: 14th International Conference on Information Technology (2017)
2. Gupta, P., Singh, G.: A novel human computer interaction aware algorithm to minimize energy consumption. In: Wireless Personal Communications (2015)
3. Serokhvostov, S., Kornushenko, A., Shustov, A., Lyapunov, S.: Numerical, Experimental and Flight Investigation of MAV Aerodynamics, et al, pp. 17–21 (2007)
4. Zhu, X., Guo, Z., Hou, Z.: Solar-powered airplanes: a historical perspective and future challenges. In: Progress in Aerospace Sciences (2014)
5. Serokhvostov, S., Chrkina, E.T.: Optimization of the trajectory and accumulator mass for the solar-powered airplane. In: International Council of the Aeronautics Sciences, (2016)
6. Nguyen, J., Lawrance, N., Fitch, R., Sukkarieh S.: Energy-constrained motion planning for Information gathering with autonomous aerial soaring." et al.

Fig. 88.1 RC Aircraft being used for this experiment

VazaZika: A Software Platform for Surveillance and Control of Mosquito-Borne Diseases

Eduardo Fernandes, Anderson Uchôa, Leonardo Sousa, Anderson Oliveira,
Rafael de Mello, Luiz Paulo Barroca, Diogo Carvalho, Alessandro Garcia,
Baldoino Fonseca, and Leopoldo Teixeira

89.1 Contextualization

The *Aedes aegypti* mosquito transmits various diseases such as Zika [1]. Mosquito-breeding sites have rapidly spread worldwide due to poor basic sanitation plus warm and humid weather [1]. Thus, various countries have started to promote public healthcare solutions aimed to support the surveillance and control of mosquito-related diseases. Unfortunately, these solutions may fall short in engaging citizens with essential tasks, especially the report of mosquito breeding sites.

In 2015, we have introduced the VazaDengue [2] platform with the purpose of collecting and managing reports of mosquito breeding sites in Brazil. We aimed to provide the public health agents with reports that help them track disease outbreaks. The platform consisted of a mobile and a web system integrated via web services. The platform offered different features: (1) citizens report the location of mosquito breeding sites through the mobile system; (2) citizens and health agents monitor the reported locations through a dynamic map provided by the mobile and the web

systems; and (3) the web system monitors social media like Twitter [3] for automatically identifying reports of mosquito breeding sites.

After deploying VazaDengue, we observed a decay in the number of new platform users and views from April 2015 (VazaDengue release) to April 2018. Such decay suggested a lack of continuous user engagement with the platform. Thus, the health agents had an insufficient number of reports to cope with disease outbreaks. We then decided to incorporate gamification [4] into VazaDengue. Gamification means applying game elements and rules to non-game contexts for engaging people [4]. We aimed to make fun the constant report of mosquito breeding sites through game elements and rules.

Gamifying the legacy platform was far from trivial. It has basically required (1) to draw inspiration from successful gamified platforms and (2) to adapt an existing gamification method [5] to support specific activities of gamifying existing platforms (e.g., revisiting requirements and architecture). This paper presents the VazaZika gamification process and conceptual model. The process has involved Brazilian public health agents and other professionals, such as epidemiologists. We aim to support software engineers in reusing knowledge from our experience with gamifying our platform.

E. Fernandes (✉) · A. Uchôa · L. Sousa · A. Oliveira · R. de Mello
A. Garcia
Informatics Department, Pontifical Catholic University of Rio de Janeiro (PUC-Rio), Rio de Janeiro, Brazil
e-mail: emfernandes@inf.puc-rio.br; auchoa@inf.puc-rio.br; lsousa@inf.puc-rio.br; aoliveira@inf.puc-rio.br; rmaiani@inf.puc-rio.br; afgarcia@inf.puc-rio.br

L. P. Barroca · D. Carvalho · B. Fonseca
Computing Institute, Federal University of Alagoas (UFAL), Maceió, Brazil
e-mail: baldoino@ic.ufal.br

L. Teixeira
Informatics Center, Federal University of Pernambuco (UFPE), Recife, Brazil
e-mail: lmt@cin.ufpe.br

89.2 Gamification Goals and Requirements

89.2.1 Defining and Prioritizing Goals

Once we evolved the VazaDengue legacy platform, we had to reason about which *gamification goals* we would like to achieve. The elicitation of these goals was based on our experience with implementing and monitoring the user engagement with those systems. The definition and prioritization

S. Latifi (ed.), *16th International Conference on Information Technology-New Generations (ITNG 2019)*,
Advances in Intelligent Systems and Computing 800,
https://doi.org/10.1007/978-3-030-14070-0_89

of goals have enabled the VazaZika development team to choose the game elements and rules that best fit these goals in the future. We have defined the gamification goals via three steps as follows.

Step 1: we have met with the public health agents aiming at understanding the VazaZika system domain, thereby answering questions such as "How do mosquito-borne diseases spread?" and "What tasks can citizens perform to support both disease surveillance and control?" Our result was an initial list of gamification goals. **Step 2:** we have ranked and refined the previously obtained list with the purpose of (1) discarding redundant goals, (2) defining more specific goals, and (3) identifying what goals should be addressed first. **Step 3:** we have discussed to what extent gamification could address each goal.

We list the VazaZika systems goals sorted by descending priorities as follows. **Goal 1:** promote a constant report of mosquito breeding sites such that tracking disease outbreaks and eliminating sites become easier for the health agents. **Goal 2:** promote such reports in all Brazilian locations. **Goal 3:** promote varied tasks in terms of purposes, difficulty, and user engagement. **Goal 4:** provide tasks to be performed individually and in teams by citizens, in order to spread the systems' user base.

89.2.2 Defining Requirements

After defining and prioritizing goals, we have performed three steps to define requirements as follows. **Step 1:** we have elicited the gamification contexts of VazaZika from the existing platform. We aimed to answer questions such as "What factors do constraint the user engagement in VazaDengue?" and "Which existing functionalities should we gamify to boost the users' engagement?". As a result, we have discarded the gamification of certain functionalities provided by VazaDengue. For instance, the report of disease cases was discarded due to a health agent demand. We provide additional details in our research companion website [6].

Step 2: we have discussed possible users of our gamified platform. We aimed to address questions such as "What are the potential users of the gamified systems?" and "How should citizens interact with our systems?" We addressed these questions by using personas [7] that help to describe user profiles based on their possible needs and expectations with a system. Profiles usually contain the user name, age, and interaction contexts [7]. We have elicited five personas for VazaZika as exemplified in the following. **Persona 1:** *Laura* is *18 years old*, she *loves playing games*, and she *lives in a community affected by several disease cases*. Our website [6] presents the full persona list.

Step 3: based on the elicited personas, as well as the existing VazaDengue systems, we have elicited the non-functional requirements for VazaZika. In total, we have defined: (1) five functional requirements, such as *The citizen can report mosquito breeding sites through text, pictures, and geolocation data*; (2) four gamification-specific requirements, such as *The citizen can perform tasks either alone or as part of a team*; and (3) six non-functional requirements, such as *The system must inter-operate through a shared communication protocol*. We present the complete requirements lists in the research companion website [6].

89.3 Conceptual Model

89.3.1 Defining Game Elements and Rules

We performed the following steps to define game elements for VazaZika. **Step 1:** by compiling a set of ten platforms that we are familiar with (e.g., Doulingo and Waze), we aimed to know: "What game elements used by familiar systems we could implement in VazaZika for engaging users?" Our goal was identifying game elements that could be perceived as successful in engaging users. We have identified 13 game elements in the platforms, which we describe in our website [6]. Points and badges are the most frequent game elements, implemented by 70% and 60% of the studied platforms. **Step 2:** as a complement to the compilation of game elements, we relied on gamification for answering this question: "What game elements VazaZika should implement to properly engage users?"

We have decided to implement 11 out of the 13 game elements elicited in Step 1 (see [6]). In a first moment, we decided not to implement both *chats* and *notifications* in the platform. However, as users start using and engaging with VazaZika, we plan to monitor users' interactions to observe whether (1) users are likely to engage through chats based on their interactions via *comments*, and if (2) adding notifications could promote a more frequent interaction with certain features that eventually become less frequently used in the long term.

89.3.2 Defining Rules and Conceptual Model

Step 1: we defined two rule categories for gamifying our platform. *Relations between the set of systems and their users (SU)* determine how the user interacts with the platform through the game elements. As an example, the VazaZika users earn points after reporting a mosquito breeding site. It aims at acknowledging the citizen so that he/she feels encouraged to report sites again. *Relations* *between a pair of game elements (EE)* determine how one element affects another. E.g., points assigned to a VazaZika user count on the user ranking. Our website [6] lists the implemented rules.

Step 2: we built the gamification conceptual model [4] for visually representing how the gamified platform should operate internally. Figure 89.1 introduces the VazaZika gamification model. The figure represents the system user, game elements, and rules. Arrows represent the gamification rules as follows. Continuous arrows represent SU rules and dotted arrows represent EE rules. Aimed to address the stakeholders' needs, we did not gamify the interactions of health agents with the platform. Thus, our model does not comprise these actions.

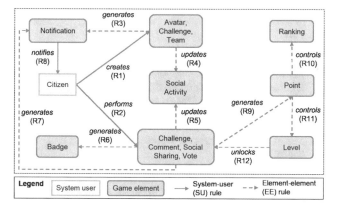

Fig. 89.1 The VazaZika gamification conceptual model

89.4 Prototyping and Implementation

89.4.1 System Prototyping

We performed five steps to define the VazaZika interface. **Step 1:** we conducted brainstorming rounds aimed to discuss the visual items of both mobile and web systems' interfaces. Our goal was answering questions like "How should we represent each game element as a visual item in the VazaZika's user interface?" and "How to organize the visual items in such a way that it makes easy to access and manage them?" **Step 2:** a few developers have drawn low-fidelity prototypes that estimate how the visual elements should look like and be organized.

89.4.2 System Implementation

Step 3: Figure 89.2a shows the main screen of the mobile system, and Fig. 89.2b shows the main screen of the web system. We highlight in the figure the visual representations of each game element implemented by VazaZika as follows: Point (A), Badge (B), Ranking (C), Social Sharing (D), Vote (E), Avatar (F), Level (G), Team (H), Social Activity (I), Challenge (J), and Comment (K).

We relied on agile development principles [8], such as short development cycles (biweekly) and iterative design, implementation, and testing. Regarding the technologies that we employed to implement the VazaZika systems, we used the following. **Mobile system:** React Framework for multi-

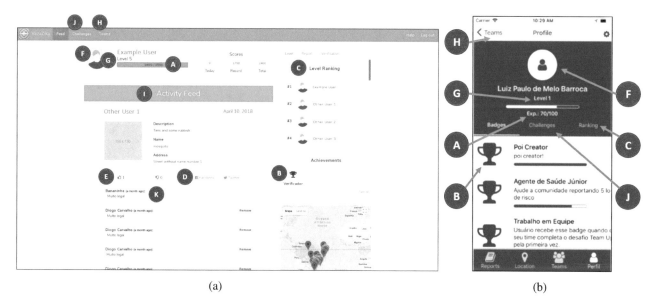

(a)   (b)

Fig. 89.2 Main screen per VazaZika system. (**a**) Mobile system. (**b**) Web system

platform interface development; SQLite for data persistence. **Web system:** PostgreSQL for data persistence; IntelliJ IDEA for development. **Both systems:** Balsamiq mockup tool; Git for system version control; JSON as our standard format for data transmission; and REST as the object transfer pattern.

89.5   Potential Knowledge Reuse

An Adapted Gamification Method: we relied on a gamification method [5] to guide the process of gamifying VazaZika. However, this method, similarly to others provided by the current literature [9, 10], focuses on the gamification of systems built from scratch. Thus, they do not consider the existing knowledge about a system (e.g., domain, requirements, and architecture). As a response to that limitation, this paper documents the gamification process, which includes activities such as reasoning about the existing system and past users (Steps 2 and 3 of Sect. 89.2.2). We expect to support developers in gamifying existing systems.

A Conceptual Gamification Model for Healthcare Systems: we have defined our conceptual model based on 11 game elements recurrently implemented by successful systems, such as Duolingo and Waze. We also carefully designed 12 rules aimed to interrelate systems and users, as well as pairs of game elements, in order to achieve well-defined gamification goals. Thus, system designers in charge of eliciting either game elements or rules to a gamified system could reuse our conceptual model to define the game elements and rules that help them achieve a satisfactory users' engagement, especially in the case of healthcare systems with a similar purpose of VazaZika (e.g., [11]).

Acknowledgements This is work is funded by Newton Fund, FAPEAL under the grant #60030 1201/2016, and CAPES/Procad project under grant #175956.

References

1. Gyawali, N., Bradbury, R., Taylor-Robinson, A.: The global spread of zika virus. Infect. Dis. Poverty **5**(1), 37 (2016)
2. Sousa, L., de Mello, R., Cedrim, D., Garcia, A., Missier, P., Uchôa, A., Oliveira, A., Romanovsky, A.: VazaDengue. Inf. Syst. (IS) **75**, 26–42 (2018)
3. Missier, P., McClean, C., Carlton, J., Cedrim, D., Silva, L., Garcia, A., Plastino, A., Romanovsky, A.: Recruiting from the network. In: 17th International Conference on Web Engineering (ICWE), pp. 437–445 (2017)
4. Werbach, K., Hunter, D.: For the Win. Wharton Digital Press, Philadelphia (2012)
5. Morschheuser, B., Hassan, L., Werder, K., Hamari, J.: How to design gamification? Inf. Softw. Technol. **95**, 219–237 (2017)
6. Fernandes, E., Uchôa, A., Sousa, L., Oliveira, A., de Mello, R., Barroca, L.P., Carvalho, D., Garcia, A., Fonseca, B., Teixeira, L.: Research companion website (2018). Available at: https://anderson-uchoa.github.io/ITNG2019/
7. Grudin, J., Pruitt, J.: Personas, participatory design and product development. In: 4th Participation and Design Conference (PDC), pp. 144–152 (2002)
8. Martin, R.: Agile Software Development. Prentice Hall, Upper Saddle River (2002)
9. Herzig, P., Ameling, M., Schill, A.: A generic platform for enterprise gamification. In: 10th Joint Working Conference on Software Architecture and 6th European Conference on Software Architecture (WICSA-ECSA), pp. 219–223 (2012)
10. Kardan, A., Arani, A.K.: A novel gamification-based architecture for web environments. In: 2nd International Conference on Web Research (ICWR), pp. 125–130 (2016)
11. Fernandes, E., Silva, M.A., Cagnin, M.I.: Sigs-s: a web application and a mobile application for social and health care data management. Braz. J. Inf. Syst. **9**(1), 81–100 (2016)

The Development of a Software System for Solving the Problem of Data Classification and Data Processing

Askar Boranbayev, Seilkhan Boranbayev, Askar Nurbekov, and Roman Taberkhan

90.1 Introduction

This work was done as part of a research grant №AP05131784 of the Ministry of Education and Science of the Republic of Kazakhstan for 2018–2020.

A recognition of a human face on an image is the main key concept in tasks such as emote recognitions and automated tracking after moving people in a camera view field. Face recognition tasks can be implemented using several approaches based on the following methods: statistical methods, graph theory, neural networks. In the developed software system, HOG, Viola-Jones and convolutional neural networks were applied. In the HOG algorithm [1], the appearance and shape of the object in the image region are described by the distribution of the intensity gradients. The implementation of descriptors is done by dividing the image into small, connected cells. For each cell, histograms of the gradient directions are calculated. The combination of histograms is a descriptor. The HOG descriptor is a good tool for finding people in images. Based on the principle of the scanning window The Viola-Jones method is one of the highly effective and popular methods for searching and classifying objects in images and video sequences in real time. The research shows that this method works well and detects facial features even when observing an object at an angle of up to 30°, reaching a recognition accuracy value of over 85%. However, at an angle of inclination greater than 30°, the probability of face detection decreases. This disadvantage makes it difficult to use the algorithm in modern production systems, taking into account their growing needs. In the PCA method faces are represented as a set (vector) of the main components of the images – "Eigenfaces". The image corresponding to each such vector has a face-like shape. PCA has two useful properties when used in face recognition. First, it can be used to reduce the dimension of the feature vectors. The second useful feature is that the PCA eliminates all statistical covariance in the vectors of the transformed objects. This means that the covariance matrix for the vectors of the transformed (learning) features will always be diagonal.

90.2 The Software System for Solving the Recognition and Classification Problem

HOG and Viola-Jones methods are used to solve the recognition and classification problem in the software system. Consider the application of the algorithm based on the HOG method for face recognition. To detect faces in an image, it is made in black and white. color data is not needed to detect faces. Then for each individual pixel in the image, its immediate surroundings. It is necessary to find out how dark the current pixel is in comparison with the immediately adjacent pixels. Then the direction in which the image becomes darker is determined [2]. If you repeat this process, each pixel will be replaced with an arrow. The image is divided into small squares of 16×16 pixels in each. The square counts how many gradient arrows show in each direction (i.e. how many arrows point up, up-right, right, etc.). Then the considered square in the image is replaced by an arrow with the direction prevailing in this square.

A. Boranbayev (✉)
Department of Computer Science, Nazarbayev University, Astana, Kazakhstan
e-mail: aboranbayev@nu.edu.kz

S. Boranbayev · A. Nurbekov · R. Taberkhan
Department of Information Systems, L.N. Gumilyov Eurasian National University, Astana, Kazakhstan
e-mail: sboranba@yandex.kz

© Springer Nature Switzerland AG 2019
S. Latifi (ed.), *16th International Conference on Information Technology-New Generations (ITNG 2019)*,
Advances in Intelligent Systems and Computing 800,
https://doi.org/10.1007/978-3-030-14070-0_90

As a result, the original image becomes a representation that shows the basic structure of the face in a simple form. Then an algorithm for estimating anthropometric points is used. There are 68 specific points (marks) on the face, the protruding part of the chin, the edges of each eye, the outer and inner edges of the eyebrows, and the like. Then the machine learning algorithm is set up to search for these 68 specific points on the face [2]. Now that you know where the eyes and mouth are, you can rotate, resize and move the image so that your eyes and mouth are centered as best you can. Only basic image transformations are done, such as rotation and scaling, which preserve parallel lines. In this case, no matter how the face is rotated, the eyes and mouth can be centered so that they are approximately in the same position in the image. This will make the accuracy of the next step much higher. The next step is, actually, the very face recognition. At this step, the main task is to train a deep convolutional neural network. The network creates 128 characteristics for each face. The learning process is valid when examining 3 faces at the same time: (1)The training image of a famous person's face is loaded; (2)Another image of the same person's face is loaded; (3)The image of the face of the new person is loaded. Further, the algorithm studies the characteristics created for each of the three images, makes the neural network adjustment so that the characteristics created for images 1 and 2 are closer to each other, and for images 2 and 3 – further. After repeating this step m times for n images of different people, the neural network is able to reliably create 128 characteristics for each person. Any 10–15 different images of the same person may well give the same characteristics. The algorithm allows you to do in the database search for an image that has characteristics that are closest to the characteristics of the desired image [3–6]. Before you begin the process of recognizing faces, you need to go through the following steps: localize faces on the image; align the face images (geometric and luminous); identify

signs; directly the recognition of faces – a comparison of the found characteristics with the standards previously laid in the database. To solve the problem of detecting faces on images and video streams in the software system, the HOG and Viola-Jones methods are used, for the recognition task, convolutional neural networks.

To carry out the work we used: Computer: Notebook HPEnvy 4, Intel Core i5-3317 U CPU@1.70GHz, 8 Gb RAM; Nvidia Geforce GT 740 M, 2Gb graphics card; Operating system: Win10, 64-bit. Language programming language Python 3.5; g ++ compiler; libraries for developing python-dev; library for scientific computing NumPy; library dlib, OpenFace, library for mathematics SciPy; library BLAS – basic subroutines of linear algebra; Git – distributed version control system.

90.3 Discussion

In order to carry out a comparative analysis of the work of HOG and Viola-Jones methods, the program has a graph showing the processing time of the image. Having analyzed the methods on a large number of images, it was found that the HOG for accuracy of detection works better, is able to detect faces at various distortions, with a head rotation of up to 40°; however, according to the detection time, Viola-Jones works faster, as illustrated in Fig. 90.1.

The Viola-Jones algorithm is trainable. For training it is necessary to have a base of positive and negative images. On positive images there are faces of people of different ages, with glasses, mustaches, etc., and on the negative – background. For successful execution of the algorithm, negative images should be larger. Since the training of classifiers is a long and complex process, the implementation of the selection stage takes place using a ready-made set of char-

Fig. 90.1 Comparison of the detection algorithms running time

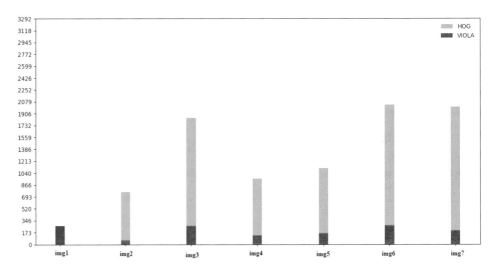

acteristics. The base is cascades from the OpenCV library: «haarcascade_frontalface_alt2.xml».

To evaluate the algorithms, a sample of 50 color images totaling more than 500 faces was applied. The experimental results show that the HOG method is more accurate than the Viola-Jones methods. The HOG method detects about 92% of the individuals represented on the test set of images, Viola-Jones – 85%. In addition, tests have shown that the HOG method significantly reduces the likelihood of false detection. To recognize faces in the image, a convolutional neural network was used. This method has certain advantages, such as the ability to recognize faces in real time from an input image or from a video stream; At the frontal position of the face and with a large scale, fully authentic recognition is performed; the software system gives positive results and when turning the face more than 20 degrees, and under poor lighting conditions.

90.4   Conclusion

The software system finds the area of the face, makes recognition of this face. Then there is a search in the database, and if the face is in the database, then there is taken available additional information (for example, name, age, etc.). A SQLite database was created, containing about 2000 images of people. Frames taken with different positions of the head and facial expressions were normalized. A set of data was created for training the neural network with variations in the form of distortions and color filters. In the program system, the existing base of famous personalities has been added.

Based on the results of the research, the Viola-Jones and HOG methods produced a good detection rate, but the Viola-Jones method works faster. Accuracy of recognition of the convolutional neural network method varies from 70–100% depending on external conditions.

Acknowledgment This work was done as part of a research grant №AP05131784 of the Ministry of Education and Science of the Republic of Kazakhstan for 2018–2020.

References

1. Extract HOG Features. [Electronic resource]: 2017 – URL: http://www.mathworks.com/help/vision/ref/extracthogfeatures.html
2. Yang, M., Ahuja, N., Kriegman, D.: Face recognition using kernel eigenfaces. Image Process.: IEEE Transactions. **1**, 37–40 (2000)
3. Boranbayev, S., Nurkas, A., Tulebayev, Y., Tashtai, B.: Method of processing big data. In: Advances in Intelligent Systems and Computing, vol. 738, pp. 757–758 (2018)
4. Boranbayev, A., Boranbayev, S., Nurusheva, A.: Analyzing methods of recognition, classification and development of a software system. In: Proceedings of Intelligent Systems Conference (IntelliSys) 2018, pp. 1055–1061. London, UK 6–7 Sept 2018
5. Boranbayev, S., Altayev, S., Boranbayev, A.: Applying the method of diverse redundancy in cloud based systems for increasing reliability. In: Proceedings of the 12th International Conference on Information Technology: New Generations (ITNG 2015), pp.796–799. Las Vegas, Nevada, USA 13–15 Apr 2015
6. Boranbayev, S., Boranbayev, A., Altayev, S., Nurbekov, A.: Mathematical model for optimal designing of reliable information systems. In: Proceedings of the 2014 IEEE 8th International Conference on Application of Information and Communication Technologies-AICT2014, pp. 123–127. Astana, Kazakhstan 15–17 Oct 2014

Comparison of Transnational Education Delivery Models

Tony de Souza-Daw, Sitalakshmi Venkatraman, Kiran Fahd, Sazia Parvin,
Logesvary Krishnasamy, Joanna Jackson, and Samuel Kaspi

91.1 Introduction

In several countries, Higher Education (HE) is still largely
operating with long-established teaching modes such as large
classroom-based delivery, which dates back centuries [1].
Historically, teaching was only provided to the gifted few,
and now formal learning has become a primary necessity
for everyone in the modern world. This has resulted in
the requirement for HE to emerge with diverse teaching
strategies in order to meet the learning style and academic
experience preferred by the ever-growing student population
[2, 3]. Together with the global economy and the advancements in technology, the education service sector of today
is flooded with heavy competition due to countless HE
providers springing up worldwide, which calls for scrutiny
of their educational standard and service quality [4, 5].

Large and well-known government-owned public HE
providers are now in direct competition with smaller and
more agile private providers that are growing quickly. In
addition, the HE funding sources from both private and
public funds are decreasing tremendously, despite the
increasing costs associated with teaching [6]. These have
dramatically changed the HE landscape in the education
service sector taking the role as a revenue generator.
This trend is on the rise with the emergence of an open

global market in education and followed by the increase
in student mobility witnessed internationally [7]. Further,
World Trade Organization's General Agreement on Trade in
Services (GATS) has facilitated in the recent establishment
of Transnational Education (TNE) partnerships worldwide
[8]. Even though TNE was introduced few decades back,
the proliferation of different forms of TNE has been
phenomenal only recently despite being recognized as a risky
venture with demonstrated failures in the past [9]. These
uncertainties of TNE form the primary motivation for this
research to understand the landscape of TNE and explore its
opportunities, challenges, different delivery models as well
as influencing success factors that would help positioning
TNE for the future.

In Sect. 91.2, we provide the basic landscape of TNE,
understanding its dimensions, opportunities, benefits as well
as challenges suggesting the need for this study. We compare the TNE delivery models in Sect. 91.6 and provide
our observations and recommendations in Sect. 91.7. Final
conclusions and future research directions are described in
Sect. 91.8.

91.2 The TNE Landscape

The original concept of TNE came from the word "transnational", which implies transactions that extend or go beyond
national borders and was developed as educational services,
teaching aid or cross-border education provided by the developed countries to the developing countries. However,
with the rapid growth of TNE in the global educational
marketplace, international regulations were established by
WTO and Revised Code of Good Practice was developed
by UNESCO and the Council of Europe [8, 10]. In line
with these international standards as well as Asia-Pacific
European Cooperation (APEC) context, TNE is defined as:

T. de Souza-Daw · S. Venkatraman · K. Fahd · S. Parvin (✉)
L. Krishnasamy · J. Jackson · S. Kaspi
Department of IT, Melbourne Polytechnic, Melbourne, VIC, Australia
e-mail: tonydesouza-daw@melbournepolytechnic.edu.au;
SitaVenkat@melbournepolytechnic.edu.au;
KiranFahad@melbournepolytechnic.edu.au;
saziaparvin@melbournepolytechnic.edu.au;
logesvarykrishnasamy@melbournepolytechnic.edu.au;
JoannaJackson@melbournepolytechnic.edu.au;
SamKaspi@melbournepolytechnic.edu.au

© Springer Nature Switzerland AG 2019
S. Latifi (ed.), *16th International Conference on Information Technology-New Generations (ITNG 2019)*,
Advances in Intelligent Systems and Computing 800,
https://doi.org/10.1007/978-3-030-14070-0_91

.... all types and modes of delivery of higher education study programs, or sets of courses of study, or educational services (including those of distance education) in which the learners are located in a country different from the one where the awarding institution is based.

In many countries such as the UK, USA Australia and India, TNE is considered as part of a wider set of cross-border activities and internationalization agenda, including establishment of international schools, research and industry partnerships [11–13]. However, in this paper we consider TNE from the primary context of offering HE programs (post-secondary education) ranging from vocational/apprenticeship education to professional awards such as certificates, diplomas and higher degrees leading to PhDs.

91.3   Opportunities/Benefits of the TNE

The transnational education (TNE) can be beneficial from different areas [14]: economic, academic, cultural as well as skills development. The opportunities exist for developing local skills, reducing capital outflow and brain drain, and putting less pressure on local education system. The impact of each area is discussed below.

Economic Impact High-quality foreign education can be achieved at lower cost through transnational education. Students can complete the desired degree at affordable, reasonable, acceptable or cheaper rate than studying the entire program abroad, which actually reduces the capital outflow. Students can continue their work and study at the same time which is being considered as a cost-effective education. As students can get their education without leaving their home country or region, it can reduce the brain drain from developing countries. The transnational education also provides an ideal opportunity for working professionals who are keen to upgrade their qualifications while working full time as they have limited scope through local education systems [15]. It allows even the working professionals to obtain higher qualifications without leaving their home country and their current employment.

Academic Impact Quality academic education can be achieved through transnational education and can be improved through the engagement of competitive HE program development among foreign and local institutions. In addition, innovative pedagogies incorporating problem-based learning and self-directed learning can be achieved from TNE [16]. TNE can offer new opportunities to practice higher academic standards of curriculum and delivery approaches. Such opportunities foster incorporation of local and international contexts of the discipline, which can enhance the teaching and learning practices to global standards. Hence, students can receive knowledge and skills with international standards of academic quality from diverse TNE programs.

Cultural/Skill Development Impact TNE provides the right platform for students to interact with fellow students and teachers from overseas with culturally diverse background, and to have the exposure to adapt themselves in a multi-cultural classroom environment as compared to their local programs. Opportunity also exists for students' academic quality improvement as the foreign and local institutions are competitively engaged in various TNE program development. Recently, universities of many host nations have emerged as international competitors in the race for talent hunts, research and innovation through TNE partnerships. Hence, there is a wider scope for students' skill development which can enhance their global career opportunities. Based on current trend in economic progress and geopolitical development, the outflow of knowledge seekers from the developing nations will compensate the inflow of students especially between Asia, Europe and North America creating an overall skill development and cultural diversity in all nations [15].

91.4   Challenges of TNE

Even though the demand for transnational education (TNE) programs offered by Australian universities has increased, especially in China [17], a range of challenges persist in the delivery of the TNE programs. Both TNE flying faculty and TNE home-based faculty face challenges in relation to the transnational education context [18, 19].

Location Challenges In transnational partnerships with relation to the geographical location of partnership, most challenges identified by the researchers are related to intercultural aspect. Culture is a broad term in TNE context; it encompasses language barriers, work culture, cultural habits, traditions, learning styles and communication styles in the partner institutions [19–22]. Similar challenges exist relating to administrative aspects that affects both the academic and operational management [20, 21]. Another challenge can be due to different perceptions of organizational status and hierarchy between members of academic staff as well as teachers and students [23]. Some studies have identified differences in learning habits and communication styles of students coming from Australian and Chinese students. It is due to different cultural backgrounds, which is perceived as a challenge by the academic staff [19, 21]. Intercultural challenges also include resistance to the western-centric programs to avoid over westernization of Asian unique civilizations like customs, languages, traditions and heritages [20, 24]. Researchers have also identified international policy challenges mainly due to the organization structure and culture [21]. These kind of issues involves procedural difficulties that emerge from incompatibilities among national and international regulations [21, 25], and additional demands or requirements from national and international funding institutions. Even after multiple changes of European Union policies, universities still need to follow national regulation in order to award a TNE degree [21, 26].

Curriculum Challenges These are internal challenges that relate to the pre-set goals and objectives, as well as the content of the TNE partnership. Challenges arise due to personal and professional approach of academic staff, weaknesses in internal arrangements like lack of responsibility, offshore staff equality and insufficient capacities of the TNE collaborator. Academic staff overburdened with their existing workload would show low level of commitment or priority or resistance to the collaboration [20–22]. The home-based academic staff would raise issues of isolation due to lack of equality, belonging, and professionalism from the provider university [20]. Challenges are also relating to the quality assurance and control of the program as many TNE programs delivered are not at the same level and standard of their source country [18, 20].

Implementation Challenges These challenges relate to rules, regulations, organizational protocols, resources and technical capacity of collaborating universities [21]. Regulation issues result from inconsistent legal and policy frameworks and barriers in establishing collaboration agreements between partner universities - for example, presence of national or institutional rules that cannot be negotiated [21, 26]. Organizational protocol issues may result from differences in organizational and administrative structures and processes, and the contextual constraints such as different time zones and academic calendars - for example, usage of different terminology and criteria for credit weightage [21, 27]. Resource challenges arise due to the limited support from finance, infrastructure and human resources. Other issues are related to the magnitude of technical infrastructure, data handling and security, availability of broadband, skilled technical support, and computational capacity to collaborate over distance. Unequal contribution by partner universities, different levels of funding, or insufficient computer or laptops for students are some of the operational challenges that can affect TNE delivery [21].

91.5 Need for the Study

HE institutions in many advanced countries are adopting TNE as a strategy to deliver higher education and other associated qualifications across geographical borders to students located either partly or wholly outside their respective countries. TNE complements several domestic HE offerings with overseas educational services, through face-to-face delivery in other countries, distance learning (including online) and blended learning approaches. From the pedagogical perspectives, such educational services offer new opportunities to practise higher academic standards of curriculum and delivery approaches to foster the local and international context of the discipline with the aim of enhancing the teaching and learning practices to global standards [28, 29]. From the perspectives of students as consumers, the TNE offers diverse opportunities for graduates to receive transcultural knowledge and skills with the aim to enhance their global career outcomes that would train them to work on both local and transnational issues of their workplace [30, 31]. These perspectives could be constrained by the mode of delivery adopted with various transnational arrangements to offer articulation, validation, franchise, branch campus, joint/dual degrees, double degrees, distance learning, and many other forms of partnerships. Overall, TNE programs are being offered through a diverse and complex range of modes of delivery with even mobility arrangements of teaching staff in providing services at local branch, satellite campuses, twinning partnerships, franchising, etc.

In recent years, with the global competition in HE and the need for self-generating revenues, universities and institutions worldwide are venturing into TNE with the main purpose of generating generous incomes. While some TNE ventures have been successful, a number of them have demonstrated failure [6, 9]. The commercial viability of TNE is based on various parameters, including the cross-border mobility of students, programs, providers or a combination of these. In this paper, we explore the TNE delivery models in the context of Australia to identify the influencing factors that have an impact on the TNE program types and their delivery modes.

91.6 Comparison of Proposed TNE Delivery Models

In the past couple of decades, many different approaches and models to TNE have evolved making the TNE landscape quite complex in practice. When the curriculum and teaching material are developed by one cultural context of a country and delivered to a different cultural context of another country, there could be issues in designing and validating such curriculum to cater to culturally different learning styles and academic experiences. Hence, TNE could be a risky venture as demonstrated by the failure of some Australian universities in their TNE ventures. It is difficult to develop a sustainable TNE venture that can adapt to the changing operational circumstances in this globalisation era. Hence, we have proposed some contemporary TNE delivery models to meet diverse requirements from both supply and demand perspectives in order to have a sustainable TNE venture. We consider five influencing factors that can impact on these models and these factors are:

- Student Affordability
- Operating Costs
- Industry Involvement
- Teaching Method
- Assessment Method

We group HE delivery modes into four categories based on the global trends and student learning styles as discussed in the previous section. These four TNE delivery modes are:

- Online
- Face-to-Face
- Mixed Mode
- Block Teaching

Most TNE programs when practically deployed would include a blend of these four delivery modes for different modules based on various influencing factors. We compare these four delivery modes based on the abovementioned five influencing factors as shown in Table 91.1. The advantage of these delivery modes for TNE programs is that they could also be adopted in domestic offering so to address any operational issues locally before considering cross border delivery.

Next, we consider three broad campus delivery types for TNE programs as follows:

- Offshore Campus (direct in-country presence)
- Twinning Campus (partnership teaching provision)
- Onshore Campus (articulation agreements)

In Table 91.2, we present the comparison of these three campus delivery types based on the first two influencing factors, namely Student Affordability and Operating Cost, while the impact by other three influencing factors will be based on the four contemporary HE program options proposed in this study. The four TNE program options are:

- Degree-Apprenticeships
- Start-up Focus Degrees
- Tailored Studies
- Multiple Major Studies

Several studies in the past have viewed TNE programs mainly from the HE provider perspectives, rather than the student perspective. The above mentioned four HE program options are proposed from student perspective for a successful TNE delivery model and Table 91.3 presents the impact of the five influencing factors on these four HE program options. It has been reported that the quality of many imported TNE programs are lower than those offered in the HE provider country [18, 20]. Our proposed TNE delivery mod-

Table 91.1 Delivery modes and their influencing factors

Factors	Delivery modes			
	Online	Face-to-face	Mixed mode	Block teaching
Student affordability	High (affordable)	Medium	Medium	Medium
Operating costs	Low Web site, servers, online material	High Building costs	High Building costs, plus online.	Same as mixed mode
Industry involvement	Low	Low	Low	Low
Teaching method	Pre-recorded lectures, remote lectures, remote labs	Classroom teaching Laboratory teaching	Classroom teaching Laboratory teaching Online support	Classroom teaching Laboratory teaching Online support
Assessment method	Must be delivered remotely (e.g. practical test with real time configurations)	Traditional Presentations/Report/ Exams/Tests/Practical	Traditional Presentations/Report/ Exams/Tests/Practical Video presentation	Traditional Presentations/Report/ Exams/Tests/Practical Video presentation

Table 91.2 Campus delivery types and their influencing factors

Factors	Campus delivery types		
	Offshore campus	Twinning campus	Onshore campus
Student affordability	Affordable (cheaper in host country)	Medium affordable (home country fee is cheaper)	High (high international fees)
Operating costs	High (building costs, councils)	Low (partner with another institute)	Medium (agent fees, traditional costs)

Table 91.3 Contemporary HE program options and their influencing factors

Factors	Contemporary HE program options			
	Degree-Apprenticeships	Start-up Focus Degrees	Tailored Studies	Multiple Major Studies
Student affordability	High Paid to study	Low Higher fees	Medium Low costs for part of the degree	High More subjects
Operating costs	Relatively low Need an assessor Classrooms are not needed	Medium Business pods rather than classrooms. Commercial Software Business Mentor	Low Normal classroom activities	Low Normal classroom activities
Industry involvement	High Entirely at Workplace	Medium Business Mentor	Low	Low
Teaching method	Work Placement Assessment Assessors	Mentor/Educator	Most efficient when taught in 3–4 week block mode	Either block mode or traditional methods.
Assessment method	On-the-job mapped to HE Learning Outcomes Educator Assessor	Real industry scenarios mapped to HE Learning Outcomes Educator Assessor	Before start of classes Formative and Summative	Before start of classes Formative and Summative

els, which can take a mix of the abovementioned delivery modes (Online, Face-to-Face, Mixed Mode and Block Teaching) with a mix of campus delivery modes (Offshore Campus, Twinning Campus and Onshore Campus) in offering our proposed HE program options (Degree-Apprenticeships, Start-up Focus Degrees, Tailored Studies and Multiple Major Studies) would facilitate a more sustainable TNE venture. Since these options could be deployed in the HE provider country itself, the quality can be ensured. Further, the five main influencing factors (Student Affordability, Operating Costs, Industry Involvement, Teaching Method and Assessment Method) would be considered a priori before making a decision on the TNE venture and the right model to be adopted.

91.7 Observations and Recommendations

HE institutions worldwide operate in a highly competitive environment and there is no doubt that HE reforms are required in this era of globalisation. While TNE programs offer opportunities for HE reforms in the global marketplace [32], HE institutions seem to find it expensive and time-consuming in understanding the partnering country's regulatory and legal requirements. Therefore, in order to achieve the targeted revenues, both the TNE provider and the host

institution could minimise costs by providing insufficient infrastructure and services or try to increase the profit by exceeding the enrolment quotas, and these could affect the quality of the TNE program. Some of the host institutions are inexperienced to evaluate and maintain the quality of the TNE program and this could result in decline or failure resulting in 'teach-out' situations or withdrawal of the TNE partnership.

While TNE offers advantage for Australian providers, there is high pressure on HE institutions in making the HE programs affordable and at the same time in offering a variety of delivery modes that are student-centric in order to cater to different student learning styles. Career focussed tailored studies and apprenticeship degree programs are a low-cost option for students wanting to get into the workforce as early as possible, and this will ease the pressure on skills migration. Start-up focus studies will support the Australian start-up business communities, which requires government and industry support. Teaching in 3–4 weeks block mode would enable much flexibility for students and staff. However, HE institutes are quite resistant to change. Overall, the most common influencing factors for various HE program options are 'Student Affordability' and 'Operating Costs'.

We recommend HE institutions to share best practices to facilitate learning from successful examples. They should explore innovative delivery models to compete internationally. The qualification standards and quality framework requirements should be communicated clearly and agreed by both the provider institution and the host institution in order to ensure high quality in their TNE delivery. Financial and sustainability risks should be avoided by adopting regular monitoring of costs, delivery and partnership. A good understanding of the associated costs and metrics for performance evaluation is essential. HE institutions should arrive at pooled knowledge and share good practices through common repositories or regional workshops that would benefit many other institutions venturing into TNE.

91.8 Conclusions and Future Research

In this global economy, several HE institutions are emerging to venture into TNE as a commercialisation strategy as well as to sustain the growing competition in the education sector. This paper presented an understanding of the opportunities and benefits of TNE as well as the challenges associated in its delivery. It identified various forms of TNE delivery and their evolution. This paper has developed a detailed understanding of the TNE landscape, describing several teaching models, delivery modes and HE program options both in traditional and in contemporary proposals. In particular, the TNE situation in the Australian context was explored and various TNE delivery models were compared. While several studies have viewed TNE from the perspective of western educational institutions, this paper has proposed innovative delivery approaches that are student-centric catering to different student needs in meeting their career outcomes. We observed that 'Student Affordability' and 'Operating Costs' are the most common influencing factors for various HE program options. Further, we considered the practicality and feasibility of delivery programs and the environment to have a focus on long term sustainability issues such as keeping the programs up-to-date in a constant changing environment, and facilitating rewards and promotions for teaching staff in a contemporary TNE model.

It is anticipated that future research will explore other curriculum issues in TNE such as transcultural differentiation, quality assurance and intercultural communication, as well as administration issues and inherent constraints that could have an impact on its delivery. A study on the TNE partnership experiences from the perspectives of both the provider institution and the host institution from a particular TNE delivery as a case study would benefit future TNE ventures. Future study will provide extensive observations from actual examples of TNE model implementation.

Acknowledgements This work has been supported by Melbourne Polytechnic Research and Scholarship Seeding Grant Program 2018.

References

1. Jackson, L.: Leaning "out" in higher education: a structural, post-colonial perspective. Policy Futures in Education. **15**(3), 295–308 (2017)
2. UK Department for Business, Innovation and Skills and Department for Education: Technical Education Reform: The Case for Change, pp. 1–28. UK Government, London. Policy Paper Ref: DFE-00163-2016 (2016)
3. New Zealand Productivity Commission: New Models of Tertiary Education: Final Report. New Zealand Government, Wellington (2017)
4. Hanushek, E.A., Schwerdt, G., Woessmann, L., Zhang, L.: General Education, Vocational Education, and Labor-Market Outcomes over the Lifecycle. J. Hum. Resour., University of Wisconsin Press. **52**(1), 48–87 (2017)
5. Mulkeen, J., Abdou, H.A., Leigh, J., Ward, P.: Degree and higher level apprenticeships: an empirical investigation of stakeholder perceptions of challenges and opportunities, Studies in Higher Education, ISSN 0307–5079, 1–14. (2017)
6. Lucianelli, G., Citro, F.: Financial conditions and financial sustainability in higher education: A literature review. In: Bolívar, M.P.R. (ed.) Financial Sustainability in Public Administration, pp. 23–53. Springer, Berlin (2017)
7. Jean Francois, E.: Building Global Education with a Local Perspective: An Introduction to Global Higher Education. Palgrave Macmillan, New York (2015)
8. The British Council.: The shape of things to come: the evolution of transnational education: data, definitions, opportunities and impacts analysis. (2013)
9. Nguyen, D.P., Vickers, M., Ly, T.M.C., Tran, M.D.: Internationalizing higher education (HE) in Vietnam: Insights from higher education leaders – an exploratory study. Educ. Train. **58**(2), 193–208 (2016)
10. Francois, E.J., et al. (eds.): Perspectives in Transnational Higher Education, pp. 3–22. Sense Publishers (2016)
11. American Council on Education: Internationalization in US Higher Education: The Student Perspective. American Council on Education, Washington, DC (2005)
12. Connelly, S., Garton, J.: Project report: enhancing Australian universities offshore QA processes – guidelines for 2+2 programs in China. Project for the Australian Vice-Chancellors' Committee, June 2005. Swinburne University of Technology, Melbourne, Australia. (2005)
13. Department for Business, Innovation and Skills: International Education – Global Growth and Prosperity: An Accompanying Analytical Narrative. United Kingdom Government, London (2013)
14. Department of Education: The Wider Benefits of Transnational Education to the UK. Robin Mellors-Bourne – Careers Research & Advisory Centre (CRAC) Ltd. (2017). https://www.qs.com/the-wider-benefits-of-international-higher-education-in-the-uk-a-report-summary/
15. Alam, F., Alam, Q., Chowdhury, H., Steiner, T.: Transnational education: Benefits, threats and challenges. Procedia Eng. **56**, 870–874 (2013). https://doi.org/10.1016/j.proeng.2013.03.209
16. British Council: The Positive Impact of Transnational Education. British Council (2013). https://www.britishcouncil.org/organisation/press/positive-impact-transnational-education
17. Huang, F.: Transnational higher education in Japan and China: a comparative study. In: Chapman, D.W., Cummings, W.K., Postiglione, G.A. (eds.) Crossing Borders in East Asian Higher Education. CERC Studies in Comparative Education, p. 27. Springer, Dordrecht (2010). https://doi.org/10.1007/978-94-007-0446-6_12
18. Phan, L.H.: Transnational Education Crossing 'Asia' and 'the West': Adjusted Desire, Transformative Mediocrity, Neo-colonial Disguise. Routledge, New York (2017)

19. Heffernan, T., Morrison, M., Basu, P., Sweeney, A.: Cultural differences, learning styles and transnational education. J. High. Educ. Policy Manag. **32**(1), 27–39 (2010)
20. O'Mahony, J.: Enhancing Student Learning and Teacher Development in Transnational Education. Higher Education Academy, York (2014)
21. Caniglia, G., Luederitz, C., Gross, M., Muhr, M., John, B., Keeler, L., von Wehrden, H., Laubichler, M., Wiek, A., Lang, D.: Transnational collaboration for sustainability in higher education: lessons from a systematic review. J. Clean. Prod. **168**, 764–779 (2017)
22. Almansour, S., Wendy, H.L.: The challenges of international collaboration: perspectives from Princess Nourah Bint Abdulrahman University. Cogent Education. **2**, 1 (2015). https://doi.org/10.1080/2331186X.2015.1118201
23. Edwards, N., Bunn, H., Morales-Mann, E., Papai, P., Davies, B.: International collaborative workshops. A 6-year partnership between Canada and China. Nurse Educ. **25**(2), 88–94 (2000)
24. Mok, K.H.: The Quest for Regional Hub of Education: Searching for New Governance and Regulatory Regimes in Singapore, Hong Kong and Malaysia, East-West. Senior Seminar on Quality Issues in the Emerging Knowledge Society, Malaysia (2009)
25. Horta, H., Patrício, M.T.: Setting-up an international science partnership program: a case study between Portuguese and US research universities. Technol. Forecast. Soc. Chang. **113**(B), 230–239 (2016)
26. Dühr, S., Cowell, R., Markus, E.: Europeanizing planning education and the enduring power of national institutions. Int. Plan. Stud. **21**(1), 1–18 (2015). https://doi.org/10.1080/13563475.2015.1114447
27. Robbert, M., Senne, L., Asgary, N.: Educating the Global IT Manager. Int. J. Technol. Knowl. Soc. **7**(1), 179–190 (2011)
28. O'Mahoney, J.: Enhancing Student Learning and Teacher Development in Transnational Education. Higher Education Academy, York (2014)
29. McNamara, J., Knight, J.: Going Global 2014: Impacts of Transnational Education on Host Countries: Academic, Cultural, Economic and Skills Impacts and Implications of Programme and Provider Mobility. British Council and DAAD, London (2014)
30. Jones, E.: Internationalization and employability: the role of intercultural experiences in the development of transferable skills. Public Money and Management. **33**, 95–104 (2013)
31. Mellors-Bourne, R., Jones, E., Woodfield, S.: Transnational Education and Employability Development. Higher Education Academy, York (2015)
32. Mellors-Bourne, R., Fielden, J., Kemp, N., Woodfield, S., Middlehurst, R.: The Value of Transnational Education to the UK. BIS Research Paper 194. Department for Business, Innovation and Skills, London (2014)

Implementation of an Acknowledgment and Signature Based Intrusion Detection System for MANETS

92

Prasanthi Sreekumari

92.1 Introduction

A Mobile Ad Hoc Network (MANET) is an interconnected system of large number of mobile nodes in which each mobile node can function as a sender, a receiver or a router [1, 2]. As shown in Fig. 92.1, due to its unique characteristics such as self-configuring, self-maintaining, dynamic changing topology, shared broadcast radio channel, MANETS are highly vulnerable for DATA traffic and CONTROL traffic attacks.

DATA traffic attack deals either drops the data packets passing through the nodes or delay the forwarding of data packets to the next node. Examples include the Black-Hole attack (ii) Gray-Hole attack and (iii) Jellyfish attack. In the case of CONTROL traffic attacks, the attacker can be launched even without having access to any cryptographic keys and subvert the functionality of the network by interrupting the flow of data packets [3]. Examples include the wormhole attack (ii) the rushing attack (iii) identity spoofing, (iv) the Sybil attack (v) the sinkhole attack, and (vi) the HELLO flood attack.

This work is mainly focuses on DATA traffic attacks, specifically detect malicious nodes in the network. Many solutions are proposed for mitigating the DATA traffic attacks in MANETS [3–8]. However, the detection of malicious nodes is still remain as a primary concern especially due to false misbehavior reports.

92.1.1 Contributions of the Paper

In this paper, we propose an acknowledgment and signature based intrusion detection system (IDS), namely MaDS (Malicious Node Detection System) for securing data packets from malicious nodes during packet transmission over MANETS.

The proposed system consists of three modes, (i) Acknowledgment (ACK), (ii) Misbehavior Report ACK Mode (MR_ACK) and (iii) Misbehavior Confirmation ACK Mode (MC_ACK). ACK modes used as a default mode to get confirmation from the destination node for receiving the data successfully. MR_ACK mode used to detect the behavior of nodes in the network. Furthermore, MC_ACK mode used to detect and confirm the malicious nodes in the network. We implemented and evaluated MaDS in NS2 and compare the result against 2ACK and EAACK in terms of packet delivery ratio.

92.1.2 Organization of the Paper

The rest of this paper is organized as follows. Section II describes the proposed intrusion detection system, MaDS. Section III presents the simulation methodology and performance results. Section IV concludes the paper.

92.2 MaDS

In this section, we describe our proposed acknowledgment and signature based intrusion detection system called MaDS in detail. Figure 92.2 shows the system architecture of MaDS. MaDS consists of three major modes, namely, (i) ACK (ii) MR_ACK and (iii) MC_ACK.

P. Sreekumari (✉)
Department of Computer Science, Grambling State University, Grambling, LA, USA
e-mail: sreekumarip@gram.edu

© Springer Nature Switzerland AG 2019
S. Latifi (ed.), *16th International Conference on Information Technology-New Generations (ITNG 2019)*,
Advances in Intelligent Systems and Computing 800,
https://doi.org/10.1007/978-3-030-14070-0_92

ACK ACK is basically an end-to-end acknowledgment scheme like other existing schemes [3–8]. In ACK mode, the sender sends a data packet to receiver through the first route and mark the packet ID and sending time. When the receiver receives the data packet, it sends back an ACK to sender along the same route. If the sender fails to receive the ACK within a predefined time period, the node change its mode to MR_ACK.

MR_ACK The MR_ACK mode is a modified version of the S-ACK mode of EAACK scheme proposed by Elhadi M. Shakshuki et al. [8]. One of the major limitations of S-ACK mode includes it cannot distinguish exactly which particular node is misbehaving node in every three consecutive nodes in the network [6]. For solving this limitation, we proposed MR_ACK scheme to detect a particular malicious node. In MR_ACK mode, when the third node receives the data packet in every three consecutive nodes, it is required to send an ACK back to the first node. If the first node does not receive this ACK from the third node within a predefined time period, first node resends the same packet and requests

an ACK to the second node in every two consecutive nodes. If the first node receives an ACK from second node within a predefined time period, then the first node reports to the source node that the third node is malicious in the route.

On the other hand, if the first node does not receive an.

ACK from the second node in the second attempt, then first node reports to the source node that the second node is malicious in the route.

MC_ACK When the source node receives malicious node report from the first node in MR_ACK mode, then the sender changes to MC_ACK mode. The main aim of MC_ACK mode is to confirm whether the malicious node report is correct or not. For confirming the malicious node report, we adopt the misbehavior report authentication scheme specified in [8]. When the sender receives an ACK from the first node by MR_ACK scheme, the sender searches its local knowledge base to find the alternative route to destination. Upon finding the new route, the sender sends data packet through the new route and marks the packet ID and sending time. If the packet is already received at destination, then the sender concludes that the report sent by the node from the previous route was not malicious. On the other hand, if the destination receives the packet for the first time, then the sender confirms that the report was correct and send alarm to other nodes in the network about the malicious nodes.

Digital Signature For protecting the data packet, we use DSA and RSA algorithms in MaDS scheme as specified in [8].

92.3 Performance Evaluation

In this section, we present simulation environment and evaluation results of MaDS with 2ACK and EAACK schemes.

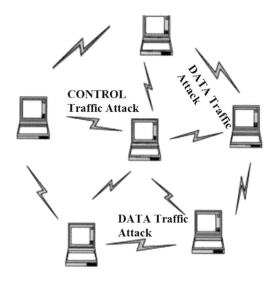

Fig. 92.1 MANET

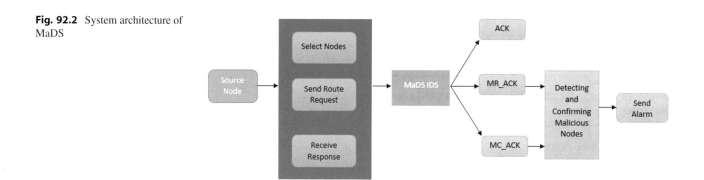

Fig. 92.2 System architecture of MaDS

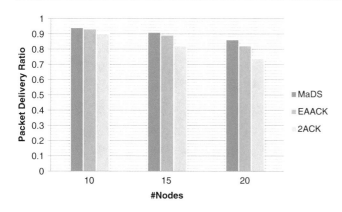

Fig. 92.3 Packet Delivery Ratio without the presence of malicious nodes

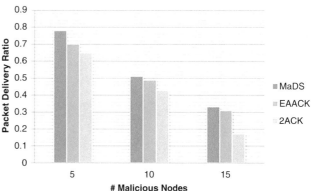

Fig. 92.4 Packet Delivery Ratio with malicious nodes

92.3.1 Simulation Setup

Simulation is conducted using the simulator NS2. We used 20 nodes in a flat space with a size of 1500 × 1500 m.

The moving speed of mobile node is limited to 10 m/s and User Datagram Protocol traffic with constant bit rate is used with a packet size of 1000 bytes. DSR routing protocol used for all schemes. We set the simulation time for 200 s. To evaluate the performance of MaDS with other schemes, we use the performance metric Packet delivery ratio (PDR) with and without the presence of malicious nodes in the network.

Figure 92.3 shows the performance of MaDS with 2ACK and EAACK in terms of packet delivery ratio using the network without having malicious nodes. We run the experiment for 10, 15 and 20 nodes. From the results, we observed that MaDS achieved highest packet delivery compared with other schemes. The main reason is, MaDS can accurately find misbehaving nodes than 2ACK and EAACK. However, when the number of nodes increases, the PDR also decreases for all schemes. Compared to 2ACK and EAACK, MaDS has better performance even the PDR is decreased.

Figure 92.4 presents the result of MaDS with 2ACK and EAACK in presence of malicious nodes. Based on the result, MaDS outperforms 2ACK and EAACK. In addition, we noted that the variation in packet delivery ratio is small compared to 2ACK and EAACK with and without the presence of malicious nodes in the network.

92.4   Conclusion

For mitigating the security issues specifically due to data traffic attacks, we designed and implemented an acknowledgment and signature based intrusion detection system for Mobile Ad-Hoc networks. From the experiments conducted by NS2, we observed that the proposed scheme MaDS was efficient to detect the malicious nodes and thereby increase the performance of the network. In future work, we check the efficiency of MaDS in terms of routing overhead and accuracy.

References

1. Sharma, S.B, Chauhan, N.: Security issues and their solutions in MANET. Published in: Futuristic Trends on Computational Analysis and Knowledge Management (ABLAZE), pp. 25–27, Feb. 2015
2. Sharma, P.K., Sharma, V.: Survey on security issues in MANET. Wormhole detection and prevention. International Conference on Computing Communication and Automation (ICCCA), pp. 637–640, 2016
3. Khalil, I.: "Mitigation of control and data traffic attacks in wireless ad-hoc and sensor networks", Proquest, 20111109
4. Marti, S., Giuli, T.J., Lai, K., Baker, M.: Mitigating routing misbehavior in mobile Ad-hoc networks. Proceedings of the 6th Annual International Conference on Mobile Computing and Networking (MobiCom'00), pp. 255–265, August 2000
5. Balakrishnan, K.; Deng, J., Varshney, V.K.: TWOACK: preventing selfishness in mobile ad hoc networks. Wireless Communications and Networking Conference, 2005 IEEE , vol. 4, no., pp. 2137–2142, 13–17 (2005)
6. Sandhiya, D., Sangeetha, K., Latha, R.S.: Adaptive Acknowledgment technique with key exchange mechanism for MANET. 2014 International Conference on Electronics and Communication Systems (ICECS) (2014)
7. Nasser, N., Chen, Y.: Enhanced intrusion detection system for discovering malicious nodes in mobile Ad-hoc networks. Communications, 2007. ICC '07. IEEE International Conference on, pp. 1154–1159, 24–28 June 2007
8. Shakshuki, E.M., Kang, N., Sheltami, T.R.: EAACK-A secure intrusion-detection system for MANETs. IEEE Trans. Ind. Electron. **60**(3), 1089–1098 (2013)

Index

Printed in the United States
By Bookmasters